Brian Bayve
October 1952

Developments in Aquaculture and Fisheries Science, 25

THE MUSSEL MYTILUS: ECOLOGY, PHYSIOLOGY, GENETICS AND CULTURE

DEVELOPMENTS IN AQUACULTURE AND FISHERIES SCIENCE

Developments in Aquaculture and Fisheries Science, 25

THE MUSSEL MYTILUS: ECOLOGY, PHYSIOLOGY, GENETICS AND CULTURE

Edited by

ELIZABETH GOSLING

Regional Technical College, Dublin Road, Galway, Ireland

ELSEVIER Amsterdam – London – New York – Tokyo 1992

ELSEVIER SCIENCE PUBLISHERS B.V.
Molenwerf 1
P.O. Box 211, 1000 AE Amsterdam, The Netherlands

ISBN: 0-444-88752-0

This book is printed on acid-free paper.

Printed in The Netherlands.

PREFACE

Mussels, of the genus *Mytilus* are among the commonest of marine molluscs and constitute an important element in the ecology of coastal waters. Mussels are nutritious and good to eat and their sedentary habit makes them suitable candidates for cultivation. Because they are sessile filter feeders and have been shown to be effective concentrators of trace toxic substances, mussels are now being widely used as biomonitoring organisms for coastal water quality. They have been extensively used as model organisms in many scientific studies and a vast body of data is available from basic physiological, biochemical, genetic and toxicological investigations.

Brian Bayne's book *Marine Mussels* was a very comprehensive review of the literature on marine mussels when it was published as an IBP volume in 1976 and it has remained until now the sole treatise on mussels. However, since 1976 there has been a wealth of published papers on mussels, not alone in the traditional areas of investigation, but also in the newer fields of nutrition, environmental monitoring, population genetics, toxicology, disease and public health. For the numerous workers involved on both the research and practical aspects of the biology of mussels it is becoming increasingly important that up-to-date and concisely written information on this experimental animal be available to them.

This book presents a thorough compilation of existing information on mussels in the form of critical review papers written by leading authorities in the field. For the numerous workers involved on both the research and practical aspects of the biology of mussels, it will serve as the standard reference text. Also, because mussels, more than any other group of marine or freshwater molluscs, have been pioneer research organisms in areas such as genetics, physiological energetics, and pollution monitoring, the information presented in this book will serve as an important starting point for those wishing to extend such studies to other species of molluscs. This book with its up-to-date information, accompanied by numerous tables and figures and an extensive bibliography of more than 2000 citations, will also provide valuable teaching material for both general as well as specialized courses on marine invertebrates.

Considerable controversy has surrounded the systematic status of many forms within the genus *Mytilus*. Much of the confusion stems from the fact that, until relatively recently, *Mytilus* taxonomy had been based solely on shell morphological characters. Protein electrophoresis along with other molecular techniques have gone quite some way in attempting to clarify the systematic status of species within the genus. For example, what have until recently been regarded as *Mytilus edulis* populations on the Pacific coasts of North America, the USSR and China have now been identified as either *Mytilus galloprovincialis* or *Mytilus trossulus*. Readers are, therefore, strongly urged to consult Chapter 1, which gives an up-to-date evaluation of *Mytilus* systematics and distribution—reviewed in the light of a very recently published paper by McDonald et al. (1991)—before preceding to other chapters.

I would like to express my gratitude to the many friends and colleagues who have helped and given advice during the preparation of the volume. In particular I would like to thank my partner Jim for his continual support and patience over the past three years. My sincere thanks to the following people who very kindly agreed to review various chapters: Andy Beaumont, David Dixon, Peter Gabbott, John McDonald, David McGrath, Raymond Seed and Catherine Thiriot-Quiévreux. Special thanks to Brian Ottway, for his invaluable comments on six of the chapters. I gratefully acknowledge the encouragement of my many colleagues in the Regional Technical College and wish to especially thank the Principal, Gay Corr and Head of School, Brian Place for permission to use the facilities of the College and School of Science respectively; thanks also to the chief librarian, Anne Walsh and to Peggy Ryan and Mary O Dea for secretarial assistance. At University College, Galway my sincere thanks to the Biochemistry Department for laser printing facilities, Noel Wilkins, Michael Couglan and Michael Laver for useful discussion, Paul Doyle for computer services and Angela Gallagher for photographic assistance. Thanks are also due to the library staff of the Queen's University, Belfast and to Paulo Prodohl for computer facilities. I gratefully acknowledge the assistance of the librarians Helene Boske and Patricia Volland–Nail at the INRA Centre de Recherche de Tours, France.

Finally, I wish to express my heartfelt appreciation of the way in which the authors of this volume worked hard to produce such excellent contributions. Without their effort the publication of this volume would not have been possible.

Elizabeth Gosling
February, 1992

CONTENTS

Chapter 7.   GENETICS OF *MYTILUS*

ELIZABETH M. GOSLING

Chapter 8.   MUSSELS AND ENVIRONMENTAL CONTAMINANTS:

BIOACCUMULATION AND PHYSIOLOGICAL ASPECTS

JOHN WIDDOWS AND PETER DONKIN

Chapter 9.   MUSSELS AND ENVIRONMENTAL CONTAMINANTS:

MOLECULAR AND CELLULAR ASPECTS

DAVID R. LIVINGSTONE AND RICHARD K. PIPE

Chapter 10. MUSSEL CULTIVATION

ROBERT W. HICKMAN

Chapter 11. MUSSELS AND PUBLIC HEALTH

SANDRA E. SHUMWAY

Chapter 12. DISEASES AND PARASITES OF MUSSELS

SUSAN M. BOWER

CONTRIBUTING AUTHORS

Brian L. Bayne Plymouth Marine Laboratory, Prospect Place, West Hoe, Plymouth, Devon PL1 3DH UK

Susan M. Bower Department of Fisheries and Oceans, Pacific Biological Station, Nanaimo, British Columbia V9R 5K6 CANADA

Peter Donkin Plymouth Marine Laboratory, Prospect Place, West Hoe, Plymouth, Devon PL1 3DH UK

Elizabeth M. Gosling Regional Technical College, Galway IRELAND

Anthony J. S. Hawkins Plymouth Marine Laboratory, Prospect Place, West Hoe, Plymouth, Devon PL1 3DH UK

Robert W. Hickman Aquaculture Research Centre, MAF Fisheries Greta Point, P.O. Box 297, Wellington NEW ZEALAND

Michael J. Kennish Institute of Marine and Coastal Sciences, New Jersey Agricultural Experiment Station, Cook College, Rutgers University, New Brunswick, New Jersey 08903 USA

David R. Livingstone NERC Plymouth Marine Laboratory, Citadel Hill, Plymouth, Devon PL1 2PB UK

Richard A. Lutz Institute of Marine and Coastal Sciences, New Jersey Agricultural Experiment Station, Cook College, Rutgers University, New Brunswick, New Jersey 08903 USA

Michel Mathieu Laboratoire de Zoologie, IBBA, Université de Caen, 14032 Caen Cedex FRANCE

Brian Morton Department of Zoology, University of Hong Kong, HONG KONG

Richard K. Pipe NERC Plymouth Marine Laboratory, Citadel Hill, Plymouth, Devon PL1 2PB UK

Raymond Seed School of Ocean Sciences, University of Wales Bangor, Menai Bridge, Gwynedd LL59 5EY UK

Sandra E. Shumway Department of Marine Resources, West Boothbay Harbor, Maine 04575 USA

Thomas H. Suchanek Division of Environmental Studies, University of California, Davis, California 95616 USA

John Widdows Plymouth Marine Laboratory, Prospect Place, West Hoe, Plymouth, Devon PL1 3DH UK

Albertus de Zwaan Delta Institute for Hydrobiological Research, Vierstraat 28, 4401 EA Yerseke THE NETHERLANDS

Chapter 1

SYSTEMATICS AND GEOGRAPHIC DISTRIBUTION OF *MYTILUS*

ELIZABETH M. GOSLING

INTRODUCTION

The family Mytilidae, to which the genus *Mytilus* Linné 1758 belongs, is believed to have its origin as far back as the Devonian era, some 400 million years ago (Soot-Ryen, 1969). The genus itself, however, is of relatively recent origin, with apparently no records older than the Pliocene (Soot-Ryen, 1955).

Mytilus is a very old name, dating back to Greek and Roman times, and is probably derived from the Greek word μυτιλοσ, mitilos, meaning sea mussel. Linnaeus (1758) in his Systema Naturae gave the genus a much broader interpretation, including in it some 22 species from many dissimilar genera e.g. *Ostrea, Anadonta, Hiatella*, in addition to true mussel species. Since then there have been numerous reviews of the mytilids, most notably those of Jukes-Brown (1905), Lamy (1936), Dodge (1952) and Soot-Ryen (1955, 1969).

Mytilus is differentiated from other genera in the family e.g. *Perna* and *Choromytilus*, by the presence of a pitted resilial ridge, several hinge teeth, the presence of an anterior adductor muscle and a more or less continuous posterior byssus and foot retractor muscle scar (Fig. 1.1).

In a comprehensive review of the genus Lamy (1936) recognized the following as distinct species of *Mytilus*. *Mytilus edulis* Linnaeus, 1758 from northern temperate latitudes, *Mytilus galloprovincialis* Lamarck, 1819 from the Mediterranean Sea, *Mytilus trossulus* Gould, 1850 from the Pacific coast of North America, *Mytilus chilensis* Hupé, 1854 from Chile, *Mytilus platensis* Orbigny, 1846 from Argentina, *Mytilus planulatus* Lamarck, 1819 from Australia and *Mytilus desolationis* Lamy, 1936 from the Kerguelen Islands. In a later review of the genus Soot-Ryen (1955), recognized *Mytilus coruscus* Gould, 1861 (=*M. crassitesta* Lischke, 1868) from Japan and China, and *Mytilus californianus* Conrad, 1837 from the Pacific coast of North America, as distinct species, but considered most of the species described by Lamy to be subspecies of *M. edulis*. Also, the New Zealand mussel *Mytilus aoteanus*, first described by Powell (1958), has since regarded as a subspecies of *M. edulis* by Fleming (1959). More recently, Scarlato and Starobogatov (1979) have

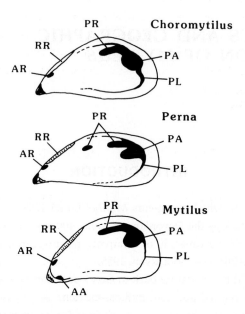

Fig. 1.1. The internal shell characteristics of the genera *Choromytilus*, *Perna* and *Mytilus*. AR: anterior retractor muscle, PR: posterior retractor muscle, RR: resilial ridge, PR: posterior retractor muscle scar, PA: posterior adductor muscle scar, PL: pallial line, AA: anterior adductor muscle scar. (After Grant et al., 1984).

described what they consider to be two new subspecies of *Mytilus* from the Pacific coast of Asia, *M. edulis kussakini* and *M. edulis zhirmunskii*.

Much of the confusion described above stems from the fact that, until relatively recently, *Mytilus* systematics has been based solely on morphological shell characteristics. Such characters are enormously plastic, being influenced by factors such as age and density of mussels, tidal level and habitat type (Seed, 1968). Clearly, systematic information that is relatively free of environmentally-induced changes is highly desirable. Over the past 20 years or so the use of biochemical techniques e.g. protein electrophoresis, have proved invaluable in quantifying genetic differences between species (see Ferguson, 1980 for details). In *Mytilus* such differences, combined with statistical techniques applied to both enzyme and morphological phenotypes, have gone some way in attempting to clarify the systematic status of species within the genus (McDonald et al., 1991).

In the first part of this chapter the reliability of the morphological characteristics, which have been traditionally used to separate the various forms of *Mytilus,* are discussed. A brief account is then presented on the contribution that electrophoresis has made in providing additional information on sytematics relationships within the genus. In order to avoid unnecessary overlap between this chapter and Chapter 7, and also because most electrophoretic investigations on mussels have not been undertaken with systematics as their prime objective, a much more detailed discussion is presented in Chapter 7. The last part of the chapter deals with the origin and present-known distribution of the various forms on a global scale—as evidenced from both morphological and electrophoretic data.

TAXONOMY OF *MYTILUS*

Morphological Shell Characteristics as Aids in *Mytilus* Taxonomy

Investigations on the taxonomy of *Mytilus* have tended to focus on the systematic relationships between *M. edulis* and the Mediterranean mussel, *M. galloprovincialis*. Since the 1860s considerable controversy has surrounded the systematic status of this mussel. While it is regarded by some as a distinct species of *Mytilus*, others consider it merely as a variety of the larger *M. edulis* complex (see Gosling, 1984 for review). *M. galloprovincialis* is believed to have diverged from *M. edulis* when the Mediterranean Sea was cut off from the Atlantic during a Pleistocene ice age, about 1–2 million years ago (Barsotti and Meluzzi, 1968). Northerly migration of *M. galloprovincialis* occurred as the ice receded, and this mussel has continued to extend its range northwards onto the Atlantic coasts of western Europe, where it is found intermixed with *M. edulis* in varying proportions (Hepper, 1957; Seed, 1972, 1974; Gosling and Wilkins, 1977, 1981; Skibinski and Beardmore, 1979).

Separation of *M. edulis* and *M. galloprovincialis* has been based primarily on external shell contours, internal features of the shell valves and the colour of the mantle edge. Although detailed descriptions of these can be found elsewhere (Verduin, 1979; Gosling, 1984 and references therein; Beaumont et al., 1989) the following is a brief account of the reported distinguishing characters.

The shell of *M. galloprovincialis* tends to be higher and flatter than in *M. edulis*, giving different transverse profiles in the two forms. The anterior end of the shell of *M. galloprovincialis* is distinctly beaked or incurved, while that of *M. edulis* has a more snub-nosed appearance (Fig. 1.2A, B). Interiorly, the anterior adductor muscle scar is small and circular in *M. galloprovincialis*, whereas in *M. edulis* it is narrow and elongated. The hinge plate is also smaller in *M. galloprovincialis*, forming a much tighter arc with its rear end more clearly delimited from the adjacent ventral edge of the valve. The hinge plate in *M. edulis* is a gently curving structure (Fig. 1.2C). The colour of the mantle edge is typically purple-violet in *M. galloprovincialis* and yellow-brown in *M. edulis*.

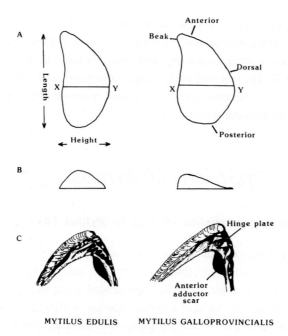

Fig. 1.2. (A) and (C): Some of the salient differences in shell morphology between *Mytilus edulis* and *Mytilus galloprovincialis*. (B): Transverse profiles through section XY of a single shell valve of *M. edulis* and *M. galloprovincialis*. (After Gosling, 1984).

Although it might appear from the foregoing that the differences between the two forms are always clearcut, this is by no means the case as will soon be made apparent.

Using these characters a number of investigators (Hepper, 1957; Lewis and Seed, 1969; Seed, 1972, 1974) have identified the *M. galloprovincialis* form on the Southwest coasts of England, and on the Atlantic coasts of Ireland and France, extending into the English Channel as far as the Cherbourg peninsula (Fig. 1.5). In these areas *M. galloprovincialis* has been found intermixed with *M. edulis* in varying proportions. In some areas, e.g. the Atlantic coasts of Ireland and North-west France, separation of the two forms of mussel has proved to be exceedingly difficult due to a considerable degree of overlap in morphological characteristics. This, together with the large number of truly intermediate forms observed, was the first indication that hybridization and introgression might be occurring in these localities (Seed, 1978).

Of the five or six morphological characters used to separate *M. edulis* and *M. galloprovincialis*, overall shell shape is the least reliable (Seed, 1972, 1974; Gosling, 1984; Beaumont et al., 1989), being influenced by environmental factors, the effects of which appear to be the same for both forms of mussel (Seed, 1978). Of the various internal shell characteristics the anterior adductor scar and hinge plate size have been generally regarded as the two most reliable in separating the two forms of mussel. While the mean adductor scar ratios (adductor scar lenght x 1000/shell length) vary from one locality to another, the values tend to be consistently lower in *M. galloprovincialis* than in *M. edulis*. The range lies between 36 and 93, and between 74 and 140, for *M. galloprovincialis* and *M. edulis* respectively (Lewis and Seed, 1969; Seed, 1972, 1974; Wilkins et al., 1983; Grant and Cherry, 1985; Lee and Morton, 1985; Beaumont et al., 1989; Sanjuan et al., 1990) and, not surprisingly, in mixed populations values overlap to a greater extent than in allopatric populations. However, even in several pure populations of *M. galloprovincialis* from the Mediterranean Sea, the percentage misidentification using this character alone was over 60% (Seed, 1972).

Like the adductor scar ratios, the hinge plate ratios (hinge plate length x 1000/shell length—although difficult to measure accurately—tend to be lower in *M. galloprovincialis* than in *M. edulis*: 50–77 in *M. galloprovincialis* and 56–115 in *M. edulis* (Lewis and Seed, 1969; Seed, 1972, 1974; Grant and Cherry, 1985; Lee and Morton, 1985; Beaumont et al., 1989; Sanjuan et al., 1990). Once again, the degree of overlap in values is greater for mixed than for pure populations of the two forms of mussel. In fact, in mixed samples from the majority of French sites sampled by Seed (1972), and at Rock in South-west England (Lewis and Seed, 1969), the frequency distributions of the hinge plate ratios were generally unimodal.

Verduin (1979) has combined adductor scar and hinge plate ratios into a single index: anterior adductor scar length/height of shell+1.5 x internal radius of the hinge plate/height of shell, which is less than 0.25 for *M. galloprovincialis*, and greater than 0.25 for *M. edulis*. The internal radius was measured by fitting the cylindrical parts of drills of known diameter into the nose of the shells. On the basis of his graphs (no data given) Verduin (1979) achieved good separation of mussels in mixed populations from South-west France. Sanjuan et al. (1990) have recently used this index, in conjunction with other shell measurements and allozyme data, to positively identify Galician mussels on the North-west coasts of Spain as *M. galloprovincialis*. Virtually all samples analysed by these authors had indices lower than 0.25. Both Verduin (1979) and Sanjuan et al. (1990) have used the index on either submerged or lower shore mussels. Difficulties will probably be encountered if this index is applied to upper shore mussels, where the shape of the anterior end of shells is often obscured by erosion, shell thickening and divergence of umbones (Lewis and Seed, 1969).

Beaumont et al. (1989) have found mantle edge colour to have good discriminatory power in separating mixed populations of *M. edulis* and *M. galloprovincialis* in South-west England (but see Lewis and Powell, 1961). However, in other areas e.g. the Atlantic coasts of France and Ireland, and even in the Mediterranean Sea, where only *M. galloprovincialis* occurs, a large percentage (20–60%) of individuals were misidentified using this character alone (Seed, 1972, 1974).

In conclusion, there is no *single* morphological character that can be reliably used to separate mixed or pure populations of the two forms of mussel (Gosling, 1984;

Beaumont et al., 1989; Koehn, 1991; McDonald et al., 1991). In some areas e.g. at Rock, good separation can be achieved using a combination of characters, such as adductor and hinge plate ratios, together with mantle edge colour. There, morphological differences are also accompanied by marked differences in breeding patterns, peacrab infestation and growth potential (Seed, 1971). However in other parts of South-west England (Lewis and Seed, 1969) and the west coast of France (Seed, 1972), only poor separation of the two mussel types is achieved using the same combination of characters. Furthermore, in areas such as the Atlantic coasts of Ireland and North-west France, where hybridization is extensive, the large number of intermediate forms makes accurate identification of *M. galloprovincialis* and *M. edulis* an almost impossible task, especially at exposed locations (Seed, 1974).

More recently, a different approach to the problem has been taken. Ferson et al. (1985) used an image analysis technique, that automatically determines the outlines of shells, to discriminate between distinct populations of *M. edulis*, and the recently rediscovered *M. trossulus* Gould, 1850 (McDonald and Koehn, 1988) in Newfoundland, Canada. Although the technique was capable of demonstrating an association between genotype and morphology, it was not able to reliably identify the population from which each specimen was collected i.e. misclassification of an individual on the basis of shell shape was about 15%, while misclassification using electrophoresis was around 7%.

McDonald et al. (1991) have used canonical variates analysis of shell traits to discriminate between three different taxa of *Mytilus* collected from sites in the northern and southern hemispheres; prior electrophoretic analysis had indicated that only one taxon was present at each of the northern hemisphere sites (see Chapter 7 for details). Eighteen morphometric characters were employed (Fig. 1.3), virtually all of which have been used by previous authors (Seed, 1972; Verduin, 1979; Beaumont et al., 1989). The analysis revealed three clusters in the northern hemisphere, corresponding to *M. edulis, M. galloprovincialis* and *M. trossulus*, with the best discrimination between *M. edulis* and *M. galloprovincialis* (Fig. 1.4). In the southern hemisphere, where, to date, *M. trossulus* has not been identified, *M. edulis* samples were morphologically intermediate between northern *M. edulis* and *M. trossulus*. In contrast, both southern and northern *M. galloprovincialis* were morphologically similar. The two characters which had previously proved to be most reliable— adductor scar and hinge plate lengths—contributed most to the canonical variates analysis, but each of these taken singly gave poor discrimination between the three different forms. A combination of all 18 characters gave good separation of the three mussel types—at least for northern hemisphere samples.

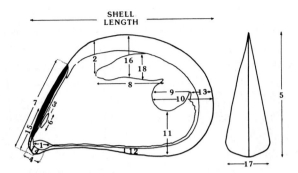

Fig. 1.3. The 18 morphometric characters used by McDonald et al. (1991) to distinguish between different forms of *Mytilus*. 1: length of the anterior adductor muscle scar. 2: distance between the anterior end of the posterior retractor muscle scar and the dorsal shell margin. 3: width of the anterior retractor muscle scar. 4: length of the hinge plate. 5: shell height. 6: length of the anterior retractor muscle scar. 7: distance between the umbo and the posterior end of the ligament. 8: length of the posterior retractor muscle scar. 9: length of the posterior adductor muscle scar. 10: distance between the anterior edge of the posterior adductor muscle scar and the posterior shell margin. 11: distance between the ventral edge of the posterior adductor muscle scar and the ventral shell margin. 12: distance between the pallial line and the ventral shell margin midway along the shell. 13: distance between the posterior edge of the posterior adductor muscle scar and the posterior shell margin. 14: number of major teeth on the hinge plate, excluding any small crenulations which may be present, especially on the posterior ventral face of the hinge plate. 15: distance between the umbo and the posterior end of the anterior retractor scar. 16: distance between the ventral edge of the posterior retractor muscle scar and the dorsal shell margin. 17: shell width. 18: width of the posterior retractor muscle scar. (After McDonald et al., 1991).

For the field taxonomist there is no single morphological character which can be used to discriminate between allopatric populations *M. edulis, M. galloprovincialis* and *M. trossulus* individuals. However, using a combination of characters (at least 6 or 7), some of which necessitates the use of a microscope, good separation can be achieved between *M. edulis* and *M. trossulus*, and *M. edulis* and *M. galloprovincialis*—but not in the case of *M. galloprovincialis* and *M. trossulus* (Fig. 1.4). In order to view this separation, fairly sophisticated computing facilities are called for. It should be emphasized that such 'pure' populations must first be identified by electrophoretic analysis. Needless to say, it is unlikely that the morphometric differences described by McDonald et al. (1991) will persist when mutivariate analysis of morphological variation is eventually applied to mussels in areas of overlap and hybridization.

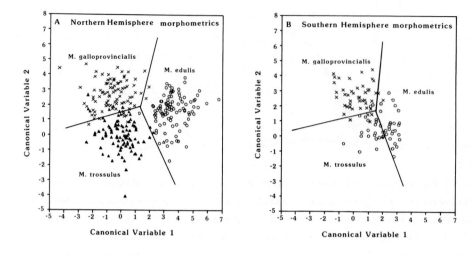

Fig. 1.4. (A) First and second canonical variates for 18 morphometric shell characters from locations in the northern hemisphere showing multivariate discrimination among the three *Mytilus* taxa. (O: *Mytilus edulis*, X: *Mytilus galloprovincialis*, ▲: *Mytilus trossulus*). To aid comparison with the southern hemisphere results, perpendicular bisectors of the lines connecting the centroids of the clusters have been drawn. (B) Scores of southern hemisphere mussels when functions from the canonical variates analysis of northern hemisphere mussels are applied to southern hemisphere samples. The lines separating the northern hemisphere clusters are repeated. (For details see McDonald et al., 1991, from which this figure was redrawn).

The Use of Allozyme Data in *Mytilus* Taxonomy

Protein electrophoresis, together with techniques such as DNA-DNA hybridization, mitochondrial DNA (mt DNA) analysis, immunology and amino acid sequencing have proved invaluable in quantifying genetic differences between different animal and plant taxa. For comparisons of closely related species, electrophoresis has proved to be a most effective technique (see Ferguson, 1980 and Murphy et al., 1990 for reviews and references).

Samples of mussels have now been analysed electrophoretically from over two hundred locations—most of these sited in the northern hemisphere. Several electrophoretic markers (protein products of gene loci) have proved useful in differentiating between different forms within the genus *Mytilus*. The best discriminating loci are mannose-6-phosphate isomerase (*Mpi*), peptidase-II (*Aap*) and glucosephosphate isomerase (*Gpi*), although it should be pointed out that none of these are truly diagnostic (see Chapter 7 for details). A locus is considered to be diagnostic if an individual can be assigned to the correct species, or form, with a probability >0.99 (Avise, 1974). These and other loci have been used—in conjunction with morphological characters (see above)—to identify populations of *M. edulis, M. galloprovincialis* and *M. trossulus* on a global scale. Generally speaking, in the northern hemisphere allopatric populations of *M. edulis, M. galloprovincialis* and *M. trossulus* are each genetically homogeneous within a particular distinct geographic region. However, when populations of a single type are compared from different geographic regions e.g. *M. edulis* from east and west sides of the North Atlantic Ocean, there are, invariably, substantial genetic differences between them.

In the southern hemisphere populations of *M. edulis* and *M. galloprovincialis* have been identified solely on electrophoretic comparisons with northern forms. On this basis McDonald et al. (1991) have suggested that the South American mussels *M. chilensis* and *M. platensis* should tentatively be included in *M. edulis*. The tentative nature of this suggestion stems from the fact that mussels on the coasts of South America do not, in fact, closely resemble northern hemisphere *M. edulis*, but appear to fall somewhere in between the morphologically (Fig. 1.4) and electrophoretically (Fig. 7.2., Chapter 7) distinguishable *M. edulis* and *M. galloprovincialis* clusters of the northern hemisphere. These authors have also suggested that *M. desolationis* from the Kerguelen Islands—considered by Blot et al. (1988) as a semispecies of *M. edulis*—should be regarded as *M. edulis*, and that the New Zealand mussel, *M. aoteanus*, and the Australian mussel, *M. planulatus*, should both be included in *M. galloprovincialis* (see details in Chapter 7). It is obvious that further sampling of mussels is necessary before a definitive statement can be made on the status of mussels in the southern hemisphere.

Where populations of two forms come into geographic contact (Figs. 1.5 and 1.6) they hybridize. The size of hybrid zones varys depending on location e.g. in North-west Europe the width of the hybrid zone between *M. edulis* and *M. galloprovincialis* is large, ~1400km, while that between North Sea *M. edulis* and Baltic *M. trossulus* is narrow, hybridization occurring over a short distance of ~150km (Väinölä and Hvilsom, 1991). Results from electrophoretic analysis has indicated that such zones are spatially complex, containing a mixture of pure, hybrid and introgressed individuals. The *M. edulis/galloprovincialis* hybrid zone in North-west Europe has been particularly well-studied (Skibinski and Beardmore, 1979). In this area the extent of introgression varies depending on location; in South-west England there is hybridization but little introgression taking place, while on the Atlantic coasts of Ireland, France and Scotland introgression can be extensive (Skibinski et al., 1983; Gosling,

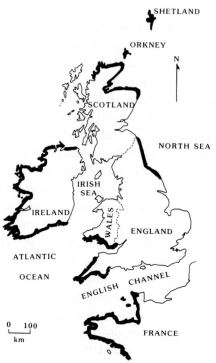

Fig. 1.5. Map showing in heavy outline the geographical areas in the British Isles and France where *Mytilus galloprovincialis* has been detected, using morphological and electrophoretic markers. (After Gosling, 1984).

1984); thus explaining the difficulty that Seed (1974) encountered in separating the two forms of mussels in this area

The question now remains: are the differences between the taxa sufficiently large to warrant full specific status for each, or should each be considered as a semispecies, ecotype or variety of *M. edulis*? There is considerable supporting evidence for the latter point of view: (1) the amount of genetic difference between any two different forms is similar to that observed for subspecies of other invertebrate taxa; (2) there is not a *single* morphological or electrophoretic character that can unequivocally assign an individual mussel to a particular taxon; (3) the occurrence of widespread hybridization in the wild, together with results on artificial hybridization, indicates that there is little evidence of genetic incompatibility; and (4), additional evidence from cytological, immunological and mtDNA studies indicates that the different forms are very closely related and do not individually merit full specific status (details and references in Chapter 7).

Many workers are in agreement on this (Skibinski et al., 1983; Gosling, 1984 and references therein; Brock, 1985; Skibinski, 1985; Hodgsen and Bernard, 1986; Edwards and Skibinski, 1987; Blot et al., 1988; Johannesson et al., 1990; Tedengren et al., 1990; Väinölä and Hvilsom, 1991). The majority of these have focused on areas of mixing where, because of hybridization and varying amounts of introgression between *M. edulis* and *M. galloprovincialis*, or between *M. edulis* and *M. trossulus*, they are impressed by the amount of genetic *similarity* between the different taxa. In areas of contact, hybridization occurs to such an extent as to contravene the present working definition of species—the biological species concept of Mayr (1970). This defines species as groups of actually or potentially interbreeding individuals reproductively isolated from other such groups and, therefore, places heavy reliance on the presence of genetically-based barriers to gene exchange between *species* pairs. Others workers hold a different viewpoint (McDonald and Koehn, 1988; Koehn, 1991; McDonald et al., 1991); these have tended to concentrate on widely separated allopatric populations of the three forms of mussel, and are therefore struck by the amount of morphological and genetic *differences* between the different taxa—differences which are maintained despite hybridization, and the massive potential for larval dispersal. These workers feel that "this genetic distinctness warrants taxonomic recognition at the species level" (McDonald et al., 1991). The difference between the two viewpoints seems to hinge on the importance placed on hybridization. Clearly the solution now lies in deciding how much hybridization is permissible before two taxa can no longer be regarded as separate species? If this can be done then we are quite some way along the road to acquiring an agreed operational definition of biological species. In the aquatic environment at least, and for *Mytilus* in particular, this is long overdue. If we cannot satisfactorly define a 'species' then it is also impossible to have a concensus on what constitues a semispecies, race, ecotype and so on.

Until then, we are still left with the problem of what to call the different taxa. In order to avoid unnecessary confusion, the most prudent scenario would seem to be to continue referring to the taxa as: *M. edulis, M. galloprovincialis* and *M. trossulus*, while at the same time recognizing—with the morphological and genetic information we have to date—that there is by no means a concensus on whether individually they merit full specific status.

What of the other *Mytilus* taxa described by Lamy and others? *M. coruscus* and *M. californianus* are readily distinguished from the other taxa by the presence of radiating ribs on the shell. To date, the specific status of *M. californianus* has never been in question. However, a recent report by Vermeij (1989) suggests that *M. californianus* and *M. coruscus* may in fact be a single species. Alternatively, some are of the opinion that *M. coruscus* may prove to be *Crenomytilus*, rather than *Mytilus* (R. Seed, personal communication, 1991). Electrophoretic analyses of *M. coruscus* populations would no doubt shed further light on its taxonomic affinities. As mentioned already, McDonald et al. (1991) suggest that the South American mussels *M. chilensis* and *M. platensis*, and also *M. desolationis*, should tentatively be included in *M. edulis*, and that the New Zealand mussel *M. aoteanus* and *M. planulatus* from Australia should be included in *M. galloprovincialis*. It is, however, debatable as to whether *M. desolationis* should be lumped with *M. edulis* since the genetic differences between the two are of the same order as those between *M. edulis* and *M. galloprovincialis* (Blot et al., 1988 and Chapter 7). McDonald et al. (1991) have also suggested that the two subspecies, *M. edulis zhurmunski* and *M. edulis kussakini*, on the basis of their described geographic distribution (Scarlato and Starobogatov, 1979), should be considered as *M. galloprovincialis* and *M. trossulus* respectively.

Thus, electrophoretic analysis has served to reduce, albeit tentatively, the original number in the genus from 12 to about five or six taxa: *M. edulis, M. galloprovincialis, M. trossulus, M. desolationis, M. californianus* and perhaps, *M. coruscus*.

GEOGRAPHIC DISTRIBUTION OF *MYTILUS*

The genus *Mytilus* is widely distributed in boreal and temperate waters of the northern and southern hemispheres (Soot-Ryen, 1955). The results from electrophoretic surveys—in particular the work of McDonald et al. (1991)—have been extremely useful in helping to map the global distribution of *Mytilus*, and this has often meant either extending or reducing the previously reported geographic range (Seed, 1976) of a particular mussel form within the genus.

Taking electrophoretic and morphological evidence into account, what follows is a brief description of the present-known distribution of *M. edulis, M. galloprovincialis, M. trossulus* and *M. californianus* (Fig. 1.6), together with some information on the historical background of each type. Because most authors cited below have accorded full specific status to each taxon, this procedure will be followed in the following section in order to avoid any unnecessary confusion.

M. edulis is widely distributed in the northern hemisphere; it occurs in European waters extending from the White Sea, U.S.S.R. (McDonald et al., 1990) as far south as the Atlantic coast of southern France (Seed, 1972, 1978; McDonald et al., 1991). The range of *M. edulis* was previously cited (Seed, 1976, 1978; Suchanek, 1985 and references therein) as extending over all of the Atlantic littoral of Europe, as far south as the North African coasts, but not into the Mediterranean Sea. A recent report by Sanjuan et al. (1990) has, however, shown that mussels in North-west Spain are in fact *M. galloprovincialis,* and they surmise that the whole of the Iberian peninsula will be found to contain *M. galloprovincialis,* rather than *M. edulis* populations as previously reported. The range of *M. edulis* was also believed to extend from the Arctic southwards to California and Japan on the Pacific coasts, and North Carolina on

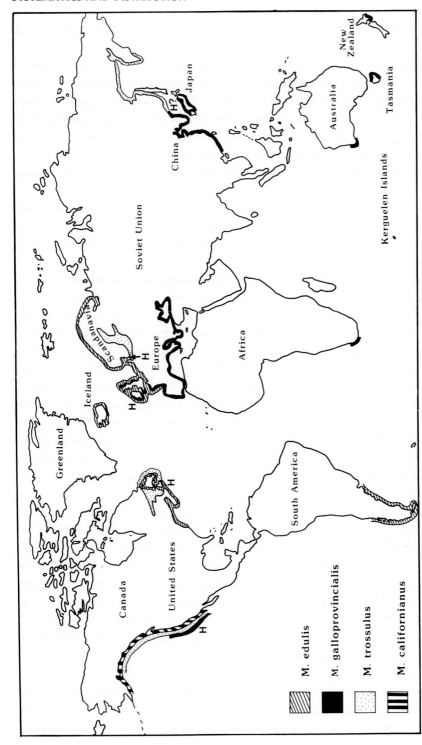

Fig. 1.6. Global distribution of the different forms of *Mytilus* based on evidence from morphological and/or genetic data of: Seed, 1976; Skibinski and Beardmore, 1979; Skibinski et al., 1983; Wilkins et al., 1983; Gosling, 1984; Koehn et al., 1984; Grant and Cherry, 1985; Lee and Morton, 1985; Suchanek, 1985; Bulnheim and Gosling, 1988; McDonald and Koehn, 1988; Varvio et al., 1988; Beaumont et al., 1989; McDonald et al., 1990, 1991; Sanjuan et al., 1990; Koehn, 1991; Väinölä and Hvilsom, 1991. (H = areas of contact and hybridization).

the Atlantic coast (Seed, 1976 and references therein). Electrophoretic evidence now indicates that *M. edulis* is absent from both Pacific coasts and has a more restricted distribution on the east coast of North America, extending from the Canadian Maritimes southwards to Cape Hatteras in North Carolina (McDonald and Koehn, 1988; Varvio et al., 1988; McDonald et al., 1990). The presence of *M. edulis* has been confirmed in Iceland (Varvio et al., 1988), but to date there is no information on whether it is present on the northern and northeastern coasts of Canada and Greenland. In the southern hemisphere McDonald et al. (1991) have suggested that mussels from South America, the Falkland and Kerguelen Islands should be included with *M. edulis*.

M. *galloprovincialis*, which is believed to have originated in the Mediterranean Sea (see below), is also found in the Black Sea, on the coasts of Spain, Portugal, the Atlantic coasts of France and the British Isles, as far north as the Shetland and Orkney Islands (see Fig. 1.5). In France, Britain and Ireland *M. galloprovincialis* is found intermixed with *M. edulis* in varying proportions, and hybridization and introgression are widespread (Gosling, 1984). In general, the extent of intergradation is greater in Ireland, Scotland and North-east England than it is in South-west England (Skibinski and Beardmore, 1979). Using electrophoretic markers and/or morphological shell characteristics, *M. galloprovincialis* has now been identified from southern California (McDonald and Koehn, 1988), Japan (Wilkins et al., 1983), Hong Kong (Lee and Morton, 1985), and northwards along the East China coast as far as Korea (McDonald et al., 1990). In the southern hemisphere populations of mussels on the west coast of Australia, Tasmania and New Zealand have been identified as *M. galloprovincialis* because of their close genetic similarity to this form of mussel in the northern hemisphere (McDonald et al., 1991). Several sample of mussels from South Africa have also been identified as *M. galloprovincialis* (Grant and Cherry, 1985; Beaumont et al., 1989).

The distribution of *M. trossulus* has been mapped solely on the basis of electrophoretic data. In the northern hemisphere it has now been identified from the Baltic Sea (Bulnheim and Gosling, 1988; Varvio et al., 1988), the east coast of the U.S.S.R. as far as the Kamchatka peninsula (McDonald et al., 1990), along the west coast of North America from central California to Alaska (McDonald and Koehn, 1988), and on the east coast in the region of the Canadian Maritimes (Koehn et al., 1984). The distribution of this form may in fact be more widespread i.e. with further sampling it may be found along the north coasts of North America and the U.S.S.R. In Europe hybridization between *M. trossulus* and *M. edulis* occurs to the east of Denmark in the Øresund—a narrow strait linking the Baltic Sea with the Kattegat (Väinölä and Hvilsom, 1991), where there is a pronounced salinity gradient. In North America the hybrid zone between these two forms appears to be in the upper regions of the Gulf of St. Lawrence (Koehn, 1991). On the west coast of North America hybridization occurs between *M. trossulus* and *M. galloprovincialis* in central California (McDonald and Koehn, 1988), while on the Pacific coast of the U.S.S.R., where contact between the two forms occurs near the North Korean and U.S.S.R. border, no evidence of hybridization has been found (Koehn, 1991). Further sampling in this area may indicate otherwise. To date no evidence of *M. trossulus* has been found in the southern hemisphere (McDonald et al., 1991).

The range of *M. californianus* extends along the Pacific coast of North America, from the Aleutian Islands in Alaska as far south as northern Mexico (Seed, 1976; Suchanek, 1985 and references therein). *M. coruscus*, is found in Japan and on the shores of the Yellow Sea and Sea of Japan (Scarlato, 1981). The range of *M. coruscus* therefore overlaps with that of *M. trossulus* and *M. galloprovincialis* in this region, but it is easily differentiated from the other two by means of shell morphological characteristics.

In summary, *M. edulis* is found in temperate latitudes in the northern (and perhaps the southern) hemisphere. *M. galloprovincialis* is also found in temperate latitudes, but its range extends into warmer waters. However, *M. trossulus* has a more northerly distribution than either *M. edulis* or *M. galloprovincialis*, and is found only in the northern hemisphere. Hybrid zones are found in areas where the ranges of any two forms overlap. In contrast to the rather disjunct distribution of *M. edulis, M. galloprovincialis* and *M. trossulus,* the range of *M. californianus* is restricted but continuous over a wide latitudinal range of about 30 degrees.

The genus *Mytilus* is of relatively recent origin with apparently no records older than two million years (Seed, 1976). Although there is virtually no information on the evolution of the genus it is generally accepted that *M. edulis* is the ancestral species from which the other forms have evolved. *M. galloprovincialis* is believed to have originated in the Mediterranean during one of the Pleistocene ice ages when the area was cut off from the Atlantic Ocean (Barsotti and Meluzzi, 1968). The warmer conditions which developed in the Mediterranean, together with the reduced contact between the Atlantic and the Mediterranean, probably favoured the process of differentiation. The present-day distribution of the *M. galloprovincialis* form in western Europe represents a northward expansion of its range since the Pleistocene era.

If this form did indeed originate in the Mediterranean how then can its present distribution outside of western Europe be explained? It is believed that *M. galloprovincialis* may have been accidentally introduced into Japan (Wilkins et al., 1983), Hong Kong (Lee and Morton, 1985), southern California (McDonald and Koehn, 1988) and South Africa (Grant and Cherry, 1985), because in each case populations of mussels in these areas bear a close morphological and electrophoretic similarity to *M. galloprovincialis* from the Mediterranean Sea. Surprisingly, however, the introduced populations show no evidence of a founder effect i.e. the level of genetic variability at the studied loci is just as great in the introduced populations as in Mediterranean mussels. This suggests that the introductions must have been large, or of a continuous nature, or both. Wilkins et al. (1983) have postulated that *M. galloprovincialis* was probably introduced to Japan about 60 years ago, since there are no references to it in the Japanese biological literature prior to the 1930s. In the case of the Hong Kong introduction, Lee and Morton (1985) believe that it occurred between 1981 and 1982, while in South Africa the introduction of *M. galloprovincialis* likely occurred in the 1960s (Grant and Cherry, 1985). McDonald and Koehn (1988) suggest that *M. galloprovincialis* was probably present in California at the turn of this century. Recent evidence of a more widespread distribution of *M. galloprovincialis* in the southern hemisphere i.e. in Australia, Tasmania and New Zealand (McDonald et al., 1991), suggests that this mussel may in fact be native to the southern hemisphere. Supporting evidence comes from the discovery of *edulis*-like fossils in all three regions (McDonald et al., 1991 and references therein). However, Grant and Cherry (1985) found no evidence of *Mytilus* fossils from South African shell-middens or raised-beach deposits dating from 10,000 and 120,000 years respectively. Koehn (1991) suggests that, since *M. galloprovincialis* is widely distributed in the southern Pacific, introductions into northern Pacific sites may have arrived via this route, rather than from the Mediterranean area. The close genetic similarity between what are believed to be introductions, and Mediterranean *M. galloprovincialis*, together with the often quite large genetic disimilarities between southern Pacific and Mediterranean *M. galloprovincialis* (see Chapter 7 for details) would argue against this. Also, if *M. galloprovincialis* is native to the southern Pacific why has it not been found in the Kerguelen Islands or South America? Alternatively, of course, it is puzzling, in view of its frequent trade in the past with the Mediterranean, that *M. galloprovincialis* was not introduced into South America. Indeed, it is striking (Fig. 1.6) that with the single exception of California, there is a very marked east-west divide in the distribution of *M. galloprovincialis*. The dilemma remains. Perhaps mtDNA data analysis, now in progress on samples from the northern and southern hemisphere (Koehn, 1991), will provide the answer.

   M. trossulus is found on both sides of the Pacific and Atlantic oceans, but is generally confined to more northerly latitudes than either *M. edulis* or *M. galloprovincialis*. Because of this, Varvio et al. (1988) have suggested a relatively ancient (1–2my), northern origin for this lineage. The Pleistocene era consisted of at least four glacial periods with interglacial periods of varying duration. At different stages during the Pleistocene ice covered Canada and parts of the U.S.A., northern Europe and Asia. As a glacial stage sets in and the climate cools a specie's favoured zone shifts equatorwards. However, among the large populations of a species left behind in the cold there will be some genotypes that can tolerate reduced temperatures and these could be the progenitors of local races persisting in higher latitudes. This may explain the origin of the *trossulus* form and why its distribution is confined to regions just south of what were once Pleistocene ice-covered areas. Koehn (1991) has suggested that *M. trossulus* may be a zoogeographical remnant i.e. that the populations we are now seeing are the remnants of what was once a much more widely distributed mussel. Morphological analysis of fossil *Mytilus* shells, along the lines of McDonald et al. (1991) may provide the answer.

M. trossulus is exceptional in that it is found in both fully marine (35°/oo) and very low salinity (6–7°/oo) habitats (Fig. 1.6). Its presence in the Baltic Sea is a relatively recent event. To colonize the Baltic *M. trossulus* must have been present in the Kattegat area when the marine connection with the Baltic—then a freshwater lake—occurred some 7000 years ago (Winterhalter et al., 1981). Why it is no longer present in northern Europe is intriguing, in view of the pattern of distribution of *M. trossulus* in other northern latitudes.

ACKNOWLEDGEMENTS

I wish to thank Andy Beaumont and Drs. Raymond Seed and John McDonald for helpful comments on earlier drafts of this chapter. I am grateful to Dr. John McDonald for permission to use figures 2, 4a and 4b from McDonald et al. (1991).

REFERENCES

Avise, J.C. 1974. The systematic value of electrophoretic data. Syst. Zool., 23: 465-481.

Barsotti, G. and Meluzzi, C., 1968. Osservazioni su *Mytilus edulis* L. e *Mytilus galloprovincialis* Lamarck. Conchiglie (Milan), 4: 50-58.

Beaumont, A.R., Seed, R. and Garcia-Martinez, P., 1989. Electrophoretic and morphometric criteria for the identification of the mussels *Mytilus edulis* and *M. galloprovincialis*. In: J. Ryland and P.A.Tyler (Editors), Proc. 23rd Eur. Mar. Biol. Symp., Swansea, U.K., 1988. Olsen and Olsen, Fredensborg, Denmark, pp. 251-258.

Blot, M., Thiriot-Quiévreux, C. and Soyer, J., 1988. Genetic relationships among populations of *Mytilus desolationis* from Kerguelen, *M. edulis* from the North Atlantic and *M. galloprovincialis* from the Mediterranean. Mar. Ecol. Prog. Ser., 44: 239-247.

Brock, V. 1985. Immuno-electrophoretic studies of genetic relations between populations of *Mytilus edulis* and *M. galloprovincialis* from the Mediterranean, Baltic, east and west Atlantic, and east Pacific. In: P.E. Gibbs (Editor), Proc. 19th Eur. Mar. Biol. Symp., Plymouth, England, 1984. Cambridge University Press, Cambridge, pp. 515-520.

Bulnheim, H.-P. and Gosling, E.M., 1988. Population genetic structure of mussels from the Baltic Sea. Helgoländer. Wiss. Meeresunters., 42: 113-129.

Conrad, T.A., 1837. Description of new marine shells from upper California, collected by Thomas Nuttall, Esq. J. Acad. Nat. Sci. (Philad.), 7: 227-268.

Dodge, H., 1952. A historical review of the mollusks of Linnaeus. Part I. The classes Loricata and Pelecypoda. Am. Mus. Hist. Bull., 100(1): 1-263.

Edwards, C.A. and Skibinski, D.O.F., 1987. Genetic variation of mitochondrial DNA in mussel (*Mytilus edulis* and *M. galloprovincialis*) populations from South West England and South Wales. Mar. Biol., 94: 547-556.

Ferguson, A., 1980. Biochemical Systematics and Evolution. Blackie, Glasgow, 194pp.

Ferson, S., Rohlf, F.J. and Koehn, R.K. 1985. Measuring shape variation of two-dimensional outlines. Syst. Zool., 34(1): 59-68.

Fleming, C.A., 1959. Notes on New Zealand Recent and Tertiary mussels (Mytilidae). Trans. R. Soc. N. Z., 87: 165-178.

Gosling, E.M., 1984. The systematic status of *Mytilus galloprovincialis* in western Europe: a review. Malacologia, 25(2): 551-568.

Gosling, E.M. and Wilkins, N.P., 1977. Phosphoglucosisomerase allele frequency data in *Mytilus edulis* from Irish coastal sites: its ecological implications. In: B.F. Keegan, P. O Céidigh and P. S. Boaden (Editors), Proc. 11th Eur. Mar. Biol. Symp., Galway, Ireland, 1976. Pergamon Press, London, pp. 297-309.

Gosling, E.M. and Wilkins, N.P., 1981. Ecological genetics of the mussels *Mytilus edulis* and *M. galloprovincialis* on Irish coasts. Mar. Ecol. Prog. Ser., 4: 221-227.

Gould, A.A., 1850. Shells from the United States Exploring Expedition. Proc. Boston Soc. Nat. Hist., 3: 343-348.

Gould, A.A., 1861. Descriptions of shells collected by the North Pacific Exploring Expedition. Proc. Boston Soc. Nat. Hist., 8: 14-40.

Grant, W.S. and Cherry M.I., 1985. *Mytilus galloprovincialis* Lmk. in southern Africa. J. Exp. Mar. Biol. Ecol., 90: 179-191.

Grant, W.S., Cherry, M.I. and Lombard, A.T. 1984. A cryptic species of *Mytilus* (Mollusca: Bivalvia) on the west coast of South Africa. S. Afr. J. Mar. Sci., 2: 149-162.

Hepper, B.T., 1957. Notes on *Mytilus galloprovincialis* Lamarck in Great Britain. J. Mar. Biol. Ass. U.K., 36: 33-40.

Hodgson, A.N. and Bernard, R.T.F., 1986. Observations on the ultrastructure of the spermatozoon of two Mytilids from the south-west coast of England. J. Mar. Biol. Ass. U.K., 66: 385-390.

Hupé, H., 1854. Moluscos. In: C. Gay (Editor), Historia fisica y politica de Chile. Zoologia, Vol. 8. C. Gay, Paris, pp. 1-407.

Johannesson, K., Kautsky, N. and Tedengren, M., 1990. Genotypic and phenotypic differences between Baltic and North Sea populations of *Mytilus edulis* evaluated through reciprocal transplantations. II. Genetic variation. Mar. Ecol. Prog. Ser., 59: 211-219.

Jukes-Brown, A.J., 1905. A review of the genera of the family Mytilidae. Malacol. Soc. Lond. Proc., 6: 211-224.

Koehn, R.K., 1991. The genetics and taxonomy of species in the genus *Mytilus*. Aquaculture, 94 (2/3): 125-145.

Koehn, R.K., Hall, J.G., Innes, D.J. and Zera, A.J., 1984. Genetic differentiation of *Mytilus edulis* in eastern North America. Mar. Biol., 79: 117-126.

Lamarck, J.B.P.A. de, 1819. Histoire naturelle des animaux sans vertèbres. Vol. 6. A.S.B. Verdiere Libraire, Paris.

Lamy, E., 1936. Révision des Mytilidae vivants du Muséum national d'Histoire naturelle de Paris. J. Conchy., 80: 66-363.

Lee, S.Y. and Morton, B.S., 1985. The introduction of the mediterranaean mussel *Mytilus galloprovincialis* into Hong Kong. Malacol. Rev., 18: 107-109.

Lewis, J.R. and Powell, H.T., 1961. The occurrence of curved and ungulate forms of the mussel *Mytilus edulis* (L.) in the British Isles, and their relationship to *Mytilus galloprovincialis* (Lmk.). Proc. Zool. Soc. Lond., 137: 583-598.

Lewis, J.R. and Seed, R., 1969. Morphological variations in *Mytilus* from S.W. England in relation to the occurrence of *Mytilus galloprovincialis* (Lmk.). Cah. Biol. Mar., 10: 231-253.

Linnaeus, C., 1758. Systema naturae per regna tria naturae. 10th Edition, Vol. 1. Regnum animale. Laurentii Salvi, Stockholm, 1384pp.

Lischke, C.E., 1868. Diagnosen neuer Meeres-Konchylien von Japan. Malakozool. Blätter 15: 218-222.

Mayr, E., 1970. Populations, Species and Evolution. Belknap Press, Cambridge, 476pp.

McDonald, J.H. and Koehn, R.K., 1988. The mussels *Mytilus galloprovincialis* and *M. trossulus* on the Pacific coast of North America. Mar. Biol., 99: 111-118.

McDonald, J.H., Koehn, R.K., Balakirev, E.S., Manchenko, G.P., Pudovkin, A.I., Sergiyevskii, S.O. and Krutovskii, K.V. 1990. Species identity of the "common mussel" inhabiting the Asiatic coasts of the Pacific Ocean. Biol. Morya, 1990 (1): 13-22.

McDonald, J.H., Seed, R. and Koehn, R.K., 1991. Allozyme and morphometric characters of three species of Mytilus in the Northern and Southern hemispheres. Mar. Biol., 111: 323-335.

Murphy, R.W., Sites, J.W. Jr., Buth, D.G. and Haufler, C.H., 1990. In: D. M. Hillis and C. Moritz (Editors), Molecular Systematics. Sinauer Associates Inc., Massachusetts, U.S.A., pp.45-126.

Orbigny, A. d', 1846. Mollusques lamellibranches. In: A. d'Orbigny (Editor), Voyage dans l'Amérique Méridionale 5(3). Bertrand and Levrault, Paris, pp. 489-758.

Powell, A.W.B., 1958. New Zealand molluscan systematics with descriptions of new species, Part 3. Rec. Auckland Inst. Mus., 5: 87-91.

Sanjuan, A., Quesada, H., Zapata, C. and Alvarez, G., 1990. On the occurrence of Mytilus galloprovincialis Lmk. on NW coasts of the Iberian Peninsula. J. Exp. Mar. Biol. Ecol., 143: 1-14.

Scarlato, O.A. 1981. Bivalve mollusks of temperate latitudes of the western portion of the Pacific Ocean. Opred. Faune SSSR, 126: 1-461.

Scarlato, O.A. and Starobogatov, Y.I., 1979. The systematic position and distribution of mussels. In: O. Scarlato (Editor), Commercial bivalve molluscan mussels and their role in the ecosystem (In Russian). Zoological Institute of the Soviet Academy of Sciences, pp. 106-111.

Seed, R., 1968. Factors influencing shell shape in the mussel Mytilus edulis. J. Mar. Biol. Ass. U.K., 48: 561-584.

Seed, R., 1971. A physiological and biochemical approach to the taxonomy of Mytilus edulis (L.) and M. galloprovincialis (Lmk.) from S.W. England. Cah. Biol. Mar., 12: 291-322.

Seed, R., 1972. Morphological variations in Mytilus from the French coasts in relation to the occurrence and distribution of Mytilus galloprovincialis (Lmk.). Cah. Biol. Mar., 13: 357-384.

Seed, R., 1974. Morphological variations in Mytilus from the Irish coasts in relation to the occurrence and distribution of Mytilus galloprovincialis (Lmk.). Cah. Biol. Mar., 15: 1-25.

Seed, R., 1976. Ecology. In: B.L. Bayne (Editor), Marine Mussels: their ecology and physiology. Cambridge University Press, Cambridge, pp. 13-56.

Seed, R., 1978. The systematics and evolution of Mytilus galloprovincialis (Lmk.). In: B. Battaglia and J.A. Beardmore (Editors), Marine Organisms: Genetics, Ecology and Evolution. Plenum Press, London, pp. 447-468.

Skibinski, D.O.F. and Beardmore, J.A., 1979. A genetic study of intergradation between Mytilus edulis and M. galloprovincialis. Experientia, 35: 1442-1444.

Skibinski, D.O.F., 1985. Mitochondrial DNA variation in Mytilus edulis L. and the Padstow mussel. J. Exp. Mar. Biol. Ecol., 92: 251-258.

Skibinski, D.O.F. and Beardmore, J.A., 1979. A genetic study of intergradation between Mytilus edulis and M. galloprovincialis. Experientia, 35: 1442-1444.

Skibinski, D.O.F., Beardmore, J.A. and Cross, T.F., 1983. Aspects of the population genetics of Mytilus (Mytilidae: Molluscs) in the British Isles. Biol. J. Linn. Soc., 19: 137-183.

Soot-Ryen, T., 1955. A report on the family Mytilidae (Pelecypoda). Allan Hancock Pacif. Exped., 20: 1-175.

Soot-Ryen, T., 1969. Family Mytilidae Rafinesque 1815. In R.C. Moore (Editor), Treatise on Invertebrate Paleontology. Part N, Vol. 1, Mollusca 6, Bivalvia. Geological Society of America and University of Kansas Press, Lawrence, pp. N271-N280.

Suchanek, T.H., 1985. Mussels and their role in structuring rocky shore communities. In: P.G. Moore and R. Seed (Editors), Ecology of Rocky Coasts. Hodder and Stoughton, Sevenoaks, Kent, pp. 70-96.

Tedengren, M., Andre, C., Johannesson, K. and Kautsky, N., 1990. Genotypic and phenotypic differences between Baltic and North Sea populations of Mytilus edulis evaluated through reciprocal transplantations. III. Physiology. Mar. Ecol. Prog. Ser., 59: 221-227.

Väinölä, R. and Hvilsom, M.M., 1991. Genetic divergence and a hybrid zone between Baltic and North Sea Mytilus populations (Mytilidae; Mollusca). Biol. J. Linn. Soc., 43(2): 127-148.

Varvio, S.-L., Koehn, R.K. and Väinölä, R., 1988. Evolutionary genetics of the Mytilus edulis complex in the North Atlantic region. Mar. Biol., 98: 51-60.

Verduin, A., 1979. Conchological evidence for the separate specific identity of Mytilus edulis L. and M. galloprovincialis Lamarck. Basteria, 43: 61-80.

Vermeij, G.J., 1989. Geographical restriction as a guide to the causes of extinction: the case of the cold northern oceans during the Neocene. Paleobiology, 15: 335-356.

Wilkins, N.P., Fujino, K. and Gosling, E.M., 1983. The Mediterranean mussel *Mytilus galloprovincialis* Lmk. in Japan. Biol. J. Linn. Soc., 20: 365-374.

Winterhalter, B., Feodén, T., Ignatius, H., Axberg, S. and Niemiströ, L. 1981. Geology of the Baltic Sea. In: A. Voipio (Editor), The Baltic Sea. Elsevier Science Publishers, B.V., Amsterdam, pp. 1-121.

Chapter 2

THE EVOLUTION AND SUCCESS OF THE HETEROMYARIAN FORM IN THE MYTILOIDA

BRIAN MORTON

INTRODUCTION

One of the most important trends in the evolution of the Bivalvia—accounting for much of the variation seen in extant and fossil species—has been the evolution of the triangular heteromyarian and circular monomyarian forms, with concomitant reduction and loss of the anterior adductor muscle, respectively.

Yonge (1953) studied the monomyarian condition in the Bivalvia but dealt only superficially with the process of heteromyarianization which must have preceded it. Early researchers, e.g. Jackson (1890), Anthony (1905) and Pelseneer (1906), recognized that the heteromyarian form was associated with the habit of byssal attachment, and Yonge (1953) demonstrated how ventral attachment has promoted the greater development of the posterior, and reduction of the anterior regions of the shell. Anthony (1905) noted, however, that not all heteromyarian bivalves are exactly alike, and divided the process of heteromyarianization into two phases: 'modiolization', where the shell is less acutely triangular, followed in some only by a second phase of 'mytilization', where a pronounced triangular form is developed.

Until relatively recently, e.g. Owen (1955), it was usual for malacologists to regard all heteromyarians as somehow related and to group them in the 'Anisomyaria'. Such an artificial scheme of classification led to confusion among some workers, who found phylogenetic significance in convergent characters. Nowhere is this better illustrated than with the Dreissenoidea and Mytiloidea, two unrelated superfamilies, both heteromyarian, but both of which Purchon and Brown (1969) considered allies, although Purchon (1987b) now recognizes their separate origins. Similarly, *Fluviolanatus* was placed in the Mytiloidea (Soot-Ryen, 1969; Waller, 1978), but probably belongs with the Arcticoidea (Morton, 1982a). Later classification schemes (Newell, 1965, 1969) have abandoned the misleading Anisomyaria, and a more meaningful approach to bivalve systematics has shown that the heteromyarian condition is of wide occurrence in the Bivalvia.

Stanley (1970) described modifications to a heteromyarian mode of life and subsequently (Stanley, 1972) elaborated on his earlier conclusions to demonstrate that both heteromyarian and monomyarian conditions result from the neotenous retention of the byssus by the various bivalve phylogenies in which they have evolved. He argued that the heteromyarian form in the Mytiloidea must have evolved, initially, in an endobyssate ancestor, cited species of *Modiolus* and *Brachidontes* as representative of such a condition, and traced an evolutionary gradient from endobyssate, modioliform genera to epibyssate, mytiliform genera. Morphological changes accompanying this trend are also seen, on a narrower scale, within the genus *Mytilus*, and Seed (1980a) has used such information to suggest that *Mytilus galloprovincialis* is of more recent origin than *Mytilus edulis*. I had arrived at the same conclusion as Stanley (1970, 1972) with regard to the freshwater Dreissenoidea (Morton, 1970), i.e. that modern epibyssate species could be derived from a more isomyarian ancestor, via the intermediary of various endobyssate taxa, e.g. *Dreissenomya* and *Congeria*. Representatives of the Mytiloidea and Dreissenoidea are gregarious and dominate marine, estuarine and freshwater environments in situations, virtually worldwide, where a byssus allows secure attachment. They attest to the success of the heteromyarian form.

This study re-examines evidence for the evolution of the heteromyarian form in the Mytiloidea, basing itself upon a study of *M. edulis* but drawing on a wealth of morphological information available for other representatives of the superfamily. The neotenous retention of the larval byssus into adult life was crucial to the evolution of the heteromyarian form but, it will be argued, other more fundamental decisions concerning the flow of water into and out of the mantle cavity must have prevised such an event(s); and thus, the success of the triangular form has its basis in the deepest history of the Bivalvia.

THE PRIMITIVE BIVALVE

Although there is debate as to whether it is a member of the Bivalvia at all (Yochelson, 1981), it is widely accepted that *Fordilla troyensis* is the oldest known bivalve mollusc (Pojeta et al., 1973) belonging to the Cycloconchoidea. These first molluscs arose in the early Cambrian and possessed a bivalved shell, a simple ligament, bivalve-like pedal muscle insertions and well-developed adductor muscle insertions. There was no shell gape; when the adductors contracted, the valve margins closed tightly (Runnegar and Pojeta, 1974). Such an early bivalve is believed to have given rise to the Babinkidae in the Ordovician (Pojeta and Runnegar, 1974; Runnegar and Pojeta, 1974), which according to McAlester (1965, 1966) was the ancestor of the modern Lucinoidea. A second Early Cambrian fossil, *Pojetaia runnegari*, from Australia, apparently possesses hinge teeth which *Fordilla* does not (Jell, 1980). Waller (1978) has shown, however, that the fossil shells of *Cycloconcha*

(Upper Ordovician) and *Actinodonta* (Silurian) have a dentition which contained certain primary cardinal teeth homologous to the teeth of the primitive heterodonts, i.e. the Crassatelloidea and Lucinoidea. Despite arguments over the status of *Fordilla* and *Pojetaia*, therefore, it is clear that by the Ordovician all of the major bivalve lineages had evolved, and these included a number of heteromyarian families, such as the Modiomorphidae, Cyrtodontidae and the Ambonychiidae (Pojeta, 1971; Pojeta and Runnegar, 1974).

By the late Palaeozoic, however, many of these bivalves had died out to be replaced in the Mesozoic by a great array of 'modern' bivalves, including numerous heteromyarian and monomyarian groups in the Caenozoic. The fact of heteromyarianism is, thus, that it is not a contemporary, advanced, phenomenon, nor is it restricted to a few, closely related, groups (the Anisomyaria), but is an ancient and widespread morphological form evolved in numerous bivalve lineages (Fig. 2.1).

Few clues are available from the fossil record to be definite about the mode of life of primitive bivalves but studies of extant representatives of, by consensus, ancient lineages have revealed much. Relevant studies include those of Allen (1958), Allen and Turner (1970) and Morton (1979) on the Lucinoidea, Allen (1968) on *Crassinella* (Crassatelloidea), Atkins (1937a), Thomas (1978) and Oliver and Allen (1980a, b) on the Arcoidea, Owen (1961) and Reid (1980) on *Solemya* (Cryptodonta), Yonge (1939) and Stasek (1965) on various protobranchs and Morton (1980a) on *Pholadomya* (Pholadomyoida).

Many contemporary authors (Atkins, 1937a; Yonge, 1953; Purchon, 1968), look to the equivalve and equilateral limopsid *Glycymeris glycymeris* for clues as to the mode of life and functional morphology of an early lamellibranch bivalve. Such a generalized primitive bivalve is illustrated in Figure 2.2A. It is equivalve and equilateral with a taxodont hinge and multivincular ligament. Anterior and posterior adductor muscles are of approximately the same size, and the animal is thus a typical isomyarian bivalve. There is a strong digging foot, no byssus (in the adult) and the ctenidia have rejection tracts located in the ventral marginal grooves of each demibranch (Atkins, 1937b; Type B). Living partially buried in sublittoral deposits, *Glycymeris* is able to orientate itself in a number of ways, although Atkins (1937a) observed that it usually lies on its side, with the posterior region pointed towards the soil-water interface. In general terms, therefore, *Glycymeris* possesses two inhalant apertures to the mantle cavity and a single exhalant aperture, although the whole of the ventral mantle margin is unfused and, in theory, water may flow in at any point. Such a pattern of circulation is seen in virtually all representatives of the Arcoida (Atkins, 1937a; Lim, 1966; Oliver and Allen, 1980a, b; Morton, 1982b).

In a number of other 'primitive' bivalve lineages, most notably the Nuculoida, Solemyoida, Lucinoidea (Allen, 1958), Crassatelloidea (Allen, 1968) and the heterodont Galeommatoidea, e.g. *Chlamydoconcha* (Morton, 1981b), an anterior in

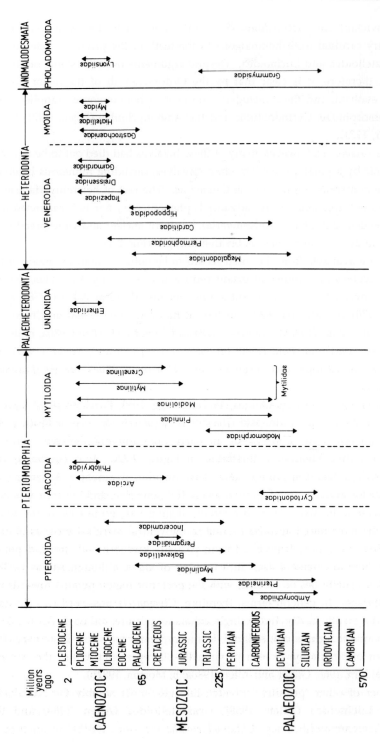

Fig. 2.1. The orders and families of bivalves possessing byssate heteromyarian representatives.

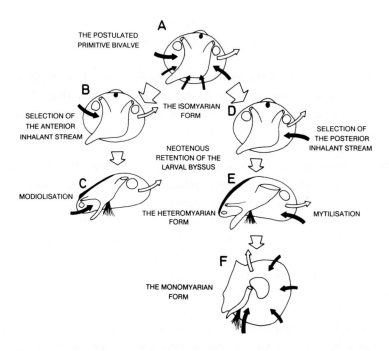

Fig. 2.2. The evolution of the heteromyarian form (and ultimately monomyarian form) from an isomyarian ancestor. (A) A postulated primitive isomyarian bivalve, e.g. representatives of the Nuculoida and Arcoida, with water capable of entering the mantle cavity from anterior and posterior directions. (B) Selection of the anterior inhalant stream by representatives of such groups as the Lucinoidea and Galeommatoidea can only result in the process of heteromyarianization leading to (C) a modioliform shell, e.g. representatives of the Arcoida. (D) Selection of the posterior inhalant stream, however, by such groups as the Mytiloidea, can result in the full expression of the heteromyarian form (E) and ultimately, the monomyarian form (F), as in representatives of the Pterioidea.

-halant aperture is retained (Fig. 2.2B). Water flow through the mantle cavity is thus in an anterior-posterior direction. Few, if any, of these bivalves are capable of evolving the heteromyarian form, since the retention and enhancement of the anterior inhalant component has led them into highly specialized, often deep-burrowing, life styles. Even though a byssus has been retained in some, allowing adoption of an epibyssate lifesyle, e.g. the Galeommatoidea and Arcoida, such a flow configuration precludes the evolution of the heteromyarian and hence the monomyarian forms, although some representatives of the Arcoida, e.g. *Barbatia*, are modioliform (Fig. 2.2C). Exceptions to this rule include the apparently monomyarian solemyid *Nucinella serrei* (Allen and Sanders, 1969). This animal is, however, unusual in that it is the posterior adductor that is lost. Similarly, *Chlamydoconcha orcutti* (Galeommatoidea) possesses only an anterior adductor (Morton, 1981b).

Philobrya munita (Arcoidea) is another exception, the possession of a posterior inhalant stream resulting, uniquely, from the loss of the anterior adductor (Morton, 1978).

The remaining bivalve lineages, however, have retained, and enhanced, the posterior component of the hitherto double inhalant stream (Fig. 2.2D). As a consequence, they have been able to exploit and enjoy the benefits of a wider range of form and lifestyles and, ultimately, to evolve the heteromyarian and monomyarian forms (Fig. 2.2E and F).

Within the Bivalvia there is thus the potential for either: (1) both anterior and posterior inhalant streams; (2) an anterior inhalant stream or (3), a posterior inhalant stream. The exhalant stream is always posterior. Onto the suite of primitive bivalves with the above-defined entrances and exits to and from the mantle cavity, the phenomenon of neoteny was to act to retain into the adult form the primitive larval attachment organ—the byssus. In the third of the above groups only, however, has the retention of the byssus had a powerful impact upon the mode of life, and thus the evolution, of the various bivalve phylogenies in which this water-circulating configuration has been adopted.

The successful settlement of the bivalve veliger larva is facilitated by the byssus, which Yonge (1962) has shown to have a wide distribution in the Bivalvia. The byssus, a series of chitinous threads secreted by a byssal gland in the foot, secures purchase, allowing the metamorphosing veliger to establish itself in a niche favoured for adult life. The evolution of the byssus must be seen as a consequence of caenogenesis and as such, has high adaptive significance. Evolved initially as a larval character, therefore, the byssus has subsequently been retained into adult life by some bivalve phylogenies (Yonge, 1962) and assumed a new significance. The retention of the byssus must also be seen as a recurrent event, since its occurrence is widespread among otherwise relatively unrelated extinct and extant bivalve lineages (Fig. 2.1). In general terms, moreover, the byssus and byssal musculature is the only larval feature that has been retained, and its retention must therefore be regarded as a consequence of partial neoteny. Neoteny, according to de Beer (1958), is of high phylogenetic significance occurring at the family level and above, which explains why the subsequent evolution of the heteromyarian form is most evident in only a few superfamilies, e.g. the Mytiloidea and Dressenoidea, among otherwise notably isomyarian orders.

Neoteny often results in the pronouncement of one organ system at the expense of others. Thus, strongly byssate bivalves do not possess a powerful digging foot; rather it is reduced and used largely as a plantar of byssal threads. In the oysters (Ostreiidae) there is extreme reduction of the foot, the shell is cemented to the substratum and the byssus is restricted to the metamorphosing larva. The byssus, retained subsequent to early phylogenetic decisions, favouring either an anterior or a posterior inhalant stream, has thus had a profound effect upon bivalve evolution. The most important effects, however, have concerned themselves with the evolution of the heteromyarian form in only some bivalve phylogenies.

FUNCTIONAL MORPHOLOGY OF *MYTILUS*

Mytilus edulis is representative of a typical heteromyarian bivalve. The anatomy has been described in detail by White (1937) and therefore, this study will investigate only those anatomical features that relate to the heteromyarian form, and update descriptive terminology not developed when White was working.

The Shell

The mytilid shell has a microstructure which is either two-layered, aragonitic and wholly nacreous in warm water representatives, or two- or three-layered aragonite and calcite in temperate species; *Mytilus* conforms to the latter plan (Taylor et al., 1969; see also p.67 Chapter 3). Often called the 'blue mussel', the species is especially well-known for its colour polymorphism (Seed, 1976), varying from brown to blue-black, with sometimes a radial colour pattern also evident. Innes and Haley (1977) present data which suggest that shell colour is determined by a single locus-two allele model, individuals homozygous for a recessive allele being blue. Newkirk (1980) indicates however, that there may be more than one locus involved in shell colour polymorphism, and both he and Mitton (1977) believe there to be a large environmental influence upon this character (see also p.368 Chapter 7). In *M. edulis*, growth rings are produced annually (Lutz, 1976). Richardson (1989) has, however, shown that microgrowth bands in the shell occur with tidal and daily periodicities, but that there is also an innate rhythm of shell deposition which is related to shell growth.

Mytilus edulis is mytiliform, with extreme reduction of the anterior but expansion of the posterior faces of the shell and ventral flattening (Fig. 2.3A). The greatest width is just ventral to the midpoint of the dorso-ventral axis of the shell (Fig. 2.3C, x-x), although the exact position varies with age. The form of *M. edulis* is, therefore, not so extreme as in other open-coast mytilids, e.g. *Septifer* (Yonge and Campbell, 1968), where ventral flattening results in a shell which is widest basally. Shell form in *M. edulis* is, however, highly variable, according to habitat and age (Seed, 1968), and some old individuals on open coasts may, therefore, match species of *Septifer* in terms of the degree of ventral flattening. Seen from the ventral surface (Fig. 2.3B), the shell margin of *M. edulis* is slightly sinusoidal.   The ligament (Fig. 2.3D), is large, planivincular, opisthodetic (i.e. posteriorly elongate), and has been described by Yonge and Campbell (1968) to comprise two layers: the outer ligament layer (Fig. 2.3E) and the posterior inner ligament layer all covered by periostracum.   Waller (1990) refers to these two ligament layers as lamellar and fibrous, respectively.

The posterior adductor muscle scar is large (Fig. 2.3E), as is that of the posterior byssal retractor muscle, which is also divided into two major components.   There is also a large posterior pedal retractor muscle scar located anterior to the byssal retractors but no anterior equivalent.   The anterior adductor muscle scar is small and

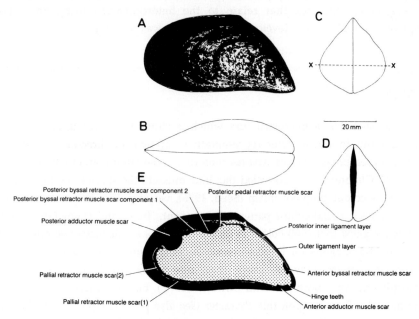

Fig. 2.3. *Mytilus edulis*. The shell as seen from (A) the right side; (B) ventrally; (C) posterior end-on with the greatest width marked by the dashed line, x–x; (D) anterior end-on, and (E) an interior view of the left shell valve.

located on the antero-ventral edge of the shell, just anterior to a weak hinge plate with approximately three denticulate hinge teeth.   The anterior byssal retractor muscle scar is located on the posterior face of the shell, below the ligament.   The pallial line is entire, with no pallial sinus, but is divided into inner and outer components, to be described later.

The Mantle

The mantle is polymorphic with respect to colour, varying from yellowish grey to chocolate brown. The mantle contains much of the gonad and the ventral mantle margin, seen in transverse section in Figure 2.4, comprises the usual three folds (Yonge, 1957, 1982): inner, middle and outer. The outer and middle folds are small, the former secreting the shell and periostracum, the latter muscular (Yonge, 1983). The inner fold is greatly enlarged, has an extensive haemocoel and can probably be inflated with blood. Between it and the general mantle surface is a deep, densely ciliated, rejectory tract for transporting pseudofaeces posteriorly. Beneath the epithelium of the inner fold is a layer of basiphilic gland cells, probably secreting mucus. Sensory papillae occur on this fold, particularly posteriorly. The pallial retractor muscle is divided into two components; the outer extends to the periostracal groove, and thus retracts the mantle margin between closing shell valves, the inner allows movement of the inner fold.

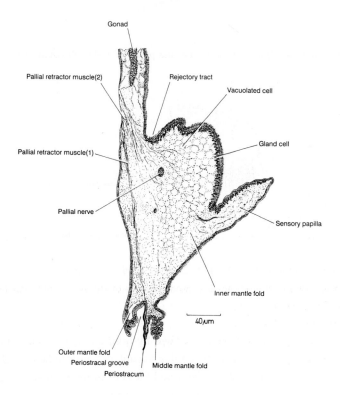

Fig. 2.4. *Mytilus edulis*. A transverse section through the left mantle margin.

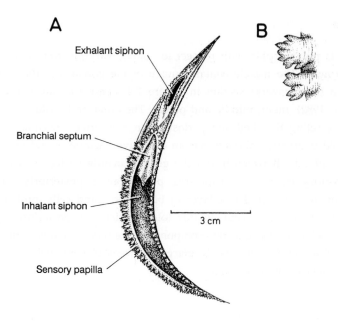

Fig. 2.5. *Mytilus edulis.* (A) The siphons as seen from the posterior right; (B) detail of two sensory papillae.

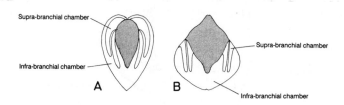

Fig. 2.6. Transverse sections through the body of (A) *Corbicula fluminea,* an isomyarian bivalve and (B) *Mytilus edulis,* a heteromyarian bivalve, showing the disposition of the supra- and infra-branchial chambers.

The periostracum of *M. edulis* has been described by Beedham (1958) who showed it to comprise three layers: a thin outer layer, a thick vacuolated middle layer and a thin internal layer. Bubel (1973a, b, c, d, e) has described the cells of the periostracal groove; the inner surface of the outer mantle fold; the outer, shell-secreting, surface of the outer fold, and the mechanism of periostracum repair.

Mantle fusion occurs only between the inhalant and exhalant apertures and is of the inner folds only (Type A; Yonge, 1957, 1982). The exhalant siphon is small and conical (Fig. 2.5A), the inhalant aperture extensive and possessing sensory papillae, typical of many other mytilids (Soot-Ryen, 1955), e.g. *Xenostrobus* (Wilson, 1967). The papillae are illustrated in greater detail in Figure 2.5B. A branchial septum occludes the upper region of the inhalant aperture.

The Ctenidia

Unlike the majority of bivalves, where a larger proportion of the mantle cavity is situated lateral to the visceral mass, e.g. *Corbicula fluminea* (Fig. 2.6A), that of the Mytiloidea, including *M. edulis*, is located largely beneath the body in the expanded ventral component of the shell (Fig. 2.6B).

The gills, more correctly referred to as ctenidia, are typical of the Mytiloidea in that they are flat, homorhabdic, nonplicate and filibranch. The ctenidia function in both respiration and feeding. Each ctenidium comprises a pair of demibranchs, inner and outer (Fig. 2.7A), which divide the pallial cavity into inhalant (infrabranchial) and exhalant (suprabranchial) chambers. Each demibranch comprises two lamellae (ascending and descending), which are held together by connective tissue junctions. The lamellae are made up of ciliated filaments through which branchial blood vessels pass. Water is driven from the inhalant to the exhalant chamber by lateral cilia arranged along the sides of the filaments; rows of laterofrontal cirri filter the water and transfer particles onto the apex of the filament, where they are transported by frontal cilia along orally-directed food acceptance tracts located in the ventral marginal food grooves of both demibranchs, the ctenidial axis and in the junctions of the ascending lamellae of the inner and outer demibranchs with the mantle and visceral mass, respectively. The ciliation (Type B(1), Atkins, 1937b)) of the ctenidial filaments of *M. edulis* has been described by Owen (1974), while Aiello and Sleigh (1972) have described the pattern of beating of the lateral cilia which create the inhalant flow through the ctenidia. Jones et al. (1990) have recently reviewed the literature on ctenidial pumping in *M. edulis*, while Bayne et al. (1976) have given a detailed account of ctenidial strucure and function. The most recent author to review bivalve filter feeding is Jørgensen (1991).

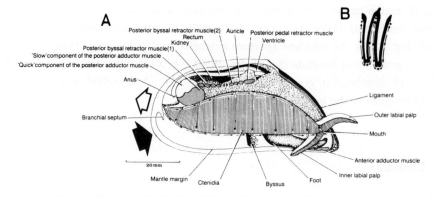

Fig. 2.7. *Mytilus edulis*. (A) The organs and ciliary currents of the mantle cavity as seen from the right side and (B) a diagrammatic section through the right ctenidium showing the ciliary currents and the food acceptance tracts (•).

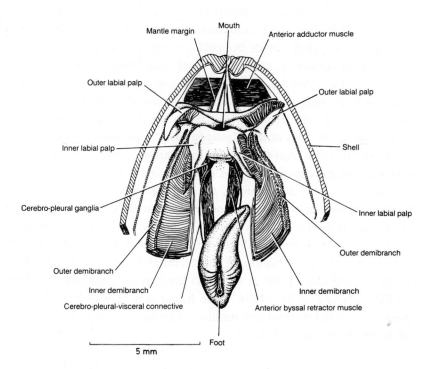

Fig. 2.8. *Mytilus edulis*. The organs of the anterior extremity of the mantle cavity as seen from the ventral surface after cutting away shell and mantle. (After Yonge and Campbell, 1968).

The Labial Palps

The ctenidial terminus of M. *edulis* ends between muscular labial palps which are relatively large and long (Fig. 2.7A), reaching posteriorly for some one third of the length of the mantle cavity. The mytilid ctenidial-labial palp junction is of Category 1 (Stasek, 1963), in that the ventral tips of the anterior filaments of the inner demibranch only are inserted unfused into a distal oral groove (Foster-Smith, 1974). The dorsal edges of the palps are united with either the mantle or the visceral mass.

In the case of the outer palps, union is with the mantle (Fig. 2.8), while in the case of the inner palps union is with the visceral mass at a point just below where the palp attaches to the ascending lamella of the inner demibranch. The sorting area of each palp is restricted to a row of ridges along the inner ventral margin (Figs. 2.7 and 2.8). Accepted particles pass over the crests of the ridges towards the mouth. Some larger particles are directed into channels between adjacent ridges and are then transported to the ventral edge of the palp. Such material is picked up by strong rejectory currents on the smooth surfaces of the inner and the outer faces of the palps and passed ventrally and posteriorly towards their tips. Much of the material arriving at the ctenidial terminus is therefore rejected, although resorting currents on the palp ridges allow for a high degree of control over the amounts either accepted or rejected; this depends on both the particulate content of the incoming water and metabolic requirements. Foster-Smith (1974) and Bayne et al. (1976) have described in more detail the functioning of the labial palps in M. *edulis*.

The Foot, Visceral Mass and Musculature

The foot (Fig. 2.9) is of the typical mytilid form, i.e. long, highly mobile and plantar. At rest, it projects into the anteriormost reaches of the mantle cavity, a small hook-like distal swelling, positioning it behind the anterior lip of the mouth. The foot, as in most bivalves, bears few ciliary tracts. The dorsal regions of the foot and the visceral mass, however, bear powerful ciliary currents, which pass unwanted material postero-dorsally and then postero-ventrally to the posterior edge of the visceral mass, where it falls onto the mantle below.

The ciliary currents of the mantle are similarly rejectory. On the general surface of each lobe, material is passed postero-ventrally to the ventral mantle margin where it accumulates in a deep rejection tract (Fig. 2.4A). Such material, in the form of a mucus-bound pseudofaecal string, is transported posteriorly towards the inhalant aperture, where it is passed dorsally and eventually rejected from the dorsal edge of the inhalant aperture, as is typical of the Mytiloidea (Morton, 1973b).

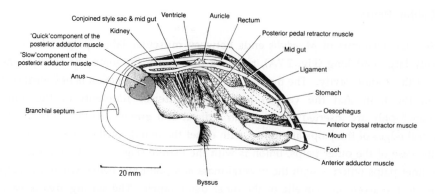

Fig. 2.9. *Mytilus edulis.* The musculature and organs of the visceral mass as seen from the right side.

The musculature of the visceral mass (Fig. 2.9) comprises the greatly enlarged and subdivided posterior byssal retractors which form two major blocks (Fig. 2.7), directly above the byssus and anterior to the posterior adductor muscle and which, in turn, is divided into 'quick' striated and 'slow' nonstriated components, for phasic and sustained adduction, respectively. Anterior to the byssal retractors is a posterior pedal retractor muscle, which unites with the byssal muscles by a number of connecting fibres. There are no anterior pedal retractor muscles, but there is a pair of anterior byssal retractor muscles, which are inserted on the posterior face of the shell beneath the ligament. The anterior byssal retractors are thus separated from the anterior adductor muscle, which is located on the antero-ventral extremity of the shell.

The Alimentary Canal

The visceral mass contains the gut. The mouth (Figs. 2.8 and 2.9) is located between outer and inner labial palps, and gives rise to a long oesophagus (Fig. 2.9) that opens into the anterior face of the stomach. The combined style sac and midgut leaves from the postero-ventral edge of the stomach and passes postero-dorsally between left and right blocks of the posterior pedal and byssal retractor muscles. Here the style sac terminates, and the now separated midgut returns on a parallel course, arches around the stomach to again turn postero-dorsally and penetrate the ventricle of the heart. It then passes over the posterior adductor muscle and terminates on its posterior face in an anus.

The stomach of *M. edulis* has been described by Graham (1949) and belongs to type III (Purchon, 1987a)(Fig. 2.10). The crystalline style projects dorsally into the stomach, revolves and apically dissolves, releasing enzymes, against a small chitinous gastric shield. On the left side of the stomach, and receiving a ventral spur of the gastric

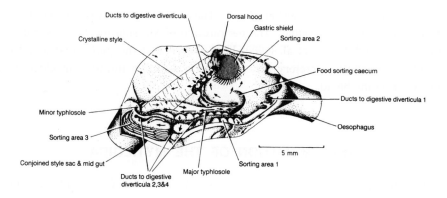

Fig. 2.10. *Mytilus edulis*. The structure and ciliary currents of the stomach as seen from the right side. (After Graham, 1949).

shield, is the left pouch into which open three or four ducts to the digestive diverticula, the site of intracellular digestion. From this pouch arises the major typhlosole which extends anteriorly and enters a large food-sorting caecum (Graham, 1949; Reid, 1965). The major typhlosole leaves this caecum and passes to the right side of the stomach, where it enlarges and, with the rejectory intestinal groove, enters the conjoined style sac and midgut. The small minor typhlosole similarly passes into the conjoined midgut and style sac from the posterior edge of the stomach. There are four openings to the digestive diverticula on the right side of the stomach, all separated from each other.

The stomach plays a final role in determining which potential food material is eventually accepted for intracellular digestion in the digestive diverticula. This is effected by three ciliated sorting areas, located close to the left pouch, food-sorting caecum and the apertures to the digestive diverticula on the right side of the stomach, respectively.

The functioning of the bivalve stomach was last reviewed by Morton (1983). The structure and functioning of the digestive diverticula have been reported upon by Owen (1955), Sumner (1969), Langton (1975) and Robinson et al. (1981), and Bayne et al. (1976) have reviewed digestion in *M. edulis*.

The Pericardium

The pericardium (Figs. 2.7 and 2.9) situated just below the dorsal edge of the shell, contains the heart, comprising a single ventricle and lateral auricles containing elements of the brown pericardial gland. The kidneys are located posterior to the pericardium above the intestine. The pericardium of *M. edulis* has been described by White (1942), and Moore et al. (1980) have shown that the pericardial gland cells possess an exocytotic excretory function, and a selective filtration mechanism for materials from the blood.

THE EVOLUTION OF THE MYTILOIDA

The Pteriomorphia includes six orders, i.e. the Arcoida, Trigonioida, Pterioida, Limoida, Ostreoida and Mytiloida (Morton, 1990), although the subclass itself may be polyphyletic particularly, for example, with regard to the Arcoida.

The extinct Modiomorphidae (Lower Ordovician-Lower Permian) possessed an opisthodetic ligament and were weakly heteromyarian, and thus modioliform. The posterior adductor muscle was larger than the anterior, although the anterior region of the shell was somewhat inflated so that the umbones were subterminal. The hinge teeth were either reduced (*Modiomorpha*) or lacking (*Byssodesma*). Pojeta and Runnegar (1974) regard the Modiomorphidae as ancestral to the Mytilidae. In *Liromytilus*, the posterior adductor occupies more than half of the posterior face of the shell. The Modiomorphidae were descendants of the most primitive pteriomorphs, the Cyrtodontidae (Pojeta, 1971; 1978; Pojeta and Runnegar, 1985), which themselves are derived from 'actinodontians' (Pojeta, 1978) or the Cycloconchoidea (Morris, 1979). If such a picture is real, then it is clear that the heteromyarian form had already appeared in the Palaeozoic, and that the subsequent radiation of the Mytiloida in the Mesozoic and Caenozoic, to dominate habitats occupied most effectively by byssally attached species, was forged in the earliest history of the order and, thus, in the deepest ancestry of the Bivalvia.

The Mytiloida

Amongst the Mytiloida, two superfamilies, the Mytiloidea and the Pinnoidea, possess heteromyarian representatives. According to Thiele (1935), the Pinnoidea represent a divergent pteriomorph lineage that has adopted an infaunal mode of life. Species of *Pinna* and *Atrina* (Fig. 2.11F) byssally attach themselves to small stones and sand grains and apparently move deeper into the sediment during development by pulling against the anchored byssus. To facilitate this, the anterior region of the shell has been both reduced and extended, resulting in a triangular form. The posterior region of the shell (fragile, easily broken, but also rapidly repaired) is greatly expanded, extending well beyond the pallial line and projects above the sediment surface. The posterior adductor muscle is large and located in the centre of the shell. The byssal retractor apparatus is dominated by the posterior element. In comparison, the anterior adductor and the anterior byssal retractor muscles are greatly reduced. The hinge line is straight and long with no teeth. Pinnids are also flattened laterally, presumably facilitating substrate penetration (Grave, 1911; Yonge, 1953). This is different from the laterally inflated epifaunal mytilids. I thus doubt that such a form could have evolved as an adaptation to life on rocky shores, and secondarily allowed a return to an endobyssate existence, as suggested by Yonge (1953, 1976). Rather, I believe that the distinctive pinnoid form to be a specific adaptation for an infaunal, endobyssate lifestyle and, though mirroring some adaptations seen in mytiloids with a similar lifestyle, to represent the result of an independant radiation.

The clearest examples of the heteromyarian form and the lifestyle which it facilitates are to be found in the Mytiloidea, and the success of this body form is reflected in the wide adaptive radiation achieved by the group. Contrary to general opinion, greatest mytilid diversity is not amongst the temperate epifaunal colonizers of the rocky intertidal, but amongst the infaunal/epifaunal occupants of soft, estuarine, tropical shores (Lee and Morton, 1985; Arnaud and Thomassin, 1990). An investigation of the Mytilidae and its three major subfamilies, the Crenellinae, Modiolinae and Mytilinae (ignoring the rock-boring Lithophaginae), gives the best insight into the evolution of the heteromyarian form (Fig. 2.11).

The Crenellinae
Crenella decussata, is an equivalve, almost equilateral, mytilid with strong marginal crenulations and an external sculpture of 50 or more radiating riblets (Fig. 2.11A). The adductor muscles are almost equal and thus isomyarian. *C. decussata* is a sublittoral colonizer of soft deposits (Soot-Ryen, 1955). *Crenella prideauxi* (Fig. 2.11B) hints at the evolution of the heteromyarian form in that the shell is markedly inequilateral with a flattened (antero-posteriorly) anterior border, effectively reducing the anterior volume of the shell, and with elongation of the posterior slope. Both species possess weak hinge plates with no teeth. Representatives of *Musculus* and *Musculista* show recognizable heteromyarian features. *Musculus discors* (Fig. 2.11C) (Soot-Ryen, 1955) is a colonizer of intertidal and subtidal algae and Merrill and Turner (1963) report that it encases itself in a byssal cocoon to protect brooded eggs. Marginal crenulations are distinct, except around the flattened antero-ventral border. The posterior crenulations and the few crenulations anterior to the umbones serve (in the absence of true hinge teeth) to align the valves correctly. The umbones are subterminal, and there is a slight inflation to the anterior slope of the shell. The ligament is opisthodetic. The reduced anterior and inflated posterior regions of the shell are connected by

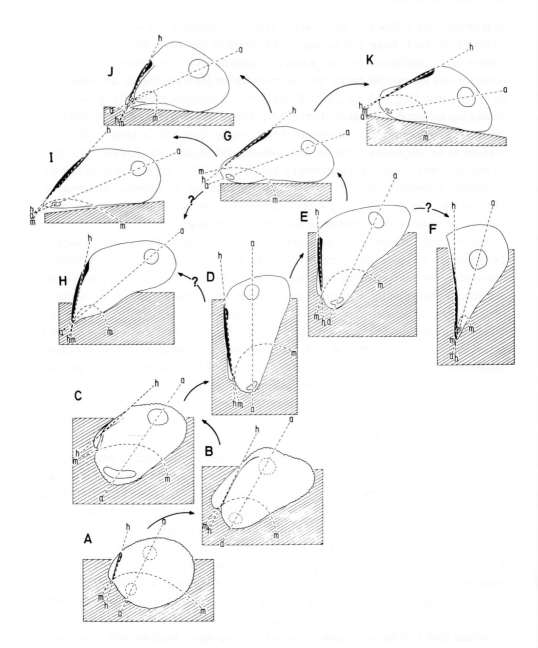

Fig. 2.11. Modern taxa of the Mytilacea used to illustrate the changes that must have occurred to obtain the heteromyarian form from an isomyarian one. (A) *Crenella decussata*, (B) *Crenella prideauxi*, (C) *Musculus discors*, (D) *Musculista senhausia*, (E) *Modiolus metcalfei*, (F) *Atrina vexillum*, (G) *Xenostrobus inconstans*, (H) *Mytella falcata*, (I) *Mytilus edulis*, (J) *Septifer virgatus*, (K) *Limnoperna fortunei*. (a = axis through the adductor muscles; h = hinge axis; m = the mid dorso-ventral axis).

correspondingly reduced and enlarged adductor muscles. *Musculista senhausia* (Fig. 2.11D) occurs in dense intertidal beds on mud flats in South-east Asia (Morton, 1973b). This is facilitated by the byssus which is formed into a protective nest that surrounds the shell, except posteriorly. In *M. senhausia*, crenulations are largely restricted to the anterior portion of the shell, just below the umbones, but are also faintly discernible just posterior to the ligament. There is a greater difference in the size of the two adductor muscles. The posterior byssal retractor muscle is large and subdivided, while the anterior is small and inserted upon the shell just under the umbones.

The Modiolinae

In representatives of *Modiolus*, shell crenulations are absent; an indication of their existence possibly being the striped pattern on the shell of some, e.g. *Modiolus striatulus* (Morton, 1977b). Species of *Modiolus* are often infaunal inhabitants of soft flats, attaching themselves to either the roots of mangroves, e.g. *M. metcalfei* (Morton, 1977a), or sea grasses, e.g. *M.* (= *Geukensia*) *demissus* and *M. americanus* (Stanley, 1970). Others, e.g. *M. modiolus* (Stanley, 1970) and *M. virgatus*, attach either to sand grains or to stones embedded in the sediment and a more advanced heteromyarian form is apparent. *M. metcalfei* possesses an almost triangular shell (Fig. 2.11E); the umbones having moved further forward and the extent of ventral, antero-posterior flattening is greater, though the shell is widest approximately midway down the dorso-ventral axis. The opisthodetic ligament often extends along approximately half the length of the shell. The posterior adductor muscle is large with the posterior byssal retractor muscle extending anteriorly from it, and divided into two muscle blocks. Such an arrangement brings it above the byssus to make attachment more secure. The anterior adductor is much reduced and the anterior byssal retractors are located, in all representatives of this genus, directly under the umbones, which have in turn moved forwards and downwards to a more terminal position.

The Mytilinae

Species of the South American *Mytella*, e.g. *M. falcata* (Fig. 2.11H) (Soot-Ryen, 1955), are similar in shell form to species of *Mytilus* and yet live, shallowly buried, in intertidal and subtidal muds. *M. falcata* has nearly terminal umbones and a short rounded anterior margin. The posterior shell margin is inflated and the ventral margin indented so that the shell is more elongate than in species of *Mytilus*.

Though similar to *Modiolus* and to the South American *Mytella* (Soot-Ryen, 1955), representatives of *Brachidontes* and *Xenostrobus* (Fig. 2.11G) (Wilson, 1967) typically inhabit estuarine rocky or oyster-covered surfaces, e.g. *B. variabilis* in Hong Kong (Morton, 1988). *X. inconstans* attaches to stones embedded in mud (Wilson,

1967). In representatives of both genera, the anterior margin of the shell is inflated but there are no true hinge teeth. The umbones are subterminal, the shell is triangular and the opisthodetic ligament long. The posterior adductor muscle is not large and the anterior adductor muscle is well formed. The posterior byssal retractor muscle is divided in *X. inconstans*, but not in either *X. pulex* or *X. securis*, and the anterior byssal retractor is located either under the umbones, e.g. *X. inconstans*, or on the posterior slope of the shell, e.g. *B. erosus*, *X. pulex* and *X. securis*. In *Brachidontes*, e.g. *B. variabilis* and *B. erosus* (Morton, 1988, 1991), the shell is strongly ribbed and marginally the valves interlock to align them correctly.

In the uniquely freshwater genus *Limnoperna*, adaptations to life on wave beaten shores are mirrored and *L. fortunei* occurs in rivers, streams and reservoirs in Southeast Asia (Morton, 1973a). *L. fortunei* (Fig. 2.11K) possesses a thin shell with a slightly concave ventral border. The posterior adductor muscle is large, the anterior small. The posterior byssal retractor is divided and the anterior byssal retractor is located on the posterior slope of the shell. The shell is also widest ventrally.

The epifaunal colonizers of hard intertidal substrata have taken the process of heteromyarianization to an extreme, and in these bivalves an important feature of the shell is extreme ventral flattening so that once byssally attached, the animal is stable in the presence of strong currents and wave action. Many such bivalves, e.g. species of *Mytilus* (Fig. 2.11I), and *Septifer* (Fig. 2.11J), are gregarious colonizers of exposed shores. Ventral flattening also tends to be emphasized in older specimens of *Mytilus galloprovincialis* (Seed, 1978), suggesting that it does have adaptive, and therefore survival value. Associated with antero-ventral flattening, the umbones have moved forward and form beaks that extend beyond the anterior end of the shell. The shell is triangular and the ligament long. The ligament of *Septifer* has been described by Yonge and Campbell (1968). The anterior adductor is reduced to minute proportions and in representatives of *Perna*, i.e. *P. viridis*, it is absent (Morton, 1987). The posterior adductor and posterior byssal retractor muscles are large, and the latter is elongated and divided into a number of components located directly above the foot. The anterior byssal retractor muscle is small and is located on the posterior slope of the shell. The anterior adductor muscle of *Septifer* is located on a septum.

Three shell axes have been identified in Figure 2.11 to illustrate the fundamental differences in shell form between endobyssate *Crenella* and epibyssate *Septifer*. In the former, the hinge (h-h) and the adductor muscle (a-a) axes are virtually parallel, as would be expected in an approximately isomyarian bivalve. In the latter, the same axes subtend an angle, reflecting the extreme reduction of the anterior region of the shell. The umbonal mid-ventral axis (m-m) divides the shell into nearly equal anterior and posterior components in *Crenella*, but the anterior component in *Septifer* is minute. The other mytilids illustrated in Figure 2.11 broadly demonstrate how changes in these three shell axes generate the heteromyarian form from an isomyarian one.

DISCUSSION

In the Arcoidea and Limopsoidea there are two inhalant streams, posterior and anterior, with the rejection of pseudofaeces taking place via the posterior. By the selection of one or other of these inhalant streams, it is possible to explain the evolution of important bivalve phylogenies. The Nuculoidea, Lucinoidea and Galeommatoidea thus possess an anterior inhalant stream only. In none of these bivalves, therefore, is it possible for the heteromyarian form to evolve, even though, as in the Galeommatoidea, a byssus is often present in the adult and some representatives of this superfamily and of the Arcoidea, e.g. *Barbatia*, are modioliform. Where the posterior inhalant stream is selected, there is the potential for the evolution of the heteromyarian form since reduction of the anterior face of the shell is possible. Furthermore, the heteromyarian form can only evolve in bivalves in which the byssus is retained by the adult. The byssus, found in most, if not all, juvenile bivalves, has been retained by a process of partial neoteny. The byssus enables the newly-settled juvenile to maintain itself on a mobile substratum until metamorphosis is complete and the adult, usually burrowing, mode of life is pursued. Dislodgement infers a dislodging agency and it is therefore likely that early bivalve lineages evolved in shallow coastal waters influenced by wave action and the tides, or both. Neoteny is believed to facilitate an escape from specialization, a return to a simpler body plan and an abrupt change of evolutionary direction. Morton (1978) has suggested, for example, that neoteny may account for the evolution of the alivincular from the duplivincular ligament, notably in the Philobryidae. The neotenous retention of the byssus, however, creates a new morphological blueprint upon which selection can work to foster adaptive radiation upon a new habitat landscape.

The retention of the byssus effectively immobilizes a bivalve with two important concomitant effects. First the pedal musculature, normally conferring mobility to the foot, is either reduced or lost. This is seen in the Mytilidae, where the anterior pedal retractors are lost and the posterior pedal retractors, though retained, are progressively reduced in more acutely heteromyarian genera, e.g. *Septifer* (Yonge and Campbell, 1968). There is, however, enhanced development of byssal retractors. Second, the foot, originally used as an organ of locomotion and digging, principally facilitates the planting of byssal threads, although locomotion is still possible. The byssus/foot thus becomes the focal point of the byssate bivalve with all other modifications having reference to this fixed point. Yonge (1953), Stanley (1972) and Oliver and Allen (1980a) argue that a return to a burrowing mode of life from an epibyssate ancestor is possible. I have reservations regarding such a hypothesis since once a character is lost, it is lost forever if the genetic factors and conditions controlling its formation are irretrievably lost, or irrevocably changed. A substitute character may appear which fulfils the same function as the old character but it is always structurally distinct and easily recognizable. I believe it unlikely that an epibyssate heteromyarian bivalve provides the blueprint for a return to a burrowing mode of life with the concomitant need for the redeployment of a pedal, as opposed to a byssal apparatus, already, in part, lost in such bivalves anyway.

In nonbyssate bivalves, orientation and derived morphological architecture relate to the sediment-water interface. Yonge and Campbell (1968) have suggested that subsequent to the evolution of the byssus the anterior inhalant stream came to be relocated posteriorly and to lie adjacent to the exhalant aperture. As discussed earlier, there is no need to 'relocate' a hitherto anterior current because the primitive bivalve probably possessed both anterior and posterior inhalant streams. It is, however, generally true that the retention of the byssus did favour bivalves in which the posterior inhalant stream was enhanced relative to the anterior because this did not, as modern bivalves demonstrate, preclude the evolution of the heteromyarian form. Byssate bivalves which have retained a dependance upon the anterior inhalant stream have been led into an evolutionary cul-de-sac, where solitary nestling is the almost only alternative. This is amply demonstrated in representatives of the Arcoida which are typically solitary and the process of heteromyarianization has proceeded only to the stage of 'modiolization'. An exception to this is *Philobrya munita*, a neotenous 'monomyarian' philobryid (Morton, 1978).

A further factor favouring a posterior inhalant stream in the ancestors of modern heteromyarian bivalves must have been the colonization by early byssate adult bivalves of either stones, plants or shells embedded in soft substrata. With a reduced head it became more important for an attached sedentary bivalve, unable to avoid potential predators by burrowing, to have an increased awareness posteriorly and from where the threat might arise. This, however, will be true of any bivalve and in all modern species the sensory functions of the head have been taken over by the mantle. An explanation for the greater development of the posterior regions of the shell and mantle in heteromyarian bivalves, therefore, must be sought elsewhere.

It has been argued by Yonge (1976) that the heteromyarian form evolved on wave-beaten rocky shores where the reduction in the anterior, and the enlargement of the posterior faces of the shell, were adaptations reducing the resistance to water flow. An abiding feature of the most important and successful heteromyarians is their gregariousness. Representatives of the Dressenoidea, e.g. *Dreissena* and *Mytilopsis* (Morton, 1969, 1981a), and the Mytiloidea, e.g. *Limnoperna, Musculista, Modiolus, Brachidontes, Xenostrobus, Mytilus* and *Septifer* and (Wilson, 1967; Yonge and

Campbell, 1968; Morton, 1973a, b, 1977a, b, 1991), occur in dense colonies often forming a 'mat'. This is as true of infaunal species as it is of epifaunal ones, especially in the tropics. Endobyssate mytilids are, moreover, typically less heteromyarian than their rocky shore or freshwater allies, and the shell is typically more elongate antero-posteriorly, with the angle between the hinge line and the line connecting the two adductors reduced. The shell is more cylindrical, i.e. modioliform, rather than mytiliform. Moreover, since the posterior byssal retractor is located further posteriorly in these bivalves than in *Septifer*, for example, contraction of this muscle pulls the animal down into the sediment—albeit only slightly. Some retreat into a byssal nest, e.g. species of *Musculus* and *Musculista* (Merrill and Turner, 1963; Morton, 1973b). The most significant adaptation of endobyssate mytilids is thus a shell in the form of a narrow elongate cone that allows large numbers of individuals to live together in dense beds. A good example of this shell form is seen in representatives of *Mytella*, e.g. *M. speciosa* (Soot-Ryen, 1955). Similarly, the extinct dreissenids, *Congeria* and *Dreissenomorpha*, were only slightly heteromyarian and, moreover, possessed a pallial sinus indicating an infaunal habitat, but with byssal attachment (as evidenced by a byssal notch). In these genera, therefore, there was the possibility of withdrawing deeper into the sediment by contraction of the posterior byssal retractors and siphonal withdrawal (Morton, 1970). The posterior margin of the shell of such infaunal heteromyarians is, therefore, only slightly inflated, when compared with their epifaunal allies. The posteriorly elongate shell keeps the inhalant and exhalant orifices in these fixed bivalves above the accretive sediment water interface. Additionally, *Mytilus galloprovincialis* becomes progressively more elongate (relative to height) with increasing size and, thus, age, further suggesting an adaptive advantage for such a shape (Seed, 1978). In the Pinnidae there is the capability for rapid regeneration of the exposed posterior region of the shell when it is damaged. The pinnid shell is thin and brittle, but this is also a feature of infaunal mytilids, e.g. *Musculista, Modiolus* and *Arcuatula* (Morton, 1973b, 1977a, b, 1980b), and can be related to the greater degree of protection afforded by an attached, infaunal, mode of life, often byssally cocooned.

Shallow burrowers often possess a shell with a pronounced sculpture of either radial or concentric ridges (Stanley, 1970). This may either give strength to the shell or it may assist in the burrowing process. In the infaunal Mytiloidea, however, which may be considered early advocates of the heteromyarian form, there is a nice example of the progressive loss of the sculpture and a trend towards a smoother shell. Isomyarian mytilids, therefore, e.g. *Crenella*, have a pronounced radial sculpture whereas in modioliform genera, e.g. *Modiolus* and *Musculista*, there are shell crenulations, the remnants of a primitive sculpture, anteriorly and posteriorly. Thin-shelled infaunal mytilids, e.g. *Musculista* and *Arcuatula*, are also slightly crenulate but retain a radial patterning. In endobyssate species, a thin shell is possible because of

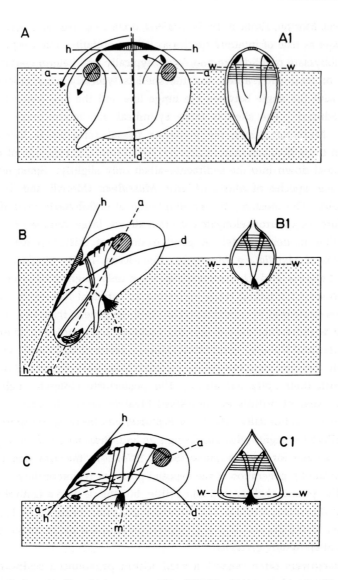

Fig. 2.12. Lateral views (A–C) and transverse sections (A1–C1) of (A) an isomyarian (B) a modioliform and (C) a mytiliform bivalve. (a = axis through the adductor muscles; d = mid dorso-ventral axis of the shell; h = hinge axis; m = the mid-dorsal and foot (byssus) axis and w = the position of greatest shell width).

the reduced need for a thick shell. Epibyssate mytilids have a much thicker shell and often, as in *Septifer, Hormomya* and *Brachidontes,* possess a distinct sculpture. In these bivalves a thicker shell is probably required to withstand the destructive (and erosive) effect of wave action. It may also serve to protect the inhabitant from shell-boring or chipping predatory snails, e.g. representatives of the Muricidae, crabs and birds (Griffiths, 1990). Representatives of the estuarine Dressenoidea also possess a thick shell, but this is significantly not the case in the freshwater mytilid *Limnoperna fortunei*, an occupant of lotic waters of larger scale.

The majority of infaunal bivalves are isomyarian (Fig. 2.12A). Infaunal heteromyarian bivalves (Fig. 2.12B) possess larger posterior adductor and posterior byssal retractor muscles and reduced anterior adductor and anterior byssal retractor muscles. This antero-posterior differential results from the enlargement of the posterior face of the shell. Most mytilids retain a pair of posterior, but not anterior pedal retractors. The byssus provided firm anchorage for the infaunal ancestors of today's epifaunal bivalves. To enhance stability around this point of attachment, the antero-ventral region of the shell valves are flattened, both laterally and antero-posteriorly. Flattening renders the animal more stable on wave-exposed shores or in lotic freshwaters, by countering an overturning force and by lowering the centre of gravity, thereby reducing the effects of drag. Such an adaptation is clearly associated with the full expression of heteromyarianism, and thus the successful colonization of dynamic, hard, aquatic habitats (Fig. 2.12C). Stanley (1970) has demonstrated that endobyssate modiolids are broadest at a point approximately half way down the dorso-ventral axis of the shell, whereas truly heteromyarian epibyssate mytilids are widest ventrally. This is especially evident in *Mytilus galloprovincialis*, in which the degree of ventral flattening increases with age (Seed, 1978).

Ventral flattening has also resulted in the umbones being developed into beaks which, with the progressive assumption of the heteromyarian form, assume a subterminal position beyond the reduced anterior edge of the shell. Ventral flattening has also been matched by hypermorphosis of the postero-dorsal regions of the shell, thus creating the triangular heteromyarian form. All heteromyarian mytilids possess a planivincular opisthodetic ligament. The evolution of this ligament structure has been from the amphidetic alivincular ligament of isomyarian bivalves, as an early consequence of the enlargement of the posterior, and reduction of the anterior faces of the shell. The great length of the ligament possibly serves to keep the shell valves correctly aligned, because an abiding feature of most heteromyarians is either the great reduction in the hinge plate and teeth (particularly anteriorly), or their loss. Sometimes, either simple denticles, e.g. in *Mytilus edulis* and species of *Brachidontes* (Morton, 1991), or marginally interlocking shell crenulations, e.g. in species of *Crenella* and *Musculista senhausia* (Morton, 1973b), function as hinge teeth, further aiding valve alignment.

Enlargement of the posterior region of the shell has also influenced the size and position of the adductor muscles. The posterior adductor is always large relative to the anterior, and the byssal retractors are similarly different. Often, the enlarged posterior byssal retractor is divided into subunits, as in *Mytilus edulis*. The reduced anterior adductor muscle varies in relative size according to the degree of heteromyarianism demonstrated by the bivalve concerned. With increasing anterior reduction of the shell, the anterior adductor is further reduced and in representatives of the genus *Perna* it has been lost. *Perna viridis* is a monomyarian which has retained the heteromyarian form (Morton, 1987). The anterior byssal retractor is always retained, but very much reduced, and with progressive heteromyarianization it loses contact with the anterior adductor and comes to lie first under the umbones, as in modiolids, and then to take up a position on the posterior face of the shell, under the ligament. A septum is found in *Septifer* (Yonge and Campbell, 1968), members of the Dressenoidea and in some philobryids (Morton, 1969; 1978). This is possibly the remnants of the hinge plate onto which the anterior adductor muscle has migrated. The anterior adductor muscle in these bivalves (but not *Philobrya*) thus extends between the left and right septa, giving a perpendicular attachment, which reduces the shearing force that must act upon any muscle contracting against its attachment surface in any other orientation.

In infaunal heteromyarian bivalves, the byssus and foot emerge from an antero-ventral position and, as noted by Stanley (1970), the elongate modiolid shell, with the posterior byssal retractor muscles located posteriorly, ensures that their contraction has the effect of pulling the bivalve down into the substratum. Conversely, the foreshortened triangular shell of epifaunal heteromyarians, with the posterior byssal retractor muscles located above the byssus, ensures that the animal is pulled down onto the substratum when this pair of muscles contract.

The changes in shell form and dimensions necessary to obtain a heteromyarian bivalve from an isomyarian one are illustrated in Figure 2.13, where reduction of the anterior and expansion of the posterior faces of the isomyarian shell (Fig. 2.13A), narrow the angles of the anterior and ventral shell demarcation lines (Fig. 2.13B) until, eventually, anterior dorso-ventral axes ($C-C_1$, $D-D_1$ and $E-E_1$) are in the same plane as ventral axes ($4-4_1$ and $5-5_1$) (Fig. 2.13D).

The line(s) of evolution which allowed colonization of rocks or plants embedded in soft deposits was, it is here contended, an essential prerequisite for the eventual successful colonization of rocky intertidal surfaces in the sea, estuaries and, ultimately, freshwater. It is significant that many estuarine mytilids have the ability, by variation in shell form, to adopt either endo- or epibyssate modes of life. One of the best examples is *Brachidontes erosus* on shores in Western Australia (Morton, 1991), where it assumes an elongate, transversely rounded, i.e. cylindrical, form infaunally, but is more triangular and ventrally flattened epifaunally. *Geukensia demissa* shows similar modifications to the shell in different components of its ecology (Seed, 1980b). Shell and habitat 'plasticity' are features of many mytilids and

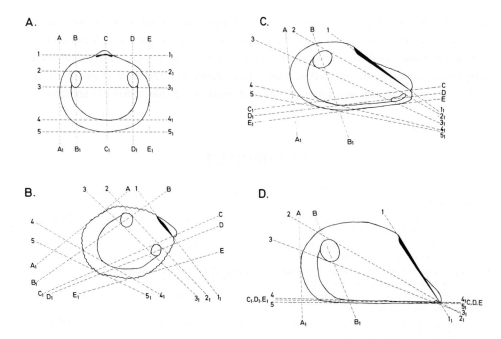

Fig. 2.13. Changes in the dorso-ventral (A–E) and anterior-posterior (1–5) shell parameters needed to produce a heteromyarian bivalve (D) from an isomyarian one (A). (A–A_1 = posterior shell edge; B–B_1 = dorso-ventral axis through the posterior adductor muscle; C–C_1 = mid dorso-ventral axis of the shell; D–D_1 = dorso-ventral axis through the anterior adductor muscle; E–E_1 = anterior shell edge; 1–1_1 = hinge axis; 2–2_1 and 3–3_1 = dorsal and ventral edges of the adductor muscles; 4–4_1 and 5–5_1 = ventral shell area beyond the pallial line).

possibly illustrate how the heteromyarian form originally evolved. The evolutionary processes hitherto discussed can be considered to have commenced in predominately infaunal lineages, so that genera such as the fossil *Dreissenomya* (Morton, 1970), and a wide range of extinct and extant mytilids, still possess features that adapt them for an endobyssate mode of life. These include a pallial sinus, an elongate, modioliform, shell, the greatest breadth mid dorso-ventrally, and a smooth, often thin, shell. Such bivalves were thus essentially preadapted for life on exposed surfaces, and an enhancement of the morphological adaptations seen in infaunal genera has led to the adoption of the true heteromyarian form and epibyssate lifestyle. Though flattened ventrally in the antero-posterior plane, early modioliform bivalves were still unstable in the lateral plane on exposed surfaces, when orientated side-on to the flow of water. Nevertheless, many species of mytilids and representatives of the Dressenoidea in freshwater are gregarious, and huge aggregations (as with their infaunal relatives) confer stability upon the population as a whole. Living in clusters, the expansion of the posterior shell face was selected for by a functional need to have free access, posteriorly, to the water above. The ventral flattening of the shell which has evolved in some of these bivalves—most notably representatives of the genus *Mytilus*—as an adaptation ensuring firmer attachment, has, however, released them from the gregarious mode of life, and some species of *Septifer*, e.g. *S. bilocularis*, are solitary.

REFERENCES

Aiello, E. and Sleigh, M.A., 1972. The metachronal wave of lateral cilia of *Mytilus edulis*. J. Cell. Biol., 54: 493-506.

Allen., J.A., 1958. On the basic form and adaptations to habitat in the Lucinacea (Eulamellibranchia). Philos. Trans. R. Soc. Lond., Ser. B., 241: 421-484.

Allen, J.A., 1968. The functional morphology of *Crassinella mactracea* (Linsley) (Bivalvia: Astartacea). Proc. Malacol. Soc. Lond., 38: 27-40.

Allen, J.A. and Sanders, H.L., 1969. *Nucinella serrei* Lamy (Bivalvia: Protobranchia), a monomyarian solemyid and possible living actinodont. Malacologia, 7: 381-396.

Allen, J.A. and Turner, J.F., 1970. The morphology of *Fimbria fimbriata* (Linné) (Bivalvia: Lucinidae). Pac. Sci., 24: 147-154.

Anthony, R., 1905. Influence de la fixation pleurothétique sur la morphologie des mollusques acédimyaires. Annals Sci. Nat. Zool. Biol. Anim., 1: 165-396.

Arnaud, P.M. and Thomassin, B.A., 1990. Habitats and morphological adaptations of mytilids (Mollusca: Bivalvia) from coastal and reefal environments in southwest Malagasia (Indian Ocean). In: B. Morton (Editor), The Bivalvia. Proceedings of a Memorial Symposium in Honour of Sir Charles Maurice Yonge, Edinburgh, 1986. Hong Kong University Press, Hong Kong, pp. 333-344.

Atkins, D., 1937a. On the ciliary mechanisms and interrelationships of lamellibranchs. I. New observations on sorting mechanisms. Q. J. Microsc. Sci., 79: 181-308.

Atkins, D., 1937b. On the ciliary mechanisms and interrelationships of lamellibranchs. III. Types of lamellibranch gills and their food currents. Q. J. Microsc. Sci., 79: 375-421.

Bayne, B.L., Widdows, J., and Thompson, R.J., 1976. Physiology. In: B.L. Bayne (Editor), Marine Mussels: their ecology and physiology. Cambridge University Press, Cambridge, pp. 121-206.

Beedham, G.E., 1958. Observations on the non-calcareous component of the shell of the Lamellibranchia. Q. J. Microsc. Sci., 99: 341-357.

Beer, G. de, 1958. Embryos and Ancestors. Oxford University Press, London, 197pp.

Bubel, A., 1973a. An electron-microscope study of periostracum formation in some marine bivalves. I. The origin of the periostracum. Mar. Biol., 20: 213-221.

Bubel, A., 1973b. An electron-microscope study of periostracum formation in some marine bivalves. II. The cells lining the periostracal groove. Mar. Biol., 20: 222-234.

Bubel, A., 1973c. An electron-microscope investigation into the distribution of polyphenols in the periostracum and cells of the inner face of the outer fold of *Mytilus edulis*. Mar. Biol., 23: 3-10.

Bubel, A., 1973d. An electron-microscope investigation of the cells lining the outer surface of the mantle in some marine molluscs. Mar. Biol., 21: 245-255.

Bubel, A., 1973e. An electron-microscope study of periostracum repair in *Mytilus edulis*. Mar. Biol., 20: 235-244.

Foster-Smith, R.L., 1974. A comparative study of the feeding mechanisms of *Mytilus edulis* (L.), *Cerastoderma edule* (L.) and *Venerupis pullastra* (Montagu) (Mollusca: Bivalvia). Ph. D. Thesis, University of Newcastle-on-Tyne, U.K.

Grave, B.H., 1911. Anatomy and physiology of the Wingshell *Atrina rigida*. Bull. Bur. Fish., Wash., 29: 409-439.

Graham, A., 1949. The molluscan stomach. Trans. R. Soc. Edinb., 61: 737-778.

Griffiths, C.L., 1990. Spatial gradients in predation pressure and their influence on the dynamics of two littoral bivalves. In: B. Morton (Editor), The Bivalvia. Proceedings of a Memorial Symposium in Honour of Sir Charles Maurice Yonge, Edinburgh, 1986. Hong Kong University Press, Hong Kong, pp. 321-332.

Innes, D.J. and Haley, L.E., 1977. Inheritance of a shell-color polymorphism in a mussel. J. Hered., 68: 203-204.

Jackson, R.T., 1890. Phylogeny of the Pelecypoda: the Aviculidae and their allies. Mem. Boston Soc. Nat. Hist., 4: 277-394.

Jell, P.A., 1980. Earliest known pelecypod on Earth—a new Early Cambrian genus from South Australia. Alcheringa, 4: 233-239.

Jones, H.D., Richards, O.G. and Hutchinson, S., 1990. The role of ctenidial abfrontal cilia in water pumping in Mytilus edulis L. J. Exp. Mar. Biol. Ecol., 143: 15-26.

Jørgensen, C.B., 1991. Bivalve filter feeding: hydrodynamics, bioenergetics, physiology and ecology. Olsen and Olsen, Fredensborg, 140pp.

Langton, R.W., 1975. Synchrony in the digestive diverticula of Mytilus edulis L. J. Mar. Biol. Ass. U.K., 55: 221-229.

Lee, S.Y. and Morton, B. 1985. The Hong Kong Mytilidae. In: B. Morton and D.Dudgeon (Editors), Proceedings of the Second International Workshop on the Malacofauna of Hong Kong and Southern China, Hong Kong, 1983. Hong Kong University Press, Hong Kong, pp. 49-76.

Lim, C.F., 1966. A comparative study on the ciliary feeding mechanisms of Anadara species from different habitats. Biol. Bull., 130: 106-117.

Lutz, R.A., 1976. Annual growth patterns in the inner shell of Mytilus edulis. J. Mar. Biol. Ass. U.K., 56: 723-731.

McAlester, A.L., 1965. Systematics, affinities, and life habits of Babinka, a transitional Ordovician lucinoid bivalve. Palaeontology (Lond.), 8: 231-246.

McAlester, A.L., 1966. Evolutionary and systematic implications of a traditional Ordovician lucinoid bivalve. Malacologia, 3: 433-439.

Merrill, A.S. and Turner, R.D., 1963. Nest building in the bivalve genera Musculus and Lima. Veliger, 6: 55-59.

Mitton, J.B., 1977. Shell colour and pattern variation in Mytilus edulis and its adaptive significance. Chesapeake Sci., 18: 387-390.

Moore, M.N., Bubel, A. and Lowe, D.M. 1980. Cytology and cytochemistry of the pericardial gland cells of Mytilus edulis and their lysosomal responses to injected horseradish peroxidase and anthracene. J. Mar. Biol. Ass. U.K., 60: 135-149.

Morris, N.J., 1979. On the origin of the Bivalvia. In: M.R. House (Editor), The origin of major invertebrate groups. Syst. Assoc. Spec. Vol., 12: 381-413.

Morton, B., 1969. Studies on the biology of Dreissena polymorpha Pall. I. General anatomy and morphology. Proc. Malacol. Soc. Lond., 38: 301-321.

Morton, B., 1970. The evolution of the heteromyarian condition in the Dreissenacea (Bivalvia). Palaeontology (Lond.), 13: 563-572.

Morton, B., 1973a. Some aspects of the biology and functional morphology of the organs of feeding and digestion of Limnoperna fortunei (Dunker) (Bivalvia: Mytilacea). Malacologia, 12: 265-281.

Morton, B., 1973b. Some aspects of the biology, population dynamics and functional morphology of Musculista senhausia Benson (Bivalvia: Mytilacea). Pac. Sci., 28: 19-33.

Morton, B., 1977a. The biology and functional morphology of Modiolus metcalfei Hanley 1844 (Bivalvia: Mytilacea) from the Singapore mangrove. Malacologia, 16: 500-518

Morton, B., 1977b. An estuarine bivalve (Modiolus striatulus) fouling raw water supply systems in West Bengal, India. J. Inst. Water Engrs. and Sci., 31: 441-452.

Morton, B., 1978. The biology and functional morphology of Philobrya munita (Bivalvia: Philobryidae). J. Zool. (Lond.), 185: 173-196.

Morton, B., 1979. The biology and functional morphology of Fimbria fimbriata (Linnaeus, 1758) (Bivalvia: Fimbriidae). Aust. Mus. Rec., 32: 389-420.

Morton, B., 1980a. The anatomy of the "living fossil" Pholadomya candida Sowerby 1823 (Mollusca: Bivalvia: Anomalodesmata). Vidensk. Medd. Dan. Naturhist. Foren., 142: 7-102.

Morton, B., 1980b. The biology and some aspects of the functional morphology of Arcuatula elegans (Mytilacea: Crenellinae). In: B. Morton (Editor), Proceedings of the First International Workshop on the Malacofauna of Hong Kong and southern China, Hong Kong, 1977. Hong Kong University Press, Hong Kong. pp. 331-345.

Morton, B., 1981a. The biology and functional morphology of *Mytilopsis sallei* (Recluz) (Bivalvia: Dreissenacea) fouling Visakhapatnam harbour, Andhra Pradesh, India. J. Molluscan Stud., 47: 25-42.

Morton, B., 1981b. The biology and functional morphology of *Chlamydoconcha orcutti* Dall with a discussion on the taxonomic status of the Chlamydoconchacea (Mollusca: Bivalvia). J. Zool. (Lond.), 195: 81-122.

Morton, B., 1982a. The biology, functional morphology and taxonomic status of *Fluviolanatus subtorta* (Bivalvia: Trapeziidae), a heteromyarian bivalve possessing "zooxanthellae". J. Malacol. Soc. Aust., 5: 113-140.

Morton, B., 1982b. Functional morphology of *Bathyarca pectunculoides* (Bivalvia: Arcacea) from a deep Norwegian fjord with a discussion of the mantle margin in the Arcoida. Sarsia, 67: 269-282.

Morton, B., 1983. Feeding and digestion in Bivalvia. In: A.S.M. Saleuddin and K.M. Wilbur (Editors), The Mollusca. Vol. 5. Physiology, Part 2. Academic Press, New York, pp. 65-147.

Morton, B., 1987. The functional morphology of the organs of the mantle cavity of *Perna viridis* (Linneaus, 1758) (Bivalvia: Mytilacea). Am. Malacol. Union (Inc.) Bull., 5: 159-164.

Morton, B., 1988. The population dynamics and reproductive cycle of *Brachidontes variabilis* (Bivalvia: Mytilidae) in a Hong Kong mangrove. Malacol. Rev., 21: 109-117.

Morton, B., 1990. Bivalvia. In: Encyclopaedia Brittanica. Vol. 24. Encyclopaedia Brittanica Inc., Chicago, pp. 321-327.

Morton, B., 1991. The biology of *Brachidontes erosus* (Bivalvia: Mytilidae) in Princess Royal Harbour, Albany, Western Australia. In: F.E. Wells, D.I. Walker, H. Kirkham and R. Lethbridge. (Editors), Proceedings of the Third International Marine Biological Workshop: The Marine Flora and Fauna of Albany, Western Australia, 1988. Western Australian Museum, Perth, pp. 693-712.

Newell, N.D., 1965. Classification of the Bivalvia. Am. Mus. Novit., No. 2206: 1-25.

Newell, N.D., 1969. Subclass Pteriomorphia Beurlen, 1944. In: R.C. Moore (Editor), Treatise on Invertebrate Paleontology. Part N, Vol. 1, Mollusca 6, Bivalvia. Geological Society of America and University of Kansas Press, Lawrence, pp. N248-N393.

Newkirk, G.F., 1980. Genetics of shell colour in *Mytilus edulis* L. and the association of growth rate with shell color. J. Exp. Mar. Biol. Ecol., 47: 89-94.

Oliver, G. and Allen, J.A., 1980a. The functional and adaptive morphology of the deep-sea species of the Arcacea (Mollusca: Bivalvia) from the Atlantic. Philos. Trans. R. Soc. Lond., Ser. B., 291: 45-76.

Oliver, G. and Allen, J.A., 1980b. The functional and adaptive morphology of the deep-sea species of the family Limopsidae (Bivalvia: Arcoida) from the Atlantic. Philos. Trans. R. Soc. Lond., Ser. B., 291: 77-125.

Owen, G., 1955. Observations on the stomach and digestive diverticula of the Lamellibranchia. I. The Anisomyaria and Eulamellibranchia. Q. J. Microsc. Sci., 96: 517-537.

Owen, G., 1961. A note on the habits and nutrition of *Solemya parkinsoni* (Protobranchia: Bivalvia). Q. J. Microsc. Sci., 102: 15-21.

Owen, G., 1972. Lysosomes, peroxisomes and bivalves. Sci. Prog., 60: 299-318.

Owen, G., 1974. Studies on the gill of *Mytilus edulis*: the eu-latero-frontal cilia. Proc. R. Soc. Lond., Ser. B., 187: 83-91.

Pelseneer, P., 1906. Mollusca. In: E.R. Lankester (Editor), A Treatise on Zoology. A. and C. Black, London, 533pp.

Pojeta, J., 1971. Review of Ordovician pelecypods. U. S. Geol. Survey Prof. Paper, 695: 1-46.

Pojeta, J., 1978. The origin and early taxonomic diversification of pelecypods. Philos. Trans. R. Soc. Lond., Ser. B., 284: 225-246.

Pojeta, J. and Runnegar B., 1974. *Fordilla troyensis* and the earliest history of pelecypod mollusks. Am. Sci., 62: 706-711.

Pojeta, J. and Runnegar, B., 1985. The early evolution of diasome molluscs. In: E.R. Trueman and M.R. Clarke (Editors), The Mollusca. Vol. 10, Evolution. Academic Press, Orlando, Florida, pp. 295-336.

Pojeta, J., Runnegar, B. and Kriz. J., 1973. *Fordilla troyensis* Barrande: the oldest known pelecypod. Science, 180: 866-868.

Purchon, R.D., 1968. The Biology of the Mollusca. Pergamon Press, Oxford, 560pp.

Purchon, R.D., 1987a. The stomach in the Bivalvia. Philos. Trans. R. Soc. Lond., Ser. B., 316: 183-276.

Purchon, R.D., 1987b. Classification and evolution of the Bivalvia: an analytical study. Philos. Trans. R. Soc. Lond., Ser. B., 316: 277-302.

Purchon, R.D. and Brown, D., 1969. Phylogenetic inter-relationships among families of bivalve molluscs. Malacologia, 9: 163-171.

Reid, R.G.B., The structure and function of the stomach in bivalve molluscs. J. Zool (Lond.), 147: 156-184.

Reid, R.G.B., 1980. Aspects of the biology of a gutless species of Solemya (Bivalvia: Protobranchia). Can. J. Zool., 58: 386-393.

Richardson, C.A., 1989. An analysis of the microgrowth bands in the shell of the common mussel Mytilus edulis. J. Mar. Biol. Ass. U.K., 69: 477-491.

Robinson, W.E., Pennington, M.R. and Langton, R.W. 1981. Variability of tubule types within the digestive glands of Mercenaria mercenaria L., Ostrea edulis L., and Mytilus edulis L. J. Exp. Mar. Biol. Ecol., 54: 265-276.

Runnegar, B. and Pojeta, J., 1974. Molluscan phylogeny: the paleontological viewpoint. Science. 1974: 311-317.

Seed, R., 1968. Factors influencing shell shape in Mytilus edulis L. J. Mar. Biol. Ass. U.K., 48: 561-584.

Seed, R., 1976. Ecology. In: B.L. Bayne (Editor), Marine mussels: their ecology and physiology. Cambridge University Press, Cambridge, pp.13-65.

Seed, R., 1978. The systematics and evolution of Mytilus galloprovincialis Lmk. In: J.A. Beardmore and B. Battaglia (Editors), Marine Organisms: Genetics, Ecology and Evolution. Plenum Press, New York, pp. 447-468.

Seed, R., 1980a. Shell growth and form in the Bivalvia. In: D.C. Rhoads and R.A. Lutz (Editors), Skeletal Growth of Aquatic Organisms. Plenum Press, New York, pp. 23-67.

Seed, R. 1980b. A note on the relationship between shell shape and life habits in Geukensia demissa and Brachidontes exustus (Mollusca: Bivalvia). J. Molluscan Stud., 46: 293-299.

Soot-Ryen, T., 1955. A report on the family Mytilidae. Allan Hancock Pacif. Exped., 20: 1-175.

Soot-Ryen, T., 1969. Superfamily Mytilacea Rafinesque, 1815. In: R.C. Moore (Editor), Treatise on Invertebrate Paleontology. Part N, Vol. 11, Mollusca 6, Bivalvia. The Geological Society of America and University of Kansas Press, Lawrence, pp. N271-N281.

Stanley, S.M., 1970. Relation of shell form to life habits in the Bivalvia (Mollusca). Geol. Soc. Amer. Mem., 125: 1-296.

Stanley, S.M., 1972. Functional morphology and evolution of byssally attached bivalve molluscs. J. Paleontol., 46: 165-212.

Stasek, C.R., 1963. Synopsis and discussion of the association of ctenidia and labial palps in the bivalved molluscs. Veliger, 6: 91-97.

Stasek, C.R., 1965. Feeding and particle-sorting in Yoldia ensifera (Bivalvia; Protobranchia) with notes on other nuculanids. Malacologia, 2: 349-366.

Sumner, A.T., 1969. The distribution of some hydrolytic enzymes in the cells of the digestive gland of certain lamellibranchs and gastropods. J. Zool. (Lond.), 158: 277-291.

Taylor, J.D., Kennedy, W.J. and Hall, A., 1969. The shell structure and mineralogy of the Bivalvia. Introduction. Nuculacea-Trigonacea. Bull. Br. Mus. (Nat. Hist.) Zool., 3: 1-125.

Thiele, J., 1935. Handbuch der systematischen Weichtierkunde. Jena, 2: 779-1154.

Thomas, R.D.K., 1978. Shell form and the ecological range of living and extinct Arcoida. Paleobiology, 4: 181-194.

Waldron, M., Packie, R.M. and Roberts, F.L., 1976. Pigment polymorphism in the blue mussel Mytilus edulis. Veliger, 19: 82-83.

Waller, T.R., 1978. Morphology, morphoclines and a new classification of the Pteriomorphia (Mollusca: Bivalvia). Philos. Trans. R. Soc. Lond., Ser. B., 284: 345-365.

Waller, T.R., 1990. The evolution of ligament systems in the Bivalvia. In: B. Morton (Editor), The Bivalvia. Proceedings of a Memorial Symposium in Honour of Sir Charles Maurice Yonge, Edinburgh, 1986. Hong Kong University Press, Hong Kong, pp. 49-71.

White, K.M., 1937. Mytilus. Liverpool Marine Biological Committee. Memoirs., 31: 1-117.

White, K.M., 1942. The pericardial cavity and the pericardial gland of the Lamellibranchia. Proc. Malacol. Soc. Lond., 25: 37-88.

Wilson, B.R., 1967. A new generic name for three recent and one fossil species of Mytilidae (Mollusca: Bivalvia) in Southern Australia with redescriptions of the species. Proc. Malacol. Soc. Lond., 37: 279-295.

Yochelson, E.L., 1981. *Fordilla troyensis* Barrande: "the oldest known pelecypod" may not be a pelecypod. J. Paleontol., 55: 113-125.

Yonge, C.M., 1939. The protobranchiate Mollusca: a functional interpretation of their structure and evolution. Philos. Trans. R. Soc. Lond., Ser. B., 230: 79-147.

Yonge, C.M., 1953. The monomyarian condition in the Lamellibranchia. Trans. R. Soc. Edinb., 62: 443-478.

Yonge, C.M., 1957. Mantle fusion in the Lamellibranchia. Pubbl. Stn. Zool. Napoli, 29: 151-171.

Yonge, C.M., 1962. On the primitive significance of the byssus in the Bivalvia and its effects in evolution. J. Mar. Biol. Ass. U.K., 42: 113-125.

Yonge, C.M., 1976. The 'mussel' form and habit. In: B.L. Bayne (Editor), Marine mussels: their ecology and physiology. Cambridge University Press, Cambridge, pp. 1-12.

Yonge, C.M., 1982. Mantle margins with a revision of siphonal types in the Bivalvia. J. Molluscan Stud., 48: 102-103.

Yonge, C.M., 1983. Symmetries and the role of the mantle margins in the bivalve Mollusca. Malacol. Rev., 16: 1-10.

Yonge, C.M. and Campbell, J.I., 1968. On the heteromyarian conditions in the Bivalvia with special reference to *Dreissena polymorpha* and certain Mytilacea. Trans. R. Soc. Edinb., 68: 21-43.

Chapter 3

ECOLOGY AND MORPHOLOGY OF LARVAL AND EARLY POSTLARVAL MUSSELS

RICHARD A. LUTZ AND MICHAEL J. KENNISH

INTRODUCTION

While the literature on *Mytilus* is extensive, most of the research on this genus has focused on the biology of adults, especially those of *Mytilus edulis* L. and *Mytilus galloprovincialis* Lmk, owing to their importance as food and as fouling organisms. Few detailed studies of the development and behaviour of early ontogenetic stages of *Mytilus* were conducted prior to the efforts of Bayne (1963, 1964a, b, 1965) in the 1960s. The reasons for the paucity of investigations on the early life history of this group, prior to the 1960s, are numerous, but certainly involve, in part, the inability of workers to correctly identify individual specimens of various bivalve species isolated from plankton samples, and problems of rearing larvae in the laboratory (Chanley and Andrews, 1971). Although data on larval mytilids are sparse, a survey of the literature on bivalve larvae indicates that veligers of *M. edulis* received more attention than those of most other bivalves (Loosanoff and Davis, 1963).

Descriptions of the reproduction, larval development, and settling behaviour of *M. edulis* date back to the early 1900s (Borisjak, 1909; Stafford, 1912). A comprehensive examination of the anatomy of this species by Field (1922) includes a brief discussion of larval characteristics. Additional studies of larval development of *Mytilus* were undertaken by Kändler (1926), Werner (1939), Thorson (1946), Rees (1950) and Chipperfield (1953). Savage (1956) reviewed some of this earlier work. Important developmental studies were carried out on *Mytilus* between 1955 and 1980 (Lubet, 1957; Bayne, 1963, 1964a, b, 1965, 1971, 1972, 1975, 1976; Loosanoff and Davis, 1963; Ockelmann, 1965; Loosanoff et al., 1966; Lubet and Le Gall, 1967; Seed, 1969, 1971, 1975, 1976; Le Gall, 1970; Bayne et al., 1975; de Schweinitz and Lutz, 1976; and Lutz and Hidu, 1979).

Since 1980, a number of studies have been undertaken to assess the endogenous and exogenous factors that affect the development and settlement of mytilid larvae (Bayne, 1983; Eyster and Pechenik, 1987). A major goal of laboratory experimentation has been to determine the relative significance of genetic and nongenetic components of larval growth and survival (Innes and Haley, 1977; Lannan, 1980). Among exogenous factors influencing larval development, temperature, salinity, and food have received the greatest attention (Riisgård et al., 1980; Jespersen and Olsen, 1982; Manahan et al., 1983; Sprung, 1984a, b, c). Evaluation of these, as well as other physicochemical factors, has been conducted both in the field (Fell and Balsamo, 1985; McGrath et al., 1988; King et al., 1989, 1990) and in the laboratory (Petersen, 1984; Eyster and Pechenik, 1987).

During the past 10 years, considerable effort has also been expended on delineating shell morphological features of mytilid larvae and postlarvae with a view towards identifying individual specimens isolated from planktonic and benthic samples, and to defining structures useful for making a variety of inferences in ecological and palaeoecological research (Jablonski and Lutz, 1980, 1983; Lutz et al., 1982; Ockelmann, 1983; Lutz, 1985a; Fuller and Lutz, 1988, 1989). Observation of the hinge apparatus, via scanning electron microscopy in particular, has facilitated identification of larval and postlarval mytilids at the species level.

Larvae of species within the genus *Mytilus*, other than those of *M. edulis*, have also received attention in the literature, albeit less frequently (Bayne, 1976). Zakhvatkina (1959) and Lubet (1973) focused their research on larvae of *M. galloprovincialis*, while Miyazaki (1935), Yoshida (1953), and Tanaka (1958) concentrated on the larval stages of *Mytilus crassitesta* (= *M. coruscus*). Yoo (1969) examined the early developmental stages of *Mytilus coruscus*. A more comprehensive treatment of the larval characteristics of the Mytilidae can be found in Chanley (1970).

This chapter provides an overview of the early life history of *Mytilus*, drawing heavily on studies conducted to date on the larval and early postlarval biology and ecology of *M. edulis*. It not only addresses fertilization and embryogenesis in mussels, but also reviews the existing knowledge of larval development, settlement, and metamorphosis. In addition, the chapter details those morphological structures of the shell which provide valuable aids for the identification of larval and postlarval stages.

FERTILIZATION, EMBRYOGENESIS AND LARVAL DEVELOPMENT

Mussels shed eggs and sperm directly from their genital ducts into the open water, where fertilization takes place. Females release oocytes in short, rod-like, yellow-orange masses that soon dissociate, whereas males discharge sperm in a steady stream (Strathmann, 1987). Hodgson and Bernard (1986) discuss the ultrastructure of the spermatozoan, while Humphries (1969) describes the fine structure of the egg. The acrosome reaction facilitates sperm penetration into the egg (Niijima and Dan, 1965). If an egg remains unfertilized for more than a few hours, its subsequent development may be arrested, particularly at unfavourable temperatures and salinities. Bayne

(1965) showed that fertilization of *M. edulis* occurs successfully at temperatures from 5 to 22°C and salinities between 15 and 40‰. *Mytilus californianus* tolerates a narrower salinity range for fertilization, and is adversely affected by salinities less than 25‰ (Young, 1941; Bayne, 1976).

Fertilized eggs usually range from 60–90μm in diameter, having a vitelline coat between 0.5 and 1.0μm thick. The initial cleavage division typically arises within one hour of fertilization (Lutz et al., 1991). Subsequent unequal, holoblastic spiral cleavage yields an embryo that begins to swim when cilia first appear after 4–5h. A ciliated trochophore stage is reached approximately 24–48h after fertilization. A larval shell begins to form shortly after the nonshelled trochophore stage. Originating from the thickened dorsal ectoderm, a shell gland secretes the first larval shell, termed the prodissoconch I (Bayne, 1976). At this point a larva possesses a straight-hinge and, in outline, exhibits a 'D'-shape with a shell length of 100 to 120μm (Jablonski and Lutz, 1980; Sprung, 1984b).

A second larval shell, the prodissoconch II, is secreted immediately subsequent to the secretion of the prodissoconch I. The mantle, rather than the shell gland, secretes the prodissoconch II, which has concentric growth lines (Millar, 1968) and other ornamentation that clearly distinguishes it from the prodissoconch I (La Barbera, 1974). Werner (1939) referred to this stage as the veliconcha—a stage which persists for several weeks (Sprung, 1984b) and is characterized by rapid larval growth from approximately 120 to 250μm in shell length, but with little increase in morphological complexity (Bayne, 1976). This planktotrophic stage is more commonly termed the 'veliger' stage. Veliger larvae have a relatively simple morphology, with a ciliated swimming organ (velum), a functional gut, and shell, providing the basic requirements for a pelagic larval phase with high dispersal capability.

The planktonic veliger actively swims by movement of the velum, a structure that first appears shortly after fertilization as a ring of long cilia around the apical plate and achieves a maximum size at a shell length of about 250–260μm. During the veliger stage, which typically persists for one to four weeks, mussel larvae actively feed in the water column, utilizing cilia on the velum to feed and swim. Gradually a rounded umbo forms in the dorsal region of the shell when the larvae attain a length of 140–150μm, marking the transition from the 'straight-hinge' to the 'umbo' stage of development. The umbo progressively extends from the hinge and shoulders in the shape of a low knob as the larvae grow larger than 210–230μm in length (Lutz, 1985b). Pigmented 'eye spots' can be seen at a shell length of 200–260μm, being most often observed when the shell attains a length of 220–230μm. However, some larvae as large as 245μm lack these structures. As metamorphosis is approached, the larvae develop a pedal organ or 'foot' which rapidly becomes functional in crawling. By the time the larvae attain lengths of 195–210μm, the foot can already be perceived in many veligers, and it becomes well-defined in mature larvae between 210 and 300μm in length. These latter individuals have the ability to actively extend the organ and at this stage of development, possessing both a velum and foot, are aptly known as 'pediveligers' capable of metamorphosing (Carriker, 1961). Bayne (1971) gives a detailed description of the gross morphology of the pediveliger larva, the life-history stage immediately preceding settlement and metamorphosis. It is a more complex phase of life than the veliconcha, and reflects the need of the individual to select a suitable substrate for a sessile existence. The general morphological features of the pediveliger larva of *M. edulis* may be summarized as follows (Bayne, 1971):

1. A large velum used in swimming and feeding
2. A foot used in crawling
3. A ciliated palp that sorts food particles
4. A mouth, oesophagus, stomach (with style sac and large digestive gland), and simple intestine
5. A thin mantle utilized in shell secretion
6. A nervous system comprised of cerebral, pedal, and visceral ganglia, together with a sensory system of statocysts, apical plate, and pigment spots
7. A few pairs of gill filaments
8. A byssus system capable of secreting simple byssal threads

If the pediveliger does not come in contact with a suitable substrate, it has the ability to delay metamorphosis for several weeks (up to 40 days at 10°C for *M. edulis*) (Bayne, 1965, 1976). During this time, the byssus system remains functional. Physical and chemical cues, together with the presence of a suitable substrate for settlement, trigger a crawling behaviour, culminating with the secretion of byssal threads by an associated series of glands in the foot of the pediveliger (Tamarin and Keller, 1972; Tamarin et al., 1974; Lane and Nott, 1975; Lane et al., 1985). The byssal threads anchor the mussel larva to the substrate, enabling the organism to commence a sessile phase of life. The entire process of crawling behaviour, secretion of byssal threads, and attachment of larvae to a substrate is termed 'settlement'. The first secretion of a byssus marks the termination of the pelagic larval life of a mussel and commencement of the process of 'metamorphosis'.

Shell deposited after metamorphosis (i.e. dissoconch) usually differs substantially from the prodissoconch in ornamentation, surface texture, microstructure, and, in the case of many species within the genus *Mytilus*, mineralogy. On the outer shell surface, an abrupt demarcation line lies at the prodissoconch-dissoconch boundary (Jablonski and Lutz, 1980). Postlarval mussels which have successfully completed settlement and metamorphosis are frequently referred to as 'plantigrades' (Bayne, 1976).

Factors Affecting Larval Development

The duration of the larval life of mussels typically ranges from about one to four weeks and is contingent upon temperature, salinity, available ration, and other factors (Bayne, 1976; Sprung, 1984a, b, c, d). While in the plankton, a larva requires a ration of approximately 30 to 60% of its own weight per day. Its growth efficiency is rather high, 60–70% net efficiency (Sprung, 1984d), and, during the larval period, an individual gains from about 0.1mg to 1.0mg in weight (Bayne, 1976). Because of the vagaries of environmental conditions, starvation, and predation by fish and invertebrates, mortality of mussel larvae approaches or exceeds 99% (Thorson, 1966; Mileikovsky, 1971; Purchon, 1977; Jørgensen, 1981).

Salinity and temperature

Cleavage and early embryonic stages of mussels have a limited tolerance to environmental change (Bayne, 1976). Hrs-Brenko (1973), conducting studies on the effects of temperature and salinity on embryonic development of M. edulis and M. galloprovincialis, recorded normal embryogenesis at 15–20°C, but not at 5 or 30°C. The lower salinity limit for normal embryogenesis ranged from 15–20°/oo at 15°C and from 20–25°/oo at 20°C, while the upper salinity limit for normal development at those temperatures was found to be between 30 and 35°/oo (Hrs-Brenko, 1973; Bayne, 1976). Bayne (1965) stated that development of the trochophore of M. edulis only proceeded normally at salinities between 30 and 40°/oo and at temperatures between 8 and 18°C. Strathmann (1987) has presented a developmental sequence from fertilization to metamorphosis at 9°C and 19–22°C for M. edulis (Table 3.1).

Hrs-Brenko (1974) indicated that the optimal temperatures and salinities for embryonic development of M. galloprovincialis are between 15 and 20°C and 27 and 40°/oo, respectively. According to her findings, temperatures below 15°C caused a delay in development to the straight-hinge stage by three or more days. The number of abnormal larvae increased substantially as the temperature and salinity deviated from the optimal range for embryonic development. At 25°C no embryonic development was apparent in any laboratory cultures of this species.

Hrs-Brenko and Calabrese (1969) investigated the combined effect of salinity and temperature on growth and survival of M. edulis larvae. They noted that the effects of these factors are significantly related only as the limits of tolerance to either factor are approached. At salinities between 15 and 40°/oo, survival of larvae was uniformly high, (70% or better) at temperatures between 5 and 20°C and was essentially 0% at 30°C. Similarly, larval growth rates were optimal at 20°C in cultures with salinities between 25 and 30°/oo, but decreased when temperatures of the cultures were lowered to 10°C or raised to 25°C. Lough (1974) and Bayne (1983) showed that larvae of M.

edulis grow normally over a wide salinity range, within which growth is dependent only on temperature.

Temperature, perhaps more than any other single factor, influences the duration of metamorphic delay (Strathmann, 1987). A strong negative correlation between temperature and maximum larval shell size has been documented for mytilids (Bayne, 1965; Lutz and Jablonski, 1978; Siddall, 1978), and has led to the suggestion that the morphometry of the prodissoconch II shell of many bivalves, including mytilids, could be employed as a relative or, in some cases, absolute paleotemperature indicator.

Food

The planktotrophic veliger larvae of *Mytilus* depend mainly on a ration of phytoplankton cells for successful growth and development (Bayne, 1983), although larval growth may be enhanced by the uptake of dissolved organic substances, especially dissolved amino acids, as well as the consumption of detritus and bacteria (Courtwright et al., 1971; Bayne, 1983; Manahan et al., 1983). Bayne (1965) successfully reared *Mytilus* larvae to metamorphosis using two algal species, *Isochrysis galbana* and *Pavlova lutheri*, with the best growth being obtained in larval cultures containing a mixture of the two algal species, rather than in those cultures containing only one species of phytoplankton. It has been noted, however, that extracellular metabolites released from algal cells may inhibit the growth of bivalve larvae (Wangersky, 1978). The inhibitory effects of such metabolites may account for Bayne's (1965) observation that the growth of *M. edulis* larvae fed with *Pavlova lutheri* cells (which support larval growth) was depressed when these cells were suspended in a medium from *Nannochloris atomus* cultures (a species which did not support larval growth).

After considerable laboratory testing of potential larval foods over the years, the nutritional values of various phytoplankton species have been determined. In general, phytoplankton devoid of a cell wall (naked flagellates) provide a better source of food than those forms with a cell wall (Bayne, 1976). Algal species found to support larval growth include *Chaetoceros calcitrans, Isochrysis galbana, Monochrysis lutheri, Tetraselmis suecica*, and *Thalassiosira pseudonana*. Other species are either a less reliable source of food for larval development (e.g. *Chlorella* spp. and *Phaeodactylum tricornutum*), or are not capable of supporting larval growth at all (e.g. *Olisthodiscus* sp.) (Bayne, 1983).

The larvae of *M. edulis* grow quite well on a diet composed of many different algal types (Bayne, 1976). However, both the size and concentration of the algal cells can influence larval growth rates. Phytoplankton cells must be small enough to be ingested by *Mytilus* larvae for growth to proceed successfully. *M. edulis* larvae filter the surrounding water most efficiently of food particles measuring 3–5µm in diameter

Table 3.1. Development sequence of *Mytilus edulis* to metamorphosis at 9 °C and 19–22 °C. D and CD cells refer to cells formed during the first spiral cleavage of a *M. edulis* embryo (for review see Rattenbury and Berg, 1954). (From Strathmann, 1987).

Temperature		Developmental Sequence
9 °C	19-22 °C	
0min	0min	Insemination
60min		First and second polar bodies emitted
		First polar lobe formation begins
	80min	First cleavage, trefoil stage
2.5h		2-cell stage, the CD cell is largest
3h		4-cell stage, the D cell is largest
8h		16- to 32-cell stage; 2d and 2D are almost of equal size, 2d forms a series of large posterior cells; 2D divides unequally, eventually producing 2 large macromeres.
	6h	Cilia appear, embryo begins to rotate.
	9–10h	Gastrulation by invagination produces a narrow tubular archenteron and small blastophore.
24h		Trochophores, uniformly ciliated except for an apical tuft, and fine nonmotile bristles on posterodorsal surface. A dorsal shell gland forms and stomodeum fuses with archenteron tip.
	22–26h	Velar cilia begin to develop.
	48h	Velar lobes develop on anterior end, apical tuft disappears, body elongates and becomes laterally compressed, shell becomes bivalved.
42h		Straight-hinge veliger stage
	72h	Velar lobes become reduced and foot appears.
66h		Prodissoconch II forming

(Jørgensen, 1981); their efficiency of retaining particles declines with decreasing or increasing particle sizes (Riisgård et al., 1980). Ultraplankton, consisting of algal cells smaller than 5-10µm, may be an ideal food source for the larvae in nature. Principally comprised of naked flagellates, the ultraplankton has a generally high food value for bivalve larvae (Davis and Guillard, 1958; Walne, 1963; Jørgensen, 1981). Moreover, the ultraplankton often accounts for well over 50% of the total phytoplankton production in estuaries and, therefore, represents a substantial nutritional source for the larvae.

Aside from the size of the algal cells, their high concentration may influence the feeding and development rates of larvae. According to Bayne (1983), the concentrations of phytoplankton that are optimal for growth of bivalve larvae in the laboratory are approximately 10–50μg mL^{-1}. These values are high relative to the mean concentrations observed in the field.

Sprung (1984b) monitored the shell growth of *M. edulis* larvae cultured in the laboratory at three different temperatures (6, 12 and 18°C) and six different food concentrations (1, 2, 5, 10, 20 and 40 *Isochrysis galbana* cells μL^{-1}). He also calculated growth, as measured by increases in larval weight, under these conditions. His findings revealed an increasing rate of larval growth with increasing concentrations of *I. galbana* up to densities of about 0.01 cells μL^{-1}, followed by a depression in growth in dense algal cultures. Larval density, meanwhile, had only minor effects on growth below 1 larva mL^{-1}. For larvae reared at 6°C at the three lowest food concentrations, growth curves (shell length vs. time) were sigmoidal. For larvae reared at all other temperatures and food concentrations, the growth curves were linear, with a maximum rate of 3.4, 8.1 and 11.8μm day^{-1} at 6, 12 and 18°C, respectively. The decline in growth rates at algal concentrations above 10 cells μL^{-1} may have resulted from: (1) the hampering of the feeding apparatus of the larvae by excessively high numbers of algal cells; and (2), the sensitivity of the larvae to algal metabolites.

Although Sprung (1984b) found no increase in the growth rates of *M. edulis* larvae at food concentrations above 0.01 cells μL^{-1}, Pechenik (unpublished data, personal communication, 1991) observed increased growth rates of *M. edulis* larvae in cultures at all food concentrations tested up to 300 cells μL^{-1}. Bayne (1965) reported optimum larval growth for *M. edulis* larvae in *I. galbana* concentrations of 0.1cells μL^{-1} (larval density = 3–10 organisms mL^{-1}). Jespersen and Olsen (1982), employing a mixture of *Isochrysis* and *Monochrysis*, recorded optimum food concentrations of 40–50 cells μL^{-1} for larval rearing of *M. edulis*, maintained at larval densities between 0.1 and 0.2 organisms mL^{-1}.

M. edulis larvae purportedly survive for protracted periods of time without food. Bayne (1976) reported that starved larvae of this species could survive for 20–30 days at 15–16°C. Older *Mytilus* larvae appear capable of surviving even longer stretches without food. Sprung (1982) documented survival of older *Mytilus* larvae for up to 150 days in sterilized seawater at 12°C. During these periods of starvation, the larvae rely on protein and lipid as energy reserves (Crisp, 1974a).

SETTLEMENT AND METAMORPHOSIS

Settlement

Bayne (1965, p.3) defines settlement as "...the descent of larvae from the plankton to the bottom substrate and the behaviour just preceding attachment." At the onset of settlement, the pediveliger slowly drops from the plankton to the seafloor, responding to light and gravity by negative phototaxis and positive geotaxis (Bayne, 1964a). The pediveliger protrudes its foot while swimming, and periodically the velum is withdrawn, whereupon the organism sinks to the bottom and begins to crawl. If the substrate is unsuitable (i.e. one not stimulating the secretion of a byssus leading to the onset of metamorphosis), the foot is withdrawn, and the larva subsequently swims off. Under more favourable substrate conditions, the larva continues to crawl for a period of time, then gradually ceases movement in response to a hierarchy of stimuli; it protrudes its foot outside the shell and quickly secretes a single byssus thread, thereby attaching the larva to the substrate (Bayne, 1965). As the pediveliger develops, its foot is of considerable importance in the selection of a settlement site. The foot contains nine kinds of glands, each having specific roles during crawling and attachment. For example, the first gland secretes a weakly acidic mucopolysaccharide, which probably facilitates pedal locomotion by ciliary gliding. The second gland contains proteinaceous vesicles, and may assist the temporary adhesion of the top of the foot during muscular pedal crawling. Several other glands are involved in the formation of the primary byssus thread (Lane and Nott, 1975). The pattern of swimming and crawling behaviour preceding attachment of mussel larvae to a substrate is consistent with that of larvae of other invertebrate groups (Thorson, 1966; Crisp, 1974b, 1984), and underscores the capability of these larvae to discriminate between different substrates at settlement. The precise stimuli from the substrate that elicit a progressive reduction in crawling, ending in byssal thread secretion and organismal attachment, have yet to be characterized. However, Cooper (1982, 1983) focusing on the role of chemical cues in the settlement and metamorphosis of planktonic pediveligers of *M. edulis*, suggested that the larvae will settle, attach, and metamorphose in response to phenolic compounds, particularly those containing catechol groups. He induced mussel larvae to settle in the laboratory by using seawater extracts of filamentous algae (*Platythamnion villosum*), or 1–10mM solutions of L-3,4–dihydroxyphenylalanine (L-DOPA).

Mytilus larvae in both the field and the laboratory attach most readily to filamentous substrates, such as bryozoans, hydroids, and filiform algae (de Blok and Geelen, 1958; Bayne, 1965; Kiseleva, 1966; Davies, 1974; Lane et al., 1985; Eyster and Pechenik, 1987). Byssal threads of previously settled mussels may also stimulate larval settlement and metamorphosis. Lane et al. (1985) describe the fine structure of drifting threads and attachment byssus threads of young postlarval mussels (*M. edulis*). Chipperfield (1953) surmises that surface discontinuities hasten the secretion of byssus threads and subsequent attachment. While metamorphosis of *M. edulis* larvae typically follows contact and subsequent attachment to filamentous substrates, water agitation may dramatically increase larval attachment to the filaments. For example, Eyster and Pechenik (1987) routinely observed a two- to eight-fold increase in the percentage attachment of mussel larvae to filamentous substrates during culture agitation. Another factor, competition between *Mytilus* species (e.g. *M. californianus* and *M. edulis*), can influence settling behaviours; *M. edulis* larvae tend to avoid adult clumps of *M. californianus*, where competition is intense and survivorship is low (Petersen, 1984). The use of artificial collectors, especially filamentous ones, deployed in the field has proven to be extremely effective in monitoring settlement of mussel larvae under natural conditions (King et al., 1990).

The absence of filamentous substrates tends to prolong the pediveliger stage and contributes to a delay of metamorphosis. As metamorphosis is delayed, the velum gradually degenerates, swimming becomes impaired, and growth in shell length is drastically reduced. The foot and gill filaments continue to grow at a very reduced rate and do not develop completely. Feeding also decreases due to disruption of feeding currents produced by cilia of the velum. Bayne (1965) recognized three pediveliger stages associated with the delay of metamorphosis of *M. edulis* larvae, which he described as the maintenance of a certain level of organization for as long a period as possible, rather than as a period of growth of new tissue. Each stage is identified by the relative size of the velum and foot and by the overall behaviour of the pediveliger. During the period of delay of metamorphosis, the larvae remain capable of attachment and further development if a suitable substrate is located. If suitable substrates are withheld from larvae in culture vessels in the laboratory, most individuals eventually die (Bayne, 1965).

When salinity, temperature, food supply, and other factors are optimal, larval development of *M. edulis* may be completed in less than 20 days (Bayne, 1965; Sprung, 1984a, b). However, growth to metamorphosis in the plankton during spring/early summer at a temperature of about 10°C normally occurs in approximately one month (Seed, 1976; Lane et al., 1985). It is not unusual for the duration of planktonic life to extend beyond two months when environmental conditions are less than optimal and suitable settlement surfaces are lacking (Bayne, 1965, 1976). The pelagic life of *Mytilus* larvae can be prolonged even more than six months due to delayed growth and metamorphosis (Lane et al., 1985).

Mature larvae usually do not initially settle on existing mussel beds, but generally attach to filamentous substrates (e.g. thecate hydroids and filamentous algae) away from adults (Petersen, 1984; Eyster and Pechenik, 1987; Strathmann, 1987; King et al., 1989, 1990). Subsequent to this primary settlement, which has been demonstrated both in the field and laboratory (Bayne, 1976; Eyster and Pechenik, 1987), newly-settled plantigrades may pass through a secondary pelagic phase, also termed a bysso-pelagic or byssus drifting phase, during which time they detach from the original settlement substrate and may attach to, and subsequently detach from, several filamentous substrates before selecting sites of permanent attachment on adult beds (see p.106–110 Chapter 4). The secretion of drifting threads, which exceed the postlarvae in length by more than two orders of magnitude, promotes the drifting capability of young postlarval mussels up to a size of approximately 2mm and may be an important strategy for their dispersal (Lane et al., 1982, 1985). Since postlarvae may require a period of two months to reach a size of 2mm, they can be transported significant distances by currents in coastal waters.

While earlier studies (Bayne, 1964b, 1976) documented the existence of the phenomenon of primary and secondary settlement of *M. edulis*, more recent work (Kautsky, 1982; Petersen, 1984; Fell and Belsamo, 1985; Eyster and Pechenik, 1987; McGrath et al., 1988) has cast doubt on its universal applicability. For example, Kautsky (1982) could not detect a secondary pelagic phase in a Baltic *M. edulis* population, while Peterson (1984) cited direct settlement of *M. californianus* plantigrades onto adult beds at Yoakam Point, Oregon, U.S.A. Fell and Balsamo (1985) discerned primary settlement of *M. edulis* on clean *Mercenaria mercenaria* cultch, but little secondary settlement on the cultch in the Thames Estuary, Long Island Sound, U.S.A. In a laboratory investigation, Eyster and Pechenik (1987) noted the induction of primary settlement and metamorphosis of *M. edulis* larvae by conspecific byssal threads, which suggested that the mussel larvae may recruit directly to the adult habitat in the field. McGrath et al. (1988), studying newly-settled *M. edulis* on an exposed rocky shore in Galway Bay, Ireland, found that mussel larvae settled directly on adult beds without an initial primary phase on filamentous substrates. It is evident, therefore, that *Mytilus* exhibits varying modes of settlement, which may not necessarily conform to the primary-secondary settlement model.

Metamorphosis

Attachment of the byssus to a substrate denotes the end of pelagic life and the onset of metamorphosis accompanied by gross morphological changes of the organism. The term 'metamorphosis' is used by Bayne (1965, p.3) "...to signify only those changes that occur between the first secretion of the byssus and the appearance of the dissoconch shell." The changes prepare the organism for the transition from a pelagic to a sessile habit.

Four gross morphological alterations can be recognized in *Mytilus* during metamorphosis: (1) secretion of byssus threads; (2) collapse and disintegration of the velum by phagocytosis; (3) formation of labial palps with the apical plate incorporated into the labial palps of the postlarva; and (4), reorientation of the mantle cavity organs (Bayne, 1971). According to Bayne (1976), 24–72h after the initial secretion of the byssus, the velum degenerates and the labial palps form. The larvae do not feed at this time but depend on stored nutrients for metabolic energy. The adult gill/palp feeding mechanism develops concomitantly with the loss of the larval feeding mechanism, and becomes functional within 48h of metamorphosis. The foot gradually migrates anteriorly and enlarges in the mantle cavity as the velum degenerates. The ctenidial filaments effectively partition the mantle cavity and form marginal interctenidial junctions that permit, for the first time, filtering of the water. By moving forward in the mantle cavity with the disappearance of the velum, the ctenidial filaments also become closely associated with the mouth and the labial palps, enhancing the feeding process. Concurrently, the labial palps also become functional, complete with ciliary sorting currents on the palps and division of the posterior mantle cavity into inhalant and exhalant regions, thus fostering the movement of food particles to the mouth.

While little or no shell growth takes place during metamorphosis, secretion of the adult dissoconch shell commences within 48h of pediveliger attachment. Different pigmentation and more conspicuous sculpturing distinguishes the dissoconch shell from the larval shell. Microstructural and mineralogical changes lend further distinction (Fuller and Lutz, 1988). Immediately subsequent to metamorphosis, a shift in the main axes of skeletal growth ensues, as manifested in a striking change of shape of the dissoconch shell. This change of shape reflects an acute posterior shift in maximum marginal growth incrementation, and reduced shell growth at the anterior margin (Bayne, 1971).

LARVAL DISPERSAL AND RECRUITMENT

The distribution of *Mytilus* populations is inextricably linked to dispersal during a planktonic larval stage. The early life-history strategy of mussels, with its associated extended planktotrophic existence, accounts for the high dispersal capability of most species within the genus *Mytilus*. In contrast to this strategy of larval development are nonplanktotrophic forms, such as lecithotrophs and 'direct developers', that spend little or no time in the plankton and, thus, have significantly lower dispersal capabilities (for review, see Jablonski and Lutz, 1983).

In most regions, *Mytilus* larvae attain peak numbers in spring and summer, with the duration of larval life depending, in part, on the available ration, temperature, salinity, and other environmental factors, mentioned earlier. The larval life of *Mytilus* consists of a precompetent period, during which time the larvae cannot be induced to metamorphose to an adult form, and a competent period, during which time the larvae respond to chemical and/or physical cues and undergo metamorphosis. The length of the precompetent period is a function of rates of growth and development of the larvae, which are controlled by both genetic and environmental factors (Day and McEdward, 1984).

When a chemical and/or physical cue is lacking, metamorphosis is delayed. However, the period of delay is limited, and it culminates either in spontaneous metamorphosis or in death of the organism (Bayne, 1965; Pechenik, 1984). While *Mytilus* larvae can delay metamorphosis and grow in the plankton until a shell length of about 350μm (Bayne, 1965; Sprung, 1984b), they typically become competent to settle at a shell length of approximately 260μm (McGrath et al., 1988). Various chemical or physical factors can induce the larvae to metamorphose. Investigations during the past decade have attempted to identify the biochemical mechanisms that normally control substrate-induced recruitment of larvae from the plankton, as well as the inductive stimuli that trigger metamorphosis (Burke, 1983, 1986; Sebens, 1983; Crisp, 1984; Morse, 1984; Pawlik, 1986; Morse et al., 1988).

Criteria used to assess larval competence for metamorphosis generally consist of observations on the presence or absence of particular morphological features (e.g. eyespots), or on the size of the larvae (Bayne, 1965; Pechenik, 1980; Sprung, 1984a, b). However, Eyster and Pechenik (1987) consider these criteria to be unreliable indicators of competence, at least in *M. edulis*. In addition to the lack of accurate predicators of metamorphic competency in mussel larvae, there is a paucity of data on the effects of environmental factors on the rate of development of larval competence for metamorophosis and, therefore, the duration of pelagic existence.

Day and McEdward (1984) maintain that larval longevity, together with water movements, regulates dispersal potential. Hence, the rate at which metamorphic competency is attained, and any subsequent delays in metamorphosis, usually affect the extent of larval dispersal. The actual degree of spread, in turn, is also influenced by larval behaviour. For instance, advective transport of planktonic larvae can be maximized or minimized by the swimming behaviour of the organisms within constraints set by the duration of the planktonic period, and strength of the water movement.

Hancock (1973) and Bayne (1976) considered the factors that most likely determine the success with which *Mytilus* larvae complete their pelagic development. These factors include: (1) suitable environmental conditions, notably temperature; (2) adequate food supply; (3) predation; (4) accidental ingestion by benthic filter-feeding adults; and (5), contact with areas and conditions favourable for settlement. Mortality of *Mytilus* during the planktonic larval period is clearly high. Large differences between fecundity and recruitment of *Mytilus* populations suggest tremendous mortality during the larval period (Thorson, 1950; Day and McEdward, 1984). Jørgensen (1981) followed a cohort of bivalve larvae, mainly *M. edulis*, during its residence in the plankton of the Isefjord, Denmark, and found a daily mortality rate of about 13%.

Although *Mytilus* larvae have relatively broad temperature and salinity tolerances, at times these factors exceed larval tolerances in nature. Extremes of these, as well as other environmental factors, however, may have less of an overall impact on the success of the larvae than biological factors, particularly predation, which has been deemed to be the single most important cause of larval mortality (Thorson, 1950). Even though physical environmental conditions may be quantitatively less significant as a direct cause of planktonic larval mortality, as suggested by Thorson (1950), they can create sublethal stresses that may indirectly affect the survival of larvae. Day and McEdward (1984) summarize the effects of sublethal stresses posed by excessive physical conditions on planktonic larvae as follows: (1) rate changes during periods of active morphogenesis can disrupt finely co-ordinated events; (2) changes in metabolic rates and feeding rates can have important long-term effects on the maintenance of energy balances; and (3), changes in the duration of pelagic life can increase the risk from other sources of mortality.

At present, quantitative data are insufficient to yield reliable conclusions on the absolute magnitude of the various environmental and biological sources of larval mortality in nature. Generalizations regarding some of these sources—temperature, salinity, starvation, predation—have been advanced (Thorson, 1950, 1966; Crisp, 1976, 1984). Much more research must be conducted on the causes of larval mortality in order to understand the risks of death so that precise models of marine life-history strategies can be formulated.

LARVAL AND POSTLARVAL SHELL MORPHOLOGY

As noted above, the larval shell secreted by mytilids is subdivided into two stages: (1) the prodissoconch I, representing the first shelled stage that develops from the nonshelled trochophore; and (2), the prodissoconch II, preceding the dissoconch or early juvenile shell, which is deposited by the mantle after metamorphosis. In many mytilids, as well as some pectinids and a few species belonging to other families, an intermediate shell, the interdissoconch, forms between the prodissoconch and dissoconch shell, and is easily delineated from the dissoconch on the shell surface

(Fuller and Lutz, 1988). The similarity in surface morphology of the prodissoconch and interdissoconch may make the larval-postlarval shell boundary difficult to discern.

Generally unornamented, the prodissoconch I has a granulated appearance under the optical microscope. When viewed under the scanning electron microscope, however, it exhibits a coarse or irregular punctate surface texture, apparently reflecting loci of shell deposition. The prodissoconch II shows a greater degree of skeletal ornamentation, most often fine surface sculpture, but generally much less ornamentation and surface texture than that of the dissoconch, from which it is usually readily distinguished. In many species, the dissoconch displays well-developed comarginal and/or radial sculpture on the exterior surface (Ansell, 1962; Carriker and Palmer, 1979; Waller, 1981; Jablonski and Lutz, 1983; Fuller and Lutz, 1988).

In addition to distinctive skeletal ornamentation, abrupt changes in mineralogy and microstructure typify transitions from the prodissoconch to the interdissoconch and early dissoconch of mytilids. The prodissoconch is always composed of aragonite, even in those taxa in which the adult shell contains calcite, and generally has a homogeneous microstructure. When present, the interdissoconch is also aragonitic and characterized by a homogenous microstructure. As in the prodissoconch shell, the surface morphology of the interdissoconch tends to be finely sculptured. In contrast, the dissoconch of mytilids may be entirely aragonitic or bimineralic (containing both aragonite and calcite) in composition, and usually consists of several shell layers with various types of microstructure (Stenzel, 1964; Taylor et al., 1969; Carriker and Palmer, 1979; Waller, 1981; Fuller and Lutz, 1988).

Table 3.2. Mineralogy and microstructure of the prodissoconch, interdissoconch, and early dissoconch in five mytilid species. (A = aragonitic, B = blocky structure, C = calcitic, H = homogeneous structure, N = nacreous structure, P = prismatic structure). (From Fuller and Lutz, 1988).

Species	Prodissoconch	Interdissoconch	Early Dissoconch
Mytilus edulis	A (H)	absent	C (P), A (N)
Ischadium recurvum	A (H)	A (H)	C (B), A (H), A (N)
Geukensia demissa	A (H)	A (H)	C (B), A (H), A (N)
Modiolus modiolus	A (H)	A (H)	C (P), A (N)
Brachidontes exustus	A (H)	A (H)	A (P), A (N)

Fuller and Lutz (1988) present a detailed summary of the shell mineralogy, microstructure, and surface morphology of the prodissoconch, interdissoconch, and early dissoconch stages of five mytilid species (i.e. *Brachidontes exustus, Geukensia demissa, Ischadium recurvum, Modiolus modiolus,* and *Mytilus edulis*) (Table 3.2). Except for *M. edulis*, all of these mytilids have an interdissoconch, which is consistent with the contention of Ockelmann (1983) that most mytilids possess this shell stage. The similarity in mineralogy, microstructure, and surface sculpture of the prodissoconch and interdissoconch in *B. exustus, G. demissa, I. recurvum,* and *M. modiolus* accounts for an inconspicuous larval-postlarval boundary. In these four species, formation of the dissoconch commences at a postsettlement stage, and is easily identified on the exterior shell surface by distinct sculpturing, and in the skeletal microstructure by a multilayered shell. When viewed in fractured sections of postlarval shell valves, a sharp boundary separates the single-layered, homogeneous microstructure of the interdissoconch from the multilayered dissoconch. This boundary is manifested on the exterior shell surface as a distinct demarcation between the interdissoconch and dissoconch.

In *M. edulis*, a demarcation resembling that in the other four species occurs in the exterior shell surface, but is located between the prodissoconch and dissoconch, since the interdissoconch is absent in this species. This boundary is strikingly evident on postlarval shells as a distinct line separating the smooth exterior surface of the prodissoconch from the comarginally ridged surface of the dissoconch. Within the shell, the boundary marks a transition from a larval shell with homogeneous microstructure to a multilayered postlarval shell comprised of a thin, aragonitic nacreous inner layer and a thick, calcitic prismatic outer layer. Formation of the prodissoconch-dissoconch boundary coincides with the time of settlement in *M. edulis* (Fuller and Lutz, 1988).

IDENTIFICATION OF LARVAL AND EARLY POSTLARVAL BIVALVES

Careful examination and interpretation of early shell morphological features of mytilids not only facilitate distinction of closely related species, but also provide insight into their ecology, biogeography, and reproductive strategy (de Schweinitz and Lutz, 1976; Lutz and Hidu, 1979; Jablonski and Lutz, 1980, 1983; Fuller and Lutz, 1988, 1989). Observations of the shell morphology of early life-history stages have been utilized for more than a century in systematic and ecological investigations of bivalve larvae (Lovén, 1848; Stafford, 1912; Odhner, 1914; Lebour, 1938; Werner, 1939; Jørgensen, 1946; Sullivan, 1948; Rees, 1950; Miyazaki, 1962; Loosanoff and Davis, 1963;

Newell and Newell, 1963; Loosanoff et al., 1966; Chanley and Andrews, 1971; Lutz and Jablonski, 1978; Fuller and Lutz, 1989). Characteristics of the shell most useful historically in the identification of the larval stages of bivalves include shell length, height, and depth, and length of the straight-hinge line (Loosanoff et al., 1966; Chanley and Andrews, 1971). The position of the larval ligament—located in the posterior region of the hinge in the Mytilidae—has also been of assistance, as have larval shape, colour, texture, and the presence or absence of a byssal notch, eyespot, or apical cilia ('apical flagellum') (Chanley and Andrews, 1971; Turner and Boyle, 1974; Culliney et al., 1975).

Examination of the larval hinge apparatus using scanning electron microscopy has enabled workers to overcome many practical identification barriers to distinguishing bivalve larval forms at the generic, or even specific, level. While the prodissoconch I possesses a weakly defined hinge structure, subsequent larval stages often have a distinct hinge. The provinculum of mytilids, which is thickened and bears numerous small rectangular teeth, forms the straight part of the hinge (Bayne, 1976).

Table 3.3. Larval hinge types of Rees (1950). (From Jablonski and Lutz, 1980).

Provinculum	No teeth	Few teeth	Many teeth
Thick		Astartacea, Hiatellacea, Mytilidae, Pholadacea and some Cardiacea, Corbulidae	Anomiacea, Mytilacea, Ostreacea, Pectinacea, Pteriacea, and some Mactracea, Veneracea
Thin	Ercinacea, Lucinacea	Some Cardiacea, Mactracea, Pandoracea, Poromyacea	Some Cardiacea, Corbulidae, Solenacea

Several features of the hinge structure are helpful in identifying larval shells; the provinculum and lateral hinge system are two such features. Denticulated structures associated with the provinculum can be diagnostic for the classification of larvae at the familial, generic, or even specific level. Shortly after metamorphosis, growth of the intra-umbonal dissoconch obscures larval dentition, limiting its usefulness as a diagnostic tool. Adult dentition that ensues has no direct relationship to larval dentition. Based on inventories of North Sea bivalves, Rees (1950) recognized five major provinculum types diagnostic at the familial or superfamilial level: (1) several or numerous rectangular teeth or crenulations in a row along a relatively thick provinculum, reminiscent of adult arcoid taxodont dentition; (2) one, two, or three strong rectangular teeth in each valve, arranged along a relatively thick provinculum; (3) a few teeth on a relatively thin provinculum; (4) right valve with spiked teeth (on a thin strip), that insert into sockets on the left valve; and (5), a thin provinculum from which teeth are entirely lacking. Superfamilies of North Sea bivalves exemplifying these five provinculum types are listed in Table 3.3. Further subdivision of the group is possible using the lateral hinge system, larval shell shape, and other characteristics.

The configuration of the lateral hinge system positioned anterior, posterior, or both, to the provinculum along the dorsal margin of the shell provides another diagnostic aid in taxonomic identification of larval shells (Rees, 1950; Cox, 1969) A component of this system is a projecting flange which, when present, usually lies on the left valve and interlocks with the right valve between the dorsal margin of the shell and an internal ridge. A solid or lamellar tooth-like projection may be found near the junction of this internal ridge and the provinculum (Rees, 1950). Unlike solid lateral teeth, that are relatively large and in contact with the outside edge of the provinculum of the opposite valve when the shell closes, lamellar lateral teeth appear thin and elongate, and they interlock with a groove bordered by the rim of the other valve. In addition to the solid or lamellar teeth, considered to be components of either the provincular or the lateral hinge system, larger 'special' teeth in various shapes occupy the juncture between the two hinge systems. The Astartacea, Mactracea, Poromyacea, and Tellinacea reportedly possess these large projections.

The shape of larval shells likewise serves a useful purpose in taxonomic classification (Werner, 1939; Rees, 1950; Yoshida, 1953; Miyazaki, 1962). The height/length and length/depth ratios, as well as the prominence of the umbo (in the prodissoconch II), are especially valuable. Loosanoff et al. (1966), Chanley and Andrews (1971), and Stephenson and Chanley (1979) give examples of the manner in which larvae can be distinguished at the species level based on shell shape.

Ligaments or 'ligament pits' are also structures that have proved useful in identifying early ontogenetic stages of bivalves. Bernard (1896), Rees (1950), Ansell (1962), Loosanoff et al. (1966), Chanley and Andrews (1971), Bayne (1976) and Le Pennec and Masson (1976) commented on the usefulness of the presence and position of these structures for classifying larvae at the superfamilial level. More recently, Lutz (1977, 1978), Lutz and Hidu (1979), Lutz and Jablonski (1979) and Jablonski and Lutz (1980) have suggested that such structures may be early postlarval features and of use in ascertaining whether or not the process of metamorphosis has commenced in a particular organism isolated from plankton or benthic samples.

Scanning electron microscopy (SEM) has significantly improved upon optical microscopy in resolving skeletal features, such as the larval hinge apparatus. Optical microscopy commonly facilitates identification of larval shells at the superfamilial, or familial, level. As mentioned above, with SEM larval specimens can be identified in many instances to the generic, or even specific, level (Lutz and Hidu, 1979; Stephenson and Chanley, 1979; Chanley and Chanley, 1980; Chanley and Dinamani, 1980; Jablonski and Lutz, 1980, 1983; Fuller and Lutz, 1988, 1989).

In addition to its usefulness in identifying bivalve specimens isolated from planktonic or benthic samples, SEM analysis of larval and early postlarval shells may be a powerful tool in a wide variety of ecological and palaeoecological studies (Jablonski and Lutz, 1980; Waller, 1981). For example, size measurements of the prodissoconch I and II, accurately and readily obtained through SEM examination, allow inferences to be drawn concerning the reproductive strategies and early developmental history of various bivalve species. The size of the initial larval shell (prodissoconch I) of planktotrophic larvae usually ranges in length from 70 to 150μm, and is generally smaller than that of lecithotrophic pelagic larvae, in which the prodissoconch I generally ranges from 135 to 230μm in length. These differences reflect variations in the amount of yolky food during early development, with lecithotrophic forms having larger diameter eggs and, consequently, larger prodissoconch I shells (Ockelmann, 1965). The prodissoconch I of nonplanktonic larvae (e.g. species with 'direct' development (Thorson, 1946, 1950; Ockelman, 1965; Mileikovsky, 1971) is even larger, ranging from about 230 to more than 500μm in length, approximating the size range of the prodissoconch II stage (200–600μm in length) of planktotrophic larvae.

As a further elaboration on the developmental strategies outlined above, planktotrophic larvae develop from small eggs released in huge numbers, with little parental investment per offspring. The prodissoconch I of planktotrophic larvae is small relative to the prodissoconch II. Nonplanktotrophic larvae, including both planktonic lecithotrophs and 'direct developers', typically arise from large eggs, with relatively few young produced per parent. In the case of lecithotrophic pelagic larvae, the prodissoconch I appears large relative to the prodissoconch II, which either lies interposed between the prodissoconch I and the dissoconch, or is absent. The diminutive size of the prodissoconch II probably reflects the brief time lecithotrophs spend in the plankton. Species with direct development, or those species that brood larvae, have the largest eggs among the Bivalvia. The prodissoconch I of these species frequently has an inflated form with irregular folds and wrinkles, while lacking the distinctive D-shape (Ockelmann, 1965; Chanley and Andrews, 1971; La Barbera, 1974). Despite these useful guidelines enabling inferences to be drawn from analyses of prodissoconch morphology, interpretations of developmental type are often ambiguous.

The identification of larval and postlarval mussels historically has been based upon analyses of a variety of shell characteristics. The 'direct method' of identification of larval bivalves entails a comparison of shell size and shape of unknown specimens with similar characteristics photographically depicted on micrographs of larvae reared in the laboratory from positively identified adult specimens (Loosanoff and Davis, 1963; Loosanoff et al., 1966; Chanley and Andrews, 1971). Also beneficial is the correlation of total shell size with appearance of an umbo, eyespot, ligament pit, or functional foot. The 'indirect method' of identification, advanced by Werner (1939) and championed by Ockelmann (1965), represents an alternate, albeit effective method of mussel larvae identification, in which much of the history of larval shell formation (i.e. size of the prodissoconch I and shape and size of the prodissoconch II) is deduced from observation of the umbonal region of the shell of young postlarval stages that have developed to point of positive species identification (Lutz, 1985a). Inherent problems associated with such 'indirect methods' have been examined by Loosanoff et al. (1966).

Over the past two decades, increased emphasis has been placed on the usefulness of structures associated with the larval and postlarval hinge apparatus for discriminating various species of mussels during early ontogenetic stages (de Schweinitz and Lutz, 1976; Lutz and Hidu, 1979; Fuller and Lutz, 1989). De Schweinitz and Lutz (1976) indicated that the veligers of *M. edulis* and *Modiolus modiolus*— larval stages extremely difficult to differentiate—could be distinguished by comparing: (1) length of the hinge line; (2) shell length of specimens between 95 and 105μm; (3) shape of the shell of umbo-stage larvae; (4) presence or absence of an eyespot in specimens less than 270μm in length; and (5), presence or absence of a functional foot in larvae less than 295μm. Differences in the length of the hinge line of the two species were readily apparent, with the hinge of straight-hinge veligers of *M. edulis* averaging 74μm in length and that of *M. modiolus* averaging 100μm. The longest hinge line encountered for *M. edulis* was 80μm, while the shortest hinge line of *M. modiolus* was 90μm. In terms of total shell length, the smallest *M. modiolus* veligers measured 105μm x 90μm, which exceeded the dimensions of the smallest *M. edulis* larva by 10μm. Eyespots occurred in *M. edulis* larvae as small as 205μm, but did not develop in *M. modiolus* larvae until they attained a size of at least 270μm. The larval foot of the pediveliger stage of *M. edulis* became functional in individuals as small as 215μm in length. However, the same development of the foot of *M. modiolus* pediveligers was not reached until a length of at least 295μm. De Schweinitz and Lutz (1976) suggested that similar type measurements and observations on the larvae of other mytilids could prove useful in distinguishing various larval stages of closely related mussel species.

While de Schweinitz and Lutz (1976) quantified larval ontogenetic changes in the external morphology of *M. edulis* and *M. modiolus*, reporting on morphometric differences at certain stages of early development of these species, identification of many larval and early postlarval stages of those two species still remained subjective. Through a detailed examination of hinge-line morphogenesis (prodissoconch I through metamorphosis), Lutz and Hidu (1979) demonstrated that larval and early postlarval specimens of these two mytilids at all stages of development may be unambiguously identified to the appropriate species based on optical microscopic examination of the hinge apparatus. For example, differences were evident between early life-history stages of *M. edulis* and *M. modiolus* after regression and quantitative comparison (analysis of covariance) of each of the following parameters: (1) larval length and provinculum length; (2) larval height and provinculum length; (3) provinculum length and number of teeth; and (4), larval length and number of teeth. Furthermore, the data presented, when combined with published descriptions of other mytilids, could be of assistance in distinguishing *M. edulis* and *M. modiolus* larvae from larval stages of other closely related species within the genus *Mytilus*.

A sequence of similar ontogenetic changes in the hinge apparatus of *M. edulis* and *M. modiolus* from the prodissoconch I stage through metamorphosis is apparent (Figs. 3.1–3.4). In particular, dentition is absent at the prodissoconch I stage. Provinculum length and complexity increase throughout larval development with progressive lateral thickening characteristic of the family Mytilidae (Rees, 1950). The number of hinge teeth in prodissoconch II and early dissoconch stages increases significantly with both total shell and provinculum length. Ultrastructural examination of anterior and posterior surfaces of the teeth of both species discloses ridged surfaces, which become increasingly obvious with shell growth.

Lutz and Hidu (1979) advocated that through inspection of the ventral surface of the hinge apparatus, an investigator can determine the point of initiation of metamorphosis in planktotrophs. Specifically, the first occurrence of a ligament pit signals the onset of metamorphosis, and the size and development of this structure infers the extent to which the metamorphic process has proceeded. The prodissoconch-dissoconch boundary is a morphological feature of value in segregating true juveniles from 'metamorphosing' postlarvae. The beginning of the dissoconch shell, in turn, establishes the termination of metamorphosis. The temporal relationship between development of the ligament pit and the prodissoconch-dissoconch boundary, therefore, contributes much information on the early development of larval mytilids.

Fuller and Lutz (1989) combined microscopy and morphometry for a comparative description of shell and hinge morphogenesis in *M. edulis*, *M. modiolus*, and four other mytilid species (i.e. *Amygdalum papyrium, Brachidontes exustus, Geukensia demissa*, and *Ischadium recurvum*) found in marine or estuarine waters of the northwestern Atlantic. Scanning electron micrographs of disarticulated valves of these species revealed differences in shell and hinge morphology, useful not only for the identification of larval and postlarval stages, but also for taxonomic classification of species within the family Mytilidae.

Shell hinge dentition is clearly an important tool for distinguishing mytilids during early ontogenetic stages. The taxonomic value of lateral hinge teeth for separation of genera within the Mytilidae has been discussed (Jukes–Brown, 1905; Siddall, 1980). Mytilids may have primary lateral teeth, secondary lateral teeth,

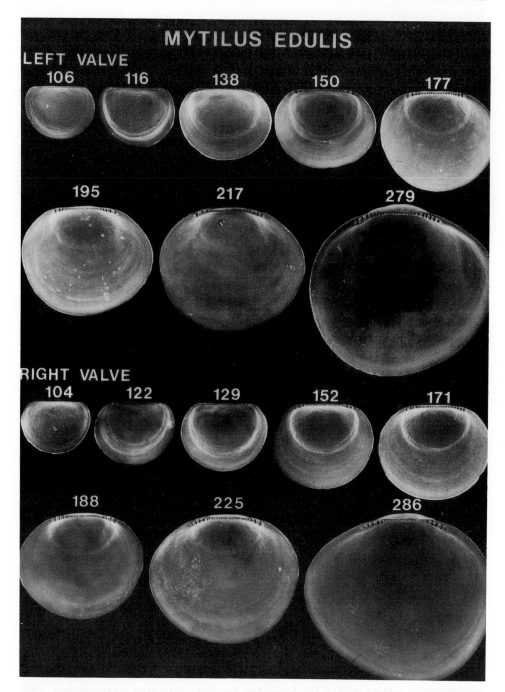

Fig. 3.1. Scanning electron micrographs of disarticulated valves of *Mytilus edulis* larvae. Numbers indicate shell length in μm. (From Fuller and Lutz, 1989).

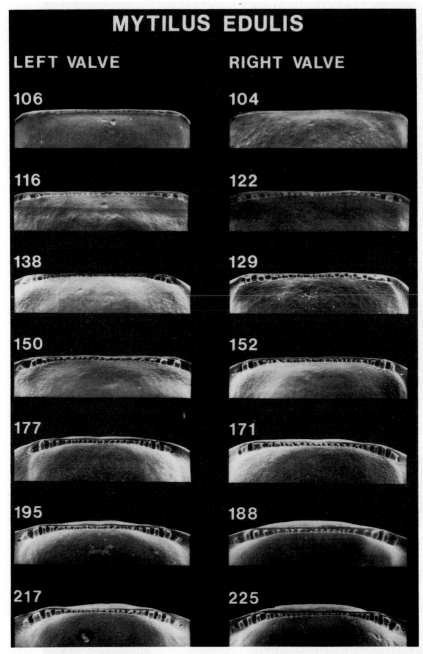

Fig. 3.2. Scanning electron micrographs of the hinge of disarticulated valves of *Mytilus edulis* larvae. Numbers indicate shell length in μm. (From Fuller and Lutz, 1989).

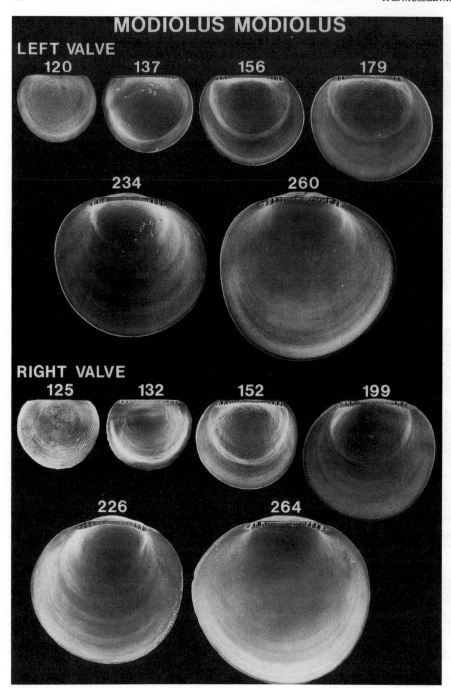

Fig. 3.3. Scanning electron micrographs of disarticulated valves of *Modiolus modiolus* larvae. Numbers indicate shell length in μm. (From Fuller and Lutz, 1989).

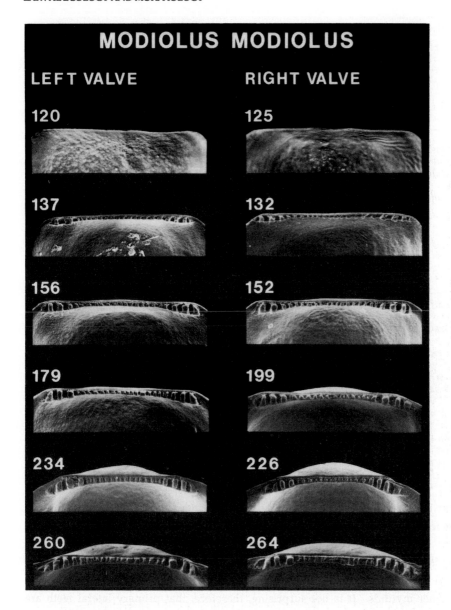

Fig. 3.4. Scanning electron micrographs of the hinge of disarticulated valves of *Modiolus modiolus* larvae. Numbers indicate shell length in μm. (From Fuller and Lutz, 1989).

Table 3.4. Presence of lateral teeth in early postlarval mytilids. (+ = present; 0 = absent; †not well-developed; blank entries signify no data). (From Fuller and Lutz, 1989).

Species	Primary Lateral	Secondary Lateral	Dysodont	Reference
Choromytilus chorus	0	+	+	Ramorino and Campos, 1983
Mytilus chilensis	0	+	+	Ramorino and Campos, 1983
Mytilus edulis	0	+	+	Cox, 1969; Le Pennec, 1980; Siddall, 1980; Redfearn et al., 1986
Mytilus galloprovincialis	0	+	+	Le Pennec and Masson, 1976; Le Pennec, 1980; Siddall, 1980
Brachidontes exustus	+	+	+	Fuller and Lutz, 1989
Brachidontes granulata	+			Ramorino and Campos, 1983
Perna canaliculus	+	+	+	Booth, 1977; Siddal, 1980; Redfearn et al., 1986
Perna perna	+	+	+	Siddall, 1980
Perna viridis	+	+	+	Siddall, 1980
Modiolus modiolus	+	0	0	Fuller and Lutz, 1989
Ischadium recurvum	0	0	+	Fuller and Lutz, 1989
Amygdalum papyrium	0	0	0	Fuller and Lutz, 1989
Geukensia demissa	0	0	0	Fuller and Lutz, 1989
Aulacomya ater	0			Ramorino and Campos, 1983
Modiolarca impacta	0			Redfearn et al., 1986
Perumytilus purpuratus	+	+		Ramorino and Campos, 1983
Semimytilus algosus	+†	+		Ramorino and Campos, 1983
Xenostrobus pulex	0			Redfearn et al., 1986

dysodont teeth, or a combination of the three. Primary lateral teeth form on the interdissoconch, whereas dysodont and secondary lateral teeth (also called posterior dysodont teeth) form on the dissoconch (Jukes-Browne, 1905; Ockelmann, 1983). Table 3.4 is a compilation of the type(s) of lateral hinge teeth observed in various early postlarval mytilids. The lack of an interdissoconch distinguishes *M. edulis* from several other mytilids. It is not surprising, therefore, that primary lateral teeth are absent in this species. *M. edulis* also differs from the other aforementioned mytilids in that its provinculum does not increase in size and complexity during postlarval stages. Jukes-Browne (1905) reported that development of the provinculum did not continue after secretion of the dissoconch in the Mytilidae. Again, lack of postlarval development of the provinculum in *M. edulis* correlates with absence of an interdissoconch in this species.

The position of the primary and secondary ligament pits likewise assists workers in identifying specimens of *M. edulis* isolated from plankton and benthic samples. Development of the primary ligament pit, which forms immediately after secretion of byssal threads, is among the earliest shell morphological changes occurring during metamorphosis of *M. edulis* larvae (Lutz and Hidu, 1979). This structure forms at settlement. The secondary ligament pit forms in late postlarval specimens at a stage after secretion of the dissoconch (Fuller and Lutz, 1989).

Employing optical and scanning electron microscopic techniques, Fuller and Lutz (1989) measured various morphological features in the larval shell of *M. edulis*. Based on their study, the earliest shells of this species ranged from 95 to 110µm long, in agreement with the size of the smallest larvae recorded by Loosanoff et al. (1966) and de Schweinitz and Lutz (1976), (94µm and 95µm, respectively). Weak dentition early in ontogeny gives way to fine, numerous teeth in later larval stages. Height of the central teeth was between 1.7 and 2.5µm, with larger teeth (up to 6.6µm high) at the anterior and posterior ends of the provinculum. Provinculum length ranged from 70 to 147µm, comparable to the range of values (71 to 133µm) reported by Lutz and Hidu (1979) for *M. edulis*. The number of teeth increased from 19 to 32. An umbo appeared at a shell length between 170 and 180µm, which was greater than the 150-µm length found by Chanley and Andrews (1971) and the 150 to 165-µm range documented by de Schweinitz and Lutz (1976) for umboned larvae. According to Fuller and Lutz (1989), the numerous, small provincular teeth in *M. edulis* larvae could be used to distinguish these larvae from those of other mytilid species.

In summary, the shell and hinge features mentioned above may be utilized for species identification of early ontogenetic stages of mussels. In addition, careful examination of certain morphological structures can shed light on ontogenetic processes. For example, observation of the ventral surface of the hinge apparatus of small mytilids may provide a means of ascertaining whether or not the process of metamorphosis has been initiated. The presence of a ligament pit would suggest that metamorphosis has commenced, while the size and development of this structure may be of assistance in determining the extent to which the process of metamorphosis has proceeded. To the degree that secretion of the dissoconch shell marks the end of metamorphosis, as proposed by Bayne (1965), the prodissoconch-dissoconch boundary provides a morphological feature useful in differentiating true juveniles from 'metamorphosing' postlarvae. On a population level, comparison of temporal relationships between development of these two morphological shell features (ligament pit and prodissoconch-dissoconch boundary) should contribute much to our understanding of ecological factors affecting metamorphosis of species within the genus *Mytilus*.

ACKNOWLEDGEMENTS

This is New Jersey Agricultural Experimental Station Publication no. F-32401-1-91 and Contribution no. 91-29 of the Institute of Marine and Coastal Sciences, Rutgers University, supported by New Jersey State funds, and the Fisheries and Aquaculture Technology Extension Center.

REFERENCES

Ansell, A. D., 1962. The functional morphology of the larva, and the post-larval development of *Venus striatula* (da Costa). J. Mar. Biol. Ass. U. K., 42: 419-443.

Bayne, B. L., 1963. Responses of *Mytilus edulis* larvae to increases in hydrostatic pressure. Nature (Lond.), 198: 406-407.

Bayne, B. L., 1964a. The responses of the larvae of *Mytilus edulis* L. to light and to gravity. Oikos, 15: 162-174.

Bayne, B. L., 1964b. Primary and secondary settlement in *Mytilus edulis* L. (Mollusca). J. Anim. Ecol., 33: 513-523.

Bayne, B. L., 1965. Growth and the delay of metamorphosis of the larvae of *Mytilus edulis* (L.). Ophelia, 2: 1-47.

Bayne, B. L., 1971. Some morphological changes that occur at the metamorphosis of the larvae of *Mytilus edulis*. In: D. J. Crisp (Editor), Proc. 4th Eur. Mar. Biol. Symp., Bangor, U.K., 1969. Cambridge University Press, London, pp. 259-280.

Bayne, B. L., 1972. Some effects of stress in the adult on the larval development of *M. edulis*. Nature (Lond.), 237: 459.

Bayne, B. L., 1975. Reproduction in bivalve molluscs under environmental stress. In: F. J. Vernberg (Editor), Physiological Ecology of Estuarine Organisms. University of South Carolina Press, Columbia, pp. 259-277.

Bayne, B. L., 1976. The biology of mussel larvae. In: B. L. Bayne (Editor), Marine Mussels: their ecology and physiology. Cambridge University Press, Cambridge, pp. 81-120.

Bayne, B. L., 1983. Physiological ecology of marine molluscan larvae. In: N. H. Verdonk, J. A. M. van den Biggelaar, and A. Tompa (Editors), The Mollusca, Vol. III, Development. Academic Press, New York, pp. 299-343.

Bayne, B. L., Gabbott, P. A., and Widdows, J., 1975. Some effects of stress in the adult on eggs and larvae of *Mytilus edulis* L. J. Mar. Biol. Ass. U. K., 55: 675-689.

Bernard, F., 1896. Deuxième note sur le developpment et la morphologie de la coquille chez des lamellibranchs. Bull. Soc. Geol. Fr., 24: 54-82.

Blok, J. W. de, and Geelen, H. J., 1958. The substratum required for the settling of mussels (*Mytilus edulis* L). Arch. Néerl. Zool., Jubilee Vol., 446-460.

Booth, J. D., 1977. Common bivalve larvae from New Zealand: Mytilacea. N. Z. J. Mar. Freshw. Res., 11: 407-440.

Borisjak, A., 1909. Pelecypoda du plankton de la Mer Noire. Bull. Scient. Fr. Belg., 42: 149-181.

Burke, R. D., 1983. The induction of metamorphosis of marine invertebrate larvae: stimulus and response. Can. J. Zool., 61: 1701-1719.

Burke, R. D., 1986. Pheromones and the gregarious settlement of marine invertebrate larvae. Bull. Mar. Sci., 39: 323-331.

Carriker, M. R., 1961. Interrelation of functional morphology, behaviour, and autecology in early stages of the bivalve *Mercenaria mercenaria*. J. Elisha Mitchell Sci. Soc., 77: 168-241.

Carriker, M. R. and Palmer, R. E., 1979. Ultrastructural morphogenesis of prodissoconch and early dissoconch valves of the oyster *Crassostrea virginica*. Proc. Natl. Shellfish. Assoc., 69: 103-128.

Chanley, P., 1970. Larval development of the hooked mussel, *Brachidontes recurvus* Rafinesque (Bivalvia:Mytilidae) including a literature review of larval characteristics of the Mytilidae. Proc. Natl. Shellfish. Assoc., 60: 86-94.

Chanley, P. and Andrews, J. D., 1971. Aids for identification of bivalve larvae of Virginia. Malacologia, 11: 45-119.

Chanley, P. and Chanley, M., 1980. Reproductive biology of *Arthritica crassiformis* and *A. bifurca*, two commensal bivalve molluscs (Liptonacea). N. Z. J. Mar. Freshw. Res., 14: 31-43.

Chanley, P. and Dinamani, P., 1980. Comparative descriptions of some oyster larvae from New Zealand and Chile, and a description of a new genus of oyster, *Tiostrea*. N. Z. J. Mar. Freshw. Res., 14: 103-120.

Chipperfield, P. N. J., 1953. Observations on the breeding and settlement of *Mytilus edulis* (L.) in British waters. J. Mar. Biol. Assoc., U. K., 32: 449-476.

Cooper, K., 1982. A model to explain the induction of settlement and metamorphosis of planktonic eyed-pediveligers of the blue mussel *Mytilus edulis* L. by chemical and tactile cues. J. Shellfish. Res., 2: 117.

Cooper, K., 1983. Potential for application of the chemical DOPA to commercial bivalve setting systems. J. Shellfish Res., 3: 110-111.

Courtright, R. C., Breese, W. P., and Krueger, H., 1971. Formulation of a synthetic seawater for bioassays with *Mytilus edulis* embryos. Water Res., 5: 877-888.

Cox, L. R., 1969. General features of the Bivalvia. In: R. C. Moore (Editor), Treatise on Invertebrate Paleontology. Part N, Vol. 1, Mollusca 6, Bivalvia. Geological Society of America and the University of Kansas Press, Lawrence, pp. 2-129.

Crisp, D. J., 1974a. Energy relations of marine invertebrate larvae. Thallasia Jugosl., 10: 103-120.

Crisp, D. J., 1974b. Factors influencing the settlement of marine invertebrate larvae. In: P. T. Grant and A. M. Mackie (Editors), Chemoreception in Marine Organisms. Academic Press, New York, pp. 177-265.

Crisp, D. J., 1976. The role of the pelagic larva. In: T. Spencer-Davis (Editor), Perspectives in Experimental Zoology. Pergamon Press, New York, pp. 145-155.

Crisp, D. J., 1984. Overview of research on marine invertebrate larvae, 1940-1980. In: J. D. Costlow and R. C. Tipper (Editors), Marine Biodeterioration: An Interdisciplinary Study. Naval Institute Press, Annapolis, Maryland, pp. 103-126.

Culliney, J. L., Boyle, P. L. and Turner, R. D., 1975. New approaches and techniques for studying bivalve larvae. In: W. L. Smith and M. H. Chanley (Editors), Culture of Marine Invertebrate Animals. Plenum Press, New York, pp. 257-271.

Davies, G., 1974. A method for monitoring the spatfall of mussels (*Mytilus edulis* L.). J. Cons. Int. Explor. Mer, 36: 27-34.

Davis, H. C. and Guillard, R. R., 1958. The relative value of ten genera of microorganisms as foods for oyster and clam larvae. Fish. Bull., U.S., 58: 293-304.

Day, R. and McEdward, L., 1984. Aspects of the physiology and ecology of pelagic larvae of marine benthic invertebrates. In: K. A. Steidinger and L. M. Walker (Editors), Marine Plankton Life Cycle Strategies. CRC Press, Boca Raton, Florida, pp. 93-120.

Eyster, L. S. and Pechenik, J. A., 1987. Attachment of *Mytilus edulis* L. larvae on algal and byssal filaments is enhanced by water agitation. J. Exp. Mar. Biol. Ecol., 114: 99-110.

Fell, P. E. and Balsamo, A. M., 1985. Recruitment of *Mytilus edulis* L. in the Thames Estuary, with evidence for differences in the time of maximal settling along the Connecticut shore. Estuaries, 8: 68-75.

Field, G. A., 1922. Biology and economic value of the sea mussel *Mytilus edulis*. Bull. U. S. Bur. Fish., 38: 127-259.

Fuller, S.C. and Lutz, R. A., 1988. Early shell mineralogy, microstructure, and surface sculpture in five mytilid species. Malacologia, 29: 363-371.

Fuller, S. C. and Lutz, R. A., 1989. Shell morphology of larval and post-larval mytilids from the northwestern Atlantic. J. Mar. Biol. Ass. U. K., 69: 181-218.

Hancock, D. A., 1973. The relationship between stock and recruitment in exploited invertebrates. J. Cons. Int. Explor. Mer, 164: 113-131.

Hodgson, A. N. and Bernard, R. T. F., 1986. Observations on the ultrastructure of the spermatozoan of two mytilids from the southwest coast of England. J. Mar. Biol. Ass. U. K., 66: 385-390.

Hrs-Brenko, M., 1973. The study of mussel larvae and their settlement in Vela Draga Bay (Pula, The Northern Adriatic Sea). Aquaculture, 2: 173-182.

Hrs-Brenko, M., 1974. Temperature and salinity requirements for embryonic development of *Mytilus galloprovincialis* Lmk. Thalassia Jugosl., 10: 131-138.

Hrs-Brenko, M. and Calabrese, A., 1969. The combined effects of salinity and temperature on larvae of the mussel *Mytilus edulis*. Mar. Biol., 4: 224-226.

Humphries, W. J., 1969. Electron microscope studies on eggs of *Mytilus edulis*. J. Ultrastruct. Res., 7: 467-487.

Innes, D. J. and Haley, L. E., 1977. Genetic aspects of larval growth under reduced salinity in *Mytilus edulis*. Biol. Bull., 153: 312-321.

Jablonski, D. J. and Lutz, R. A., 1980. Molluscan larval shell morphology: ecological and paleoecological applications. In: D.C. Rhoads, and R.A. Lutz (Editors), Skeletal Growth of Aquatic Organisms. Plenum Press, New York, pp. 323-377.

Jablonski, D. J. and Lutz, R. A., 1983. Larval ecology of marine benthic invertebrates: paleobiological implications. Biol. Rev., 58: 21-89.

Jespersen, H. and Olsen, K., 1982. Bioenergetics in veliger larvae of *Mytilus edulis* L. Ophelia, 21: 101-113.

Jørgensen, C. B., 1946. Reproduction and larval development of Danish marine bottom invertebrates. 9. Lamellibranchia. Meddr. Kommn. Danm. Fisk. -øg Havunders., Ser. : Plankton, 4: 277-311.

Jørgensen, C. B., 1981. Mortality, growth, and grazing impact on a cohort of bivalve larvae, *Mytilus edulis* L., Ophelia, 20: 185-192.

Jukes-Browne, A. J., 1905. A review of the genera of the family Mytilidae. Proc. Malacol. Soc. Lond., 6: 211-224.

Kändler, R., 1926. Muschellarven aus dem Helgoländer Plankton. Wissenschaftliche Meeresuntersuchungen der Kommission sur Wissenshaftlichen Untersuchung der Deutschen Meere, Abt. Helgoland, 16: 1-9.

Kautsky, N., 1982. Quantitative studies on gonad cycle, fecundity, reproductive output and recruitment in a Baltic *Mytilus edulis* population. Mar. Biol., 68: 143-160.

King, P. A., McGrath, D., and Gosling, E. M., 1989. Reproduction and settlement of *Mytilus edulis* on an exposed rocky shore in Galway Bay, west coast of Ireland. J. Mar. Biol. Ass. U. K., 69: 355-365.

King, P. A., McGrath, D., and Britton, W., 1990. The use of artificial substrates in monitoring mussel (*Mytilus edulis* L.) settlement on an exposed rocky shore in the west of Ireland. J. Mar. Biol. Ass. U. K., 70: 371-380.

Kiseleva, G. A., 1966. Factors stimulating larval metamorphosis of the lamellibranch *Brachyodontes lineatus* (Gmelin). Zool. Zh., 45: 1571-1573.

La Barbera, M., 1974. Larval and postlarval development of five species of Miocene bivalves (Mollusca). J. Paleontol., 48: 256-277.

Lane, D. J. W. and Nott, J. A., 1975. A study of the morphology, fine structure, and histochemistry of the foot of the pediveliger of *Mytilus edulis* L. J. Mar. Biol. Ass. U. K., 55: 477-495.

Lane, D. J. W., Nott, J. A., and Crisp, D. J., 1982. Enlarged stem glands in the foot of the postlarval mussel, *Mytilus edulis*: adaptation for byssopelagic migration. J. Mar. Biol. Ass. U. K., 62: 809-818.

Lane, D. J. W., Beaumont, A. R., and Hunter, J. R., 1985. Byssus drifting and the drifting threads of the young postlarval mussel *Mytilus edulis*. Mar. Biol., 84: 301-308.

Lannan, J. E., 1980. Broodstock management of *Crassostrea gigas*. I. Genetic and environmental variation in survival in the larval rearing system. Aquaculture, 21: 323-336.

Lebour, M. V., 1938. Notes on the breeding of some lamellibranchs from Plymouth and their larvae. J. Mar. Biol. Ass. U. K., 23: 119-144.

Le Gall, P., 1970. Étude des moulières Normandes; renouvellement, croissance. Vie Milieu, 21B: 545-590.

Le Pennec, M., 1980. The larval and postlarval hinge of some families of bivalve molluscs. J. Mar. Biol. Ass. U.K., 60: 601-617.

Le Pennec, M. and Masson, M., 1976. Morphogenèse de la coquille de *Mytilus galloprovincialis* (Lmk.) élevé au laboratoire. Cah. Biol. Mar., 17: 113-118.

Loosanoff, V. L. and Davis, H. C., 1963. Rearing of bivalve molluscs. Adv. Mar. Biol., 1: 1-136.

Loosanoff, V. L., Davis, H. C., and Chanley, P. E., 1966. Dimensions and shapes of some marine bivalve molluscs. Malacologia, 4: 351-435.

Lough, R. G., 1974. A re-evaluation of the combined effects of temperature and salinity on survival and growth of *Mytilus edulis* larvae using response surface techniques. Proc. Natl. Shellfish Assoc., 64: 73-76.

Lovén, S., 1848. Bidrag till Kannedomen on Utvecklingen af Mollusca Acephala, Lamellibranchiata. K. Sven Vetenskapsakad. Handl., 4: 299-435.

Lubet, P., 1957. Cycle sexuel de *Mytilus edulis* L. et de *Mytilus galloprovincialis* Lmk. dans le Bassin d'Arcachon (Gironde). Année Biol., 33: 19-29.

Lubet, P., 1973. Exposé synoptique des données biologique sur la moule *Mytilus galloprovincialis* (Lamarck 1819). Synopsis FAO sur les pêches No. 88. (SAST-Moule, 3, 16(10), 028, 08, pag. var.) FAO, Rome, 51pp.

Lubet, P. and Le Gall, P., 1967. Observations sur le cycle sexuel de *Mytilus edulis* L. à Luc-s-Mer. Bull. Soc. Linn. Normandie, 10: 303-307.

Lutz, R. A., 1977. Shell morphology of larval bivalves and its use in ecological and paleoecological studies (Abstract). Geol. Soc. Am. Bull. 9: 1079.

Lutz, R. A., 1978. A comparison of hinge line morphogenesis in larval shells of *Mytilus edulis* and *Modiolus modiolus*. Proc. Natl. Shellfish. Assoc., 66: 83.

Lutz, R. A., 1985a. Identification of bivalve larvae and postlarvae: a review of recent advances. Am. Malacol. Union Inc. Bull., Spec. Ed. 1, pp. 59-78,

Lutz, R. A., 1985b. Mussel aquaculture in the United States. In: J. V. Huner and E. E. Brown (Editors), Crustacean and Mollusk Aquaculture in the United States. AVI Publishers, Westport, Connecticut, pp. 311-363.

Lutz, R. A. and Jablonski D., 1978. Classification of bivalve larvae and early postlarvae using scanning electron microscopy. Am. Zool., 18: 647.

Lutz, R. A. and Hidu, H., 1979. Hinge morphogenesis in the shells of larval and early post-larval mussels (*Mytilus edulis* L. and *Modiolus modiolus* L.). J. Mar. Biol. Ass. U. K., 59: 111-121.

Lutz, R. A. and Jablonski D., 1979. Micro- and ultramorphology of larval bivalve shells: ecological, paleoecological, and paleoclimatic implications. Proc. Natl. Shellfish. Assoc., 69: 197-198.

Lutz, R., Goodsell, J., Castagna, M., Chapman, S., Newell, C., Hidu, H., Mann, R., Jablonski D., Kennedy, V., Siddall, S., Goldberg, R., Beattie, H., Falmagne, C., Chestnut, A., and Partridge, A., 1982. Preliminary observations on the usefulness of hinge structures for identification of bivalve larvae. J. Shellfish. Res., 2: 65-70.

Lutz, R. A., Chalermwat, K., Figueras, A., Gustafson, R. G., and Newell, C. 1991. Mussel aquaculture in marine and estuarine environments throughout the world. In: Menzel, W. (Editor), Culture of Estuarine and Marine Bivalve Mollusks in Temperate and Tropical Regions. CRC Press, Inc., Boca Raton, Florida, pp. 57-97.

Manahan, D. T., Wright, S. H., and Stephens, G. C., 1983. Simultaneous determination of net uptake of 16 amino acids by a marine bivalve. Am. J. Physiol., 244: 832-838.

McGrath, D., King. P. A., and Gosling, E. M., 1988. Evidence for the direct settlement of *Mytilus edulis* L. larvae on adult mussel beds. Mar. Ecol. Prog. Ser., 47: 103-106.

Mileikovsky, S. A., 1971. Types of larval development in marine bottom invertebrates, their distribution and ecological significance: a reevaluation. Mar. Biol., 10: 193-213.

Millar, R. H., 1968. Growth lines in the larvae and adults of bivalve molluscs. Nature (Lond.), 217: 683.

Miyazaki, I., 1935. On the development of some marine bivalves, with special reference to the shelled larvae. J. Imp. Fish. Inst. (Jpn.), 31: 1-10.

Miyazaki, I., 1962. On the identification of lamellibranch larvae. Bull. Jpn. Soc. Sci. Fish., 28: 955-966.

Morse, D. E., 1984. Biochemical control of larval recruitment and marine fouling. In: J. D. Costlow and R. C. Tipper (Editors), Marine Biodeterioration: An Interdisciplinary Study. Naval Institute Press, Annapolis, Maryland, pp. 134-140.

Morse, D. E., Hooker, N., Morse, A. N. C., and Jensen, R. A., 1988. Control of larval metamorphosis and recruitment in sympatric agariciid corals. J. Exp. Mar. Biol. Ecol., 116: 193-217.

Newell, G. E. and Newell, R. C., 1963. Marine Plankton. Hutchinson Educational Ltd., London, 207pp.

Niijima, L. and Dan, J., 1965. The acrosome reaction in *Mytilus edulis* L. 1. Fine structure of the intact acrosome. 2. Stages in the reaction observed in supernumerary and calcium-treated spermatozoa. J. Cell Biol., 25: 243-259.

Ockelmann, K. W., 1965. Developmental types in marine bivalves and their distribution along the Atlantic coast of Europe. In: L. R. Cox and J. F. Peake (Editors), Proceedings of the First European Malacological Congress, London, 1962. Conchological Society of Great Britain and Iceland and the Malacological Society of London, pp. 25-35.

Ockelmann, K. W., 1983. Descriptions of mytilid species and definition of the Dacrydiinae n. subfam. (Mytilacea-Bivalvia). Ophelia, 22: 81-123.

Odhner, N. H., 1914. Notizen über die fauna der Adria bei Rovigno. Berträge zur kenntnis der marine molluskenfauna von Rovigno in Istrein. Zool. Anz., 44: 156-170.

Pawlik, J. R., 1986. Chemical induction of larval settlement and metamorphosis in the reef-building tube worm *Phragmatopoma californica* (Sabellariidae, Polychaeta). Mar. Biol., 91: 59-68.

Pechenik, J. A., 1980. Growth and energy balance during the larval lives of three prosobranch gastropods. J. Exp. Mar. Biol. Ecol., 44: 1-28.

Pechenik, J. A., 1984. The relationship between temperature, growth rate, and duration of planktonic life for larvae of the gastropod *Crepidula* fornicata (L.). J. Exp. Mar. Biol. Ecol., 74: 241-257.

Petersen, J. H., 1984. Larval settlement behavior in competing species: *Mytilus californianus* Conrad and *M. edulis* L. J. Exp. Mar. Biol. Ecol., 82: 147-159.

Purchon, R. D., 1977. The Biology of the Mollusca. Pergamon Press, Oxford, 560pp.

Ramorino, L. and Campos, B., 1983. Larvas y postlarvas de Mytilidae de Chile (Mollusca: Bivalvia). Rev. Biol. Mar., 19: 143-192.

Rattenbury, J.C. and Berg, W.E., 1954. Embryonic segregation during early development of *Mytilus edulis*. J. Morph. 95: 393-414.

Redfearn, P., Chanley, P., and Chanley, M., 1986. Larval shell development of four species of New Zealand mussels: (Bivalvia, Mytilacea). N. Z. J. Mar. Freshw. Res., 20: 157-172.

Rees, C. B., 1950. The identification and classification of lamellibranch larvae. Hull Bull. Mar. Ecol., 3: 73-104.

Riisgård, H. U., Randløv, A., and Kristensen, P. S., 1980. Rates of water processing, oxygen consumption, and efficiency of particle retention in veligers and young post-metamorphic *Mytilus edulis*. Ophelia, 19: 37-47.

Savage, R. E., 1956. The great spatfall of mussels in the River Conway Estuary in spring 1940. Fish. Invest. Minist. Agric. Fish. Food, Lond., Ser II, 20: 1-21.

Schweinitz, E. H. de, and Lutz, R. A., 1976. Larval development of the northern horse mussel, *Modiolus modiolus* (L.), including a comparison with the larvae of *Mytilus edulis* L. as an aid in planktonic identification. Biol. Bull., 150: 348-360.

Sebens, K. P., 1983. Settlement and metamorphosis of a temperate soft-coral larva (*Alcyonium siderium* Verrill): induction by crustose algae. Biol. Bull., 165: 286-304.

Seed, R., 1969. The ecology of *Mytilus edulis* L. (Lamellibranchiata) on exposed rocky shores. 1. Breeding and settlement. Oecologia (Berl.), 3: 277-316.

Seed, R., 1971. A physiological and biochemical approach to the taxonomy of *Mytilus edulis* L. and *M. galloprovincialis* Lmk. from south-west England. Cah. Biol. Mar., 12: 291-322.

Seed, R., 1975. Reproduction in *Mytilus* (Mollusca: Bivalvia) in European waters. Pubbl. Stn. Zool. Napoli, 39:317-334.

Seed, R., 1976. Ecology. In: B. L. Bayne (Editor), Marine Mussels: their ecology and physiology. Cambridge University Press, Cambridge, pp. 13-65.

Siddall, S. E., 1978. The development of the hinge line in tropical mussel larvae of the genus *Perna*. Proc. Natl. Shellfish. Assoc., 68: 86.

Siddall, S. E., 1980. A clarification of the genus *Perna* (Mytilidae). Bull. Mar. Sci., 30: 858-870.

Sprung, M., 1982. Untersuchungen zum energiebudget der larven der miesmuschel, *Mytilus edulis*. L. Ph.D. Thesis, University of Kiel, Kiel, Germany.

Sprung, M., 1984a. Physiological energetics of mussel larvae (*Mytilus edulis*). II. Food uptake. Mar. Ecol. Prog. Ser., 17: 295-305.

Sprung, M., 1984b. Physiological energetics of mussel larvae (*Mytilus edulis*). I. Shell growth and biomass. Mar. Ecol. Prog. Ser., 17: 295-305.

Sprung, M., 1984c. Physiological energetics of mussel larvae (*Mytilus edulis*). III. Respiration. Mar. Ecol. Prog. Ser., 18: 171-178.

Sprung, M., 1984d. Physiological energetics of mussel larvae (*Mytilus edulis*). IV. Efficiencies. Mar. Ecol. Prog. Ser., 18: 179-186.

Stafford, J., 1912. On the recognition of bivalve larvae in plankton collections. Contrib. Can. Biol. Fish., 1906-1910: 221-242.

Stenzel, H. B., 1964. Oysters: composition of the larval shell. Science, 145: 155-156.

Stephenson, R. L. and Chanley, P. E., 1979. Larval development of the cockle *Chione stutchburyi* (Bivalvia: Veneridae) reared in the laboratory. N. Z. J. Zool., 6: 553-560.

Strathmann, M. F., 1987. Reproduction and Development of Marine Invertebrates of the northern Pacific coast. University of Washington Press, Seattle, 670pp.

Sullivan, C. M., 1948. Bivalve larvae of Malpeque Bay, P. E. I. Fish. Res. Board Can., Bull. No. 77, 36pp.

Tamarin, A. and Keller, P. J., 1972. An ultrastructural study of the byssal thread forming system in *Mytilus*. J. Ultrastruct. Res., 40: 401-416.

Tamarin, A., Lewis, P., and Askey, J., 1974. Specialized cilia of the byssus attachment plaque forming region in *Mytilus californianus*. J. Morphol., 142: 321-327.

Tanaka, Y., 1958. Studies on molluscan larvae. Venus, 20: 207-219.

Taylor, J. D., Kennedy, W. J., and Hall, A., 1969. The shell structure and mineralogy of the Bivalvia: Introduction. Nuculacea-Trigonacea. Bull. Br. Mus. (Nat. Hist.) Zool., 3: 1-125.

Thorson, G., 1946. Reproduction and larval development of Danish marine bottom invertebrates. Meddr. Kommn. Danm. Fisk. -øg Havunders. Ser. : Plankton, 4: 1-523.

Thorson, G., 1950. Reproductive and larval ecology of marine bottom invertebrates. Biol. Rev., 25: 1-45.

Thorson, G., 1966. Some factors influencing the recruitment and establishment of marine benthic communities. Neth. J. Sea Res., 3: 267-293.

Turner, R. D., and Boyle, P. J., 1974. Studies of bivalve larvae using the scanning electron microscope and critical point drying. Am. Malacol. Union Inc. Bull., 40: 59-65.

Waller, T. R., 1981. Functional morphology and development of veliger larvae of the European oyster, *Ostrea edulis* Linné. Smithson. Contrib. Zool., No. 328, 70pp.

Walne, P. R., 1963. Observations on the food value of seven species of algae to the larvae of *Ostrea edulis*. I. Feeding Experiments. J. Mar. Biol. Ass. U. K., 43: 767-784.

Wangersky, P. J., 1978. Production of dissolved organic matter. In: O. Kinne (Editor), Marine Ecology, Vol. 4. John Wiley, Chichester, New York, pp. 115-220.

Werner, B., 1939. Über die entwicklung und artunterscheidung von muschellarven des Nordseeplanktons, unter besonderer beruchsichtigung der schalenentwicklung. Zool. Jahrb. Abt. Anat. Ontog., 66: 1-54.

Yoo, S. K., 1969. Food and growth of the larvae of certain important bivalves. Bull. Pusan Fish. Coll. (Nat. Sci.), 9: 65-87.

Yoshida, H., 1953. Studies on larvae and young shells of industrial bivalves in Japan. J. Shimonoseki Coll. Fish., 3: 1-106.

Young, R. T., 1941. The distribution of the mussel *Mytilus californianus* in relation to the salinity of its environment. Ecology, 22: 379-386.

Zakhvatkina, K. A., 1959. Larvae of bivalve molluscs of the Sevastopol region of the Black Sea. Tr. Sevastop. Biol. Stn. Im. A. D. Kovalenskago Akad. Nauk. Ukr. SSR., 11: 108-152.

Chapter 4

POPULATION AND COMMUNITY ECOLOGY OF *MYTILUS*

RAYMOND SEED AND THOMAS H. SUCHANEK

INTRODUCTION

The Mytilidae is a family of considerable antiquity dating back to the Devonian era (Moore, 1983) and includes many important byssally attached genera such as *Choromytilus*, *Perna*, *Modiolus* and *Aulacomya*, as well as *Mytilus* itself. The development of byssal attachment threads by adult mussels, probably as a result of their neotenous retention by the byssate postlarval stages of some earlier burrowing taxa, coupled with the associated evolution of a heteromyarian wedge-shaped shell (the 'typical' mussel form) has enabled mytilid mussels to successfully exploit hard or semiconsolidated substrata (Yonge, 1976; Seed, 1990, and Chapter 2) and to dominate rocky shore habitats on all continents. Their worldwide success as dominant space occupiers, however, is perhaps most pronounced on flat or gently shelving wave-exposed shores in temperate latitudes (Lewis, 1964; Suchanek, 1985).

Mussels belonging to the genus *Mytilus* are widely distributed throughout the cooler waters of both the northern and southern hemispheres and have proved to be model organisms for various physiological, biochemical and genetic investigations (Chapters 5, 6 and 7). They are also important economically as food and fouling organisms (Chapter 10) and as biomonitors of coastal water quality (Chapters 8 and 9). In historic and prehistoric times *Mytilus* shells were used extensively for tools by native North American Indians as well as by the Pilgrim settlers (Miller, 1980). With few exceptions, in most exposed or moderately wave-exposed locations in temperate zone habitats *Mytilus* spp. form the foundation for a variety of diverse hard-shore communities. In this chapter we shall briefly examine selected aspects of the population and community ecology of mussels belonging to this successful and widely distributed genus. Despite recent changes in the taxonomy and known distribution patterns of *Mytilus* worldwide (McDonald and Koehn, 1988; McDonald et al., 1991; see also Chapters 1 and 7) we have retained the specific names as used by authors in the primary citations in order to avoid any unnecessary confusion.

DISTRIBUTIONAL AND ZONATIONAL PATTERNS

On a worldwide basis, mussels of the family Mytilidae form the foundation (both in terms of strict percent cover, as well as organic production) for most exposed rocky shore communities within the temperate zone (Suchanek, 1985). The genus *Mytilus* is the most diverse and widely distributed of the Mytilidae, its representatives typically occupying significant space on intertidal sites on most major continents. A relatively small-sized member of the family that has received most attention, *Mytilus edulis* has the widest distributional patterns of the genus, and is typically quite eurytopic, with abilities to withstand wide fluctuations in salinity, desiccation, temperature and oxygen tension. As a result it often occupies, or has the capability of occupying, a broad variety of microhabitats. This allows it to extend its zonational range from the high intertidal to subtidal regions, its salinity range from estuaries to fully oceanic sites, and its climatic regime from mild, subtropical locations to ice-scoured and frequently frozen habitats. Other members of the genus (e.g. *Mytilus californianus* and *Mytilus galloprovincialis*), are usually more restricted to narrower bands and sites controlled by both physical and biological factors.

Although sometimes found in abundance subtidally (Newcombe, 1935; Paine, 1976b; Tursi et al., 1985), *Mytilus* species typically occur in intertidal habitats; this limited distribution appears mostly controlled by biological factors of predation and competition, rather than an inability to survive the conditions found in subtidal habitats. At Tatoosh Island, Washington, U.S.A., when given proper refuges from seastar predation by *Pisaster ochraceus*, *M. edulis* settles and survives in relatively high frequency on low intertidal and/or subtidal substrata, such as the frond crotches of brown algae (e.g. *Lessoniopsis*), with densities ranging from 5–80 mussels plant[-1] (Suchanek, unpublished results). *M. californianus*, a typically dominant intertidal mussel in Washington, is usually limited by *Pisaster* predation in the low intertidal (Paine 1974, 1976a), but can be found occupying vast areas subtidally to depths of nearly 30m on seamounts (Scagel, 1970; Chan, 1973; Paine 1976b; Suchanek, unpublished results). In these unique habitats, *Pisaster* is in extremely low abundance (ca. 0–0.017m[-2]), resulting in *M. californianus* cover from ca. 28–100% (Paine, 1976b).

Although suitable for the development of healthy mussel beds, many moderately exposed habitats in Europe, as well as North America, display a distinct absence of dense *M. edulis* populations, but the reasons behind this phenomenon are likely to have diverse origins. One possible explanation is provided from results of research at Lough Ine, South-west Ireland (Ebling et al., 1964), where the absence of mussels from areas of moderate exposure is apparently due to intense crab predation. In contrast, *Mytilus* in sheltered areas experience a size-refuge from predation, and the absence of crabs in exposed habitats provides a spatial refuge. Within the San Juan Archipelago,

Washington state, U.S.A., *M. edulis* is also scarce in habitats of intermediate exposure, probably because of log damage as well as dogwhelk predation (Dayton, 1971).

Subtidal populations often occur on seamounts (Paine, 1976b), dock pilings (Suchanek, 1978) and offshore oil platforms (Page and Hubbard, 1987). Typically, the unique aspects of these populations include the continuous growth and lack of predators, allowing individual *Mytilus* to attain large sizes, often in relatively short periods of time. Subtidal *Mytilus* aggregations that reach 120cm in thickness have been reported from oil platforms off the coast of California (Simpson, 1977). Growth rates for these continually submerged individuals have also been documented as some of the highest on record, yielding ca. 50mm mussels in 6–8mo (Page and Hubbard, 1987).

Factors Limiting Upper and Lower Limits

Upper limits

Upper distributional limits for *Mytilus* spp. are typically rather constant over long periods of time (see Suchanek, 1985 for review and data for *M. edulis* and *M. californianus*). Physiological intolerance to temperature extremes and desiccation represent the most important factors in the determination of upper limits for *Mytilus* populations in rocky intertidal sites. *M. edulis* is known to withstand extreme cold and even freezing (Williams, 1970; Aarset, 1982). The presence of nucleating agents in the haemolymph during winter probably insures that freezing (which occurs in extracellular compartments; Kanwisher, 1959, 1966) takes place at a few degrees below zero, thereby preventing intracellular freezing and subsequent injury (Aunaas, 1982; Aunaas et al., 1988). In most temperate zone habitats *M. edulis* is only subjected to lethal low temperatures periodically, if at all. However, in some locations such as the St. Lawrence Estuary, temperatures typically reach –30 to –35°C each winter (Bourget, 1983). Williams (1970) has shown that *M. edulis* can survive even after tissue temperatures declined to –10°C . Furthermore, in 24h laboratory tests using large *M. edulis* (>3cm), Bourget (1983) found that the median lethal temperature (MLT) was as low as –16°C, whereas for juveniles (<1.5cm), the MLT was considerably higher (–12.5°C). If exposed for only 16h, comparably sized *M. edulis* had minimum MLTs of –20°C and –12.5°C respectively. However, probably of more relevance to indigenous *M. edulis* populations are cyclic exposures to 'sublethal' temperatures (e.g. –8°C every 12.4h), with significant damage, which may lead to death even after three to four such cyclic exposures (Bourget, 1983).

M. californianus is much more stenothermal than *M. edulis* and appears unable to tolerate freezing conditions. Suchanek (1985) reported a December 1983 winter freeze at Tatoosh Island that caused substantial mussel mortality and significantly lowered the upper distributional limit of *M. californianus*. Five years later (January 1989) another significant freeze has lowered the upper limit of the *M. californianus* zone by ca. 0.27 vertical meters, as measured at a nearby mainland site, Shi-Shi (Suchanek, unpublished data). In the upper 1.65m of this zone, *M. californianus* cover has been reduced from nearly 90% to ca. 25%, whereas in the remaining 1.31m, mussel cover has been reduced to ca. 50%. In fact, freezing is the most likely factor in preventing *M. californianus* from dominating most intertidal sites north of Sitka, Alaska, where *M. edulis* replaces it as the dominant space occupier. As a result, *M. californianus* is usually restricted to intertidal pools, crevices, or subtidal habitats from Sitka northward to the Aleutian Islands. Paine (1986) has shown that the upper distributional limits of both *M. edulis and M. californianus* do not appear to be affected by El Niño episodes over a 14-year period. However, Paine did find that the intensity of mussel bed matrix disruption by storm-related phenomena did show temporal correlation with El Niño periodicity, but the ultimate causes of such events are speculative (see below for other disturbance data).

The detrimental effects of extreme high temperatures on setting upper limits for *Mytilus* spp. has also been well-documented. For nearly all mytilid species studied so far, high temperatures usually interact additively or synergistically with desiccation to control upper zonational limits. Occasional, sudden and massive mortalities at the upper limit of intertidal mussel bands are often correlated with prolonged periods of unusually high temperatures and associated desiccation stress (Suchanek, 1978, 1985; Tsuchiya, 1983). For instance, in northern Japan an extreme period of hot days (air temperature ca. 34°C, resulting in mussel tissue temperatures >40°C) caused mortality for about 50% of intertidal *M. edulis* within one hour, with mortality occurring over approximately the upper 75% of the intertidal range for *M. edulis* (Tsuchiya, 1983). Interestingly, another mussel, *Septifer (Mytilisepta) virgatus*, living at the same site above the *M. edulis* zone, suffered only limited (16.8%) mortality.

In consistently warmer climates, behavioural activities probably modify the abilities of some *Mytilus* species to circumvent the effects of desiccation and/or high temperature. British *M. edulis* have an upper sustained thermal tolerance limit of about 29°C (Read and Cumming, 1967; Almada-Villela et al., 1982), but can probably withstand somewhat higher temperatures for short periods of time (Cawthorne, 1979). European *M. edulis* very likely never experience temperatures greater than about 25°C, but the Indian mussel, *Perna* (formerly *Mytilus) viridis*, is predictably exposed to annual mean temperatures of ca. 27.3°C year-round, with deviations of only about 1.4°C (Davenport, 1983). Both species close their valves in response to aerial exposure (as well as to lowered salinity), but *P. viridis* also first takes a bubble of air into the mantle cavity, which probably increases its tolerance to desiccation in such high temperature environments, especially in high intertidal sites (Davenport, 1983). If similar behavioural differences exist between *Mytilus* and *Septifer* (see above), this may also help to explain their significant differences in survivability at high temperatures.

Finally, recruitment or movement into cracks, crevices or pools obviously affords much better protection from the physical effects of both temperature and desiccation, although those habitats can also protect mussels from the influence of storm waves and wave-driven logs. Observations on the effect of such microhabitats in increasing survival in mussel populations have been reviewed in Suchanek (1985).

In some cases, competition with other fauna or flora can also influence upper distributional patterns for *Mytilus* spp. For example, in very exposed rocky intertidal sites along the U.S. west coast, individuals of *M. californianus*, at the upper limit of their intertidal distribution, appear to be negatively affected by the sea palm *Postelsia palmaeformis*. During a typical growth season at Bodega Bay, California, about 5% or more of the *Postelsia* population is overgrowing *Mytilus* at the upper tidal range (Suchanek, unpublished results). Although the mechanisms involved are not completely clear, *Postelsia* could adversely affect *M. californianus* by several potential mechanisms: (1) overgrowth of valves and feeding apertures by sea palm holdfasts, with subsequent reduction of food intake for *Mytilus*, resulting in elimination of the mussels, (2) overgrowth of the mussels by holdfasts with subsequent dislodgement during storms, again resulting in elimination of the mussels or (3), proprietary settlement by *Postelsia* in high disturbance gaps, preventing *Mytilus* from attaining space in those intertidal sites (Suchanek, unpublished results), although Paine (1979, 1984) states that mussels eventually reclaim this lost space.

Lower limits

Lower zonational limits for many sessile fauna, including mussels, have been shown to be under strong influence from biological factors, especially predators (Connell, 1972; Paine, 1974). For over five decades seastars have been recognized as the most important predators establishing *Mytilus* lower limits (Newcombe, 1935; Kitching et al., 1959; Paris, 1960; Ebling et al., 1964; Kitching and Ebling, 1967; Paine, 1974; Menge, 1983; and see review by Suchanek, 1985). On the east coast of England predatory seastars (*Asterias rubens*) and dogwhelks (*Nucella lapillus*) control the lower limits of *M. edulis* beds, essentially eliminating these mussels from the lower intertidal zone (Seed, 1969b). In Ireland, crabs (*Carcinus* and *Liocarcinus*), dogwhelks (*Nucella*) and seastars (*Marthasterias*) most likely control mussel zonation (Kitching and Ebling, 1967). Along New England shores strong evidence shows that the lower limit of *M. edulis* is controlled by a suite of consumers: *Nucella*, two seastars of the genus *Asterias*, and three crabs (*Carcinus* sp. and *Cancer* spp.) (Menge, 1983 and see Suchanek, 1985). At Shi-Shi in Washington state, U.S.A. *M. edulis* settles

Fig. 4.1. Photograph from Shi-Shi, Washington state, North America, illustrating *Pisaster* predation on *Mytilus* beds. Central core bed is composed of *Mytilus californianus* (central left and fouled with light coloured barnacles, mostly *Semibalanus cariosus*), surrounded by a dense ring of *Mytilus edulis*. After locating these recently recruited *M. edulis*, the *Pisaster* herd at the perimeter consumed over 90% of the bed within one week. The central *M. californianus* core is relatively protected from such predation by virtue of its location on the highest portion of an intertidal mound.

unpredictably in patches of open space, typically just below the intertidal distribution of *M. californianus* (Suchanek, unpublished results). These individuals grow quickly to reproductive size but are soon 'discovered' and consumed by 'herds' of roaming *Pisaster ochraceus* seastars, which can eliminate large beds of *M. edulis* within a few days (Fig. 4.1). Physiological intolerance to desiccation is believed to be the most likely factor that limits the upshore extension of most predators controlling the lower limits of *Mytilus* beds. As such, the upper intertidal zones represent an effective refuge for *Mytilus*, although this incurs costs of lowered fitness for these mussels in terms of reduced reproductive output (see Suchanek, 1981).

The lower limits of other *Mytilus* species appear to be controlled in much the same way as that of *M. edulis*. In an elegant removal experiment, Paine (1974) showed causality by systematically removing the predatory seastar *Pisaster* from a stretch of Tatoosh Island shoreline in Washington state, which resulted in a downward vertical extension of the *M. californianus* band by an average of 0.84m. In other experimental removals, Paine (1971) working with the seastar *Stichaster* in New Zealand, and Paine et al. (1985) working with the seastar *Heliaster* in Chile, showed similar trends in altering the lower limits of the indigenous mussels *Perna* and *Perumytilus* respectively. Typically, when seastar predators are then allowed to re-enter the system, the lower limit returns to the previous state. However, if the mussels are able to grow beyond the size which seastars are capable of consuming (Paine, 1976a; Paine et al., 1985; and see p.128–132 on size-limited predation), an altered state in which mussels exist below their typical lower limit may persist for up to 10 to 30 years for *Mytilus* and *Perna* respectively. A removal of *Pisaster* over a six and a half year period from a region in central California, where sea otters occur, did not result in the predicted lowering of the mussel zone, and in some transects resulted in an upward movement of the mussel zone (VanBlaricom, 1988), although other factors such as variable recruitment may be responsible for these inconsistent results.

Physical factors also occasionally fix lower limits to mussel zonation patterns. In California (Cimberg, 1975; Littler et al., 1983) sand burial periodically can limit the lower distribution of *M. californianus*, and Daly and Mathieson (1977) report similar results for *M. edulis* populations in New Hampshire, U.S.A.

Competition from mussels or other sessile organisms clearly sets the lower limits for *Mytilus* in at least some sites. Along the west coast of North America, *M. californianus* dominates *M. edulis* in most exposed rocky intertidal habitats (Suchanek, 1978, 1981). Whether this effect is mediated by the 'brute force' of crushing its congener (as suggested by Harger, 1972), or some other mechanism, is still unclear. *M. edulis* can certainly withstand the rigours of extremely wave-battered shores, including the maintenance of byssal thread attachments under the most high-energy conditions, since it occupies a predictable band above the *M. californianus* zone (Suchanek, 1978). In addition, it temporarily occupies disturbance gaps within the *M. californianus* zone, and is commonly found in the holdfasts and stipe crotches of the brown alga *Lessoniopsis*, well below the *M. californianus* zone. In Washington, as disturbance gaps heal, *M. edulis* is eventually replaced by *M. californianus*. In Alaska, however, where *M. californianus* is eliminated from most intertidal sites by freezing, *M. edulis* is free to dominate the majority of intertidal space at tidal heights more typical of *M. californianus* from more southerly locations. The lower limits of *M. edulis* in Alaska are also determined by dogwhelk and seastar predation (see above), but where predators are eliminated (e.g. sites near glacial melt waters, where salinity is extremely low as in Glacier Bay, Alaska), *M. edulis* dominates intertidal and subtidal space, with a vertical extent of 5.5m or more (Suchanek, unpublished results).

It should be noted that in all of the studies discussed above on similarities and differences in competitive abilities, physiological adaptations, and distributions of '*M. edulis*' at various geographic locations along the west coast of North America, the assumption is that the small blue or blue-black *M. edulis*-looking morphotype, identified in the field by previous investigators, is indeed *M. edulis*. New information on the systematics of *Mytilus* (McDonald et al., 1991; see also Chapters 1 and 7) may be especially critical in understanding unusual differences in both distribution and competitive abilities between mussels from different geographic sites.

For instance, Harger (1972) reports that *M. edulis* is a species adapted to calm waters and protected embayments and is not found on the exposed outer coast of California, where the larger *M. californianus* is dominant. Suchanek (1978, 1985), however, reported that *M. edulis* maintains a predictably discrete and dominant zone in extremely wave-exposed rocky shores in the high intertidal zone in Washington, and usually in the mid to high intertidal zone in Alaska. Are these contradictory reports indicative of different ecological niches involving habitat preferences, physiological tolerances and/or competitive abilities for two or more species of *Mytilus*? Considerably more ecological, physiological and genetic data are needed before we can answer these questions.

How or whether we must now alter our current interpretations of species-species interactions involving a universal '*M. edulis*' from these locations along the Pacific coast of North America is uncertain. More understanding of the identity of the *M. edulis* morphotypes and genotypes is needed before we can truly be confident that the similarities or differences we report are real and meaningful characteristics attributed to a single species. See Chapters 1 and 7 for a more detailed discussion.

REPRODUCTION

Most marine organisms have a geographical range over which they survive and a narrower range over which they breed successfully. Investigations of natural reproductive cycles are therefore central, not only to studies of population dynamics, but also to our understanding of biogeography and speciation. In the case of commercial bivalves such as *Mytilus*, which are known to exhibit a significant loss of condition following spawning, they will also indicate when the population can be harvested most effectively. Studies of natural reproductive cycles are most valuable when carried out over several years since the onset and duration of both gametogenesis and spawning in *Mytilus* can exhibit considerable temporal and spatial variation. The reproductive cycle comprises the entire sequence of events from activation of the gonad, through gametogenesis to the release of ripe gametes (= spawning), and the subsequent recession of the gonad. Typically, it can be broadly divided into a reproductive period which starts with the initiation of gametogenesis and culminates in the emission of gametes, and a reproductively quiescent period in which the energy stores required to fuel gametogenesis are accumulated (see Gabbott, 1983). The reproductive period is characterized by one or more gametogenic cycles each of which is followed by the release of gametes when the reproductive follicles become partially or completely emptied. Like many temperate water bivalves, most *Mytilus* populations exhibit a seasonal pattern of reproduction. In some populations, however, mussels may slowly dribble gametes more or less continuously over an extended part of the year, a pattern more typically associated with populations in less cyclical environments.

Anatomy and Histology of the Gonad

Apart from a few hermaphrodites the sexes in *Mytilus* are separate and most populations contain approximately equal numbers of males and females (Seed, 1976; Sunila, 1981; Kautsky, 1982a; Brousseau, 1983; Sprung, 1983), although these cannot be distinguished on external characters. The colour of the reproductive tissue of *M. edulis* varies considerably, but typically females are orange coloured while males are creamy-white; the mantle tissues of recently spent mussels are thin and transparent, often with reddish-brown blotches. A simple and convenient colorimetric method for determining the sex of *M. edulis* has recently been described by Jabbar and Davies (1987).

M. edulis can become sexually mature in its first year but the size at which this occurs depends largely on local growth rates (Seed, 1976; Sprung, 1983 and references therein). Most gametes are generated within the extensive mantle folds, though small amounts of the reproductive tissue also extend into the visceral mass and mesosoma. Paired gonoducts, which discharge into the mantle cavity on papillae situated between the mesosoma and the inner gill lamellae, lead into five major canals with convoluted walls forming longitudinal ciliated ridges. These lead in turn into a series of smaller canals which have part of their walls of ciliated columnar epithelium. Each of these fine ducts eventually terminates in a genital follicle. Oogonia and spermatogonia are budded off from the germinal epithelium of these follicles. Early oocytes are connected to the epithelium by a broad stalk but this gradually becomes more slender and finally ruptures to leave the mature ova free within the follicular cavity. Spermatogonia give rise, in turn, to concentric bands of spermatocytes, spermatids and spermatozoa, the latter converging towards the centre of the follicles in the form of dense lamellae.

Methods Used for Assesssing the Reproductive Cycle

Several methods have been used to assess the course of the reproductive cycle. These may involve direct observations of spawning in natural or laboratory populations as well as the macro- and microscopic appearance of the gonad throughout the year. Observations of spawning in field populations provide the most reliable evidence of natural spawning though these data may be difficult or impossible to obtain.

Alternatively, the reproductive period can be inferred from the appearance of larvae in the plankton, or the recruitment of juvenile mussels (= spat) to the populations. Data obtained using these indirect methods, however, are generally less reliable since larvae and spat may have been transported by currents over considerable distances from parental stocks which have experienced quite different environmental conditions. However, they can serve as valuable checks on data obtained by more direct methods.

The most reliable and detailed information regarding the annual reproductive cycle is that obtained from histological preparations of mussels sampled at regular intervals throughout the year. Several schemes have been used to classify the reproductive condition in *Mytilus*. While some of these simply rely on the general macroscopic features such as the colour, texture or thickness of the gonad, others are based on microscopic appearance of squashes or thin, stained sections of mantle tissue (Lubet, 1959; Wilson and Seed, 1974; Seed and Brown, 1977; Sunila, 1981; Kautsky, 1982a; Fell and Balsamo, 1985; King et al., 1989). From such preparations various stages in the reproductive cycle (developing, ripe, spawning and spent) can be recognized. Developing and spawning stages are usually further subdivided, thus resulting in several arbitrary stages into which any individual mussel can be assigned. The general reproductive condition of the population can then be assessed by calculating a mean gonad index. This is obtained by multiplying the number of mussels in each stage by the numerical ranking of that stage and dividing the resulting value by the total number of mussels in the sample. Gametogenesis leads to an increase in this index, while a decrease in the index denotes spawning.

One of the major disadvantages with such arbitrary classifications of gonad condition is that they are rather subjective, and do not fully recognize the occurrence of intermediate stages of development; the gonad index is thus a nominal rather than an interval measurement. Moreover, they provide no information on mantle nutritive storage cells. Consequently, some workers (Bayne et al., 1978; Lowe et al., 1982; Newell et al., 1982; Rodhouse et al., 1984a; Hawkins et al., 1985) have preferred to use more quantitative stereological methods. These enable the changes in the volume fractions of different components within the gonad (e.g. gametes, storage cells) throughout the course of gametogenesis to be established from point counts on test grids applied to random thin sections of mantle tissue. Rodhouse et al. (1984a), however, found that there was generally good agreement between the gonad index and gamete volume fraction (i.e. the proportion of mantle tissue that consists of follicles containing developing and ripe gametes) in *M. edulis* from Killary Harbour, western Ireland (Fig. 4.2). Stereology thus offers a simple technique whereby the main events in the reproductive cycle, including periods of energy storage and utilization, can be quantified and related to environmental variables. The technique has also

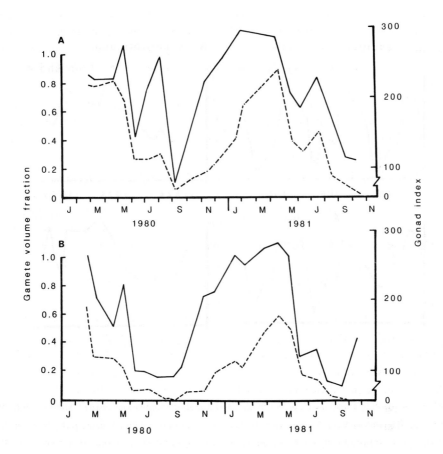

Fig. 4.2. Annual cycle of gamete volume fraction (dashed line) and gonad index (solid line) in (A) cultivated, and (B) wild *Mytilus edulis* from Killary Harbour, Ireland. (After Rodhouse et al., 1984a).

proved useful in quantifying the effects of various pollutants on gonad development in *M. edulis* (see p.452 Chapter 9).

Gonad indices have sometimes been expressed as a function of gonad weight and shell length (Suchanek, 1981). However, some caution is required when using gonad or mantle weights as indicators of gamete production since these tissues also contain variable amounts of stored nutrient reserves such as glycogen. Thompson (1984a) used the DNA content of male mussels to establish the course of the gametogenic cycle in a subarctic population of *M. edulis* in Newfoundland, Canada. Other workers (Wilson and Seed, 1974; Seed, 1975; Wilson, 1988) have assessed reproductive

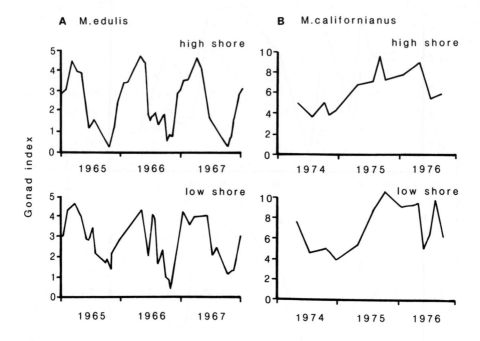

Fig. 4.3. Comparison of gonad development in high and low shore populations of (A) *Mytilus edulis* from Filey Brigg, England and (B) *Mytilus californianus* from Tatoosh Island, Washington state, North America; note the different gonad indices ranging from completely spent (0) to fully ripe (5 and 10 in *M. edulis* and *M. californianus* respectively). (After Seed, 1975 and Suchanek, 1981).

condition in terms of the density and size of oocytes in mantle sections. More recently, image analysis has been used in the case of oysters (Wilson and Simons, 1985), and could also presumably be applied to *Mytilus*.

The Annual Cycle

Figure 4.3A illustrates the annual fluctuations in the gonad index of *M. edulis* from high and low intertidal sites in North-east England, over a three-year period. The gonad index used varies from 0 (completely spawned out) to 5 (fully ripe).

Redevelopment of the resting gonad commences during October or November and gametogenesis then proceeds throughout the winter so that by early spring the gonads of most mussels are morphologically ripe. During the spring months a period of partial spawning is followed by rapid gametogenesis until by early summer the gonads are again fully ripe. This second period of gametogenic activity is more evident among mussels in the more favourable feeding conditions of the lower intertidal zone. Additional, less intensive, spawnings may occur throughout the summer until by late August or September the gonad index reaches its lowest value, as most mussels enter their reproductively quiescent phase. Throughout late August, September and October the mantle tissue becomes progressively thickened with varying amounts of nutrient reserves, which are used subsequently to fuel gametogenesis throughout winter when food supplies may be more limiting. A significant feature of the annual cycle of the populations described above is the actual duration of the spawning period, with some gamete release apparently occurring throughout much of the year (see also Fell and Balsamo, 1985), albeit with seasonal peaks during the spring and summer. Neither body size nor tidal level appeared to have any marked or consistent effects on the timing of spawning (Seed, 1975; but see Campbell, 1969; Kautsky, 1982a). Male mussels, however, were generally at a rather more advanced stage of gametogenesis than females at any particular time (Seed, 1969a), possibly because sperm can be produced at a faster rate than ova, which have large yolky reserves. Newell et al. (1982) also noted that male mussels had a higher gamete volume fraction than females during the course of maturation.

Protracted reproductive periods resulting from repeated spawnings, particularly during the spring and summer, have also been reported for natural *M. edulis* populations elsewhere (Wilson and Seed, 1974; Seed and Brown, 1977; Briggs, 1978; Sunila, 1981; Lowe et al., 1982; Brousseau, 1983; Fell and Balsamo, 1985; McKenzie, 1986; King et al., 1989), and appear to be characteristic of many cultivated mussels growing under particularly favourable nutrient conditions (Lutz et al., 1980; Rodhouse et al., 1984a; Zhang, 1984; Wilson, 1987; Wallace, 1990). Other populations, however, may exhibit a single short spawning period lasting only a few weeks (Chipperfield, 1953; Kautsky, 1982a; Newell et al., 1982), suggesting that food is perhaps more limiting at other times of the year.

Bayne (1976) classifies bivalve reproductive strategies according to the relationship between spawning and storage cycles. Conservative species utilize nutrient stores accumulated during the summer and autumn for gametogenesis during the winter, and are thus partially buffered from adverse environmental changes. In opportunistic species, however, gametogenesis is more closely linked with the prevailing food supply, thus enabling them to capitalize on periods when feeding conditions are especially favourable. *M. edulis* is apparently quite flexible and can exhibit both types of strategy (Lowe et al., 1982; Rodhouse et al., 1984a). In some populations it clearly follows a conservative strategy, spawning in the early spring, thus enabling the developing larvae to exploit the spring phytoplankton bloom, which is so characteristic of coastal waters in temperate latitudes. In many populations, however, the spring spawning is followed by further opportunistic spawnings, which derive from energy resources accumulated concurrently with gametogenesis. Populations grown under particularly favourable culture conditions may be entirely opportunistic in their reproductive strategy (Rodhouse et al., 1984a). Mussels which spawn late in the year, when food supply may be at, or even below, maintenance levels, can be at considerable reproductive risk since winter stress may kill those with insufficient nutrient reserves. Late spawnings are to be anticipated, therefore, only when energy reserves surplus to those required for basal metabolism and necessary gamete production over winter occur. *Mytilus*, therefore, seems to have a remarkable ability to adjust its reproductive strategy according to the prevailing environmental conditions.

In addition to the annual variations in the timing of both gametogenesis and spawning which can occur within any particular mussel population, reproductive cycles in *Mytilus* also vary geographically. The reproductive cycles of several mytilids from various parts of their geographical range are extensively reviewed by Seed (1976) and Suchanek (1985). Marked variations also occur between coexisting populations of different species. Along the west coast of North America, for instance, *M. edulis* spawns over a relatively restricted period during the late autumn or early winter (October to February), whereas in the same region *M. californianus* apparently dribbles gametes continuously throughout the year though never spawning out completely (Fig. 4.3B). These differences seem to be related to other aspects of the life-history strategies of these two closely related mussels (Suchanek, 1981), and are thus probably important in facilitating their continued coexistence in many west coast habitats. Differences have also been reported in the reproductive cycles of coexisting populations of *M. edulis* and *M. galloprovincialis* from South-west England (Seed, 1971; Gardner and Skibinski, 1990).

Factors Controlling the Reproductive Cycle

Although the reproductive cycle in *Mytilus* has been well-documented, we still have only a partial understanding of the complex interactions between those exogenous (e.g. temperature, food, salinity) and endogenous (e.g. nutrient reserves, hormonal cycles, genotype) factors that determine the initiation and duration of gametogenesis and spawning (Newell et al., 1982; see also Chapter 6). The proximate cues responsible for triggering spawning are probably of only brief duration and may be quite different from those which control the growth and maturation of gametes, and which will thus operate over a more extended time-scale.

Many attempts have been made to determine the key factor(s) which synchronize the reproductive cycle to the prevailing environmental conditions. Of these, sea temperature, which varies seasonally and latitudinally in a moderately uniform manner, has perhaps received most attention, and the concept of a causal relationship between this factor, reproduction and geographical distribution has become widely established as a general zoogeographical principal. Bayne (1975) described a linear relationship between the rate of gametogenesis in *M. edulis* and the rate of temperature change measured as 'day-degrees' (see p.117); the duration of spawning, by contrast, was more variable and seemed to be related to nutritional status and fecundity. Gamete formation in *Mytilus*, however, is evidently not inhibited by low temperature, since differentiation in some populations can proceed even when the water temperature is close to zero (Kautsky, 1982a; Thompson, 1984a, b).

Reproductive cycles in *Mytilus* vary latitudinally both in terms of their onset and duration, with mussels from the warmer, more southerly, waters of the northern hemisphere generally reproducing earlier in the year than conspecifics further north. Moreover, the cyclical nature of gametogenesis and cycles of storage and utilization of reserves are typically less pronounced in more southerly populations (Gabbott, 1975). In Britain *M. edulis* populations on the west coast reproduce earlier in the year than those on the colder east coast (Seed, 1975), while in South-west England the Mediterranean mussel, *M. galloprovincialis* spawns several weeks later than *M. edulis* when the sea water temperature for that locality is at its maximum (Seed, 1971). Such observations suggest that sea temperature acts as a principle factor in controlling the broader aspects of the annual cycle of *Mytilus*.

In a study of several populations of *M. edulis* along the eastern seaboard of North America, however, Newell et al. (1982) found no discernible latitudinal trends in reproduction. Two populations on Long Island, at the same latitude and experiencing the same temperature regime, exhibited the greatest temporal differences in gametogenesis, with summer reproductive maxima separated by an interval of three months. Newell et al. (1982) attributed these differences to temporal and quantitative variations in the energy content of the available food. Again, working on Long Island, Fell and Balsamo (1985) also concluded that temperature was not a major factor determining the time of the reproductive period in *M. edulis*. Bayne and Worrall (1980) showed that gamete production in *M. edulis* is initiated by a rise in temperature only if sufficient nutrients are available, either as energy reserves or as recently ingested food. Some authors (Lubet and Aloui, 1987) have suggested that a 'temperature window' may exist outside which gametogenesis declines, or does not occur, but inside which the reproductive strategy will depend to a large extent on food availability. This window presumably will vary according to the temperature range normally experienced by any particular population and to which it will therefore be adapted.

Food abundance certainly appears to be the primary factor controlling gonad growth in *M. edulis* in the Baltic (Kautsky, 1982a). No food storage occurs in these mussels since only during the spring phytoplankton bloom is there sufficient food to allow gametogenesis to proceed. Food shortage outside this bloom period presumably explains the absence of any secondary spawnings in these mussels. However, when caged mussels were maintained under more favourable feeding conditions, they ripened in January when the gonad index of the natural population was still close to its minimal value. Pieters et al. (1979) also indicated that a close relationship exists between food availability and gametogenesis in *M. edulis*. In coastal waters adjacent to industrial areas the gametogenic cycle of mussels may be subjected to stresses imposed by the input of toxic metals (Myint and Tyler, 1982). Reproductive failure has been reported in *M. edulis* during brown tide conditions (Tracey, 1988).

Thus, superimposed upon the overall effects of latitude, and therefore temperature, on the reproductive cycle are variations due to habitat-specific differences in the time and duration of maximum food availability. Any factor which results in altered food availability, or the ability of mussels to assimilate this food, will alter the nutrient storage cycle, and thus the timing of gametogenic events (Newell et al., 1982).

Rising, falling and fluctuating temperatures have all been reported to stimulate spawning in *Mytilus* (Chipperfield, 1953; Campbell, 1969; Wilson and Seed, 1974; Kennedy, 1977; Hines, 1979; Kautsky, 1982a; Wilson, 1987). Some workers have suggested that spawning occurs only within a critical temperature range (Zhang et al., 1980; Sprung, 1983), while others (Young, 1946; Sunila, 1981) have found little or no evidence that spawning is induced by temperature or temperature change. Thermal limits may be set by long-term average temperatures. Elvin and Gonor (1979), for example, speculate that in *M. californianus* average temperatures may set threshold levels for nerves responsive to thermal shock, and that rapid thermal changes then trigger the release of neurosecretions from cerebral ganglia and associated spawning events.

Physical stimulation caused by jarring or scraping the shell and/or by pulling or cutting the byssus threads will often cause ripe *M. edulis* to spawn (Suchanek, 1978; Lutz et al., 1980; Wilson, 1987). These are precisely the environmental cues received during periods of rough weather, which could also signal the presence of storm-generated patches of bare rock onto which mussels can settle. This pattern apparently correlates well with spawning and subsequent recruitment events for *M. edulis* on the west coast of North America (Suchanek, 1985). Several workers (Battle, 1932; Chipperfield, 1953; Wilson, 1987) have correlated spawning with phases of the moon and tidal fluctuations, while salinity changes may also initiate spawning in some mussels (Parulekar et al., 1982; but see Fell and Balsamo, 1985). Currently, the relative contribution of these and other exogenous factors to spawning is uncertain. Whatever factors are involved in initiating spawning, the presence of gametes in the water stimulates other ripe mussels to spawn, thereby enhancing the chances of fertilization.

Thus, from the extensive literature, a wide range of exogenous factors have been suggested as controls for both gametogenesis and spawning in *Mytilus*. Of these, temperature and food supply seem to be particularly important. However, these and other factors probably interact with endogenous factors (see also Chapter 6) in a complex manner to control the initiation of the gametogenic cycle and synchronize spawning. In field populations, a major difficulty in demonstrating a simple causal relationship between single environmental variables and complex processes such as reproduction, is that many of these variables (e.g. temperature, food supply, salinity, light) often covary or interact, sometimes even synergistically. Data, particularly on European mussels, has shown that *Mytilus* has a remarkable ability to vary its reproductive cycle in response to annual fluctuations in exogenous conditions. The precise pattern which the reproductive cycle takes probably depends on external factors which time endogenous events in such a way that maximizes survival and reproductive success. It seems likely, therefore, that *Mytilus* does not exhibit a single reproductive strategy, but rather exhibits a variety of patterns depending on the particular environmental regime (Newell et al., 1982).

Reproductive Output

Estimates of reproductive output or fecundity in *Mytilus* have usually been obtained either directly, by inducing mussels to spawn in the laboratory and then counting or weighing the gametes released, or indirectly from allometric equations relating weight loss on spawning to dry body weight or shell length (Thompson, 1979; Bayne and Worrall, 1980; Kautsky, 1982a; Sprung, 1983; Rodhouse et al., 1984b). Reproductive output in mussels, as in many other bivalves, can account for a substantial proportion of both total production and standing crop (Griffiths and Griffiths, 1987). It can also represent a significant energy subsidy to the pelagic system. Kautsky (1982a), for instance, estimated that the reproductive output of Baltic *M. edulis* was equivalent to half the zooplankton production, and was thus an important food source for herring larvae and carnivorous zooplankton. Individual female *M. edulis* (ca. 7cm shell length) can produce around 7–8 x 10^6 eggs during a complete spawning, while even larger individuals can produce as many as 40 x 10^6 eggs (Thompson, 1979).

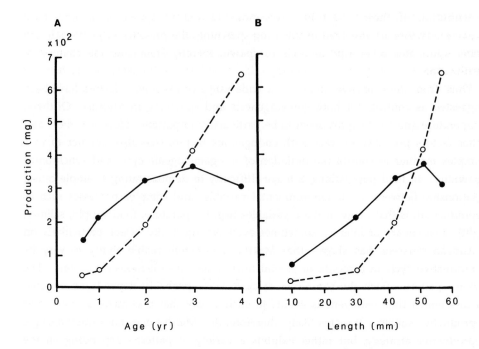

Fig. 4.4. Somatic (solid circles) and gamete (open circles) production as a function of age and shell length in low shore *Mytilus edulis* from Long Island, New York, North America. (After Rodhouse et al., 1986).

Energy which is surplus to metabolic requirements can be utilized for somatic growth and/or for gamete production (see p.122 and Chapter 5). However, the proportion of this energy surplus allocated to reproduction (= reproductive effort) varies according to body size or age (Bayne and Worrall, 1980; Kautsky, 1982a; Sprung, 1983; Thompson, 1984b; Rodhouse et al., 1986). Reproductive effort should not be confused with reproductive value which is the average expected lifetime fecundity (see for e.g. Bayne et al., 1983; Thompson, 1984b). Young mussels grow rapidly and convert little or no energy into reproduction, but with increasing size there is a gradual transition from somatic growth to reproduction, so that in the largest mussels most production (sometimes >90%) is channelled into gamete synthesis (Fig. 4.4). In these large individuals reproductive tissues may sometimes account for over 50% of the soft body weight (Thompson, 1979; Kautsky, 1982a). The reproductive effort of mussel populations will thus depend on their size (age) structure; those dominated by smaller, younger individuals will have a lower reproductive effort than populations consisting mainly of larger mussels, where gamete production may approach the total annual production. Okamura (1986) found that isolated individuals of *M. edulis*, or those growing in small clumps (6–9 mussels), grew more rapidly and had a greater reproductive effort than those in larger clumps (21–28 mussels). Moreover, mussels in the centre of groups had reduced growth rates compared with those situated at the edge, whose growth and reproductive output was similar to isolated individuals. In natural populations most mussels occur within a matrix of very large groups where growth rate (see also p.121–122) and reproductive effort will therefore be substantially reduced.

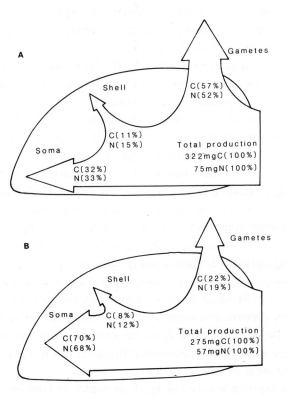

Fig. 4.5. Allocation of carbon and nitrogen in (A) wild and (B) cultivated *Mytilus edulis* from Killary Harbour, Ireland, once total cumulative production in the two populations is approximately equal. (After Rodhouse et al., 1984b).

Reproductive output is influenced by environmental variables such as temperature, food supply and tidal exposure, since these will broadly determine levels of net production. Bayne et al. (1983) reported a ten-fold difference between the maximum and minimum values for egg production, reproductive effort and reproductive value in *M. edulis* from six contrasted sites on the English and Welsh coasts. Fecundity may also vary from year to year, suggesting that the proportion of energy allocated to reproduction is adjusted according to the available food ration (Thompson, 1979). Bayne and Worrall (1980) compared growth and reproductive output in *M. edulis* from two contrasting sites near Plymouth, England. The Lynher population received a richer food supply than the population at Cattewater, which was located near the outfall of a small electricity generating station, and was thus subjected to additional temperature stress during winter and spring. Consequently, the Lynher population had a greater overall production, spawned twice each year and expended up to 60% of its total production on reproduction. At Cattewater, production was lower, there was only a single spawning each year, and reproductive output accounted for 26% of total production.

M. edulis, however, appears to be able to buffer its reproductive effort from the full effects of environmental stress. Thus, while temperature extremes and desiccation in the upper intertidal zone produce higher metabolic costs, energy seems to be shunted away from somatic growth, rather than from gamete production (see Suchanek, 1985, and references therein). Sprung (1983) showed that although intertidal mussels at Helgoland produced significantly smaller eggs than those grown subtidally, (but see Bayne et al., 1983), they reached sexual maturity at a smaller body size and actually had a higher egg output relative to shell length. In Killary Harbour, western Ireland, naturally occurring intertidal mussels allocated a greater proportion of their energy budget to reproduction compared with mussels cultivated on ropes, which channelled more of their energy into somatic growth (Fig. 4.5). Such observations, perhaps rather surprisingly, suggest that *M. edulis* actually reduces its proportional allocation of resources to gamete production in the presence of environmental amelioration. Reproductive effort in *M. edulis*, however, varies with body size and age, and mussels from different populations, even those of similar shell length, may vary considerably in terms of age. Bayne (1976) found that even when *M. edulis* had been starved for 30 days it still continued to produce gametes, although it had to significantly deplete its energy reserves in order to do so.

Baltic mussels experience very low levels of predation (and low Ca^{2+} levels) and tend therefore to produce relatively thin shells with small adductor muscles. Accordingly, these mussels are able to allocate a considerably greater proportion of their energy budget to reproduction compared with mussels of similar size from fully marine areas (Kautsky et al., 1990). It is now known, however, that Baltic and North Sea *Mytilus* are genetically differentiated (see Chapters 1 and 7 and references therein), and that reproductive output in *Mytilus* also appears to be related to genotype (Hilbish and Zimmerman, 1988; Gardner and Skibinski, 1990). Rodhouse et al. (1986) found a positive correlation between fecundity and multiple locus heterozygosity, at least amongst larger mussels, which had grown beyond the size at which gamete production started to exceed somatic production.

Condition

Condition indices in which the amount of flesh is related to the quantity of shell have been used extensively for many years, both in scientific research and in the commercial fishery. Many different methods exist for measuring condition (Lutz, 1980; Aldrich and Crowley, 1986; Davenport and Chen, 1987). A commonly used index, however, is that in which dry flesh weight is expressed as a proportion of the internal cavity volume of the shell (i.e. whole volume less the volume occupied by the actual shell valves). Methods utilizing wet flesh weight or volume are less sensitive due largely to the difficulty in standardizing the degree of wetness. Condition indices can be measured either for individual mussels or for whole populations. However, in view of the large amount of natural variability in condition, individual measurements are preferred whenever condition is investigated in relation to biological aspects, such as mortality or parasitic infection. The use of grouped data can be justified for population comparisons when the differences between sample means are more relevant than the degree of individual variation.

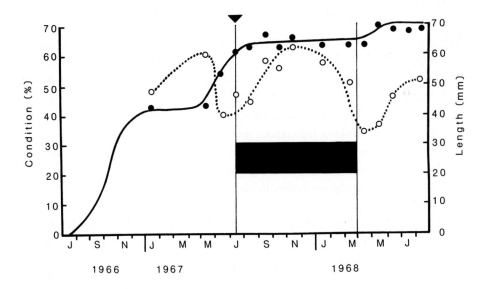

Fig. 4.6. Growth in shell length (solid circles) and condition, wet flesh volume as a percentage of the internal volume of the shell (open circles) of *Mytilus edulis* from Linne Mhuirich, Scotland; arrowhead denotes a mean length of 60mm, black bar the suggested time of harvesting. (After Mason, 1976).

Condition indices in *Mytilus* vary according to body size (Baird, 1958), season (Mason, 1976; Dix and Ferguson, 1984; Rodhouse et al., 1984a), level of parasitic infection (Kent, 1979; Thiesen, 1987; see also Chapter 12) and with local environmental conditions, especially the availability of food and degree of aerial exposure (Baird, 1966; Seed, 1980; Yamada, 1989). Seasonal changes are due to a complex interaction of those factors such as temperature, food supply and salinity which are thought to influence somatic growth and reproductive development. Figure 4.6 illustrates how the condition index of *M. edulis* in Linne Mhuirich, Scotland, increases during the autumn and winter months, when shell growth has virtually ceased. The subsequent steep decline in condition coincides with the main spring spawning period of these particular mussels. Marked differences in condition can also occur between mussels grown on the seabed and those grown in suspended culture on rafts (Fréchette and Bourget, 1985). Beatty and Aldrich (1989) also reported a significant increase in condition when mussels were only slightly elevated above the sea bottom. Such differences presumably reflect the better quality of digestible food and the generally lower concentration of sediment present in the water column, compared with the water at or immediately above the seabed itself.

SETTLEMENT AND RECRUITMENT

Following spawning and external fertilization, developing mussel larvae spend a variable period of time as part of the temporary or meroplankton when they are passively drifted by water currents, often over considerable distances. In temperate waters *Mytilus* larvae are generally abundant throughout the spring and summer months though several studies (Seed, 1969a; Rodhouse et al., 1985) have recorded *M. edulis* larvae in the plankton throughout much of the year. Such extended periods of larval abundance are due, at least in part, to the protracted spawning periods which are known to characterize many mussel populations.

Because of the ongoing confusion in the literature with respect to the use of the terms settlement versus recruitment, the term settlement shall refer here to the process whereby larval individuals come in contact with, and permanently attach to, the substratum, a process which clearly includes metamorphosis (see also p.63–64 Chapter 3). Recruitment, on the other hand, shall be defined as the process of successful colonization after some specified period of time, during which some postsettlement mortality will generally have occurred. Because it is often difficult, or even impossible, to assess settlement in field populations, settlement is often inferred from recruitment data measured in days or even weeks after settlement has actually taken place.

In view of the difficulties inherent in following larval cohorts in the field, most larval life spans have been determined either directly from laboratory cultures or inferred from the time difference between spawning and the subsequent recruitment of juvenile mussels within the same geographical region (Suchanek, 1985). Estimates obtained using the latter approach, however, can be confounded by the immigration of larvae from distant mussel stocks in which the temporal spawning pattern may have been quite different. Although the larvae of *M. edulis* become competent to settle at a shell length of approximately 260μm they are able to delay their metamorphosis and remain within the plankton until they have grown to around 350–400μm (Bayne, 1965; Sprung, 1984). Fuller and Lutz (1988) found that in field populations the mean size of *M. edulis* postlarvae at the time of their settlement was 300–350μm (see larval settlement section Chapter 3). Any delay in metamorphosis, however, is accompanied by a decreased ability to discriminate between different settlement surfaces. Thus, although 2–4 weeks seems to be the normal duration of planktonic life for *M. edulis*, this can vary according to temperature, food supply and the availability of a suitable settlement surface, so that 10 weeks or more may elapse between fertilization and the settlement of postlarval mussels (= plantigrades).

Much of the early literature fails to distinguish between the primary settlement of early plantigrades (250–400μm) on filamentous surfaces and the secondary settlement of later plantigrades (>500μm) onto established mussel beds. The association between recently settled *M. edulis* and filamentous substrata has long been recognized (see Seed 1976 and references therein), but the significance of such observations was not fully appreciated until de Blok and Geelen (1958) showed that early plantigrades settled on filamentous surfaces such as hydroids and various algal species from which they subsequently disappeared. Early experiments with various substrata indicated a distinct preference by early plantigrades for filamentous surfaces but that this preference subsequently changed. The suitability of the substratum seemed to be related to its general surface texture rather than to any chemical attraction (but see Cooper, 1981). Maas Geesteranus (1942) showed that young mussels would settle on most types of substrata provided these were firm and had a roughened or discontinuous surface. He also showed that plantigrades would attach and detach themselves many times before finally settling on the established mussel bed. He argued that mussel beds attract further recruits by virtue of their surface texture and that the byssal threads of the mussels themselves seemed to be important in this respect. *Mytilus* species are now known to settle on a wide variety of filamentous substrata, including the byssal filaments of conspecific adults (Petraitis, 1978; Suchanek, 1981; Hosomi, 1984; Eyster and Pechenik, 1987), filamentous algae (Paine, 1974; Suchanek, 1978; Petersen, 1984a, b; King et al., 1990) and fibrous ropes used in the mytiliculture industry (Mason, 1976; Lutz, 1980).

Bayne (1964) demonstrated that mussels pass successively from the plankton to sites of temporary attachment on filamentous algae, and from these, via a secondary pelagic phase, to sites of more permanent attachment on adult beds. This secondary pelagic phase has been termed bysso-pelagic migration or byssus drifting, and is facilitated by the secretion of long fine byssus-like threads (Sigurdsson et al., 1976; de Blok and Tan-Mass, 1977; Board, 1983; Lane et al., 1985). Young postlarval mussels retain this ability to drift, up to a size of about 2–2.5mm. Growth to this size may take several months and a significant part of this time could, therefore, be spent undergoing repeated phases of bysso-pelagic migration. The primary attachment phase seems to be a natural prelude to final settlement and may have considerable adaptive value. An initial attachment period away from the established mussel bed effectively reduces intraspecific competition, and also prevents the small, vulnerable postlarval stages from entering the strong inhalant currents of larger mussels.

The relatively marked seasonal abundance of early plantigrades on filamentous algae reported by Bayne (1964) and King et al. (1989) was not observed by Seed (1969a) for wave-exposed rocky shores in North-east England, where high densities of these mussels persisted throughout much of the year. Filamentous algae, together with other algae such as *Corallina* and *Mastocarpus* (= *Gigartina*) appeared to provide an extensive pool of young mussels, many of which could be migrating onto the adult beds more or less at any time of the year, thus accounting for the sporadic and often unpredictable pulses of recruitment that characterize many *Mytilus* populations (Seed, 1969a; Dare, 1976; Lewis, 1977). Early plantigrades normally remain on their primary attachment sites until they are approximately 1–2mm in shell length, though some may actually remain there until they are twice this size. The time taken to achieve this size will depend on individual growth rates. However, many autumn-spawned plantigrades which settle late in the year, when conditions for growth are becoming increasingly unfavourable, frequently overwinter on the algae, leaving only with the onset of more favourable conditions during the following spring. While migration from primary attachment sites to the adult habitat appears to be due to changes in the ecological requirements of the plantigrades, many mussels will also be liberated involuntarily by the seasonal die-back of their host algae, or through the action of winter storms.

More recently evidence has emerged which indicates that in some populations early plantigrades of *M. edulis* may settle directly onto adult beds, without an initial growth phase on filamentous substrata as postulated by the primary-secondary settlement model (McGrath et al., 1988; King et al., 1990). The absence of mussels >400μm in plankton samples from Norwegian fjords has also been cited as evidence for the absence of a secondary pelagic phase (Bøhle, 1971). Kautsky (1982a) similarly found no evidence for this phase in his study of Baltic mussels. Direct settlement of plantigrades onto adult beds has also been reported for *M. californianus* (Petersen,

1984a,b). The existence of varying modes of settlement may thus provide yet another example of the plasticity in the biology of *Mytilus*. Whether the mode observed on any particular shore is genetically-based or is a response to varying environmental conditions still has to be determined (McGrath et al., 1988).

The onset, duration and intensity of settlement and recruitment exhibits considerable spatial and temporal variation (see reviews by Seed, 1976; Suchanek, 1985). Seasonal patterns have normally been determined either by using artificial surfaces or by noting the relative abundance of plantigrades in samples of filamentous algae and/or established mussel beds. A wide range of artificial substrata has been used to collect settling plantigrades (see King et al., 1990); a major advantage of these collectors is that they present a constant surface area of relatively uniform textural composition and, when routinely deployed, they enable recruitment to be quantified over fixed intervals of time. Field studies have clearly demonstrated that smooth surfaces are generally unattractive to prospecting plantigrades and that maximum settlement occurs on roughened, scarred or fibrous substrata (Seed, 1969b; Dare et al., 1983; King et al., 1990).

When collected regularly throughout the year samples of filamentous algae and/or established mussel populations can provide useful semi-quantitative information concerning recruitment. Such data, however, provide only an estimate of the total number of plantigrades present at any given time; differences between successive samples therefore, represent a balance between the numbers of larvae settling and the subsequent loss of plantigrades through postsettlement mortality and growth out of the size categories being sampled. Furthermore, it is often difficult or even impossible to separate recently recruited mussels from the large numbers of small (<2.0mm) competitively suppressed individuals which characterize many *Mytilus* populations year round (Seed, 1969b; Kautsky, 1982b). These small, slow-growing mussels apparently become recruited to the breeding stock only after they have been effectively released from intense intraspecific competition, as larger mussels are lost from the population (Kautsky, 1982a,b). In this way the population is effectively stabilized and maintained at or near the carrying capacity of the area with respect to the available food and space. Only if settlement fails for several successive years will this appreciably affect the size of such populations. Kautsky (1982a) thus draws an interesting distinction between the initial recruitment of juveniles to the population, and subsequent recruitment from the persistent pool of small competitively suppressed mussels to the breeding stock.

Mytilus is highly gregarious and dense settlements often occur around the edges and in between individual mussels in established populations. Hosomi (1984) has shown that recruitment of *M. galloprovincialis* in Osaka Bay, Japan, was proportional to the density or biomass of the adult population, because recuitment occurred only amongst the byssal threads of adult mussels. Such gregarious behaviour is probably adaptive since *Mytilus* occurs predominantly in the intertidal or shallow subtidal zones and will therefore be subjected to mechanical forces of water movement especially on wave-exposed coasts. The reduced surface area exposed to such forces by mussels living in dense clusters, together with the mutual support afforded by neighbouring individuals, makes clumps of mussels better able to withstand these forces than isolated individuals (Harger, 1972; Paine, 1974). Mussels living in clumps may also be less vulnerable to predators. Plantigrades, once settled, provide loci for further recruitment and so the colony gradually extends. Juvenile mussels reach established beds by their ability to attach and detach themselves from unsuitable surfaces until a favourable habitat is encountered, usually in cracks or crevices in the rock surface or amongst the matrix of byssal threads provided by conspecifics. Once established, mussel beds will tend to increase in size both through gregarious settlement and growth of individual mussels. The competitive dominance of mytilid mussels on many rocky shores is probably, at least partly attributable to their ability to crawl extensively over the substratum even for some time after their initial settlement.

SOMATIC GROWTH

Growth has been extensively documented in *Mytilus* partly because of its commercial and ecological importance and partly because its growth history, as in many bivalves, is permanently recorded in the shell as a series of growth checks, a feature which makes these animals especially amenable to growth studies. Growth is usually assessed in one of two ways; either the size of the whole organism is related to age (= absolute growth), or the rate of growth of one size variable is related to that of another variable (= allometric growth). Reviews of growth in bivalves generally, and in *Mytilus* in particular, are provided respectively by Seed (1980) and Seed and Richardson (1990).

Methods for Estimating Absolute Growth

Although growth is most appropriately measured as the rate of change in biomass, in bivalves this can only be accurately determined once the animal has been removed from its shell; this is because live weight is strongly affected by variations in shell shape and thickness and by the amount of water retained in the mantle cavity (Griffiths and Griffiths, 1987). Consequently, shell length is a more commonly used indicator of size, and this, in turn, can then be related to weight, volume or even energy content by one or more allometric functions (p.121–122).

Analysis of size frequency distribution

Where recruitment to the population is seasonal, individual year classes can often be identified as distinct modes in plots of size frequency distributions. Changes in the position of these modes over time enable the mean growth rate of each year class to be estimated. In *Mytilus* this method has limited application because extended periods of recruitment and variable individual growth rates usually result in an inevitable merging of age classes (Seed, 1976; Kautsky, 1982b; Craeymeersch et al., 1986). Occasionally, such overlapping distributions can be adequately resolved by using various graphical or mathematical techniques (Grant, 1989). However, even when size frequency analysis is used (Bayne and Worrall, 1980; Rodhouse et al., 1984a) it provides only a measure of the average growth of mussels within the population, and such estimates may have been substantially modified by size-specific natural mortality.

Use of growth checks on or within the shell

Surface growth rings on the shells of many bivalves, including mussels, have been used extensively in age determination. These rings are produced during periods of suspended shell growth, and may be associated with various environmental factors including seasonal changes in temperature or food availability, prolonged stormy weather, or even with the annual reproductive cycle. Consequently, they cannot be assumed to be annual in origin, and even when annual rings are present their use in age determination can be confounded by other nonannual growth checks. Earlier rings may also be worn away due to shell abrasion, and in older, slow growing mussels, rings at the posterior shell margin become closely packed and difficult to resolve.

The shell of *Mytilus* consists of three layers, a thin outer periostracum, a middle prismatic layer and an inner nacreous layer. The periostracum and prismatic layers are secreted by the mantle epithelium around the margins of the shell, while the nacreous layer is deposited by the general outer surface of the mantle, and thus effectively thickens and strengthens the shell. Acetate peel replicas of polished and etched longitudinal shell sections reveal a series of distinct growth bands within the middle and inner layers. Nacreous lines are formed annually (Lutz, 1976), whereas microgrowth bands within the prismatic shell have a tidal periodicity, and can thus be used to detect short-term as well as longer term variations in individual growth rates (Richardson, 1989; Richardson et al., 1990).

Other direct measurements of shell growth

Successive measurements of marked or caged mussels can provide valuable records of the effects of size, season and environmental conditions on growth (Seed, 1976; Kautsky, 1982b; Page and Hubbard, 1987), providing that caging itself does not influence growth rate through its effects on water movement and food supply. Precise measurements of linear growth have been obtained using a sensitive laser diffraction technique first developed by Strømgren (1975). This method has been used in several laboratory-based studies to measure the effects of temperature (Almada-Villela et al., 1982), salinity (Gruffydd et al., 1984), photoperiod (Strømgren, 1976a, b), algal diets (Strømgren and Cary, 1984) and heavy metals (Redpath, 1985; see also Chapter 9) on the shell growth of *M. edulis*. More recently, a photographic technique which involves digitizing negative images of shell outlines has been used by Davenport and Glasspool (1987). Though less sensitive than the laser method, this technique allows the outlines and projected areas of shells to be calculated; the recorded images thus contain valuable information regarding changes in shell shape as well as size.

Estimates from physiological measurements

An alternative approach to the study of mussel growth is that based on the energy balance equation. Here the growth potential or 'scope for growth' is estimated from physiological measurements of the various components of the energy budget (Navarro and Winter, 1982; Widdows et al., 1984; Thompson, 1984a; see also p.405–406 Chapter 8). When integrated over time and applied to individuals of different size these data can be used to derive an average growth curve. Scope for growth does not differentiate between energy that is used for somatic growth or reproductive output, but has considerable practical advantages in that it can be assessed over short-term laboratory experiments. It thus provides an excellent method for quantifying the responses of individual mussels to changing environmental conditions such as food supply, temperature, salinity and contaminants. In *Mytilus* a close correspondence between growth rates estimated from energy budgets, and actual growth rates measured by more direct methods, has been reported by several workers (Bayne and Worrall, 1980; Riisgård and Randløv, 1981; Hamburger et al.,1983).

Quantitative expressions of growth

A habitat by virtue of its resource limiting environmental conditions imposes a maximum size beyond which further growth proceeds only slowly, if at all. The maximum attainable size (L_∞) under any set of environmental conditions can be approximated using the Ford–Walford plot, where length at age t + 1 years is plotted against length at t years (Cerrato, 1980). Maximum size is given where the line of best fit intercepts a point of zero growth when $L_t = L_t + 1$. This parameter is basic to many growth equations, two of which, the von Bertalanffy and the Gompertz have been widely used to describe and compare growth rates in *Mytilus* (Bayne and Worrall, 1980; Rodhouse et al., 1984a; Thompson, 1984b; Craeymeersch et al., 1986). Sigmoidal growth curves of the Gompertz type appear to be more characteristic of slower growing populations; faster growing populations generally lack any obvious inflexion point and for these the von Bertalanffy equation may provide a better description of growth (Griffiths and Griffiths, 1987). In the von Bertalanffy equation:

$$L_t = L_\infty \ [1 - e^{-k(t - t_0)}]$$

where k is the growth constant reflecting the rate at which maximum size (L_∞) is approached and t_0 is another constant representing time when $L_t = 0$. The Gompertz equation is similar but uses the logarithms of length:

$$\log_{10} L_t = \log_{10} L_\infty \ [\ 1 - e^{-k^1(t - t_1)}]$$

where k^1 is the rate constant and t_1 a constant representing time when $L_t = 1$.

Such equations do not reflect growth variations caused by seasonal fluctuations in temperature. Consequently, the product which integrates temperature and time (= day degrees) is sometimes incorporated as an independent variable (Ursin, 1963; Rodhouse et al., 1984a). Both equations assume that growth is determinate and that some maximum attainable size exists for any given population. Yet growth in many bivalves, including mussels, may not always be determinate, at least over their realized life span and may not, therefore, cease at any fixed adult size (Seed, 1980; Gardner and Thomas, 1987). Some workers (Davenport et al., 1984) have used polynomial expressions to describe shell growth in preference to the more commonly used growth equations (see also Chapter 5).

Variability in growth rate

Growth rate in *Mytilus* varies according to size, age and environmental conditions. Even mussels of similar size and age grown under apparently identical conditions can exhibit widely different rates and it is now known that growth variation is at least partially determined by genotype (p.121 and Chapters 5 and 7). However, variations attributable to this source are probably minor compared with those resulting from environmental factors. For example, under heavy settlement, growth in *M. edulis* individuals can vary as much as 10-fold (Trevelyan, 1991). Under optimal conditions *M. edulis* can attain lengths of 60–80mm within two years, whereas in the marginal

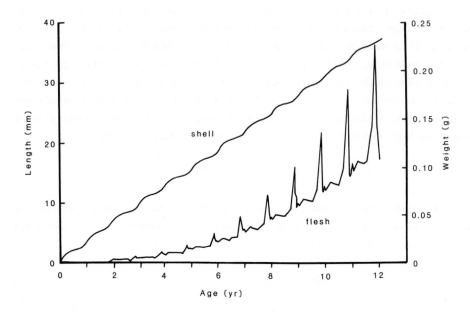

Fig. 4.7. Seasonal patterns of shell and flesh (dry weight) growth in *Mytilus edulis* from the Baltic. (After Kautsky, 1982b).

conditions of the high intertidal zone growth is substantially reduced, and mussels may reach lengths of only 20–30mm after 15–20 years (Seed, 1976). *M. californianus* can achieve faster growth rates and attain a much larger body size (up to 250mm) than *M. edulis* (Paine, 1976a; Suchanek, 1981; Yamada, 1989). Various environmental factors influence growth rate in *Mytilus*, and these are generally such that in temperate waters shell growth is rapid during the spring and summer, and slow or absent during the colder months. Flesh weight, by contrast, exhibits pronounced seasonal peaks associated with the annual reproductive cycle. Thus the pattern of growth in temperate water populations consists of alternating increments in shell length during the spring and summer and flesh weight during winter (Fig. 4.7).

Environmental Modulators of Growth

Temperature
Almada Villela et al. (1982) examined the effect of several constant temperatures on *M. edulis* and found that between 3°C and 20°C linear growth increased logarithmically; above 20°C growth declined sharply, while at lower temperatures (3

and 5°C) it proceeded only very slowly. These workers further noted that while a cyclical temperature range occasionally produced better growth than either of the temperatures at the extremes of the experimental regime, no constant pattern emerged. The absence of any adverse effects of fluctuating temperature and the ability to acclimate to temperature changes, at least over part of their physiological range, indicates that *Mytilus* is well-adapted to life in the constantly changing environmental conditions usually associated with coastal and estuarine waters.

A relationship between growth and temperature is clearly demonstrated when shell length is plotted against age in day degrees. However, growth rates expressed in these terms are not always consistent, which suggests that factors other than temperature (e.g. food supply) are probably involved (Wilson, 1977; Thompson, 1984b). A major difficulty in correlating single environmental variables with growth rate in field populations is that the correlation, although obvious, may not be causal. Page and Hubbard (1987) concluded that the temporal patterns of growth in *M. edulis* on a production platform off the Californian coast were determined mainly by variations in phytoplankton, and that temperature could be virtually eliminated as an important growth regulator over the range (10–18°C) normally experienced by these mussels. Similarly, in a study of mussels in western Sweden, Loo and Rosenberg (1983) found that low temperatures (<5°C) did not seem to limit growth whenever these coincided with the spring phytoplankton bloom. Physiological studies on *M. edulis* have also demonstrated that between 10 and 20°C water temperature has little effect on scope for growth (Bayne et al., 1976).

Salinity

Brackish estuaries and lagoons are favourable habitats for mussel growth but this probably reflects the increased food levels in these environments rather than any beneficial effects of reduced salinity. Indeed, lowered salinity may have a detrimental effect on growth and can even be lethal to mussels under extreme conditions (Almada-Villela, 1984; Gruffydd et al., 1984). *M. edulis*, however, can survive considerably reduced salinities and will even grow as dwarfed individuals in the inner Baltic, where salinities can be as low as 4–5°/oo (Kautsky, 1982b). Results from reciprocal transplant experiments suggest that differences in growth rate and maximum size between North Sea and Baltic mussels are mainly due to physiological adaptations to environmental salinity (Kautsky et al., 1990).

Bøhle (1972) found that at various steady state salinities mussels gradually acclimated to lowered salinity levels. Acclimation to fluctuating salinities, however, did not occur to the same extent, presumably because the period of exposure to the lowest salinities was too brief. *M. edulis* can effectively isolate itself from low salinity by closing its valves and maintaining a relatively high osmotic concentration within the mantle fluid (Davenport, 1979; Aunaas et al., 1988). However, since feeding is suspended while the valves remain closed, growth rate will inevitably be depressed.

The influence of salinity on growth may be due to reduced metabolic efficiency. Tedengren and Kautsky (1986), for example, found that at ambient salinities the oxygen to nitrogen ratio was consistently lower in Baltic mussels than in those from the North Sea. Since a lowered O:N ratio is energetically unfavourable (Bayne et al., 1985), this would contribute to the lower growth rate and smaller maximum size of Baltic mussels.

Food supply and tidal exposure

Probably the single most important factor in determining growth rate is food supply, since this provides the necessary energy to sustain growth. Mussels are efficient filter feeders removing particles down to 2–3μm with 80–100% efficiency (Møhlenberg and Riisgård, 1977). The total amount of particulate material present in suspension (= seston) contains several potentially utilizable food types. These include bacteria, phytoplankton, fine organic detritus and material of inorganic origin, though the precise nutritional contribution that each of these makes to the diet varies seasonally, and among mussels of different size (Rodhouse et al., 1984a; Page and Hubbard, 1987). Dissolved organic matter may also contribute to the energy intake of *Mytilus* (Manahan et al., 1983; Siebers and Winkler, 1984).

Several authors have identified seasonal and regional variations in both the quantity and quality of utilizable food as important determinants of mussel growth (Ceccherelli and Rossi, 1984; Fréchette and Bourget, 1987). Field growth rates often exceed those recorded in the laboratory, irrespective of food supply and temperature. Kiørboe et al. (1981) suggest that this may be due to the stimulatory effects of resuspended bottom material which, as well as serving as an additional food supply, may also enable the mussel to exploit its full clearance potential. Despite the broad correlation between growth rate and particulate food, mussels can buffer their shell growth during short-term temporal variations in food availability by utilizing glycogen reserves accumulated prior to and during gametogenesis (Bayne et al., 1983).

Mussels only feed when they are submerged. At some point along the intertidal gradient, therefore, the energy required for metabolism during aerial exposure will exceed that available during the feeding period. Figure 4.8 illustrates the progressive reduction in growth rates with increasing tidal elevation in the black mussel *Choromytilus meridionalis*, as well as the exceedingly rapid growth rates under the most favourable conditions. Baird (1966) estimated that the point of zero growth in *M. edulis* was approximately 55% aerial exposure though this will presumably vary

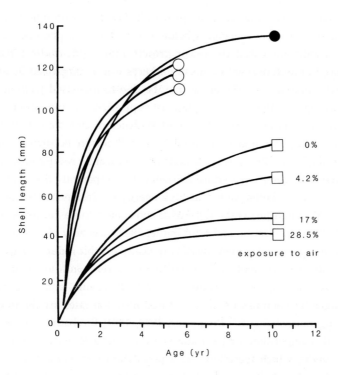

Fig. 4.8. Growth rates of South African *Choromytilus meridionalis* at different sites and tidal levels; Table Bay, offshore (solid circles), Saldanha Bay, raft culture (open circles) and False Bay (open squares). (After Griffiths and Hockey, 1987).

according to local conditions, such as the degree of wave splash. While growth declines with tidal exposure (see also Yamada (1989) for *M. californianus*), life expectancy often increases, since predation pressure in the upper shore is substantially reduced. However, when old slow growing mussels are transferred to more benign conditions downshore they are able to grow rapidly (Seed, 1968), indicating that the potential for growth remains intact for many years, even among mussels that have been prevented from exploiting this potential, owing to unfavourable environmental conditions. The relationship between growth rate and water depth reported by several workers (Rodhouse et al., 1984a; Page and Hubbard, 1987) is generally thought to reflect variations in food availability within the water column.

Other factors

Light seems to have a detrimental effect on growth in *Mytilus*. Continuous darkness, reduced levels of irradiance, wavelengths below 600–700nm and photoperiods of 7h or less, all significantly increased the linear growth rate in *M. edulis* (Strømgren, 1976a, b). Enhanced growth during periods of darkness was accompanied by increased defaecation, which suggests that the effect may be due to increased feeding activity (Nielsen and Strømgren, 1985). This view is supported by the observation of Ameyaw–Akumfi and Naylor (1987) who found evidence of circadian rhythmicity in shell gaping of *M. edulis*, with a greater duration of shell closure occurring during the hours of expected daylight. Strong wave action can significantly reduce growth rate in *M. edulis*, presumably by reducing feeding efficiency; this apparently does not occur in *M. californianus*, a species better suited to high-energy environments (Harger, 1970; Suchanek, 1981). Although pea crabs were once considered to be harmless commensals (Wells, 1928, 1940; MacGinitie and MacGinitie, 1949) it is now clear that they are parasitic on *Mytilus* species. Infested mussels typically exhibit reduced growth rates and shell shape distortions, and are generally in poorer condition compared to noninfested mussels (Seed, 1969c; Anderson, 1975; Bierbaum and Ferson, 1986). Reduced food intake by the mussel and/or reduced filtration rates are the most likely factors in lowering mussel fitness. (Pregenzer, 1979, 1981; Bierbaum and Shumway, 1988). Intraspecific competition for food and space can lead to extreme variations in growth rate. *M. edulis*, which recruited into a population consisting of one-year-old mussels, grew at less than half the rate of those recruiting onto an adjacent bare rock surface (Seed, 1969b). In populations consisting of several age classes even greater growth reductions occur, as the majority of small mussels trapped amongst the byssal threads of larger conspecifics are at a severe competitive disadvantage (see also Dare and Edwards, 1976; Kautsky, 1982b). Environmental contaminants such as tributyltin, heavy metals and petroleum hydrocarbons can all cause significant reductions in growth rate, often at exceedingly low concentrations (for references see Seed and Richardson, 1990).

Several environmental factors can thus modulate growth in *Mytilus*. Of these perhaps the availability of a suitable food resource is the most important since, without this, sustained growth cannot occur. Given adequate food several factors, particularly temperature, salinity and aerial exposure may interact, sometimes synergistically, resulting in various rates and seasonal patterns of growth. Such interactions, and the tendency for some variables to covary, makes it extremely difficult to identify the precise influence of any single factor on growth in natural mussel populations.

The importance of genotype

The relationship between growth rate and genotype has been demonstrated in several bivalves including *Mytilus* (see p.352–357 Chapter 7). Particular attention has focussed on the strong positive correlation between individual growth rate and the degree of heterozygosity measured at several polymorphic enzyme loci (Koehn and Gaffney, 1984; Zouros et al., 1988). Several recent studies have revealed that more heterozygous *M. edulis* individuals have lower energy requirements for maintenance metabolism, and a higher efficiency for protein synthesis (Diehl et al., 1986; Hawkins et al., 1986). The higher energy status of these individuals may thus be reflected in faster somatic growth in juvenile mussels (Koehn and Gaffney, 1984), or in higher fecundity and production in those which have attained reproductive size (Rodhouse et al., 1986). In addition to growing faster, more heterozygous mussels also tend to achieve more uniform average growth rates (Koehn and Gaffney, 1984).

Allometric growth

So far we have considered how linear growth rate can be modulated by environmental and genetic factors. However, mussels, like most organisms, exhibit progressive changes in their relative proportions with increasing body size. The relationship between any two size variables (x and y) can be expressed by the allometric equation $y = ax^b$ where a and b are constants. The exponent or growth coefficient b represents the relative growth rate of the two variables, while a is the value of y when x is unity. In its linearized logarithmic form this becomes $\log y = \log a + b \log x$. The slope (b) and intercept (A) of such transformed data are estimated by regression analysis (Brown et al., 1976; Aldrich and Crowley, 1986). Changes in relative proportions may simply be associated with the maintenance of physiologically favourable surface area to volume ratios as body size increases; alternatively, they may reflect adaptive responses to changing environmental conditions (see p.191–193 Chapter 5).

Of the various environmental factors that are known to influence shell shape in bivalves, population density (= crowding) seems to be particularly important in the case of *Mytilus* (Seed, 1968, 1973, 1978; Brown et al., 1976). The shells of densely packed mussels are proportionately more elongate with higher length to height ratios than those from less crowded conditions. This effect, which is exaggerated in older individuals, presumably has adaptive value since the posterior feeding currents will be effectively elevated above neighbouring conspecifics. Such ontogenetic and phenotypic variations in the allometric relationships of *Mytilus* and other bivalves are discussed in more detail elsewhere (Seed, 1980) and will not, therefore, be considered further in this account. The allometric equation has also found extensive use in physiological investigations, and in studies of mussel production for estimating flesh weights from measurements of shell length. Here dry flesh weight is usually regressed against length for population samples taken at regular intervals throughout the year. These regressions are then compared by covariance analysis and used to estimate the mean weight of mussels of standard length. Temporal variations in these adjusted mean weights are then presumed to reflect changes in productivity, fecundity or physiological condition (Bayne and Worrall, 1980; Kautsky, 1982b; Rodhouse et al., 1984a, b; Thompson, 1984a). However, this procedure, which assumes that changes in the covariate (length) are trivial, has recently been criticized by Hilbish (1986), who showed that shell and soft tissues in *M. edulis* can exhibit markedly different seasonal growth patterns (but see Beatty and Aldrich, 1989). Where such uncoupling of growth occurs, seasonal variations in flesh weight adjusted to mussels of a standard length should perhaps be interpreted with caution.

PRODUCTION

Production represents the net gain in body energy and occurs when the energy content of the absorbed ration exceeds metabolic requirements (see details in Chapter 5). This energy surplus can then be utilized for somatic and/or gamete production. Although reproductive output is thus an essential component of total production, in many studies it is often ignored, and commonly used definitions of production take into account only somatic growth. The partitioning of surplus energy between somatic growth and gamete production and its relationship to age is of fundamental importance within the context of life history strategies (Seed and Brown, 1978; Thompson, 1979). If energy intake falls below the maintenance ration then body reserves are utilized and negative growth (= degrowth) may occur. Production of individual mussels can be measured directly in the field or under controlled laboratory conditions, as actual rates of somatic growth and reproductive output; alternatively, it can be estimated indirectly as 'scope for growth' (Griffiths and Griffiths, 1987; see also p.114 and Chapters 5 and 8).

The methods by which production is calculated for individual mussels can also be applied to populations or even to whole communities (Asmus, 1987), provided that certain group attributes, such as size (= age) distributions and changes in population density, are taken into account. The data thus obtained are then generally expressed in terms of ash free dry weight (AFDW) or energy flux per unit area of habitat (kJ m^{-2} yr^{-1}). Estimates of production are extremely sensitive to variations in the size (age) composition of the population as well as to changing environmental conditions. Increase in body size, for instance, is usually accompanied by a transition from somatic production to gamete production (Thompson, 1984b; Rodhouse et al., 1985; Craeymeersch et al., 1986), while growth efficiency declines as total production fails to keep pace with the increase of respiratory energy expenditure. Consequently, production is often estimated separately for the different age or size classes represented within the population.

Although there have been relatively few studies of production in natural mussel populations, available data indicate that production levels in *Mytilus* can be extremely high and may rival even those reported for other highly productive systems such as tropical rain forests and kelp beds (Whittaker, 1975; Leigh et al., 1987). In Morecambe Bay, England, for instance, production by two year classes of *M. edulis* (1968 and 1969) amounted to 62.89 x 10³ and 86.40 x 10³ kJ m⁻² yr⁻¹ respectively, some 2.5 to 3 times their maximum standing crop (Dare, 1976; see also p.137). Most production occurred in the first year following settlement and although these mussels survived into their third year, production had virtually ceased after sixteen months (Fig. 4.9) due to the high rates of mortality, mainly from physical factors, that characterize this particular population. Production of organic material in the shell was 32–34% of organic flesh production; thus, while the organic component of the shell of *M. edulis* is small (probably <5% of total shell weight), production is high by virtue of the large bulk of shell in the population (see also Gardner and Thomas, 1987). Dare (1976) draws attention to the large amount of production that is made available to decomposers and to the detrital food chain in mussel populations of this sort. This contrasts with

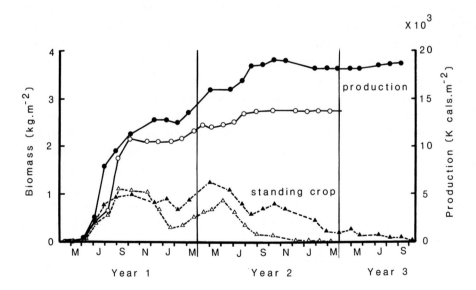

Fig. 4.9. Standing crop (dry biomass) and production of dry flesh in the 1968 (open symbols) and 1969 (solid symbols) year classes of *Mytilus edulis* from the low shore, Morecambe Bay, England. (After Dare, 1976).

other populations such as those in the Ythan Estuary, Scotland, in which predation apparently accounts for most of the annual mussel production (Milne and Dunnet, 1972; Baird and Milne, 1981).

The Morecambe Bay population studied by Dare is perhaps rather unusual in that it consisted entirely of young, fast-growing mussels with a considerable rate of turnover. In populations where the biomass tends to be dominated by older, slower growing individuals production levels are usually much lower with typical values generally falling within the range of 2.0 x 10^3 and 14.5 x 10^3 kJ m^{-2} yr^{-1} (Milne and Dunnet, 1972; Rodhouse et al., 1985; Asmus, 1987). High levels of production, comparable to those reported by Dare (1976), however, have been recorded in populations of *M. galloprovincialis* (Ceccherelli and Rossi, 1984; Hosomi, 1985) and *M. californianus* (Leigh et al., 1987), as well as in populations of *M. edulis* elsewhere (Dare and Edwards, 1976). In Killary Harbour, western Ireland, production levels of cultured mussels exceeded those of wild shore mussels by an order of magnitude (Rodhouse et al., 1985). The shore population, however, was dominated by larger, older mussels (40–60mm) with a high reproductive output. Consequently, these mussels contributed significantly to the larval population of this particular inlet. Carbon flow per unit area in a Swedish mussel culture described by Loo and Rosenberg (1983) was approximately ten times lower (ca. 13.5kg m^{-1}) than in Killary Harbour (Rodhouse et al., 1985). Production levels in *Mytilus* vary seasonally due to the timing of events such as the spring phytoplankton bloom and the onset of spawning. Other factors including temperature will also presumably contribute to the strength of the seasonal effect. Variations amongst mussels grown under similar environmental conditions strongly suggest that genetic differences may also partially determine the level of secondary production (Mallet and Carver, 1989 and Chapter 7).

Flow-through plastic tunnels have recently been used to determine the uptake and release of materials in the tidal waters passing over natural mussel beds (Siebers and Winkler, 1984; Dame and Dankers, 1988; Prins and Smaal, 1990). Such studies elegantly demonstrate that intertidal beds of *Mytilus* are capable of removing substantial amounts of particulate material from the water column, transforming some of this into biomass, and releasing some constituents as dissolved waste products. Dense assemblages of filter feeding organisms like mussels can thus function as central processors of estuarine and coastal materials, and by regenerating nutrients could, therefore, play a central role in controlling levels of eutrophication and primary production in inshore waters.

MORTALITY

Survival and Longevity

Survivorship curves for *Mytilus* have generally been constructed either from changes in the relative abundance of mussels (e.g. numbers m^{-2}) belonging to different year classes represented within the population, or by following the survival of marked cohorts of approximately similar size or age. Unfortunately, individual year groups in mussel populations are not always clearly defined, although their overlapping size frequencies can sometimes be adequately resolved (see p.113). Estimates of mortality obtained from changes in population density can also be confounded by the immigration or emigration of mussels between successive sampling dates. The demography of bivalve populations has been extensively reviewed by Cerrato (1980).

Figure 4.10 illustrates the survival of *M. edulis* settlements at two low intertidal sites in Morecambe Bay, England, and shows that few mussels in these populations survived beyond their second or third year (see also Emmett et al., 1987). By following the survival of marked cohorts at three tidal levels on an exposed rocky shore on the North-east coast of England, Seed (1969b) similarly found that mortality in the lower shore was severe, mainly as a result of intense predation. At higher tidal elevations, however, reduced levels of predation led to enhanced survival, although growth rates were considerably reduced. Thus, while lowshore populations consisted almost entirely of mussels that were under three years old, highshore populations often contained twenty or more year classes (Seed, 1969b). Senility, therefore, is unlikely to be a major source of mortality in *Mytilus* under most ecological conditions. Long-lived specimens of *M. edulis* (18–24 years) have also been reported by other workers (Thiesen, 1973), while *M. californianus* appears to experience even greater longevity, with some individuals possibly surviving for 50–100 years in certain relatively undisturbed populations (Suchanek, 1981).

Thiesen (1968) estimated the mortality (cause unspecified) of *M. edulis in* a fishery in the Danish Wadden Sea by observing the amounts and size of mussels laid and refished on the different mussel banks. He concluded that mortality was size-dependent, with the mean annual mortality varying between 68% and 34% in mussels of 25mm and 50mm shell length, respectively. These figures contrast with values of between 74% and 98% for the short-lived but fast-growing Morecambe Bay population studied by Dare (1976). Several workers (Seed, 1969b; Ceccherelli and Rossi, 1984) have suggested that faster growing mussels may be less long-lived because they will attain the size limit imposed by the environment much more rapidly than those living in habitats where growth rates are much slower. Mortality, like growth and reproduction, may also be genetically determined (see Chapter 7 and references therein).

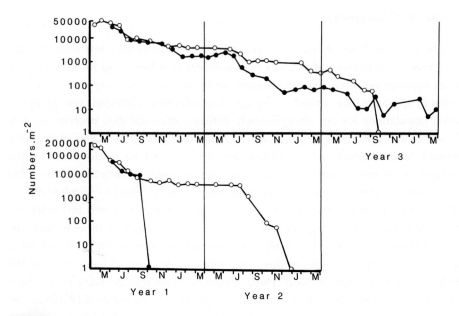

Fig. 4.10. Survival curves for the 1968 (solid circles) and 1969 (open circles) year classes of *Mytilus edulis* at two low shore sites in Morecambe Bay, England. (After Dare, 1976).

Causes of Natural Mortality

Physical factors

While *Mytilus* is highly adaptable and especially tolerant of a wide range of environmental conditions, extremes of physical factors such as storms, temperature and desiccation, and excessive deposition of silt are all known to cause mortality in mussels. These factors will vary seasonally and their combined and/or synergistic effects can occasionally result in spectacular mass mortalities.

Mortality from high temperatures and/or desiccation is well-documented (Suchanek, 1978; Peterson, 1979; Tsuchiya, 1983) and physiological intolerance to desiccation is probably the single most important factor determining the upper limits of mussel zonation (p.89–91). Juveniles are especially vulnerable and thus generally tend to settle gregariously into established mussel clumps, where they are presumably afforded somewhat greater protection. Low temperatures can be equally damaging. In Alaska, littoral *M. californianus* exposed to freezing conditions suffered 100% mortality, whereas *M. edulis*, a more eurythermal species, remained alive even after spending several months frozen solid in ground ice (Suchanek, 1985). Ice crushing or scouring, however, is known to result in heavy mortalities of *M. edulis* populations in boreal regions such as Baffin Island (Stephenson and Stephenson, 1972), Glacier Bay, Alaska (Suchanek, 1985) and the Gulf of St. Lawrence (Bergeron and Bourget, 1986).

Storm generated waves and wave-driven logs can also cause extensive mussel mortality (Dayton, 1971; Lubchenco and Menge, 1978; Witman and Suchanek, 1984; Suchanek, 1985; Denny, 1987 and see section: Disturbance and Recovery, below). Storms are known to act in a density-dependent manner on *M. californianus* (Harger and Landenberger, 1971), and since *M. edulis* has a less robust shell and weaker byssal attachment than its congener, the effects of storm damage in mixed populations depends on the relative abundance of the two species (Harger, 1970, 1972). *M. edulis*, however, is known to be able to adjust its attachment strength according to the prevailing conditions, particularly the degree of water movement (Price, 1982; Young, 1985).

Excessive levels of silt and inorganic detritus (Ceccherelli and Rossi, 1984) and biodeposits (= mussel mud) produced by the mussels themselves (Tsuchiya, 1980) can also be extremely damaging. Thus, while *M. californianus* can attach strongly, and is otherwise well-adapted to waveswept environments, it does not crawl as effectively as *M. edulis*. Consequently, it suffers burial and heavy mortality from excessive siltation in more protected embayments. Suffocation by excessive biodeposits and subsequent destruction of the beds by waves and tidal scour were also identified as major causes of mortality in *M. edulis* in Morecambe Bay (Dare, 1976) and the Menai Strait in North Wales (Dare and Edwards, 1976). Pollution probably only becomes a significant mortality factor when mussels are stressed or perhaps weakened by disease (Sunila and Lindström, 1985).

Biological factors

The natural enemies of mussels fall into four main categories: predators, parasites, pathogens and competitors for food and space. Parasites and pathogens are discussed in detail in Chapter 12 and with the exception of pea crab infestation, already dealt with on p.120, will not, therefore, be considered further in this section. Organisms which bore into the shell (e.g. *Cliona*, *Polydora*) may cause mortality indirectly by weakening the shell structure, thus increasing the mussel's vulnerability to predators. Blooms of toxic algae can occasionally result in heavy mortalities (Tracey, 1988) although, like many parasitic infections, their effects appear to be mainly sublethal. High levels of natural mortality have also been reported in some populations of *M. edulis* at times of metabolic stress (Emmett et al., 1987). This appears to be related to the relatively poor condition of postspawned mussels when nutrient reserves in the mantle are at their lowest levels, and is therefore often most severe among larger mussels which have the highest reproductive effort (Worrall and Widdows, 1984). Alternatively, it may be associated with high temperatures (>20°C), particularly at times when food is in short supply (Incze et al., 1980).

Predation is undoubtedly the single most important source of natural mortality in *Mytilus*. Moreover, many mussel predators such as crabs (Jubb et al., 1983; ap Rheinallt, 1986), starfish (Menge, 1972; O'Neill et al., 1983), gastropod molluscs (Hughes and Dunkin, 1984; Hughes and Burrows, 1990) and shorebirds (Incze et al., 1980; Durrell and Goss Custard, 1984; Feare and Summers, 1985; Meire and Ervynck, 1986; Bustnes and Erikstad, 1990; Raffaelli et al., 1990) are known to forage selectively on specific size ranges of *Mytilus*. Consequently, these predators have the potential to influence population size structure as well as overall abundance and local distribution patterns (p.91–94). However, coexistence is often facilitated by virtue of temporal (= size) and spatial refuges, where the impact of the predator is significantly reduced (Seed, 1969b, 1992; Dayton, 1971; Paine, 1974, 1976a; Bayne and Scullard, 1978; Elner, 1978; Campbell, 1983). Because mussels can effectively escape predation by growing out of the size range normally taken by any particular predator, the length of time for which they remain vulnerable will thus depend on growth rate, and this in turn is a function of geographic location and tidal elevation.

The dogwhelk *Nucella* (=*Thais*) *lapillus* is a widely distributed littoral predator in northern Europe and along the Atlantic coast of North America. It is especially abundant on wave-exposed shores, where it feeds extensively on both barnacles and mussels (Seed, 1969b; Menge, 1983). The occurrence of dogwhelks on mussel beds, however, is highly seasonal (Tongiorgi et al., 1981) and during the colder months adult whelks aggregate in crevices and pools as part of their breeding cycle (Feare, 1971a). Emergence from winter aggregations occurs in spring, and over the summer months large numbers of dogwhelks can often be found foraging on *M. edulis* in the middle and lower shore (Fig. 4.11). Feeding in *N. lapillus*, however, may be severely curtailed during periods of desiccation or strong wave action (Hughes and Burrows, 1990). Dogwhelks are also major predators of *M. californianus* along the west coast of North America (Paine, 1974; Suchanek, 1978; Palmer, 1983). However, when given a choice, both *Nucella canaliculata* and *N. emarginata* showed a strong preference for *M. edulis* (Harger, 1972; Suchanek, 1978, 1981). The precise mechanism for such selection is uncertain but is presumably based upon the increased time or energy expended on drilling a thicker shell (containing an extra prismatic layer in *M. californianus*) and/or the potential lower calorific value of *M. californianus* tissue (Suchanek, 1981; Palmer, 1983). Whatever the reason, laboratory experiments have shown that *N. canaliculata* fed on *M. edulis* grow at a significantly faster rate than those fed on *M. californianus* (Palmer, 1983). Mussels attacked by dogwhelks are easily

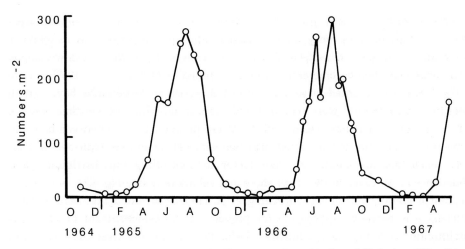

Fig. 4.11. Seasonal abundance of *Nucella lapillus* on low shore *Mytilus edulis* beds at Filey Brigg, England. (After Seed, 1969b).

identified by the small hole drilled through the shell. Most mussels are drilled through the thinnest part of the shell (Seed, 1969b), or above the underlying digestive gland, which is rich in glycogen and is easily digested (Hughes and Dunkin, 1984). Dogwhelks take from several hours to well over a day to handle prey and usually move net distances of <0.2m between meals (Hughes and Drewett, 1985). Feeding rates vary from about 0.1–0.6 mussels whelk[-1] day[-1] depending on shell thickness and temperature (Stickle et al., 1985). Seed (1969b) found large numbers of drilled mussels in the low intertidal zone on the North-east coast of England and attributed the absence of *Mytilus* near deep crevices and pools to locally intense predation as dogwhelks entered and left their dense winter aggregations. Similar high levels of dogwhelk predation have been reported among low-shore populations of *M. edulis* in Alaska, where the average percentage of drilled shells ranged from 61% at protected sites, to 95% on more wave-exposed areas (Suchanek, 1978).

In addition to dogwhelks, several other gastropods such as *Ocenebra, Urosalpinx, Acanthina, Ceratostoma* and *Jaton* are also known to feed on mussels (see Seed, 1976 for references). *Mytilus*, however, is not entirely defenseless as shown by its ability to immobilize predatory gastropods by means of its byssal threads. This method of defense, in which several individual mussels may actually cooperate in subduing the predator, has been observed in response to both dogwhelks (Wayne, 1980; Petraitis, 1987) and oysterdrills (Carricker, 1981). Gastropods are not considered to be important predators of mussels in the Danish Wadden Sea (Thiesen, 1968), nor in Morecambe Bay, England, (Dare, 1976). Similarly, *Thais clavigera* was not identified as an important predator of mussels by Hosomi (1984) in his detailed study of *M. galloprovincialis* in Japan.

Starfish are major mussel predators in many areas. *Asterias rubens* is often present at low densities on most rocky shores in northern Europe but periodically its numbers rise dramatically so that it effectively blankets much of the middle and lower shore. Such areas may then become almost totally denuded of *Mytilus* (Seed, 1969b). Dare (1976, 1982) also recorded large invasions of *A. rubens* in Morecambe Bay; starfish densities up to 450 m^{-2} were recorded and the swarm, which at one time covered 2.25ha of ground, may have cleared up to 4000t of first year mussels between June and September. Such swarms of starfish are often very patchy and unpredictable in their occurrence, but are clearly a major factor in controlling the distribution and abundance of *M. edulis* in the lower shore and sublittorally (see Fig. 4.1). As a result of transplantation experiments, Kitching et al. (1959) concluded that *Marthasterias glacialis* was partially responsible for preventing the establishment of *M. edulis* sublittorally in Lough Ine, South-west Ireland, while on the west coast of North America *Pisaster ochraceus* effectively controls the distribution of *M. californianus* on the lower shore. Continued removal of this starfish results in the encroachment by the mussels into areas not previously occupied, eventually producing a virtual monoculture of mussels occupying the primary substratum (Paine, 1974). Asteroid starfish have also been identified as important predators of *M. edulis* on both the Pacific (Suchanek, 1978; Paine, 1980) and Atlantic (Peterson, 1979; Menge, 1983) coasts of North America. Interestingly, a close association between *M. edulis* and the anemone *Metridium senile* appears to afford the mussel significant protection against its asteroid predator *Asterias forbesii* (Kaplan, 1984).

In the low-salinity waters of both the Baltic and Glacier Bay, Alaska, however, the scarcity of mussel predators has effectively allowed *M. edulis* to become the dominant space occupying organism down to depths of 30m and 3m respectively (Kautsky, 1981, 1982b; Suchanek, 1985). Moreover, in view of the low predation pressures (and presumably the low Ca^{2+} levels) experienced by Baltic mussels, selection has favoured individuals with thinner shells and smaller adductor muscles, but with a higher fecundity (Kautsky et al., 1990). Consequently, when these mussels are transplanted to fully marine sites in the North Sea, they are more readily attacked and easily pulled open by *Asterias*. Hancock (1965) showed experimentally that mussels with larger adductor muscles were less vulnerable to starfish predation.

In the low intertidal zone of exposed rocky shores of the Pacific North-west Paine (1976a) has documented size-limited predation on *M. californianus* by the seastar *Pisaster ochraceus*. From a series of laboratory choice experiments, in which *P. ochraceus* selected medium-sized *M. californianus*, rejecting both small- and large-sized individuals, McClintock and Robnett (1986) conclude that this species is maximizing energy intake, and minimizing time spent foraging and handling. However, Paine's (1976a) field observations do not support this evidence, since *P. ochraceus* chose *M. californianus* of all sizes, especially small individuals.

Several other reports have shown size-limited predation on mussels: Elner (1978) for *Carcinus maenas* on *M. edulis* (maximum size taken = 70mm); Campbell (1983) for *Asterias forbesii* on *M. edulis* (maximum size taken = 70mm); Briscoe and Sebens (1988) for *Strongylocentrotus droebachiensis* on *M. edulis* (maximum size = 16mm). With unlimited food supply and reduced predation, even *M. edulis* (typically viewed as a relatively small species) can attain sizes (up to 140mm in length at Bodega Harbor, California) well beyond most predators' capabilities (see Suchanek 1978).

Among the many avian predators of mussels, oystercatchers (*Haematopus* spp.) and eider ducks (*Somateria* spp.) are both known to feed extensively on *Mytilus* (Dunthorn, 1971; Heppleston, 1971; Incze et al., 1980; Zwarts and Drent, 1981; Swennen et al., 1983; Goss-Custard and Durrell, 1987; Bustnes and Erikstad, 1990; Raffaelli et al., 1990), and immense flocks of these birds can sometimes be responsible for heavy mortalities, particularly on commercial mussel beds in wave-protected environments. In some years flocks of eider (>4000 birds) in the Ythan Estuary, North-east Scotland, can account for most of the surplus mussel production (Milne and Dunnet, 1972; Baird and Milne, 1981). More than 60% of the adult eider diet is represented by mussels. Raffaelli et al. (1990) showed that over a 60-day period a flock of 500 eiders removed approximately 4500 mussels m^{-2} mostly from the preferred (10–25mm) size range. When feeding on mussels eiders remove entire mussel clumps along with the focal prey item, thus generating bare patches within the mussel bed. Eiders can, therefore, have a significant impact on the population dynamics of *M. edulis*, not only through direct predation, but also as a result of increased mortality of the large numbers of mussels shaken from clumps (Raffaelli et al., 1990).

Mytilus is often the principal food supply of oystercatchers during the winter months and mussel production appears to be the major factor limiting the density of overwintering flocks in certain areas (Craeymeersch et al., 1986). In the East Scheldt, Holland, oystercatchers consumed about 40% of the annual mussel production (Meire and Ervynck, 1986). Mussels encrusted with barnacles, however, were rarely taken and there appeared to be strong selection against thicker-shelled mussels. In experiments using mussels as prey, Leopold et al. (1989) showed that oystercatchers tended to select mussels that were easiest to open. Drinnan (1958) estimated that oystercatchers in the Conwy Estuary, North Wales, could ingest their own body weight of wet shellfish per day, with individual birds consuming up to 574 mussels (average length 25.7mm) or 186 mussels (37.5mm) during each low-tide period. By the end of the winter period a significant proportion of the larger mussels had been removed from the population. On more wave-exposed shores small numbers of oystercatchers feed mainly on limpets and dogwhelks and consequently fewer mussels are taken in these habitats (Feare, 1971b). Using exclusion cages, Marsh (1986) showed that birds (black oystercatchers, surf birds and gulls) significantly reduced recruitment of juvenile *M. californianus* and *M. edulis*. Moreover, on surfaces previously lacking mussels, clumps of *Mytilus* became established within the exclosures but not in the control plots. Other birds that are known to feed on littoral mussels include scooters, sandpipers, knot, turnstones and even crows (Dare and Edwards, 1976; Feare and Summers, 1985; Yamada, 1989; Whiteley et al., 1990).

Crabs, particularly *Carcinus* and *Cancer*, can include large numbers of *Mytilus* in their diet (Kitching et al., 1959; Perkins, 1967; Walne and Dean, 1972; Elner, 1981; Menge, 1983; Ceccherelli and Rossi, 1984; Jensen and Jensen, 1985; Davidson, 1986; Ameyaw-Akumfi and Hughes, 1987; Gardner and Thomas, 1987). Many of the more recent studies of crab predation have focussed on predator preferences, the mechanics of shell crushing and energy maximization (Elner, 1978; Elner and Hughes, 1978; Jubb et al., 1983; Cunningham and Hughes, 1984; ap Rheinallt, 1986). Mortality from crab predation is generally most intense in the lower shore and sublittoral zone, where crabs are particularly abundant, and where they are able to forage for longer periods of time. Ebling et al. (1964) reported extensive crab predation in Lough Ine, South-west Ireland, and tentatively attributed the absence of *M. edulis* sublittorally in many localities to this cause. In the Menai Strait, North Wales, Davies et al. (1980) found that *M. edulis* protected by crabproof fences survived well, whereas unprotected control plots soon became completely denuded of live mussels.

Small mussels are especially vulnerable since these can be easily crushed by virtually all size ranges of crabs, whereas larger mussels are available only to larger crabs with strong chelae. Crabs will often actively select smaller mussels even when larger mussels, which they are capable of opening, are freely available. Vulnerability to crab predation will, therefore, generally decline as mussels increase in body size. Adult *Carcinus maenas* varies in colour from green to red depending at least partially on the length of intermoult. Red and green varieties are now known to have distinct physiological, ecological and behavioural characteristics (Reid and Aldrich, 1989; Kaiser et al., 1990). Green crabs are physiologically more tolerant of extreme conditions, but red crabs are structurally stronger and able to exploit a wider range of prey. In laboratory experiments red crabs exhibited a significant preference for larger *M. edulis* and usually dominated green crabs in aggressive disputes over prey. Harger (1972) showed that both *Cancer antennarius* and *Pachygrapsus crassipes* had a preference for *M. edulis* over *M. californianus*. Predation rates were such that mussels required 6–8 weeks from the time of settlement before they were large enough to escape predation by these crabs, and Harger (1972) concluded that, in order to survive on most rocky shores inhabited by crabs, mussels would have to settle at densities in excess of 10000 m^{-2}. When these two mytilids co-occurred the thicker-shelled *M. californianus* was afforded some protection by the presence of its more vulnerable congener.

One predator, whose profound effects on the abundance and distribution of at least two *Mytilus* species has not been recognized widely, is the sea otter *Enhydra lutris* (VanBlaricom, 1988). In central California, sea otters create numerous discrete gaps in

M. californianus beds (VanBlaricom, 1988), similar to those formed by log or wave damage (Paine and Levin, 1981), by removing clumps of mussels (of all size classes), which are then sorted and consumed on the sea surface by pounding the mussels on a flat stone on the sea otter's chest, or against other mussels. From these clumps sea otters typically consume individual *M. californianus* from 40–120mm length. Yet even those not selected for consumption and discarded by the sea otters will most certainly experience mortality from other benthic predators, burial on the seabed or eventual stranding in upper intertidal regions as a result of wave action. For the size range of mussels (up to 150mm length) observed in this region there was therefore no effective (size) refuge from sea otter predation (*sensu* Paine, 1976a). Thus, although sea otters are selective in terms of size preference the result of their foraging activities affects all size classes of mussels. Otters usually dive repeatedly in one region during high tide, typically increasing the area of an individual gap by this method, although some observations have been made of sea otters climbing up onto the exposed intertidal zone to forage on mussels (Harrold and Hardin, 1986; and see VanBlaricom, 1988). Although most foraging occurs during the period from January to June, there is tremendous variability in sea otter foraging rates; this variability may be linked to differences in prey quality, or even differences in individual foraging preferences and behaviour (VanBlaricom, 1988).

From data gathered before the 1989 Exxon Valdez oil spill, *M. edulis* comprised up to 40% of the diet of sea otters from Green Island, Prince William Sound, Alaska (Estes et al., 1981; VanBlaricom, 1987, 1988). As with *M. californianus*, clumps of *M. edulis* are removed, then sorted and crushed with the canines and/or consumed whole at the sea surface without the aid of tools. No observations were made of predation during low tide. In contrast to the foraging method used on *M. californianus*, sea otters remove *M. edulis* clumps independently of one another, thereby creating a more patchy landscape with smaller foraging gaps in the mussel beds, except in regions of extreme predation pressure, where vast areas are denuded. Since the largest *M. edulis* found were ca. 90mm, these mussels also could not attain a refuge in size from sea otter predation.

In southern California, nocturnal spiny lobsters (*Panulirus interruptus*) exert a significant influence on the population dynamics of intertidal *Mytilus* spp. (Robles, 1987). Despite continuous recruitment of *Mytilus* into these habitats, no established mussel beds are present at these sites. Here whelks, fish and lobsters consume *Mytilus* but only the lobsters specialize on the mussels. It is not surprising that these lobsters can have such a dramatic impact on mussels since it takes them less than 1.5min on average to consume a small (<20mm) *M. californianus*. While intertidal mussel-foraging by lobsters occurs throughout the year, lobster densities and resultant impacts on mussels during the autumn/winter period are less than half of those during spring and summer (Robles, 1987). American lobsters (*Homarus americanus*) in

Nova Scotia also have a dramatic influence on *M. edulis* populations (Elner and Campbell, 1987), especially at night (Lawton, 1987). Analyses of stomach contents from both adults and juveniles reveal that mussels (*Mytilus edulis and Modiolus modiolus* combined) comprise 9.6–16.2% of the lobster diet, although Elner and Campbell (1987) argue that *M. edulis* is the more important prey species.

Fish are well known predators of mussels. In Morecambe Bay, England, Dare (1976) found that the stomachs of 15 flounders, *Platichthys flesus* (23–38cm in length), contained the umbones of an average of ca. 150 young mussels (2–570 fish⁻¹). All sizes of mussels from 1 to 15mm were eaten, corresponding to the size range of this population. Eight plaice, *Pleuronectes platessa* of similar size had each eaten 105 mussels on average (30–175), while dabs, *Limanda limanda*, were also reported to feed on mussel spat in this region. Edwards et al. (1982) suggest that fish, especially the cunner (*Tautogolabrus*) may play a significant role in controlling the vertical distribution of *M. edulis* in New England.

Apart from the major groups of mussel predators already mentioned, mammals, including seals and walrus, and even turtles are also reported to feed on mussels (Seed, 1976; Hurd et al., 1979; Suchanek, 1985), while the grazing activities of sea urchins (Briscoe and Sebens, 1988) and limpets (Seed, 1969b; Connell, 1972) may account for some mortality of small, recently recruited mussels, particularly in the low shore. Polychaete worms (Hosomi, 1980) and the polyclad *Stylochus mediterraneus* (Galleni et al., 1977) are also predators of *Mytilus*.

Although sea urchins are known to consume kelp, their omnivorous habit is now being more appreciated, and their consumption of and control over mussel populations is more evident. From observations and field and laboratory experiments in the Gulf of Maine and the St. Lawrence Estuary, it is clear that the sea urchin *Strongylocentrotus droebachiensis* can essentially eliminate *M. edulis* from subtidal habitats (Himmelman et al., 1983; Briscoe and Sebens, 1988). *S. droebachiensis* is a facultative specialist that typically selects *M. edulis* when algal resources are depleted, but its effect is probably not as significant as that of seastars or predatory fish (Briscoe and Sebens, 1988).

It is clear, therefore, that similar suites of predators operate on *Mytilus* in different parts of its geographical range. The impact of many of these predators is often highly seasonal. Crabs and dogwhelks, for example, are generally most active during the spring and summer months, whereas the impact of many avian predators such as oystercatchers occurs mainly during the colder winter months, when huge flocks may be temporarily resident in the coastal zone. Figure 4.12 summarizes the major sources of mortality of *M. edulis* in Morecambe Bay, England, and shows how these mussels become vulnerable to different sources of mortality as they grow, and how this mortality is related to season.

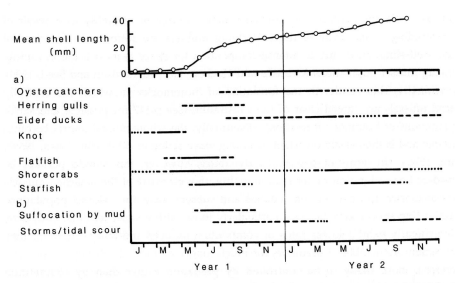

Fig. 4.12. The major identified sources of *Mytilus edulis* mortality from (a) predators and (b) biophysical factors in Morecambe Bay, England, and their effective periods relative to the seasonal growth rate of the mussel population; maximum effects denoted by heavy lines, slight effects by dotted lines. (After Dare, 1976).

Mussels are the supreme competitors for space amongst the ground-covering organisms on horizontal or gently sloping rocky shores (Lewis, 1977) and only rarely do we see other species displacing them from mid to lowshore sites under these conditions. On steeper rock faces mussels are generally less effective spatial competitors and consequently tend to be replaced by barnacles or algae (Lewis, 1964; Paine, 1974; Menge, 1976). Occasionally, however, when two mytilid species co-occur, as do *M. californianus and M. edulis* along much of the west coast of North America, interspecific competition can often result in the partial exclusion of one species (Harger, 1972; Suchanek, 1985). Nevertheless, these mussels exhibit very different life history strategies which facilitate their widespread coexistence (Suchanek, 1981).

In a fascinating account of an introduced species, one mytilid (*Mytilaster minimus*) found in the Mediterranean Sea may now be experiencing competitive interference from a second mytilid species (*Brachidontes variabilis*) that migrated through the Suez Canal (Safriel and Sasson-Frostig, 1988). The two species have virtually identical habitat requirements, although different densities of each are found under different conditions of wave exposure. Results of field experiments indicate that adult *B. variabilis* disproportionately inhibit both survivorship and mean shell length of *M. minimus* recruits, but environmental patchiness probably permits coexistence (Safriel and Sasson-Frostig, 1988).

Intraspecific competition can also be a major source of mortality as a result of overcrowding when many of the underlying mussels are suffocated or starved of food. Self-elimination due to intense competition for space is most acute in rapidly growing mussel populations (Griffiths and Hockey, 1987; Richardson and Seed, 1990). Occasionally, this results in the formation of 'hummocks' in which the centrally placed mussels are forced clear of the substratum (see p.147 for possible alternative explanations of hummock formation). Eventually, the whole mussel matrix becomes unstable and is then easily detached by strong wave action or tidal scour (Seed, 1969b; Dare, 1976). In terms of population dynamics, however, this should perhaps be considered as emigration rather than mortality, though many of the detached mussels are transported to unfavourable habitats and subsequently die. Mussel populations experiencing slower growth rates, particularly those with a small terminal body size, will generally exhibit lower rates of competition-induced mortality, although they will remain vulnerable to predators for much longer periods. Such populations are, therefore, more likely to be controlled by predation rather than by competition (Griffiths and Hockey, 1987). In the Exe Estuary, England, McGrorty et al. (1990) showed that the mortality of *M. edulis* during their first winter was strongly and positively density-dependent. This provided a powerful regulating mechanism by which the adult population was effectively stabilized despite wide annual variations (up to x17) in the density of juvenile mussels.

Fouling organisms are increasingly being recognized as significant sources of mortality in littoral and sublittoral mussel populations (Paine, 1979; Suchanek, 1985). Mortality usually occurs when mussels are dislodged as a result of the increased weight or shearing stresses in the form of drag or lift imposed on them by the fouling organisms, especially by barnacles and seaweeds (Witman and Suchanek, 1984; Denny, 1987). Feeding currents may also be restricted or even totally occluded if the valve openings are overgrown. Even without causing outright mortality, fouling organisms may lower the fitness of mussels by reducing body tissue and/or gamete development (Paine, 1976). The mechanism responsible for such lowered fitness is probably related to reduced food intake, or to the greater expenditure of energy on maintenance and the production of byssus for more secure anchorage (see also p.143–144). Fouling, especially by ascidians, is often a major problem on rope grown mussels (Mason, 1972 and see p.487 Chapter 10). Fouling organisms, however, are not always detrimental to their hosts. A recent period of freezing temperatures along the Pacific coast of North America resulted in mass mortality of *M. californianus*. However, mussels, which were insulated with the epiphytic alga *Endocladia muricata*, survived the freeze, whereas those without the epiphyte died (Brosnan, 1990).

THE *MYTILUS* BED COMMUNITY

The *Mytilus* Matrix and Associated Fauna

Mytilus aggregations are highly productive assemblages. *Mytilus californianus*, the largest *Mytilus* species from the North American continent, typically has a range of mussel abundance of 459–11,098 individuals m^{-2} (mean 5,795 ± SD 3,451) (Suchanek, 1979), a calculated mussel meat standing crop of 1.3–6.5kg m^{-2} dry wt (energy = 29–144J mm^{-2}) and a mussel meat productivity of ca. 0.2–1.9kg m^{-2} yr^{-1} dry wt (energy = 3.8–42J mm^{-2} yr^{-1}), rivalling the productivity of some of the most productive rainforests (Leigh et al., 1987). This compares with *M. edulis* communities from the Bay of Fundy with mussel abundance of 700–4,000 individuals m^{-2} (Newcombe, 1935), Narragansett Bay, Rhode Island, where whole mussels represent ca. 77% (11kg m^{-2}) of the total community dry weight and the mussel meat represents ca. 82% (1.5kg m^{-2}) of the total community meat dry weight (Nixon et al., 1971); Morecambe Bay, England, where mussels have a maximum standing crop value of 1.5kg m^{-2} AFDW (Dare, 1976), and the northern Wadden Sea (North Sea) with numerical abundance values of ca. 1,631 ± 151 individuals m^{-2}, a range of standing crop values for whole mussels of 0.9–1.4kg m^{-2} AFDW (which is ca. 25 times higher than other communities in the Wadden Sea), and a range of whole mussel productivity values from 0.2–0.6kg m^{-2} yr^{-1} AFDW, 93% of which is due to the mussels themselves, and 7% to the organisms found in association with the mussels (Asmus, 1987). In comparison, Craeymeersch et al. (1986) found mean densities of *M. edulis* in the southwestern Netherlands of 1,590 m^{-2} and annual productivity of mussel flesh of 0.16kg m^{-2} yr^{-1} AFDW, whereas Kautsky (1982a) found densities of 36,000–158,000 m^{-2} (see Thompson, 1984b) in subtidal Baltic Sea populations. These extremely productive populations most likely result from a significant lack of predatory pressures on *Mytilus* in totally submerged populations (Kautsky, 1981). *M. galloprovincialis* communities from Italy show similar trends, with numerical abundance values for mussels of 333–11,536 m^{-2} (extrapolated from 600cm^2 samples) which comprise ca. 4.7–91.2% of the benthic community biomass at these sites (Tursi et al., 1985). Comparable values for the closely related mytilid, *Perna perna*, yield mussel numerical abundance values of 633–1587 m^{-2} (extrapolated from 100cm^2 samples), and standing crop estimates for whole mussels of 0.4–1.4kg m^{-2} AFDW, of which *Perna* alone represents 82–94% (Jacobi, 1987a).

    Mytilus assemblages have long served as the focal point for numerous empirical and theoretical studies on population and community ecology, especially the importance of predation and disturbance events on structuring marine intertidal communities. In the past, most attention has been focussed on *Mytilus* population dynamics *per se*, although many investigators have referred to the mussel populations as *Mytilus* communities or mussel communities, probably because they represent neatly identifiable entities and dominant biomass structural components within specific regions of the intertidal zone (Shelford et al., 1935; Hewatt, 1937; Rigg and Miller, 1949). Paine's (1966, 1969) seminal works on the regulation of *Mytilus* 'community' diversity by predation, and Dayton's (1971) analysis of physical disturbance on *Mytilus* populations, have stimulated a wide variety of studies on processes responsible for regulating community structure in rocky intertidal systems, as well as many other habitats. However, these studies have dealt almost exclusively with the dominant biomass component (mussels) and/or other primary space occupiers, with little reference to the diverse biological community found associated with the mussels. Hewatt (1935), Newcombe (1935) and Ricketts and Calvin (1939) were among the first to address taxonomic complexity within *Mytilus* beds, but not until recently has the full extent of this diversity been appreciated and more fully documented (see Suchanek, 1979, 1980, 1985; Kanter, 1978, 1980, and Paine and Suchanek, 1983 for *M. californianus* beds; Tsuchiya, 1979 and Tsuchiya and Nishihira, 1985, 1986 for *M. edulis* beds; Tsuchiya and Bellan-Santini, 1989 for *M. galloprovincialis* beds).

As *Mytilus* beds age and grow, they increase not only their biological component, the living mussels, but they also enlarge their physical component, producing structurally complex entities that are capable of harbouring a diverse assemblage of associated fauna and flora. Bed thickness, connectedness between individuals and sediment loading are all increased within the bed, changing dramatically the microhabitats under, between and around the mussels. Although only the bottom layer of mussels is attached to the primary substratum, individuals in subsequent layers solidify the bed structure by interconnecting byssal threads to many neighbouring individuals. This usually forms a dense structure with numerous interstices, which provides refuge for a myriad of associated fauna. Thus, *Mytilus* beds are composed primarily of three components: (1) a physical matrix of interconnected living and dead mussel shells, which may occur as a monolayer or multilayer up to five or six mussel layers deep, (2) a bottom layer of accumulated sediments, mussel faeces and pseudofaeces, organic detritus and shell debris, and (3) a taxonomically diverse assemblage of associated fauna and flora (Suchanek, 1979).

The data given below on associated fauna and flora pertain primarily to characteristics of naturally occurring mussel beds, although there are also data on associated organisms in cultured *Mytilus* populations (Plessing, 1981; Mattsson and Linden, 1983).

Considerable progress has been made in understanding the development of diverse *Mytilus* communities (Suchanek, 1979, unpublished results; Tsuchiya and Nishihira, 1985, 1986). Unless indicated, the information presented below has been gleaned from the works cited in the preceding sentence. In general, mussel bed thickness and structural complexity increase with mussel bed age. Sediments and debris accumulate in direct proportion to the thickness of the beds. Further mussel recruitment promotes even greater physical and/or chemical changes and as more layers are added, existing microhabitats are modified and new ones created. Observations by Newcombe (1935) indicate that for a M. *edulis* bed on soft substrata, sediments beneath the bed become more anoxic with increased mussel cover, thereby eliminating some infaunal bivalve species such as *Mya arenaria*. Nixon et al. (1971) found that M. *edulis* beds in Rhode Island contained ca. 14.4kg m^{-2} dry wt of trapped sediment with a mean organic content of 3.86%. For M. *californianus* beds ca. 24cm thick from Tatoosh Island, Washington, dry weight of sediment can exceed 80kg m^{-2} (Suchanek, 1979).

Mussel beds can attain considerable thickness, with intertidal M. *edulis* beds reaching 10cm thickness (Nixon et al., 1971), subtidal M. *edulis* beds being reported as thick as ca. 120cm (Simpson, 1977), and intertidal M. *californianus* beds attaining a maximum thickness of ca. 40cm (Suchanek, unpublished results). In M. *californianus* beds, mussel shells also increase the ratio of shell surface area to rock surface area by multipliers of ca. 5 (in monolayered beds) to 30 (in multilayered beds), providing considerable increased substrata for settlement of associated fauna and flora (Suchanek, unpublished results). Within the interstices of this matrix light, temperature and wave action are diminished, whereas sedimentation and relative humidity are increased. Figure 4.13 illustrates the significant changes in microhabitat regimes as both temperature and relative humidity are altered at various depths within M. *californianus* beds on the west coast of North America at Tatoosh Island, Washington and Bodega Bay, California (from Suchanek, unpublished results). On a sunny day, temperatures at the base of a 25cm thick M. *californianus* bed can be 5-13°C cooler than at the surface, whereas relative humidity can increase by ca. 15%, affording greater protection to species with more restricted physiological tolerances to these parameters.

Community diversity within mussel beds is surprisingly high. Species richness (S) and the Shannon-Wiener diversity index (H') increase with both mussel bed age and thickness. M. *edulis* from Northern Ireland had at least 34 associated species within the interstices of the bed, of which nearly half are Crustacea (Briggs, 1982). For M. *edulis* beds in Japan, Tsuchiya and Nishihira (1985, 1986) found that S was greatest in older and larger beds, and H' and Margalef's equitability index (J') were higher at the perimeter and central portions of adult beds than in young beds or old beds. Although 69 species representing eight phyla were found in this community, 98% of the individuals encountered were accounted for by three phyla: Annelida (46%), Arthropoda (39%), Mollusca (13%), and no algae were reported. Surprisingly, Asmus (1987) found no correlation of increased biomass of associates with increased M. *edulis* density in the northern Wadden Sea. Also surprising is the low species richness (12 taxa), and no significant increase of species richness of associated fauna inside M.

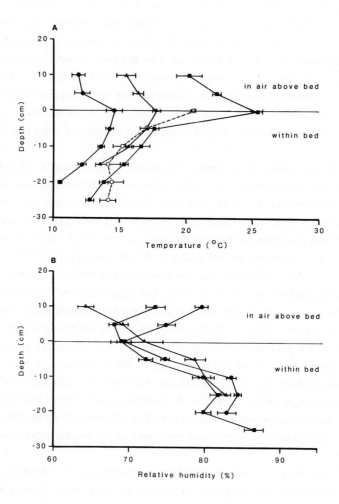

Fig. 4.13. Variations in (A) temperature and (B) relative humidity above and within dense beds of *Mytilus californianus* from three sites in Bodega Bay, California, North America (closed symbols), and one site at Tatoosh Island, Washington state, North America (open symbols); central symbols are mean values (typically N = 5), error bars are ± 1 SE.

edulis beds, as compared with those found outside mussel beds in Maine, U.S.A., documented by Commito (1987).

For *M. californianus* in Washington, Suchanek (1979, 1980) sampled mussel beds over a variety of thicknesses, tidal heights and exposure conditions. He found that S (but not H' or J') increased with mussel bed age, and decreased with tidal height at most study sites to a maximum of about 135 taxa for any one sample (ca. 30 x 30cm) yielding a total community S of over 300 taxa representing three divisions of macro-algae, 12 phyla of invertebrates and 3 species of bony fish. Nearly 90% was accounted for by four phyla: Arthropoda (38%, with barnacles alone comprising 24%), Mollusca (35%), Bryozoa (9%) and Annelida (7%). All sites showed a significant increase in S, H' and J' with increasing mussel bed thickness. Species richness showed a significant inverse correlation with tidal height and only suggestive (nonsignificant) positive trends with increasing wave exposure. For samples of ca. 0.10m², S ranged from ca. 25 species at upper intertidal low-wave exposure sites, to 135 species at low intertidal high-wave exposure sites. As a comparison, *Perna perna* beds from Brazil also harbour a diverse fauna, with Crustacea (70%), Polychaeta (6%) and Cnidaria (6%) making up the most numerically abundant taxa, although no quantitative species richness or diversity values were given (Jacobi, 1987a).

Selected taxa specialize on specific microhabitats within the mussel matrix. Some prefer the organically rich sediments at the base (bivalves, sipunculids and some poly-chaetes); some prefer the interior facies of old broken mussel shells (bryozoans and hydroids); and some only the surface mussels (algae and most barnacles), although other species are distributed throughout the entire vertical strata (sea cucumbers, anemones, boring clionid sponges, crabs, nemerteans and errant polychaetes) (Suchanek, 1979). While there is some overlap between groups, associated fauna and flora can be divided into three major categories based on their interaction with, or relative position within, the mussel matrix (Suchanek, 1979). 'Epizoans', typically ranging from 17–33% of the associated species, are sessile forms that use the mussel shells directly as a substratum; they either grow on or bore into mussel shells. Examples include all attached algae, barnacles and hydroids. 'Mobile fauna' (58–74% of associates) may include representatives from many trophic and/or taxonomic groups, but at least have the ability to move freely throughout the mussel matrix. Examples include porcellanid crabs, numerous amphipods and isopods and free-roving gastropods. 'Infauna' (5–21% of associates) are those organisms that are associated with, and usually directly dependent upon, the sediment and/or accumulated detritus at the base of the mussel bed, and are species typical of most soft sediment environments. Examples include sipunculids, sediment dwelling polychaetes and ophiuroids. Furthermore, while some sediment contributes to habitat heterogeneity and the promotion of diversity for some taxa, an excess in mussel beds has been found to adversely affect some amphipod assemlages (Tsuchiya and Nishihira, 1985; Jacobi, 1987b).

The specialization that derives from the microhabitat differences encountered by fauna and flora within different regions of the mussel matrix results in a predictable vertical stratification. For example, because algae need high levels of incident radiation, they are always found on the upper surfaces of the mussel matrix. The common sea cucumber found in *M. californianus* beds, *Cucumaria pseudocurata*,

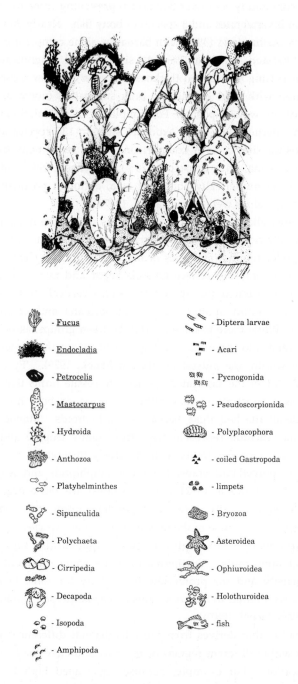

 - <u>Fucus</u>

 - <u>Endocladia</u>

 - <u>Petrocelis</u>

 - <u>Mastocarpus</u>

 - Hydroida

 - Anthozoa

 - Platyhelminthes

 - Sipunculida

 - Polychaeta

 - Cirripedia

 - Decapoda

 - Isopoda

 - Amphipoda

 - Diptera larvae

 - Acari

 - Pycnogonida

 - Pseudoscorpionida

 - Polyplacophora

 - coiled Gastropoda

 - limpets

 - Bryozoa

 - Asteroidea

 - Ophiuroidea

 - Holothuroidea

 - fish

Fig. 4.14. Diagrammatic cross section through a dense *Mytilus californianus* bed from Tatoosh Island, Washington state, North America, showing vertical stratification of typically abundant fauna and flora.

from data obtained in laboratory tanks, has the ability to move throughout the bed, but typically chooses sites in the mid to lower region, and specializes in the consumption of *Mytilus* faeces and pseudofaeces (Suchanek, unpublished results). Figure 4.14 is a diagrammatic cross-section through a *M. californianus* bed showing the vertically stratified nature of these typical associated fauna and flora (from Suchanek, 1979). The specific nature of preferred site selection within a mussel matrix by associated fauna has been shown even on the scale of individual mussel shells. Laihonen and Furman (1986) reported site-specific settlement of a barnacle epibiont, primarily near the siphonal aperture, on *M. edulis* from the Baltic Sea.

Mytilus Dislodgement by Biota

Habitat energy regimes can influence greatly the byssal attachment strength of intertidal mussels, and therefore play a critical role in protection from dislodgement. Byssal attachment strength of *M. californianus* is significantly greater than that of its smaller congener *M. edulis* and is significantly greater at the edges of a bed than at the periphery (Witman and Suchanek, 1984). Other information on differences in attachment strengths between *M. galloprovincialis* and *M. edulis* is provided in Skibinski and Gardner (1991). Attachment strength for *M. edulis* from exposed habitats is ca. 15 times higher than in protected habitats (Witman and Suchanek, 1984), and in experiments where *M. californianus* were held down under plastic covered mesh in exposed rocky intertidal habitats, those mussels that were held down firmly seldom or never laid down byssal threads (Suchanek, unpublished results). However, those that were loose under the plastic developed numerous byssal attachments, presumably in order to stabilize their positions. These data indicate that *Mytilus* detects and responds to movement by wave energy or drag forces by the production of increased numbers of byssal threads.

Despite increased production of byssal threads, excessive fouling by epibionts can increase significantly the risk that mussels will be dislodged from the primary substratum (Dayton, 1971, 1973; Paine, 1979; Witman and Suchanek, 1984; Witman, 1987). In general, low-profile epizoans (e.g. sponges, bryozoans, encrusting coralline algae) that form thin encrustations and do not increase the vertical relief of the mussel, do not place mussels at much risk of dislodgement, but those epibionts (e.g. macroalgae, kelp, barnacles) that increase the effective height of mussels above the mussel bed plane will lead to increased drag and lift amplification and higher risk of dislodgement (Witman and Suchanek, 1984; Denny, 1987; Suchanek, unpublished results). Another generalization is hypothesized for encrusting epizoans (e.g. tunicates, sponges) which might occlude the feeding gape of mussels, and thereby cause death, dislodgement or lowered fitness through a weakened state due to starvation (Paine, 1976b; Suchanek, unpublished results).

Symbiosis

At least four types of symbioses, commensalism (+0), amensalism (-0), parasitism (+-) and mutualism (++), have been reported between mussels and their associates. Barnacle settling patterns on live *M. edulis*, result in significantly faster barnacle growth rates as a result of enhanced food resources, indicating a commensal relationship (Laihonen and Furman, 1986). Over 76% of those barnacles settling on live mussels chose the region closest to the siphonal aperture and, during a 10-week period, grew an average of 4mm larger (rostro-carinal length) than those settling on dead mussel shells. No significant difference in growth rates was observed between mussels with and without barnacle fouling. It is interesting to note that in many cases *M. edulis* is able to clean itself by sweeping its prehensile foot over the dorsal part of the shell (Thiesen, 1972), whereas *M. californianus* lacks that capability. Although fouling does occur on *M. edulis*, this may help to explain why Tsuchiya and Nishihira (1985, 1986) reported no algae fouling *M. edulis* in their study of associated organisms (see above).

In southern California, U.S.A., facultative red algal epiphytes have a negative effect on both the survivorship and reproductive capabilities of *M. californianus* and provide virtually no protection to the mussels (Dittman and Robles, 1991). In an epiphyte removal experiment, overgrown mussels had significantly lower growth rates and gonad weights than comparable individuals that were cleaned of their epiphytes.

In Oregon, U.S.A., the red alga *Endocladia muricata* can have a variable influence on *Mytilus*, depending on the environmental conditions (D. Brosnan, 1990 and personal communication, 1990). Under heavy wave action *M. californianus* individuals that are heavily fouled with *Endocladia* are dislodged at a significantly higher frequency than those without fouling. *Endocladia* also likely interferes with mussel feeding and growth, resulting in a clearly amensal relationship. However, as decribed already (p.136) mussels that are fouled by *Endocladia* are also insulated from extremes in temperature, and survive drastic freezes at significantly higher rates than those without such epiphytic growth, producing a commensal (or possibly mutualistic) relationship.

The lack of shell cleaning ability by *M. californianus* may help to explain why associated grazers (e.g. limpets, snails and chitons) are especially important in reducing excessive epibiotic fouling on this species, and why a mussel-grazer mutualism may be essential in maintaining community stability for this assemblage of mussels and associates (Suchanek, 1979). *M. californianus* beds harbour high densities of mobile grazers, ca. 3–16 times that of the surrounding open rock substratum. Several factors contribute to these increased densities since *M. californianus* beds provide significant benefits to grazers: (1) increased surface area on mussel shells increases grazer food resources; (2) modified microhabitat regimes within the interstices of the mussel matrix provide protection from extremes in physical variables such as relative humidity and temperature, and (3) the mussel bed matrix provides mobile grazers with a spatial refuge from visual predators. Grazers, in turn, provide significant benefit to the mussels: (1) they remove potentially harmful epibiota, reducing the risk that mussels might be fouled extensively and then dislodged when they encounter high-wave conditions, and (2) grazers enhance the speed of mussel bed recovery from disturbance by nearly seven-fold. As a result, grazers stabilize the physical mussel matrix, within which a diverse community of associates has developed. In so doing, this mutualistic association also enhances the stability of the entire mussel bed community (Suchanek, 1979, 1985).

Grazers not only reduce algal cover on the mussels themselves, but often produce distinctive patterns (browse zones) within and around *Mytilus* beds that indicate the intensity and extent of grazing (Suchanek, 1978; Sousa, 1984). Figure 4.15 illustrates the nature of browse zones at the perimeter (upper) and within disturbance gaps (lower) in *M. californianus* beds from Shi-Shi, Washington state, showing the relative distance travelled by molluscan grazers foraging out from the mussel matrix. Similar patterns are found in *Perumytilus purpuratus* beds in Chile (Suchanek, unpublished results). In a series of experiments Suchanek (1979) erected experimental barriers that prevented the movement of grazing molluscs at the exterior and interior edges of *M. californianus* beds at Tatoosh Island and Shi-Shi, Washington, and demonstrated that browse zones are indeed created by the foraging activities of molluscan grazers. Since the mussel bed matrix provides refuge from visual predators such as birds, mobile grazers typically remain within the interstices of the bed during low tide. When *M. californianus* beds are submerged grazers usually migrate out ca. 20cm from the edges of the bed, creating a halo effect of reduced algal cover (Suchanek, 1978, 1979, and Fig. 4.15 upper). When disturbance gaps are formed in the central portions of these same *M. californianus* beds, grazers also forage ca. 20cm into the gaps to consume algae within that zone (Fig. 4.15 lower). This effectively alters the successional sequences and replacement order of gap colonizing species (see section: Recovery from disturbance, p.149ff).

Fig. 4.15. Browse zones of bare space created by molluscan grazers associated with *Mytilus californianus* beds in Washington state, North America. External browse zones, about 20 to 30cm into the surrounding *Fucus* cover (upper), surround two isolated *M. californianus* beds at Shi-Shi. Internal browse zones (lower) extend inwards about 20cm into the *Porphyra* covered central region of a natural disturbance gap at Tatoosh Island. The smaller disturbance gap in (lower) has a diameter of about 20cm and therefore is devoid of all algal growth (see text).

DISTURBANCE AND RECOVERY OF *MYTILUS* BEDS

Disturbance

The integrity and physical stability of *Mytilus* beds depends directly on the attachment strength of their byssal threads. In general, mussels appear to respond to increased wave force by the production of greater numbers of byssal threads. *M. edulis* produce more byssus during stormy winter periods than in calm summer periods (Price, 1980, 1982) and *M. californianus* produce stronger (more numerous) attachments at the edges of beds than in the central regions (Witman and Suchanek, 1984). However, when byssal thread attachments are disrupted, individual or groups of mussels within the matrix are dislodged and disturbance gaps are formed. These gaps within mussel beds may be initiated by either physical factors (Dayton, 1971; Levin and Paine, 1974, 1975; Paine and Levin, 1981; Sousa, 1985; Denny, 1987), or biological processes (Paine, 1966, 1969, 1974, 1979; Dayton, 1971, 1973; Suchanek, 1978; Witman and Suchanek, 1984; Paine et. al., 1985; VanBlaricom, 1988). Wave action (Dayton, 1971; Paine and Levin, 1981), epizoism (Denny, 1987), or predation (Van Blaricom, 1988) are likely to lead to subsequent expansion of these disturbance gaps.

The mechanism of initial gap formation in *Mytilus* beds may involve a single process or a combination of processes: log battering, wave action, fouling, and hummocking. Log damage seems to occur more frequently in regions of heavy logging (Dayton, 1971). Wave action most likely involves a combination of drag (Witman and Suchanek, 1984) and fluid dynamic lift (Denny, 1987) from breaking waves, especially if mussels are significantly fouled by epibiota. Heavy fouling, particularly by the brown algae *Fucus and Postelsia* and the barnacle *Semibalanus cariosus*, is especially common in regions where mobile predators or grazers are absent, or in low density. Heavy fouling by these three taxa is common on dislodged *M. californianus* that have been washed ashore along outer coast beaches (Witman and Suchanek, 1984; Suchanek, unpublished results). Hummocking, the phenomenon where mussel clumps are raised above the ambient level of the bed because they lack byssal attachments to the primary substratum but remain attached to each other, is a common feature of some *M. californianus* beds in Washington.

Hummocks may be formed by intense intraspecific pressures as young mussels grow larger within the matrix of the bed forcing other individuals upwards without making firm byssal attachments to the primary substratum. Or, they may also be formed by biological agents such as decapods, either purposefully or mistakenly nipping at *Mytilus* 'feet' as the mussels attempt to extend them to secure byssal attachments. Hummocks are often connected to other hummocks by small tunnels with the same characteristics, and are typically inhabited by vast numbers of crabs such as *Petrolisthes* spp. (Fig. 4.16). Because of the weak attachment of these hummocks to

Fig. 4.16. Diagram of hummocks in *Mytilus californianus* beds at Tatoosh Island, Washington state, North America. Note that such hummocks are typically connected by tunnels or passageways and populated with high numbers of porcellanid crabs.

the primary substratum, they may serve as initiation sites in the formation of disturbance gaps (Denny, 1987).

Gap formation in *Mytilus californianus* mussel cover resulting from wave or log damage occurs most frequently during the winter season and can disrupt up to 65% of some mussel beds in one season; studies over six winters (199 site-years of data) showed typical removal rates of ca. 0.4–5.4% of the mussel cover per month, with 89% of the disturbances affecting <25% of the beds each winter (Paine and Levin, 1981). The remaining observations span the range from 25–65% and might, more appropriately, be termed catastrophes, not an event to which *M. californianus* beds are typically exposed each winter. Gaps which occur in summer are smaller and may form 5–10 times slower (Paine and Levin, 1981). The initial size of disturbance gaps from wave or log damage in *M. californianus* beds ranges from the size of an individual mussel to areas of ca. 57.5m² and the distribution of gap sizes approximates a log-normal (Dayton, 1971; Suchanek, 1979; Paine and Levin, 1981). Subsequent enlargement may result in a nearly 5000% increase in gap size (Dayton, 1971), most likely as a result of weaker byssal thread attachments in the central portions of the beds, where the gap was initiated, than at the edges (Witman and Suchanek, 1984). Only 1.6% of such disturbance gaps subsequently enlarged when they were formed during or just prior to summer months (when wave action is minimal), but over 21% of comparable gaps enlarged when they were formed during or just prior to winter months (Paine, 1989).

Biological disturbances that disrupt the *Mytilus* matrix, such as predation by crabs and seastars or epizoism by algae (e.g. *Fucus* or *Postelsia*), usually occur on the scale of individual or few mussels, although the actions of the predatory seastar *Pisaster* can result in larger gap formation in *M. californianus* beds. This is done by the removal of one or several mussels from the matrix, which can then undergo enlargement as described above.

Sea otters also cause disturbance gaps in both *M. edulis* and *M. californianus* beds. At one site in central California, sea otter predation removed about 20% of the *M. californianus* cover, where the mean size of gaps was $0.25m^2$ (range $= 0.03$–$1.34m^2$) (VanBlaricom, 1988). In this study sea otters created more gaps than did wave/log damage. The size-frequency distribution of these gaps is quite similar to those produced by wave or log damage in Washington state (Paine and Levin, 1981), although wave or log damage produced more smaller and more larger gaps in mussel beds than did sea otters. In Alaska, a major immigration of male sea otters is believed to have been responsible for a large-scale *M. edulis* mortality event (100% in some regions) in Prince William Sound during the winter of 1979–80, but the mussel beds recovered quickly and returned to moderately dense coverage in 1984 (VanBlaricom, 1988).

Twenty six percent of sea otter generated gaps enlarged before they healed, which in some cases took over two years (VanBlaricom, 1988). This increased the chance that various subsections of the gap might be set to different successional ages, and therefore might experience different successional processes; thus, making the patch mosaic of primary space occupiers even more complex in regions inhabited by sea otters, than in areas where gaps are formed from wave or log damage.

Within the context of the variable disturbance events described above, *Mytilus* beds have been characterized as having 'mosaic' patterns composed of disturbance gaps with a wide variety of alternative primary space occupiers, most of which are found in gaps of varying ages and varying successional stages.

Recovery from disturbance

Nearly complete recovery from physical and/or biological disturbance is a characteristic common to *Mytilus* beds throughout the world. Most *Mytilus* assemblages are fairly dominant and persistent features of the habitats in which they occur, and their superior ability to regain lost space has been interpreted as a deterministic process (e.g. Paine, 1974, 1984). Parameters that influence this reclamation process are: size of initial disturbance gap, season of disturbance, height of mussel bed on the shore, angle of substratum, age of bed and intensity of larval settlement.

Paine and Levin (1981) provide substantial documentation on the basic recovery process for mid intertidal *M. californianus* beds. In their study, recovery of very small gaps was almost instantaneous (0.2cm day^{-1}) due to leaning or collapse of the adjacent mussel matrix, especially if the surrounding matrix was thick. For intermediate sized gaps ($<3.0m^2$) their recovery was slower (0.05cm day^{-1}) and more dependent on lateral movement of mussels from the edges of the disturbance gap. For very large gaps mussels may settle directly from the plankton onto the primary substratum or onto filamentous substrata within the gap. Alternatively, mussel recolonization into large gaps can be enhanced by two other mechanisms. Firstly, *M. californianus* larvae tend to settle most heavily onto conspecific byssal threads; therefore, byssal threads of adult mussels that have been dislodged from other regions and reattach within the gap, provide foci for mussel larvae settling from the plankton (Suchanek, 1978, 1981), thus enhancing gap healing. Secondly, mussels dislodged from the primary substratum leave remnant byssal threads, which remain in the gap for several months after a major disturbance, and these threads can act as similar foci for mussel larval settlement/recruitment (Suchanek, unpublished results; see also p.108–112).

A predictable series of biological events accompany the recolonization of disturbance gaps in mid intertidal *M. californianus* beds in Washington (Suchanek, 1979; Paine and Levin, 1981). Remnant byssal threads typically remain on the primary substratum for about 10mo. During this period, the substratum is first colonized macroscopically by diatoms and filamentous algae, as well as by barnacles and *M. edulis*. The latter may persist as a fugitive in these gaps up to three years from the initial disturbance event, with coverage sometimes reaching 70–80% of the gap at extrapolated densities of ca. 12,400 ± 5990 individuals m^{-2} (Suchanek, 1978, and unpublished results). This allows ample time for *M. edulis* to attain reproductive size and spawn (Suchanek 1978, 1981). During this phase the gap substratum is also colonized by numerous algae and balanomorph and gooseneck barnacles, during which time *Nucella* dogwhelks, at mean densities of ca. 50–75 m^{-2}, consume a majority of the *M. edulis* (Suchanek, 1978, 1981; Paine and Levin, 1981). Typically, ca. 85 ± 14% of the dead *M. edulis* shells surveyed within the boundaries of these gaps show clearly identifiable *Nucella* drill holes (Suchanek, 1978). Similar processes appear to occur in *M. californianus* gaps in California, but the *M. edulis* are consumed at such a rapid rate that they likely never reach reproductive size (Sousa, 1984).

M. californianus first colonizes these gaps about 20–26mo after the initial disturbance, and rapidly increases its coverage of the substratum to over 80% after about 36mo, with a matrix about 14cm thick, consisting of 2–3 mussel layers. After 60–80mo the substratum is completely reclaimed by *M. californianus*, with a 15–20cm thick matrix comprising 3–4 mussel layers. Figure 4.17 illustrates this sequence of recolonization events at a site on Tatoosh Island, Washington state over 15 years. After complete coverage of the substratum the mussel bed continues to increase in thickness to a maximum of about 40cm with 5–6 mussel layers.

Seasonality is also critical to the sequence of recovery events that follow. In addition to an increased rate of gap enlargement during winter seasons, as compared with summer seasons (see above), the rate of recovery is also significantly reduced during winter months (Suchanek, unpublished results). Figure 4.18 shows the maximum monthly rate of gap closure (cm^2 mo^{-1}) for 55 relatively small (ca. 900cm^2)

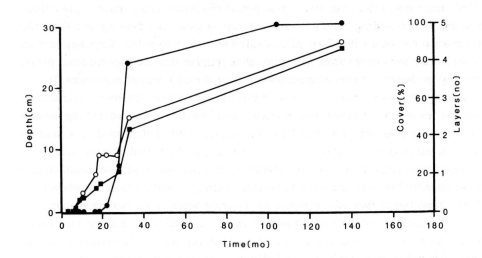

Fig. 4.17. Changes in mussel bed depth (solid squares), number of mussel layers (open circles) and mussel cover (solid circles) during the re-establishment of a *Mytilus californianus* bed at Tatoosh Island, Washington state, North America, following a storm which completely denuded the substratum in November 1975 (time = 0).

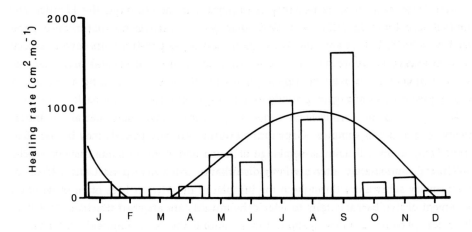

Fig. 4.18. Maximum monthly rate of disturbance gap closure (= healing rate) for 55 artificially created gaps in *Mytilus californianus* beds at Tatoosh Island and Shi-Shi, Washington state, North America, during 1974–1978. Data were pooled for all tidal heights; fourth order polynomial curve fitted to data points.

artificially created gaps in *M. californianus* beds at Tatoosh Island and Shi-Shi, Washington monitored over a four year period (Suchanek and Duggins, unpublished results). This healing rate/time relationship is described best by the 4th order polynomial $y = 980.3 - 1048.1\,x + 321.2\,x^2 - 31.4\,x^3 + 0.9\,x^4$, $R^2 = 0.62$. Gaps healed much faster during the months from May through September than during any other period. Healing in these gaps was accomplished mostly through mussel movement (leaning and re-attachment of byssal threads), rather than by larval settlement/recruitment. This series of experiments also revealed that the initial period of gap healing (immediately after gap initiation) has the highest rate of closure during summer months, especially for gaps that occur in thick, multilayered mussel beds. The mechanism(s) responsible for this phenomenon is unclear. Since mussels tend to produce more byssal threads during periods of stronger wave action (see above), one might hypothesize that winter would be a period when more healing would occur through the extension and production of more byssal threads, thereby resulting in lateral movement; but the data do not support this prediction. Alternatively, mussels may discriminate between rough and calm sea conditions by changing the nature of byssal thread production. In rough winter conditions they might produce many byssal threads in their immediate vicinity in an attempt to become as rigidly attached to the underlying substratum as possible. During calmer summer conditions they might extend their foot to attach byssal threads further away in order to reclaim substratum lost during disturbance events.

Patch size as well as seasonality is important in determining the identity and timing of colonizing fauna and flora that occupy recovering disturbance gaps. Suchanek (1978, 1981), Paine and Levin (1981) and Sousa (1984) discuss several aspects of competitively subordinate taxa (including *M. edulis* and several barnacles and algae) that colonize in winter—a period when the probability of finding a colonizable gap is greatest. Winter gaps provide excellent opportunities for barnacle larvae (e.g. *Balanus glandula* and *Semibalanus cariosus*) and certain algal taxa that are abundant in spring, but not in summer. Gaps formed during summer periods may be colonized initially by a very different assemblage of fauna and flora; in some cases barnacles colonizing immediately after a winter disturbance patch may persist and remain on the primary substratum throughout the lifespan of the *Mytilus* bed, often providing a basement of live barnacle tests to which the mussels are attached (Suchanek, 1979; D. Brosnan, personal communication, 1990). Along the New England and Alaskan shorelines, the colonization of *Balanus* in disturbed areas appears to be more than an incidental event in the recovery process as barnacle tests act as significant recruitment sites for *M. edulis* (Menge, 1976; Suchanek and Duggins, unpublished results).

Since gap size also determines the ability of mobile grazers to reach and consume colonizers in the interior regions of gaps in *M. californianus* beds (see above and Suchanek, 1978, 1979; Sousa, 1984) size is intimately related to the overall recovery process, and especially the interaction between alternative space occupiers and mussels. Floral and faunal assemblages affected by grazers within the browse zone (typically ca. 10–20cm width—see Fig. 4.15 lower) are completely different than those within the central regions of gaps (usually more than 40cm diameter) (Suchanek, 1978, 1979; Sousa, 1984). For instance, during the early stages of gap colonization, diatoms and the red alga *Porphyra* occupy 80–90% of the primary substratum within the central portions of large gaps, but are nearly or completely absent within the browse zones (Suchanek, unpublished results). Many other algae and some sessile fauna are similarly but not as dramatically affected. However, other genera (e.g. *Alaria*, *Hedophyllum* and *Petrocelis*) appear to benefit from the grazing pressures exerted in the browse zone, probably because these genera have poorer competitive abilities and/or better antigrazer defenses. Moreover, some genera (e.g. *Halosaccion*) may benefit from grazing pressures during early stages of gap recolonization but be excluded completely from the browse zone during later stages (Suchanek, unpublished results). One of the most significant benefits that grazers can impart to *M. californianus* populations is an enhanced rate of recovery from disturbance. *M. californianus* beds at Shi-Shi, Washington, which had grazers experimentally removed, recovered lost space seven times slower than beds with their full complement of grazers (Suchanek, 1979).

Angle of substratum is important in determining not only the relative cover of *Mytilus*, but also the rate of recovery from disturbance. In regions of lower wave intensity (e.g. the relatively protected fjords within Glacier Bay, Alaska, or on any protected dock or piling) *Mytilus* spp. very effectively colonize vertical substrata. However, in regions of intense wave action *Mytilus* is not at all effective in maintaining a persistent and dominant coverage on vertical or nearly vertical slopes. In New England *M. edulis* dominates horizontal and inclined substrata, but does not compete very effectively with barnacles on vertical substrata (Menge, 1976). For *M. edulis*, the combination of gravity and wave action appears to dislodge individuals from the substratum; for *M. californianus* competition from the stalked barnacle *Pollicipes* also seems to be a contributing factor in its inability to dominate vertical slopes. In terms of recovery rate, on more steeply angled slopes the leaning component of gap healing (Paine and Levin, 1981) contributes more to gap recovery than on horizontal substrata, especially on the uphill side of the gap.

Recovery time from a major disturbance is variable for different species of *Mytilus*, but for most *Mytilus* assemblages it is a long-term process. *M. edulis* appears to colonize and recover from perturbation more quickly than other species. On a regional scale its recruitment typically shows temporal predictability, but spatial predictability is low and is dependent upon available areas of disturbed primary substratum. In Washington, for example, where *M. edulis* is typically a fugitive in the mid to low intertidal zone, it commonly recruits into disturbance gaps within the dominant *M. cali-*

fornianus cover, although not all gaps are colonized heavily (Suchanek, 1978). At Washington sites, where *M. edulis* is eliminated from the substratum by extremes in temperature, such as the very high intertidal zone where it is dominant, it recovers lost space typically within 1–3 years (data extracted from Paine, 1986). Recovery of *M. edulis* in New England is enhanced by the presence of barnacles on the primary substratum (Menge, 1976). Substratum with prior barnacle cover yielded nearly 100% recovery by *M. edulis*, whereas substratum without barnacles was virtually devoid of mussels. In Mutsu Bay, Japan, an artificial removal of *M. edulis* from a rocky shore resulted in complete recovery within three years (Hoshiai, 1964; Tsuchiya, 1983). In artificially created disturbance gaps (50 x 50cm) in Torch Bay, an Alaskan fjord, *M. edulis* populations had not completely recovered (range = 5–90% recovery) after almost seven years (Suchanek and Duggins, unpublished results). Here, exposure dictated recovery rates with gaps at intermediate exposure sites healing about twice as fast as those at very exposed sites.

In several studies after 3–5 years of data collection, very little recovery had been noted in disturbed *M. californianus* assemblages (Hewatt, 1935; Cimberg, 1975; Sousa, 1984), even though mussel larval settlement in this species occurs continuously throughout the year (Suchanek, 1981). An eight-year data-set by Castenholz (1967) also showed no recovery within that time period. Paine and Levin (1981) provide excellent data and a model of gap healing in *M. californianus* beds, and show that recovery from disturbance in mid intertidal mussel beds, measured as a cycling time (or rotation period) for these beds ranges from ca. 8–35 years depending on location. That is, mussel beds or portions of mussel beds destroyed by some disturbance event will recolonize and return to their original condition within this time period. While these results are consistent with other studies for mid intertidal horizontal *M. californianus* beds (Suchanek, unpublished results), some long-term data sets show variable agreement with these predictions for both vertical slopes and high intertidal mussel beds on horizontal platforms.

In a 21-year data set from a large intertidal stack at Trinidad Head, California (Cimberg, 1975 and personal communication, 1990), *M. californianus* showed minimal or no recovery after both artificial and natural mussel removals on a near vertical slope (Fig. 4.19). In 1968 *M. californianus* dominated a 30–40cm band around the stack (Fig. 4.19left). That year all biota was removed from a 20cm vertical swath on both the seaward and leeward sides of the stack, and the site was monitored for long-term changes. In 1972 the remainder of the mussel band was eliminated by heavy *Pisaster* predation (Fig. 4.19centre) and showed no recovery through 1989 (see example in Fig. 4.19right). Since the stack is completely surrounded by sand, it is unclear how *Pisaster* colonized it, but it could have been the result of a massive larval settlement.

A 16-year data set at Shi-Shi, Washington from a *M. californianus* bed on horizontal substrata shows similar results (Suchanek and Duggins, unpublished

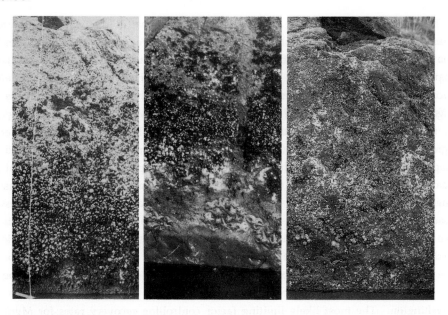

Fig. 4.19. Serial photographs of an intertidal stack at Trinidad Head, California, North America. In 1968 *Mytilus californianus* occupied a continuous band around the bottom portion of the stack (left). In 1972 *Pisaster* predation resulted in the elimination of all *M. californianus* from the stack (centre). Also note in (centre) a centrally located 20cm vertical swath denuded in 1968. As of 1989, *M. californianus* has not recolonized this site (note condition in (right) taken in 1986). All data and photographs from R.L. Cimberg (personal communication, 1990).

results). From 1974 through 1976 artificial disturbance gaps (ca. 30 x 30cm) were created in the high intertidal mussel bed region, but where *M. californianus* still occupied 100% of the substratum as a monolayer. Healing rates of these gaps were variable and several enlarged their area, but none healed completely. Because of the slow recovery rates, and the fact that some had continued to enlarge, the mathematically calculated mean projected healing times of these gaps (n = 6) for data through 1988 was 600 ± 1139 years. It is highly unlikely that recovery would take this long since a massive larval settlement/recruitment event would probably result in complete recolonization of this intertidal region. However, during the winter of 1988/1989 a severe freeze virtually eliminated the entire upper vertical 0.3m of the *M. californianus* beds at this site (also see discussion on p.89–91). During the last visit to this site (August 1990) the mussel beds surrounding these gaps and contributing to gap healing were nonexistent. Therefore, the former gaps have technically enlarged infinitely in size so that the region is one entire 'gap' and the recovery process for this entire mussel bed must start again from virtually bare rock, and it is unclear how long this may take.

The *Mytilus* beds at Shi-Shi are located in a region that is relatively less exposed than comparable beds at Tatoosh Island. Differences between and within these sites in the rates of recovery for disturbance gaps in intertidal mussel beds suggest that both exposure and tidal height play a significant role in the relative 'stability' (i.e. rate of recovery) of the dominant space occupiers. In the Washington study sites, gaps at high intertidal and less exposed sites recover much more slowly than comparable gaps at mid and low intertidal and more exposed sites. These results are in general agreement with those of Lewis (1977), who reported that mussel assemblages found in upper intertidal zones and in the most sheltered sites, experience the least amount of change per unit time. Lewis (1977) presents these data as evidence for increased 'stability' (i.e. lack of change). However, if some disturbance disrupted these mussel assemblages, it is highly likely that the rate of change back to the original condition (i.e. the rate of recovery) would be much slower at those sites than at lower intertidal and more exposed sites. Therefore, without entering into a semantic discussion about the term 'stability', the dynamics (or at least relative rates of change) in *M. edulis* beds in North-west Europe are comparable to those in *M. californianus* beds in Washington. The most likely limiting factor controlling recovery rates for *Mytilus* beds in upper intertidal sites in Washington is larval settlement and subsequent recruitment by mussels. In Europe, upper intertidal habitats also contribute greatly to long-term mussel bed 'stability' by virtue of their role as refuges from predation by *Nucella and Asterias* (Lewis, 1977).

Paine (1984) has used *M. californianus*, and the processes that allow it to return to dominant space occupier status, as a clear example of ecological determinism. While this principle is accurate for those mussel beds at mid to lower exposed rocky intertidal sites, those mussel beds on vertical substrata, or at high intertidal sites, are much more variable and may display alternative stable states (*sensu* Sutherland, 1974) for long periods of time. The examples given above for long-term data sets at Trinidad Head, California (Cimberg, 1975 and personal communication, 1990) and Shi-Shi, Washington (Suchanek and Duggins, unpublished results) provide evidence for these alternative states. Data from Trinidad Head indicate that *M. californianus* was replaced as the dominant primary space occupier on almost vertically sloped substrata, primarily by the barnacles *Semibalanus cariosus* and *Chthamalus dalli*, over a period of at least 17–21 years (R.L. Cimberg, personal communication, 1990). Data from Shi-Shi over a 16-year period indicate that on high intertidal horizontal substrata, *M. californianus*, once removed, can be replaced by a variety of fauna and flora, primarily *Semibalanus cariosus* and the brown alga *Fucus gardneri*; the mathematically projected healing times exceed many hundreds of years (Suchanek and Duggins, unpublished results). It is with great interest that these disturbance sites will continue to be monitored for as long as possible.

REFERENCES

Aarset, A.V., 1982. Freezing tolerance in intertidal invertebrates (A review). Comp. Biochem. Physiol., 73A: 571-580.

Aldrich, J.C., and Crowley, M., 1986. Condition and variability in Mytilus edulis L. from different habitats in Ireland. Aquaculture, 52: 273-286.

Almada-Villela, P.C., 1984. The effects of reduced salinity on the shell growth of small Mytilus edulis. J. Mar. Biol. Ass. U.K., 64: 171-182.

Almada-Villela, P. C., Davenport, J., and Gruffydd, L.L.D., 1982. The effects of temperature on the shell growth of young Mytilus edulis L. J. Exp. Mar. Biol. Ecol., 59: 275-288.

Ameyaw-Akumfi, C. and Hughes, R.N., 1987. Behaviour of Carcinus maenas feeding on large Mytilus edulis. How do they assess the optimal diet? Mar. Ecol. Prog. Ser., 38: 213-216.

Ameyaw-Akumfi, C. and Naylor, E., 1987. Temporal patterns of shell-gape in Mytilus edulis. Mar. Biol., 95: 237-242.

Anderson, G.L., 1975. The effects of intertidal height and the parasitic crustacean Fabia subquadrata Dana on the nutrition and reproductive capacity of the California sea mussel Mytilus californianus Conrad. Veliger, 17: 299-306.

Asmus, H., 1987. Secondary production of an intertidal mussel bed community related to its storage and turnover compartments. Mar. Ecol. Prog. Ser., 39: 251-266.

Aunaas, T., 1982. Nucleating agents in the haemolymph of an intertidal mollusc tolerant to freezing. Experientia, 38: 1456-1457.

Aunaas, T., Denstad, J.-P. and Zachariassen, K.E., 1988. Ecophysiological importance of the isolation response of hibernating blue mussels (Mytilus edulis). Mar. Biol., 98: 415-419.

Baird, R.H., 1958. Measurement of condition in mussels and oysters. J. Cons., 23: 249-257.

Baird, R.H., 1966. Factors affecting the growth and condition of mussels (Mytilus edulis). Fish. Invest. Minist. Agric. Fish. Food, Lond., Ser II, 25: 1-33.

Baird, D. and Milne, H., 1981. Energy flow in the Ythan estuary, Aberdeenshire, Scotland. Est. Coast. Shelf Sci., 13: 455-472.

Battle, H., 1932. Rhythmical sexual maturity and spawning of certain bivalve mollusks. Contrib. Can. Biol. Fish., 7: 257-276.

Bayne, B.L., 1964. Primary and secondary settlement in Mytilus edulis L. (Mollusca). J. Anim. Ecol., 33: 513-523.

Bayne, B.L., 1965. Growth and delay of metamorphosis of the larvae of Mytilus edulis (L). Ophelia, 2: 1-47.

Bayne, B.L., 1975. Reproduction in bivalve molluscs under environmental stress. In: F.J. Vernberg (Editor), Physiological Ecology of Estuarine Organisms. University of South Carolina Press, Columbia, pp. 259-277.

Bayne, B.L., 1976. Aspects of reproduction in bivalve molluscs. In: M. Wiley (Editor), Estuarine Processes. Vol. 1. Uses, stresses and adaptation to the estuary. Academic Press, New York, pp. 432-448.

Bayne, B.L. and Scullard, C., 1978. Rates of feeding by Thais (Nucella) lapillus (L.). J. Exp. Mar. Biol. Ecol., 32: 97-111.

Bayne, B.L. and Worrall, C.M., 1980. Growth and production of mussels (Mytilus edulis) from two populations. Mar. Ecol. Prog. Ser., 3: 317-328.

Bayne, B.L., Widdows, J., and Thompson, R.J., 1976. Physiological integrations. In: B.L. Bayne (Editor), Marine Mussels: their ecology and physiology. Cambridge University Press, Cambridge, pp. 261-299.

Bayne, B.L., Holland, D.L., Moore, M.N., Lowe, D.M., Widdows, J., 1978. Further studies on the effects of stress in the adult eggs of Mytilus edulis. J. Mar. Biol. Ass. U. K., 58: 825-841.

Bayne, B.L., Salkeld, P.N. and Worrall, C.M., 1983. Reproductive effort and value in different populations of the marine mussel, Mytilus edulis. L. Oecologia (Berl.), 59: 18-26.

Bayne, B.L., Brown, D.A., Burns, K., Dixon, D.R., Ivanovici, A., Livingstone, D.R., Lowe, D.M., Moore, M.N., Stebbing, A.R.D. and Widdows, J., 1985. The Effects of Stress and Pollution on Marine Animals. Praeger Press, New York, 384pp.

Beatty, N., and Aldrich, J.C., 1989. Effects of changes in microhabitat on the morphology and condition of Mytilus edulis L. In: J.C. Aldrich (Editor), Phenotypic Responses and Individuality in Aquatic Ectotherms. Japaga, Ashford, Ireland, pp. 41-54.

Bergeron, P. and Bourget, E., 1986. Shore topography and spatial partitioning of crevice refuges by sessile epibenthos in an ice disturbed environment. Mar. Ecol. Prog. Ser., 28: 129-145.

Bierbaum, R.M. and Ferson, S., 1986. Do symbiotic pea crabs decrease growth rate in mussels? Biol. Bull., 170: 51-61.

Bierbaum, R.M., and Shumway, S.E., 1988. Filtration and oxygen consumption in mussels, Mytilus edulis, with and without pea crabs, Pinnotheres maculatus. Estuaries, 11: 264-271.

Blok, J.W. de, and Geelen, H.J., 1958. The substratum required for the settling of mussels (Mytilus edulis L.). Arch. Néerl. Zool., Jubilee Vol., pp. 446-460.

Blok, J.W. de, and Tan-Mass, M., 1977. Function of byssus threads in young postlarval Mytilus. Nature (Lond.), 267: 558.

Board, P., 1983. The settlement of post larval Mytilus edulis (settlement of post larval mussels). J. Molluscan Stud., 49: 53-60.

Bøhle, B., 1971. Settlement of mussel larvae Mytilus edulis on suspended collectors in Norwegian waters. In: D.J. Crisp (Editor), Proc. 4th Eur. Mar. Biol. Symp. Bangor, U.K., 1969. Cambridge University Press, Cambridge pp. 63-69.

Bøhle, B., 1972. Effects of adaptation to reduced salinity on the filtration activity and growth of mussels (Mytilus edulis). J. Exp. Mar. Biol. Ecol., 10: 41-49.

Bourget, E., 1983. Seasonal variations of cold tolerance in intertidal mollusks and their relation to environmental conditions in the St. Lawrence Estuary. Can. J. Zool., 61: 1193-1201.

Briggs, R.P., 1978. Lough Foyle mussels. Annual Report of the Fisheries Research Laboratory D.A.N.I., pp. 14-17.

Briggs, R.P., 1982. Community structure and growth of Mytilus edulis L. in Lough Foyle. Proc. R. Ir. Acad., 82: 245-259.

Briscoe, C.S. and Sebens, K.P., 1988. Omnivory in Strongylocentrotus droebachiensis (Muller) (Echinodermata: Echinoidea): predation on subtidal mussels. J. Exp. Mar. Biol. Ecol., 115: 1-24.

Brosnan, D.M., 1990. Fairweather friends: relationships between plants and animals can change as environmental conditions alter (Abstract). Plant-Animal Interactions in the Marine Benthos; Systematics Association Symposium, Liverpool, 1990.

Brousseau, D.J., 1983. Aspects of reproduction of the blue mussel, Mytilus edulis (Pelecypoda, Mytilidae) in Long Island Sound. Fish. Bull., 81: 733-739.

Brown, R.A., Seed, R. and O'Connor, R.J., 1976. A comparison of relative growth in Cerastoderma (=Cardium) edule, Modiolus modiolus, and Mytilus edulis (Mollusca: Bivalvia). J. Zool. (Lond.), 179: 297-315.

Bustnes, J.O. and Erikstad, K.E., 1990. Size selection of common mussels, Mytilus edulis, by common eider, Somateria mollissima, energy maximization or shell weight minimization. Can. J. Zool., 68: 2280-2283.

Campbell, D.B., 1983. Determination of the foraging strategy of Asterias forbesii (Echinodermata: Asteroidea). Ph.D. Thesis, University of Rhode Island, U.S.A.

Campbell, S.A., 1969. Seasonal cycle in the carotenoid content in Mytilus edulis. Mar. Biol., 4: 227-232.

Carriker, M.R., 1981. Shell penetration and feeding by naticacean and muricacean predatory gastropods: A synthesis. Malacologia, 20: 403-422.

Castenholz, R.W., 1967. Stability and stresses in intertidal populations. In: T.A. Olson and F.J. Burgess (Editors), Pollution and Marine Ecology. Interscience Publishers, New York, pp. 15-28.

Cawthorne, D.F., 1979. Some effects of fluctuating temperature and salinity upon cirripedes. Ph.D. Thesis, University of Wales, U.K.

Ceccherelli, V.U. and Rossi, R., 1984. Settlement, growth and production of the mussel Mytilus galloprovincialis. Mar. Ecol. Prog. Ser., 16: 173-184.

Cerrato, R.M., 1980. Demographic analysis of bivalve populations. In: D.C, Rhoads and R. A. Lutz (Editors), Skeletal Growth of Aquatic Organisms. Plenum Press, New York, pp. 417-468.

Chan, G.L., 1973. Subtidal mussel beds in Baja California, with a new record size for Mytilus californianus. Veliger, 16: 239-240.

Chipperfield, P.N.J., 1953. Observations on the breeding and settlement of Mytilus edulis (L.) in British waters. J. Mar. Biol. Ass. U. K., 32: 449-476.

Cimberg, R.L., 1975. Zonation, species diversity and redevelopment in the rocky intertidal near Trinidad, northern California. M. Sc. Thesis, Humboldt State University, California, U.S.A.

Commito, J.A., 1987. Adult-larval interactions: predictions, mussels and cocoons. Est. Coast. Shelf Sci., 25: 599-606.

Connell, J.H., 1972. Community interactions on marine rocky intertidal shores. Ann. Rev. Ecol. Syst., 3: 169-192.

Cooper, K., 1981. A model to explain the induction of settlement and metamorphosis of planktonic eyed-pediveligers of the blue mussel Mytilus edulis L. by chemical and tactile cues (Abstract). J. Shellfish Res., 2: 117.

Craeymeersch, J.A., Herman, P.M.J. and Meire, P.M., 1986. Secondary production of an intertidal mussel (Mytilus edulis L.) population in the Eastern Scheldt (S.W. Netherlands). Hydrobiologia, 133: 107-115.

Cunningham, P.N. and Hughes, R.N., 1984. Learning of predatory skills by shore crabs Carcinus maenas feeding on mussels and dogwhelks. Mar. Ecol. Prog. Ser., 16: 21-26.

Daly, M.A. and Mathieson, A.C., 1977. The effects of sand movement on intertidal seaweeds and selected invertebrates at Bound Rock, New Hampshire. Mar. Biol., 43: 269-293.

Dame, R.F. and Dankers, N., 1988. Uptake and release of materials by a Wadden Sea mussel bed. J. Exp. Mar. Biol. Ecol., 118: 207-216.

Dare, P.J., 1976. Settlement, growth and production of the mussel, Mytilus edulis L., in Morecambe Bay, England. Fish. Invest. Minist. Agric. Fish. Food Lond., Ser. II., 28: 1-25.

Dare, P.J., 1982. Notes on the swarming behaviour and population density of Asterias rubens L. (Echinodermata: Asteroidea) feeding on the mussel Mytilus edulis. J. Cons., 40: 112-118.

Dare, P.J. and Edwards, D.B., 1976. Experiments on the survival, growth, and yield of relaid seed mussels (Mytilus edulis L.) in the Menai Strait, North Wales. J. Cons., 37: 16-28.

Dare, P.J., Davies, G. and Edwards, D.B., 1983. Predation on juvenile Pacific oysters (Crassostrea gigas Thunberg) and mussels (Mytilus edulis L.) by shore crabs (Carcinus maenas (L.)). Minist. Agric. Fish. Food, Fish. Tech. Rep. 73: 1-15.

Davenport, J., 1979. The isolation response of mussels (Mytilus edulis L.) exposed to falling sea water concentrations. J. Mar. Biol. Ass. U. K., 59: 124-132.

Davenport, J., 1983. A comparison of some aspects of the behaviour and physiology of the Indian mussel Perna (= Mytilus) viridis and the common mussel Mytilus edulis L. J. Molluscan Stud., 49: 21-26.

Davenport, J. and Chen, X., 1987. A comparison of methods for the assessment of condition in the mussel (Mytilus edulis L.). J. Molluscan Stud., 53: 293-297.

Davenport, J. and Glasspool, A.F., 1987. A photographic technique for the measurement of short term shell growth in bivalve molluscs. J. Molluscan Stud., 53: 299-303.

Davenport, J., Davenport, J. and Davies, G., 1984. A preliminary assessment of growth rates of mussels from the Falkland Islands (Mytilus chilensis Hupé and Aulacomya ater (Molina)). J. Cons., 41: 154-158.

Davidson, R.J., 1986. Mussel selection by the paddle crab Ovalipes catharus (White): evidence of flexible foraging behaviour. J. Exp. Mar. Biol. Ecol., 102: 281-299.

Davies, G., Dare, D.B. and Edwards, D.B., 1980. Fenced enclosures for the protection of seed mussels (Mytilus edulis L.) from predation by shore crabs (Carcinus maenas (L.)). Minist. Agric. Fish. Food, Fish. Res.Tech. Rep., 56: 1-14.

Dayton, P.K., 1971. Competition, disturbance, and community organisation: the provision and subsequent utilization of space in a rocky intertidal community. Ecol. Monogr., 41: 351-389.

Dayton, P.K., 1973. Dispersion, dispersal, and persistence of the annual intertidal alga, Postelsia palmaeformis Ruprecht. Ecology, 54: 433-438.

Denny, M., 1987. Lift as a mechanism of patch initiation in mussel beds. J. Exp. Mar. Biol. Ecol., 113: 231-245.

Diehl, W.J., Gaffney, P.M. and Koehn, R.K., 1986. Physiological and genetic aspects of growth in the mussel Mytilus edulis L. Oxygen consumption, growth and weight loss. Physiol. Zool., 59: 201-211.

Dittman, D. and Robles, C., 1991. Effect of algal epiphytes on the mussel Mytilus californianus. Ecology, 72: 286-296.

Dix, T.G. and Ferguson, A., 1984. Cycles of reproduction and condition in Tasmanian blue mussels Mytilus edulis planulatus. Aust. J. Mar. Freshw. Res., 35: 307-313.

Drinnan, R.E., 1958. The winter feeding of the oystercatcher (Haematopus ostralegus) on the edible mussel (Mytilus edulis) in the Conway Estuary, North Wales. Fish. Invest. Minist. Agric. Fish. Food, Lond., Ser II, 22: 1-15.

Dunthorn, A.A., 1971. The predation of cultivated mussels by eider. Bird Study, 18: 107-112.

Durrell, S.E.A. and Goss-Custard, J.D., 1984. Prey selection within a size-class of mussels. Anim. Behav., 32: 1197-1203.

Ebling, F.J., Kitching, J.A., Muntz, L. and Taylor, C.M., 1964. The ecology of Lough Ine XIII. Experimental observations of the destruction of *Mytilus edulis* and *Nucella lapillus* by crabs. J. Anim. Ecol., 33: 73-82.

Edwards, D.C., Conover, D.O. and Sutter, F., 1982. Mobile predators and the structure of marine intertidal communities. Ecology, 63: 1175-1180.

Elner, R.W., 1978. The mechanics of predation by the shore crab *Carcinus maenas* (L.) on the edible mussel, *Mytilus edulis* L. Oecologia (Berl.), 36: 333-344.

Elner, R.W., 1981. Diet of green crab *Carcinus maenas* (L.) from Port Herbert, Southwestern Nova Scotia. J. Shellfish Res., 1: 89-94.

Elner, R.W., and Hughes, R.N., 1978. Energy maximization in the diet of the shore crab, *Carcinus maenas*. J. Anim. Ecol., 47: 103-116.

Elner, R.W. and Campbell, A., 1987. Natural diets of lobster *Homarus americanus* from barren ground and macroalgal habitats off southwestern Nova Scotia, Canada. Mar. Ecol. Prog. Ser., 37: 131-140.

Elvin, D.W., and Gonor, J.J., 1979. The thermal regime of an intertidal *Mytilus californianus* Conrad population on the central Oregon coast. J. Exp. Mar. Biol. Ecol., 39: 265-279.

Emmett, B., Thompson, K. and Popham, J.D., 1987. The reproduction and energy storage cycles of two populations of *Mytilus edulis* (Linné) from British Columbia. J. Shellfish Res., 6: 29-36.

Estes, J.A., Jameson, R.J. and Johnson, A.M., 1981. Food selection and some foraging tactics of sea otters. In: J.A. Chapman and D. Pursley (Editors), Proceedings of the Worldwide Fur Bearers Conference Number 2. World Fur Bearers. Frostburg, Maryland, U.S.A., 1980, pp. 606-641.

Eyster, L.S. and Pechenik, J.A., 1987. Attachment of *Mytilus edulis* L. larvae on algal and byssal filaments is enhanced by water agitation. J. Exp. Mar. Biol. Ecol., 114: 99-110.

Feare, C.J., 1971a. The adaptive significance of aggregation behaviour in the dogwhelk *Nucella lapillus* (L.). Oecologia (Berl.), 7: 117-126.

Feare, C.J., 1971b. Predation of limpets and dogwhelks by oystercatchers. Bird Study, 18: 121-129.

Feare, C.J. and Summers, R.W., 1985. Birds as predators on rocky shores. In: P.G. Moore and R. Seed (Editors), The Ecology of Rocky Coasts. Hodder and Stoughton, Sevenoaks, U.K., pp. 249-264.

Fell, P.E. and Balsamo, A.M., 1985. Recruitment of *Mytilus edulis* L. in the Thames Estuary, with evidence for differences in the time of maximal settling along the Connecticut shore. Estuaries, 8: 68-75.

Fréchette, M. and Bourget, E., 1985. Energy flow between the pelagic and benthic zones: Factors controlling particulate organic matter available to an intertidal mussel bed. Can. J. Fish. Aquat. Sci., 42: 1158-1165.

Fréchette, M. and Bourget, E., 1987. Significance of small scale spatio-temporal heterogeneity on phytoplankton abundance for energy flow in *Mytilus edulis*. Mar. Biol., 94: 231-240.

Fuller, S.C. and Lutz, R.A., 1988. Early shell mineralogy, microstructure, and shell sculpture in live mytilid species. Malacologia, 29: 363-371.

Gabbott, P.A., 1975. Storage cycles in marine bivalve molluscs: a hypothesis concerning the relationship between glycogen metabolism and gametogenesis. In: H. Barnes (Editor), Proc. 9th Eur. Mar. Biol. Symp., Oban, Scotland, 1974. Aberdeen University Press, Aberdeen, pp. 191-211.

Gabbott, P.A., 1983. Development and seasonal metabolic activities in marine molluscs. In: P.W. Hochachka (Editor), The Mollusca 2. Environmental Biochemistry and Physiology. Academic Press, London, pp. 165-217.

Galleni, L., Ferrero, E., Salghetti, V., Tongiorgi, P. and Salvadego, P., 1977. Ulteriori osservazioni sulla predazione di *Stylochus mediterraneus* (Turbellaria, Polycladida) sui Mytili e suo orientamento chemiotattico. Atti IX Congr. Soc. Ital. di Biol. Mar., Lacco Ameno d'Ischia, 1977, pp. 259-261.

Gardner, J.P.A. and Thomas, M.L.H., 1987. Growth, mortality and production of organic matter by a rocky intertidal population of *Mytilus edulis* in the Quoddy Region of the Bay of Fundy. Mar. Ecol. Prog. Ser., 39: 31-36.

Gardner, J.P.A. and Skibinski, D.O.F., 1990. Genotype-dependent fecundity and temporal variation of spawning in hybrid mussel (*Mytilus*) populations. Mar. Biol., 105: 153-162.

Gardner, J.P.A. and Skibinski, D.O.F., 1991. Biological and physical factors influencing genotype-dependant mortality in hybrid mussel populations. Mar. Ecol. Prog. Ser., 71(3): 235-244.

Goss-Custard, J.D. and Durrell, S.E.A., 1987. Age related effects of oystercatchers, *Haematopus ostralegus*, feeding on mussels, *Mytilus edulis*. II. Aggression. J. Anim. Ecol., 56: 537-548.

Grant, A., 1989. The use of graphical methods to estimate demographic parameters. J. Mar. Biol. Ass. U. K., 69: 367-371.

Griffiths, C.L. and Griffiths, R.J., 1987. Bivalvia, In: T.J. Pandian and F.J. Vernberg (Editors), Animal Energetics Vol. 2. Bivalvia through Reptilia. Academic Press, California, pp. 1-88.

Griffiths, C.L. and Hockey, P.A.R., 1987. A model describing the interactive role of predation, competition and tidal elevation in structuring mussel populations. In: A.I.L. Payne, J.A. Gulland and K.H. Brink (Editors), The Benguela and Comparable Ecosystems. S. Afr. J. Mar. Sci., 5: 547-556.

Gruffydd, L.L.D., Huxley, R. and Crisp, D.J., 1984. The reduction in growth of *Mytilus edulis* in fluctuating salinity regimes measured using laser diffraction patterns and the exaggeration of this effect by using tap water as the diluting medium. J. Mar. Biol. Ass. U. K., 64: 401-409.

Hamburger, K., Møhlenberg, F., Randløv, A. and Riisgård, H.U., 1983. Size, oxygen consumption and growth in the mussel *Mytilus edulis*. Mar. Biol., 75: 303-306.

Hancock, D.A., 1965. Adductor muscle size in Danish and British mussels in relation to starfish predation. Ophelia, 2: 253-267.

Harger, J.R.E., 1970. The effect of wave impact on some aspects of the biology of sea mussels. Veliger, 12: 401-414.

Harger, J.R.E., 1972. Competitive co-existence: maintenance of interacting associations of the sea mussels *Mytilus edulis* and *Mytilus californianus*. Veliger, 14: 387-410.

Harger, J.R.E. and Landenberger, D.E., 1971. The effect of storms as a density dependent mortality factor on populations of sea mussels. Veliger, 14: 195-201.

Harrold, C. and Hardin, D., 1986. Foraging on land by the California sea otter, *Enhydra lutris*. Mar. Mamm. Sci., 2: 309-313.

Hawkins, A.J.S., Salkeld, P.N., Bayne, B.L., Gneiger, E. and Lowe, D.M., 1985. Feeding and resource allocation in the mussel *Mytilus edulis*: evidence for time-averaged optimization. Mar. Ecol. Prog. Ser., 20: 273-287.

Hawkins, A.J.S., Bayne, B.L. and Day, A.J., 1986. Protein turnover, physiological energetics and heterozygosity in the blue mussel, *Mytilus edulis*: the basis of variable age-specific growth. Proc. R. Soc. Lond., Ser. B, 229: 161-176.

Heppleston, P.B., 1971. The feeding ecology of oystercatchers (*Haematopus ostralegus* L.) in winter in northern Scotland. J. Anim. Ecol., 40: 651-672.

Hewatt, W.G., 1935. Ecological succession in the *Mytilus californianus* habitat as observed in Monterey Bay, California. Ecology, 16: 244-251.

Hewatt, W.G., 1937. Ecological studies on selected marine intertidal communities of Monterey Bay, California. Am. Midl. Nat., 18: 161-206.

Hilbish, T.J., 1986. Growth trajectories of shell and soft tissue in bivalves: seasonal variation in *Mytilus edulis* L. J. Exp. Mar. Biol. Ecol., 96: 103-113.

Hilbish, T.J. and Zimmermann, K.M.,1988. Genetic and nutritional control of the gametogenic cycle in *Mytilus edulis*. Mar. Biol., 98: 223-228.

Himmelman, J.H., Cardinal, A. and Bourget, E., 1983. Community development following removal of urchins, *Strongylocentrotus droebachiensis*, from the rocky subtidal zone of the St. Lawrence Estuary, eastern Canada. Oecologia (Berl.), 59: 27-39.

Hines, A.H., 1979. Effects of a thermal discharge on reproductive cycles in *Mytilus edulis* and *Mytilus californianus*. Fish. Bull., 77: 498-503.

Hoshiai, T., 1964. Synecological study on intertidal communities. V. The interrelation between *Septifer virgatus* and *Mytilus edulis*. Bull. Mar. Biol. Stn. Asamushi, Tokyo, 12: 37-41.

Hosomi, A., 1980. Studies on the spat recruitment and age structure in the population of the mussel, *Mytilus galloprovincialis* Lamarck, with special reference to the cause of the extinction of population. Venus, 39: 155-166.

Hosomi, A., 1984. Ecological studies on the mussel *Mytilus galloprovincialis* (Lamarck): rise and fall of recruited cohorts in adult mussel bed. Venus, 43: 157-171.

Hosomi, A., 1985. The production, daily production, biomass and turnover rate of the mussel *Mytilus galloprovincialis*. Venus, 44: 270-277.

Hughes, R.N. and Dunkin, S. B. de, 1984. Behavioural components of prey selection by dogwhelks, *Nucella lapillus* (L.) feeding on mussels, *Mytilus edulis* L. in the laboratory. J. Exp. Mar. Biol. Ecol., 77: 45-68.

Hughes, R.N. and Drewett, D., 1985. A comparison of the foraging behaviour of dogwhelks, *Nucella lapillus* (L.) feeding on barnacles or mussels on the shore. J. Molluscan Stud., 51: 73-77.

Hughes, R.N. and Burrows, M.T., 1990. Energy maximization in the natural foraging behaviour of the dogwhelk *Nucella lapillus* (L.). In: M. Barnes and R.N. Gibson (Editors), Trophic Relationships in the Marine Environment. Aberdeen University Press, Aberdeen. pp. 517-527.

Hurd, L.E., Smedes, G.W. and Dean, T.A., 1979. An ecological study of a natural population of diamondback terrapins (*Malaclemys t. terrapin*) in a Delaware salt marsh. Estuaries, 2: 28-33.

Incze, L.S., Lutz, R.A. and Watling, L., 1980. Relationship between effects of environmental temperature and seston on growth and mortality of *Mytilus edulis* in a temperate northern estuary. Mar. Biol., 57: 147-156.

Jabbar, A. and Davies, J.L., 1987. A simple and convenient biochemical method for sex identification in the marine mussel *Mytilus edulis* L. J. Exp. Mar. Biol. Ecol., 107: 39-44.

Jacobi, C.M., 1987a. The invertebrate fauna associated with intertidal beds of the brown mussel *Perna perna* (L.) from Santos, Brazil. Stud. Neotrop. Fauna Environ., 22: 57-72.

Jacobi, C.M., 1987b. Spatial and temporal distribution of Amphipoda associated with mussel beds from the Bay of Santos (Brazil). Mar. Ecol. Prog. Ser., 35: 51-58.

Jensen, K.T. and Jensen, J.N., 1985. The importance of some epibenthic predators on the density of juvenile benthic macrofauna in the Danish Wadden Sea. J. Exp. Mar. Biol. Ecol., 89:157-174.

Jubb, C.A., Hughes, R.N. and Rheinallt, T. ap, 1983. Behavioural mechanisms of size selection by crabs, *Carcinus maenas* (L.) feeding on mussels, *Mytilus edulis* L. J. Exp. Mar. Biol. Ecol., 66: 81-87.

Kaiser, M.J., Hughes, R.N. and Reid, D.G., 1990. Chelal morphometry, prey-size selection and aggressive competition in green and red forms of *Carcinus maenas* (L.). J. Exp. Mar. Biol. Ecol., 140: 121-134.

Kanter, R.G., 1978. Structure and diversity in *Mytilus californianus* (Mollusca: Bivalvia) communities. Ph.D. Thesis, University of Southern California, U.S.A.

Kanter, R.G., 1980. Biogeographic patterns in mussel community distribution from the southern California Bight. In: D.M. Power (Editor), The California Islands, Proceedings of a Multidisciplinary Symposium. Santa Barbara Museum of Natural History, Santa Barbara, California, 1979, pp. 341-355.

Kanwisher, J.W., 1959. Histology and metabolism of frozen intertidal animals. Biol. Bull., 116: 258-264.

Kanwisher, J.W., 1966. Freezing in intertidal animals. In: H.T. Meryman (Editor), Cryobiology. Academic Press, New York. pp. 487-494.

Kaplan, S.W., 1984. The association between the sea anemone *Metridium senile* (L.) and the mussel *Mytilus edulis* (L.): reduced predation by the starfish *Asterias forbesii* (Desor). J. Exp. Mar. Biol. Ecol., 79: 1550-1557.

Kautsky, N., 1981. On the role of the blue mussel (*Mytilus edulis* L.) in a Baltic coastal ecosystem and the fate of the organic matter produced by the mussels. Kieler Meeresforsch. Sonderh., 5: 454-461.

Kautsky, N., 1982a. Quantitative studies on the gonad cycle, fecundity, reproductive output and recruitment in a Baltic *Mytilus edulis* population. Mar. Biol., 68: 143-160.

Kautsky, N., 1982b. Growth and size structure in a Baltic *Mytilus edulis* population. Mar. Biol., 68: 117-133.

Kautsky, N., Johannesson, J. and Tedengren, M., 1990. Genotypic and phenotypic differences between Baltic and North Sea populations of *Mytilus edulis* evaluated through reciprocal transplantations. 1. Growth and morphology. Mar. Ecol. Prog. Ser., 59: 203-210.

Kennedy, V.S., 1977. Reproduction in *Mytilus edulis aoteanus* and *Aulacomya maoriana* (Mollusca: Bivalvia) from Taylors Mistake, New Zealand. N. Z. J. Mar. Freshw. Res., 11: 255-267.

Kent, R.M.L., 1979. The influence of heavy infestations of *Polydora ciliata* on the flesh content of *Mytilus edulis*. J. Mar. Biol. Ass. U. K., 59: 289-297.

King, P.A., McGrath, D. and Gosling, E.M., 1989. Reproduction and settlement of *Mytilus edulis* on an exposed rocky shore in Galway Bay, west coast of Ireland. J. Mar. Biol. Ass. U. K., 69: 355-365.

King, P.A., McGrath, D. and Britton, W., 1990. The use of artificial substrates in monitoring mussel (*Mytilus edulis* L.) settlement on an exposed rocky shore in the west of Ireland. J. Mar. Biol. Ass. U. K., 70: 371-380.

Kiørboe, T., Møhlenberg, F. and Nøhr, O., 1981. Effect of suspended bottom material on growth and energetics in *Mytilus edulis*. Mar. Biol., 61: 283-288.

Kitching, J.A. and Ebling, F.J., 1967. Ecological studies at Lough Ine. Adv. Ecol. Res., 4: 198-291.

Kitching, J.A., Sloane, J.F. and Ebling, F.J., 1959. The ecology of Lough Ine VIII. Mussels and their predators. J. Anim. Ecol., 28: 331-341.

Koehn, R.K. and Gaffney, P.M., 1984. Genetic heterozygosity and growth rate in *Mytilus edulis*. Mar. Biol., 82: 1-7.

Laihonen, P. and Furman, E.R., 1986. The site of settlement indicates commensalism between blue mussel and its epibiont. Oecologia (Berl.), 71: 38-40.

Lane, D.J.W., Beaumont, A.R. and Hunter, J.R., 1985. Byssus drifting and the drifting threads of the young post-larval mussel *Mytilus edulis*. Mar. Biol., 84: 301-308.

Lawton, P., 1987. Diel activity and foraging behavior of juvenile American lobsters, *Homarus americanus*. Can. J. Fish. Aquat. Sci., 44: 1195-1205.

Leigh, E.G., Paine, R.T., Quinn, J.F. and Suchanek, T.H., 1987. Wave energy and intertidal productivity. Proc. Natl. Acad. Sci. (U.S.A.), 84: 1314-1318.

Leopold, M.F., Swennen, C. and Bruijn, L.L.M. de, 1989. Experiments on selection of feeding site and food size in oystercatchers, *Haematopus ostralegus*, of different social status. Neth. J. Sea Res., 23: 333-346.

Levin, S.A. and Paine, R.T., 1974. Disturbance, patch formation, and community structure. Proc. Natl. Acad. Sci. (U.S.A.), 71: 2744-2747.

Levin, S.A. and Paine, R.T., 1975. The role of disturbance in models of community structure. Ecosystem Analysis and Prediction. Society for Industrial and Applied Mathematics, Philadelphia, Pennsylvania, U.S.A., pp. 56-67.

Lewis, J.R., 1964. The Ecology of Rocky Shores. English Universities Press, London, 323pp.

Lewis, J.R., 1977. The role of physical and biological factors in the distribution and stability of rocky shore communities. In: B.F. Keegan, P. O Céidigh and P. J.S. Boaden (Editors), Biology of Benthic Organisms. Proc. 11th Eur. Mar. Biol. Symp., Galway, Ireland, 1976. Pergamon Press, London, pp. 417-424.

Littler, M.M., Martz, D.R. and Littler, D.S., 1983. Effects of recurrent sand deposition on rocky intertidal organisms: importance of substrate heterogeneity in a fluctuating environment. Mar. Ecol. Prog. Ser., 11: 129-139.

Loo, L.-O. and Rosenberg, R., 1983. *Mytilus edulis* culture, growth and production in western Sweden. Aquaculture, 35: 137-150.

Lowe, D.M., Moore, M.N. and Bayne, B.L., 1982. Aspects of gametogenesis in the marine mussel *Mytilus edulis* L. J. Mar. Biol. Ass. U. K., 62: 133-145.

Lubchenco, J.L. and Menge, B.A., 1978. Community development and persistence in a low rocky intertidal zone. Ecol. Monogr., 59: 67-94.

Lubet, P., 1959. Recherches sur le cycle sexuel et l'émission des gametes chez les Mytilidae et les Pectinidae (Moll. Bivalves). Rev. Trav. Off. (Scient. Tech.) de Pêch. Marit., 23: 387-548.

Lubet, P. and Aloui, N., 1987. Limites letales thermiques et action de la temperature sur les gametogeneses et l'activité neurosecretrice chez la moule (*Mytilus edulis* et *M. galloprovincialis*, Mollusque Bivalve). Haliotis, 16: 309-316.

Lutz, R.A., 1976. Annual growth layers in the shell of *Mytilus edulis*. J. Mar. Biol. Ass. U. K., 56: 723-731.

Lutz, R.A., 1980. Mussel Culture and Harvest: A North American Perspective. Elsevier Science Publishers, B.V., Amsterdam, 305pp.

Lutz, R.A., Incze, L.S., Porter, B. and Stotz, J.K., 1980. Seasonal variation in the condition of raft-cultivated mussels (*Mytilus edulis* L.). Proc. World Maricul. Soc., 11: 262-268.

Maas Geesteranus, R.A., 1942. On the formation of banks of *Mytilus edulis*. Arch. Néerl. Zool., 6: 283-325.

MacGinitie, G. and MacGinitie, N., 1949. Natural history of marine animals. McGraw-Hill, New York, 473pp.

Mallet, A.L. and Carver, C.E.A., 1989. Growth, mortality, and secondary production in natural populations of the blue mussel, *Mytilus edulis*. Can. J. Fish. Aquat. Sci., 46: 1154-1159.

Manahan, D.T., Wright, S.J. and Stephens, G.C., 1983. Simultaneous determination of net uptake of 16 amino acids by a marine bivalve. Am. J. Physiol., 244: 832-838.

Marsh, C.P., 1986. Rocky intertidal community organization: the impact of avian predators on mussel recruitment. Ecology, 67: 771-786.

Mason, J., 1972. The cultivation of the European mussel, *Mytilus edulis* Linnaeus. Oceanogr. Mar. Biol. Annu. Rev., 10: 437-460.

Mason, J., 1976. Cultivation. In: B.L. Bayne (Editor), Marine Mussels: their ecology and physiology. Cambridge University Press, Cambridge, pp. 385-410.

Mattsson, J. and Linden, O., 1983. Benthic macrofauna succession under mussels, *Mytilus edulis* L. (Bivalvia), cultured on hanging long-lines. Sarsia, 68: 97-102.

McClintock, J.B. and Robnett, Jr., T.J., 1986. Size selective predation by the asteroid *Pisaster ochraceus* on the bivalve *Mytilus californianus*: a cost-benefit analysis. Pubbl. Stn. Zool. Napoli, Mar. Ecol., 7: 321-332.

McDonald, J.H. and Koehn, R.K., 1988. The mussels *Mytilus galloprovincialis* and *M. trossulus* on the Pacific coast of North America. Mar. Biol., 99: 111-118.

McDonald, J.H., Seed, R. and Koehn, R.K., 1991. Allozymes and morphometric characters of three species of *Mytilus* in the Northern and Southern hemispheres. Mar. Biol., 111: 323-335.

McGrath, D., King, P.A. and Gosling, E.M., 1988. Evidence for the direct settlement of *Mytilus edulis* larvae onto adult mussel beds. Mar. Ecol. Prog. Ser., 47: 103-106.

McGrorty, S., Clarke, R.T., Reading, C.J. and Goss-Custard, J.D., 1990. Population dynamics of the mussel *Mytilus edulis*: density changes and regulation of the population in the Exe Estuary Devon. Mar. Ecol. Prog. Ser., 67: 157-169.

McKenzie, J.D., 1986. The reproductive cycle of *Mytilus edulis* L. from Lough Foyle. Ir. Nat. J., 22: 13-16.

Meire, P.M. and Ervynck, A., 1986. Are oystercatchers (*Haematopus ostralegus*) selecting the most profitable mussels (*Mytilus edulis*)? Anim. Behav., 34: 1427-1435.

Menge, B.A., 1972. Competition for food between two intertidal starfish species and its effect on body size and feeding. Ecology, 53: 635-644.

Menge, B.A., 1976. Organization of the New England rocky intertidal community: role of predation, competition, and environmental heterogeneity. Ecol. Monogr., 46: 355-393.

Menge, B.A., 1983. Components of predation intensity in the low zone of the New England rocky intertidal region. Oecologia (Berl.), 58: 141-155.

Miller, B.A., 1980. Historical review of U.S. mussel culture and harvest. In: R.A. Lutz (Editor), Mussel Culture and Harvest: A North American Perspective. Elsevier Science Publishers, B.V., Amsterdam, pp. 18-37.

Milne, H. and Dunnet, G.M., 1972. Standing crop, productivity and trophic relations of the fauna of the Ythan estuary. In: R.S.K. Barnes and J. Green (Editors), The Estuarine Environment. Applied Science Publishers, London, pp. 88-106.

Møhlenberg, F. and Riisgård, H.U., 1977. Efficiency of particle retention in thirteen species of suspension feeding bivalves. Ophelia, 17: 239-246.

Moore, R.C., 1983. Treatise on Invertebrate Paleontology. Geological Society of America and University of Kansas Press, Lawrence, pp. 1953-1983.

Myint, U.M. and Tyler, P.A., 1982. Effects of temperature, nutritive and metal stressors on the reproductive biology of *Mytilus edulis*. Mar. Biol., 67: 209-223.

Navarro, J.M. and Winter, J.E., 1982. Ingestion rate, assimilation efficiency and energy balance in *Mytilus chilensis* in relation to body size and different algal concentrations. Mar. Biol., 67: 255-266.

Newcombe, C.L., 1935. A study of the community relationships of the sea mussel, *Mytilus edulis* L. Ecology, 16: 234-243.

Newell, R.I.E., Hilbish, T.J., Koehn, R.K. and Newell, C.J., 1982. Temporal variation in the reproductive cycle of *Mytilus edulis* L. (Bivalvia, Mytilidae) from localities on the east coast of the U.S.A. Biol. Bull., 162: 299-310.

Nielsen, V.M. and Strømgren, T., 1985. The effect of light on the shell length, growth, and defecation rate of *Mytilus edulis* (L.). Aquaculture, 47: 205-211.

Nixon, S.W., Oviatt, C.A., Rogers, C. and Taylor, K., 1971. Mass and metabolism of a mussel bed. Oecologia (Berl.), 8: 21-30.

Okamura, B., 1986. Group living and the effects of spatial position in aggregations of *Mytilus edulis*. Oecologia (Berl.), 69: 341-347.

O'Neill, S.M., Sutterlin, A.M. and Aggett, D., 1983. The effect of size-selective feeding by starfish (*Asterias vulgaris*) on the production of mussels (*Mytilus edulis*) cultured on nets. Aquaculture, 35: 211-220.

Page, H.M. and Hubbard, D.M., 1987. Temporal and spatial patterns of growth in mussels *Mytilus edulis* on an offshore platform: relationships to water temperature and food availability. J. Exp. Mar. Biol. Ecol., 111: 159-179.

Paine, R.T., 1966. Food web complexity and species diversity. Am. Nat., 100: 65-75.

Paine, R.T., 1969. A note on trophic complexity and community stability. Am. Nat., 103: 91-93.

Paine, R.T., 1971. A short-term experimental investigation of resource partitioning in a New Zealand rocky intertidal habitat. Ecology, 52: 1096-1106.

Paine, R.T., 1974. Intertidal community structure: experimental studies on the relationship between a dominant competitor and its principal predator. Oecologia (Berl.), 15: 93-120.

Paine, R.T., 1976a. Size-limited predation: an observational and experimental approach with the Mytilus-Pisaster interaction. Ecology, 57: 858-873.

Paine, R.T., 1976b. Biological observations on a subtidal Mytilus californianus bed. Veliger, 19: 125-130.

Paine, R.T., 1979. Disaster, catastrophe, and local persistence of the sea palm Postelsia palmaeformis. Science, 205: 685-687.

Paine, R.T., 1980. Food webs: linkage, interaction strength, and community infrastructure. J. Anim. Ecol., 49: 667-685.

Paine, R.T., 1984. Ecological determinism in the competition for space. Ecology, 65: 1339-1348.

Paine, R.T., 1986. Benthic community-water column coupling during the 1982-1983 El Niño. Are community changes at high latitudes attributable to cause or coincidence? Limnol. Oceanogr., 31: 351-360.

Paine, R.T., 1989. On commercial exploitation of the sea mussel, Mytilus californianus. N.W. Environ. J., 5: 89-97.

Paine, R.T. and Levin, S.A., 1981. Intertidal landscapes: disturbance and the dynamics of pattern. Ecol. Monogr., 51: 145-178.

Paine, R.T. and Suchanek, T.H., 1983. Convergence of ecological processes between independently evolved competitive dominants: a tunicate-mussel comparison. Evolution, 37: 821-831.

Paine, R.T., Castilla, J.C. and Cancino, J., 1985. Perturbation and recovery patterns of starfish dominated intertidal assemblages in Chile, New Zealand and Washington State. Am. Nat., 125: 679-691.

Palmer, A.R., 1983. Growth rate as a measure of food value in thaidid gastropods: assumptions and implications from prey morphology and distribution. J. Exp. Mar. Biol. Ecol., 73: 95-124.

Paris, O.H., 1960. Some quantitative aspects of predation by muricid snails on mussels in Washington Sound. Veliger, 2: 41-47.

Parulekar, A.H., Dalal, S.F., Ansari, Z.A. and Harkantra, S.N., 1982. Environmental physiology of raft grown mussels Perna viridis in Goa, India. Aquaculture, 29: 83-94.

Perkins, E.J., 1967. Some aspects of the biology of Carcinus maenas (L.). Trans. Dumfries. Galloway Nat. Hist. Antiq. Soc., Ser. 3, 44: 47-56.

Petersen, J.H., 1984a. Establishment of mussel beds: attachment behaviour and distribution of recently settled mussels (Mytilus californianus). Veliger, 27: 7-13.

Petersen, J.H., 1984b. Larval settlement behavior in competing species: Mytilus californianus Conrad and M. edulis L. J. Exp. Mar. Biol. Ecol., 82: 147-159.

Peterson, C.H., 1979. The importance of predation and competition in organizing the intertidal epifaunal communities of Barnegat Inlet, New Jersey. Oecologia (Berl.), 39: 1-24.

Petraitis, P.S., 1978. Distributional patterns in juvenile Mytilus edulis and Mytilus californianus. Veliger, 21: 288-292.

Petraitis, P.S., 1987. Immobilization of the predatory gastropod, Nucella lapillus by its prey, Mytilus edulis. Biol. Bull., 172: 307-314.

Pieters, H., Kluytmans, J.H., Zurburg, W. and Zandee, D.I., 1979. The influence of seasonal changes on energy metabolism in Mytilus edulis (L.). Growth rate and biochemical composition in relation to environmental parameters and spawning. In: E. Naylor and R.G. Hartnoll (Editors), Cyclic Phenomena in Marine Plants and Animals. Pergamon Press, New York, pp. 285-291.

Plessing, T.V., 1981. Desarrollo de las comunidades de mitilidos, Yaldad, Chiloe, Chile. Archos. Biol. Med. Exp., 14: 287.

Pregenzer, C., 1979. The effect of Pinnotheres hickmani on pumping rates in Mytilus edulis. Aust. J. Mar. Freshw. Res., 30: 547-550.

Pregenzer, C., 1981. The effect of Pinnotheres hickmani on the meat yield (condition) of Mytilus edulis measured several ways. Veliger, 23: 250-253.

Price, H.A., 1980. Seasonal variation in the strength of byssal attachment of the common mussel, Mytilus edulis. L. J. Mar. Biol. Ass. U. K., 60: 1035-1037.

Price, H.A., 1982. An analysis of factors determining seasonal variation in the byssal attachment strength of Mytilus edulis. J. Mar. Biol. Ass. U. K., 62: 147-155.

Prins, T.C. and Smaal, A.C., 1990. Benthic-pelagic coupling: the release of inorganic nutrients by an intertidal bed of *Mytilus edulis*. In: M. Barnes and R.N. Gibson (Editors), Trophic Relationships in the Marine Environment. Aberdeen University Press, Aberdeen, pp. 89-103.

Raffaelli, D., Falcy, V. and Galbraith, C., 1990. Eider predation and the dynamics of mussel bed communities. In: M. Barnes and R.N. Gibson (Editors),Trophic Relationships in the Marine Environment. Aberdeen University Press, Aberdeen, pp. 157-169.

Read, K.R.H. and Cumming, K.B., 1967. Thermal tolerance of the bivalve molluscs *Modiolus modiolus* L., *Mytilus edulis* L. and *Brachidontes demissus* (Dillwyn). Comp. Biochem. Physiol., 22: 149-155.

Redpath, K.J., 1985. Growth inhibition and recovery in mussels (*Mytilus edulis*) exposed to low copper concentrations. J. Mar. Biol. Ass. U. K., 65: 421-431.

Reid, D.G. and Aldrich, J.C., 1989. Variations in response to environmental hypoxia of different colour forms of the shore crab, *Carcinus maenas*. Comp. Biochem. Physiol., 92A: 535-539.

Rheinallt, T. ap, 1986. Size selection by the crab *Liocarcinus puber* feeding on mussels *Mytilus edulis* and shore crabs *Carcinus maenas*: the importance of mechanical factors. Mar. Ecol. Prog. Ser., 29: 45-53.

Richardson, C.A., 1989. An analysis of the growth bands in the shell of the common mussel *Mytilus edulis*. J. Mar. Biol. Ass. U. K., 69: 477-491.

Richardson, C.A. and Seed, R., 1990. Predictions of mussel (*Mytilus edulis*) biomass on an offshore platform from single population samples. Biofouling, 2: 289-297.

Richardson, C.A., Seed, R. and Naylor, E., 1990. Use of internal growth bands for measuring individual and population growth rates in *Mytilus edulis* from offshore production platforms. Mar. Ecol. Prog. Ser., 66: 259-265.

Ricketts, E.F. and Calvin, J., 1939. Between Pacific Tides. Stanford University Press, Stanford, 320pp.

Rigg, G.R. and Miller, R.C., 1949. Intertidal plant and animal zonation in the vicinity of Neah Bay, Washington. Proc. Calif. Acad. Sci., 26: 323-357.

Riisgård, H.U. and Randløv, A., 1981. Energy budgets, growth and filtration rates in *Mytilus edulis* at different algal concentrations. Mar. Biol., 61: 227-234.

Robles, C., 1987. Predator foraging characteristics and prey population structure on a sheltered shore. Ecology, 68: 1502-1514.

Rodhouse, P.G., Roden, C.M., Burnell, G.M., Hensey, M.P., McMahon, T., Ottway, B. and Ryan, T.H., 1984a. Food resource, gametogenesis, and growth of *Mytilus edulis* on the shore and in suspended culture: Killary Harbour, Ireland. J. Mar. Biol. Ass. U. K., 64: 513-529.

Rodhouse, P.G., Roden, C.M., Hensey, M.P. and Ryan, T.H., 1984b. Resource allocation in *Mytilus edulis* on the shore and in suspended culture. Mar. Biol., 84: 27-34.

Rodhouse, P.G., Roden, C.M., Hensey, M.P. and Ryan, T.H., 1985. Production of mussels, *Mytilus edulis*, in a suspended culture and estimates of carbon and nitrogen flow: Killary Harbour, Ireland. J. Mar. Biol. Ass. U. K., 65: 55-68.

Rodhouse, P.G., McDonald, J.H., Newell, R.I.E. and Koehn, R.K., 1986. Gamete production, somatic growth, and multiple-locus enzyme heterozygosity in *Mytilus edulis*. Mar. Biol., 90: 209-214.

Safriel, U.N. and Sasson-Frostig, Z., 1988. Can colonizing mussel outcompete indigenous mussel? J. Exp. Mar. Biol. Ecol., 117: 211-226.

Scagel, R.F., 1970. Benthic algae of Bowie Seamount. Syesis, 3: 15-16.

Seed, R., 1968. Factors influencing shell shape in the mussel *Mytilus edulis*. J. Mar. Biol. Ass. U. K., 48: 561-584.

Seed, R., 1969a. The ecology of *Mytilus edulis* L. (Lamellibranchiata) on exposed rocky shores 1. Breeding and settlement. Oecologia (Berl.), 3: 277-316.

Seed, R., 1969b. The ecology of *Mytilus edulis* L. (Lamellibranchiata) on exposed rocky shores 2. Growth and mortality. Oecologia (Berl.), 3: 317-350.

Seed, R., 1969c. The influence of the pea crab *Pinnotheres pisum* Penn. in the two types of *Mytilus* (Mollusca: Bivalvia) from Padstow, south-west England. J. Zool. (Lond.), 158: 413-420.

Seed, R., 1971. A physiological and biochemical approach to the taxonomy of *Mytilus edulis* L. and *M. galloprovincialis* Lmk. from south-west England. Cah. Biol. Mar., 12: 291-322.

Seed, R., 1973. Absolute and allometric growth in the mussel *Mytilus edulis* L. (Mollusca: Bivalvia). Proc. Malacol. Soc. Lond., 40: 343-357.

Seed, R., 1975. Reproduction in *Mytilus edulis* L. (Mollusca: Bivalvia) in European waters. Pubbl. Stn. Zool. Napoli, 39: 317-334.

Seed, R., 1976. Ecology. In: B.L. Bayne (Editor), Marine Mussels: their ecology and physiology. Cambridge University Press, Cambridge, pp. 13-65.

Seed, R., 1978. The systematics and evolution of *Mytilus galloprovincialis* Lmk., In: B. Battaglia and J. Beardmore (Editors), Marine Organisms: Genetics, Ecology and Evolution. Plenum Press, New York, pp. 447-468.

Seed, R., 1980. Shell growth and form in the Bivalvia. In: D.C. Rhoads and R.A. Lutz (Editors), Skeletal Growth of Aquatic Organisms. Plenum Press, New York, pp. 23-67.

Seed, R., 1990. Taxonomic and evolutionary relationships within the genus *Mytilus*. In: B. Morton (Editor), Proceedings of a Symposium in Honour of Sir Charles Maurice Yonge, Edinburgh, 1986. Hong Kong University Press, Hong Kong, pp. 93-107.

Seed, R., 1992. Crabs as predators of marine bivalve molluscs. Asian Mar. Biol., in press.

Seed, R. and Brown, R.A., 1977. A comparison of the reproductive cycles of *Modiolus modiolus* (L.), *Cerastoderma* (=*Cardium*) *edule* (L.) and *Mytilus edulis* in Strangford Lough, Northern Ireland. Oecologia (Berl.), 30: 173-188.

Seed, R. and Brown, R.A., 1978. Growth as a strategy for survival in two marine bivalves, *Cerastoderma edule* (L.) and *Modiolus modiolus* (L.). J. Anim. Ecol., 47: 283-292.

Seed, R. and Richardson, C.A., 1990. *Mytilus* growth and its environmental responsiveness. In: G.B. Stefano (Editor), The Neurobiology of *Mytilus edulis*. Manchester University Press, Manchester, pp. 1-37.

Shelford, V.E., Weese, A.O., Rice, L.A., Rasmussen D.I. and MacLean, A., 1935. Some marine biotic communities of the Pacific coast of North America. Ecol. Monogr., 5: 249-354.

Siebers, D. and Winkler, A., 1984. Amino acid uptake by mussels, *Mytilus edulis*, from natural sea water in a flow through system. Helgolander Wiss. Meeresunters., 38: 189-199.

Sigurdsson, J.B., Titman, C.W. and Davies, P.A., 1976. The dispersal of young post-larval bivalve molluscs by byssus threads. Nature (Lond.), 262: 386-387.

Simpson, R.A., 1977. The biology of two offshore oil platforms. Institute of Marine Resources, University of California, I.M.R. Ref. 76-13.

Sousa, W.P., 1984. Intertidal mosaics: patch size, propagule availability, and spatially variable patterns of succession. Ecology, 65: 1918-1935.

Sousa, W.P., 1985. Disturbance and patch dynamics on rocky intertidal shores. In: S.T.A. Pickett and P.S. White (Editors), The Ecology of Natural Disturbance and Patch Dynamics. Academic Press, New York, pp. 101-124.

Sprung, M., 1983. Reproduction and fecundity of the mussel *Mytilus edulis* at Helgoland (North Sea). Helgolander Wiss. Meeresunters., 36: 243-255.

Sprung, M., 1984. Physiological energetics of mussel larvae *Mytilus edulis* 1. Shell growth and biomass. Mar. Ecol. Prog. Ser., 17: 283-293.

Stephenson, T.A., and Stephenson, A., 1972. Life Between Tidemarks on Rocky Shores. W.H. Freeman and Co., San Francisco, 425pp.

Stickle, W.B., Moore, M.N. and Bayne, B.L., 1985. Effects of temperature, salinity, and aerial exposure on predation and lysosomal stability of the dogwhelk *Thais* (*Nucella*) *lapillus* (L.). J. Exp. Mar. Biol. Ecol., 93: 235-258.

Strømgren, T., 1975. Linear measurements of growth of shells using laser diffraction. Limnol. Oceanogr., 20: 845-848.

Strømgren, T., 1976a. Growth patterns of *Mytilus edulis* in relation to individual variation, light conditions, feeding and starvation. Sarsia, 60: 25-40.

Strømgren, T., 1976b. Length growth of *Mytilus edulis* (Bivalvia) in relation to photoperiod, irradiance and spectral distribution of light. Sarsia, 61: 31-40.

Strømgren, T. and Cary, C., 1984. Growth in length of *Mytilus edulis* L. fed on different algal diets. J. Exp. Mar. Biol. Ecol., 76: 23-34.

Suchanek, T.H., 1978. The ecology of *Mytilus edulis* L. in exposed rocky intertidal communities. J. Exp. Mar. Biol. Ecol., 31: 105-120.

Suchanek, T.H., 1979. The *Mytilus californianus* community: Studies on the composition, structure, organization, and dynamics of a mussel bed. Ph.D. Thesis, University of Washington, U.S.A.

Suchanek, T.H., 1980. Diversity in natural and artificial mussel bed communities of *Mytilus californianus*. Am. Zool., 20: 807.

Suchanek, T.H., 1981. The role of disturbance in the evolution of life history strategies in the intertidal mussels *Mytilus edulis* and *Mytilus californianus*. Oecologia (Berl.), 50: 143-152.

Suchanek, T.H., 1985. Mussels and their role in structuring rocky shore communities. In: P.G. Moore and R. Seed (Editors), The Ecology of Rocky Coasts. Hodder and Stoughton, Sevenoaks, U.K., pp. 70-96.

Sunila, I., 1981. Reproduction of *Mytilus edulis* L. (Bivalvia) in a brackish water area, the Gulf of Finland. Ann. Zool. Fenn., 48: 121-128.

Sunila, I. and Lindstrøm, R., 1985. Survival, growth, and shell deformities of copper and cadmium-exposed mussels (*Mytilus edulis* L.) in brackish water. Est. Coast. Shelf Sci., 21: 555-565.

Sutherland, J.P., 1974. Multiple stable points in natural communities. Am. Nat., 108: 859-873.

Swennen, C., Bruijn, L.L.M. de, Duiven, P., Leopold, M.F. and Marteijn, E.C.L., 1983. Differences in bill form of the oystercatcher *Haematopus ostralegus*: a dynamic adaptation to specific foraging techniques. Neth. J. Sea Res., 17: 57-83.

Tedengren, M. and Kautsky, N., 1986. Comparative studies of the physiology and its probable effect on size in blue mussels (*Mytilus edulis*) from the North Sea and Northern Baltic proper. Ophelia, 25: 147-155.

Theisen, B.F., 1968. Growth and mortality of culture mussels in the Danish Wadden Sea. Medd. Dan. Fisk. Havunders., 6: 47-78.

Theisen, B.F., 1972. Shell cleaning and deposit feeding in *Mytilus edulis* L. (Bivalvia). Ophelia, 10: 49-55.

Theisen, B.F., 1973. The growth of *Mytilus edulis* L. (Bivalvia) from Disko and Thule district, Greenland. Ophelia, 12: 59-77.

Theisen, B.F., 1987. Infestation of *Mytilus edulis* by *Mytilicola intestinalis*. Ophelia, 27: 77-86.

Thompson, R.J., 1979. Fecundity and reproductive effort of the blue mussel (*Mytilus edulis*), the sea urchin (*Strongylocentrotus droebachiensis*) and the snow crab (*Chionectes opilio*) from populations in Nova Scotia and Newfoundland. J. Fish. Res. Board Can., 36: 955-964.

Thompson, R.J., 1984a. The reproductive cycle and physiological ecology of the mussel *Mytilus edulis* in a subarctic, non-estuarine environment. Mar. Biol., 79: 277-288.

Thompson, R.J., 1984b. Production, reproductive effort, reproductive value and reproductive cost in a population of the blue mussel *Mytilus edulis* from a subarctic environment. Mar. Ecol. Prog. Ser., 16: 249-257.

Tongiorgi, P., Nardi, P., Galleni, L., Nigro, M. and Salghetti, 1981. Feeding habits of *Ocinebrina edwardsi* (Mollusca: Prosobranchia) a common mussel drill of the Italian coasts. Pubbl. Stn. Zool. Napoli, Mar. Ecol., 2: 169-180.

Tracey, G.A., 1988. Feeding reduction, reproductive failure, and mortality in *Mytilus edulis* during the 1985 'Brown tide' in Narragansett Bay, Rhode Island. Mar. Ecol. Prog. Ser., 50: 73-81.

Trevelyan, G.A., 1991. Aquacultural ecology of hatchery-produced juvenile bay mussels, *Mytilus edulis* L. Ph.D. Thesis, University of California, U.S.A.

Tsuchiya, M., 1979. Quantitative survey of intertidal organisms on rocky shores in Mutsu Bay, with special reference to the influence of wave action. Bull. Mar. Biol. Stn. Asamushi, Tokyo, 16: 69-86.

Tsuchiya, M., 1980. Biodeposit production by the mussel *Mytilus edulis* L. on rocky shores. J. Exp. Mar. Biol. Ecol., 47: 203-222.

Tsuchiya, M., 1983. Mass mortality in a population of the mussel *Mytilus edulis* L. caused by high temperature on rocky shores. J. Exp. Mar. Biol. Ecol., 66: 101-111.

Tsuchiya, M. and Nishihira, M., 1985. Islands of *Mytilus* as habitat for small intertidal animals: effect of island size on community structure. Mar. Ecol. Prog. Ser., 25: 71-81.

Tsuchiya, M. and Nishihira, M., 1986. Islands of *Mytilus* as habitat for small intertidal animals: effect of *Mytilus* age structure on the species composition of the associated fauna and community organization. Mar. Ecol. Prog. Ser., 31: 171-178.

Tsuchiya, M. and Bellan-Santini, D., 1989. Vertical distribution of shallow rocky shore organisms and community structure of mussel beds (*Mytilus galloprovincialis*) along the coast of Marseille, France. Mesogée, 49: 91-110.

Tursi, A.,Matarrese, A., Liaci, L.S., Cecere, E., Montanaro, C. and Chieppa, M., 1985. Struttura della popolazione di *Mytilus galloprovincialis* Lamarck presente nei branchi naturali del Mar Piccolo di Taranto. Quad. Ist. Ric. Pesca Marit., 4: 183-203.

Ursin, E., 1963. On the incorporation of temperature in the von Bertalanffy growth equation. Medd. Dan. Fisk. Havunders., 4: 1-16.

VanBlaricom, G.R., 1987. Regulation of mussel population structure in Prince William Sound, Alaska. Natl. Geog. Res., 3: 501-510.

VanBlaricom, G.R., 1988. Effects of foraging by sea otters on mussel-dominated intertidal communities. In: G.R. VanBlaricom and J.A. Estes (Editors), The Community Ecology of Sea Otters. Ecological Studies 65: Analysis and Synthesis. Springer-Verlag, New York. pp. 48-91.

Wallace, S., 1990. Spawning patterns in populations of cultured and wild mussels (Mytilus edulis L.) in Bantry Bay, south-west Ireland. Ph.D. Thesis, National University of Ireland, Cork.

Walne, P.R. and Dean, G.D., 1972. Experiments on predation by the shore crab Carcinus maenas L. on Mytilus and Mercenaria. J. Cons., 34: 190-199.

Wayne, T.A., 1980. Antipredatory behaviour of the mussel Mytilus edulis. Am. Zool., 20: 789.

Wells, W., 1928. Pinnotheridae of Puget Sound. Publ. Puget Sound Biol. Stn., Univ. Wash., 6: 283-314.

Wells, W., 1940. Ecological studies on the pinnotherid crabs of Puget Sound. Univ. Wash. Publ. Oceanogr., 2: 19-50.

Whiteley, J.D., Pritchard, J.S. and Slater, P.J.B., 1990. Strategies of mussel dropping in Carrion Crows Corvus c. corone. Bird Study, 37: 12-17.

Whittaker, R.H., 1975. Communities and Ecosystems. MacMillan Publishing Company, New York, 385pp.

Widdows, J., Donkin, P., Salkeld, P.N., Clearly, J.J., Lowe, D.M., Evans, S.V. and Thompson, P.E., 1984. Relative importance of environmental factors in determining physiological differences between two populations of mussels (Mytilus edulis). Mar. Ecol. Prog. Ser., 17: 33-47.

Williams, R.J., 1970. Freezing tolerance in Mytilus edulis. Comp. Biochem. Physiol., 35: 145-161.

Wilson, J.H., 1977. The growth of Mytilus edulis from Carlingford Lough. Ir. Fish. Invest. Ser. B (Mar.), 17: 1-15.

Wilson, J.H., 1987. Mussels: The problem of early spawning. Aquacult. Ir., 30: 20-21.

Wilson, J.H., 1988. Distribution of oyster Ostrea edulis, mussel Mytilus edulis and anomiid larvae in Bertraghboy Bay, Co. Galway. Ir. Fish. Invest. Ser. B (Mar.), 31: 1-11.

Wilson, J.H. and Seed, R., 1974. Reproduction in Mytilus edulis L. (Mollusca: Bivalvia) in Carlingford Lough, Northern Ireland. Ir. Fish. Invest. Ser. B (Mar.), 15: 1-30.

Wilson, J.H. and Simons, J., 1985. Gametogenesis and breeding of Ostrea edulis in the west of Ireland. Aquaculture, 46: 307-321.

Witman, J.D., 1987. Subtidal coexistence: storms, grazing, mutualism, and the zonation of kelps and mussels. Ecol. Monogr., 57: 167-187.

Witman, J.D., and Suchanek, T.H., 1984. Mussels in flow: drag and dislodgement by epizoans. Mar. Ecol. Prog. Ser., 16: 259-268.

Worrall, C.M. and Widdows, J., 1984. Investigation of factors influencing mortality in Mytilus edulis L. Mar. Biol. Lett., 5: 85-97.

Yamada, S.B., 1989. Mytilus californianus, a new aquaculture species. Aquaculture, 81: 275-284.

Yonge, C.M., 1976. The "mussel" form and habit. In: B.L. Bayne (Editor), Marine Mussels: their ecology and physiology. Cambridge University Press, Cambridge, pp 1-12.

Young, A., 1985. Byssus thread formation by the mussel Mytilus edulis: effects of environmental forces. Mar. Ecol. Prog. Ser., 24: 261-271.

Young, R.T., 1946. Spawning and settling season of the mussel Mytilus californianus. Ecology, 27: 354-363.

Zhang, F., 1984. Mussel culture in China. Aquaculture, 39: 1-10.

Zhang, F., He, Y., Liu, X., Li, S., Ma, J. and Lou, Z., 1980. The breeding seasons of mussel Mytilus edulis in Jiaszhou Bay, Shandong Province, China (Chinese with English Abstract). Oceanogr. Limnol. Sin., 11: 341-350.

Zouros, E., Romero-Dorey, M. and Mallet, A.L., 1988. Heterozygosity and growth in marine bivalves: further data and possible explanations. Evolution, 42: 1332-1341.

Zwarts, L. and Drent, R.H., 1981. Prey depletion and the regulation of predator density: oystercatchers (Haematopus ostralegus) feeding on mussels (Mytilus edulis). In: N.V. Jones and W.J. Wolff (Editors), Feeding and Survival Strategies of Estuarine Organisms. Plenum Press, New York, pp. 193-216.

Chapter 5

PHYSIOLOGICAL INTERRELATIONS, AND THE REGULATION OF PRODUCTION

ANTHONY J.S. HAWKINS AND BRIAN L. BAYNE

INTRODUCTION

Production, defined as the accretion of somatic tissues and/or gametes, has been extensively studied in mussels. This reflects their widespread distributions throughout coastal waters of each hemisphere, and stems from the need to resolve: ecological consequences of the interrelationships between mussels and their environment (Bayne and Hawkins, 1992), the importance of mussels in sustenance fisheries, in mariculture, and as fouling organisms (Mason, 1976; Lutz, 1980), and the widespread use of mussels as biological indicators of environmental pollution (Widdows and Donkin, 1989 and Chapters 8 and 9).

Such studies in the genera *Mytilus, Choromytilus, Aulacomya, Geukensia, Arca* and *Perna* have shown that mussels tolerate a wide range of environmental conditions, facilitated by a remarkable plasticity of physiological response. This plasticity is demonstrated by rates of growth and/or reproduction that not only differ between species, but which also vary dramatically between populations of the same species, between individuals of the same population, and with seasonal changes in the natural environment (Bayne and Worrall, 1980; Hawkins et al., 1985, 1989a; Seed and Richardson, 1990; Anwar et al., 1990; Jørgensen, 1991).

This chapter considers processes and interrelations effecting such variability, first presenting a mechanistic consideration of the physiology of production, and then addressing the regulation of production. In so doing, it has not been possible to review all relevent literature. Instead, points are illustrated with selected examples that are pertinent to current understanding, and the reader is referred to previous comprehensive reviews of physiological energetics and growth in mussels and other bivalves (see Bayne, 1976a; Bayne and Newell, 1983; Griffiths and Griffiths, 1987; Seed and Richardson, 1990).

PHYSIOLOGY OF PRODUCTION

The physiology of production may be subdivided into processes whereby mussels acquire nutrients, and processes which influence subsequent utilization of those nutrients.

Nutrient Acquisition

Particulate

Mussels acquire nutrients using mucociliary mechanisms on the ctenidia and labial palps to filter and ingest a variety of suspended particulates which include bacteria, phytoplankton, detritus and microzooplankton (for reviews see Bayne and Newell, 1983; Griffiths and Griffiths, 1987; Jørgensen, 1991; and also Jones et al., 1990 and Chapter 2). Particles are then transported along ciliated grooves to the labial palps. At low levels of abundance, all filtered material is conveyed from the palps for ingestion through the mouth. However, as seston availability increases above threshold concentrations that correspond with saturation of the mussel's digestive capacity, ingested ration remains relatively constant, and the 'surplus' filtered material is bound with mucus to form pseudofaeces destined for pre-ingestive rejection (see p.201–206). Ingested material is subjected to extracellular digestion throughout the gut, supplemented by intracellular digestion of selected particles within tubules of the digestive gland, and followed by continuing absorption of organics throughout the intestine (Hawkins et al., 1986b), before egestion as faeces (for review see Morton, 1983).

The ctenidia of mussels often retain suspended bacteria smaller than 1μm, but with efficiencies of less than 20 to 30% (for reviews see Kemp et al., 1990; Langdon and Newell, 1990; Prieur et al., 1990). These efficiencies improve progressively with increasing particle size, reaching 100% for seston larger than 4μm (Lucas et al., 1987; Riisgård, 1988). However, species such as the saltmarsh mussel, *Geukensia demissa*, (Wright et al., 1982) and the brown mussel *Perna perna* (Schleyer, 1981), living in habitats in which the bacterial biomass is characteristically high, display high retention efficiencies for particles <0.6μm, and thus appear specifically adapted to exploit the bacterial size-fraction. In addition, an increasing body of work indicates that efficiencies of retention not only depend on particle size, but also upon shape, motility, density and chemical cues such as algal ectocrines (Bayne et al., 1977a; Newell et al., 1989; Ward and Targett, 1989; Kemp et al., 1990; Prieur et al., 1990). Similar factors may be involved during the preferential rejection of nutritionally undesirable food particles within pseudofaeces. *Mytilus edulis*, for example, produces pseudofaeces in which ratios of inorganics to both chlorophyl *a* (Kiørboe and

Møhlenburg, 1981) and organics (Bayne et al., 1989) may be more than double those in the available seston.

Differential processing also occurs within the gut. Mussels synthesize a wide variety of digestive carbohydrases, proteases, lipases, esterases and phosphatases, also showing the ability to lyse and assimilate ingested bacteria (Morton 1983; Lucas and Newell, 1984; McHenery and Birkbeck, 1985; Muir et al., 1986; Seiderer et al., 1987; Prieur et al., 1990; Teo and Sabapathy, 1990). Nevertheless, there is clear evidence in mussels of differential absorption from specific algae within mixed cell suspensions, resilient species even remaining viable following their egestion and resuspension from faeces (Fielding, 1987). These differences may be associated with different rates of gut passage. Foster–Smith (1975), for example, showed that alumina particles passed through the gut of M. edulis more quickly than algal cells. Similarly, results of a series of experiments, using analytical flow cytometry, have indicated that transit times in M. edulis of 0.3g dry tissue weight were independent of particle size, but differed markedly according to algal species over a range of less than 0.25h for Dunaliella to more that 1.0h for Chlorella (Hawkins, Bayne and Klumpp, unpublished data).

Mussels, therefore, show considerable variability in the relative utilization of different suspended particulates. This variability stems collectively from preferential retention on the ctenidia, pre-ingestive rejection within pseudofaeces, and differential processing within the gut. Consequently, mussels may impose qualitative as well as quantitative changes on suspended particulates within the natural environment, and the relative nutritional significances of different suspended particulates cannot be inferred from their available concentrations.

Studies relating utilization of different components of the available seston to metabolic requirements of mussels in their natural environment have shown that phytoplankton and organically-rich detrital particles generally represent the main sources of nutrition (for review see Bayne and Hawkins, 1992). Nevertheless, mussel growth may be enhanced following addition of a small proportion of silt to the diet (Winter 1976; Kiørboe et al., 1980, 1981). This stems from the volumetric constraints that are inherent in digestion (Taghon and Jumars, 1984); enhanced growth results because, relative to certain phytoplankton or detritus, the nonrefractory organic matter carried by silt represents a greater fraction of total particle volume (Bayne et al., 1987). In addition, under circumstances when phytoplankton and detrital organics are not abundant, it has been estimated that unattached bacteria and the bacteria bound to other particles may then be collectively utilized to satisfy from 70–73% and 8–30% of the metabolic requirements for nitrogen and carbon, respectively (for reviews see Langdon and Newell, 1990; Prieur et al., 1990). Clearly, bacteria may satisfy significant proportions of the requirements for nitrogen, since bacterial nitrogen contents per unit carbon are more than twice those in phytoplankton (Seiderer and Newell, 1985).

Similar studies have shown that cellulose within natural detrital material satisfies less than 15% of the mussel's total carbon requirements (Stuart et al., 1982; Fielding et al., 1986; Kreeger et al., 1988; Langdon and Newell, 1990). Nevertheless, vascular plant detritus may contribute indirectly, by acting as a carbon source for cellulolytic bacteria and perhaps nanozooplankton (Langdon and Newell, 1990). As suggested by Crosby et al. (1990), the significance of this indirect contribution may be much greater, given higher bacterial abundance and/or greater proportional attachment of bacteria to larger particles, thereby increasing the retention and ingestion of bacteria.

The foregoing examples serve to emphasize the rather obvious point that importance of individual particle-types to mussel nutrition is partially dependent on their relative abundances, which may show considerable seasonal and spatial variability (Widdows et al., 1979; Seiderer and Newell, 1985; Berg and Newell, 1986; Fielding and Davis, 1989).

Dissolved

Biologists have long speculated that the vast quantities of dissolved organic material (DOM) present in seawater may be available to marine organisms for utilization as a source of nutrition (for reviews see Gorham, 1990 and Manahan, 1990). While early critics argued that measured removal could be an artefact deriving from bacterial contamination or exchange diffusion, more recent work under bacteria-free conditions, using high performance liquid-chromatography (HPLC) and amperometric detection, has confirmed the net uptake by marine mussels both of amino acids (Manahan et al., 1982) and sugars (Colwell and Manahan, 1988; Welborn and Manahan, 1988). In *Mytilus californianus*, active transport systems are well-adapted for the net accumulation of amino acids against chemical gradients in excess of one million to one (Wright et al., 1989). Total uptake depends upon the concentration and composition of available amino acids (Jørgensen, 1982; Manahan et al., 1983), external salinity and sodium concentration (Wright et al., 1989), as well as the rate of seawater flow across each mussel (Siebers and Winkler, 1984).

Considering the contribution of accumulated DOM to metabolic demands, then FAA uptake by *Mytilus* spp. has been calculated to represent between as much as 34 and 40% of the requirements for growth and reproduction, respectively (Stephens, 1981; Manahan et al., 1983). However, these authors were relating rates of uptake measured to metabolic requirements that had been determined in separate studies by other researchers working on different populations of mussels. When all relevant parameters have been measured in the same mussels, then the data indicate that uptake of dissolved amino acids and sugars from natural seawater may represent a much lower percentage (13–24%) of measured energy expenditure (Siebers and Winkler, 1984; Gorham, 1990).

Similarly, Melaouah (1990) reported that measured uptake of an artificial mixture of ^{14}C-labelled amino acids represented less than 10% of the energy expenditure in *M. edulis* larvae. Embryos and larvae of the oyster *Crassostrea gigas* cultured in particle-free seawater nevertheless showed net gain in biomass during embryogenesis and early larval development when, as growth proceeded, they also increased their ability to obtain DOM in direct proportion to the increase in their metabolic demand (Manahan, 1990). It seems likely that, for mussels in coastal waters, uptake of DOM will prove of greatest ecological significance during times when other sources of nutrition are least abundant and/or nutritionally incomplete. However, to fully understand the ecological significance of dissolved uptake, more information is needed on natural concentrations of DOM, with simultaneous measures of uptake, metabolism and growth in the same individuals (Gorham, 1990).

Whereas most filter-feeding bivalves rely through heterotrophic processes on organic materials accumulated from seawater, autotrophic pathways have been described in the gills of certain species (for reviews see Fiala-Medioni and Felbeck, 1990 and Prieur et al., 1990). By providing organic carbon from reduced CO_2, these pathways may either provide an additional supply to heterotrophic processes, or they may represent the only way to colonize environments which are poor in organic particulates, but which are rich in dissolved materials (Page et al., 1990). For example, *Bathymodiolus thermophilus* and other undescribed mussels living in deep-sea environments obtain the majority or all of their nutrient requirements both from uptake of dissolved amino acids and chemoautotrophic processes. This is known to involve symbiotic sulphur-oxidizing bacterial associations in hydrothermal environments, as well as methanotrophic bacterial associations in the vicinity of hydrocarbon-seeps (Childress et al., 1986; Fiala-Medioni et al., 1986; Brooks et al., 1987; Fisher et al., 1987).

Nutrient Utilization

Costs of living
Metabolism provides the power that is essential for life processes. To fuel metabolic processes, nutrients are acquired for conversion to kinetic energy by means of fission and oxidation, but with direct costs incurred through the dissipation of energy as heat, and indirect costs that include losses of assimilated materials (Fig. 5.1).

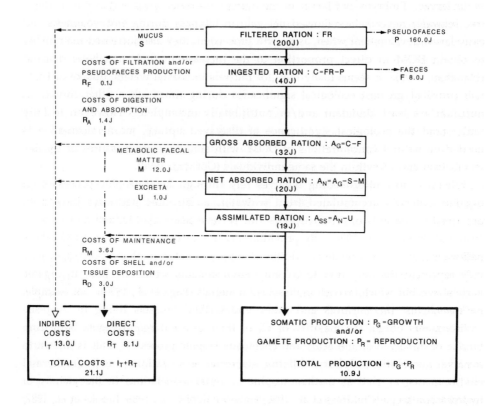

Fig. 5.1. Flow diagram showing components of energy acquisition, energy expenditure and production, each expressed as J h^{-1} in a typical mussel of 1g dry tissue weight, feeding with a clearance rate of 2L h^{-1} at seston concentrations of 10mg L^{-1}, of which 50% is organic matter with an energy content of 20J mg^{-1}. Gross and net absorption efficiencies are assumed to be 80% and 50%, respectively. Note (a) the absence of quantitative estimates of energy either acquired as dissolved nutrients, lost as mucus, or deposited as byssus threads; and (b), that energy values given in brackets are independent of direct costs.

Direct costs:

Direct costs represent as much as 29% of the energy retained in tissues by *M. edulis* under normal conditions of growth (Hawkins et al., 1989b; Widdows and Hawkins, 1989). The generation and loss of energy as heat may result both from normal aerobic metabolism, entailing the catabolism of glucose by the Krebs cycle, and also from anaerobic processes that enable mussels to tolerate oxygen depletion during extended periods of either environmental or physiological hypoxia (for reviews see Bayne and Newell, 1983; Shick et al., 1983 and Chapter 6).

Determination of the total rate of heat dissipation from living organisms by means of calorimetry is a nonspecific measure of total metabolic activity and all direct costs. Alternatively, specific components of metabolism are measured indirectly either by biochemical determinations addressing the utilization of metabolic substrates and accumulation of metabolic end products, or by respirometric quantifications of O_2 uptake and CO_2 production. Such indirect measures of direct costs commonly rely on oxycalorific equivalents, estimating heat loss from O_2 uptake. Oxycaloric equivalents of O_2 uptake are nevertheless appropriate in the majority of physiological studies, and these range from -440 to $-480kJ$ mol^{-1} O_2, depending on the relative balance of substrates metabolized to satisfy catabolic requirements (Gnaiger, 1983). Only under conditions of physiological and environmental hypoxia or anoxia, will the generalized oxycalorific equivalent be inappropriate, requiring more detailed measures of CO_2 production, biochemical fluxes, and/or calorimetry (Widdows, 1987, 1989).

Indirect costs:

Indirect costs result from losses of endogenous substances that are either rejected as mucus within pseudofaeces, egested as metabolic faecal matter, or excreted as metabolic end products (Fig. 5.1).

The amount of mucus rejected as pseudofaeces has yet to be documented for any bivalve. Yet, mussels may produce very considerable quantities of pseudofaeces when quantities of suspended material retained by the ctenidia exceed the capacity for ingestion (Jørgensen, 1991; see also p.201–206). Indeed, Prins and Smaal (1989) have recorded net losses of carbon from M. edulis at high particle concentrations exceeding about 40mg L^{-1}, which they attributed to losses of mucus in pseudofaeces, and which imply very significant indirect costs of filtration in turbid environments.

Collective egesta comprising fragments of digestive epithelium, enzymes and other substances secreted into but not reabsorbed from the gut are termed metabolic faecal losses (MFL). Such losses were first identified in mussels from analyses of nitrogen content (Hawkins et al., 1983) and isotope depuration (Famme and Kofoed, 1982; Hawkins and Bayne, 1984) within faeces, and have subsequently been shown to comprise more than 60% of egested organic matter (Hawkins and Bayne, 1985; Hawkins et al., 1990). Associated computations have shown that MFL varies with season and physiological condition, and represents between 25 and 89% of absorbed nitrogen, amounting to losses between 0.73 and 2.45g N per 100g of total dry mass ingested (Hawkins and Bayne, 1985).

These losses were considerably greater than values observed (<0.22g N per 100g) in other invertebrates and fish (for reviews see Hawkins and Bayne, 1985 and Steffens, 1989). In contrast to these organisms, in which digestion is predominantly extracellular, bivalves utilize intracellular digestion—a process which involves rejection of cellular contents within residual bodies exocytosed during cycles of digestion and reconstitution (Platt, 1971). It seems likely then, that such high MFL from mussels reflects additional costs associated with intracellular digestion, which must be viewed as one component of the total costs of feeding (Bayne and Hawkins, 1990; see also p.180–184).

Costs associated with MFL have previously been 'hidden' in studies which relied solely upon measures, either of gross or of net efficiencies, with which digested material is absorbed across the epithelium of the gut. The former takes no account of MFL, whereas estimates of net absorption efficiency account for, but do not quantify, this loss to the organism. By measuring both net and gross efficiencies simultaneously, one can quantify the influence of MFL as shown in Fig. 5.2 for *M. edulis* fed mixed diets of differing organic content. In this instance, net absorption efficiencies were measured by comparing the organic and inorganic content of the food and faeces, quantifying inorganics either as ash (Conover, 1969) or as biogenic silica (Tande and Slagstad, 1985). Alternatively, gross absorption efficiencies were measured by labelling the algal diet with ^{14}C and computing the ratio $[(^{14}C_{ingested} - ^{14}C_{defaecated})/^{14}C_{ingested}]$ (cf. Hawkins and Bayne, 1984). Both sets of results show that gross absorption efficiencies remained independent of dietary organic content, whereas net absorption efficiencies were significantly lower, declining exponentially (Bayne et al., 1987) with reducing food quality, and reaching negative values when dietary organic content reached less than 10 to 20% (Fig. 5.2).

Presumably, the decline in net absorption efficiencies with increasing dietary ash content reflects continuing digestive 'investment' so that, irrespective of dietary composition, MFL is relatively constant per unit of the total food ingested. Certainly, these findings illustrate a distinct uncoupling between digestion and net absorption. This uncoupling has also been evidenced by previous observations that, with increasing dietary ash content, efficiencies with which chlorophyll was degraded remained unchanged, despite a coincident decline in net efficiencies for absorption of the total products of digestion (Hawkins et al., 1986b). MFL therefore represents a significant cost of digestion. Although representing an essentially constant percentage of total metabolic expenditure in mussels of different size under standard conditions (Hawkins et al., 1990), costs deriving from MFL need not be fixed, so that the rate of 'digestive investment' should be considered as a potential component effecting compensatory adjustments in feeding behaviour (Willows, 1992).

High performance liquid chromatography (HPLC) now enables measurements to be made of the net fluxes of metabolites during a single passage of natural seawater over the mussel gill. Problems associated with previous spectrophotometric analyses of water, in which animals had been incubated for extended periods, are thereby

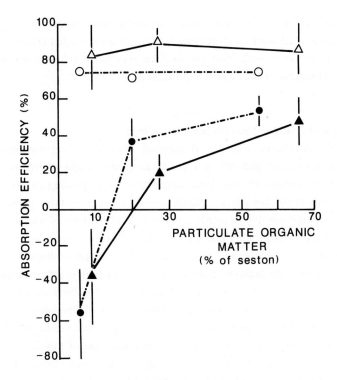

Fig. 5.2. Gross and net absorption efficiencies (open and closed symbols, respectively) in relation to dietary organic content in *Mytilus edulis*. Data are means ± SD (n = 5) from Bayne and Hawkins (1990), and represent the results of two separate experiments in which net absorption was measured as described in the text by quantifying inorganics either as ash (triangles) or as biogenic silica (circles).

avoided. Using HPLC, it has been shown that ammonia is the only nitrogenous waste product excreted in *M. edulis* (for discussions see Manahan et al., 1983 and Griffiths and Griffiths, 1987). This indirect loss normally represents less than 10% of the total energy dissipated from adult *Mytilus* in healthy physiological condition (Bayne and Newell, 1983; Hawkins and Bayne, 1985). Accordingly, excretion has often been ignored, or estimated by difference, in studies of production. However, recent investigations have shown that losses of ammonia from juvenile *M. edulis* (10mg dry tissue) acclimated to a submaintenance ration were energetically equivalent to as much as 63% of the heat dissipation measured calorimetrically. Such losses reflect the catabolism of endogenous proteins needed to support metabolic requirements, given low energy intake and the relative absence of prestored carbon-rich reserves in small mussels (Hawkins et al., 1989b). Alternatively, losses of ammonia may increase exponentially with increased dietary protein content. Enhanced excretion under such circumstances is predominantly of exogenous (ingested) nitrogen, and results from increased oxidation which is required to avoid deleterious consequences of the imbalance associated with an increasing surplus of amino acids (Hawkins and Bayne, 1991). Therefore, in juvenile mussels, and for specific diets, excretion may represent losses of considerable physiological significance.

Metabolic partitioning of total energy expenditure
All metabolic reactions, whether concerned with obtaining nutrients, molecular repair and replacement, growth, or reproduction, are closely interrelated. Therefore, subdivision of metabolic reactions according to physiological function is not consistent with their dynamics. Nevertheless, the concepts of compensation and adaptation presuppose the potential for variable utilization of assimilated nutrients and energy (Calow, 1985; Sibly and Calow, 1986). Such subdivision is therefore useful in quantifying relative metabolic allocations to different physiological functions.

Positive associations between rates of heat loss (metabolic energy expenditure) and growth suggest that metabolism may be statistically resolved into two components (Close et al., 1983; Kiørboe et al., 1985, 1987). One component represents energy required to maintain normal processes at equilibrium, and varies in positive relation with body mass (see p.191–195). The other is a function of the dietary intake above that required for maintenance, comprising the well-established heat increment following feeding. This is generally referred to as the 'specific dynamic action' (SDA), and stems from the component costs of food collection, digestion/absorption, and growth.

These component costs have rarely been convincingly defined and quantified, and opinions differ as to their relative significances (Jobling, 1983; Gaffney and Diehl, 1986). To establish the separate costs of maintenance, feeding, digestion/absorption and growth in *M. edulis*, growth measured directly in terms of the total energy content of soft tissues and shell has recently been compared with coincident calorimetric measures of heat dissipation, thus obviating uncertainties associated with digestibility, assimilation and the conversion factors required for indirect estimates of nutrient balance (Hawkins et al., 1989b; Widdows and Hawkins, 1989). Findings are illustrated in Fig. 5.3 for mussels of 10mg dry tissue that were administered experimental rations of similar biochemical composition, but which ranged in quantity up to a maximum availability when ingestion was more than seven times more than would have been required to maintain body mass at energy equilibrium.

The minimum rate of energy expenditure required for maintenance processes during starvation was determined as 3.11J day^{-1}, representing the rate of heat dissipation at zero ingestion (Hawkins et al., 1989b). Maintenance functions are many and varied, ranging from service processes which include work by the kidney and

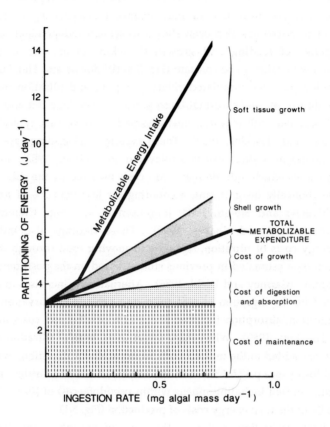

Fig. 5.3. Partitioning of metabolizable energy intake into growth (soft tissue and shell) and metabolic costs according to rate of ingestion by *Mytilus edulis* of 10mg dry tissue weight. Metabolizable energy intake was computed as the sum of heat loss and total energy retention, both measured directly, and where the y-intercept of 3.11J day^{-1} represents the minimum rate of energy expenditure required for maintenance processes at zero ingestion (Hawkins et al., 1989b; Widdows and Hawkins, 1989).

heart, to cellular maintenance effected by processes such as ion transport, lipid resynthesis and protein turnover (Milligan and Summers, 1986). Protein turnover is the continuous degradation and replacement of cellular proteins that is required for repair, and to enable the metabolic adjustments necessary for regulation and adaptation. The proportions of total maintenance energy expenditure which are represented by theoretical costs of the synthesis effecting protein turnover, as well as constant linear increases of about 11.4J total heat loss per mg^{-1} protein synthesis, are remarkably consistent between mussels and a wide variety of other animals

(Hawkins, 1991). This means that whole-body protein turnover represents a general index of the energy requirements for maintenance (Hawkins et al., 1989b). In addition, protein turnover has also been shown to remain independent of the wide experimental range of feeding and growth (Hawkins et al., loc. cit.), so that maintenance costs were taken to be constant (Fig. 5.3) (Widdows and Hawkins, 1989).

Inert suspended particles stimulated ciliary pumping and filtration mechanisms following starvation in *M. edulis,* so that these particles were ingested and defecated, but were not associated with elevated heat dissipation above that typical of unfed mussels (Widdows and Hawkins, 1989). These experimental findings are consistent with earlier estimates in *Mytilus chilensis* (Navarro and Winter, 1982), and confirm theoretical calculations which indicate that direct costs incurred by the ciliary pump in bivalves are metabolically insignificant, accounting for less than 3% of total energy expenditure (Silvester and Sleigh, 1984; Jørgensen et al., 1986; Clemmesen and Jørgensen, 1987; Bernard and Noakes, 1990). These findings also indicate that 'mechanical' energy costs of filtration, sorting and moving food through the gut appear to have been over estimated in previous studies based on the postprandial rise in oxygen consumption (Gaffney and Diehl, 1986). Alternatively, increases in heat dissipation stimulated by the ingestion of alga following prolonged starvation indicated that costs of digestion/absorption incurred about 17% of all direct costs in *M. edulis* (Widdows and Hawkins, 1989). When these direct costs, both of filtration and digestion/absorption, are added to the indirect costs of intracellular digestion, evidenced as metabolic faecal losses (see p.177–178), then the combined costs of nutrient acquisition represent a minimum (not including mucus within pseudofaeces) of [(0.1 + 1.4 + 12.0) ÷ 21.1] x 100 = 64% of the total energy costs of production (Fig. 5.1).

Remaining heat dissipation is due to the costs of growth, computed as total metabolic expenditure minus the summed costs of maintenance, filtration and digestion/absorption. Estimated costs of growth increased from zero, when the body mass was at energy equilibrium, to as much as 34% of total metabolic expenditure at the highest ration representing 7.3% of body mass day^{-1} (Fig. 5.3). However, these costs remain poorly understood. Just as in other animals (Close et al., 1983; Kiørboe et al., 1987), there was a large discrepancy between empirical and theoretical estimates of the energy costs associated with measured depositions of protein and nonprotein substances (Hawkins et al., 1989b). Part of this discrepancy may represent costs of integrated processes that are necessarily incurred coincident with tissue deposition, but which are not directly involved with synthetic pathways. In particular, there is a need to direct more attention away from the energetics of production, viewed solely in terms of synthesis and deposition, towards energetics of mediator molecules, which facilitate the control necessary for physiological regulation (Reeds et al., 1985; Millward and Rivers, 1989).

The net result is an energy surplus deposited as byssus, shell and soft tissue. Perhaps, because it is not widely appreciated that byssal threads are mostly protein, and because organic matter in most bivalve shells amounts to less than 5% by weight, these materials have mainly been ignored in studies of production. However, research in a variety of mussel species (Hawkins and Bayne, 1985 and references therein) has shown that both the shell and byssus threads may represent significant fractions of total production. Although not resolving the production of byssus from that of soft tissue, the findings illustrated in Figure 5.3 indicate that shell growth in *M. edulis* increased in direct proportion with ration, continuing at submaintenance rations, despite the net loss of soft tissues, and representing a major component exceeding 20% of total energy deposition.

Whatever the material, total energy deposition increased with food intake to comprise the largest fraction (58%) of metabolizable energy at the highest ration (Fig. 5.3). Conversely, total metabolic expenditure dominated the components of net energy balance at lower rations. Within this expenditure, costs of maintenance represented the single largest component. Even at the highest experimental ration, maintenance accounted for more than 50% of total metabolizable expenditure, and for all of the metabolizable energy available during complete starvation (Fig. 5.3). It is of physiological significance then, that weight-specific costs of maintenance are not fixed, varying between and within individual mussels. A genetic basis for maintenance energy expenditure has been established from numerous studies of selective breeding in mammals (Payne, 1984; van Steenbergen, 1987), as well as from associations with mean heterozygosity in mussels (Hawkins et al., 1986, 1989a). In addition, and independent of age-related changes, weight-specific rates of protein turnover and of maintenance expenditure undergo large seasonal adjustments (Hawkins et al., 1989b, discussing the results of Hawkins, 1985), and are responsive to environmental change (Bayne and Newell, 1983; Hawkins et al., 1987b).

Through their effects on metabolic partitioning, differences in protein/maintenance metabolism have major impacts on energy status (Reeds et al., 1985; van Steenbergen, 1987; Waterlow, 1988). For example, Figure 5.4 illustrates the simulated results of a 20% reduction in energy requirements for maintenance in *M. edulis*. Findings predict that total cumulative production was more than doubled over 10 years, leading both to larger somatic size and to higher reproductive output (R.I. Willows, personal communication, 1989). These predictions are consistent with experimental observations whereby individual mussels with lower genotype-dependent maintenance expenditure show faster somatic growth when immature, a higher reproductive output, and a greater seasonal capacity for production (see p.187–191).

It is important, also, that costs of digestion/absorption and growth are not considered as being solely dependent upon ingested ration per unit body size. While

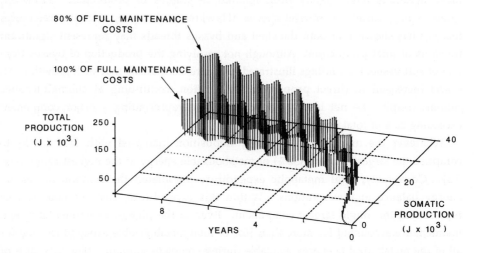

Fig. 5.4. Simulated effects of a 20% reduction in energy requirements for maintenance upon performance over 10 years subsequent to settlement in *Mytilus edulis*. The simulation has been developed by Dr R. I. Willows (Plymouth Marine Laboratory), using Michaelis-Menton type functions to partition energy between growth, reproduction and storage, depending upon the amount of storage already accumulated and the current energy balance, and applying allometric relations to determine how the allocation of material changes with mussel size.

Figure 5.3 illustrates relations for *M. edulis* under defined conditions, there is widespread recognition of potential trade-offs, dependent upon the variable allocation of ingested nutrients between these different physiological functions (Calow, 1985; Sibley and Calow, 1986; Willows, 1992). Weiser (1989) has recently developed the concept of 'sequential energy allocation', whereby metabolizable energy is utilized by poikilotherms according to priorities of demand, so that a given function may be suppressed in favour of another function of higher 'priority', thus leading to temporal changes in metabolic partitioning. We will return to this variability in later sections, emphasizing the consequences both of genetic differences and compensatory adjustments in energy allocation (see p.188–191 and p.209–211).

Differential utilization

Assimilated nutrients are either catabolized to provide the kinetic energy that powers life processes, or are deposited as lipid, carbohydrate or protein. However, considerable variability exists both in the comparative efficiencies with which assimilated nu-trients are incorporated within tissues, and in the subsequent relative utilization of deposited substrates. At the simplest level of interpretation, lipids and carbohydrates serve mainly as fuels, whereas proteins comprise the primary structural and func-tional elements of the body. Nevertheless, dietary protein is also degraded as a source of metabolizable energy. In addition, both the turnover and net deposition of protein are energetically expensive, and deposited protein may ultimately be used as an en-ergy reserve.

To address the nature of these interrelations in M. *edulis*, Hawkins and Bayne (1991) have further analysed data from above-mentioned experiments (p.180–184 and Fig. 5.3), investigating effects of food intake on protein metabolism, substrate deposition and the metabolic partitioning of heat losses (Hawkins et al., 1989b; Widdows and Hawkins, 1989). Net growth efficiencies for protein and energy, absorbed from a diet of pure phytoplankton, are illustrated in Figure 5.5a, and show monotonic increases with ration towards an asymptotic maximum of approximately 80% for protein. Such high efficiencies approximate the theoretical maximum for conversion of absorbed nutrients to body tissue (Calow and Townsend, 1981). Alternatively, those efficiencies reaching a lower maximum of about 60% for absorbed energy correspond with the mean net growth efficiency of 62% calculated for a variety of mussels by Griffiths and Griffiths (1987).

These findings also confirm previous observations whereby, irrespective of seasonal changes, growth efficiencies for carbon were consistently less than for nitrogen (Bayne and Widdows, 1978; Hawkins, 1985; Hawkins and Bayne, 1985). In general terms, relatively greater conservation of absorbed nitrogen reflects the incorporation of amino acids within proteins, whereas carbon-rich substrates are mainly oxidized for the energy needed to deposit, renew and replace those proteins and other substrates. For these reasons, elemental turnover of carbon is faster than for nitrogen (Hawkins, 1985), and mussels are seen to catabolize carbohydrates as the 'preferred' respiratory substrate (Gabbott and Bayne, 1973; Hawkins et al., 1985). Such conservation also results in reducing rates of protein loss at higher rates of ingestion (Fig. 5.5b; see also Bayne and Newell, 1983). Given unchanging rates of protein breakdown, this sparing of endogenous protein is enabled by increased proportional recycling whereby up to as much as 74% of the products of protein breakdown are used to build new proteins (Fig. 5.5c) (Hawkins et al., 1989b). Differential recycling of the products of protein breakdown therefore represents a compensatory variable that affects the relative catabolism of metabolic substrates.

Nitrogen has historically been assumed to be the most likely element limiting heterotrophic growth in the sea (Roman, 1983). However, similar to fin-fishes (Bowen, 1987), but in contrast to mammals, mussels do not require more protein either for maintenance or growth (Hawkins and Bayne, 1991). Furthermore,

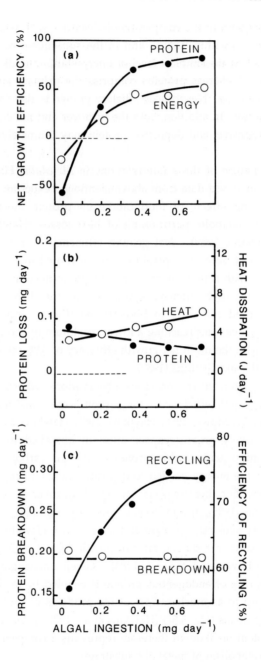

Fig. 5.5. Metabolic responses to different ration levels in *Mytilus edulis* of 10mg dry tissue weight: (a) net growth efficiencies for protein and energy; (b) protein loss and heat dissipation; and (c) protein breakdown and efficiency of protein recycling. Data of Hawkins et al. (1989b).

comparison of maintenance requirements for absorbed carbon and absorbed nitrogen in *M. edulis* from an open-shore population, feeding mainly upon phytoplankton, has shown that maintenance C:N ratios vary seasonally from at least 83:1 during winter to as low as 12:1 during summer (Hawkins and Bayne, 1985), but with an average from all available data of 16:1 (Bayne and Hawkins, 1992). This average compares well with the ratio of 17:1 calculated by Russell-Hunter (1970) for animals generally. Phytoplankton alone have mean C:N ratios of between only 5 and 8:1 (Fenchell and Jørgensen, 1976) which indicates that *Mytilus* is more likely to be growth-limited (Hawkins and Bayne, 1985; Fielding and Davis, 1989) by metabolizable energy than by amino-nitrogen. In addition, this helps to explain why detritus, with C:N ratios that commonly range between 10 and as much as 35:1 (Russell-Hunter, 1970) may represent a nutritionally significant component of the diet ingested by mussels in their natural habitat (see p.175–180).

REGULATION OF PRODUCTION

Production is controlled by the complex interplay between endogenous influences that are inherent to the organism, and exogenous influences deriving from the external environment. The manner in which temporal and spatial differences in production are influenced by exogenous influences such as food availability, temperature, and salinity are reviewed comprehensively elsewhere (Bayne, 1976a; Bayne and Newell, 1983; Griffiths and Griffiths, 1987; Seed and Richardson, 1990; and Chapter 4). The present section describes how endogenous factors determine the maximal capacity for production, which although genetically determined (see p.188– 191), may alter during an individual's lifetime according to size (p.191–195) and physiological status (p.195–198). Any direct ('immediate') response to environmental change is therefore dependent upon endogenous factors, with possible interaction between environmental factors (p.198–200). In addition, compensatory mechanisms operate to help maintain production at least partially independent of most of the variation experienced by mussels in their natural environment (p.200–208). Long-term reductions in production only lead to stress, mortality and associated evolutionary responses (resulting from natural selection) when environmental influences exceed homeostatic capacities under relatively extreme conditions (p.208– 210).

Endogenous Factors

Genotype

Changes in growth and mortality following reciprocal transfers between populations have indicated that genetic constitution may explain significant proportions (up to 28%) of the variances in production and/or survival in *Mytilus* (Dickie et al., 1984; Widdows et al., 1984; Mallet et al., 1987; Blot and Thiriot-Quiévreux, 1989; Mallet and Carver, 1989; Kautsky et al., 1990; Mallet et al., 1990). Alternatively, high heritabilities have indicated an impressive potential for directional genetic selection, resulting in enhanced rates of growth that increased by as much as 35% per generation (Mallet et al., 1986; Strømgren and Nielsen, 1989). Genetic manipulation, particularly by cytogenetic induction of polyploidy, has also been shown to enhance somatic growth in the larvae of *M. edulis* (Yamamoto and Sugawara, 1988; Beaumont and Kelly, 1989) and other shellfish species.

Whilst confirming a genetic basis to production, these studies do not enhance the mechanistic understanding of genotype-dependent physiology. Such mechanisms have been addressed studying individual multilocus genotypes. Similar to other shellfish species, marine mussels exhibit high levels of genetic variability, measured as enzyme polymorphisms, both on a micro- and macrogeographic scale (for reviews see Hawkins, 1988 and Chapter 7). These enzyme polymorphisms are statistically associated with significant proportions of marked differences in growth and reproduction, such as are commonly observed between individuals of the same population. Specifically, and as widely reported among other animal species (for reviews see Mitton and Grant, 1984; Koehn, 1991), independent laboratories have confirmed that greater heterozygosity, computed for individual *M. edulis* as the mean number of electrophoretically-measured enzyme loci that are polymorphic, is associated both with lower energy demands of maintenance metabolism (Diehl et al., 1986; Hawkins et al., 1986a, 1989a) and faster growth rates (Koehn and Gaffney, 1984; Hawkins et al., 1986a; Gentili and Beaumont, 1988; Gaffney, 1990; and Chapter 7 for review).

This suggests a general mechanism whereby genotype affects energy metabolism and physiological performance. Investigating this mechanism, then higher energy requirements in more homozygous *M. edulis* derive, at least in part, from faster rates of whole-body protein turnover (Hawkins et al., 1986a, 1989a). Indeed, findings to date show that genotype-dependent differences between rates of whole-body protein turnover may act as indicators of individual fitness; individuals with greater intensities of whole-body protein turnover and higher associated costs being relatively disadvantaged, when compared in terms of the metabolizable energy remaining available for other processes (Hawkins, 1991).

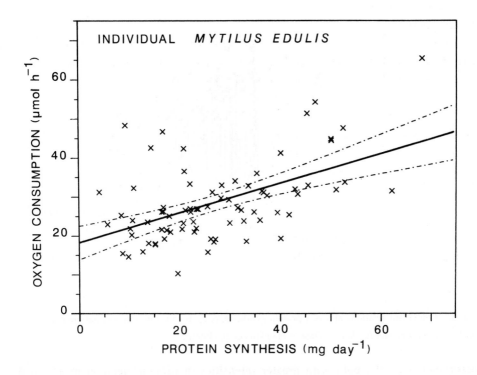

Fig. 5.6. Oxygen consumption in relation to whole-body protein synthesis within individual *Mytilus edulis*. The line ± 95% CI was fitted by least squares. (From Hawkins et al., 1989a).

Protein turnover is integral to maintenance metabolism, effecting the degradation and replacement of cellular proteins, and thereby facilitating the metabolic adjustments that are required for regulation and adaptation (Hawkins, 1991). Per unit of the net deposition of protein that is evidenced as growth, more intense protein turnover involves faster rates both of protein synthesis and protein breakdown. Both the synthesis and breakdown of protein are ATP-dependent; costs of protein turnover account for a major proportion of total metabolic expenditure (Hawkins, 1991). Further, associated costs of integrated maintenance functions are necessarily increased coincident with protein turnover, so that energy expenditure may correlate with measures of whole-body protein synthesis (Hawkins et al., 1989b, and references therein). Such a correlation is illustrated for 87 individual *M. edulis* in Figure 5.6. Energy expenditure, measured as oxygen consumption in these same 87 individuals, was significantly negatively correlated with mean individual heterozygosity (Hawkins et al., 1989a). Thus, less heterozygous individuals tend to have higher energy costs of maintenance, which, as illustrated in Figure 5.7, are statistically

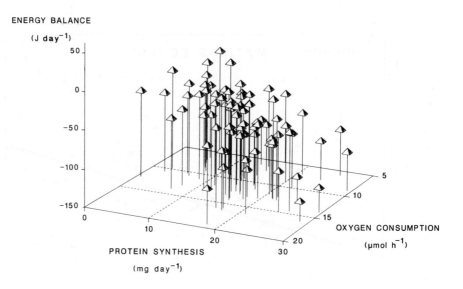

Fig. 5.7. Coincident relations between protein synthesis, oxygen consumption and net energy balance within individual *Mytilus edulis*. Data of Hawkins et al. (1989a).

associated (P<0.001) both with greater intensities of protein metabolism and with reduced energy balance/growth.

These physiological consequences of genotype are consistent with the synthesis of enzymes and other proteins representing the most direct expression of genetic information, and also with the observation (p.183 and Fig. 5.4) that even small differences in maintenance metabolism have major impacts on production. However, compensatory re-utilization of amino acids, which derive from the breakdown involved in protein turnover, ensures relatively constant proportional depositions of the nitrogen absorbed (Hawkins et al., 1987b, 1989a). Such constancy was also observed between fast- and slow-growing size classes from one cohort of *M. edulis* (Hawkins et al., 1986a). Therefore, and despite associated differences in the intensity of whole-body protein turnover, phenotypic effects associated with mean individual heterozygosity derive, most obviously, from different maintenance requirements for energy, rather than for amino nitrogen.

Given these relations, mean individual heterozygosity in *Mytilus* has also been shown to correlate positively with other energy-dependent traits such as feeding rate (Hawkins et al., 1986a), reproductive output (Rodhouse et al., 1986), viability (Diehl and Koehn, 1985), and the seasonal capacity for production (Hawkins et al., 1989a). Collective findings therefore confirm that genotype-dependent differences in protein/maintenance metabolism result in important functional consequences assessed in terms of physiological energetics and performance.

It is important to appreciate, however, that mean heterozygosity typically explains a small percentage (<16%) of the individual variance in growth rate. Further, the relationship is not always statistically evident, and appears to depend on age and reproductive stage, environmental influences, and whether genetic background effects are nonrandom (Beaumont et al., 1983; Koehn and Gaffney, 1984; Gentili and Beaumont, 1988; Gosling, 1989; Koehn, 1991). Also, we have yet to resolve the biochemical mechanism by which heterozygosity varies with maintenance requirements. One possibility, termed the 'associated-overdominance hypothesis', is that studied alleles are acting as indicators of heterozygosity at linked loci affecting metabolism (Zouros et al., 1988). Alternatively, experimental findings from a major study in the marine bivalve, *Mulinia lateralis*, suggest that phenotypic effects of heterozygosity on growth are locus-specific, independent of heterozygosity *per se*, and related to the function of the enzyme either in protein catabolism or glycolysis (Koehn et al., 1988). Koehn's findings complement our own observations, that processes associated with protein turnover incur the energetic consequences which underly heterozygosity relations. It seems, therefore, that heterozygosity relations may depend upon the metabolic significances of scored allele products (Koehn et al., 1988; Koehn, 1991). See Chapter 7 for a more detailed discussion of this topic.

While there is a clear genetic base to production in *Mytilus*, it is important to appreciate that physiological consequences of genotype are modulated according to environmental circumstances, an example of which will be discussed later (p.209–210).

Body size
Unlike the situation in mammals, growth over the realized life span of mussels does not cease upon reaching a maximum size at a particular age. Instead, after a rapid increase following settlement, growth in body size continues within nonlimiting environments, but at reducing rates (for reviews see Bayne and Newell, 1983; Bayne et al., 1983 and Fig. 5.8). Comparing standard growth curves, Theisen (1973) concluded that the sigmoidal Gompertz equation accurately describes growth in shell length for young *Mytilus*, whereas the von Bertalanffy equation, illustrated in Figure 5.8, is best suited for larger mussels (see also Chapter 4). Ultimately, however, because both of these equations assume a maximum determinable size for any given population, some workers have also used polynomial expressions to describe mussel growth (Davenport et al., 1984; Seed and Richardson, 1990).

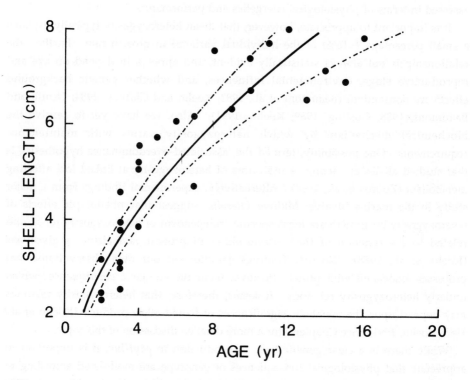

Fig. 5.8. Shell length in relation to age in *Mytilus edulis* from the Lyhner Estuary, southwest England. The line ± 95% CI was fitted by least squares according to the von Bertalanffy model of growth (see text). Unpublished results of Mr. C. M. Worrall (Plymouth Marine Laboratory).

To help understand these deviations from standard growth curves, organ sizes and physiological rates (y) have normally been related to body size (x, the dry weight of soft tissues) by the allometric equation:

$$y = a.x^b,$$

where a and b are fitted parameters. For aerobic energy expenditure in *Mytilus*, whether measured as oxygen uptake or as heat loss, the power coefficient of body mass b approximates 0.75 (Bayne, 1976a; Bayne and Newell, 1983; Widdows, 1987). This indicates that, as for animals in general, metabolism is limited by the surface area available for oxygen diffusion, declining on a weight-specific basis with increasing size. It should be noted, however, that this 'text-book' relationship only applies beyond the velichoncha stage, since mass-specific metabolism in *M. edulis* undergoes a dramatic increase throughout earlier larval development (Sprung and Widdows, 1986). Furthermore, in contrast to aerobic energy expenditure, anoxic heat dissipation varies in direct proportion to body mass (Widdows, 1987).

Nevertheless, under normoxic conditions, b-values in the majority of bivalves approximate 0.75 for metabolic rate and which is significantly higher than values averaging 0.44 ± 0.12 (± 2SE, n=10) for ingestion rate, determined over size-ranges that include large adults (Bayne and Newell, 1983). Absorption efficiencies generally remain independent of mussel size (Griffiths and Griffiths, 1987). Thus, and despite higher pseudofaeces thresholds in larger *M. edulis* (Widdows et al., 1979), reducing weight-specific energy intake is a primary physiological constraint on lifetime production.

Considering the components of feeding that limit intake, weight exponents for gill area (0.72 ± 0.08), palp area (0.68 ± 0.19) and gut content (0.68 ± 0.16) are similar to that of 0.75 for energy expenditure, whereas values both for ingestion rate (0.35 ± 0.12) and gut passage time (0.34 ± 0.12) are significantly reduced (Hawkins et al., 1990 and references therein). Thus, lifetime production is not constrained by morphological features of ingestive and digestive capacity which, given nonlimiting food availability, impose the critical limits to maximum rate of growth (Bayne and Hawkins, 1990). Instead, these findings collectively identify limitations to production as being linked with markedly reduced rate-functions with which mussels filter particles from the water column and pass them through their digestive system.

Winter (1973) postulated that such reduced feeding activity may stem from slower cilial activity in larger individuals. Whatever the exact cause, metabolic rates decline less rapidly than ingestion rates, so that per unit ingestion, larger mussels need to off-set greater metabolic expenditure, and grow fastest at higher rations than do smaller mussels (Thompson and Bayne, 1974; Stuart, 1982). An associated consequence is that net growth efficiency [(net balance/absorption) x 100] decreases with increasing mussel size (Fig. 5.9). This reduction has only previously been documented for energy (for reviews see Bayne and Newell, 1983; and Griffiths and Griffiths, 1987). More recently, it has been shown that weight exponents (± 2SE) of maintenance requirements computed separately for energy (0.69 ± 0.13; Bayne and Hawkins, 1992) and amino-nitrogen (0.77 ± 0.19; Hawkins and Bayne, 1991) are statistically similar in *M. edulis*, which suggests that net conversions of energy and protein are equally dependent on mussel size.

In addition to the influence of diminishing food intake upon production, there is evidence in larger mussels of decreased metabolic efficiency *per se*. This may be assessed as the gross metabolic efficiency with which a measured increment in absorption between two ration levels is used to effect increased net balance (Hawkins and Bayne, 1985). On this basis, data in table 2 of Thompson and Bayne (1974), which describes net energy balance at two ration levels, each below the requirements for energy equilibrium, indicate that gross metabolic efficiencies for maintenance

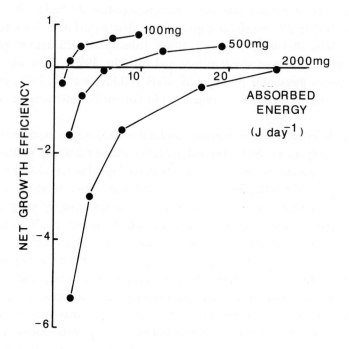

Fig. 5.9. Net growth efficiency (net energy balance/net energy absorbed) in relation to absorbed energy and body size (mg total dry tissue) in *Mytilus edulis*. (After Thompson and Bayne, 1974).

decreased with increasing animal size from 55 to 15% in *M. edulis* of 0.25 to 2.00g dry flesh weight. Gross metabolic efficiencies therefore decline with growth; helping to explain why, as shown in papers reviewed by Griffiths and Griffiths (1987), peak growth efficiency is generally a decreasing function of mussel size (Fig. 5.9).

Finally, and in addition to slowing ingestion and reducing metabolic efficiency, somatic growth is limited by increasing proportional allocations of total production to the synthesis of gametes that are subsequently expelled during periods of spawning (for reviews see Bayne and Newell, 1983; Bayne et al., 1983; Griffiths and Griffiths, 1987). Juvenile mussels utilize virtually all metabolizable energy available above the requirements of maintenance for somatic growth. However, once sexual maturity is reached, at sizes that may vary substantially both between populations and species, then reproductive effort (reproductive production/total production) is a rapidly increasing function of growth (Bayne et al., 1983; Griffiths and Griffiths, 1987 and Chapter 4). Figure 5.10 illustrates steep rises in reproductive effort with greater body size in *M. edulis*, indicating that the largest individuals may expel more than 90% of annual production upon spawning, as seen in many other studies among

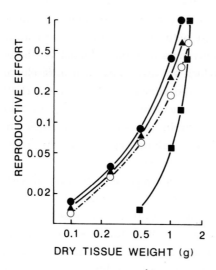

Fig. 5.10. Reproductive effort (production of gametes/total production) related to dry tissue weight in *Mytilus edulis* from four different populations in Britain. (After Bayne et al., 1983).

mussels generally (Bayne and Newell, 1983; Griffiths and Griffiths, 1987). Clearly, increasing proportional allocation of available energy to reproduction represents the main 'brake' on growth.

There are, nevertheless, large differences between populations (Fig. 5.10) and species in the proportional allocations of available energy to somatic growth or reproduction; these differences being modulated according to both endogenous and environmental influences (Bayne and Newell, 1983; Bayne et al., 1983). This reflects an adaptive potential for compensatory adjustments in the relative deposition of different products (p.200–208), and helps to explain why standard growth curves may be inappropriate for *Mytilus* and other bivalves.

Physiological status

Until the last decade, it had generally been assumed that physiological variability was the direct result of environmental changes in temperature, available food and other factors; the individual maximizing nutrient gain under all circumstances, when endogenous factors affect only the allocation of nutrients, rather than efficiencies of utilization. This view is still held by Jørgensen (1991). However, physiological rates may vary in the same individual fully-acclimated under identical environmental conditions. Short-term rhythmicity, for example, is widespread among bivalves (for review see Bayne and Newell, 1983). In *Mytilus*, cycles of digestion, absorption and excretion undergo seasonally variable periodicities of less than eight hours, that bear no consistent relationship with exogenous influences, and which reflect co-ordination in the component activities of nutrient aquisition and utilization (Hawkins et al., 1983, and references therein). Ultimately, these cycles stem from sequential cellular changes which are associated with intracellular digestion within diverticulae that ramify from the stomach. Endogenous rhythms with similar periodicities that correlate with coincident rates of growth are also evidenced by the deposition of visible bands within the shells of mussels and other shellfish continuously immersed in the laboratory, or in the natural environment (Richardson, 1989, and references therein).

Alternatively, and particularly in temperate latitudes, mussels undergo annual reproductive cycles that are associated with marked seasonal changes in gross biochemical composition and physiological rate processes (Bayne and Newell, 1983; Gabbott, 1983; Newell and Thompson, 1984; Rodhouse et al., 1984; Thompson 1984b; Hawkins et al., 1985; Griffiths and Griffiths, 1987; Mallet and Carver, 1989; see also Chapter 6). Coincidentally, changing relationships recorded between physiological rates and environmental factors that include the concentration of suspended food particles indicate concurrent adjustments in physiological status. For example, despite similar availabilities of the same unicellular alga, rates of ingestion by *M. edulis*, acclimated in the laboratory during winter, were less than half those measured under identical conditions during summer (Hawkins and Bayne, 1984). Hummel (1985) has described similar seasonality in physiological response for the filter-feeding bivalve, *Macoma balthica*; ingestion of natural seston during the spring remained low at particle concentrations which were subsequently associated with much faster rates of feeding.

These observations suggest that net energy balance in mussels and other shellfish is not maintained at a continuous steady maximum. Instead, they support the hypothesis which predicts that feeding behaviour may only be optimized in 'some time-averaged sense' (Doyle, 1979). Simply increasing available food from low levels need not lead to faster growth, at least during certain times of the year. Unless accompanied by an enhanced rate of digestion, increased ingestion would actually reduce net energy gain (Bayne et al., 1988). On this basis, Hawkins et al. (1985) suggested that endogenous factors may inhibit short-term adjustments in response to changing food environment during months when, on average, the high ash content of food commonly limits growth. This presumably helps to minimize metabolically wasteful costs incurred by adjustments in feeding and/or the activity of digestive enzymes. Certainly, the induction of digestive enzymes in *Choromytilus meridionalis* is temperature dependent (Seiderer and Newell, 1979; and see p.206–207).

Table 5.1. Net protein balances, gross metabolic efficiencies (G_2) and absorbed maintenance requirements (MR) for carbon and the protein equivalents of amino-nitrogen in 45 to 57mm shell-length *Mytilus edulis* acclimated to seasonally-standardized laboratory diets in March, June and October 1981. Also presented are the relative contributions of protein (W_P) to the total mass of catabolic substrates, calculated for RQ values of 0.95/0.85 (cf. Gnaiger, 1983). Data are synthesized from Hawkins (1985), Hawkins and Bayne (1985) and Hawkins et al. (1985).

	Month		
	March	June	October
Net protein balance (mg day^{-1}g^{-1} dry tissue protein)	−0.3	+0.35	+0.37
Whole body protein turnover (mg protein day^{-1}g^{-1} dry tissue protein)	0.29	0.05	0.01
G2 : carbon	+4	+83	+89
G2 : protein (%)[a]	+72	+91	+87
MR : carbon	213	28	35
MR : protein (mg day^{-1}g^{-1}dry tissue protein)	2.9	2.0	0.9
W_P	0.33/0.44	0.19/0.26	0.10/0.14

[a] Note a distinction here from the 'gross efficiency' of Thompson and Bayne (1974) and others.

Endogenous regulation of the seasonal balance between acquisition and utilization of nutrients was confirmed in the above-mentioned experiments of Hawkins and Bayne (1984) by changing gross efficiencies with which *M. edulis* utilized absorbed energy to increase net energy balance (Table 5.1) (for overview see Hawkins and Bayne, 1991). Indeed, irrespective of standardized experimental food availability, gross efficiencies for carbon/energy varied from 89% in October (autumn) to as little as 4% in March (winter). A small proportion of this variability may have reflected changing costs associated with the seasonal deposition of different biochemical substrates. It is also possible that gross efficiencies (computed on the basis of absorbed energy alone) were lower during winter due to reduced subsidization of metabolic requirements from prestored reserves (Hawkins et al., 1985). What is certain, however, is that rates of whole-body protein turnover were 30 times faster per unit tissue protein in March than in October (Hawkins, 1985). Therefore, energy costs associated with more intense protein metabolism at least partially accounted for the coincident reduction in gross efficiency, together with increased maintenance requirements for carbon that were about six times higher in March than in October (Table 5.1) (Hawkins and Bayne, 1985).

Recycling of the products of protein breakdown directly to protein synthesis meant that the consequences of variable protein turnover were far greater when expressed in terms of conversion efficiencies and maintenance requirements for energy than for the protein equivalents of amino-nitrogen (Table 5.1) (Hawkins and Bayne, 1991). Presumably, faster protein turnover during March, when protein comprised at least 33% of all catabolized substrates (Table 5.1), was incurred as a regulated response to the need for enhanced metabolism of stored or structural proteins, effecting mobilization while enabling the preferential catabolism of non-essential amino acids (Hawkins and Bayne, 1991). This infers that the relative utilization of absorbed nutrients is not only influenced by the variable subsidization of metabolic requirements from prestored reserves, but also by increased energy requirements for protein turnover as such subsidization diminishes.

Endogenous regulation of protein turnover according to requirements for the mobilization of amino acids from stored or structural proteins for use as fuels in response to starvation or injury, is consistent with observations that the same hormones affect both protein degradation and energy metabolism in fish and in mammals (Millward, 1989). Active metabolic control is also evidenced by seasonal alterations in enzyme activity, which are sufficient to effect significant differences in glycolytic flux (Churchill, 1987), as well as by seasonal changes in neurobiology (Stefano and Stefano, 1990).

There is, therefore, considerable variability in physiological status, which is effected both by endogenous and by exogenous influences. Short-term rhythms inherent to *Mytilus* and other shellfish mean that much of the considerable variability apparent between samples, and even replicates, may be avoided by undertaking physiological measurements over periods of time that are sufficient to provide integrated averages. Certainly, these rhythms, together with natural variations in food availablity, associated with seasonal cycles of substrate storage and reproduction, rule out the possibility of defining any 'normal' physiological state. For these reasons, and the need to account for genotypic differences (see p.188–191), a range of values, both in absolute and in qualitative terms, is most appropriate for defining nutrient requirements within any given population of mussels (Hawkins and Bayne, 1991).

Responses to Exogenous Influences

Interrelations affecting the direct response to environmental change

The temporal instabilities of numerous environmental factors, of which the more important include food availability, temperature and salinity, induce associated changes in the biochemistry, physiology and production of bivalves generally (see Bayne, 1976a; Newell, 1979; Griffiths and Griffiths, 1987 and Seed and Richardson, 1990 for reviews). Compensatory adjustments may help to maintain production relatively independent of environmental change (p.200–208). However, before such compensation can occur, the direct ('immediate') effects of environmental change may alter physiological steady-state, with associated consequences that are detrimental to performance. It is therefore important to understand the basis both of inter- and intra-individual variability in the direct response to a given environmental change.

As discussed already (p.188), genetically less heterozygous individuals generally have higher weight-specific energy requirements for maintenance and slower rates of production. There are at least three means by which more intense maintenance metabolism may also reduce relative performance during the immediate response to changing conditions. Firstly, an individual with higher maintenance energy expenditure may display greater physiological variability, maintaining net growth throughout a narrow range of food abundance, and suffering negative energy balance under circumstances that deviate less from the optimum (Koehn and Bayne, 1989). Secondly, although not yet confirmed in other organisms, it has been suggested that reduced maintenance requirements in more heterozygous tiger salamanders are associated with higher potential increments between nonactive and maximal expenditures, and thus a greater 'scope for activity' (Mitton et al., 1986). Thirdly, faster intensities of maintenance metabolism, measured directly as whole-body protein turnover, appear to amplify metabolic responses, resulting in greater thermal sensitivity, measured as Q_{10} values for both protein synthesis and oxygen uptake (Hawkins et al., 1987b). As a result, when mussels were exposed to increased temperatures that were high enough to induce stress, individuals with more intense maintenance metabolism, measured at the temperature of prior acclimation, showed the greatest increases in metabolic expenditure. This resulted in their reduced survival relative to individuals with slower maintenance metabolism (Hawkins et al., 1987b). For what may be any or all of the above reasons, individuals with greater mean heterozygosity are better at resisting environmental change (Blot and Thiriot-Quiévreux, 1989; Scott and Koehn, 1990), as also evidenced both by higher developmental stability (Mitton and Koehn, 1985), and correlations linking enzyme polymorphism with ecological heterogeneity (Hawkins, 1988).

Apart from genotype, different responses to environmental change between individual mussels, that occur despite prior acclimation to the same conditions, also depend on body size and physiological condition. Weight exponents for energy expenditure are generally higher than for nutrient acquisition (p.191–195). Therefore, given that environmental changes result in reduced absorption and/or increased energy expenditure, smaller individuals remain in positive energy or nutrient balance over a wider environmental range, thus maintaining net production for longer periods during times of extreme perturbation (Bayne, 1985). For the same reasons, it seems possible that seasonal cycles of growth and reproduction may change with age. Certainly, differential fecundity and temporal variation of spawning is known to stem from genotype-dependent differences in physiological response, as has been described for specific allozyme variants (Hilbish and Zimmerman, 1988; and see p.209–211), and between hybrid populations of *M. edulis/M. galloprovincialis* (Gardner and Skibinski, 1990).

The influence of physiological condition is indicated by seasonally-dependent variations in the acute response to environmental change. In particular, when available in sufficient quantity, carbon-rich reserves are hydrolyzed to sustain metabolic expenditure relatively independent of nutrient availability (Hawkins et al., 1985). This means that, relative to the net balance for energy/carbon, nitrogen balance is more independent of starvation during summer and autumn than in winter (Gabbott and Bayne, 1973; Bayne and Scullard, 1977; Widdows, 1978; Hawkins and Bayne, 1985). Such differences are presumably linked with coincident seasonal variation observed both in thermal sensitivity (Newell and Pye, 1970) and thermal tolerance (Worrall and Widdows, 1984).

The combined role of endogenous factors in modulating individual responses will depend on the nature, amplitude and frequency of environmental change. In addition, however, direct responses and ultimately survival vary according to potentially synergistic interactions between separate environmental variables (Newell, 1979; Bayne and Newell, 1983; Griffiths and Griffiths, 1987). There have been relatively few multifactorial analyses of environmental influences in mussels, but each serves to emphasize both the natural complexity and ecological relevance of associated interrelations (Bayne 1973a, b; Bayne et al., 1973; Bayne and Scullard, 1977; Incze et al., 1980; Hawkins et al., 1987a, His et al., 1989; Clarke and Griffiths, 1990; Pechenick et al., 1990).

Effective compensation for environmental change

Jørgensen (1991) chooses to view the mussel as an autonomous unit, incapable of regulation, so that any temporal variation of 'automatized' physiological processes occurs solely in direct response to environmental factors (see also p.195). Yet a compelling body of evidence confirms that mussels and other animals act as homeostatic systems, responding to the 'direct' effects of environmental change by modulating their biochemistry, physiology and/or morphology in order to compensate, at least in part, for any reduction in performance (for reviews see Bayne, 1976a, Newell and Branch, 1980; Bayne and Newell, 1983; Griffiths and Griffiths, 1987). Following the direct ('immediate') consequences of environmental change, active adjustments over a period of acclimation or compensation, lasting days or even weeks, slow to a new steady-state. If rate of production at this new steady-state is similar to that before the preceding environmental change, then compensation has been complete. Alternatively, if newly-acclimated production remains less than at the previous steady-state, then compensation is incomplete, leading to impaired performance and possibly death (Kinne, 1971; Vernberg and Vernberg, 1972; Prosser, 1973; Calow, 1976, 1989). The present section will describe examples of complete compensation in *Mytilus* to changes in food, temperature and salinity.

Availability and composition of food:
Among sessile suspension feeders, seston availability is generally the single most important exogenous variable acting to regulate production. Both the quantity and composition of available seston may fluctuate greatly within coastal waters (Bayne and Widdows, 1978; Widdows et al., 1979; Incze et al., 1980; Rodhouse et al., 1984; Smaal et al., 1986; Bayne et al., 1987; Prins and Smaal, 1989). Over intervals that may range from tidal to seasonal, quantity commonly varies from <3 to >100 total dry mg L^{-1}, of which between 5 and 80% by mass may be organic, with a C:N ratio that may change from <4 to >26 (Bayne and Hawkins, 1990).

Despite such variability, mussels maintain relatively constant rates of nutrient acquisition by balancing rates of ingestion against elements of digestive physiology to effect constancies both of organic gut content and absorption efficiency (Navarro and Winter, 1982; Bayne et al., 1989; Bayne and Hawkins, 1990, 1992). The so-called 'functional response' to variations in the quantity of available food has been thoroughly studied in *M. edulis* (for reviews see Winter, 1978; Bayne and Newell, 1983; Griffiths and Griffiths, 1987 and Bayne and Hawkins, 1990). Findings collectively indicate that from minimum seston concentrations, rates of ingestion increase with food availability, but stabilize upon reaching a threshold concentration at which pseudofaeces are first produced, coincident with filling of the digestive tract. At higher concentrations rates of water pumping and filtration may decrease (Widdows et al., 1979; but see Prins and Smaal, 1989), and the production of pseudofaeces increases, thereby maintaining ingestion at a constant maximal rate over wide ranges of food availability (p.172–174).

Given decreasing availability of seston below concentrations enabling maximal ingestion, then at least two interrelations help to sustain absorption. Firstly, an inverse relationship may exist between filtration rate and particle retention efficiency (Wilson and Seed, 1974). Secondly, reduced ingestion is normally associated with longer residence times of food in the the digestive diverticulae (Hawkins and Bayne, 1984), as well as in the digestive system as a whole (Bayne et al., 1984, 1987). This increases the potential for assimilation from ingested particles, as confirmed by positive correlations both between absorption efficiency and gut passage time (Bayne et al., 1984, 1987), and between absorption efficiency and ingestion rate (Widdows, 1978; Hawkins and Bayne, 1984).

In addition to the limitation of production by total food availability, the nutritional value of seston is frequently reduced by increasing proportions of inorganic silt, which act to dilute the accompanying organic matter (Widdows et al., 1979; Prins and Smaal, 1989). To some extent, the potential reduction in rate of absorption is offset by mechanisms which result in pre-ingestive selection. At concentrations below the threshold for pseudofaeces production, Newell et al. (1989) have described how *M. edulis* preferentially retains algal cells, relative to inorganic particles of equivalent size, on the gills. Nevertheless, due primarily to metabolic faecal losses (see p.177–180), net absorption efficiency, measured within true egested faeces, declines exponentially to negative values when the organic content of available seston falls below 10 to 25% of total dry mass (Fig. 5.2) (Hawkins et al., 1986b; Bayne et al., 1987; Navarro et al., 1991). At higher concentrations, of the seston retained on the gills, mussels may preferentially ingest organically-rich particles, rejecting less desirable food items in pseudofaeces (Kiørboe and Møhlenberg, 1981; Bayne et al., 1989 and p.172–174).

Mechanisms resulting in selective retention on the gill are poorly understood, but are likely to depend directly upon particle physics and/or chemistry (p.172–174). Similarly, particle characteristics also determine their differential entanglement within mucus strings carried across the gill filaments, before rejection in pseudofaeces. However, such entanglement is ultimately dependent upon chance encounter, so that rates and net efficiencies of rejection are modulated by the relative abundance of mucus. This is only secreted when particle retention on the gills exceeds the capacity for ingestion, and is then produced at rates that increase in positive relation with the concentration of available seston (Jørgensen, 1991). The secretion of mucus therefore represents a regulatory adjustment, which not only prevents clogging of the gills, but also helps to maximize organic ingestion.

Clearly, through their influences both on particle retention and pre-ingestive rejection, separate but interrelated adjustments that occur in direct ('immediate') response to the food environment help to maintain organic ingestion at maximal capacity. These adjustments are clearly beneficial. Yet to maximize absorption over the natural range of seston concentrations, such regulation must ultimately be both of organic ingestion and absorption efficiency.

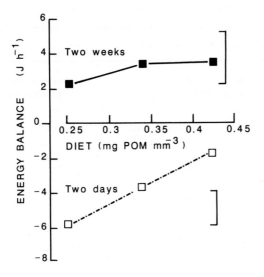

Fig. 5.11. Net energy balances two days and two weeks following transfer of *Mytilus edulis* from the natural environment to an experimental range of diets that varied in quality, measured as organic content per unit volume of particulate material. Bars represent the least-significant difference between the mean energy balances. (After Bayne et al., 1987).

The active regulation of absorption efficiency has been confirmed by experimental manipulations of laboratory diet. Indirect evidence first came from an experiment showing that *M. edulis* was able to express similar net energy balances following laboratory acclimation over two weeks to each of three diets (Bayne et al., 1987) (Fig. 5.11). All three diets were similar in terms of the volume of particles available per unit volume of seawater. This volume concentration was below the threshold for pseudofaeces production, and represented a reduction in total food available within the natural environment from which the mussels had been collected. In addition, each diet differed in the relative proportions of alga and silt, ranging between 10 and 50% organic matter, that corresponded with 0.25 and 0.43mg particulate organic matter mm[-3] (Fig. 5.11). Discussing these findings, Bayne and Hawkins (1990) described how adjustments to the lowest quality diet included elevated rates of ingestion, but unchanging gut passage times, so that total gut content increased, helping to maintain organic gut content independent of changes either in food quality or food quantity (cf. Bayne et al., 1989). In addition, and despite doubled rates of ingestion, efficiency of absorption was maintained, and even enhanced, accounting for up to more than 40%

Table 5.2. Maltose liberation by α-amylase and laminarinase from eluates of styles and digestive glands of *M. edulis* collected directly from the natural environment, and following acclimation to two ration levels (see text). Data (mean ± SE, n = 5) is expressed as mg maltose liberated per mg total protein in the eluates at 15.0 ± 1.0°C.

Condition	Specific α-amylase activity (mg maltose mg protein^{-1} h^{-1})		Specific laminarinase activity (mg maltose mg protein^{-1} h^{-1})	
	Style	Digestive gland	Style	Digestive gland
Natural environment	4.45 ±0.83	4.46 ±0.56	2.90 ±0.57	1.36 ±0.19
High ration	4.69 ±0.83	2.46 ±0.56	2.52 ±0.57	1.15 ±0.19
Low ration	1.12 ±0.93	1.43 ±0.62	2.18 ±0.77	0.64 ±0.21

of the compensatory increase in net energy balance (Bayne et al., 1987). Subsequent measurements in *M. edulis* acclimated to different concentrations of seston below the threshold for pseudofaeces production also showed that absorption efficiencies were independent of ration, irrespective of associated differences both in ingestion rate and gut passage time (Bayne et al., 1989; Bayne and Hawkins, 1990). Indirect evidence for the regulation of absorption by compensatory adjustments in the quantity and/or specific activities of digestive enzymes comes from work linking digestive ability with dietary composition (Seiderer and Newell, 1979; Stuart et al., 1982; Lucas and Newell, 1984; Langdon and Newell, 1990). More convincingly, however, Table 5.2 summarizes results of an experiment (Bayne and Hawkins, unpublished data) in which *M. edulis* were collected from the wild and fed the alga *Phaeodactylum tricornutum* for seven days at high and low rations (7.8 x 10^7 and 1.0 x 10^7 cells mussels^{-1} day^{-1}, respectively), but which each represented a reduction in food availability relative to the natural environment. Activities of carbohydrase enzymes were then assayed within the style and digestive gland according to Fielding (1987), and compared with those from mussels collected in the wild. Findings show enzyme activities both for α-amylase and laminarinase that were consistently reduced at the low ration, down to levels as low as 25% of those in mussels from their natural environment (Table 5.2). Bayne and Hawkins (1990) have reported reduced activity of carbohydrase enzymes following acclimation of *M. edulis* over two weeks to an increased dietary ash content.

Apart from physiological and enzymatic compensations to the food quantity and composition, reciprocal transplantations of *M. edulis* between the Wadden Sea and

North Sea have shown that, in response to changing concentrations of suspended matter, individual mussels are able to effect phenotypic adjustment in the morphology of their feeding apparatus (Essink et al., 1989). Theisen (1982) had earlier postulated that morphometric differences observed between these habitats were genetically determined. However, Essink et al. (loc. cit.) have shown that the relative dimensions of gills and/or labial palps in M. *edulis* may alter within three months following transplantation to environments with either increased or decreased seston availability.

To summarize, individual mussels effect a complementary set of interrelated compensatory adjustments which occur over different time scales according to separate levels of biological organization, and help to maintain energy balance independent of wide variations both in the quantity and organic content of available seston. Considering the direct physiological responses, findings suggest that at all but the lowest (environmentally unrealistic) concentrations of food, filtration and retention tend towards maximal physiological capacities, only decreasing upon saturation of the digestive system at increased seston concentrations. Simultaneous regulatory adjustments in rate of pseudofaeces production help both to protect gill function and to maximize organic ingestion. Any change in dietary ash content incurs associated adjustments in rate of gut passage and/or total gut content that help to maintain a constant volume of organic matter within the digestive tract. Compensatory adjustments in the quantity and/or nature of digestive enzymes, which may not be complete for days or even weeks, also help to maintain absorption efficiency independent of gut processing rate. These collective compensations therefore provide the means to help maximize net energy balance within any given food environment. Both environmental and endogenous influences may, however, constrain each of the above processes at rates or efficiencies below their maximal physical potentials. In addition, those potentials may themselves be modulated by morphological compensations occuring over periods of months. Somatic compensations of this nature may not only help to maintain the flexibility of potential response to further challenges, but are also less 'expensive' than physiological and biochemical compensations involving integrations across many interdependent processes (cf. Bateson, 1963).

Further to acclimatory adjustments regulating the acquisition of nutrients, there is widespread evidence in fish and mammals of compensatory metabolic changes in response to differences in the molecular composition of absorbed organics (Tacon and Cowey, 1985; Reeds, 1988). In mussels, Bayne, Hawkins, Uriarte and Farias (unpublished data) compared physiological responses following acclimation of M. *edulis* over 17 days to rations that were available at similar concentrations (4–7 x 10^{-3} cells mL^{-1}), but which differed significantly in protein content. Figure 5.12 illustrates how protein loss increased with dietary protein content. Although associated with declining metabolic efficiencies with which absorbed amino nitrogen is retained as protein, such enhanced excretion is regulatory, representing oxidation of the increasing surplus of exogenous amino acids in order to avoid the deleterious effects of amino acid imbalance (cf. Harper et al., 1970). Lower efficiencies of protein deposition can also be expected following reduction in total energy intake, consequent upon the 'protein-sparing' effect of dietary energy (Hawkins and Bayne, 1991).

Temperature:

Temperature has a direct effect upon metabolic rate in poikilotherms, so that compensatory acclimation has long been recognized as important for maintaining production within thermally unstable environments (Weiser, 1973; Newell, 1979). *M. edulis* achieves at least partial thermal independence by effecting complete or near complete acclimation of routine oxygen consumption, filtration rate, net energy balance and shell growth over two to three weeks following transfer to constant temperatures between 5 and 20°C (Widdows and Bayne, 1971; Bayne et al., 1973, 1977b; Widdows, 1973; Thompson and Newell, 1985; Hawkins et al., 1987b; Nielson, 1988).

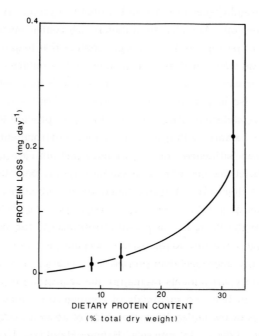

Fig. 5.12. Protein loss (± 95% CI) in relation to dietary protein content in *Mytilus edulis* of 1g dry tissue weight. The curve was fitted by least squares, as described in Hawkins and Bayne (1991).

Given short-term fluctuations, such as occur over a tidal cycle, M. *edulis* may also adapt by reducing the amplitude of associated physiological responses, thereby enhancing thermal independance within the range of fluctuating temperature; and enabling the mussel to function over short periods at higher temperatures than were possible under constant laboratory conditions (Widdows, 1976). Huppert and Laudien (1980) described how changing temperatures also induce elevated heat resistance in the gill epithelium. More recently, compensatory reductions in aerial respiration, following exposure to elevated temperature, have been shown to help improve net energy balance in a high intertidal population of the ribbed mussel, *Geukensia demissa* (Wilbur and Hilbish, 1989).

The biochemical basis of compensation for temperature in mussels is known to involve acclimation of the rate with which vesicles containing the monoamine 5-hydroxytryptamine are transported along axons to the nerve terminal (Stefano and Aiello, 1990). In addition, several studies have described seasonal and/or temperature-related regulation of enzyme reaction rates. This regulation may be quantitative, resulting in changed concentration of the same enzyme, and thus in altered specific activity (Livingstone, 1981; Churchill, 1987; see also p.251–253 Chapter 6). Alternatively, this regulation may be qualitative, leading to induction of thermo-isozymes, that differ in structure, but which have the same function (Livingstone and Bayne, 1974; Seiderer and Newell, 1979; Ramos-Martinez and Torres, 1985).

Salinity:
Changes in external salinity may disrupt the steady-state balance between influx and efflux of cellular water and salts. To help reduce the rate of associated changes in cell volume, mussels and other bivalves respond immediately by closing the shell, and then effect iso-osmotic intracellular regulation by adjusting the intracellular concentration of ions, amino acids and other small molecules to maintain the volume of cells relatively constant (for review see Lange, 1972). Thus, following a sudden change in salinity, physiological rates of feeding and metabolism are initially depressed, but gradually recover as the mussel regains normal function coincident with regulatory changes in intracellular concentrations of osmotic effectors over the next two or more days (for reviews see Bayne, 1976a, 1985; also see p.251–253 Chapter 6).

Although able to compensate for fluctuations in temperature, M. *edulis* appears relatively limited in its ability to effect compensatory physiological adjustments that act to regulate net energy balance in response to short-term fluctuations in salinity (Widdows, 1985). To some extent, this inability is offset by slow osmotic equilibration, so that the amplitude of extracellular osmotic changes is lower than the amplitude of

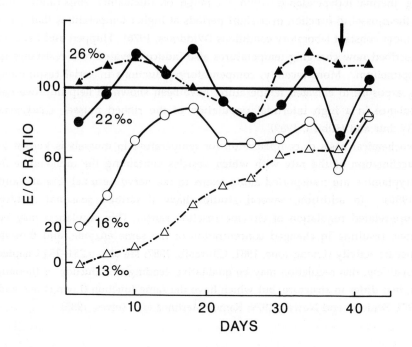

Fig. 5.13. Mean specific rate of shell growth expressed as a percentage of that in the controls (E/C Ratio) over time following transfer of *Mytilus edulis* to several experimental salinities. The vertical arrow indicates when mussels were transferred back to the control salinity of 32°/oo. (After Almada-Villela, 1984).

fluctuations within the external medium (Livingstone et al., 1979). Certainly, there is slow acclimation in response to single acute changes. For example, when compared 14 days after transfer from 30°/oo to a range of reduced salinities, *Choromytilus chorus* had only regained the previous (acclimated) net energy balance at salinities higher than 24°/oo (Navarro, 1988). Alternatively, when monitored until all compensation was complete, which required up to 44 days, *M. edulis* transferred from a control salinity of 32°/oo was seen to regulate shell growth independent of a wider range of salinities—as low as 13°/oo—when the period required for complete compensation varied in positive relation with the size of initial salinity change (Fig. 5.13) (Almada-Villela, 1984).

Stress, mortality and natural selection

Homeostatic capacity in *Mytilus* may vary seasonally, as seen to date by the relative magnitude of responses to reduced salinity (Livingstone et al., 1979), and by different periods required for shell growth to attain a new steady-state following increased temperatures (Nielson, 1988). Such variation, contrasted with environmental fluctuations, often results in transient or longer term periods when exogenous variables exceed the homeostatic capacity of an animal. Any acclimation is then incomplete, and the associated reduction in net energy gain signifies a 'stressed' condition (Bayne, 1985; Widdows and Donkin, 1989). The coincident decline in fitness may be evidenced as reduced tolerance of further environmental change, smaller maximum body size, and changes in age-specific gametogenesis. Reproductive output varies from year to year, as well as between populations of *M. edulis*, suggesting that gamete production changes according to environmental suitability (Thompson, 1979, 1984a; Bayne and Worrall, 1980; Bayne et al., 1983). Bayne (1985), however, pointed out that lifetime egg production in mussels would be maximized under suboptimal conditions if, associated with a reduction in adult body size, reproductive effort is increased. This helps to explain observations of enhanced reproductive effort following experimental starvation (Bayne, 1976b) and increased aerial exposure (Griffiths, 1981; Rodhouse et al., 1984).

Evidence therefore indicates that reproductive output is maintained at the relative expense of somatic tissues under moderately unfavourable conditions. Only when the stress becomes greater, is reproductive effort reduced, helping to preserve structural integrity (Fig. 5.14) (Bayne et al., 1983; Bayne, 1986). These adjustments are clearly compensatory, and confer added flexibility within a variety of phenotypic responses to environmental change, by enabling the individual to maximize reproductive output under favourable conditions, and to prolong survival during periods of extreme stress.

Ultimately, however, extended stress exceeding the mussels' homeostatic capacity will result in death. Such mortality is differential, with variable periods of survival according to body size (Wallis, 1975; Bayne et al., 1977b; Worrall and Widdows, 1984) and condition (Hawkins et al., 1987b). In addition, it is especially important to recognize the influence of genotype (Hvilsom, 1983; Hawkins et al., 1989c; Beaumont et al., 1988, 1989, 1990), that confers the potential for evolutionary responses resulting from natural selection over one or more generations. Unequivocal confirmation of natural selection requires evidence linking genotype via components of fitness to the environment. It is therefore necessary to identify appropriate functional differences in the biochemistry and physiology of relevant allozyme variants. Such evidence is extremely limited among animals generally. In mussels, this has nevertheless been achieved for leucine aminopeptidase (*Lap*), which degrades cellular proteins to amino acids (for review see Koehn and Hilbish, 1987 and Chapter 7). The catalytic efficiencies

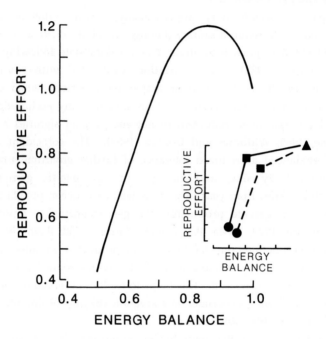

Fig. 5.14. Reproductive effort in relation to energy balance, each scaled to the predicted maxima from a model described by Bayne (1986). The inset, from the same model and data of Bayne (1986), illustrates relative values measured in two age classes (dashed line: 7yr; solid line: 8yr) from three populations of *Mytilus edulis*, and indicates that reproductive effort may be increasingly suppressed when growth is depressed.

of allozymes at this locus differ significantly; *M. edulis* genotypes with *Lap^{94}* have markedly greater specific activity than those without the *Lap^{94}* allele, and this results in faster adjustment and stabilization of the pool of free amino acids used to regulate cell volume in response to increased salinity. However, as part of the acclimation to reduced salinity, genotypes with the *Lap^{94}* allele excrete greater excesses of amino acids, which may derive solely from the breakdown of tissue proteins (Hawkins and Hilbish, 1992).

Sublethal effects of what may be small but frequent changes in salinity have been evidenced as seasonally reduced protein balances among *M. edulis* with the *Lap^{94}* allele, resulting in significantly delayed gametogenesis and reduced output of gametes, compared with individuals which do not possess this allele (Hilbish and Zimmerman, 1988). Moreover, genetic adaptation has been demonstrated by selection against the *Lap^{94}* allele among postlarval *M. edulis* reared in the laboratory at low salinity (Beaumont et al., 1989), as well as by widespread geographical correlations between environmental salinity and *Lap* genotype (Koehn et al., 1976; Theisen, 1978; Gartner-Kepkay et al., 1983; Garthwaite, 1986; McDonald and Siebenaller, 1989).

SYNOPSIS

Production is controlled according to the complex interplay between endogenous processes that are inherent to the organism, and exogenous influences deriving from the external environment.   Exogenous influences have been widely studied in isolation, food availability being recognized as of the greatest general significance to temporal and spatial differences in production.   There remains a need for controlled multifactorial analyses of environmental influences, which may interact both with one another and with endogenous influences to affect growth and ultimately survival.   In general, however, studies to date have shown that mussels demonstrate a remarkable variety of physiological responses, enabling survival over a wide range of environmental conditions.

Endogenous influences determine maximal capacity for production, which although influenced by genotype, may alter during an individual's lifetime according to size, morphological changes, and physiological status.   Within this capacity, intra-individual variation in production is consequent upon endogenous modulation of processes affecting both the acquisition and subsequent relative utilization of nutrients.   Acquisition may be modulated by changes in filtration rate, particle retention, pre-ingestive rejection, gut content and associated gut passage time, the quantity and/or nature of digestive enzymes, and morphology of the feeding apparatus.   Variable utilization is evidenced as changes in conversion efficiency, that may either derive from adjustments in the degree of metabolic subsidization from prestored reserves, the relative partitioning of available resources between maintenance metabolism and production, and/or the comparative deposition of different biochemical substrates.

This collective variability affords the flexibility whereby maintenance and production are shown to be regulated processes.   Mussels and other marine invertebrates respond to the 'direct' effects of environmental change by modulating their biochemistry, physiology and/or morphology in order to compensate, at least in part, for any reduction in performance.   Such compensation may result from a heirarchy of responses occurring over different overlapping time scales. Phenotypic responses range from direct behavioural reactions, through biochemical and/or physiological compensation over days or weeks, to morphological changes over periods of months, to adaptive adjustments of seasonal reproductive effort under conditions of incomplete compensation, and culminating with mortality. Genetic responses confer additional flexibility evidenced within populations as adaptation resulting from natural selection over one or more generations. To understand the consequences of environmental change and the processes affecting physiological evolution, it is therefore essential to resolve both the nature and time course of coincident responses.

Evaluation of both exogenous and endogenous factors as determinants of production is to address complex, subtle and dynamic interactions between the mussel and its environment. Such evaluation may be facilitated by formulating quantitative expectations based on what may previously be understood of the processes involved. Indeed, the validity of modelling acquisition and/or expenditure of energy has recently been confirmed by simulations that reasonably approximate responses observed in natural populations of mussels (Ross and Nisbet, 1990; Willows, 1992) and finfish (for review see Steffens, 1989). Given further development and refinement, the main future challenge will be to integrate simulations of individual growth and reproduction within 'structured population models' (Metz and Diekmann, 1986; Nisbet et al., 1989), with a view to predicting population dynamics, species distribution and ecological impact.

REFERENCES

Almada-Villela, P.C., 1984. The effects of reduced salinity on the shell growth of small *Mytilus edulis*. J. Mar. Biol. Ass. U.K., 64: 171-182.

Anwar, N.A., Richardson, C.A. and Seed, R., 1990. Age determination, growth rate and population structure of the horse mussel, *Modiolus modiolus*. J. Mar. Biol. Ass. U.K., 70: 441-457.

Bateson, G., 1963. The role of somatic change in evolution. Evolution, 17: 529-533.

Bayne, B.L., 1973a. Physiological changes in *Mytilus edulis* L. induced by temperature and nutritive stress. J. Mar. Biol. Ass. U.K., 53: 39-58.

Bayne, B.L., 1973b. The responses of three species of bivalve mollusc to declining oxygen tension at reduced salinity. Comp. Biochem. Physiol., 45A: 793-806.

Bayne, B.L. (Editor), 1976a. Marine mussels: their ecology and physiology. Cambridge University Press, Cambridge, 506pp.

Bayne, B.L., 1976b. Aspects of reproduction in bivalve molluscs. In: M. Wiley (Editor), Estuarine Processes. Vol. 1. Uses, stresses and adaptation to the estuary. Academic Press, New York, pp. 432-448.

Bayne, B.L., 1985. Responses to environmental stress: tolerance, resistance and adaptation. In: J.S. Gray and M.E. Christiansen, (Editors), Proc. 18th Eur. Mar. Biol. Symp., Oslo, Norway, 1983. John Wiley, New York, pp. 331-349.

Bayne, B.L., 1986. Measuring the effects of pollution at the cellular and organism level. In: G. Kullenberg, (Editor), The role of the oceans as a waste disposal option. D. Reidel, Netherlands, pp. 617-634.

Bayne, B.L. and Scullard, C., 1977. Rates of nitrogen excretion by species of *Mytilus* (Bivalvia: Mollusca). J. Mar. Biol. Ass. U.K., 57: 355-369.

Bayne, B.L. and Widdows, J., 1978. The physiological ecology of two populations of *Mytilus edulis* L. Oecologia, 37: 137-162.

Bayne, B.L. and Worrall, C.M., 1980. Growth and production of mussels *Mytilus edulis* from two populations. Mar. Ecol. Prog. Ser., 3: 317-328.

Bayne, B.L. and Newell, R.C., 1983. Physiological energetics of marine molluscs. In: K.M. Wilbur and A.S. Saleuddin (Editors), The Mollusca, Vol. 4. Academic Press, New York, pp. 407-515.

Bayne, B.L. and Hawkins, A.J.S., 1990. Filter-feeding in bivalve molluscs: controls on energy balance. In: J. Mellinger (Editor), Animal nutrition and transport processes, Vol. 1. Nutrition in wild and domestic animals. Karger, Basel, pp. 70-83.

Bayne, B.L. and Hawkins, A.J.S., 1992. Ecological and physiological aspects of herbivory in benthic suspension-feeding molluscs. In: D.M. John, S.J. Hawkins and J.H. Price (Editors), Plant-animal interaction in the marine benthos. Syst. Assoc. Symp. Ser., Oxford University Press, Oxford, in press.

Bayne, B.L., Thompson, R.J. and Widdows, J., 1973. Some effects of temperature and food on the rate of oxygen consumption by Mytilus edulis L. In: W. Wieser (Editor), Effects of temperature on ectothermic organisms. Springer-Verlag, Berlin, pp. 189-193.

Bayne, B.L., Widdows, J. and Newell, R.I.E., 1977a. Physiological measurements on estuarine bivalve molluscs in the field. In: B.F. Keegan, P. O Céidigh and P. S. Boaden (Editors), Proc. 11th Eur. Mar. Biol. Symp., Galway, Ireland, 1976. Pergamon Press, London, pp. 57-68.

Bayne, B.L., Widdows, J. and Worrall, C.M., 1977b. Some temperature relationships in the physiology of two ecologically distinct bivalve populations. In: F.J. Vernberg, A. Calabrase, F.P. Thurberg and W. Vernberg (Editors), Physiological responses of marine biota to pollutants. Academic Press, New York, pp. 379-400.

Bayne, B.L., Salkeld, P.N. and Worrall, C.M., 1983. Reproductive effort and value in different populations of the marine mussel, Mytilus edulis L. Oecologia, 59: 18-26.

Bayne, B.L., Klumpp, D.W. and Clarke, K.R., 1984. Aspects of feeding, including estimates of gut residence time, in three mytilid species (Bivalvia, Mollusca) at two contrasting sites in the Cape Peninsula, South Africa. Oecologia, 64: 26-33.

Bayne, B.L., Hawkins, A.J.S. and Navarro, E., 1987. Feeding and digestion by the common mussel Mytilus edulis L. (Bivalvia: Mollusca) in mixtures of silt and algal cells at low concentrations. J. Exp. Mar. Biol. Ecol., 111: 1-22.

Bayne, B.L., Hawkins, A.J.S. and Navarro, E., 1988. Feeding and digestion in suspension-feeding bivalve molluscs: the relevance of physiological compensations. Am. Zool., 28: 147-159.

Bayne, B.L., Hawkins, A.J.S., Navarro, E. and Iglesias, J.I.P., 1989. The effects of seston concentration on feeding, digestion and growth in the mussel Mytilus edulis. Mar. Ecol. Prog. Ser., 55: 47-54.

Beaumont, A.R. and Kelly, K.S. 1989. Production and growth of triploid Mytilus edulis larvae. J. Exp. Mar. Biol. Ecol., 132: 69-84.

Beaumont, A.R., Beveridge, C.M. and Budd, M.D., 1983. Selection and heterozygosity within single families of the mussel Mytilus edulis (L.). Mar. Biol. Lett., 4: 151-151.

Beaumont, A.R., Beveridge, C.M., Barnet, E.A., Budd, M.D. and Smyth-Chamosa, M., 1988. Genetic studies of laboratory reared Mytilus edulis. I. Genotype specific selection in relation to salinity. Heredity, 61: 389-400.

Beaumont, A.R., Beveridge, C.M., Barnet, E.A. and Budd, M.D., 1989. Genetic studies of laboratory reared Mytilus edulis. II. Selection at the leucine amino peptidase (Lap) locus. Heredity, 62: 169-176.

Beaumont, A.R., Beveridge, C.M., Barnet, E.A. and Budd, M.D., 1990. Genetic studies of laboratory reared Mytilus edulis. III. Scored loci act as markers for genotype-specific mortalities which are unrelated to temperature. Mar. Biol., 106: 227-233.

Berg, J.A. and Newell, R.I.E., 1986. Temporal and spatial variations in the composition of seston available to the suspension feeder Crassostrea virginica. Est. Coast. Shelf Sci., 23: 375-386.

Bernard, F.R. and Noakes, D.J., 1990. Pumping rates, water pressures, and oxygen use in eight species of marine bivalve molluscs from British Columbia. Can. J. Fish. Aquat. Sci., 47: 1302-1306.

Blot, M. and Thiriot-Quiévreux, C., 1989. Multiple locus fitness in a transfer of adult Mytilus desolationis (Mollusca: Bivalvia). In: J. Ryland and P.A.Tyler (Editors), Proc. 23rd Eur. Mar. Biol. Symp., Swansea, U.K., 1988. Olsen and Olsen, Fredensborg, Denmark, pp. 259-264.

Bowen, S.H., 1987. Dietary protein requirements of fishes—a reassessment. Can. J. Fish. Aquat. Sci., 44: 1995-2001.

Brooks, J.M., Kennicut, I.I.M.C., Fisher, C.R., Macko, S.A., Cole, K., Childress, J.J., Bidigare, R.R. and Vetter, R.D., 1987. Deep-sea hydrocarbon seep communities: evidence for energy and nutritional carbon sources. Science, 238: 1138-1142.

Calow, P., 1976. Biological machines: a cybernetic approach to life. Edward Arnold Ltd., London, 134pp.

Calow, P., 1985. Adaptive aspects of energy allocation. In: P. Tytler and P. Calow (Editors), Fish energetics: new perspectives. Croom Helm, London, pp. 13-31.

Calow, P., 1989. Are individual production rates optimized by natural selection? In: W. Wieser and E. Gnaiger (Editors), Energy transformations in cells and organisms. Georg Thieme Verlag, New York, pp. 264-269.

Calow, P. and Townsend, C.R., 1981. Resource utilization in growth. In: C.R. Townsend and P. Calow (Editors), Physiological ecology: an evolutionary approach to resource use. Blackwell Scientific Publications, London, pp. 220-244.

Childress, J.J., Fisher, C.R., Brooks, J.M., Kennicut, M.C., Bidigare, R. and Anderson, A.E., 1986. A methanotrophic marine molluscan (Bivalvia, Mytilidae) symbiosis: mussels fueled by gas. Science, 233: 1306-1308.

Churchill, H.M., 1987. Enzymic aspects of the seasonal regulation of metabolism in the common mussel *Mytilus edulis* L. Ph.D. Thesis, University of Exeter, U.K.

Clarke, B.C. and Griffiths, C.L., 1990. Ecological energetics of mussels *Choromytilus meridionalis* under simulated intertidal rock pool conditions. J. Exp. Mar. Biol. Ecol., 137: 63-77.

Clemmesen, B. and Jørgensen, C.B., 1987. Energetic costs and efficiencies of ciliary filter feeding. Mar. Biol., 94: 445-449.

Close, W.H., Berschauer, F. and Heavens, R.P., 1983. The influence of protein: energy value of the ration and level of feed intake on the energy and nitrogen metabolism of the growing pig. Br. J. Nutr., 49: 255-269.

Colwell, S.J. and Manahan, D.T., 1988. A comparison of the transport rate by marine invertebrate larvae of monsaccharides and alanine from seawater. Am. Zool., 28: 131A.

Conover, R.J., 1969. Assimilation of organic matter by zooplankton. Limnol. Oceanogr., 11: 338-345.

Crosby, M.P., Newell, R.I.E. and Langdon, C.J., 1990. Bacterial mediation in the utilization of carbon and nitrogen from detrital complexes by the American oyster, *Crossostrea virginica*. Limnol. Oceanogr., 35: 625-635.

Davenport, J., Davenport, J. and Davies, G., 1984. A preliminary assessment of growth rates of mussels from the Falkland Islands (*Mytilus chilensis* Hupé and *Aulacomya ater* (Molina)). J. Cons. Int. Explor. Mer, 41: 154-158.

Dickie, L.M., Boudreau, P.R. and Freeman, K.R., 1984. Influences of stock and site on growth and mortality in blue mussel *Mytilus edulis*. Can. J. Fish. Aquat. Sci., 41: 134-140.

Diehl, W.J. and Koehn, R.K., 1985. Multiple-locus heterozygosity, mortality and growth in a cohort of *Mytilus edulis*. Mar. Biol., 88: 265-271.

Diehl, W.J., Gaffney, P.M. and Koehn, R.K., 1986. Physiological and genetic aspects of growth in the mussel *Mytilus edulis*. 1. Oxygen consumption, growth and weight loss. Physiol. Zool., 59: 201-211.

Doyle, R.W., 1979. Ingestion rate of a selective deposit feeder in a complex mixture of particles: testing the energy-optimisation hypothesis. Limnol. Oceanogr., 24: 867-874.

Essink, K., Tydeman, P., de, Koning, F. and Kleef, H.L., 1989. On the adaptation of the mussel *Mytilus edulis* L. to different environmental suspended matter concentrations. In: R.Z. Klekowski, E. Styczynska-Jurewicz and L. Falkowski (Editors), Proc. 21st Eur. Mar. Biol. Symp., Gdansk, Poland, 1988. Ossolineum, Gdansk, pp. 41-51.

Famme, P. and Kofoed, L.H., 1982. Rates of carbon release and oxygen uptake by the mussel, *Mytilus edulis* L., in response to starvation and oxygen. Mar. Biol. Lett., 3: 241-256.

Fenchel, T. and Jørgensen, B.B., 1976. Detritus food chains of aquatic ecosystems and the role of bacteria. Adv. Microb. Ecol., 1: 1-49.

Fiala-Medioni, A., Alayse, A.M. and Cahet, G., 1986. Evidence of in situ uptake and incorporation of bicarbonate and amino acids by hydrothermal vent mussels. J. Exp. Mar. Biol. Ecol., 96: 191-198.

Fiala-Medioni, A. and Felbeck, H., 1990. Autotrophic processes in invertebrate nutrition: bacterial symbosis in bivalve molluscs. Comp. Physiol., 5: 49-69.

Fielding, P.J. and Davis, C.L., 1989. Carbon and nitrogen resources available to kelp bed filter feeders in an upwelling environment. Mar. Ecol. Prog. Ser., 55: 181-189.

Fielding, P.J., Harris, J.M., Lucas, M.I. and Cook, P.A., 1986. Implications for the assessment of crystalline style activity in bivalves when using the Bernfeld and Nelson-Somogyi assays for reducing sugars. J. Exp. Mar. Biol. Ecol., 101: 269-284.

Fisher, C.R., Childress, J.J., Oremland, R.S. and Bidigare, R.R., 1987. The importance of methane and thiosulfate in the metabolism of the bacterial symbionts of two deep-sea mussels. Mar. Biol., 96: 59-71.

Foster-Smith, R.L., 1975. The effect of concentration of suspension and inert material on the assimilation of algae by three bivalves. J. Mar. Biol. Ass. U.K., 55: 411-418.

Gabbott, P.A., 1983. Development and seasonal metabolic activities in marine molluscs. In: P.W. Hochachka (Editor), The Mollusca, Vol. 2. Academic Press, New York, pp. 165-217.

Gabbott, P.A. and Bayne, B.L., 1973. Biochemical effects of temperature and nutritive stress on Mytilus edulis L. J. Mar. Biol. Ass. U.K., 53: 269-286.

Gaffney, P.M., 1990. Enzyme heterozygosity, growth rate, and variability in Mytilus edulis: another look. Evolution, 44: 204-210.

Gaffney, P.M. and Diehl, W.J., 1986. Growth, condition and specific dynamic action in the mussel Mytilus edulis recovering from starvation. Mar. Biol., 93: 401-409.

Gardner, J.P.A. and Skibinski, D.O.F., 1990. Genotype-dependent fecundity and temporal variation of spawning in hybrid (Mytilus) populations. Mar. Biol., 105: 153-162.

Garthwaite, R., 1986. The genetics of California populations of Geukensia demissa (Dillwyn) (Mollusca): further evidence on the selective importance of leucine aminopeptidase variation in salinity acclimation. Biol. J. Linn. Soc., 28: 343-358.

Gartner-Kepkay, K.E., Zouros, E., Dickie, L.M. and Freeman, K.R., 1983. Genetic differentiation in the face of gene flow: a study of mussel populations from a single Nova Scotia embayment. Can. J. Fish. Aquat. Sci., 40: 443-451.

Gentili, M.R. and Beaumont, A.R., 1988. Environmental stress, heterozygosity and growth rate in Mytilus edulis. J. Exp. Mar. Biol. Ecol., 120: 145-153.

Gnaiger, E., 1983. Calculation of energetic and biochemical equivalents of respiratory oxygen consumption. In: E. Gnaiger and H. Forstner (Editors), Polarographic oxygen sensors: aquatic and physiological applications. Springer-Verlag, Berlin, pp. 337-345.

Gorham, W.T., 1990. Uptake of dilute nutrients by marine invertebrates. In: J. Mellinger (Editor), Animal nutrition and transport processes. Vol. 1. Nutrition in wild and domestic animals. Karger, Basel, pp. 70-83.

Gosling, E.M., 1989. Genetic heterozygosity and growth rate in a cohort of Mytilus edulis from the Irish coast. Mar. Biol., 100: 211-215.

Griffiths, C.L. and Griffiths, R.J., 1987. Bivalvia. In: T.J. Pandian and F.J. Vernberg, (Editors), Animal energetics. Vol. 2: Bivalvia through Reptilia. Academic Press, New York, pp. 1-88.

Griffiths, R.J., 1981. Production and energy flow in relation to age and shore level in the bivalve Choromytilus meridionalis (Kr.). Est. Coast. Shelf Sci., 13: 477-493.

Harper, A.E., Benevenga, N.J. and Wohlhueter, R.M., 1970. Effects of ingestion of disproportionate amounts of amino acids. Physiol. Rev., 50: 428-558.

Hawkins, A.J.S., 1985. Relationships between the synthesis and breakdown of protein, dietary absorption and turnovers of nitrogen and carbon in the blue mussel, Mytilus edulis L. Oecologia, 66: 42-49.

Hawkins, A.J.S., 1988. Genetic variations in the physiology of marine shellfish. In: J.C. Iturrondobeitia (Editor), Actas del Congreso de Biologia Ambiental, II Congreso Mundial Vasco, Vol. 1. Servicio Editorial de la Universidad del Pais Vasco, Bilbao, pp. 121-131.

Hawkins, A.J.S., 1991. Protein turnover: a functional appraisal. Funct. Ecol., 5(2): 222-233.

Hawkins, A.J.S. and Bayne, B.L., 1984. Seasonal variation in the balance between physiological mechanisms of feeding and digestion in Mytilus edulis (Bivalvia: Mollusca). Mar. Biol., 82: 233-240.

Hawkins, A.J.S. and Bayne, B.L., 1985. Seasonal variation in the relative utilization of carbon and nitrogen by the mussel, Mytilus edulis: budgets, conversion efficiencies and maintenance requirements. Mar. Ecol. Prog. Ser., 25: 181-188.

Hawkins, A.J.S. and Bayne, B.L., 1991. Nutrition of marine mussels: factors influencing the relative utilizations of protein and energy. Aquaculture, 94(2/3): 177-196.

Hawkins, A.J.S. and Hilbish, T.J., 1992. Towards an understanding of the true costs of cell volume regulation: protein metabolism during hyperosmotic adjustment. J. Mar. Biol. Ass. U.K., in press.

Hawkins, A.J.S., Bayne, B.L. and Clarke, K.R., 1983. Co-ordinated rhythms of digestion, absorption and excretion in Mytilus edulis (Bivalvia: Mollusca). Mar. Biol., 74: 41-48.

Hawkins, A.J.S., Salkeld, P.N., Bayne, B.L., Gnaiger, E. and Lowe, D.M., 1985. Feeding and resource allocation in the mussel *Mytilus edulis*: evidence for time-averaged optimization. Mar. Ecol. Prog. Ser., 20: 273-287.

Hawkins, A.J.S., Bayne, B.L. and Day, A.J., 1986a. Protein turnover, physiological energetics and heterozygosity in the blue mussel, *Mytilus edulis*: the basis of variable age-specific growth. Proc. R. Soc. Lond., 229: 161-176.

Hawkins, A.J.S., Bayne, B.L., Mantoura, R.F.C., Llewellyn, C.A. and Navarro, E., 1986b. Chlorophyll degradation and absorption throughout the digestive system of the blue mussel *Mytilus edulis* L. J. Exp. Mar. Biol. Ecol., 96: 213-223.

Hawkins, A.J.S., Menon, N.R., Damodaran, R. and Bayne, B.L., 1987a. Metabolic responses of the mussels *Perna viridis* and *Perna perna* to declining oxygen tension at different salinities. Comp. Biochem. Physiol., 88: 691-694.

Hawkins, A.J.S., Wilson, I.A. and Bayne, B.L., 1987b. Thermal responses reflect protein turnover in *Mytilus edulis*. Funct. Ecol., 1: 339-351.

Hawkins, A.J.S., Bayne, B.L., Day, A.J. Rusin, J. and Worrall, C.M., 1989a. Genotype-dependent interrelations between energy metabolism, protein metabolism and fitness. In: J. Ryland and P.A.Tyler (Editors), Proc. 23rd Eur. Mar. Biol. Symp., Swansea, U.K., 1988. Olsen and Olsen, Fredensborg, Denmark, pp. 283-292.

Hawkins, A.J.S., Widdows, J. and Bayne, B.L., 1989b. The relevance of whole-body protein metabolism to measured costs of maintenance and growth in *Mytilus edulis*. Physiol. Zool., 62: 745-763.

Hawkins, A.J.S., Rusin, J., Bayne, B.L. and Day, A.J., 1989c. The metabolic/physiological basis of genotype-dependent mortality during copper exposure in *Mytilus edulis*. Mar. Environ. Res., 28: 253-257.

Hawkins, A.J.S., Navarro, E. and Iglesias, J.I.P., 1990. Comparative allometries of gut content, gut passage time and metabolic faecal loss in *Mytilus edulis* and *Cerastoderma edule*. Mar. Biol., 105: 197-204.

Hilbish, T.J. and Zimmerman, K.M., 1988. Genetic and nutritional control of the gametogenic cycle in *Mytilus edulis*. Mar. Biol., 98: 223-228.

His, E., Robert, R. and Dinet, A., 1989. Combined effects of temperature and salinity on fed and starved larvae of the Mediterranean mussel *Mytilus galloprovincialis* and the Japanese oyster *Crassostrea gigas*. Mar. Biol., 100: 455-463.

Hummel, H., 1985. Food intake of *Macoma balthica* (Mollusca) in relation to seasonal changes in its potential food on a tidal flat in the Dutch Wadden Sea. Neth. J. Sea Res., 19: 52-76.

Huppert, H.W. and Laudien, H., 1980. Influence of pretreatment with constant and changing temperatures on heat and freezing resistance in gill-epithelium of the mussel *Mytilus edulis*. Mar. Ecol. Prog. Ser., 3: 113-120.

Hvilsom, M.M., 1983. Copper-induced differential mortality in the mussel *Mytilus edulis*. Mar. Biol., 76: 291-295.

Incze, L.A., Lutz, R.A. and Watling, L., 1980. Relationships between effects of environmental temperature and seston on growth and mortality of *Mytilus edulis* in a temperate northern estuary. Mar. Biol., 57: 147-156.

Jobling, M., 1983. Towards an explanation of specific dynamic action (SDA). J. Fish Biol., 23: 549-555.

Jones, H.D., Richards, O.G. and Hutchinson, S., 1990. The role of ctenidial abfrontal cilia in water pumping in *Mytilus edulis* L. J. Exp. Mar. Biol. Ecol., 143: 15-26.

Jørgensen, C.B., 1982. Uptake of dissolved amino acids from natural seawater in the mussel *Mytilus edulis*. Ophelia, 21: 215-221.

Jørgensen, C.B., 1991. Bivalve filter feeding: hydrodynamics, bioenergetics, physiology and ecology. Olsen and Olsen, Fredensborg, 140pp.

Jørgensen, C.B., Møhlenberg, F. and Sten-Knudsen, O., 1986. Nature of relation between ventilation and oxygen consumption in filter feeders. Mar. Ecol. Prog. Ser., 29: 73-88.

Kautsky, N., Johannesson, K. and Tedengren, M., 1990. Genotypic and phenotypic differences between Baltic and North Sea populations of *Mytilus edulis* evaluated through reciprocal transplantations. I. Growth and morphology. Mar. Ecol. Prog. Ser., 59: 203-210.

Kemp, P.F., Newell, S.Y. and Krambeck, C., 1990. Effects of filter-feeding by the ribbed mussel *Geukensia demissa* on the water-column microbiota of a *Spartina alterniflora* saltmarsh. Mar. Ecol. Prog. Ser., 59: 119-131.

Kinne, O., 1971. Salinity: invertebrates. In: O. Kinne (Editor), Marine ecology. Wiley-Interscience, London, pp. 821-996.

Kiørboe, T. and Møhlenberg, F., 1981. Particle selection in suspension-feeding bivalves. Mar. Ecol. Prog. Ser., 5: 291-296.

Kiørboe, T., Møhlenberg, F. and Hamburger, K., 1985. Bioenergetics of the planktonic copepod Acartia tonsa: relation between feeding, egg production and respiration, and composition of specific dynamic action. Mar. Ecol. Prog. Ser., 26: 85-97.

Kiørboe, T., Møhlenberg, F. and Nohr, O., 1980. Feeding, particle selection and carbon absorption in Mytilus edulis in different mixtures of algae and resuspended bottom material. Ophelia, 19: 193-205.

Kiørboe, T., Møhlenberg, F. and Nohr, O., 1981. Effect of suspended bottom material on growth and energetics in Mytilus edulis. Mar. Biol., 61: 283-288.

Kiørboe, T., Munk, P. and Richardson, K., 1987. Respiration and growth of larval herring Clupea harengus: relation between specific dynamic action and growth efficiency. Mar. Ecol. Prog. Ser., 40: 1-10.

Koehn, R.K., 1991. The genetics and taxonomy of species in the genus Mytilus. Aquaculture, 94(2/3): 125-145.

Koehn, R.K. and Gaffney, P.M., 1984. Genetic heterozygosity and growth rate in Mytilus edulis. Mar. Biol., 82: 1-7.

Koehn, R.K. and Hilbish, T.J., 1987. The adaptive importance of genetic variation. Am. Sci., 75: 134-141.

Koehn, R.K. and Bayne, B.L., 1989. Towards a physiological and energetic understanding of the energetics of the stress response. Biol. J. Linn. Soc., 37: 157-171.

Koehn, R.K., Milkman, R. and Mitton, J.B., 1976. Population genetics of marine pelecypods. IV. Selection, migration and genetic differentiation in the blue mussel Mytilus edulis. Evolution, 30: 2-32.

Koehn, R.K., Diehl, W.J. and Scott, T.M., 1988. The differential contribution by individual enzymes of glycolysis and protein catabolism to the relationship between heterozygosity and growth rate in the coot clam, Mulinia lateralis. Genetics, 118: 121-130.

Kreeger, D.A., Langdon, C.J. and Newell, R.I.E., 1988. Utilization of refractory cellulosic carbon derived from Spartina alterniflora by the ribbed mussel Geukensia demissa. Mar. Ecol. Prog. Ser., 42: 171-179.

Langdon, C.J. and Newell, R.I.E., 1990. Utilization of detritus and bacteria as food sources by two bivalve suspension-feeders, the oyster Crassostrea virginica and the mussel Geukensia demissa. Mar. Ecol. Prog. Ser., 58: 299-310.

Lange, R., 1972. Some recent work on osmotic, ionic and volume regulation in marine animals. Oceanogr. Mar. Biol. Annu. Rev., 10: 97-136.

Livingstone, D., 1981. Induction of enzymes as a mechanism for the seasonal control of metabolism in marine invertebrates: glucose 6-phosphate dehydrogenases from the mantle and hepatopancreas of the common mussel Mytilus edulis. Comp. Biochem. Physiol., 69B: 147-156.

Livingstone, D. and Bayne, B.L., 1974. Pyruvate kinase from the mantle tissue of Mytilus edulis. Comp. Biochem. Physiol., 48B: 481-497.

Livingstone, D.R., Widdows, J. and Fieth, P., 1979. Aspects of nitrogen metabolism of the common mussel Mytilus edulis: adaptation to abrupt and fluctuating changes in salinity. Mar. Biol., 53: 41-55.

Lucas, M.I. and Newell, R. C., 1984. Utilization of saltmarsh grass detritus by two estuarine bivalves: carbohydrase activity of crystalline style enzymes of the oyster Crassostrea virginica (Gmelin) and the mussel Geukensia demissa (Dillwyn). Mar. Biol. Lett., 5: 275-290.

Lucas, M.I., Newell, R.C., Shumway, S.E., Seiderer, L.J. and Bally, R., 1987. Particle clearance and yield in relation to bacterioplankton and suspended particulate availability in estuarine and open coast populations of the mussel Mytilus edulis. Mar. Ecol. Prog. Ser., 36: 215-224.

Lutz, R.A., 1980. Mussel Culture and Harvest: A North American Perspective. Elsevier Science Publishers, B.V., Amsterdam, 305pp.

Mallet, A.L. and Carver, C.E.A., 1989. Growth, mortality, and secondary production in natural populations of the blue mussel, Mytilus edulis. Can. J. Fish. Aquat. Sci., 46: 1154-1159.

Mallet, A.L., Freeman, K.R. and Dickie, L.M., 1986. The genetics of production characters in the blue mussel Mytilus edulis. I. A preliminary analysis. Aquaculture, 57: 133-140.

Mallet, A.L., Carver, C.E.A., Coffen, S.S. and Freeman, K.R., 1987. Winter growth of the blue mussel Mytilus edulis L.: importance of stock and site. J. Exp. Mar. Biol. Ecol., 108: 217-228.

Mallet, A.L., Carver, C.E.A. and Freeman, K.R., 1990. Summer mortality of the blue mussel in eastern Canada: spatial, temporal, stock and age variation. Mar. Ecol. Prog. Ser., 67: 35-41.

Manahan, D.T., 1990. Adaptations by invertebrate larvae for nutrient acquisition from seawater. Am. Zool., 30: 147-160.

Manahan, D.T., Wright, S.W., Stephens, G.C. and Rice, M.A., 1982. Transport of dissolved amino acids by the mussel *Mytilus edulis*: demonstration of net uptake from natural seawater. Science, 215: 1253-1255.

Manahan, D.T., Wright, S.H. and Stephens, G.C., 1983. Simultaneous determination of net uptake of 16 amino acids by a marine bivalve. Am. J. Physiol., 244: 832-838.

Mason, J., 1976. Cultivation. In: Bayne, B.L. (Editor), Marine mussels: their ecology and physiology. Cambridge University Press, Cambridge, pp. 385-410.

McDonald, J.H. and Siebenaller, J.F., 1989. Similar geographic variation at the *Lap* locus in the mussels *Mytilus trossulus* and *M. edulis*. Evolution, 43: 228-231.

McHenery, J.G. and Birkbeck, T.H., 1985. Uptake and processing of cultured micro-organisms by bivalves. J. Exp. Mar. Biol. Ecol., 90: 145-163.

Melaouah, N., 1990. Absorption et metabolisation de substances organiques dissoutes au cours du developpement larvaire de *Mytilus edulis* L. (Bivalves). Oceanol. Acta, 13: 245-255.

Metz, J.A.J. and Diekmann, O., 1986. The dynamics of physiologically structured populations. Lecture notes in biomathematics. Vol. 68. Springer-Verlag, Heidelberg, 511pp.

Milligan, L.P. and Summers, M., 1986. The biological basis of maintenance and its relevance to assessing responses to nutrients. Proc. Nutr. Soc., 45: 185-193.

Millward, D.J., 1989. The nutritional regulation of muscle growth and protein turnover. Aquaculture, 79: 1-28.

Millward, D.J. and Rivers, J.P.W., 1989. The need for indispensable amino acids: the concept of the anabolic drive. Diabete. and Metab., 5: 191-211.

Mitton, J.B. and Grant, M.C., 1984. Associations among protein heterozygosity, growth rate and developmental homeostasis. Annu. Rev. Ecol. Syst., 15: 479-499.

Mitton, J.B. and Koehn, R.K., 1985. Shell shape variation in the blue mussel, *Mytilus edulis* L. and its association with enzyme heterozygosity. J. Exp. Mar. Biol. Ecol., 90: 73-80.

Mitton, J.B., Carey, C. and Kocher, J.D., 1986. The relation of enzyme heterozygosity to standard and active oxygen consumption and body size of tiger salamanders, *Ambystoma tigrinum*. Physiol. Zool., 59: 574-582.

Morton, B.S., 1983. Feeding and digestion in Bivalvia. In: A.S.M. Saleuddin and K.M. Wilbur (Editors), The Mollusca, Vol. 5. Physiology, Part 2. Academic Press, New York, pp. 65-147.

Muir, D.G., Seiderer, L.J., Davis, C.L., Painting, S.J. and Robb, F.T., 1986. Filtration, lysis and absorption of bacteria by mussels *Choromytilus meridionalis* collected under upwelling and downwelling conditions. S. Afr. J. Mar. Sci., 4: 169-179.

Navarro, E., Iglesias, J.I.P., Perez Camacho, A., Labarta, U. and Beiras, R., 1991. The physiological energetics of mussels (*Mytilus galloprovincialis* Lmk.) from different cultivation rafts in the Ria de Arosa (Galicia, Spain). Aquaculture, 94 (2/3): 197-212.

Navarro, J.M., 1988. The effects of salinity on the physiological ecology of *Choromytilus chorus* (Molina, 1782) (Bivalvia: Mytilidae). J. Exp. Mar. Biol. Ecol., 122: 19-33.

Navarro, J.M. and Winter, J.E., 1982. Ingestion rate, assimilation efficiency and energy balance in *Mytilus chilensis* in relation to body size and different algal concentrations. Mar. Biol., 67: 255-266.

Newell, C.R., Shumway, S.E., Cucci, T.L. and Selvin, R., 1989. The effects of natural seston particle size and type on feeding rates, feeding selectivity and food resource availability for the mussel *Mytilus edulis* Linnaeus, 1758 at bottom culture sites in Maine. J. Shellfish Res., 8: 187-196.

Newell, R.C., 1979. Biology of intertidal animals. Marine Ecological Surveys Ltd., Faversham, 781pp.

Newell, R.C. and Pye, V.I., 1970. Seasonal changes in the effect of temperature on the oxygen consumption of the winkle *Littorina littoria* L. and the mussel *Mytilus edulis*. Comp. Biochem. Physiol., 34A: 367-383.

Newell, R.C. and Branch, G.M., 1980. The effects of temperature on the maintenance of metabolic energy in marine invertebrates. Adv. Mar. Biol., 17: 329-396.

Newell, R.I.E. and Thompson, R.J., 1984. Reduced clearance rates associated with spawning in the mussel, *Mytilus edulis* (L.) (Bivalvia, Mytilidae). Mar. Biol. Lett., 5: 21-33.

Nielson, M.V., 1988. The effect of temperature on the shell-length growth of juvenile *Mytilus edulis* L. J. Exp. Mar. Biol. Ecol., 123: 227-234.

Nisbet, R.M., Gurney, W.S.C., Murdoch, W.W. and McCauley, E., 1989. Structured population models: a tool for linking effects at individual and population level. Biol. J. Linn. Soc., 37: 79-99.

Page, H.M., Fisher, C.R. and Childress, J.J., 1990. Role of filter-feeding in the nutritional biology of a deep-sea mussel with methanotrophic symbionts. Mar. Biol., 104: 251-257.

Payne, P.R., 1984. Variability of nutrient requirements. In: A. Velasquez, H. Bourges and I.P. Montfort (Editors), Genetic factors in nutrition. Academic Press, New York, pp. 177-187.

Pechenik, J.A., Eyster, L.S., Widdows, J. and Bayne, B.L., 1990. The influence of food concentration and temperature on the growth and morphological differentiation of blue mussel *Mytilus edulis* L. larvae. J. Exp. Mar. Biol. Ecol., 136: 47-64.

Platt, A.M., 1971. Studies on the digestive diverticulae of *Mytilus edulis*. Ph.D. Thesis, Queens University, Belfast, U.K.

Prieur, D., Mevel, G., Nicolas, J.L., Plusquellec, A. and Vigneulle, M., 1990. Interactions between bivalve molluscs and bacteria in the marine environment. Oceangr. Mar. Biol. Annu. Rev., 28: 277-352.

Prins, T.C. and Smaal, A.C., 1989. Carbon and nitrogen budgets of the mussel *Mytilus edulis* L. and the cockle *Cerastoderma edule* (L.) in relation to food quality. Scient. Mar., 53: 477-482.

Prosser, C.L., 1973. Comparative animal physiology. W.B. Saunders Company, Philadelphia, 966pp.

Ramos-Martinez, J.I. and Torres, A.M.R., 1985. Glutothione reductase of the mantle tissue from sea mussel *Mytilus edulis* L. I. Purification and characterisation of two seasonal enzymatic forms. Comp. Biochem. Physiol., 80B: 355-360.

Reeds, P.J., 1988. Regulation of protein metabolism. In: J.F. Quirke and H. Schmid (Editors), Control and regulation of animal growth. Pudoc, Wageningen, pp. 254-344.

Reeds, P.J., Fuller, M.F. and Nicholson, B.A., 1985. Metabolic basis of energy expenditure with particular reference to protein. In: J.S. Garrow and D. Halliday (Editors), Substrate and energy metabolism in man. Libbey, London, pp. 46-57.

Richardson, C.A., 1989. An analysis of the microgrowth bands in the shell of the common mussel *Mytilus edulis*. J. Mar. Biol. Ass. U.K., 69: 477-491.

Riisgård, H.U., 1988. Efficiency of particle retention and filtration rate in 6 species of the northeast American bivalves. Mar. Ecol. Prog. Ser., 45: 217-223.

Rodhouse, P.G., Roden, C.M., Burnell, G.M., Hensey, M.P., McMahon, T., Ottway, B. and Ryan, T.H., 1984. Food resource, gametogenesis and growth of *Mytilus edulis* on the shore and in suspended culture: Killary Harbour, Ireland. J. Mar. Biol. Ass. U.K., 64: 513-529.

Rodhouse, P.G., McDonald, J.H., Newell, R.I.E. and Koehn, R.K., 1986. Gamete production, somatic growth and multiple-locus enzyme heterozygosity in *Mytilus edulis*. Mar. Biol., 90: 209-214.

Roman, M.R., 1983. Nitrogenous nutrition in marine invertebrates. In: E.J. Carpenter and D.G. Capone (Editors), Nitrogen in the marine environment. Academic Press, London, pp. 347-384.

Ross, A.H. and Nisbet, R.M., 1990. Dynamic models of growth and reproduction of the mussel *Mytilus edulis* L. Funct. Ecol., 4: 777-787.

Russel-Hunter, W.D., 1970. Aquatic productivity. Macmillan, London, 306pp.

Schleyer, M.H., 1981. Microorganisms and detritus in the water column of a subtidal reef of Natal. Mar. Ecol. Prog. Ser., 4: 307-320.

Scott, T.M. and Koehn, R.K., 1990. The effect of environmental stress on the relationship of heterozygosity to growth rate in the coot clam *Mulinia lateralis* (Say). J. Exp. Mar. Biol. Ecol., 135: 109-116.

Seed, R. and Richardson, C.A., 1990. Mytilus growth and its environmental responsiveness. In: G.B. Stefano (Editors), Neurobiology of *Mytilus edulis*. Manchester University Press, Manchester, pp. 1-37.

Seiderer, L.J. and Newell, R.C., 1979. Adjustments of the activity of α-amylase extracted from the style of the black mussel *Choromytilus meridionalis* (Krauss) in response to thermal acclimation. J. Exp. Mar. Biol. Ecol., 39: 79-86.

Seiderer, L.J. and Newell, R.C., 1985. Relative significance of phytoplankton, bacteria and plant detritus as carbon and nitrogen resources for the kelp bed filter-feeder *Choromytilus meridionalis*. Mar. Ecol. Prog. Ser., 22: 127-139.

Seiderer, L.J., Newell, R.C., Schultes, K., Robb, F.T. and Turley, C.M., 1987. Novel bacteriolytic activity associated with the style microflora of the mussel *Mytilus edulis* (L.). J. Exp. Mar. Biol. Ecol., 110: 213-22.

Shick, J.M., De Zwaan, A. and Bont, A.M.T., 1983. Anoxic metabolic rate in the mussel *Mytilus edulis* L. estimated by simultaneous direct calorimetry and biochemical analysis. Physiol. Zool., 56: 56-63.

Sibly, R.M. and Calow, P., 1986. Physiological ecology of animals: an evolutionary approach. Blackwell Scientific Publications, Oxford, 179pp.

Siebers, D. and Winkler, A., 1984. Amino acid uptake by mussels, *Mytilus edulis*, from natural seawater in a flow-through system. Helgoländer. Wiss. Meeresunters., 38: 189-199.

Silvester, N.R. and Sleigh, M.A., 1984. Hydrodynamic aspects of particle capture by *Mytilus*. J. Mar. Biol. Ass. U.K., 64: 859-879.

Smaal, A.C., Verhagen, J.H.G., Coosen, J. and Haas, H.A., 1986. Interaction between seston quantity and quality and benthic suspension feeders in the Oosterschelde, The Netherlands. Ophelia, 26: 385-399.

Sprung, M. and Widdows, J., 1986. Rate of heat dissipation by gametes and larval stages of *Mytilus edulis*. Mar. Biol., 91: 41-45.

Steenbergen, E.J. van, 1987. Genetic variation of energy metabolism in mice. In: M.W.A.Verstegen and A.M. Henken (Editors), Energy metabolism in farm animals. Martinus Nijhoff, Boston, pp. 467-477.

Stefano, G.B. and Aiello, E., 1990. Thermal acclimation during monoamine axonal transport in *Mytilus edulis*: pharmacological characteristics. In: G.B. Stefano (Editor), Neurobiology of *Mytilus edulis*. Manchester University Press, Manchester, pp. 175-188.

Stefano, J.M. and Stefano, C.B., 1990. Neural regulation of seasonality and rhythmicity in *Mytilus edulis*. In: G.B. Stefano (Editor), Neurobiology of *Mytilus edulis*. Manchester University Press, Manchester, pp. 164-174.

Steffens, W., 1989. Principles of fish nutrition. Ellis Horwood Ltd., Chichester, 384pp.

Stephens, G.C., 1981. The trophic role of dissolved organic material. In: A.R. Longhurst (Editors), Analysis of marine ecosystems. Academic Press, London, pp. 271-291.

Strømgren, T. and Nielsen, M.V., 1989. Heritability of growth in larvae and juveniles of *Mytilus edulis*. Aquaculture, 80: 1-6.

Stuart, V.R., 1982. Absorbed ration, respiratory cost and resultant scope for growth in the mussel *Aulacomya ater* (Molina) fed on a diet of kelp detritus of different ages. Mar. Biol. Lett., 3: 289-306.

Stuart, V.R., Field, J.G. and Newell, R.C., 1982. Evidence for absorption of kelp detritus by the ribbed mussel *Aulacomya ater* using a new ^{51}Cr-labelled microsphere technique. Mar. Ecol. Prog. Ser., 9: 263-271.

Tacon, A.G.J. and Cowey, C.B., 1985. Protein and amino acid requirements. In: P. Tytler and P. Calow (Editors), Fish energetics: new perspectives. Croom Helm, London, pp. 155-183.

Taghon, G.L. and Jumars, P.A., 1984. Variable ingestion rate and its role in optimal foraging behaviour of marine deposit feeders. Ecology, 65: 549-558.

Tande, K.S. and Slagstad, D., 1985. Assimilation efficiency in herbivorous aquatic organisms—the potential of the ratio method using ^{14}C and biogenic silica as markers. Limnol. Oceanogr., 30: 1093-1099.

Teo, L.H. and Sabapathy, U., 1990. Preliminary report on the digestive enzymes present in the digestive gland of *Perna viridis*. Mar. Biol., 106: 403-407.

Theisen, B.F., 1973. The growth of *Mytilus edulis* L. (Bivalvia) from Disko and Thule district, Greenland. Ophelia, 12: 59-77.

Theisen, B.F., 1978. Allozyme clines and evidence of strong selection in three loci in *Mytilus edulis* L. (Bivalvia) from Danish waters. Ophelia, 17: 135-142.

Theisen, B.F., 1982. Variation in size of gills, labial palps and adductor muscle in *Mytilus edulis* L. (Bivalvia) from Danish waters. Ophelia, 21: 49-63.

Thompson, R.J., 1979. Fecundity and reproductive effort in the blue mussel (*Mytilus edulis*), the sea urchin (*Strongylocentrotus droebachiensis*), and the snow crab (*Chionoecetes opilio*) from populations in Nova Scotia and Newfoundland. J. Fish. Res. Board Can., 36: 955-964.

Thompson, R.J., 1984a. Production, reproductive effort, reproductive value and reproductive cost in a population of the blue mussel *Mytilus edulis* from a subarctic environment. Mar. Ecol. Prog. Ser., 16: 249-257.

Thompson, R.J., 1984b. The reproductive cycle and physiological ecology of the mussel *Mytilus edulis* in a subarctic, non-estuarine environment. Mar. Biol., 79: 277-288.

Thompson, R.J. and Bayne, B.L., 1974. Some relationships between growth, metabolism and food in the mussel *Mytilus edulis*. Mar. Biol., 27: 317-326.

Thompson, R.J. and Newell, R.I.E., 1985. Physiological responses to temperature in two latitudinally separated populations of the mussel *Mytilus edulis*. In: P.E. Gibbs (Editors), Proc. 19th Eur. Mar. Biol. Symp., Plymouth, England, 1984. Cambridge University Press, Cambridge, pp. 481-495.

Vernberg, W.B. and Vernberg, F.J., 1972. Environmental physiology of marine animals. Springer-Verlag, Berlin, 346pp.

Wallis, R.L., 1975. Thermal tolerance of *Mytilus edulis* of eastern Australia. Mar. Biol., 30: 183-191.

Ward, J.E. and Targett, N.M., 1989. Influence of marine microalgal metabolites on the feeding behaviour of the blue mussel *Mytilus edulis*. Mar. Biol., 101: 313-321.

Waterlow, J.C., 1988. The variability of energy metabolism in man. In: K. Blaxter and I. MacDonald, (Editors), Comparative Nutrition. Libbey, London, pp. 133-139.

Weiser, W., 1973. Temperature relations of ectotherms: a speculative view. In: W. Weiser (Editor), Effects of temperature on ectothermic organisms. Springer-Verlag, Berlin, pp. 1-24.

Weiser, W., 1989. Energy allocation by addition and compensation: an old principle revisited. In: W. Weiser and E. Gnaiger (Editors), Energy transformations in cells and organisms. Georg Thieme Verlag, Stuttgart, pp. 98-105.

Welborn, J.R. and Manahan, D.T., 1988. The use of amperometric detection for the study of sugar uptake from seawater in molluscan larvae. Am. Zool., 28: 130A.

Widdows, J., 1973. Effect of temperature and food on the heart beat, ventilation rate and oxygen uptake of *Mytilus edulis*. Mar. Biol., 20: 269-276.

Widdows, J., 1976. Physiological adaptation of *Mytilus edulis* to cyclic temperatures. J. Comp. Physiol., 105: 115-128.

Widdows, J., 1978. Combined effects of body size, food concentration and season on the physiology of *Mytilus edulis*. J. Mar. Biol. Ass. U.K., 58: 109-124.

Widdows, J., 1985. The effects of fluctuating and abrupt changes in salinity on the performance of *Mytilus edulis* L.. In: J.S. Gray and M.E. Christiansen, (Editors), Proc. 18th Eur. Mar. Biol. Symp., Oslo, Norway, 1984. John Wiley, New York, pp. 555-566.

Widdows, J., 1987. Application of calormetric methods in ecological studies. In: A.M. James (Editor), Thermal and energetic studies of cellular biological systems. Wright, Bristol, pp. 182-215.

Widdows, J., 1989. Calorimetric and energetic studies of marine bivalves. In: W. Weiser and E. Gnaiger (Editors), Energy transformations in cells and organisms. Georg Thieme Verlag, Stuttgart, pp. 145-154.

Widdows, J. and Bayne, B.L., 1971. Temperature acclimation of *Mytilus edulis* with reference to its energy budget. J. Mar. Biol. Ass. U.K., 51: 827-843.

Widdows, J. and Donkin, P., 1989. The application of combined tissue residue chemistry and physiological measurements of mussels (*Mytilus edulis*) for the assessment of environmental pollution. Hydrobiologia, 188/189: 455-461.

Widdows, J. and Hawkins, A.J.S., 1989. Partitioning of rate of heat dissipation by *Mytilus edulis* into maintenance, feeding and growth components. Physiol. Zool., 62: 764-784.

Widdows, J., Fieth, P. and Worrall, C.M., 1979. Relationships between seston, available food and feeding activity in the common mussel *Mytilus edulis*. Mar. Biol., 50: 195-207.

Widdows, J., Donkin, P., Salkeld, P.N., Cleary, J.J., Lowe, D.M., Evans, S.V. and Thompson, P.E., 1984. Relative importance of environmental factors in determining physiological differences between two populations of mussels (*Mytilus edulis*). Mar. Ecol. Prog. Ser., 17: 33-47.

Wilbur, A.E. and Hilbish, T.J., 1989. Physiological energetics of the ribbed mussel *Geukensia demissa* (Dillwyn) in response to increased temperature. J. Exp. Mar. Biol. Ecol., 131: 161-170.

Willows, R.I., 1992. Optimal digestive investment: a model for filter-feeders experiencing variable diets. Limnol. Oceanogr., in press

Wilson, J.H. and Seed, R., 1974. Laboratory experiments on pumping and filtration in *Mytilus edulis* L. using suspensions of colloidal graphite. Ir. Fish. Invest. Ser. B (Mar.), 14: 1-20.

Winter, J.E., 1973. The filtration rate of *Mytilus edulis* and its dependence on algal concentration, measured by a continuous automatic recording apparatus. Mar. Biol., 22: 317-328.

Winter, J.E., 1976. Feeding experiments with *Mytilus edulis* L. at small laboratory scale. II. The influence of suspended silt in addition to algal suspensions on growth. In: G. Persoone and E. Jaspers (Editors),

Vol. 1. Proc. 10th Eur. Mar. Biol. Symp., Ostend, Belgium, 1975. Universa Press, Wetteren, pp. 583-600.

Winter, J.E., 1978. A review of the knowledge of suspension-feeding in lamellibranchiate bivalves, with special reference to artificial aquaculture systems. Aquaculture, 13: 1-33.

Worrall, C.M. and Widdows, J., 1984. Investigation of factors influencing mortality in *Mytilus edulis* L. Mar. Biol. Lett., 5: 85-97.

Wright, R.T., Coffin, R.B., Ersing, C.P. and Pearson, D., 1982. Field and laboratory measurements of bivalve filtration of natural marine bacterioplankton. Limnol. Oceanogr., 27: 91-98.

Wright, S.H., Moon, D.A. and Silva, A.L., 1989. Intracellular Na^+ and the control of amino acid fluxes in the integumental epithelium of a marine bivalve. J. Exp. Zool., 142: 293-310.

Yamamoto, S. and Sugawara, Y., 1988. Induced triploidy in the mussel, *Mytilus edulis* by temperature shock. Aquaculture, 72: 21-29.

Zouros, E., Romero-Dorey, M. and Mallet, A.L., 1988. Heterozygosity and growth in marine bivalves: further data and possible explanations. Evolution, 42: 1332-1341.

Chapter 6

CELLULAR BIOCHEMISTRY AND ENDOCRINOLOGY

ALBERTUS DE ZWAAN AND MICHEL MATHIEU

CELLULAR ENERGY METABOLISM IN THE MYTILIDAE: AN OVERVIEW

The Aerobic/Anaerobic Transition

Mytilidae belong to the faculative anaerobes i.e. they can live either aerobically or anaerobically but prefer to use oxygen (respiration) when it is available, as it allows a much more economical use of fuel molecules. Mytilidae are also called euryoxic, because they tolerate a wide range of oxygen concentrations, including zero concentration. Adult mytilidae have a high tolerance of anoxia and show high LT_{50} values e.g. 35 days for *Mytilus edulis* at 10°C, and 15 days for *Mytilus galloprovincialis* at 20°C have been reported (Theede et al., 1969; de Zwaan et al., 1991b). *Mytilus* spp. live in intertidal and shallow waters which are characterized by large fluctuations in temperature, salinity and oxygen availability. All of those changes can induce an anaerobic type of energy metabolism.

Metabolic depression and anaerobiosis are clearly implicated as key factors in freezing temperature survival. Low temperature induces a switch to anaerobic metabolism. *Modiolus demissus* held anaerobically had a greater resistance to cold than mussels kept aerobically at the same temperature (for reviews see Aarset, 1982, and Storey and Storey, 1988). The anaerobic end-product strombine has been identified as a cryoprotectant in *M. edulis* (Loomis et al., 1988). In the subarctic region only a few species survive the winter in the upper littoral zone and *M. edulis* is among the most freeze-tolerant species. The mantle cavity of *M. edulis* contains about ten times as much water as the tissues. When mussels freeze during periods of low temperatures, the latent heat of fusion tends to keep the animals at a temperature which is above ambient temperature. The latent heat of fusion of seawater may, therefore, act as a heat store preventing the interior of the mussel from reaching ambient temperature during low tide (Aarset, 1982).

Isolation behaviour is also utilized to minimize contact with low external salinities. Mussels have peripheral salinity receptors on the tentacles of the inhalant siphon and they can tightly adduct their shell valves as seawater concentrations fall (Davenport, 1985).

Behaviours to conserve water during intertidal exposure at the same time impair respiratory gas exchange. The Mytilidae possess poor morphological adaptations for aerial respiration. The lack of an oxygen-carrying pigment, together with a poor open circulatory system, makes oxygen storage and transportation to tissues impossible. Simultaneous calorimetry and respirometry experiments have established that *M. edulis* is largely anaerobic during aerial exposure (Widdows and Shick, 1985). The compensation for reduced feeding time involves energy conservation; there is little evidence for energy supplementation, such as increases in feeding rate or absorption efficiency (Shick et al., 1988). During anaerobiosis there is an almost complete reduction in the rates of energetic processes such as digestion and absorption of food and growth. The cost of digestion and absorption of food is in *M. edulis* about 17% of the total metabolic energy expenditure, and the cost of growth ranges from 0-30% (Widdows and Hawkins, 1989; also see details in Chapter 5). There will be reduced muscular activity of e.g. the heart, foot, byssus gland, shell, whereas the adductor muscle, once contracted, is not in an active state as a result of the so-called 'catch' phenomenon. During catch the muscle is able to maintain tension or shortening for a long period, although the metabolic rate is reduced almost to resting levels (Zange et al., 1989). Consequently, in sessile bivalves, the ATP turnover is less than 10% of the resting aerobic rates; for *M. edulis* a value of 4% is reported by Widdows (1987) and for *M. galloprovincialis* a value of 4.5% by de Zwaan et al. (1991b). The relationship between the anoxic state, relative to the normoxic rate of energy expenditure and anoxia tolerance, is shown in interspecific comparisons with adult bivalves (Widdows et al., 1989; Brooks et al., 1991).

Increasing pollution of estuarine and coastal habitats with organic matter and nutrient salts strongly stimulates primary production. Eutrophication, therefore, causes an increased organic load on the bottom, which causes increased respiration. Decomposition of sulphur-containing detritus by bacteria results in the release of dissolved sulphide. The main known effect of H_2S is similar to that of free cyanide. Both compounds inhibit aerobic metabolism by blocking cytochrome aa_3. Furthermore, it may block the oxygen-binding site on the haemoglobin molecule. Eutrophication, therefore, causes seasonal cycles of severe lack of oxygen and the release of sulphide in and near the bottom sediments. A high tolerance of anoxia is, therefore, a prerequisite for survival of *Mytilus* populations in eutrophicated biotopes.

Anoxia tolerance is a characteristic of the adult sessile bivalve; tolerance increases with larval development. Total energy metabolism is sustained mainly by aerobic metabolism down to pO_2 values of 2 or 4kPa for early larval stages and juveniles, respectively. The ability to conserve energy expenditure under anoxic conditions appears to be related to anoxia tolerance. Prodissoconch larvae maintain feeding activity under anoxic conditions (Widdows et al., 1989). Adult bivalves lacking red blood cells can switch off aerobic metabolism under conditions of severe hypoxia (Mangum and Winkle, 1973). The critical pO_2 of *M. galloprovincialis* for the induction of fermentative pathways of ATP production is about 8kPa. A graded increase in the output of anaerobic products (succinate, alanine) occurs at oxygen tensions below 8kPa and reaches a maximum at about 4kPa. This suggests that fermentation pathways are maximally activated at all oxygen tensions below 4kPa (de Zwaan et al., 1991b). Metabolic depression and activation of anaerobic metabolism develop simultaneously (van den Thillart et al., 1992).

The above conditions, caused by external physical factors, induce so-called environmental or organism-level anaerobiosis. Because of its coupling to energy conservation it is also known as a low-power output mode of anaerobic metabolism. Metabolic rate depression has a much larger impact on long-term survival than the other biochemical adaptations to anoxia tolerance such as: (1) maintenance of high reserves of glycogen and aspartate (de Zwaan and Putzer, 1985), and (2) the use of alternative pathways of fermentative ATP production to enhance anaerobic ATP yield (see below).

Anaerobic energy metabolism is also involved in providing ATP when energy demand exceeds the capacity of aerobic energy production. This may be due to exercise (functional anaerobiosis, or high-power output mode of anaerobic catabolism), or restoration of normal physiological functioning, following exercise or environmental anaerobiosis (recovery anaerobiosis). The high-power output mode is linked to classical glycolysis, although in marine invertebrates, including the Mytilidae, it usually does not terminate in lactate but in so-called opines. The increase in anoxia tolerance with larval development appears to go along with a shift from functional anaerobiosis to environmental anaerobiosis (Widdows et al., 1989).

The type of contraction of the posterior adductor muscle (PAM) occurring when shell valves are closed during aerial exposure, and the shell valve adductions during re-immersion, rely on tonic and phasic contractions, respectively. The PAM and anterior byssus retractor muscle (ABRM) are composed of smooth muscle fibres (opaque, smooth or catch part) and (irregular-oblique) striated muscle fibres (phasic or transluscent part). The smooth part is responsible for keeping the valves closed and the phasic part for rapid closure of the shell. The tonically contracted PAM reaches the catch state, where it relaxes very slowly, whereas phasic contraction is characterized by rapid relaxation. Catch is the ability to maintain tension at a considerably reduced ATP cost compared to phasic tension development (for literature see Sailer et al., 1990, and Watabe and Hartshorne, 1990). Functional anaerobiosis under aerobic conditions only appears to be involved in repeated phasic contractions. This is indicated by a continuous breakdown of arginine phosphate, a decreasing ATP level and the accumulation of octopine (Zange et al., 1989).

Patterns of Energy Metabolism

This chapter deals mainly with the metabolism of carbohydrates and lipids. For protein metabolism we refer to Chapter 5 and for amino acid metabolism to Bishop et al. (1983).

There is ample evidence to show that the current view of mammalian bioenergetics also applies to bivalves. Most of the oxygen consumption of bivalves is mitochondrial and linked to oxidative catabolism. Glucose, triglycerides, free fatty acids and a number of amino acids serve as energy sources. They all can provide acetyl-CoA, which is the direct fuel of the TCA cycle. In this cycle the carbon of the acetyl group is released as CO_2 and hydrogen reduces the NAD^+ and FAD^+ coenzymes. These reduced coenzymes donate electrons to the electron transport (respiratory) chain until finally oxygen is reduced to water. This electron transfer is coupled to the synthesis of ATP and is known as oxidative phosphorylation. The processes involved in aerobic degradation of carbohydrates are depicted in Fig. 6.1A.

Degradation of glucose and some amino acids leads to the formation of pyruvate, which subsequently can be converted to acetyl-CoA by the pyruvate dehydrogenase complex. The latter is located in the mitochondrial matrix and has been studied for the mytilid *Modiolus demissus* by Paynter et al. (1985a). The constituent enzymes of the TCA cycle are also located within the matrix of the mitochondria, with the exception of succinate dehydrogenase, which is located on the inner surface of the inner mitochondrial membrane (Addink and Veenhof, 1975). Citrate synthase, 2-oxoglutarate dehydrogenase, succinate dehydrogenase and isocitrate dehydrogenase catalyze nonequilibrium steps. These enzymes have been studied in some detail in bivalves, because of their importance in regulating the carbon flow rate of the cycle (Ryan and King, 1962; Alp et al., 1976; Philip et al., 1976; Head, 1980a, b; Head and Gabbott, 1980, 1981; Kargbo and Swift, 1983; Ellington, 1985; Ruiz et al., 1985; Karam et al., 1987). A transhydrogenase ($NADPH^+ + NAD^+ \rightarrow NADP^+ + NADH$) is associated with the mitochondrial membrane (Zandee et al., 1980a). The electron transport chain forms a part of the inner mitochondrial membrane.

The inner mitochondrial membrane shows selective permeability; only small, neutral or lipophilic molecules can pass this barrier by diffusion. It poses an impenetrable barrier to the coenzyme NAD(H), whereas transport of tricarboxylic and

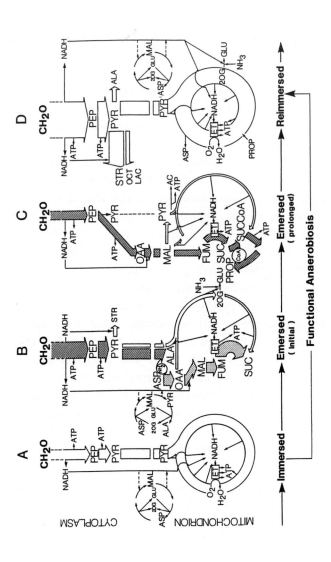

Fig. 6.1. Schematic diagrams illustrating transitions in the pathway of glycogen (and aspartate) degradation in relation to increased energy demand (exercise or functional anaerobiosis—see text for explanation) and cycles of normoxia and anoxia (environmental aerobiosis—see text for explanation). B and C represent the transition from the initial stage to the steady state stage of anaerobiosis. For the outside panels A and D, other fuels may be involved besides (or instead of) carbohydrate (not indicated).

The width of the arrows are an indication of the relative carbon flux through the pathways. Abbreviations: Ac, acetate; ALA, alanine; ASP, aspartate; CH_2O and Glc_n, glycogen; ET, electron transfer chain; FUM, fumarate; GLU, glutamate; LAC, lactate; MAL, malate; OAA and OXA, oxaloacetate; OCT, octopine; 2-OG and 2-OGLU, 2-oxoglutarate; PEP, phosphoenolpyruvate; PROP, propionate; PYR, pyruvate; STR, strombine; SUC, succinate; SUCCoA, succinyl CoA.

dicarboxylic intermediates depend on the presence of specific carriers in the inner membrane (see LaNoue and Schoolwerth, 1979 for a general review).

The process of β-oxidation (degradation of fatty acids) and the TCA cycle generate reducing agents (NADH, FADH) inside the mitochondrion, but glycolysis and oxidation of fermentative products derived from pyruvate generate reducing agents (NADH) in the cytosol. So-called shuttle systems oxidize cytosolic NADH, and transfer the reducing equivalents to the mitochondrial electron transport chain (see Dawson (1979) for a review of the mammalian system, and Zammit and Newsholme (1976) for shuttle system enzymes of marine invertebrates; see also Fig. 6.1A and D).

It is generally accepted that under anaerobic conditions carbohydrates are the sole or main source of energy. Unlike proteins and lipids, glucose can be metabolized without the help of the TCA cycle in such a way that energy can be conserved as ATP (substrate-linked phosphorylation) without changing the redox state of the cell. The classic example is glycolytic fermentation in skeletal muscle (Embden–Meyerhof–Parnas pathway). This linear pathway involves one oxidation, catalyzed by glyceraldehyde-3-phosphate dehydrogenase (GAPDH), and one reduction, catalyzed by lactate dehydrogenase (LDH). The 1:1 organization of the GAPDH/LDH reactions maintains a constant redox state in the cell. Both the hydrogen donor (G3P) and the hydrogen acceptor (pyruvate) are formed in the cytoplasm. The intermediates of the coupled redox reactions are not directly connected in the overall pathway, and therefore, the coenzyme NAD^+ mediates the hydrogen transport.

The reaction steps of the Embden–Meyerhof–Parnas pathway in which glucose-6-phosphate is converted to pyruvate have been well-documentated for *M. edulis*, both in terms of enzyme activities (Crabtree and Newsholme, 1972; Churchill and Livingstone, 1989), and intermediates (Beis and Newsholme, 1975; Ebberink and de Zwaan, 1980). Fermentations consist of both linear and branched pathways (Livingstone, 1983). During environmental anoxia marine mussels accumulate a variety of organic acids such as succinate, propionate, acetate and alanine, in addition to small amounts of CO_2 and pyruvate derivatives such as iminocarboxylic acids (opines) and lactate. Some of the end-products are formed in the cytoplasm (opines, lactate and possibly alanine), and others in the mitochondrion (acetate, succinate, propionate, CO_2 and alanine).

Anaerobic metabolism can rely entirely on carbohydrate, because electron acceptor (e.g. pyruvate, fumarate, 2-oxoglutarate) and electron donor molecules can be formed, both in the cytoplasm and mitochondrion. Since the inner mitochondrial membrane of *M. edulis* tissues is impermeable to $NAD^+(H)$ (Zaba et al., 1978), a barrier exists against the coenzyme that couples the redox reactions in the overall pathway. This implies that redox balance will be maintained on both sides of the barrier, or that shuttle systems must be operating. Depending on the phase of anaerobiosis, and the particular tissue, either the first or both mechanisms are involved.

In mammals the mitochondria loose their function in energy production under anaerobic conditions. In vivo investigations with radiolabelled metabolites strongly indicate that in M. *edulis* (de Zwaan et al., 1975a; Wijsman et al., 1977) the TCA cycle remains partially or totally functional under anaerobic conditions. The formation of succinate in anaerobic mitochondria occurs via a NADH-dependent reduction of fumarate, and this step needs to be balanced by the regeneration of NADH. Starting in 1968 with a paper by Stokes and Awapara a large number of fermentation schemes for bivalves were proposed in the five years following. This fruitful period of scientific research has been well-documentated by de Zwaan (1977). Common to most schemes is that the route leading to succinate accumulation (the glucose-succinate pathway) proceeds via classical glycolysis up to the stage of phosphoenolpyruvate (PEP); this then carboxylates to form oxaloacetate.

AEROBIC METABOLISM AT THE ORGANISM, THE TISSUE AND SUBCELLULAR LEVEL

Whole Animals

Evidence that energy metabolism will at least be partially aerobic, when oxygen uptake is possible, follows from the simple observation that mussels incubated in anoxic seawater show considerably higher concentrations of established fermentative compounds and lower concentrations of the potential anaerobic fuel aspartate, than mussels taken from aerated seawater. Moreover, volatile fatty acids are only released to the incubation seawater when this is deprived of oxygen (Kluytmans et al., 1975, 1978). Combustion of glucose into CO_2 and H_2O has been unequivocally proven with the use of U-^{14}C labelled glucose (de Zwaan et al., 1975a). In aerobically incubated M. *edulis* 20% of the radiolabel was recovered as exhaled CO_2 (another 3% as dissolved or bound CO_2), whereas under anoxic conditions CO_2 accounted for only about 1% of the radioactivity recovered. In agreement with the above was the observed shift in total incorporation of radiolabel in the organic acid fraction from 11.1% (aerobic) to 56.6% during anoxia.

The integrated operation of the TCA cycle and the respiratory chain in M. *edulis* was shown by the conversion of U-^{14}C-glutamate (de Zwaan et al., 1975a) and 2,3-^{14}C-succinate (Wijsman et al., 1977) into aspartate (and other compounds). With glutamate, the conversion rate in the presence of oxygen was three times higher and release of ^{14}C as CO_2 was enhanced by a factor of four.

Recently, dissolved sulphide and cyanide have been added to aerated seawater containing the bivalves *Mytilus galloprovincialis* and *Scapharca inaequivalvis* (de Zwaan et al., 1991a). The metabolic response was compared to that induced by incubation in anoxic seawater. That both poisons must have blocked the electron transfer chain in spite of the fact that sufficient oxygen was present in the seawater to allow full respiration, follows from the accumulation of typical anaerobic end-products and the utilization of aspartate (Fig. 6.2).

Separate Organs

Gill tissue

Beating of the lateral cilia of *M. edulis* which create the inhalant and exhalant water currents, seems to be particularly dependent on the presence of oxygen. The activity is inhibited by the inhibitors of cytochrome aa_3, cyanide and azide, the uncoupler of oxidative phosphorylation, dinitrophenol (DNP), and the competitive inhibitor of succinate dehydrogenase, malonate (Usuki, 1956a, b). Succinate is an established aerobic substrate in the gills of *Modiolus demissus* and *M. edulis*. This substrate stimulated oxygen consumption in intact gill pieces and in whole tissue homogenates of both species, which could be counteracted by malonate (Malanga and Aiello, 1972). By contrast, pyruvate, acetate or malate did not produce any significant increase in the rate of oxygen consumption in gills of either species. It has been shown, histochemically, that localization of succinate dehydrogenase is almost exclusively in the lateral ciliated cells of *Modiolus* and *Mytilus* gills (Bouffard and Aiello, 1969). Burchem et al. (1984) measured oxygen consumption in isolated gill tissue of *M. demissus* in the presence and absence of DNP and azide. The controls consumed oxygen linearly for at least 60min. DNP stimulated oxygen consumption, and azide blocked tissue oxygen consumption. These results support the conclusion that most of the tissue oxygen consumption is mitochondrial, that the mitochondria are probably tightly coupled in vivo, and that the gills possess 'normal' (mammalian type) mitochondria with no obvious unusual respiratory properties.

Heart tissue

No data are available for *Mytilus* but an analysis of energy metabolism in the ventricle of four other marine bivalves, with considerable diversity of life-style, revealed similarities (Ellington, 1985). A dominant role of carbohydrates in bivalve ventricles is illustrated by the presence of abundant glycogen particles. These particles are often concentrated in the vicinity of mitochondria. Glycogen content in bivalve

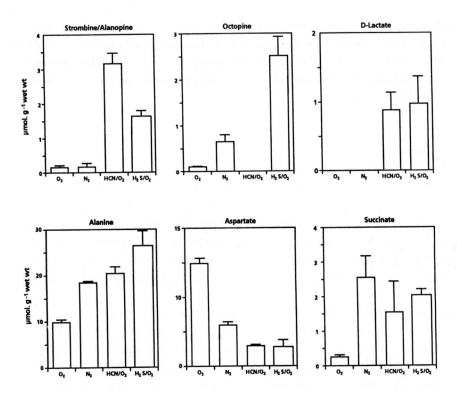

Fig. 6.2. Concentrations of the anaerobic fuel aspartate, and fermentative products, in adductor muscle of *Mytilus galloprovincialis* after 12h incubation under four different conditions: O_2, oxygenated seawater; N_2, N_2-bubbled, degassed seawater; HCN/O_2, oxygenated seawater with cyanide present ($\leq$1 mmol L^{-1}); H_2S/O_2, oxygenated seawater with sulfide present ($\leq$1 mmol L^{-1}). Each bar represents the mean $\pm$ SD (n = 3). (A. de Zwaan and P. Cortesi, unpublished data).

ventricles is extremely high—up to 22% of the total dry weight of the tissue (Ellington, 1985).

Integration of glycolysis, TCA cycle and electron transport chain in oxidative energy provision for mechanical work has been demonstrated in isolated heart of the bivalve *Tapes watlingi* (Jamieson and de Rome, 1979) using various metabolic inhibitors of oxidative phosphorylation. Cyanide caused abrupt cessation of respiration and contraction, azide produced a slower effect, while rotenone had a slight effect. If, however, antimycin A was added after rotenone, respiration ceased. DNP dramatically increased the rate of oxygen consumption. These results illustrate that the heart mitochondria possess coupled respiration in vivo. With nitrogen substituted for oxygen, and in the presence of the glycolytic inhibitors 2-deoxyglucose and iodoacetate, it was shown that (anaerobic) glycolysis is able to replace oxidative phosphorylation in providing energy for contraction. It was also shown that noncarbohydrate substrates add to aerobic energy production in order to obtain maximum levels of activity.

Mantle tissue

In mantle tissue of *M. edulis* the role of carbohydrates in catabolism depends very much on the season and/or which type of cells are considered; these aspects will be covered later on p.254–257 and p.284–287. Ibarguren et al. (1990) investigated seasonal changes in glycolytic activity in the mantle, hepatopancreas and posterior adductor muscle (PAM) of *M. galloprovincialis*, using fructose 2,6-P_2 as a marker of the state of the glycolytic pathway. In mantle and hepatopancreas glycolytic activity was maximal in winter and at a minimum in summer, while small fluctuations took place in the PAM.

   Chen and Awapara (1969) incubated mantle of the bivalve *Rangia cuneata* with U-^{14}C-glucose in a medium saturated with oxygen, and observed that radiolabel was almost quantitatively incorporated into alanine. They considered this as strong evidence that pyruvate is poorly oxidized, or not oxidized at all, under aerobic conditions. Later studies by Zaba and Davies (1981, 1984) on mantle of *M. edulis* came to the same conclusion. These authors reported oxygen consumption rates in sliced mantle preparations which lie in the range 2.1–7.0μmol h^{-1} g^{-1} wet wt, values comparable with those for the intact animal. They proposed, like Chen and Awapara (1969), that the 'anaerobic' pathway of phosphoenolpyruvate metabolism (see below) may also be used by mantle in the presence of oxygen as a consequence of the low activity of pyruvate dehydrogenase (Addink and Veenhof, 1975). Despite the relatively low rates of incorporation of ^{14}C from carbohydrate precursors into CO_2, Zaba and Davies (1981) established that oxidative metabolism in mantle slices was active under the conditions of incubation. The mean rate of CO_2 production from four replicate incubations was 5.6μmol h^{-1} g^{-1} wet wt, which is consistent with the above quoted oxygen consumption rates. It appears, therefore, that the TCA cycle is operative in mantle, but that a very small proportion of its fuel may originate from exogenous glucose. Most likely, the CO_2 production and oxygen consumption were due to the aerobic metabolism of other substrates, such as proteins and lipids (Zaba and Davies, 1981).

Oxidative Metabolism in the Mitochondrion

Zaba (1983) measured oxygen uptake on crude homogenates of gill and mantle of *M. edulis* after addition of ADP, Pi, succinate and glutamate and observed two unusual features of mitochondrial respiration. Firstly, the rate of respiration was dependent on the oxygen tension and secondly, it was relatively insensitive to 1mM cyanide. The first point suggests a low affinity of the terminal oxygenase for oxygen. The author concluded that if a convential electron transport-linked oxidase (cytochrome aa_3) is present, it clearly is not the only type of oxidative activity in the tissues. In order to test the possibility that part of the measured oxygen uptake was the result of the activity of oxidases, other than that of a mitochondrial electron transport oxidase, the action of two additional inhibitors was studied. Neither a xanthine oxidase inhibitor nor a monoamino oxidase inhibitor had any effect on the oxygen uptake in gill mitochondria. Although the author points to the fact that the possibility cannot be excluded that other mitochondrial oxidases are involved, he speculated also about the presence of an alternative or modified electron transport system. From an earlier study of mitochondrial preparations from *M. edulis* mantle (and other tissues) Zaba et al. (1978) tentatively concluded that cytochromes of the *b*- and *c*-type were present, but no aa_3 cytochrome. However, at room temperature it is not possible to positively identify the cytochrome peaks. Coupled mitochondria were obtained from PAM, mantle and digestive gland with respiratory control ratios (RCR) of 1.4, 3.0 and 1.6, respectively (RCR is the ratio of state 3 over state 4 respiration; metabolic states according to the terminology of Chance and Williams, 1956). Succinate, glutamate and malate plus pyruvate had a clear stimulatory effect on respiration in the presence of ADP and Pi (state 3; see Chance and Williams, 1956). Electron micrographs showed both broken and intact mitochondria. Physical disruption was further indicated by the use of exogenous NADH as substrate. NADH appeared to be indeed able to stimulate respiration, but the rate was substantially increased by sonification of the preparation. In both cases stimulation by ADP and Pi failed. This indicates that for mussel mitochondria, as for mammalian preparations, NADH is a nonpenetrant, and also that disrupted mitochondria fail to show coupling. Akberali and Earnshaw (1982) confirmed the characteristics observed by Zaba et al. (1978) for mitochondria from mantle and digestive gland of *M. edulis*. P/O ratios could not be established in mitochondrial preparations of Zaba et al. (1978) due to the fact that respiration remained stimulated after addition of ADP. Disruption most likely resulted in induction of an ATP-ase activity, which continuously regenerated ADP from the ATP formed, resulting in a continuation of state 3. By slightly modifying the isolation method de Zwaan and Holwerda (unpublished data) extended the studies with mantle tissue mitochondria of *M. edulis*. Using the 3000g mantle fraction (cellular debris and nuclear fraction removed) they showed, by the effects of ADP, KCN and DNP on respiration, that at least 84% of oxygen consumption could be accounted for by mitochondrial respiration, and that this relied upon the classical respiration chain, including the terminal complex IV, cytochrome aa_3. When mitochondria were separated from the cytosol by further centrifugation at 20,000g and resuspended, oxygen consumption in the presence of ADP and Pi was completely blocked by concentrations of cyanide as low as 30µM, irrespective of whether succinate, or glutamate plus malate, were used as electron donors (Holwerda and de Zwaan, 1979). Rotenone (1µM) and amytal (5mM), which both act on NADH dehydrogenase (complex I), blocked respiration with malate plus glutamate present by 71% and 76%, respectively, but displayed only slight effects with succinate as substrate. This agrees with the mammalian concept that for the oxidation of succinate, complex I of the electron transport chain is bypassed. Malonate (8.4mM) inhibited succinate oxidation by 86%. Antimycin A (0.1µM), which inhibits electron transport in the span from cytochrome *b* to *c* (ubiquinone dehydrogenase, complex III), blocked respiration by over 80% with both types of substrates. Cytochrome *c* is not integrated with the four respiratory complexes, which are tightly bound to the inner membrane, but provides a 'mobile' link between complex III and IV. Mitochondrial particles of mantle showed about a five-fold increase in oxygen consumption rate when cytochrome *c* was added (de Zwaan and Holwerda, unpublished data). These results unequivocally prove that mantle mitochondria in *M. edulis* possess a mammalian-type respiratory chain which accounts for all oxygen consumed. These findings contrast strongly with the observed insensitivity to cyanide reported by Zaba (1983) for mantle mitochondria from the same species.

By adding the substrate analogue of ATP, adenylimidodiphosphate, to inhibit possible ATP hydrolysis by endogenous ATPase activity, de Zwaan and Holwerda (unpublished data) indeed observed a transition from state 3 to 4 after addition of ADP to mantle mitochondria of *M. edulis* (compare incubation b and c of Table 6.1). A normal stoichiometry was obtained for oxygen consumption and phosphorylation (P/O ratio of approximately 3 in the case of incubation c of Table 6.1). Moreover, a clear adenylate kinase activity (ATP + AMP ↔ 2 ADP) appeared to be present in the mitochondrial preparations (see incubation e of Table 6.1). This in turn can be suppressed by adding di-adenosyl pentaphosphate (incubation f of Table 6.1). After addition of ADP to mitochondria the endogenous activity of both enzymes may regenerate ADP from ATP in the case of endogenous AMP, which may counteract the return to state 4.

In the above studies on *M. edulis* a mixture of sucrose and mannitol has been used to isolate the mitochondria. As pointed out by Burcham et al. (1984), in such preparations a mannitol oxidation using molecular oxygen may occur. When applying isolation media containing D-mannitol, Burcham et al. (1984) also obtained rapid rates of oxygen consumption in the absence of added substrate in *M. demissus* mitochondria. However, the complete cyanide inhibition of oxygen utilization observed by Holwerda and de Zwaan (1979) excludes the possibility that this type of oxidation was associated with their mitochondrial preparations, but might indeed have been responsible for part of the oxygen consumption in the studies by Zaba (1983). Omission of KCl in the isolation buffer provided mitochondria from *M. demissus* which showed a transition to a reduced respiratory rate (Burcham et al., 1984). These mitochondria could oxidize glutamate, proline, succinate and malate, and with most of these substrates showed a high degree of respiratory control with P/O ratios characteristic of the particular substrate. Of the substrates tested, only pyruvate failed to generate state 3 respiration. Pyruvate lacked a 'sparking' effect with malate as substrate. These results suggest the absence of a rapid oxidation of exogenously supplied pyruvate by the mitochondria.

In order to obtain information about which fuels are providing substrates for the TCA cycle, as well as their relative importance, a number of potential substrates have been added to mitochondrial suspensions in order to determine their effect on oxygen consumption rate. Lipid substrates have only been used in bivalve studies on ventricle (Ballantyne and Storey, 1983) and hepatopancreas of *Mercenaria mercenaria* (Ballantyne and Moon, 1985) and hepatopancreas of *Mytilus edulis* (Ballantyne and Storey, 1984). These studies indicate that mitochondria can be tissue-specific regarding type and preference of substrate. The heart was unable to use lipid as palmitoyl-L-carnitine which, on the contrary, was the preferred substrate of hepatopancreas mitochondria. The heart showed a high capacity to use amino acids. Besides proline, L-ornithine, L-aspartate, L-glutamate and L-arginine were suitable substrates. Interspecific differences for particular tissues seem to exist for pyruvate and proline; proline, the preferred substrate of the ventricle of *M. mercenaria* could not be oxidized by hepatopancreas mitochondria. Proline also appears to be oxidized by gill mitochondria of *Modiolus demissus*, but not by mitochondria of the hepatopancreas, mantle and PAM of *M. edulis* (Zaba et al., 1978), and not by the gill of *Crassostrea virginica* (Burcham et al., 1983). *M. mercenaria* ventricle and hepatopancreas showed state 3 respiration comparable with glutamate with a P/O ratio of near 3. However, like *M. demissus* gill, the hepatopancreas of *M. edulis* failed to oxidize pyruvate (Ballantyne and Moon, 1985). Rates of oxidation of malate and 2-oxoglutarate in the mitochondria of the ventricle and hepatopancreas of *M. mercenaria*, and the capability to oxidize α-glycerophosphate, imply the existence of a malate-aspartate shuttle and α-glycerophosphate cycle for cytosolic redox balance (Ballantyne and Storey, 1983, 1984).

Table 6.1. Effect of addition of substrate, ADP, ATP, AMP and the inhibitors of ATPase, adenoylimidodiphosphate (AMP-PNP) and adenylate kinase, diadenosylpentaphosphate (Ap$_5$A) on the oxygen consumption rate of mitochondria isolated from mantle of *Mytilus edulis*. (de Zwaan and Holwerda , unpublished data).

	Additions	Oxygen consumption (% control)	[1]RCR
a	None added	100	
	malate (6mM) + glutamate (6mM)	448	
	ibid + ADP (0.8mM)	829	1.85
b	malate (6mM) + glutamate (6mM)	530	
	ibid + ATP (0.8mM)	855	1.61
c	malate (6mM) + glutamate (6mM) + AMP-PNP (0.4mM)	387	
	ibid + ADP (0.4mM)		
	at start, state 3	910	2.35
	after return to state 4	415	1.07
	+ ADP (0.8mM)	934	2.25
d	malate (6mM) + glutamate (6mM) + AMP-PNP (0.4mM)	398	
	ibid + ATP (0.8mM)	452	1.14
	+ ATP (2.0mM)	477	1.20
e	malate (6mM) + glutamate (6mM)	369	
	ibid + AMP-PNP (0.4mM) + ATP (0.15mM)	380	1.03
	+ AMP (0.2mM)	779	2.11
f	malate (6mM) + glutamate (6mM) + AMP-PNP (0.4mM) + Ap$_5$A (0.25mM)	390	
	ibid + ATP (0.8mM)	413	1.06
	+ AMP (0.2mM)	471	1.21

[1]RCR is the ratio of state 3 over state 4 respiration (Chance and Williams, 1956)

The amino acids cysteine and ornithine were suitable substrates for the hepatopancreas of *M. edulis*, but alanine, glycine, proline, asparagine, arginine, aspartate and citrulline were not oxidized at detectable levels. This was also the case for α-glycerophosphate, fumarate and oxaloacetate. The mitochondria preferred lipid substrates and accepted a range of fatty acid chain lengths (Ballantyne and Moon, 1985). Zandee and Kruytwagen (cited in Kluytmans et al., 1985) measured a high acivity of α-hydroxyacyl-CoA-dehydrogenase in muscle, mantle, gill and especially hepatopancreas. The hepatopancreas, therefore, appears to be well-equipped for fatty acid oxidation.

ANAEROBIC METABOLISM

Pathways

In *M. edulis*, as in marine euryoxic invertebrates in general (Bryant, 1991), there is a great metabolic diversity in anaerobic energy metabolism during environmental anaerobiosis. The accumulation pattern of end-products varies from tissue to tissue (Kluytmans et al., 1977; Zurburg and Kluytmans, 1980; Zurburg and Ebberink, 1981), the region within a particular organ (Greenfield and Crenshaw, 1981), and with season (Kluytmans et al., 1980). Moreover, it has been established that, within the time course of anaerobiosis, gradual transitions take place in carbon flow (Kluytmans et al., 1977). In order to illustrate the latter, a metabolic map has been designed in which panels represent metabolic schemes in a sequence in which they may occur in upper intertidal mussels over the tidal cycle (Fig. 6.1). The map, which was first published in 1983 (de Zwaan, 1983), has undergone essential revision, especially concerning the initial or transition stage of anaerobiosis (Fig. 6.1B). The metabolic changes designated in a certain panel, when compared with its preceding one, do not have to be synchronized; for example, panels B and C suggest that the succinate conversion into propionate is coupled to the conversion of PEP away from alanine to oxaloacetate. It is, however, only intended to indicate that both changes take time to evolve during the anaerobic transition. During prolonged anoxia succinate is converted into propionate (Fig. 6.1C) and length of the lag period is inversely related to the incubation temperature. Excretion of volatile fatty acids is observed (Kluytmans et al., 1977, 1978). Schulz et al. (1984) proposed a role for acetyl-CoA transferase in determining the length of the lag time; they observed a striking similarity between the pH optimum of CoA transferase and that of the overall propionate synthesis in vitro from succinate in intact mantle mitochondria.

For PAM the transition stage of fermentation may last up to 10h or even longer (Ebberink et al., 1979; de Zwaan et al., 1983a), before going over to the anaerobic steady state phase (Fig. 6.1C). In the transition stage some strombine is formed, as well as succinate and alanine (de Zwaan and Zurburg, 1981). Succinate is derived from aspartate and the other end-products from glucose; as time proceeds, glycogen is converted into succinate and/or propionate (Fig. 6.1C). This change is initiated by the formation of oxaloacetate from PEP by the action of PEP-carboxykinase (de Zwaan et al., 1983a), an enzyme generally found in euryoxic invertebrate muscles, as opposed to vertebrate muscle (de Zwaan and de Bont, 1975). Oxaloacetate is converted into succinate via malate and fumarate. In this route, the glycolytically reduced NADH is oxidized by the malate dehydrogenase reaction (see de Zwaan, 1977 for review). Malate, therefore, is generally considered as the main carbohydrate-derived anaerobic fuel for the mitochondrion, compared to pyruvate during aerobic catabolism. In agreement with this it has been observed that oxygen stimulates the utilization of pyruvate, but reduces that of malate (Table 6.2B).

Anaerobic Substrate Utilization in Mitochondria

In a simplified scheme, with solely succinate or propionate accumulating, it can be shown that the redox state of the mitochondrion does not change when two equivalents of malate are oxidized to succinate (via malic enzyme and part of the TCA cycle running in a clockwise direction) and five are reduced (via fumarate) to succinate and/or propionate (Gnaiger, 1977, 1983; de Zwaan, 1983). The stoichiometry is as follows:

$$7 \text{ malate} \rightarrow 6 \text{ succinate} + 4CO_2 + 3H_2O \tag{1}$$
$$7 \text{ malate} \rightarrow 6 \text{ propionate} + 10CO_2 + 3 H_2O \tag{2}$$

In order to study anaerobic substrate utilization, and the pathway along which these substrates are metabolized, potential substrates were incubated with mitochondria isolated from mantle tissue by de Zwaan and co-workers. In addition, arsenite (an inhibitor of pyruvate dehydrogenase and 2-oxoglutarate dehydrogenase) and monofluoroacetate (an inhibitor of aconitase) were added, to give additional information on the involvement of specific reaction steps. Only the carbon flow schemes for malate and pyruvate plus alanine have been published so far (de Zwaan et al., 1981 and de Zwaan, 1991, respectively); relevant results for these incubations and for other substrates are summarized in Table 6.2.

Table 6.2. (A) The anaerobic transformation of substrates by isolated mitochondria from the mantle of *Mytilus edulis*. Changes (δ) are shown in main metabolite concentrations after 3h incubation. (B) The utilization of pyruvate and malate after 1h incubation under both aerobic ($+O_2$) and anaerobic ($-O_2$) conditions. The data are expressed in μmol mL^{-1} mitochondrial suspension (1mL suspension corresponds to 2g wet wt and approximately 15mg mitochondrial protein). The procedure for malate, described in de Zwaan et al. (1981), is also applied for the other incubations. (de Zwaan, Veenhof and Holwerda, unpublished data) FAc, monofluoroacetate: As, arsenite: n.c., no change; other abbreviations as in Figs. 6.1 and 6.3.

Substrate added	δ Added substrate(s)		δ SUC	δ ALA	δ Others
A. 6mM					
(1) PYR	−1.14		0.51	0.37	0.38 (Ac)
(2) PYR + As	−0.74		0.39	0.32	0.14 (2-OGLU)
(3) PYR + FAc	−1.12		0.35	0.34	0.10 (CIT)
(4) PYR + ASP	−2.66	−1.37	1.03	1.57	1.02 (Ac)
(5) PYR + ASP +As	−1.96	−1.18	0.61	1.02	0.17 (2-OGLU)
(6) 2-OGLU + FAc	−0.99		0.43	−0.04	0.43 (GLU) n.c. (CIT)
(7) 2-OGLU + ASP + FAc	−4.12	−3.28	2.59	0.76	2.50 (GLU) 0.19 (CIT)
(8) MAL	−5.48		3.36	0.34	0.29 (PYR) 0.65 (Ac) n.c. (CIT)
(9) MAL + As	−5.10		2.52	0.37	2.24 (PYR) n.c. (Ac) n.c. (CIT) 0.21 (2-OGLU)
(10) MAL + FAc	−5.74		3.64	0.30	0.37 (PYR) 0.69 (Ac) 0.56 (CIT)
(11) MAL + PYR	−5.87	−1.41	4.59	0.38	
(12) MAL + PYR + As	−5.90	+ 2.73	3.10	0.39	0.20 (2-OGLU)
(13) MAL + PYR + FAc	−6.14	− 1.09	4.19	0.31	0.63 (CIT)
B. 2.5mM					
PYR −O_2	−0.27				
PYR +O_2	−0.64				
MAL −O_2	−1.70				
MAL +O_2	−1.31				

About 85% of the added malate was metabolized, in the absence of oxygen, by the mitochondria under the conditions shown (see de Zwaan et al., 1981). Succinate accounted for the larger part of the end-products (Table 6.2, incubation 8). It was entirely formed from malate via the fumarate reductase step. The volatile fatty acids acetate and propionate, were also produced in substantial amounts. Propionate was formed from malate mainly via initial conversion into pyruvate and acetyl-CoA, followed by the TCA cycle reactions from citrate to succinyl-CoA. There was also a flow towards alanine and acetate. The results strongly support the concept that the so-called intramitochondrial dismutation of malate (part of the malate pool is reduced and the other part oxidized) is responsible for both the accumulation of succinate and the maintenance of redox balance (mitochondrial parts in panels B and C of Fig. 6.1). The following overall reaction for the anaerobic utilization of malate can be derived from the carbon-flow scheme:

$$7 \text{ malate } (+ 0.14 \text{ glutamate } + 0.07 \text{ aspartate } + 0.21 \text{ NH}_3) \rightarrow$$
$$4.27 \text{ succinate } + 0.77 \text{ propionate } + 4.76 \text{ CO}_2 (+ 0.42 \text{ alanine } + 0.77 \text{ acetate}$$
$$+ 0.28 \text{ fumarate } + 0.35 \text{ pyruvate}) \tag{3}$$

This equation approaches the theoretical equations (1) and (2) above very closely. Different reaction steps contributed to redox balance. Reduction of fumarate was the only reaction utilizing NADH. Of the reducing power, 51.6% was generated in the conversion of malate into pyruvate and oxaloacetate. Pyruvate dehydrogenase accounted for a further 26.1% and the remainder came from the isocitrate dehydrogenase and 2-oxoglutarate dehydrogenase steps (de Zwaan et al., 1981).

Simultaneous addition of pyruvate resulted in total utilization of malate, probably due to an increased capacity to condense oxaloacetate with acetyl-CoA (Table 6.2, reaction 11). The accumulation of citrate in incubations involving monofluoroacetate is evidence for the involvement of the citrate synthase step in anaerobic metabolism. Pyruvate and 2-oxoglutarate were metabolized to similar extents—about 20% the rate of malate utilization. Interestingly, their degradation rates increased dramatically when aspartate was added simultaneously (Table 6.2, incubations 1 versus 4, and 6 versus 7). In these incubations oxidation was supplemented by amination via the activity of glutamate-oxaloacetate-(GOT) and glutamate-pyruvate-transaminase (GPT). Aspartate addition strongly promotes the formation of alanine. In vivo the cytoplasm contains high concentrations of free amino acids which can be transferred to the mitochondrion (Livingstone et al., 1979). Incubation 4, therefore, mimics the transition phase depicted in Fig 6.1B.

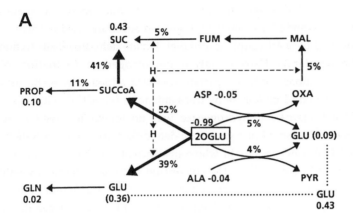

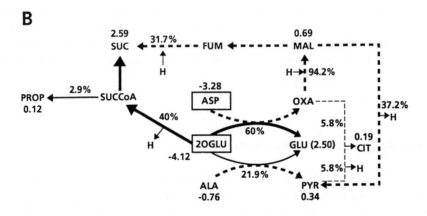

Fig. 6.3. Carbon flow schemes for the anaerobic transformation of 2-oxoglutarate (A), and 2-oxoglutarate plus aspartate (B), in the presence of the aconitase inhibitor monofluoroacetate, by isolated mitochondria from the mantle of *Mytilus edulis*. Figures near substrates and intermediates show the change in concentration in μmol mL^{-1} mitochondrial suspension after 3h incubation (see also incubations 6 and 7 of Table 6.2). Figures along the arrows represent the substrate flow through the reaction steps as a percentage of the total 2-oxoglutarate (solid lines), or aspartate (dotted lines in (B)) turnover.

Abbreviations: see legend to Figure 1; CIT, citrate; --->H, concomitant reduction of coenzyme; H --->, concomitant oxidation of coenzyme.

The various incubations show a high degree of metabolic flexibility in the way redox balance can be maintained within the mitochondrion. When adding only 2-oxoglutarate (Fig. 6.3A), there was an equimolar accumulation of succinate and glutamate, together accounting for about 90% of utilized substrate. Redox balance was almost entirely obtained by a 1:1 coupling of the glutamate dehydrogenase (reductive amination of 2-oxoglutarate)/2-oxoglutarate dehydrogenase (oxidation of 2-oxoglutarate) redox reactions. When aspartate was also added (Fig. 6.3B), most 2-oxoglutarate was aminated to glutamate by transamination (via GOT) with aspartate, with the result that glutamate dehydrogenase was no longer involved in redox control. These observations show that glutamate dehydrogenase can contribute to the reoxidation of NADH during periods in which insufficient fumarate is generated to feed the fumarate reductase complex. The conversion of 2-oxoglutarate, by either a redox reaction (involving oxidation or reduction), or by transamination (thus creating, indirectly, oxaloacetate for the reductive pathway to succinate), can therefore, play a vital role in redox homeostasis of the mitochondrion. During the transition from aerobiosis to anaerobiosis the NADH/NAD$^+$ ratio will, at reducing tissue oxygen tension, gradually rise while pyruvate is still a main cytosolic end-product of glycolysis. During this period reduction by glutamate dehydrogenase of 2-oxoglutarate formed in the mitochondrion will occur. This is in accord with the small increases in glutamate levels found during the initial stage of anaerobiosis (de Zwaan, 1977; Wijsman et al., 1977; see also Fig. 6.1B).

Intracellular Location of Malate Formation

Collicutt (1975) showed that about 63% of radioactive U-^{14}C-aspartate added to oyster ventricle was incorporated into malate and succinate, and about 35% into pyruvate, alanine and (the later identified) alanopine. Fifty three per cent ^{14}C-glucose was metabolized mainly to ^{14}C-alanine compared with 3% percent to succinate. On the basis of these results Collicutt and Hochachka (1977) proposed a scheme in which the glucose-alanine pathway and the aspartate-succinate pathway were coupled to the transfer, via the activity of GOT and GPT, of nitrogen from aspartate to pyruvate (Fig. 6.1B). Additional support for this concept came from studies in which inhibitors of the transaminases, glycolysis and PEP-carboxykinase were added to oyster ventricle (Foreman and Ellington, 1983), adductor muscle of *M. edulis* (de Zwaan et al., 1982) and whole *M. edulis* (de Zwaan et al., 1983a). Moreover, the absence of an inhibitory effect with hadacidin showed that the purine nucleotide cycle is not involved in the inverse correlation of changes in the levels of aspartate and alanine (de Zwaan et al., 1982).

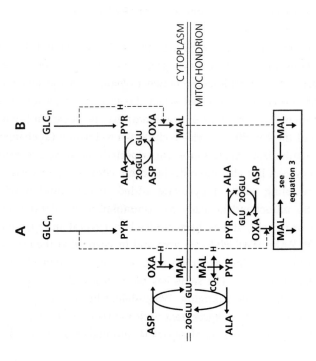

(A) PYR+2ASP+NADH+H⁺ → 2ALA+CO₂+MAL+NAD⁺+H₂O   (B) PYR+ASP+NADH+H⁺ → ALA+MAL+NAD⁺+H₂O

Fig. 6.4. Two concepts to connect the glycogen-alanine and aspartate-succinate pathways by redox coupling and transaminase reactions. In (A) there is a topographic separation between the operation of opine dehydrogenase (cytosol) and pyruvate glutamate transaminase (mitochondrion), thus preventing competition for pyruvate within the same cell compartment. The overall reaction at the bottom of the figure illustrates the stoichiometric differences between the two concepts. Abbreviations: see legend to Figure 6.1.

Until recently it was generally assumed that the enzymes which are responsible for the coupling of both pathways would operate entirely in the cytosol. This would ensure both redox balance in this compartment by the GAPDH/malate dehydrogenase (MDH) couple, and the provision of malate to the mitochondrion. Nonetheless, the whole set of reactions can equally well be carried out within the mitochondria (Paynter et al., 1984b, 1985b). This gives the same result, assuming that the reducing power created by the GAPDH reaction can be (in)directly passed to the mitochondrion. In the following section we will show how this may occur. The concerted action of GPT, GOT and MDH results in the following overall reaction:

$$\text{pyruvate} + \text{aspartate} + \text{NADH} + \text{H}^+ \rightarrow \text{alanine} + \text{malate} + \text{NAD}^+ \tag{4}$$

The right hand part (B) of Fig. 6.4 illustrates the location in the cytosol and the left hand part (A) in the mitochondrion. Since a major part of malate is converted into succinate (equation 3) a near 1:1:1 stoichiometry between changes in the levels of alanine, aspartate and succinate is to be expected when the reaction steps involved operate in the cytosol (equation 4). A number of arguments are, however, in favour of a mitochondrial location of the coupling. These include the following: (a) changes in the actual aspartate (or alanine)/succinate ratios are, especially in muscle, greater than unity; (b) not only the cytosol but also the mitochondrion contains the necessary enzymes to catalyze the coupling; and (c) pyruvate added in combination with aspartate to isolated mantle mitochondria leads to accumulation of alanine and succinate. As to the first argument, it has been noticed in several independent experiments carried out with PAM of *M. edulis* (in vivo and in vitro) that the ratio of changes in the concentrations of alanine, aspartate and succinate is often close to 1:1:0.5 (de Zwaan et al., 1982, 1983a, b). The same has been observed for muscular tissues (ventricle, foot) of other molluscs (Ellington, 1981; Meinardus and Gäde, 1981; Gäde and Ellington, 1983). This ratio is in agreement with the redox coupling of the glycogen-alanine pathway with the conversion of aspartate into malate as depicted in Fig. 6.4A.

Studies with isotopes (Collicutt and Hochachka, 1977), metabolic inhibitors (de Zwaan et al., 1982), and the observation of malic enzyme activity in the mitochondria of bivalves and gastropods (de Zwaan, 1977; de Zwaan et al., 1981; Ellington, 1985; Paynter et al., 1985b), have revealed that not only is the nitrogen of aspartate incorporated into alanine, but so also is a substantial part of the carbon. Incorporation of aspartate carbon into alanine probably involves an initial, and then a final transamination step and the decarboxylation of malate to pyruvate, resulting in the overall reaction (see Fig. 6.4A):

$$aspartate \rightarrow CO_2 + alanine \tag{5}$$

In the studies cited above concentrations of amino acids were determined with L-specific enzymes. However, D-isomers of alanine and aspartate have also been detected in the Mytilidae. In *M. edulis* and *Modiolus capex* no D-alanine was detected but D-aspartate was found in concentrations approaching those of the L-isomer. The metabolic role of D-amino acids in marine invertebrates is still unknown (Felbeck and Wiley, 1987).

As far as the second argument is concerned, it appears that GPT has, in general, a higher activity than MDH in mitochondria than in the cytosol (Paynter et al., 1984a, 1985b). In a study of the subcellular distribution of aminotransferases in the gill tissue of four bivalves it was found that *C. virginica* and *M. edulis* showed both mitochondrial and cytosolic activity for GPT, *Mercenaria mercenaria* had trace activity of GPT in the cytosol, whereas in *Modiolus demissus* it was strictly mitochondrial. GOT showed distinct mitochondrial and cytosolic isozymes in all animals (Paynter et al., 1984b). Earlier studies on anaerobic metabolism of *M. demissus* have shown that the gill tissue behaves like that of other euryoxic bivalves, namely by accumulation of succinate and utilization of aspartate (Baginski and Pierce, 1978). Paynter et al. (1984a), therefore, conclude that, at least in *M. demissus*, alanine production and the coupling to aspartate turnover must be mitochondrial.

The third argument concerns direct evidence for the presence of a coupling between aspartate utilization and alanine formation within mitochondria. When an equimolar amount of aspartate and pyruvate were added simultaneously to mantle mitochondria, about 45% of pyruvate was metabolized, half of which appeared as alanine (incubation 4 of Table 6.2). Assuming coupling of pyruvate metabolism (giving alanine) and aspartate (giving succinate and propionate) by GPT, GOT and MDH, a flow diagram for metabolite conversion can be constructed which showed reasonable balance for hydrogen, nitrogen and carbon (de Zwaan, 1991). Since the incubation was carried out with resuspended mitochondria the possibility (in vivo) of utilizing reducing equivalents generated in the cytosol by GAPDH was excluded. Reducing equivalents for the observed reduction of fumarate, therefore, had to be generated in the mitochondria itself and appeared to be mainly obtained by the oxidation of pyruvate via acetyl-CoA to acetate. In vivo acetate is only a minor end-product when compared to propionate (Kluytmans et al., 1975, 1977, 1978). Only close to death does there appear to be high acetate accumulation and excretion in *M. edulis* (van den Thillart and de Vries, 1985; Hummel et al., 1989).

Transfer of Reducing Equivalents from the Cytoplasm to the Mitochondrion

A major problem that arises when the alanine and malate accumulation occurs in the mitochondria is that the direct redox link between GAPDH and MDH is lost. In aerobically functioning tissues an identical problem is solved by the malate aspartate (or other) shuttle (s), which transfer hydrogen from the cytosol to the mitochondrial respiratory chain (Fig. 6.1A and D). When intracellular conditions are anoxic, the mitochondrial MDH will reverse its catalytic direction from oxaloacetate towards malate and, instead, catalyzes the second step in the conversion of aspartate into succinate. This blocks the participation of MDH in the classical malate-aspartate shuttle. This reversal will be due to a rise in the $NADH/NAD^+$ ratio which, at the same time, reduces the activity of pyruvate dehydrogenase (Paynter et al., 1985a). This, in turn, increases the oxaloacetate available for MDH. Since, in bivalves, malic enzyme operates in vivo in the direction of oxidative decarboxylating (de Zwaan, 1977; de Zwaan et al., 1981) it can take over the function displayed under aerobic conditions by MDH in the mitochondrial part of the malate-aspartate shuttle. The adaption of the malate-aspartate shuttle to operate under anaerobic conditions can be accomplished by the replacement of GOT and the mitochondrial MDH by GPT and malic enzyme, respectively. Instead of a shuttle system in which aspartate is cycled, a net conversion of aspartate into alanine and CO_2 is obtained (equation 5).

The physiological relevance may be, therefore, that this reaction sequence can functionally replace the malate-aspartate shuttle in transferring cytoplasmic hydrogen from the cytosol through the inner mitochondrial membrane, in order to supply reducing equivalents for the mitochondrial production of malate from aspartate. The coupling of the aspartate-succinate pathway with the glycogen-alanine pathway, and the connection with the 'aspartate dependent hydrogen translocation' is depicted in Figs. 6.1B and 6.4A. The scheme gives a good explanation as to why aspartate utilization should be almost double that of succinate production (see overall reaction equation at the bottom of Fig. 6.4A).

When, after a period of anaerobiosis, the aspartate pool is greatly diminished and succinate carbon is also delivered from carbohydrate (Fig. 6.1A and B), the discrepancy between the stoichiometry of aspartate and succinate will be less than two, and at some stage succinate production will start to exceed aspartate utilization. In in vivo studies with *M. edulis* de Zwaan et al. (1983a) observed that even after 18h following the onset of anaerobiosis (aerial exposure) the decrease in aspartate in PAM exceeded the increase in succinate (no propionate formed) by a factor of 2. By contrast, for other tissues, the discrepancy between the two metabolites was already reversed within the first 6h. This indicates that the transition stage is much more important in muscle tissue than in other tissues.

In an in-depth phylogenetic survey study including about 200 species of several phyla, Livingstone et al. (1983, 1990) have determined whether the presence of certain pathways can be correlated with the specific activities of particular enzymes involved in these pathways. The occurrence of an aspartate-malate shuttle predicts a relationship between MDH and GOT. This was observed for the vertebrates and most other phyla. The pathways depicted in Fig. 6.4 also predict a relationship between MDH and GOT, *and* between GOT and GPT. This was observed only for nonvertebrate phyla.

When aerobic conditions return, MDH can again catalyze the oxidation of malate, despite an unfavourable dissociation constant. This is due to both a rapid removal of oxaloacetate and the reoxidation of NADH. Thus, the malate-aspartate shuttle will again take over the transfer of reducing equivalents from the cytosol to the mitochondrion.

Anaerobic ATP Generation Steps in the Mitochondrion

The mitochondrial reduction of fumarate to succinate by the fumarate reductase complex is theoretically sufficiently exergonic to enable the phosphorylation of a mole of ADP (Gnaiger, 1977; Fig. 6.1B and C, mitochondrial part). The reaction is as follows:

fumarate + NADH + H$^+$ → succinate + NAD$^+$
standard free energy of hydrolysis: –67.70 kJ mol^{-1}

Holwerda and de Zwaan (1979 and 1980) have examined the NADH-fumarate reductase enzyme system, using intact mitochondria and submitochondrial particles from the mantle tissue of *M. edulis*. On incubating intact mitochondria anaerobically with malate as substrate, the enzyme activity was found to be well above the in vitro rate of succinate production by *M. edulis* (de Zwaan, 1977). The enzyme activity was stimulated by ADP plus inorganic phosphate. ATP was formed with a P/fumarate ratio of 0.46, indicating a theoretical yield of one mole ATP per mole of fumarate reduced. Oligomycin and various uncouplers inhibited the synthesis of ATP. These results indicate that synthesis of ATP is coupled to the transport of electrons from NADH to fumarate. The energy-conserving formation of succinate from malate or fumarate has also been reported for many microorganisms and other lower Metazoa (Behm, 1991).

Kinetics of the NADH-fumarate reductase reaction in *M. edulis* were studied with cyanide-poisoned submitochondrial particles (Holwerda and de Zwaan, 1980; Zandee et al., 1980a). The low K_m values established for NADH and fumarate in *M. edulis* enable the oxidation of low concentrations of NADH by low concentrations of fumarate. Experiments with inhibitors (rotenone, amytal and malonate) showed the involvement of two flavoproteins, NADH-quinone reductase and quinone-fumarate reductase (or complexes I and II of the electron transport chain) in electron transport between NADH and fumarate.

NADPH appeared to be a poor substrate for the fumarate reductase reaction. However, when added together with NAD^+, maximum reductase activity was obtained, indicating the presence of a transhydrogenase associated with the submitochondrial particles, thus explaining why malic enzyme, which is $NADP^+$-dependent in bivalves (see de Zwaan, 1977), can generate reducing equivalents for the fumarate reductase system (see previous section).

The pathway for propionate production in mantle mitochondria of *M. edulis* has been studied by Schulz et al. (1982) and Schulz and Kluytmans (1983). Propionate is formed in a cyclic manner in a pathway through which succinyl-CoA is converted to succinate in four steps catalyzed by methylmalonyl-CoA isomerase, methylmalonyl-CoA racemase, propionyl-CoA carboxylase and acyl (propionyl)-CoA transferase, thus closing the cycle. In the propionyl-CoA carboxylase step ATP is formed by substrate-linked phosphorylation of ADP (Schulz et al., 1983; Fig. 6.1B and C, mitochondrial part; for a review on comparative aspects of propionate metabolism see Halarnkar and Blomquist, 1989).

Mitochondrial participation in carbohydrate fermentation can, therefore, add two extra sites of ATP generation (in the route fumarate-propionate) to the cytosolic substrate-linked phosphorylations. By comparing overall equations for carbohydrate fermentations it appears that the maximum energetic advantage obtained with propionate fermentation, in terms of extra mitochondrial ATP yield, is 2.14 times higher than that of simple lactate fermentation (Gnaiger, 1977; de Zwaan, 1983).

Anaerobic Pyruvate Metabolism in the Cytoplasm

Pyruvate oxidoreductases

Six end-products of anaerobic pyruvate metabolism have been identified in molluscs. These are alanine, lactate, octopine, α- and β-alanopine, strombine and tauropine. The imino acid derivatives octopine, α- and β-alanopine, strombine and tauropine are formed by reductive condensation of pyruvate with the amino acids arginine, α- and β-alanine, glycine and taurine, respectively. These condensation products are collectively known as opines (Schell et al., 1979). Some molluscan tissues may exhibit all enzyme activities for the production of the listed pyruvate derivatives; e.g. the adductor muscle of the marine bivalve *Scapharca inaequivalvis* (Sato et al., 1987).

The PAM of *M. edulis* possesses at least the enzymes to form the first five pyruvate derivatives. The discovery of the opines has been reviewed by de Zwaan and Dando (1984).

In a phylogenetic survey for pyruvate reductase activity by Livingstone et al. (1983, 1990) the opine dehydrogenases showed the highest activities in muscular tissues. They were generally absent in nonmarine species. The adductor muscles of most marine bivalves contain both alanopine and strombine (Sato et al., 1982a; Storey et al., 1982). They show similar opine dehydrogenase activity, with either glycine or alanine as the amino acid substrate (Dando et al., 1981; Sato et al., 1982b; Livingstone et al., 1983). It is apparent that bivalve adductor muscles predominantly contain the broadly specific strombine dehydrogenase. A distinct alanopine dehydrogenase is found in the mantle, gill, midgut gland and foot of *M. edulis* (Dando, 1981; Dando et al., 1981). Subcellular fractionation studies indicate that 98% of the activity of the opine dehydrogenases is present in the cytosol (Dando, 1981; Fields and Hochachka, 1981). It has been noted by several authors that the presence of these enzymes in a tissue is related to the concentration of the appropriate free amino acid (Ellington, 1979; Barrett and Butterworth, 1981; Dando et al., 1981; Fields and Quinn, 1981; Siegmund and Grieshaber, 1983; Kreutzer et al., 1989).

With the discovery of the opine dehydrogenases it is now apparent that muscular tissue in *M. edulis* has a considerable potential for the direct cytoplasmic reduction of pyruvate. The ratio of pyruvate kinase to total pyruvate reductase activity is similar for the PAM of *M. edulis* and for vertebrate skeletal muscle (de Zwaan and Zurburg, 1981). It is still unclear why a tissue such as the adductor muscle of *M. edulis* should have different pyruvate oxidoreductases. Fields and Quinn (1981) used computer simulations to compare the effect of the different enzymes on the redox ratio ($NADH/NAD^+$), on the assumption that substrates and products were maintained at instantaneous equilibrium. The redox ratio increased as the dehydrogenase products, lactate, malate or octopine, accumulated. Dehydrogenases with a large pool (relative to the K_m) of the non-NADH cosubstrate(s) would be the most effective at maintaining a low redox ratio, especially if the cosubstrate was continuously replenished. Thus, MDH works well at maintaining the redox ratio during anoxia when aspartate is converted to succinate. Octopine dehydrogenase is favoured when phosphoarginine hydrolysis increases the free arginine pool, and alanopine and strombine formation occurs in tissues with large free amino acid pools of alanine and glycine, respectively. Various studies have shown that both the succinate pathway, and the pathways terminating in the reduction of pyruvate, co-exist in the same tissue (de Zwaan and Zurburg, 1981; de Zwaan et al., 1983a, b).

Physiological role of opine formation
The physiological role played by the opine dehydrogenases during environmental and functional hypoxia has been extensively reviewed (Gäde, 1980, 1983; Livingstone et al., 1983; de Zwaan and Dando, 1984; Gäde and Grieshaber, 1986). The opine pathways appear to operate when there is an increased glycolytic flux. Livingstone (1982) and de Zwaan and Putzer (1985) have calculated the actual rates of energy production by the opine and succinate pathway in different molluscs. For bivalves the rate of energy production (μmoles ATP equivalents g^{-1} min^{-1}) of the former pathway was in the order of 0.1–1.0, compared to a range of 0.005–0.010 in the latter. It has been argued by Livingstone et al. (1981) that, depending on the organism, the role of the opine pathway is to maintain resting (aerobic) rates of energy production in anoxic situations, or to increase them during exercise in a manner analogous to the lactate pathway in vertebrates.

Muscular activity:
Few studies have been made on the formation of alanopine and strombine during muscular activity. In a study by Shick et al. (1986) *M. edulis* were subjected to an 8h exposure/4h immersion regime and the number of valve adductions were counted. Significant correlations were obtained between the number of adductions during recovery and strombine formation. Exceptionally active mussels accumulated more strombine in their PAM and had a higher anaerobic contribution to muscular activity during recovery (Shick et al., 1986).

During aerial exposure the PAM of *M. edulis* also forms strombine, but little or no alanopine, lactate or octopine (de Zwaan and Zurburg, 1981; Zurburg et al., 1982; Kreutzer et al., 1989). However, on reimmersion after a small lagtime, a much faster rate of formation of the pyruvate derivatives strombine, octopine and lactate occurs. The accumulation of strombine was much less important in other organs (Zurburg et al., 1982; de Zwaan et al., 1983a). The short delay in the accumulation of pyruvate derivatives probably coincides with the initial suppression of shell valve movements during recovery (Shick et al., 1986). Once the adductor muscle starts to produce shell valve movements, the energy demand appears to be too high to be met by respiration alone. Recovery from anaerobic metabolism has been extensively reviewed (de Zwaan, 1977, 1983; Ellington, 1983a; with respect to *M. edulis* we refer especially to de Zwaan et al., 1983b; Shick et al., 1983, 1986; Widdows and Shick, 1985). The prompt oxidation of succinate during reimmersion (de Zwaan et al., 1983b) may serve to replenish phosphoarginine levels, since the phosphagen increases before the anaerobic end-products begin to accumulate (Gnaiger, unpublished data, cited in Shick et al., 1986).

The anaerobic energy metabolism of posterior adductor muscle and anterior byssus retractor muscle (ABRM) preparations of *M. edulis* has been extensively studied both at normoxia and hypoxia (Zange et al., 1989). Active contraction induced by acetylcholine, proceeds to catch when the neurotransmitter is subsequently washed out. Catch can be abolished by addition of serotonin. Both active state and catch or tonic contraction have been energetically characterized, and compared to repeated phasic contraction. Hypoxia did not change the contractility during phasic contractions. Under hyoxia energy was provided to the resting control muscles by means of phosphoarginine and ATP depletion, by octopine and succinate formation and, in addition, by some accumulation of alanopine. Under normoxia, during the active state, muscles depleted phosphoarginine, which was replenished during catch. Anaerobic glycolytic ATP production did not occur. Under hypoxic conditions octopine and succinate accumulated, but only during the initial 30min of catch. Both the recovery of phosphagen and the reduction in anaerobic metabolic rate indicated a drop in energy requirements during catch.

Additional ATP turnover was only found during the period of shortening, i.e. when mechanical work was performed. The increased energy demand, during repeated phasic contractions elicited in aerated media containing serotonin, was not only indicated by a continuous breakdown of the phosphoarginine, but also by decreasing ATP levels and the accumulation of octopine.

Noticeably, ABRM and PAM of *M. edulis* appear to accumulate a different opine (octopine and strombine, respectively) during mechanical work. Bivalve PAM contains a broadly specific strombine dehydrogenase with similar activities for both glycine and alanine as substrates. ABRM also shows dehydrogenase activity with both amino acids, although a much higher activity is found for octopine dehydrogenase (de Zwaan and Zurburg, 1981).

Temperature and salinity:
Another possible role suggested for opine formation is in the osmotic control of cell volume (Zandee et al., 1980a; Dando et al., 1981). This appears unlikely, however, since the reported opine concentrations in aerobic tissues are low. Preliminary experiments with *M. edulis* exposed to different salinity regimes have failed to show any marked change in the concentrations of alanopine or strombine (Dando, unpublished data, cited in de Zwaan and Dando, 1984). Opine formation during glycolysis differs from lactate formation in that the end-product will not change the intracellular osmolarity, and this could be important during long-term anaerobiosis in osmoconformers (Dando et al., 1981).

The potential for strombine formation by anaerobic PAM, however, appears to be strongly influenced by salinity and temperature. In a study by de Zwaan and Putzer

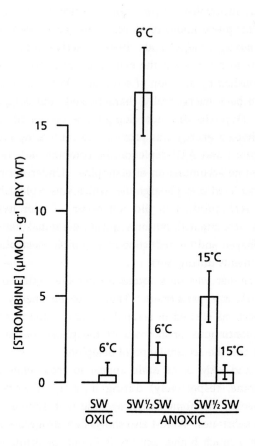

Fig. 6.5. Levels of strombine after 72h oxic or anoxic exposure in seawater in adductor muscle of *Mytilus edulis*. The mussels were acclimated to aerated seawater ot 6 °C and 32°/oo. Transfer to anoxic seawater was combined with concurrent changes in salinity (32°/oo (SW) → 16°/oo (1/2 SW)) and temperature (6 °C → 15 °C). (A. de Zwaan and V.M. Putzer, unpublished data).

(unpublished results) mussels acclimated to 6 °C and 32°/oo (ambient field conditions) were subjected to anoxic conditions combined with concurrent changes in salinity and temperature. Both acute temperature increase and hyposalinity shock blocked strombine formation under anoxic conditions (Fig. 6.5). The high accumulation of strombine in full strength seawater at low temperature is interesting in the light of the proposed function of strombine (and taurine) as cryoprotectants in intertidal *M. edulis* (Loomis et al., 1988). Exposure to anaerobic conditions has been shown to increase the freeze tolerance of *M. edulis* (Theede, 1972) and *M. demissus* (Murphy, 1977). Loomis et al. (1988) have proposed that taurine and strombine act as cryoprotectants for membranes by directly interacting with them.

Oxygen tension:
De Zwaan et al. (1991b) studied metabolic responses of *Mytilus galloprovincialis* to a series of low oxygen tensions. A graded increase in the output of anaerobic products (succinate, alanine) occured at oxygen tensions below 2.4kPa and reached a maximum at 4.5kPa. In PAM the highest net accumulation of lactate + octopine + strombine/alanopine occurred in the animals exposed to an oxygen tension of 4.5kPa. The net accumulation of these end-products decreased at both higher and lower oxygen tensions and in complete anoxia the total was only 40% of the value at 4.5kPa. The results suggest that the involvement of imino acids (opines) and lactate as products of fermentative metabolism is greatest in hypoxia. In complete anoxia, and at very low oxygen tensions that are functionally anoxic to the animal, these products are of much less importance to overall ATP-generation. It appears, then, that the compounds are produced when aerobic ATP production needs to be supplemented in an attempt to maintain high and/or constant levels of ATP use. This indicates that lactate and the imino acids are products of hypoxia stress and not of anoxia. Their accumulation may in general indicate a situation in which metabolic rate depression as a function of anaerobiosis is disturbed (stress). This is in accord with the experiments, described earlier, in which cyanide or sulfide blocked the respiratory chain of *M. galloprovincialis*. These poisons caused stronger changes in the levels of the pyruvate derivatives than incubation in deoxygenated seawater (de Zwaan and Cortesi, unpublished data; see Fig. 6.2).

Reoxidation of Cytosolic NADH: Transfer into the Mitochondrion versus Oxidation by Pyruvate Reductase

More than other tissues, muscle is involved in both environmental and functional anaerobiosis. Environmental anaerobiosis requires fuel conserving pathways (high ATP yield/unit fuel), while functional anaerobiosis requires a high ATP output pathway (high ATP output/unit time). In both pathways pyruvate will be converted into an end-product; in environmental anaerobiosis this is achieved by transamination into alanine, and in functional anaerobiosis by reduction or reductive condensation with an amino acid into lactate or an opine, respectively (de Zwaan and van den Thillart, 1985). When aerobic conditions prevail, pyruvate is directed towards the TCA cycle in spite of the fact that the pyruvate oxidoreductases are also present in the cytosol (Fig. 6.1A). These act mainly during functional anaerobiosis, hypoxia and reimmersion. It seems, therefore, that during aerobiosis and environmental anaerobiosis, cytosolic MDH overrules the pyruvate oxidoreductases for oxidizing glycolytic NADH (Fig. 6.1A and B). During aerobic glycolytic flux, reoxidation of NADH generated by the GAPDH step, will be entirely compensated for by the malate-aspartate shuttle. This can be maintained during environmental anaerobiosis because of the absence of a Pasteur effect (see below) and the possibility of modifying the shuttle for anaerobic functioning (see p.246). When, during phasic contractions, hypoxia or recovery, the rate of reduction of NAD^+ is higher, owing to a 'flare-up' of glycolysis, then the reoxidation rate of NADH associated with the hydrogen translocation system, the surplus NADH, remains available for the pyruvate oxidoreductases to form lactate or opines.

De Zwaan et al. (1983a) observed that mussels stressed by keeping the shell valves tightly closed with a rubber band, were handicapped in suppressing metabolism. Recently it was established that this treatment reduced anoxic survival time to 73% of that of control animals (de Zwaan, 1989). The forced closure resulted in a much higher energy expenditure. Strombine and, to a lesser extent octopine, accumulated in much higher amounts than in untreated anaerobic control groups. This is an illustration that the ratio of the formation of succinate (plus propionate) over pyruvate derivatives is determined by the glycolytic flux. This agrees with the above proposed competitive model for reoxidation of cytosolic NADH between hydrogen translocation to the mitochondrion and cytosolic pyruvate reduction by the action of pyruvate oxidoreductases. Further support for this model has been obtained from in vitro and in vivo studies with PAM and ventricle in which generation of glycolytic NADH and utilization of aspartate were reduced by inhibitors (de Zwaan et al., 1982 and 1983a; Foreman and Ellington, 1983). For an explanation we refer to de Zwaan and Putzer (1985).

THE PENTOSE PHOSPHATE PATHWAY

The primary function of the pentose phosphate pathway (or hexose monophosphate shunt) is to provide reducing power required for extramitochondrial reduction reactions, and the ribose-5-phosphate necessary for nucleotide and nucleic acid synthesis. During the multicyclical processes of the pentose phosphate pathway, three molecules of glucose-6-phosphate produce three molecules of CO_2 and three 5-carbon residues. These 5-carbon residues are transformed to regenerate two molecules of glucose-6-phosphate and one molecule of glyceraldehyde-3-phosphate. As two

Table 6.3. Pentose phosphate pathway enzymes identified in *Mytilus* spp.

Enzyme	Species	Reference
Glucose-6-phosphate dehydrogenase 6-Phosphogluconate dehydrogenase Transketolase Transaldolase	*Mytilus californianus*	Bennett and Nakada, 1968
Glucose-6-phosphate dehydrogenase 6-Phosphogluconate dehydrogenase	*Mytilus edulis*	Gabbott and Head, 1980
Glucose-6-phosphate dehydrogenase	*Mytilus edulis*	Livingstone, 1981
6-Phosphogluconate dehydrogenase	*Mytilus galloprovincialis*	Garcia Martin et al., 1984

molecules of glyceraldehyde-3-phosphate can reform a molecule of glucose-6-phosphate phosphate, the pentose phosphate pathway can completely oxidize glucose, as can the Embden–Meyerhof–Parnas pathway. Both pathways result in the reduction of a nicotinamide coenzyme. In the pentose phosphate pathway $NADP^+$ is the election acceptor while NAD^+ is the electron acceptor in the Embden–Meyerhof–Parnas pathway.

The enzymes involved in the pentose phosphate pathway are extramitochondrial. The dehydrogenation of glucose-6-phosphate to form 6-phosphogluconate is catalyzed by $NADP^+$-dependent glucose-6-phosphate dehydrogenase (G6PDH) and gluconolactone hydrolase. The next oxidation step is catalyzed by 6-phosphogluconate dehydrogenase (6PGDH) with $NADP^+$ as hydrogen acceptor to form ribulose-5-phosphate, which is transformed to ribose-5-phosphate. The pathway may continue further by the action of an epimerase, a transketolase and a transaldolase. The complete oxidation of glucose to CO_2 via the pentose phosphate pathway requires enzymes capable of converting glyceraldhyde-3-phosphate to glucose-6-phosphate (which permit the Embden–Meyerhof–Parnas pathway to function in reverse). The overall equation is:

$$\text{glucose-6-phosphate} + 12NADP^+ + 7H_2O \rightarrow 6CO_2 + 12NADPH + 12H^+ + Pi$$

In *M. edulis*, it is clear that the hexose phosphates are associated with the large stock of glycogen found in the storage tissues. The hexose phosphates are formed from glycogen, either by the action of glycogen phosphorylase, or indirectly by amyloglucosidase and hexokinase. These hexose phosphates are equally available to the Embden–Meyerhof–Parnas pathway and the pentose phosphate pathway, the latter often being associated with biosynthetic requirements. Several enzymes associated with the pentose phosphate pathway have been identified in mussels (Table 6.3). Rodrigez-Segade et al. (1979, 1980) have demonstrated that the inhibition of G6PDH in mussel digestive gland by $NADP^+$ could be countered by the oxidized form of glutathion (GSSG); this enzyme would also appear to be regulated by Mg^{2+} ions and by a proteinaceous cofactor that has been purified.

The mantle of the mussel is the site of extensive transformations during the annual sexual cycle. Nutrient reserves, especially glycogen, laid down in summer are utilized in autumn and winter for the formation of gametes. Gametogenesis involves extensive accumulation of lipids in the germ cells (Gabbott, 1975; see also Neuroendocrine regulation of storage tissue metabolism, p.284ff, this chapter).

This conversion of glycogen into lipids, confirmed by the increase of triglycerides in the mantle (Waldock and Holland, 1979) and by the studies of Zaba and Davies (1979, 1981) on the utilization of ^{14}C-glucose, indicates that glucose is oxidized, at least partially, by the pentose phosphate pathway. This underlines the importance of G-6-PDH in the supply of NADPH required for lipogenesis. In mantle tissue, the specific activity of G-6-PDH is higher in females than in males (Cienfuegos et al., 1977; Livingstone, 1981) and undergoes seasonal variations. This is also the case for 6-phosphogluconate deshydrogenase (6-P-GDH) (Gabbott and Head, 1980). In *M. edulis*, the maximal activity of these two enzymes occurs in summer, probably due to the increase in food intake and modifications in carbohydrate metabolism of the mantle and digestive gland. During this period there is an increase in the digestive gland lipid and carbohydrate content (Thompson et al., 1974) and this synthetic activity is accompanied by an increased demand for NADPH. On the other hand, the apparent K_m of G-6-PDH for G-6-P and $NADP^+$ only shows slight seasonal variations (Livingstone, 1981).

The regulation of the pentose phosphate pathway in the mussel is important in mantle tissue, which is the site of both nutrient storage and gametogenesis. This is especially so in females, where the reproductive effort necessitates the accumulation of lipids and the synthesis of nucleic acids (Gabbott, 1975; Waldock and Holland, 1979). Zaba and Davies (1981) estimated total glucose utilization by measuring quantities of tritium recovered in metabolites following administration of [2–3H]-glucose. Using data on the conversion of [1-^{14}C] and [6-^{14}C]glucose to $^{14}CO_2$ these authors calculated a C1/C6 ratio of 27:58 for values of glucose utilization ranging from 1.04 to 1.84µmol h^{-1}

g^{-1} wet wt. The application of the method of Katz and Wood (1963) gave an estimation of the contribution of the pentose phosphate pathway of between 0.8 and 1.8% of total exogenous glucose utilization, or, in other words, 8.4–32.0nmol h^{-1} g^{-1} wet wt. However, as pointed out by Zaba and Davies (1981), the method employed tends to underestimate the contribution of the pentose phosphate pathway, and a more realistic value would be 1.1–2.6%. These results demonstrate that the pentose phosphate pathway is indeed active in *M. edulis* (as reflected by a high C1/C6 ratio in CO_2 given off), but that its contribution to glycogen metabolism is rather low (C1/C6 ratio in glycogen and amino acids close to unity). In addition, there is a discrepancy between the activity of G-6-PDH and the metabolic flux attributed to the pentose phosphate pathway (Zaba and Davies, 1981). It is, therefore, possible that the demand for NADPH is fulfilled by other means, such as, for example, NADP-dependent isocitrate dehydrogenase, (Gabbott and Head, 1980). Nonetheless, it is apparent that a nonoxidative pentose phosphate pathway is functional, and can supply pentoses required for nucleic acid synthesis.

Substrates present in the mantle are ultimately derived from the circulatory system, either directly or from specialized cells of the nutrient storage tissue. Glucose, in particular, may be derived from the free metabolite in the haemolymph, or from the hydrolysis of stored glycogen. The relative importance of these two potential sources of glucose in the mussel has been studied by Livingstone and Clarke (1983) and found to vary seasonally. During periods of abundant food supply (summer and autumn in the English Channel, U.K.) extracellular free glucose is the principal source. In winter, however, when intracellular glycogen content is high and the food supply less abundant, glycogen becomes the major source of glucose. This hypothesis is supported by annual variations of haemolymph glucose content (Livingstone and Clarke, 1983). With an extracellular glucose concentration of 5mM, phosphorylation of glucose is possible throughout the year (Zaba et al., 1981). This concentration is much higher than that actually encountered in the haemolymph of *M. edulis* (≤1mM: Bayne, 1973; Whittle, 1982; Livingstone and Clarke, 1983). Despite this fact, however, the incorporation and phosphorylation of glucose by *M. edulis* tissues has been demonstrated (Zaba et al., 1981) with summer values of 1.69μmol h^{-1} g^{-1} wet wt and winter values of 0.78μmol h^{-1} g^{-1} wet wt. The analysis of the radioactivity of glucose-6-phosphate shows that extracellular glucose is the sole source of hexose phosphate during the summer, while stored glycogen accounts for 40% of hexose phosphate production in winter.

METABOLIC RATE

Estimation of Metabolic Rate

Various methods are available to estimate the metabolic rate (also known as energy demand, or ATP turnover rate). Under aerobic conditions it can be derived directly from the rate of oxygen consumption using oxycaloric coefficients (indirect calorimetry) to convert oxygen uptake to energy demand. Anoxic metabolic rates can be calculated from the accumulation of glycolytic end-products and changes in ATP and phosphoarginine. This biochemical method has often been applied for *Mytilus* spp. (de Zwaan and Wijsman, 1976; Ebberink et al., 1979; Zurburg and Ebberink, 1981; Shick et al., 1983; de Zwaan et al., 1991b). When energy metabolism has simultaneous aerobic and anaerobic components, during, for example, aerial exposure (Widdows et al., 1979; Widdows and Shick, 1985), or during hypoxia (de Zwaan et al., 1991b; van den Thillart et al., 1992), energy demand can be measured by both oxygen uptake and the biochemical method, or alternatively, by direct calorimetry. Heat production is an unspecific measure of the enthalpy changes, accompanying all metabolic reactions of an organism and can be measured continuously. Direct calorimetry, and the application of thermodynamics in the study of energy metabolism of euryoxic invertebrates, has been introduced and strongly promoted by Gnaiger (Gnaiger 1977, 1983).

Aerobic heat production in *M. edulis* measured directly at 15°C in oxygenated seawater appears to be in good agreement with that calculated from the rate of oxygen consumption (Famme et al., 1981; Shick and Widdows, 1981). Under anaerobic conditions Gnaiger (1980) calculated, using mainly biochemical data from the literature, that there was a large discrepancy between direct and indirect (biochemical) calorimetric measurements in *M. edulis* and other euryoxic invertebrates. Simultaneous measurements by Shick et al. (1983) of the anoxic metabolic rate in *M. edulis*, by direct (heat production) and indirect calorimetry, qualitatively confirmed Gnaiger's suggestion of an 'excess' anoxic heat production, which cannot be accounted for by the stoichiometric equations of glycogen fermentation. However, individual variability was sufficiently large that the average 35% discrepancy up to 12h of anoxia was not statistically significant, and either method provided a reasonable estimate of anoxic energy demand. During 48h of anoxia the discrepancy increased to 63% and was statistically significant. The authors suggest that excess heat production during prolonged anoxia may have included enthalpy changes owing to dissolution of calcium carbonate in the shell by the acidic end-products of anaerobiosis, tissue autolysis after lysosomal destabilization, and 'anoxic endogenous oxidation'. The first argument has been disputed by Hardewig et al. (1991). They, for the first time, obtained a close agreement on energy expenditure derived from direct and indirect heat production for an intertidal invertebrate. The concept of 'anoxic endogenous oxidation' has been discussed in previous reviews by de Zwaan (1977, 1983) and due to a lack of new supporting evidence in euryoxic invertebrates, will not be dealt with here.

Despite the discrepancies between 'indirect' and 'direct' calorimetry, all studies have shown that metabolic rate is dramatically reduced in anoxia. Metabolic rate depression starts immediately the oxygen concentration of the seawater begins to fall, but, at this time, respiration still overrules fermentation in meeting the cellular energy demand (van den Thillart et al., 1992). The final anoxic metabolic rate is only a fraction of the rate in oxygenated seawater. Comparing biochemical data with oxygen uptake values from the literature, de Zwaan and Wijsman (1976) calculated the anoxic energy demand in *M. edulis* to be 5.4% of the aerobic value. Biochemical and oxygen uptake studies for *M. galloprovincialis* revealed that after 12h of exposure to anoxic water the anoxic metabolic rate was about 4.5% of the normoxic resting rate (de Zwaan et al., 1991b). In biochemical and oxygen uptake studies carried out simultaneously Widdows et al. (1979) calculated that the total ATP production in *M. edulis* after 6h of air exposure represented 12% of the aerobic ATP production in water over a similar period. During the first 6h of aerial exposure, anaerobic energy production appeared to be about 60% of total ATP production. More recently, Widdows and Shick (1985) studied physiological responses of *M. edulis* to intertidal conditions which included measurements of heat dissipation and oxygen uptake. Mussels acclimated to intertidal and subtidal regimes had markedly different rates of total heat dissipation and oxygen uptake when exposed to air. Specimens which continuously feed under subtidal conditions, when subjected to acute air exposure for 5h, had a significantly higher heat dissipation than intertidal mussels, but no measurable oxygen uptake; the intertidal mussels, on the other hand, had a rate of oxygen uptake, which represented 40% of total heat dissipation in air. Similarly, Shick et al. (1983) estimated an aerobic contribution of 28% of total energy metabolism in air for *M. edulis* .

Pamatmat (1980) reported that the anaerobic rate of heat dissipation by *M. edulis* exposed to air was 7–18% of the aerobic aquatic heat dissipation. Comparable studies by Shick et al. (1983) revealed values of 7–10% for periods of up to 24h. Heat production values for *M. edulis* in anoxic seawater by Hammen (1980) and Famme et al. (1981) appeared to be considerably higher than the above values. This agrees with the much higher ATP turnover rate calculated from biochemical data for *M. edulis* kept in anoxic seawater when compared to air (Zurburg and Ebberink, 1981). Biochemical analyses and direct calorimetry have revealed that the ATP consumption with time (up to 10–12h) gradually drops to low values (Ebberink et al., 1979; Pamatmat, 1980; Zurburg and Ebberink, 1981; Shick et al., 1983). For *M. edulis* PAM energy expenditure was reduced by a factor of 5.3 in the first 12h of shell valve closure and remained constant for the next 12h (Ebberink et al., 1979). During the transition stage of anaerobiosis (Fig. 6.1B) transphosphorylation of phosphoarginine and ADP may be as important as catabolism in contributing to the anerobic power output. In the transition stage the glycolytic flux has not yet reached its lower limit of the anoxic steady state phase (Fig. 6. 1C) which, besides succinate, results in alanine as well as some opine production (de Zwaan and Putzer, 1985). The initial glycolytic rate during long-term shell valve closure (catch) appears to be inversely related to the quantity of stored phosphoarginine. The latter shows a strong seasonal variation (Zurburg and Ebberink, 1981).

To obtain the same amount of ATP, fermentation leading to succinate and propionate production requires six to seven times more glycogen than aerobic metabolism. The reduction in metabolic rate, however, compensates for, or even overrides, the reduction in the yield of ATP as compared with respiration. A Pasteur effect (increased carbon flow through the glycolytic pathway to compensate for low efficiency in ATP generation by fermentation) is therefore avoided, or even reversed, during environmental anaerobiosis.

Regulation of Metabolic Rate

Metabolic arrest, the ability to greatly reduce ATP requirements in the anoxic state, compared to the basal rate of ATP utilization under aerobic conditions, is probably the most effective adaptation to extend survival in the absence of environmental oxygen. In order to be a viable survival strategy, mechanisms for both reducing cellular energy expenditure and balancing the rates of ATP-producing and ATP-utilizing reactions in anoxia must be present. The necessity for strict control over glycolysis is obvious, since this is the primary pathway of ATP production during anaerobiosis. The basic kinetic properties and the effect of modulators on glycolytic enzymes have been reviewed by de Zwaan (1977, 1983) and will not, therefore, be dealt with here. For glycogen phosphorylase and hexokinase we refer to Ebberink and Salimans (1982) and Livingstone and Clarke (1983), respectively. Some basic kinetic properties of all glycolytic enzymes of mantle and PAM of *M. edulis* are presented by Churchill and Livingstone (1989). Storey (1986b) deals with aspartate activation of pyruvate kinase (PK) in anoxia-tolerant molluscs, including PK from hepatopancreas of *M. edulis*. Effects of fructose 2,6-biphosphate on glycolytic flux regulation in *M. galloprovincialis* are presented in Villamarin et al. (1990a) and Ibarguren et al. (1990), and the effect of pH on phosphofructokinase (PFK) from the same species in Villamarin et al. (1990b). Additional information on a comparative study of kinetics of PK and PEPCK in *M.*

edulis and *M. galloprovincialis* can be found in de Vooys (1980) and de Vooys and Holwerda (1986). Studies with monoclonal antibodies have confirmed conclusions based on kinetic data that a L-type PK isozyme is present in various tissues of *M. galloprovincialis* (Papadopoulos et al., 1990).

Recent studies have revealed three mechanisms of glycolytic rate depression that operate in the tissues of marine molluscs during aerobic-anaerobic transition. These are: (1) reversible phosphorylation control over the activity state of regulatory enzymes, (2) association-dissociation of enzymes from complexes bound to the subcellular particulate fraction, and (3) changes in the concentration of fructose 2,6-biphosphate, a potent activator of PFK (Storey, 1985a, 1986a; Storey and Storey, 1990).

The first mechanism appears to be most important in the Mytilidae and has received much attention. The aerobic-anaerobic transition may result in a reduction in the percentage of glycogen phosphorylase in the phosphorylated, active *a* form and appears to induce a stable modification of the properties of PFK and PK, which is consistent with a less active form of both enzymes being present in the anoxic state. One way of assessing the importance of enzyme covalent modification to metabolic rate depression and anoxia survival is to compare the enzymic responses to anoxia by bivalve species with different anoxia tolerances. This approach has been recently followed in an in-depth analysis of the physiological and metabolic reasons for the steep decline in populations of the commercially important bivalve species *Venus gallina* and *Mytilus galloprovincialis* in the North Adriatic Sea, at the same time that populations of the introduced Indo-Pacific clam *Scapharca inaequivalvis* are soaring. *V. gallina* shows very poor tolerance of anoxia, with LT_{50} (50% survival) values of only 4 days at 17–18°C, compared with nearly 20 days for the two other species (Brooks et al., 1991; de Zwaan et al., 1991b). Thus, *M. galloprovincialis* and *S. inaequivalvis*, both with anoxic metabolic rates about 4.5% that of aerobic values, are clearly better prepared for long-term anoxia survival than is *V. gallina*, with a corresponding value of 13%. Furthermore, only *M. galloprovincialis* and *S. inaequivalvis* showed evidence of specific molecular mechanisms for the co-ordinated reduction of the rates of various metabolic processes in anoxia. Anoxia-induced post-translational modifications of key cellular enzymes (including glycogen phosphorylase, PFK and PK), producing less active forms, occurred in both *M. galloprovincialis* and *S. inaequivalvis* but not (or to a far lesser extent) in *V. gallina*. Thus, anoxia led to a significant decrease in the percentage of phosphorylase in the active form in the PAM of *M. galloprovincialis* as well as a 3.2-fold decrease in the I_{50} (50% inhibition) for Mg.ATP of PFK (de Zwaan et al., 1991b). Similar changes in phosphorylase and PFK (as well as significant alterations of other PFK kinetic parameters) were noted for *S. inaequivalvis* foot muscle after anoxic exposure, but anoxia had no effect on the properties of either enzyme from *V. gallina* foot (Brooks et al., 1991; van den Thillart et al., 1992).

Ebberink and Salimans (1982) determined the levels of glycogen phosphorylase in vivo in the PAM of *M. edulis* after various periods of valve closure. In contrast to *M. galloprovincialis*, there was no detectable conversion of phosphorylase *a* to *b*. They also found no conversion of inactive phosphorylase kinase to the active form.

The second mechanism has been established for whelk muscle during environmental anoxia (Plaxton and Storey, 1986). The aerobic/anaerobic transition leads to a decrease in the association of glycolytic enzymes with the particulate fraction; the probable result being a decrease in cellular organization, thereby promoting a decrease in glycolytic flux. This mechanism has not been investigated in bivalves. The third mechanism has been studied, to some extent, in a number of bivalve species, including *M. edulis* (Storey, 1985b). Fructose-2,6 biphosphate appears to play a significant role in decreasing glycolytic rate (at least in soft tissues) during anoxia by depressing carbohydrate availability for anabolic purposes.

Depression of metabolic rate by covalent modification

Stable modification of enzymes of glycolysis is an integral part of the survival strategy in anoxia-induced metabolic arrest. That this modification is due to the actions of protein kinases or phosphatases on PFK and PK has been suggested from the effects of in vitro treatments of tissues (or homogenates) with agents that mimic the changes in enzyme properties seen during the aerobic/anaerobic transition.

The discovery that (de)phosphorylation is probably involved in the regulation of PK activity in the PAM stemmed from the observation of electrophoretically distinguishable forms by Siebenaller (1979). Studies by Holwerda and co-workers (Holwerda et al., 1981, 1983) offered evidence for the involvement of a cAMP-dependent protein kinase in covalent modification during environmental anoxia and temperature adjustment (for a detailed description see de Zwaan and Dando, 1984). Holwerda et al. (1984) followed the process of phosphorylation and dephosphorylation in the mantle and PAM of *M. edulis*. During exposure to air there appears to be a shift towards phosphorylated PK in both organs, whereas the opposite is observed during re-immersion. The changes were more pronounced in the mantle than in the PAM. This was re-established in a more recent study by Holwerda et al. (1989) on the time-dependent decrease of PK activity in four tissues of *M. edulis* during aerial exposure. By the use of excised gill Holwerda et al. (1989) showed that the anoxic state *per se* of the isolated organ could trigger enzyme modification. It is, therefore, unlikely that inactivation will be induced by hormonal action. Anaerobiosis probably brings about a change in the intracellular milieu, resulting directly in the modification of PK.

Phosphorylation can probably occur both via a Ca^{2+}/calmodulin and a cAMP-dependent mechanism. Calmodulin and a calmodulin-dependent protein kinase have recently been isolated and characterized from PAM of *M. edulis* (Sailer et al., 1990). Gill flaps incubated in aerated medium in the presence of the calcium ionophore A-23187 showed a decrease in the activity ratio for PK, whereas incubation in aerated Ca-free seawater, with or without the ionophore resulted in unalterated activity ratios (Holwerda et al., 1989). Further evidence was obtained from incubation of the supernatant fraction of gill homogenate. Here the PK activity ratio fell to 85% of the control (not incubated) value after 1.5h of incubation. The decrease could be prevented by the presence of the calmodulin inhibitor trifluorperazine or EGTA (Ca^{2+}-chelator).

The decreasing effect of ATP plus cAMP (indicative of a cAMP-dependent mechanism) on enzyme activity has been compared in gill and muscle homogenates in *M. edulis* (Holwerda et al., 1989). The response was considerably stronger in muscle than in gill. The cAMP-dependent mechanism was dependent on a factor present in the low molecular weight fraction of a Sephadex G-25 gel-filtrate of centrifuged (50,000g x 20min) homogenate. This essential factor was different from the calcium ion.

Inactivation of PK by phosphorylation also helps to control the PEP-branchpoint, by re-routing glycolytic carbon away from PK and into the PEP carboxykinase reaction, and hence into succinate synthesis (see de Zwaan, 1977 and de Zwaan and Dando, 1984 for details on the control of the PEP-branchpoint).

Covalent enzyme modification also appears to be involved in the control of PFK. PFK isolated from ABRM of *M. edulis* undergoes a stable modification that changes its kinetic properties in response to anoxic exposure (Michaelidis and Storey, 1990). Affinity for both substrates (Fru-6-P, ATP) decreased in anoxia, and the kinetic constants for Pi activation, and for PEP and ATP inhibition, changed significantly. This probably converts the enzyme to a less active enzyme form. In vitro incubation of an ABRM-PFK preparation with protein kinase second messengers indicated that changes in the phosphorylation state are responsible for the anoxia-induced modification of the enzyme. A cAMP-dependent protein kinase appears to be present in ABRM homogenates that can phosphorylate PFK and, in doing so, activates the enzyme by increasing its affinity for Fru-6-P. Alkaline phosphatase decreased both the maximal activity of the enzyme and its affinity for Fru-6-P. These results, therefore, strongly indicate that the effect of phosphorylation is enzyme activation. Furthermore, the results suggest that the aerobic versus anoxic forms of ABRM-PFK represent the phosphorylated and dephosphorylated enzyme, respectively.

Also, results from studies on hepatopancreas and PAM of *M. edulis* (Michaelidis and Storey, 1991) indicated that the depression of glycolytic rate that occurs with the transition to anoxia is facilitated by reversible phosphorylation control over the activity of PFK. This study again showed that the properties of PFK of both tissues alter as a result of anoxia exposure. In particular, PFK affinity for Fru-6-P, and the effects of activators on the enzyme, were decreased in anoxia. These stable alterations are consistent with the presence of a less active enzyme form during anaerobiosis. Again, the results indicate that the aerobic versus anoxic forms of PFK represent the phosphorylated and dephosphorylated enzyme, respectively. This mechanism is opposite to that of PK inactivation in anoxia; as shown by Holwerda (see above), the anoxic form of this enzyme is clearly the phosphoprotein. The results for PAM-PFK also differed distinctly from those obtained by de Zwaan et al. (1991b) for PAM in *M. galloprovincialis* in that: (1) the anoxic enzyme form of *M. edulis* showed reduced maximum activity, but remained unchanged in *M. galloprovincialis*, and (2) the decreased response to ATP inhibition of PFK from *M. edulis* PAM was opposite to that of PFK from *M. galloprovincialis*. In the latter species the I_{50} for ATP dropped strongly in anoxia, the value being only 32% of the control aerobic value after 24h. In both species PFK from anoxic muscle showed a reduced affinity for Fru-6-P.

Depression of metabolic rate by acidification

Little is known about factors triggering the modification of enzymes involved in metabolic rate depression. It appears that anoxia-tolerant bivalves in general are more vulnerable to 'forced' anoxia caused by blocking the respiratory chain with cyanide or hydrogen sulfide in aerated seawater, than to anoxia caused by the absence of oxygen in the seawater (environmental anoxia). In PAM of both *M. galloprovincialis* (see Fig. 6.2) and *S. inaequivalvis* (de Zwaan et al., 1991a), as well as the red blood cells of the latter species, the anaerobic pathway leading to pyruvate derivatives was clearly operating at higher rates in the presence of these poisons. This resulted in a decreased tolerance to anoxia. Possibly the stimulus to trigger the process(es) leading to metabolic arrest will only be evoked when oxygen tension in the blood and the tissues drop gradually, as is the case with environmental anoxia, but not when the respiratory chain is specifically and promptly blocked. In agreement with this assumption it has been observed that in *S. inaequivalvis* the increased sensitivity for PFK inhibition by ATP (I_{50}) starts immediately the oxygen consumption rates drop, due to a reduced oxygen concentration in the ambient seawater. Oxygen consumption and I_{50} for ATP appear to be strongly integrated over the whole range of oxygen tensions from oxygen saturated seawater to anoxic water (van den Thillart et al., 1992).

A slightly alkaline intracellular pH (pHi) seems important for many cellular functions. There is a correlation between increased pH and, for example, glycolytic activity, cell motility and membrane transport (Nuccitelli and Heiple, 1982). Studies on the direct effect of a pH decrease on the kinetic properties of PFK and PK showed a decreased affinity for substrate, an increased sensitivity towards inhibitors and a decreased sensitivity towards activators (see de Zwaan, 1977 and 1983 for a summary; Holwerda et al., 1983; Villamarin et al., 1990b). It may be that metabolic acidosis during anaerobiosis has a dual effect on metabolic rate depression by direct action on enzyme catalytic activity, and by triggering the modification of enzymes by (de)phosphorylation. In their search for the prime stimulus of enzyme deactivation, Holwerda et al. (1989) considered cellular acidosis as a possibility. Intracellular pHi was manipulated by incubating gills in the presence of the weak organic acid 5,5-dimethyloxazolidine-2,4-dione (DOMO) at lowered pH. The activity ratio of PK decreased slowly but steadily.

The development of metabolic acidosis due to fermentative cellular processes has been reported for several molluscs, including *M. edulis* for both the intra- and extracellular compartment. This subject has been dealt with in previous reviews (de Zwaan, 1977; de Zwaan and van den Thillart, 1985) but will be continued here with the discussion of additional relevant literature.

A theoretical consideration of metabolic acidosis in relation to anaerobiosis has been given by Hochachka and Mommsen (1983) and Pörtner (1987). Changes in pO_2, pCO_2, pH, Ca^{2+} and NH_4^+ of the hemolymph of *M. edulis* during hypoxia (and recovery) are presented in Jokumsen and Fyhn (1982), de Zwaan et al. (1983b) and Shick et al. (1986). Acid-base balance in *M. edulis* has been very well-studied both for hemolymph and intracellular fluid in relation to shell valve closure and respiratory acidosis (hypercapnia) (Booth et al., 1984; Lindinger et al., 1984; Walsh et al., 1984). In these studies pHi was established by the DMO/inulin method. During 8h exposure in air PAM pHi decreased from 7.38 to 7.12, foot pHi from 7.42 to 6.92 and mantle pHi from 7.15 to 6.27. These pHi decreases were inversely correlated with tissue buffering capacity. Referring to pH profiles for PK and PEPCK (de Zwaan et al., 1975b and de Zwaan and de Bont, 1975, respectively), these authors concluded that their observed pHi change in the PAM after 8h air exposure would have virtually no effect on PK and PEPCK activity (Walsh et al., 1984). The relatively small drop of pHi in PAM is probably due to the important contribution made by the hydrolysis of phosphoarginine to ATP in the transition phase of environmental anaerobiosis, which results in the release of the base arginine (Pörtner, 1987). For PAM of *M. edulis*, ATP equivalents derived from phosphoarginine, ATP and metabolic fermentation as a percentage of total ATP turnover, were, during the first 4h of anaerobiosis 60, 9 and 31%, respectively, as compared to 0, 0 and 100 percent after 12h (de Zwaan and Putzer, 1985). Acidosis, therefore, probably does not occur during the first hours of aerial exposure (as opposed to tissues with low or no phosphoarginine).

Also, Ellington (1983b) observed a relatively small degree of intracellular acidification in in vitro preparations of M. *edulis* PAM using phosphorus nuclear magnetic resonance (NMR) spectrometry. According to the author this probably reflects low rates of energy metabolism in this muscle, although he does point out that the degree of acidification may be sufficient to have important regulatory effects on energy metabolism in catch muscle.

More recently, Zange et al. (1990) established a pHi of resting ABRM in M. *edulis* of 7.4, determined both by DOMO and NMR-spectrometry. The pHi was established during a contraction-catch-relaxation cycle. During catch the pHi was not significantly different from its value at rest.

LIPID METABOLISM

The biochemistry of lipid metabolism in the bivalves has been the subject of a number of studies, largely due to the important position held by this group in marine food chains. Indeed, these filter-feeding animals feed abundantly on phytoplankton which, therefore, plays a role in the accumulation of lipids formed during primary production. Another important consideration in the lipid biology of bivalve molluscs is its link with pollution (see Chapter 9). From a physiological viewpoint, lipids constitute an important form of nutrient storage. This is used to fuel gametogenesis and is also the principle form of nutrient storage in the oocytes themselves, destined to cover the energetic requirements of the embryos and early larval stages (reviewed by Holland, 1978). Other aspects include the now evident role of steroid hormones in the regulation of reproduction and growth (see p.276ff). The role of prostaglandins in the regulation of gamete emission in bivalve molluscs has been studied in the scallop, *Patinopecten yessoensis* (Matsutani and Nomura, 1986; Osada et al., 1989).

The ever increasing perfection of the techniques used to study lipids has served to increase our knowledge of the lipid biochemistry of the mussel. This is especially so for the detection and identification of lipid substances in tissues (see Joseph, 1982 for review). Recent results have, thus, tended to overshadow those from earlier studies (see, for example, Gardner and Riley, 1972 for a critical review of lipid extraction methods). Lipid metabolism and especially its regulation remain, however, an area where our understanding is still fragmentary.

Lipid Distribution

There exists a notewordy heterogeneity in the lipid composition of different bivalve species. According to Mane and Nagabhushanam (1975) sedentary species, including *M. edulis*, which are occasionally obliged to survive periods of anaerobiosis, have a lower lipid content than other species. Indeed, during anaerobiosis, glycogen would appear to be a better form of energy reserve. Even under these conditions, however, the role of lipids cannot be ignored (van der Horst, 1974; Oudejans and van der Horst, 1974; Zs. Nagy and Galli, 1977). Taking into account the fact that the lipid content of mussels increases at a much lower rate than carbohydrate and protein content during growth, Zandee et al. (1980a) suggest that it is the lipids that constitute the principal source of energy for growth.

The lipid content of the mussel undergoes only weak seasonal variations. In *Mytilus platensis*, for example, the lipid content varies from 0.9 to 1.8% of the wet weight (DeMoreno et al., 1980). Differences also exist in the lipid content of various organs of the mussel. The mantle and the digestive gland generally display the highest neutral lipid content, probably reflecting the storage capacities of these tissues. For populations of mussels in the English Channel, lipids are conserved for gametogenesis between autumn and spring, whereas carbohydrates and proteins are used for both energy production and gametogenesis. The highest values observed in the gonad coincide with gametogenetic activity and oocyte maturation: gamete release is reflected by a sharp fall in lipid concentration (literature cited in Pieters et al., 1980). In addition, the transformation of stored carbohydrates in the nutrient storage tissue into lipids stored in the gametes has been suggested by Gabbott (1975). The relative proportions of polar and nonpolar lipids in whole tissues also undergo seasonal variations (DeMoreno et al., 1980).

Fatty Acids

Numerous authors have described the fatty acid composition of the mussel (Rodegker and Nezenzel, 1964; Bannantyne and Thomas, 1969; Cerma et al., 1970; Calzolari et al., 1971; Gardner and Riley, 1972; Paradis and Ackman, 1977; DeMoreno et al., 1980). With the exception of a few minor differences of opinion, most authors agree that the predominant fatty acids are 16 : 0 (palmitic acid), 16 : 1 (palmitoleic acid), 18 : 1 (oleic acid), as well as the long-chain polyunsaturated fatty acids 20 : 5 ω 3 (eicosapentaenoic acid) and 22 : 6 ω 3. Palmitic acid is the major final product of fatty acid synthetase activity in animal tissues, and constitutes the precursor for *de novo* synthesis of saturated and unsaturated long-chain fatty acids. In fish and higher vertebrates, fatty acids produced by the elongation and desaturation of palmitate have a ω 9 configuration. Indeed, these organisms are incapable of synthesizing ω 3 (linolenic acid) and ω 9 (linoleic acid) series fatty acids. As these fatty acids are essential requirements for growth, they must be furnished in the diet. Kanazawa et al. (1979) have shown that whereas freshwater invertebrates are capable of actively transforming linolenic acid to 20 : 5 ω 3 and 22 : 6 ω 3, this is not the case for marine invertebrates. It is, therefore evident that feeding constitutes the major source of long-chain polyunsaturated fatty acids for marine mussels.

One of the principal characteristics of the fatty acid composition of the mussel is the diversity of molecules encountered. The fatty acid distribution varies between species, and even within the same species can vary geographically and seasonally. These variations result principally from the influence of food supply and temperature. For example, observed variations in the 20 : 5 ω 3 and 22 : 6 ω 3 contents can be attibuted to diatoms, dinoflagellates and bacteria on which the mussel feeds. This direct influence of the food supply on fatty acid composition has been clearly demonstrated for oyster species by Watanabe and Ackman (1974). Temperature influences the degree of unsaturation of fatty acids. A decrease in temperature results in an increase in the degree of unsaturation (literature cited in DeMoreno et al., 1980).

The fact that lipid content varies relatively little in *Mytilus* should not overshadow the important role of fatty acids, particularly as stuctural components of cell membranes, and also in general metabolism. The total lipid content of the mussel represents about 10% (maximum) of the dry weight (Kluytmans et al., 1985). Of this, 10% (= 1% of dry wt) represents structural lipids (membranes), and the remaining 90% can be attributed to storage lipids. In starved mussels, or mussels that have recently spawned, the carbohydrate and protein concentrations increase very rapidly after feeding. After feeding if, on the other hand, animals are in normal condition, the accumulation of lipids is low, even though the planktonic algal food is lipid-rich. It would, thus, appear that the ingested lipids are rapidly metabolized, whereas the proteins and glycogen are stored (Kluytmans et al., 1985). Moreover, Ballantyne (1984) has shown that lipids are the preferential substrate of digestive gland mitochondria and are oxidized in greater quantity than pyruvate or other metabolites.

In *Mytilus* the 16 : 0 and 18 : 0 fatty acid content diminish from April to January for English Channel populations and increase again between January and spring. This increase coincides with a decrease in glycogen content (Gabbott, 1983). The increase in the concentrations of saturated fatty acids or 16 : 0, 16 : 1, 18 : 0 and 18 : 1 mono-unsaturated fatty acids corresponds to *de novo* synthesis prior to spawning, undoubtedly linked to the conversion of glycogen into lipids within the developing eggs (see Gabbott, 1975). The fact that the lipid content of the mantle tissue remains relatively stable during autumn and winter when food supply is minimal, whereas the protein and carbohydrate contents undergo a steady decline, is evidence for lipid biosynthesis (Kluytmans et al., 1975). A study of two enzymes characteristic of *de novo* fatty acid synthesis (acetyl CoA carboxylase and fatty acid synthetase) has shown that their activities in the mantle and digestive gland of *M. edulis* are negligible (Kluytmans at al., 1985). This result is in apparent contradiction with the measured increase in mantle lipid concentration (which does not coincide with a decrease in other tissues) during the winter when the food supply is poor. Measurements made during the prespawning period (April–May), however, have shown increased activities of these enzymes in female mantle, and very much lower activities in male mantle. Lubet et al. (1986) have confirmed that the measured lipid content of the female mantle was twice that observed for the males. In addition, these authors have demonstrated that the 16 : 0 / 16 : 1 ratio of saturated to monosaturated differs between males and females, apparently reflecting differences in the activity of the desaturase. Therefore, the biosynthesis of fatty acids in the mussel is a cyclic phenomenon, linked to the reproductive cycle, and especially to the production of female gametes and to vitellogenesis. It is likely that this biosynthesis is under hormonal control (see later).

Sterols and Steroids

A number of sterols and steroids have been identified in the bivalves (see reviews by Bergman, 1949, 1962; Austin, 1970; Idler et Wiseman, 1972; Voogt, 1972; Goad, 1976). They are all derived from a common nucleus: cholestane and their synthesis takes place in three stages: (1) formation of squalene from acetyl CoA; (2) cycling of squalene to form lanosterol and (3) demethylization of lanosterol to give zymosterol, the precursor of all of the other sterols.

The bivalves contain a complex mixture of Δ5 sterols which differ in the length of their side-chain and in the degree of unsaturation of the nucleus and side chain. Up to 13 different sterols occur in any one species (Idler et Wiseman, 1972). Nine sterols have been identified in *M. edulis* by a number of authors (Toyama and Tanaka, 1956; Takanashi and Mitsuhashi, 1968; Idler and Wiseman, 1971; Teshima and Kanazawa, 1972), although there is a lack of agreement as to their proportional concentrations:

22-*trans*-24 norcholesta 5-22-dien-3β-ol

22-*cis* cholesta 5-22-dien-3β-ol

22-*trans* cholesta 5-22-dien-3β-ol

cholesterol

brassicasterol

demosterol

22-23 dihydrobrassicasterol (campesterol)

poriferasterol

β-sitosterol

The predominant sterols are often the C_{27} sterols, often in association with C_{26}, C_{28}, C_{29} and C_{30}. The most abundant sterol is cholesterol which represents between 41.5 and 58.7% of the sterol content. Some of these sterols, especially β-sitosterol seem to vary according to locality. These variations have been attributed to the diet of the animals. Indeed, according to Goad (1978), bivalve sterols are of multiple origin. They can be produced by *de novo* biosynthesis from acetyl CoA via mevalonic acid and squalene, or they may be derived from the absorption and assimilation of dietary sterols. The assimilated sterols may be modified within the mollusc. Symbiotic organisms, especially algae, may be another source of sterols. The study of sterol biosynthetic mechanisms has been hampered by a number of problems linked to the dilution of radioactive precursors. In addition, Goad (1978) has pointed out that there is a difference between the ability to synthetize sterols and the necessity to do so.

It is generally accepted, however, that the C_{27} sterols are synthetized *de novo* whereas the C_{26}, C_{28}, C_{29} and C_{30} sterols are of dietary origin. It would appear that the molluscs are capable of synthesizing sterols, but that their actual sterol production is low. As far as *Mytilus* is concerned, some results would appear to be contradictory. For example, Fagerlund and Idler (1960, 1961a, b) have demonstrated the formation of ^{14}C-cholesterol from ^{14}C-acetate, and [14]C-24-methylenecholesterol from ^{14}C-cholesterol in *M. californianus*, whereas Walton and Pennock (1972) were unable to show sterol synthesis from ^{14}C-mevalonate in *M. edulis*. In the latter study, however, the radiolabel was located in substances closely related to sterols, such as ubiquinone, dolichol, farnesol and geranylfarnesol. Although Voogt (1975a, b) considers that there is no strong evidence that mussels synthesize sterols, Teshima and Kanazawa (1974) have shown that the administration of ^{14}C-mevalonic acid leads to the synthesis of radiolabelled squalene and sterols (cholesterol, 24-methylene-cholesterol and 22-dehydrocholesterol), showing that the mussel has the capacity to alkylate C_{24}. On the other hand, however, Teshima et al. (1979) have shown that the mussel is incapable of converting β-sitosterol into C_{27} sterols, even though its conversion into 5α-stigmastan-3β-ol is possible, suggesting that the mussel lacks the capacity to dealkylate C_{29} sterols.

As far as steroid hormones are concerned, the results to date are still fragmentary. The presence of steroids has been demonstrated in the tissues of a number of bivalve species (see reviews by Reis-Henriques et al., 1990 and by Lubet and Mathieu, 1991). Using thin-layer chromatography and mass spectrometry, Reis-Henriques et al. (1990) have recently demonstrated the presence of progesterone, androstenedione, testosterone, 5α-dihydrotestosterone, estradiol 17β, androsterone and 3α-androstanediol in

M. edulis. The pathways of steroid biosynthesis in the mussel have been studied by de Longcamp et al. (1974), who have measured the activities of 3β-hydrosteroid dehydrogenase/Δ5 isomerase, C_{17-20} lyase, 17β-hydrosteroid-dehydrogenase and 5α-reductase. These enzymatic activities, however, function with a low yield, except when the substrates furnished are œstrogens. Equally, of all the steroids identified in *M. edulis,* it is progesterone, followed by androstendione that are present in the highest concentrations, and these vary seasonally regardless of gender (Reis-Henriques et al., 1990). The exact role of the identified steroid hormones in the bivalves remains to be elucidated. According to Reis-Henriques and Coimbra (1990) progesterone undergoes similar variations in concentration in male and female *M. edulis,* suggesting that the hormone has a similar role in both sexes. The peaks of progesterone concentration occur in July and October, coinciding in the study area (Portugal) with the two periods of intense gametogenesis, and suggesting, perhaps, that progesterone, or one of its derivatives, plays a role in gamete multiplication. The cells involved in steroid synthesis in bivalves have yet to be localized. The accessory cells of the gonadic tubules, described by Pipe (1987) and Mathieu (1987b), which contain numerous lipid inclusions, might be likely candidates.

Prostaglandins

The presence of prostaglandins has been demonstrated in a number of animals and plants (for marine invertebrates, see Nomura and Ogata, 1976; Srivastava and Mustafa, 1984). The presence of PG-E2 in *M. edulis* has been demonstrated by Nomura and Ogata (1976), although its concentration is very low. This prostaglandin has also been described in the mussel *Modiolus demissus* by Freas and Grollman (1980). Srivastava and Mustafa (1984) have confirmed that the mussel is capable of synthesizing PG-E2, and also PG-F2α and PG-D2, using arachidonic acid as substrate. As is the case in mammals, the conversion of arachidonic acid into prostaglandins requires the presence of oxygen (Srivastava and Mustafa, 1985).

In invertebrates the prostaglandins are important in the regulation of ionic fluxes, thermoregulation, regulation of reproduction and spawning, and in mechanisms of cell aggregation (Stanley-Samuelson, 1987). The induction of spawning by the addition of hydrogen peroxide to the water, or by UV irradiation of the water (producing H_2O_2), which in both cases stimulates peroxidation (Uki and Kikuchi, 1974; Morse et al., 1977), has led to the hypothesis that the prostaglandins might stimulate gamete release. Although the administration of PG-F2α has no direct effect on gamete emission by the scallop *Pecten yessoensis* (Matsutani and Nomura, 1986)– serotonin being the stimulatory agent (Matsutani and Nomura, 1982). The annual cycle of PG-E2 and PG-F2α concentrations in the gonad suggests that these molecules are involved in the regulation of spawning in females (Osada et al., 1989). This hypothesis is supported by the fact that the induction of spawning is inhibited, in the females of this species, by the administration of aspirin, an inhibitor of the cyclooxygenase activity of prostaglandin endoperoxidase synthetase (Matsutani and Nomura, 1986).

METABOLISM AND GAMETOGENESIS: NEUROENDOCRINE REGULATION

Endocrine Organs and Neurosecretory cells

The presence of neurohormones in molluscs was first demonstrated by Scharrer (1935). It was not until 1955, however, that Gabe described for the first time, neurosecretory cells in bivalves. As far as mussels are concerned, a number of descriptive studies of the different types of neurosecretory cells have been published (Lubet, 1955, 1959; Tsuneki, 1974; Illanes-Bucher, 1979; Illanes-Bucher and Lubet, 1980). The nervous system of *M. edulis* comprises three pairs of ganglia: a pair of cerebropleural ganglia situated symmetrically along the anterior foot retractor muscles; a pair of visceral ganglia situated on the ventral surface of the posterior adductor muscles and a pair of pedal ganglia that are fused together and situated at the base of the foot, buried in the mass of the anterior byssus retractor muscles (Fig. 6.6). The ganglia are small, their longest dimension never exceeding 2mm, but each contains numerous neurosecretory cells. No specific spatial organisation of the nervous cells is evident in *M. edulis*.

The authors cited above distinguish four distincts types of neurosecretory cell in *M. edulis*. This classification is based upon histological and histochemical data, the criteria employed being the shape, size and localization of the cells, as well as their annual cycle of neurosecretory activity. Using the nomenclature suggested by Illanes-Bucher and Lubet (1980) the a1 cells are the most numerous. These cells are pyriform or elongate and measure between 6 and 15μm in length. The a2 cells are more or less spherical with a diameter of between 20 and 30μm. The a3 cells are of irregular shape and measure 20 to 30μm in diameter. Finally, the a4 cells are spherical with an irregular contour and a diameter of 12 to 15μm. The contents of all of these cells have a high affinity for basic stains (especially fuchsin paraldehyde and thionone paraldehyde) following oxidation with potassium permanganate. Although, the use of these signaletic stains has permitted the visualization of axonal migration of

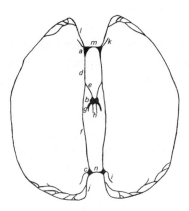

Fig. 6. 6. A schematic representation of the nervous system in *Mytilus edulis*. a: cerebral ganglia; b: pedal ganglia; c: visceral ganglia; d: cerebrovisceral pedal connective; e: cerebropedal connective; f: cerebrovisceral connective; g: pedal nerve; h: byssal retractor nerve; i: branchial nerve; j: posterior pallial nerve; k: anterior pallial nerve; l: buccal nerve; m: cerebral commissure; n: visceral commissure.

secretory products, no specific neurohaemal structure, into which neurohormones are secreted, has been identified. According to Illanes-Bucher (1979) 75% of the neurosecretory cells in *M. edulis* are found in the cerebropleural ganglia. This percentage corresponds to 1500–4000 active cells per ganglion, depending on the season. This number is high when compared to other molluscan nervous systems (pulmonates, for example). In *Mytilus grayanus* the neuroendocrine cells represent around 10% of the nervous cells of the mussel, estimated to number 20,000–45,000 per ganglion (Motavkine and Varaksine, 1989). In this species the majority of the neurosecretory cells are of the a1 variety (83%), and 92% are found in the cerebropleural ganglia.

More recently, cytochemical data have been complemented by results from immunocytochemical studies. The neurosecretory cells are neurones that possess an endocrine function by producing peptides. The latter are liberated into the circulatory system, often following axonal transport. It is currently considered that the neuropeptides are widely distributed within the animal kingdom and much experimental evidence is available to support this view. The use of antibodies raised against vertebrate and, to a lesser extent, invertebrate peptide hormones has become common, and has permitted the identification of a number of immunologically related molecules in invertebrates.

Although such studies can only provide limited data concerning the molecular structure of the neuropeptides, they provide a very useful tool for the characterization and mapping of neurosecretory cells, that complements the morphological data. An excellent example of this is the precise cartography of the neurosecretory cells of the pond snail *Lymnaea stagnalis* performed by Schot et al. (1981) and Boer and van Minnen (1988) using 21 different antibodies. As far as the mussel is concerned, a few immunocytochemical studies have refined the cytochemical data cited above. An enkephalin-like molecule has been located in the pedal ganglia of *M. edulis* (Martin et al., 1986) and a met-enkephalin-like peptide in all three types of ganglia (Andersen et al., 1986). The latter authors also describe a β-endorphin-related molecule in the pedal ganglia. Analogous molecules have been located in the digestive tract (Andersen et al., 1986) and an αMelano-Stimulating

Hormone-like (αMSH-like) molecule demonstrated in the pedal (Andersen et al., 1986) and cerebropleural ganglia (Mathieu and van Minnen, 1989). In their study of the cerebropleural ganglia, Mathieu and van Minnen (1989) have also described a cholecystokinin-gastrin-like (CCK-gastrin-like) molecule and a somatostatin-like molecule. The above studies all involved antibodies raised against vertebrates peptides. Data on substances related to invertebrate peptides are also available. An FMRFamide (Phe-Met-Arg-Phe-amide) has been located in the pedal (Andersen et al., 1986) and in the cerebropleural ganglia (Mathieu and van Minnen, 1989). Three antibodies raised against *Lymnaea stagnalis* peptides have also given positive results with the cerebropleural ganglia of *M. edulis*. These peptides are αCDCP (αCaudo-Dorsal-Cells-Peptide), a peptide involved in the spawning of *Lymnaea*; APGWamide (Ala-Pro-Gly-Trp-amide), that regulates the activity of the male genital tract (Mathieu and van Minnen, 1989); and MIPc (Molluscan-Insulin related-Peptide-C), a fragment of an insulin-like molecule produced by *Lymnaea* Light Green Cells (LGC) (van Minnen and Mathieu, unpublished results).

The significance of these immunocytochemical studies is not necessarily the same for all of the peptides. As far as the biochemical nature of the molecules is concerned, only the enkephalins have been purified and sequenced (Leung and Stefano, 1983); these peptides have the same primary structure in the mussel as in vertebrates. No other peptide identified in the mussel by immunological techniques, has to date been sequenced.

Nerve cells showing an immunoreactivity with anti-somatostatin are numerous in *Mytilus*. The majority are a1 type cells, as described by Illanes-Bucher (1979), although a few belong to other morphologically different neurosecretory cell categories. Somatostatin-like molecules have been identified in a variety of other molluscan species (*Lymnaea stagnalis*: Schot et al., 1981; *Physa* sp. : Grimm-Jørgensen, 1983; *Helix aspersa*: Assaka et al., 1987). Among all of these pulmonate molluscs, the somatostatin-like molecule appears to be implicated in the control of somatic growth and, according to the above authors, could act as a growth hormone. In other pulmonate molluscs (*Helisoma*: Bulloch, 1987; *Physella*: Grimm-Jørgensen, 1987) somatostatin activates in vitro neurite outgrowth. In the mussel, a somatostatin-like molecule has been partially purified from an extract of the cerebral ganglia (Toullec, 1989). The apparent molecular weight was close to 2kDA, similar to that of vertebrate somatostatin. This molecule has not, however, been shown to affect somatic growth in vitro. It would seem more likely that somatostatin plays a role in interneuronal communication and hence, in the regulation of certain neurosecretory cells i.e. it appears to act as a parahormone. The same is probably true for the enkephalins.

As far as FMRF-amide is concerned, closely related molecules have been identified in a number of molluscs (Greenberg et al., 1988), and their physiological role as cardioaccelerators has been demonstrated. In the mussel, FMRFamide-containing nerve cells have been shown to be motor neurons by Illanes-Bucher (1979), and it appears likely that the FMRFamide-like molecule also plays a cardioexcitatory role in this species.

The antibodies raised against CCK and αMSH react with a very limited number of neurosecretory cells in well-defined locations. These cells are all of the same type. This suggests that these cells are responsible for the production of neurohormones, the physiological roles of which remain to be elucidated.

The immunocytochemical data obtained with antibodies raised against *Lymnaea* hormones pose different problems, as far as their interpretation is concerned. These antibodies were raised against purified hormones of known physiological function, from a phylogenetically closely-related mollusc. That they cross-react with small, homogeneous groups of mussel neurosecretory cells is, therefore, not surprising, but for many of them, their physiological role remains obscure. This is particularly true of the egg-laying hormone, and of the peptide that regulates the male genital tract. Indeed, the usual target cells for these hormones in *Lymnaea* are absent in *Mytilus*. The positive results obtained with *Lymnaea* antiMIPc are very promising in relation to growth regulation. In *Lymnaea* the LGCs that produce MIP have for some time been known to be responsible for the production of growth hormone (Geraerts, 1976). Recently, the use of molecular biological techniques has shown that these cells express genes coding for 6 pre-proinsulin-like molecules (Smit et al., 1988). Such pre-prohormone molecules possess a signal sequence that is presumably involved in the secretory process. The prohormone may be a simple precursor of the hormone, or be comprised of several identical or nonidentical peptides. These peptides are derived from the prohormone by the processing of the latter by endoproteolytic enzymes recognizing specific hydrolytic sites. In the light of recently published results, especially those concerning the egg-laying hormone of *Aplysia* (Scheller et al., 1982;

Mahon et al., 1985), FRMFamide precursors of *Aplysia* (Schaefer et al., 1985) and pre-proinsulin of *Lymnaea*, the above model for biosynthesis and post-translational processing of neuropeptides can be considered to be common among the molluscs.

Neuroendocrine Regulation of Growth

In the vertebrates, several hormones are known to stimulate growth e.g. growth hormone (GH), insulin and insulin-like hormones, thyroid hormones, androgenic steroids; and the mechanisms by which growth is regulated are complex. In the molluscs, somatic growth is linked to that of the shell, whether the shell is internal or external. Lubet (1971) was the first to demonstrate the role played by the nervous ganglia in the control of molluscan growth, and the existence of a molluscan growth hormone, by the ablation of nervous tissue in the slipper limpet *Crepidula fornicata*. For *M. edulis* the majority of studies have centered upon the effects of external factors on growth. Thus, the effects of temperature, nutrition, age and population structure have been demonstrated. Studies of the effects of internal factors, i.e. endocrine control, are fewer. This has largely been due to experimental difficulties linked to the anatomical and biological particularities of this species. Several lines of research are currently being pursued in order to identify bivalve growth hormones. The first strategy involves the purification of active molecules from endocrine organ extracts (nervous ganglia). These molecules are identified at each stage of the purification by bioassay. For *M. edulis*, the bioassay employed involves the use of enzymatically dispersed cell suspensions cultivated in vitro in a liquid culture medium. This method circumvents the problems of response time, and of the increased quantities of hormone required in whole animal experiments, and alleviates problems of heterogeneity implicit in tissue culture experiments. Such a bioassay has led to the partial purification of a low molecular weight cerebral factor (around 1kDa) that stimulates the in vitro incorporation of radiolabelled amino acids and thymidine into the macromolecular fraction in somatic cells. The incorporation of amino acids is indicative of protein synthesis and, thus, of cell growth. The incorporation of thymidine is a marker of DNA synthesis, and thus of cellular multiplication. The active molecule can therefore, be considered to be a growth factor (Toullec et al., 1989). A similar molecule has been purified from *Lymnaea* by Ebberink and Joosse (1985). In both cases, the active molecule is very hydrophilic. Recently, it has been shown that in extracts deproteinized by gel chromatography (excluding molecules >2.5kDa), the active factor is soluble in acetic acid (1N) and in certain organic solvents (acetonitrile and methanol) without loss of activity (Robbins and Mathieu, unpublished data).

The loss of activity of cerebral extracts following tryptic digestion suggests that the active molecule is a peptide.

The second strategy consists in using oligonucleotide probes constructed from knowledge of the structure of peptide hormones purified from other species, or from gene sequences derived from other species. This molecular biological approach is currently being used to identify and study insulin-like genes in the mussel genome using probes constructed from known sequences of *Lymnaea* MIP

Neuroendocrine Control of Gametogenesis

Mussels of the genus *Mytilus* undergo an annual reproductive cycle punctuated by a rest period that is often short, but which varies according to latitude. This resting phase is followed by a period of gametogenetic activity—lasting a large part of the year—which leads to a single or several phases of gamete emission. In the latter case, the emptying of the gonadic tubules from the mantle is followed by a period of intense gametogenesis that serves to reconstitute the gonad. The sequences involved in the annual reproductive cycle of *Mytilus* spp. have been described elsewhere (Chipperfield, 1953; Lubet, 1959; Seed, 1975; Lowe et al., 1982; see also Chapter 4).

Spermatogenesis

Spermatogenesis in *Mytilus* is well-documented (Field, 1922; Franzen, 1955; Lubet, 1959; Niijima and Dan, 1965; Longo and Dornfield, 1967) and would appear to be similar to that described for other molluscs (Bloch and Hew, 1960; de Jong-Brink et al., 1977). Sixty four spermatogonia are derived, by succesive mitotic divisions, from each primary spermatogonium. These spermatogonia differentiate into first order spermatocytes, which then undergo meiosis to become successively 128 second order spermatocytes, followed by 256 spermatids. Finally, the spermatids differentiate into spermatozoa.

Throughout spermatogenesis, the cells are linked by the synchronous development of cytoplasmic bridges. In *M. edulis*, the spermatids are linked in groups of a least four cells (Longo and Dornfield, 1967). The duration of spermatogenesis has been estimated by the incorporation of radiolabelled thymidine (a DNA precursor) followed by autoradiography. In *M. californianus*, only the spermatogonia and the first order spermatocytes are labelled after 10h of incorporation (Kelley et al., 1982). In *M. edulis* , the first appearance of radiolabel in the first order spermatocytes is very rapid, occuring after only 1.5h of incorporation (Mathieu, 1987a; Fig. 6.7). In this case the second order spermatocytes showed no incorporation of radiolabel, even after 6h of incubation. The first radiolabelled spermatozoa are observed after 10 days and become abundant around day 16 (Kelley et al., 1982). The labelled spermatozoa migrate progressively towards the centre of the gonadic follicles.

Oogenesis

Oogenesis follows a similar pattern to spermatogenesis. Primary oogonia, derived from stem cells, undergo successive mitoses to give secondary oogonia that enter the first division of meiosis to become first order oocytes. These oocytes are blocked in the diplotene phase of the first meiotic division and undergo previtellogenesis, during which there is limited growth, and synthesis of large quantities of RNA. A portion of this RNA is stocked, either in the germinal vesicle or in the cytoplasm, to be used in the first stages of development. The oocytes then enter a period of vitellogenesis. They store large quantities of vitellus, mainly lipids, and grow rapidly, reaching 70μm in diameter in *M. edulis*.

The formation of vitelline reserves in invertebrates has recently been reviewed by Wourms (1987). As far as *M. edulis* is concerned, the ultrastructural changes during oogenesis have been studied in detail by Pipe (1987), and indicate that during the initial phase of development, the oocyte absorbs macromolecules by endocytosis. This phase is followed by the intracellular biosynthesis of macromolecules, especially lipids and proteins, and to a lesser extent glycogen. Substrates for the biosynthesis are furnished by the degradation of cells forming the storage tissue. The utilization of stored carbohydrate reserves, represented in large part by glycogen stored in specialized glycogen cells (= vesicular connective tissue cells), and lipids (especially triglycerides) has been discussed in detail by Zandee et al. (1980b) and Gabbott (1983). Pipe (1987) has described the important role played by intragonadic follicle cells. These follicle cells are of heterogeneous shape, but are characterized by the presence of lysosomes, a well-developed rough endoplasmic reticulum, numerous mitochondria and lipid inclusions. The follicle cells are closely linked to the oocytes by desmosome-like 'gap-junctions' that allow the transfer of ions and small molecules. According to Pipe (1987) the follicle cells of *M. edulis* have the capacity to absorb, transform and synthesize material. The hypothesis that they play a role in the resorption of material liberated during oocyte lysis (Pipe and Moore, 1985) was not confirmed by the later study of Pipe (1987).

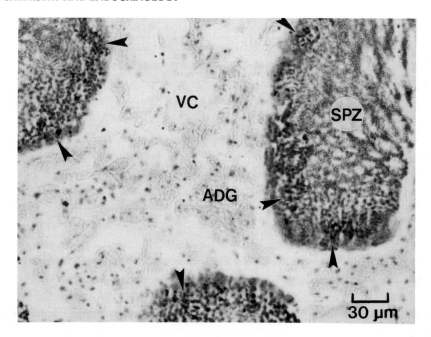

Fig. 6. 7. Autoradiograph of male *Mytilus edulis* mantle after two hours ³H-thymidine labelling. SPZ, spermatozoa; VC, vesicular cell; ADG, adipogranular cell. Arrows, labeled spermatogonia and primary spermatocytes.

Hormonal regulation of gametogenesis

The idea that the gametogenesis of the mussel is under hormonal control is not new and was originally founded on observations that the annual cycle of ganglionic neurorosecretory activity coincided with the annual reproductive cycle. The neurosecretory activities of *M. galloprovincialis* (Lubet, 1959), *Mytilus perna* (Umiji, 1969), *Crenomytilus grayanus* (Motavkine and Varaskine, 1989) and female *M. edulis* (Lubet et al., 1986) have been shown to be low during the sexual rest phase, and to progressively increase in synchrony with the developing gonad, to reach a maximum shortly before the emission of gametes (Fig. 6.8). Direct experimental evidence to support the view that the nervous ganglia might control gametogenesis in the mussel was not immediately forthcoming. This was largely due to the fact that ablation of nervous ganglia in the mussel is not easy, and only the cerebral ganglia can be destroyed without killing the mussel. In addition, the large number of neurosecretory cells (several thousands for each ganglion) and their small size (15–17µm for the most abundant category), renders the selective destruction of particular cells impossible. Finally, the open circulatory system comprising open vessels and sinuses with a large volume of haemolymph circulating at low pressure, provides

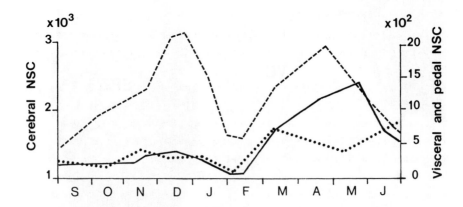

Fig. 6.8. Variation of active neurosecretory cell numbers in *Mytilus edulis* ganglia during the annual reproductive cycle. Cerebral ganglia: - - - - - ; Pedal ganglia: ——— ; Visceral ganglia: ···········

only slow and irregular diffusion of any injected substances such as endocrine organ extracts .

Ablation of endocrine organs:
Though the cerebropleural ganglia contain 75% of the total number of neurosecretory cells of *M. edulis* and 98% of the a1-type cells, whose annual secretory cycle reflects that of the annual reproductive cycle (Illanes-Bucher, 1979), their bilateral ablation does not appear to inhibit gametogenesis (Lubet, 1959, 1966). As no regeneration of these organs was apparent, Lubet proposed that other organs, such as the visceral ganglia, may replace by vicariance, the missing cerebropleural ganglia.

Tissue culture experiments:
The use of tissue culture techniques has provided evidence for the stimulatory role played by the nervous ganglia in the gametogenesis of *M. edulis*. Mantle explants, when cultured in isolation on a solid gel-based medium, undergo a number of perturbations as far as the gonadic tissue is concerned. These perturbations underline the different levels of neuroendocrine control exerted by the ganglia. In explants of male animals the results of culture in isolation include an absence of gonial mitosis, spermatogonial anomalies, clumping (often in groups of four), necrosis of first order spermatocytes and the nonformation of second order spermatocytes. The spermatozoa are apparently normal. No ovogonial mitosis is present in the explants of female mussels cultured in isolation. In addition, a number of anomalies are observed in the previtellogenic and vitellogenic oocytes. These anomalies include abnormal stacking

of the endoplasmic reticulum, cytoplasmic vacuolization, disappearance of lipids globules and oocyte lysis (Lubet et al., 1986). The majority of these abnormalities can be prevented if the explants are cultured in association with the two autologous cerebropleural ganglia. In particular, the ganglia permit the restoration of gonial mitosis in the mantle explants.

The role of the visceral ganglia is less clear. Thoughout most of the year, their presence is insufficient to maintain gametogenesis in the mantle explants. At the end of the gametogenetic cycle, however, their effects are similar to those of the cerebropleural ganglia. Various experiments in which mantle explants were cultured with heterogenous ganglia have demonstrated that the active factors are neither sex-specific nor species-specific in bivalves (Mathieu, 1987b). It is at present impossible to determine the exact number of hormonal factors involved in the regulation of gametogenesis in mussels. Equally, it is not known whether these factors act directly on target cells or whether they act indirectly via another cell type which, once stimulated, releases the true gonadotropins.

Dispersed cell suspensions:

If in vitro tissue culture techniques provide valuable qualitative results, they are less suited to the quantification of the effects of hormonal extracts in the mussel. This is due to the uneven diffusion of the active substances and to the particularly heterogeous structure of the mantle tissue that renders comparative studies very difficult. These difficulties have been overcome by the use of dispersed cell suspensions, involving the mechanical or enzymatic dissociation of target tissues (Lenoir and Mathieu, 1986). The culture of these cells in a liquid nutritive medium has led to the development of a number of bioassays based upon the incorporation of radiolabelled metabolites, or on changes in enzymic activities, and has thus permitted the quantitative study of a number of hormonal substances.

In their studies of factors responsible for the stimulation of gonial mitosis, Mathieu et al. (1988) employed mantle tissue dissociated with the proteolytic enzyme pronase and two bioassays for DNA synthesis. The first of these involved the incorporation of ^{3}H-thymidine. Preliminary, autoradiographic studies had shown that the oogonia were the only cell type to incorporate the radiotracer. The second bioassay involved the measurement of aspartate transcarbamylase (ATCase) activity. ATCase is a key enzyme in *de novo* pyrimidine biosynthesis and its activity has been closely correlated with active cell multiplication in *M. edulis* (Figs. 6.9 and 6.10) (Mathieu, 1985, 1987a) and numerous other organisms (reviewed by Mathieu, 1985).

Using these techniques, it has been demonstrated that acid extracts of cerebropleural ganglia stimulate in vitro gonial mitosis in *M. edulis*. The active

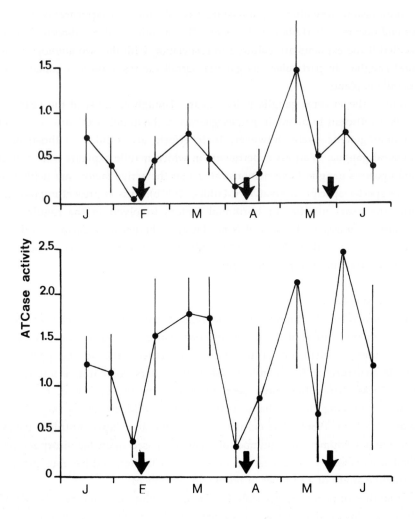

Fig. 6.9. ATCase activity levels in male and female gonads of *Mytilus edulis* during the reproductive period. Results are expressed as nmol per min per mg of protein. Spawning periods are indicated by arrows. Male, upper graph; Female, lower graph.

factor has been shown to be of a peptide nature, to have a molecular weight <5000Da, and to be relatively thermostable. A molecule possessing similar characteristics has been identified in the haemolymph (Mathieu et al., 1988). It has not been possible to work with male mussels since the enzymic dissociation produces a viscous gel associated with the high DNA concentrations inherent in the male gonad (also see Peek and Gabbott, 1989).

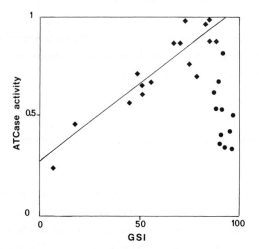

Fig. 6.10. Correlation between ATCase activity and gonadosomatic index (GSI) in mussel mantle tissue. ♦, stages I, II and II(A1); •, stages III (A2).

In addition to this oogonial-mitosis *stimulating* factor, there is also evidence for a factor that *inhibits* oogonial mitosis. Mathieu et al. (1982) have demonstrated a sharp decline in mantle ATCase activity immediately prior to gamete release. This can be interpreted as an inhibition of mitosis at maturity, and is true of both males and females. Furthermore, in a study of the parasitic castration of *M. edulis* by a trematode, Coustau et al. (1991) have shown that mantle extracts of both mature and immature mussels inhibit ^{3}H-thymidine incorporation by dispersed mantle tissue, whereas only haemolymph from mature mussels produces a similar effect. These results indicate that a factor capable of blocking gonial mitosis is present in the mantle tissue but is only released at sexual maturity. The nature and origin of this molecule have yet to be established, as do the factors leading to its liberation.

Transport of hormonal factors:
The results obtained from tissue-culture and cell dispersion techniques suggest that the neurohormones that act upon the gonad do so by diffusion, and are therefore secreted by the neurosecretory cells into the general circulation. Motavkine and Varaskine (1989), however, have described numerous axonal terminals in the gonad wall of *Patinopecten yessoensis* and have hypothesized that in bivalves, neurohormones may reach the close proximity of target cells by a process of axonal transport. The coexistence of these two processes is probable.

Although the role of the nervous ganglia in the functioning of gametogenesis has been clearly demonstrated, it remains to be established whether the neurohormones act directly on their target cells, or whether they act via a relay system involving other cell types in the mantle tissue; these intermediary cells, once stimulated release the hormone active on the target cells. In both tissue culture and dispersed cell techniques all the cell types of the mantle are present together.

Neuroendocrine Regulation of Storage Tissue Metabolism

Annual storage cycle

The mantle tissue of *M. edulis* undergoes seasonal variations in biochemical composition (de Zwaan and Zandee, 1972) that reflect changes in its cellular composition. During sexual maturation the gonad, present in the form of ramified tubules, progressively invades the somatic storage tissues. The functioning of this storage tissue can be split into two distinct phases: (a), a phase of nutrient storage during the summer, corresponding to the period of sexual rest and coincident with an abundant food supply, and (b) a phase of nutrient mobilization during the autumn and winter.

The volume occupied by the storage tissue is maximal in summer and decreases during autumn and winter as nutrient reserves are utilized to fuel the reproductive effort (Lowe et al., 1982; Worral and Widdows, 1984; Pipe, 1985, 1987). This decrease continues until the first emission of gametes in spring. This seasonal cycle involves adipogranular cells (ADG) that store protein, lipids and glycogen, and the vesicular connective tissue (VCT) cells that store large quantities of glycogen. The analysis of seasonal changes in the mantle volume occupied by these cells between 1977 and 1980 (Zaba et al., 1981) has shown that the developmental cycle of the two cell types is subtly different. During the summer of 1978 the development of the two cell types was synchronous but during the summers of 1977 and 1979, however, the VCT cells attained maximal development in June, whereas the ADG cells reached this point later in September (Fig. 6.11). In addition, in 1979 the maximal volume attained by the VCT cells was more than 30% in excess of values normally observed, while that of the ADG cells was more than 30% less than in preceding years. These variations were reflected in the maximal glycogen levels of the mantle tissue: 43.7%, 44.8% and 53.3% dry weight for the years 1977, 1978 and 1979, respectively (Gabbott and Whittle, 1986). These observations, which clearly demonstrate the influence of external factors such as temperature, food availability (Chapter 4) and contaminants (see p.452–453 Chapter 9) on the development of the storage tissue during the summer, have been confirmed

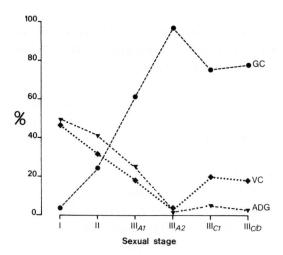

Fig. 6.11. Comparative development of gonadic and storage tissues in *Mytilus edulis* mantle during a sexual cycle. The results are expressed as percentage of volume occupied by adipogranular cells (ADG), vesicular cells (VC) and gonad (GC).

by the further studies of Lowe et al. (1982), who demonstrated differences in the developmental cycles of storage tissue from two geographically separated British populations of *M. edulis*, one of which was the population studied by Zaba et al. (1981). The repercussions on the quantity of stored reserves for gametogenesis have been studied by Pipe (1985). These mainly concern oocyte development. It is proposed that oocytes which develop early in the season do so at the expense of metabolites liberated by the ADG and VCT cells. Later in the season, when this source of metabolites is exhausted, vitellogenesis may be fueled directly by circulating metabolites originating from the transformation of ingested food. If the abundance of food remains low during this period, the oocytes contain less glycogen.

Studies concerning the metabolism of storage tissue in the mussel have attempted to demonstrate that, on the one hand, the two cell types involved differ in function and, on the other hand, that the various stored metabolites, viz. lipids, carbohydrates and proteins, are subject to differences in the way that they are metabolically regulated. The cyclical nature of the metabolism of the storage cells and the coincidences with that of the sexual cycle, suggest that their function may be under hormonal control. The origin and nature of hormonal factors affecting the metabolism of the storage cells have therefore, been studied (see later).

Glycogen metabolism

It is unquestionable that glycogen constitutes the dominant energy reserve fueling gametogenesis, even though Zaba and Davies (1984) have shown that it is not uniquely employed as a source of energy and metabolic precursors for the various synthetic reactions involved in the reproductive effort. When the food supply is abundant, the maintenace of mantle metabolism is assured by a rapid transfer of metabolites from the digestive gland (Bayne et al., 1975); glycogen in the storage tissue is, thus, preserved for use as an energy source for gametogenesis. Moreover, Zaba and Davies (1981) have demonstrated that, in summer, glucose-6-phosphate is derived almost exclusively from exogenous glucose. If the supply of exogenous glucose is insufficient (in winter or during periods of limited food supply) the stored glycogen reserves may be used. The mobilization of glycogen reserves may occur by its transfer into cytoplasmic vesicles (Bayne et al., 1982; Pipe, 1987) or by the appearance of localized zones of lysis (Lubet et al., 1976). The degradation of glycogen to glucose-6-phosphate may follow one of two possible metabolic pathways: (a) phosphorolysis catalyzed by glycogen phosphorylase and phosphoglucomutase, or (b), hydrolysis, followed by the phosphorylation of glucose, involving amyloglucosidase and hexokinase. The existence of both pathways has been demonstrated in *M. edulis* (Alemany and Rosell-Perez, 1973; Zaba, 1981).

Nonhormonal regulation of glycogen metabolism

Glycogen metabolism of the mussel is subjected to hormonal and nonhormonal control. In the first case the effectors are neuroendocrine factors. In the case of nonhormonal control, the principal factor is the circulating free glucose concentration. In both cases, regulation may be at the level of glucose transport into the cells, or at the level of the metabolic pathways involved in the synthesis and degradation of glycogen within the cells.

As is the case in the mammals, the activities of glycogen synthase and glycogen phosphorylase in the mussel are regulated by a chain of phosphorylation-dephosphorylation reactions. The phosphorylated form of glycogen phosphorylase is active, whereas that of glycogen synthase is inactive. Cook and Gabbott (1978) identified the two forms of glycogen synthetase in *M. edulis*. In the series of papers cited below Gabbott and co-workers used the earlier designation of glycogen synthetase rather than glyggogen synthase. The active I form is independent of glucose-6-phosphate concentration, whereas the inactive D form is glucose-6-phosphate dependent. As shown by Gabbott and Whittle (1986) it is the conversion of the D-form to the I-form that is responsible for glycogen synthesis during the summer period of sexual rest. Measured activities of the I (active) form during May and June were sufficient to account for the quantities of glycogen stored during this period. The total glycogen synthetase activity (I + D forms) in the mantle peaked in May–June after the spawning period. At this time, the total activity is two or three times greater than that observed in winter. The I-form shows greatest seasonal fluctuations, with activity in June about 10 times greater than in December. In a companion paper Whittle and Gabbott (1986) showed that in vitro culture of mantle slices in the presence of 2–10mM glucose increased the I activity of glycogen synthetase, but had no effect on the total (I + D) activity. In the presence of 5mM glucose, the glycogen synthetase activity (I-form) of winter mussels increased to reach values typical for the summer months. The effect of glucose was observed after 3h of incubation and was temperature-dependent. The authors concluded that both factors, increased glucose levels due to feeding and higher seasonal temperatures, contribute to the activation of glycogen synthetase in the summer (Whittle and Gabbott, 1986). During the autumn and winter, at which time there is intense gametogenesis, the activity of glycogen synthetase is low. A weak, but measurable, activity of the I-form remains, however. Gabbott (1983) interprets this low activity as a potential for the recycling of glucose carbon during gametogenesis.

Adipogranular cells

ADG cells have been described in a number of species of bivalve molluscs (List, 1902; Kollman, 1908; Daniel, 1923; Froutin, 1937; Lubet, 1959; Herlin-Houtteville, 1974). Stereological, ultrastructural and histochemical studies have provided an insight into certain aspects of their structure and function. The ADG cells are large (25μm) with an excentric nucleus. They contain glycogen, in the form of α and β particles, and numerous electron-dense granules that are surrounded by a membrane and which are most probably protein in nature (Herlin-Houtteville, 1974; Bayne et al., 1982). Recent biochemical analyses have, indeed, confirmed that ADG cells contain abnormally high levels of protein (75–85% dry wt; Gabbott and Peek, 1991). The cells are characterized by an extensive rough endoplasmic reticulum, numerous free ribosomes and small vesicles associated with the Golgi apparatus, again indicative of the abundant synthesis of protein. Mitochondria are rare and the cytoplasm contains large lipid inclusions. According to Herlin-Houtteville (1974), ADG cells arise by de-differentiation of granulocytes, at the end of the sexual cycle and during the sexual rest phase. The young ADG cells first accumulate lipids, in part derived from the degeneration of any remaining gametes. Proteins are then synthesized and glycogen storage occurs as a final phase.

According to Bayne et al. (1982), the regression of the ADG cells corresponds to a mechanism of autophagy. In particular, glycogen appears to be sequestered in simple or double membrane-bound vesicles. These membranes are characteristic of the formation of autophagic vacuoles (autophagosomes). The content of the vesicles undergo transformations that can be observed in ultrastructural studies. The ADG cells contain increasing numbers of autophagosomes, for the most part containing β-glycogen particles, as regression proceeds. The transformation of these phagosomes into autolysosomes results in the degeneration of the ADG cells which accompanies gametogenesis. Several studies have corroborated the role of lysosomes in the degradation of the ADG cells. The seasonal variations of lysosomal hydrolases (β-N-acetylhexosaminidase, β-glucuronidase) observed in histochemical studies are such that increasing activities coincide with ADG-cell regression (Lowe et al., 1982).

In order to gain a better insight into the biochemical functions of the ADG cells, Peek and Gabbott (1989) developed a method of isolating them following enzymic dissociation of the mantle tissue. The dispersed cells were separated according to their buoyant densities on a gradient of Percoll; cells with a density <1.08g mL^{-1} were discarded, the remaining suspension consisting almost exclusively of ADG cells. The incorporation of ^{3}H glucose and ^{14}C-glucose into the glycogen of these cells occurred at rates of 80 and 120nmol glucose h^{-1} g^{-1} wet wt, respectively, at an extracellular glucose concentration of 0.5mM. The rate of $^{14}CO_2$ production by the ADG cells was very low (18nmol glucose h^{-1} g^{-1} wet wt) and equal to only 6% of the total glucose consumed. The kinetics of glucose utilization, as estimated by the detritiation of 2-^{3}H glucose, show an apparent K_m of 0.56 ± 0.09mM and a V_{max} of 632 ± 48nmol h^{-1} g^{-1} wet wt. In addition, the technique for isolating ADG cells has permitted Peek et al. (1989) to study the seasonal distribution of enzymically dispersed cells from the mantle of *M. edulis* and, later, Peek and Gabbott (1990) to localize various lysosomal hydrolase activities, and to compare seasonal variations of their activities in ADG cells with those in the oocytes. Acid phosphatase and cathepsin L appear to play an important role in oocyte maturation, whereas β-N-acetyl-glucosaminidase, β-glucuronidase and cathepsin B participate in the degradation of the ADG cells.

Hormonal control of ADG cells

The liberation of metabolites by ADG cells appears to be under hormonal control. In mantle tissue culture experiments, it has been shown that the cerebropleural ganglia stimulate the degradation of the storage tissue in addition to stimulating gametogenesis. The visceral ganglia, on the other hand, would seem to have the reverse effect (Mathieu and Lubet, 1980; Lubet, 1981; Lubet and Mathieu, 1982; Whittle et al., 1983). The precise nature of this regulation, and of the factors involved, remain to be established. The idea that the follicle cells are stimulated by a neurohormone originating in the cerebropleural ganglia and, in turn, secrete a steroid hormone has been proposed by Peek et al. (1989). Moreover, de Longcamp et al. (1974) have identified the presence of testosterone, oestrogens and enzyme systems involved in steroid synthesis in the mantle of *M. edulis*. More recently, Reis-Henrique et al. (1990) have confirmed the findings of de Longcamp et al. (1974), and have identified the presence of progesterone, androstenedione, testosterone, 5α-dihydrosterone, 17β-oestradiol and oestrone using gas-liquid chromatography coupled with mass spectrometry. Steroid hormones may well act upon the metabolism of ADG cells, since Moore et al. (1978) have demonstrated the induction of lysosomal labilization in the digestive cells of *M. edulis* by oestradiol and progesterone.

Vesicular connective tissue cells (glycogen cells)

Cells specialized in the storage of glycogen have been described in a number of molluscan species and under a variety of names. In the mussel the term vesicular connective tissue (VCT) cell is frequently applied, although the term glycogen cell, used by Joosse and Geraerts (1983) to describe morphologically similar cells in the gastropod *Lymnaea stagnalis*, is simpler and descriptive and has been employed by Lenoir et al. (1989) to describe the cells from *M. edulis*. In *Mytilus* these cells are observed in optical microscopy as large, homogenous plaques. The nucleus is small, spherical, peripheral and often not observed. The quasi-totality of the cell is filled with glycogen granules and occasional lipid droplets are observed (Herlin-Houtteville, 1974).

This very particular structure renders the glycogen cells extremely fragile and it has proved impossible to purify them from the mantle tissue where a number of cell types coexist. Lenoir et al. (1989), however, developed a method of purifying the cells (to >95% homogeneity) from enzymically dissociated labial palps. The glycogen cells are the only storage cells in the palps. The technique involves density gradient centrifugation on Percoll and velocity sedimentation in culture medium containing bovine serum albumin. The purified glycogen cells have a density between 1.09 and 1.20g mL^{-1}. A study of their metabolism has shown that they produce very low quantities of $^{14}CO_2$ from U-^{14}C-glucose (less than 6%) comparable to the CO_2 production by mantle slices (Zaba and Davies, 1980; Zaba et al., 1981) and purified ADG cells (Peek and Gabbott, 1989). In addition, it has been shown for mantle slices, by comparisons of 1-^{14}C-glucose and 6-^{14}C-glucose utilization, that the majority of the CO_2 produced comes from the pentose phosphate pathway, via 6-phosphogluconate dehydrogenase (Zaba and Davies, 1981). The oxidation of glucose by the tricarboxylic acid cycle is negligible. There is, however, a marked discrepancy between the activity of glucose-6-phosphate dehydrogenase and the metabolic flux supported by the pentose phosphate pathway (see review by Zaba and Davies, 1984, and p.254–257).

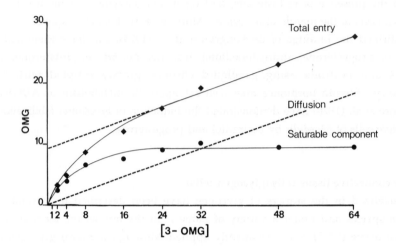

Fig. 6.12. Study of 3-*o*-MeG (3-orthomethyl glucose) zero-trans-entry in glycogen cells. Effect of 3-*o*-MeG concentration [3-*o*-MeG]. Results are expressed as nmoles of 3-*o*-MeG h^{-1} in 10^6 glycogen cells; means and standard deviations of four samples.

Total entry (♦) is the expression of experimental assays. Diffusion curve is the extrapolation of the linear part of the total entry curve. Saturable component (•) is obtained by difference between Total entry curve and Diffusion curve.

These authors have also pointed out that the aerobic metabolism of the mussel mantle tissue is atypical in that the oxidation of carbohydrates is incomplete and that amino acids may form a carbon 'sink' (Zaba and Davies, 1984). Furthermore, Famme and Kofoed (1982) have shown that organic carbon compounds are excreted by intact mussels, regardless of the pO_2.

Finally, despite the fact that the glycogen cells are highly specialized for the storage and mobilization of glycogen, Lenoir et al. (1989) have reported that the incorporation of ^{14}C into glycogen is not the predominant fate of exogenous added ^{14}C-glucose. In experiments performed in February, 29.7%, 30.2% and 21.8% of the ^{14}C were found in glycogen, amino acids and organic acids, respectively. These results agree with those of Zaba et al. (1981) for mantle slices.

Certain aspects of the glycogen metabolism of glycogen cells and ADG cells can be compared. The kinetics of glucose transport (zero-trans entry) has been studied in the two cell types, using a nonmetabolized analogue of glucose, viz. 3-*o*-methyl glucose (ADG cells: Ruiz-Ruano and Gabbott, unpublished data; glycogen cells: Lenoir, 1989). The mechanisms of glucose transport appear to be similar in the two types of storage tissue cells. Firstly, glucose transport is very slow in both cell types, reaching equilibrium after 90min in ADG cells and 120min for glycogen cells. For comparison, in vertebrate cells equilibrium is obtained within a few seconds to a minute. This slow transport would appear incompatible with any mechanism for short-term glycemic regulation in the mussel. Secondly, in both cases transport is a facilitated process that can be divided into two components: a nonsaturable diffusion component, and a saturable component that dominates at increased extracellular glucose concentrations (Fig. 6.12). The membrane receptors that ensure this transport have no affinity for L-glucose in either cell type. It is apparent, however, that the rate of entry of glucose into glycogen cells is greater than that into ADG cells (V_{max} = 3.55µmoles h^{-1} g^{-1} wet wt for glycogen cells and 0.07–0.08µmoles h^{-1} g^{-1} wet wt for ADG cells). In both cases the K_m (7mM for glycogen cells and 1.3–3.3mM for ADG cells) is higher than the level of glycemia observed in the mussel (0–1.5mM: Livingstone and Clarke, 1983) suggesting that it is the extracellular glucose concentration that regulates glucose entry into the storage tissue cells. The fact that the values of K_m for glucose utilization by mantle slices (Zaba and Davies, 1980) and purified ADG cells (Peek and Gabbott, 1989) are both within the limits of glycemia for the mussel suggests that the transport of glucose into the storage-tissue cells is not a limiting factor in the synthesis of glycogen. It is also noteworthy that glucose transport functions well within its maximal capacities, and is therefore capable of responding to transient increases in haemolymph glucose concentration resulting from feeding activities.

In other respects, it is apparent that the circulating glucose concentration, linked to feeding, plays an important role in the synthesis of glycogen. In *M. edulis* the incorporation of ^{14}C-glucose into glycogen of isolated glycogen cells, is linear over 5h, and shows saturation kinetics with respect to extracellular glucose concentration (Lenoir et al., 1989) identical to those observed in *Lymnaea* glycogen cells (Hemminga et al., 1985). These results are in agreement with those obtained for the oyster, *Ostrea edulis*, by L–Fando et al. (1972), who suggested the existence of two possible limiting factors: glucose transport, and the activity of hexokinase that catalyzes the phosphorylation of glucose to glucose-6-phosphate which is metabolized by the cell.

The results of Ruiz-Ruano and Gabbott, and those of Lenoir (see above) suggest that the former possibility can be excluded as a rate-limiting factor. The second hypothesis, however, is supported by the results of Livingstone and Clarke (1983). The activity of *M. edulis* mantle hexokinase varies with the season at both saturating (20mM) and nonsaturating (0.5mM) concentrations of glucose. The hexokinase activity increases in spring and diminishes during the summer to reach its lowest values in autumn and winter. The increase in the specific activity of hexokinase in the spring coincides with a decrease in the free glucose concentration of the mantle and corresponds, according to Livingstone and Clarke, to the degradation of intracellular glycogen. In addition, maximum hexokinase activity coincides with the optimal rate of glucose utilization in the mantle, measured by Zaba et al. (1981). In summer and autumn, the hexokinase activity is minimal but shows the highest values of K_m. The enzyme is in its least saturable state and can therefore respond to increases in circulating glucose concentration linked to feeding activities.

The quantity of glycogen stored in the cells also plays a role in the regulation of glucose incorporation into glycogen. This feedback phenomenon, previously described by L–Fando et al. (1972) for *Ostrea edulis*, has been demonstrated in *M. galloprovincialis* by Gomorosova (1976) and in *M. edulis* by Gabbott and Whittle (1986).

Hormonal regulation of glycogen metabolism

Gabbott and Whittle (1986) have noted that the stimulatory effect of glucose on glycogen synthetase activity is more intense in vivo than in vitro for a given concentration of glucose. The glucose concentration required to produce a stimulatory effect in vitro is 2–5mM, well in excess of observed haemolymph glucose concentrations (0.5–1.5mM; Livingstone and Clarke, 1983). This suggests that there exists a factor that acts in synergy with the activation of glycogen synthesis by glucose concentration. With the use of purified glycogen cells, Robbins et al. (1991) have identified a factor in cerebropleural ganglia extracts of *M. edulis* that stimulates glycogen synthesis in vitro. This factor is acid soluble, has hydrophobic properties and a molecular weight of about 1.5kDa.

Using the same techniques, Robbins et al. (1990) have partially characterized a second factor that stimulates glycogen mobilization in purified glycogen cells. The stimulation of glycogenolysis can be provoked by acid extracts of cerebral ganglia and by haemolymph serum. The same neuroendocrine factor inhibits glycogen synthesis, but does not appear to act upon zero-trans glucose entry into the glycogen cells. The molecular weight is 20–30kDa and the factor acts in a dose-dependent manner, even with short incubation times.

Although the activation of gonial mitosis and the degradation of glycogen cells are synchronous and complementary, the neuroendocrine factors involved are distinct. Both factors originate in the cerebropleural ganglia, but their molecular weights are very different. The site of action of the gonial-mitosis-stimulating factor is, as yet, unknown, but the use of purified glycogen cells shows that the glycogen mobilizing factor acts directly on the target cells.

ACKNOWLEDGEMENTS

Michel Mathieu is grateful to Ian Robbins for critical reviews of the typescript and help with English translation. Both authors would like to express their thanks to very valuable criticism by Dr. P.A. Gabbott. This is contribution No. 561 of the Delta Institute for Hydrobiological Research, Yerseke, the Netherlands.

REFERENCES

Aarset, A.V., 1982. Freezing tolerance in intertidal invertebrates. Comp. Biochem. Physiol., 73A: 571-580.

Addink, A.D.F. and Veenhoff, P.R., 1975. Regulation of mitochondrial matrix enzymes in M. edulis (L). In: H. Barnes (Editor), Proc. 9th Eur. Mar. Biol. Symp., Oban, Scotland, 1984. Aberdeen University Press, Aberdeen, pp. 109-119.

Akberali, H.B. and Earnshaw, M.J., 1982. Studies of the effects of zinc on the respiration of mitochondria from different tissues in the bivalve mollusc M. edulis (L.). Comp. Biochem. Physiol., 72C: 149-152.

Alemany, M. and Rosel-Perez, M., 1973. Two different amylase activities in the sea mussel, Mytilus edulis L. Rev. Esp. Fisiol., 29: 217-222.

Alp, P.R., Newsholme, E.A. and Zammit, V.A., 1976. Activities of citrate synthase and NAD$^+$-linked and NADP$^+$-linked isocitrate dehydrogenase in muscle from vertebrates and invertebrates. Biochem. J., 154: 689-700.

Andersen, A.C., L'Hermite, A., Ferrand, R. and Dubois, M., 1986. Immunohistological detection of methionine-enkephaline-like and endorphin-like material in the digestive tract and in the nervous system of the mussel: Mytilus edulis L. Gen. Comp. Endocrinol., 62: 111-119.

Assaka, L., Marchand, C.R., Strosser, M.T. and van Dorselaer, A., 1987. Detection of somatostatin-14 peptide in the hepatopancreas of the snail Helix aspersa. C. R. Soc. Biol. (Paris), 181(2): 187-193.

Austin, J., 1970. The sterols of marine invertebrates plants. In: M.H. Briggs (Editor), Advances in steroïd biochemistry and pharmacology, 1. Academic Press, New York, pp. 73-96.

Baginski, R.M. and Pierce Jr., S.K., 1978. A comparison of amino acid accumulation during high salinity adaptation with anaerobic metabolism in the ribbed mussel, Modiolus demissus. J. Exp. Zool., 203: 419-428.

Ballantyne, J. S., 1984. Characterization of mitochondria from the hepatopancreas of Mytilus edulis. Proc. 1st Int. Congr. Comp. Biochem. Physiol. 127A:

Ballantyne, J.S. and Storey, K.B., 1983. Mitochondria from the ventricle of the marine clam, Mercenaria mercenaria: substrate preferences and effects of pH and salt concentration on proline oxidation. Comp. Biochem. Physiol., 76B: 133-138.

Ballantyne, J.S. and Storey, K.B., 1984. Mitochondria from the hepatopancreas of the marine clam Mercenaria mercenaria: substrate preferences and salt and pH effects on the oxidation of palmitoyl-L-carnitine and succinate. J. Exp. Zool., 230: 165-174.

Ballantyne, J.S. and Moon, T.W., 1985. Hepatopancreas mitochondria from Mytilus edulis: substrate preferences and effects of pH and osmolarity. Mar. Biol., 87: 239-244.

Bannantyne, W.R. and Thomas, J., 1969. Fatty acid composition of New Zealand shellfish lipids. N. Z. J. Sci., 12: 207-212.

Barrett, J. and Butterworth, P.E., 1981. A novel amino acid-linked dehydrogenase in the sponge Halichondria panicea (Pallas). Comp. Biochem. Physiol., 70B: 141-146.

Bayne, B.L., 1973. Physiological changes in Mytilus edulis L. induced by temperature and nutritive stress. J. Mar. Biol. Ass. U. K., 53: 39-58.

Bayne, B. L., Gabbott, P. A. and Widdows, J., 1975. Some effects of stress in the adult, on the eggs and larvae of Mytilus edulis L. J. Mar. Biol. Ass. U.K., 55: 675-690.

Bayne, B. L., Bubel, A., Gabbott, P. A., Livingstone, D. R., Lowe, D. M. and Moore, M. N., 1982. Glycogen utilization and gametogenesis in *Mytilus edulis* L. Mar. Biol. Lett., 3: 89-105.

Behm, C.A., 1991. Fumarate reductase and the evolution of electron transport systems. In: C. Bryant (Editor), Metazoan life without oxygen. Chapman and Hall, London, pp. 88-108.

Beis, I. and Newsholme, E.A., 1975. The contents of adenine nucleotides, phosphagens and some glycolytic intermediates in resting muscles from vertebrates and invertebrates. Biochem. J., 152: 23-32.

Bennett, R. and Nakada, H.I., 1968. Comparative carbohydrate metabolism of marine molluscs. I. The intermediary metabolism of *Mytilus californianus* and *Haliotis rufescens*. Comp. Biochem. Physiol., 24B: 787-797.

Bergmann, W., 1949. Marine products. XXV. Comparative biochemical studies on the lipids and sterols of marine invertebrates. J. Mar. Res. 8: 137-176.

Bergmann, W., 1962. Sterols, their structure and distribution. In: M. Florkin and S. Mason (Editors), Comparative Biochemistry, 3. Academic Press, New York, pp. 103-162.

Bishop, S. H. Ellis, L.L. and Burcham, J.M., 1983. Amino acids in molluscs. In: P.W. Hochachka (Editor), The Mollusca, Vol I: Metabolic Biochemistry and Molecular Biomechanics. Academic Press, New York, pp. 243-327.

Bloch, D. and Hew, H., 1960. Schedule of spermatogenesis in the pulmonate snail *Helix aspersa*, with special reference to histone transition. J. Biophys. Biochem. Cytol., 7: 515-531.

Boer, H.H. and Minnen, J. van, 1988. Immunocytochemistry and hormonal peptides in molluscs: optical and electron microscopy and the use of monoclonal antibodies. In: M.C. Thorndyke and G.J. Goldsworthy (Editors), Neurohormones in Invertebrates, immunocytochemistry and ultrastructure. Cambridge University Press, Cambridge, pp. 19-41.

Booth, C.E., McDonald, D.G. and Walsh, P.J., 1984. Acid-base balance in the sea mussel, *Mytilus edulis*. I. Effects of hypoxia and air-exposure on hemolymph acid-base status. Mar. Biol. Lett., 5: 347-358.

Bouffard, T. and Aiello, E.L., 1969. Histochemical localization of succinic dehydrogenase in bivalve gill. Am. Zool., 9: 582.

Brooks, S.P.J., Zwaan, A. de, Thillart, G. van den, Cattani, O., Cortesi, P. and Storey, K.B., 1991. Differential survival of *Venus gallina* and *Scapharca inaequivalvis* during anoxic stress: covalent modification of phosphofructokinase and glycogen phosphorylase during anoxia. J. Comp Physiol., B, 161: 207-212.

Bryant, C., 1991. Metazoan life without oxygen. Chapman and Hall, London, 291pp.

Bulloch, A.G.M., 1987. Somatostatin enhances neurite outgrowth and electrical coupling of regenerating neurons in *Helisoma*. Brain Res., 412(1): 6-17.

Burcham, J.M., Paynter, K.T. and Bishop, S.H., 1983. Coupled mitochondria from oyster gill tissue. Mar. Biol. Lett., 4: 349-356.

Burcham, J.M., Ritchie, A. and Bishop, S.H., 1984. Preparation and some respiratory properties of coupled mitochondria from ribbed mussel (*Modiolus demissus*) gill tissue. J. Exp. Zool., 229: 55-67.

Calzolari, C., Cerma, E. and Stancher, B., 1971. Applicazione della gas-cromatographia nella determinazione degli acidi grassi di alcuni gasteropodi e lamellibranchi dell'alto adriatico durante un ciclo annuale. Riv. Ital. Sostanze Grasse, 48: 605-616.

Cerma, E., Stancher, B. and Baradel, P., 1970. Molluscs of the upper Adriatic sea (chemical composition of some of the gastropods and the lamellibranchs). Rass. Chim., 22: 39-43.

Chance, B. and Williams, G.R., 1956. The respiratory chain and oxidative phosphorylation. Adv. Enzymol., 17: 65-134.

Chen, C. and Awapara, J., 1969. Effect of oxygen on the end-products of glycolysis in *Rangia cuneata*. Comp. Biochem. Physiol., 3: 395-401.

Chipperfield, P.N.J., 1953. Observations on the breeding and settlement of *Mytilus edulis* L. in British waters. J. Mar. Biol. Ass. U.K., 32: 449-476.

Churchill, H.M. and Livingstone, D.R., 1989. Kinetic studies of the glycolytic enzymes from the mantle and posterior adductor muscle of the common mussel, *Mytilus edulis* L., and use of activity ratio (V_m/v) as an indicator of apparent K_m. Comp. Biochem. Physiol., 94B: 299-314.

Cienfuegos, J.E., Garcia Martin, L.O. and Carrion, A., 1977. Seasonal variations in the activity of enzymes of the pentose pathway in mussel mantle (*Mytilus edulis* L.). 11th FEBS Meeting Copenhagen, August 1977.

Collicutt, J.M., 1975. Anaerobic metabolism in the oyster heart. M.Sc. Thesis, University of British Columbia, Canada.

Collicutt, J.M. and Hochachka, P.W., 1977. The anaerobic oyster heart: coupling of glucose and aspartate fermentation. J. Comp. Physiol., 115: 147-157.

Cook, P. A. and Gabbott, P. A., 1978. Glycogen synthetase in the sea mussel *Mytilus edulis* L. I. Purification, interconversion and kinetic properties of the I and D forms. Comp. Biochem. Physiol., 60B: 419-421.

Coustau, C., Renaud, F., Delay, B., Robbins, I. and Mathieu, M., 1991. Mechanisms involved in parasitic castration: in vitro effects of the trematode *Prosorhynchus squamatus* on the gamatogenesis and the nutrient storage metabolism of the marine bivalve *Mytilus edulis*. Exp. Zool., 73(1): 36-43.

Crabtree, B. and Newsholme, E.A., 1972. The activities of phosphorylase, hexokinase, phosphofructokinase, lactate dehydrogenase and the glycerol 3-phosphate dehydrogenases in muscles from vertebrates and invertebrates. Biochem. J., 126: 49-58.

Dando, P.R., 1981. Strombine [N-(carboxymethyl)-D-alanine] dehydrogenase and alanopine [meso-N)-(1-carboxyethyl)-alanine] dehydrogenase from the mussel *Mytilus edulis* L. Biochem. Soc. Trans., 9: 297-298.

Dando, P.R., Storey, K.B., Hochachka, P.W. and Storey, J.M., 1981. Multiple dehydrogenases in marine molluscs: electrophoretic analysis of alanopine dehydrogenase, strombine dehydrogenase, octopine dehydrogenase and lactate dehydrogenase. Mar. Biol. Lett., 2: 249-257.

Daniel, R.J., 1923. Seasonal changes in the chemical composition of the mussel (*Mytilus edulis*). Trans. Liverpool Biol. Soc., 37: 85-106.

Davenport, J., 1985. Osmotic control in marine animals. In: M.S. Laverack (Editor), Physiological adaptations of marine animals, Vol. 39, Symp. Soc. Exp. Biol. The Company of Biologists Ltd., University of Cambridge, pp. 207-245.

Dawson, A.G., 1979. Oxidation of cytosolic NADH formed during aerobic metabolism in mammalian cells. Trends Biochem. Sci., 4(8): 171-176.

DeMoreno, J.E.A., Pollero, R.J. and Moreno, V.J., 1980. Lipids and fatty acids of the mussel (*Mytilus platensis* d'Orbigny) from South Atlantic waters. J. Exp. Mar. Biol. Ecol., 48: 263-276.

Ebberink, R.H.M. and Zwaan, A. de, 1980. Control of glycolysis in the posterior adductor muscle of the sea mussel *Mytilus edulis*. J. Comp. Physiol., 137: 165-171.

Ebberink, R.H.M. and Salimans, M., 1982. Control of glycogen phosphorylase activity in the posterior adductor muscle of the sea mussel *Mytilus edulis*. J. Comp. Physiol., 148: 27-33.

Ebberink, R.H.M. and Joosse, J., 1985. Molecular properties of various snail peptides from brain and gut. Peptides (N.Y.), 6: 451-457.

Ebberink, R.H.M., Zurburg, W. and Zandee, D.I., 1979. The energy demand of the posterior adductor muscle of *Mytilus edulis* in catch during exposure to air. Mar. Biol. Lett., 1: 23-31.

Ellington, W.R., 1979. Octopine dehydrogenase in the basilar muscle of the sea anemone *Metridium senile*. Comp. Biochem. Physiol., 63B: 349-354.

Ellington, W.R., 1981. Energy metabolism during hypoxia in the isolated, perfused ventricle of the whelk, *Busycon contrarium* Conrad. J. Comp. Physiol., 142: 457-464.

Ellington, W.R., 1983a. The recovery from anaerobic metabolism in invertebrates. J. Exp. Zool., 228: 431-444.

Ellington, W.R., 1983b. The extent of intracellular acidification during anoxia in the catch muscles of two bivalve molluscs. J. Exp. Zool., 227: 313-317.

Ellington, W.R., 1985. Cardiac energy metabolism in relation to work demand and habitat in bivalve and gastropod mollusks. In: R. Gilles (Editor), Circulation, respiration and metabolism. Springer-Verlag, Berlin, pp. 356-376.

Fagerlund, U.H.M. and Idler, D.R., 1960. Marine sterols-VI. Sterol biosynthesis in molluscs and echinoderms. Can. J. Biochem. Physiol., 38: 997-1002.

Fagerlund, U.H.M. and Idler, D.R., 1961a. Marine sterols - VIII. In vivo transformation of the sterol side chain by a clam. Can. J. Biochem. Physiol., 39: 505-509.

Fagerlund, U.H.M. and Idler, D.R., 1961b. Marine sterols-IX. Biosynthesis of 24-methylenecholesterol in clams. Can. J. Biochem. Physiol., 39: 1347-1355.

Famme, P. and Kofoed, L. H., 1982. Rates of carbon release and oxygen uptake by the mussel *Mytilus edulis* in response to starvation and oxygen. Mar. Biol. Lett., 3: 241-256.

Famme, P., Knudsen, J. and Hansen, E.S., 1981. The effect of oxygen on the aerobic-anaerobic metabolism of the marine bivalve, *Mytilus edulis* L. Mar. Biol. Lett., 2: 345-351.

Felbeck, H. and Wiley, S., 1987. Free D-amino acids in the tissues of marine bivalves. Biol. Bull., 173: 252-259.

Field, I. A., 1922. Biology and economic value of the sea mussel *Mytilus edulis*. Bull. U.S. Bur. Fish. Wash., 38: 127-259.

Fields, J.H.A. and Hochachka, P.W., 1981. Purification and properties of alanopine dehydrogenase from the adductor muscle of the oyster *Crassostrea gigas* (Mollusca, Bivalvia). Eur. J. Biochem., 114: 615-622.

Fields, J.H.A. and Quinn, J.F., 1981. Some theoretical considerations on cytosolic redox balance during anaerobiosis in marine invertebrates. J. Theor. Biol., 88: 35-45.

Foreman, R.A. and Ellington, W.R., 1983. Effects of inhibitors and substrate supplementation on anaerobic energy metabolism in the ventricle of the oyster, *Crassostrea virginica*. Comp. Biochem. Physiol., 74B: 543-547.

Franzen, A., 1955. Comparative morphological investigations into the spermiogenesis among Mollusca. Zool. Bidr. Upps., 31: 355-382.

Freas, W. and Grollman, S., 1980. Ionic and osmotic influences on prostaglandin release from the gill tissue of the marine mussel, *Modiolus demissus*. J. Exp. Biol., 84: 169-185.

Froutin, G. H., 1937. Contribution à l'étude du tissu conjonctif des Mollusques et plus particulièrement des Lamellibranches et des Gastéropodes. Thèse Sciences, Paris.

Gabbott, P.A., 1975. Storages cycles in marine bivalve molluscs: a hypothesis concerning the relationship between glycogen metabolism and gametogenesis. In: H. Barnes (Editor), Proc. 9th Eur. Mar. Biol. Symp., Oban, Scotland, 1984. Aberdeen University Press, Aberdeen, pp. 191-211.

Gabbott, P.A., 1983. Developmental and seasonal metabolic activities in marine molluscs. In: P.W. Hochachka (Editor), The Mollusca, Environmental biochemistry and physiology, 2. Academic Press, New York, pp. 165-217.

Gabbott, P.A. and Head, J.H., 1980. Seasonal changes in the specific activities of the pentose phosphate pathway enzymes, G6PDH and 6PGDH and NADP-dependent isocitrate dehydrogenase in the bivalves *Mytilus edulis*, *Ostrea edulis* and *Crassostrea gigas*. Comp. Biochem. Physiol., 66B: 279-284.

Gabbott, P. A. and Whittle, M. A., 1986. Glycogen synthetase in the sea mussel *Mytilus edulis* L. II: Seasonal changes in glycogen content and glycogen synthetase activity in the mantle tissue. Comp. Biochem. Physiol., 83B: 197-207.

Gabbott, P.A. and Peek, K., 1991. Cellular biochestry of the mantle tissue of the mussel *Mytilus edulis* L. Aquaculture, 94(2/3): 165-176.

Gabe, M., 1955. Particularités histologiques des cellules neurosécrétrices chez quelques lamellibranches. C. R. Acad. Sci., Paris, 240: 1810-1812.

Gäde, G., 1980. Biological role of octopine formation in marine molluscs. Mar. Biol. Lett., 1: 121-135.

Gäde, G., 1983. Energy metabolism of arthropods and mollusks during environmental and functional anaerobiosis. J. Exp. Zool., 228: 415-429.

Gäde, G. and Ellington, W.R., 1983. The anaerobic molluscan heart: adaptation to environmental anoxia. Comparison with energy metabolism in vertebrate hearts. Comp. Biochem. Physiol., 76A: 615-620.

Gäde, G. and Grieshaber, M.K., 1986. Pyruvate reductases catalyze the formation of lactate and opines in anaerobic invertebrates. Comp. Biochem. Physiol., 83B: 255-272.

Garcia Martin, L.O., Abad, M., Sanchez, J.L. and Galarza, A., 1984. Purification and properties of 6-phosphogluconate dehydrogenase from *Mytilus galloprovincialis* digestive gland. Comp. Biochem. Physiol., 79B: 599-606.

Gardner, D. and Riley, J.P., 1972. The component fatty acids of the lipids of some species of marine and freshwater molluscs. J. Mar. Biol. Ass. U.K., 52: 827-838.

Geraerts, W.P.M., 1976. Control of growth by the neurosecretory hormone of the light green cells in the freshwater snail *Lymnaea stagnalis*. Gen. Comp. Endocrinol., 29: 61-71.

Gnaiger, E., 1977. Thermodynamic considerations of invertebrate anoxibiosis. In: I. Lamprecht and B. Schaarschmidt (Editors), Applications of calorimetry in life sciences. Walter de Gruyter, Berlin, pp. 281-303.

Gnaiger, E., 1980. Energetics of invertebrate anoxibiosis: direct calorimetry in aquatic oligochaetes. FEBS Lett., 112: 239-242.

Gnaiger, E., 1983. Heat dissipation and energetic efficiency in animal anoxibiosis: economy contra power. J. Exp. Zool., 228: 471-490.

Goad, L.J., 1976. The sterols of marine algae and invertebrate animals. In: D.C. Malins and J.R. Sargent (Editors), Biochemical and Biophysical Perspectives in Marine Biology, 3. Academic Press, New York, 213 pp.

Goad, L.J., 1978. The sterols of marine invertebrates: Composition, biosynthesis and metabolites. In: P.J. Scheuer (Editor), Marine natural products, Academic Press, New York, pp. 75-172.

Goromosova, S.A., 1976. Glycogen synthetase and fructose diphosphatase activity in the tissue of mollusks (Mytilus galloprovincialis) and crustaceans (Balanus improvisus, Carcinus maenas). J. Evol. Biochem. Physiol. (Engl. Trans.), 12: 331-334.

Greenberg, M.J., Payza, K., Nachman, R.J., Holman, G.M. and Price, D.A., 1988. Relationships between the FMRFamide-related peptides and other peptide families. Peptides (N.Y.), 9: 125-135.

Greenfield, E. and Crenshaw, M.A., 1981. Variations in the rate of anaerobic succinate accumulation within the central and marginal regions of an euryoxic bivalve mantle. Mar. Ecol., 2: 353-362.

Grimm-Jørgensen, Y., 1983. Possible physiological roles of thyrotropin releasing hormone and a somatostatin-like peptide in gastropods. In: J. Lever et H. H. Boer (Editors), Molluscan Neuroendocrinology. North–Holland, Amsterdam, pp. 21-28.

Grimm-Jørgensen, Y., 1987. Somatostatin and calcitonin stimulate neurite regeneration of molluscan neurons in vitro. Brain Res., 403(1): 121-126.

Halarnkar, P.P. and Blomquist, G.J., 1989. Comparative aspects of propionate metabolism. Comp. Biochem. Physiol., 92B: 227-231.

Hammen, C.S., 1980. Total energy metabolism of marine bivalve mollusks in anaerobic and aerobic states. Comp. Biochem. Physiol., 67A: 617-621.

Hardewig, I., Addink, A.D.F., Grieshaber, M.K., Pörtner, H.D. and Thillart, G. van den, 1991. Metabolic rates at different oxygen levels determined by direct and indirect calorimetry in the oxyconformer Sipunculus nudus. J. Exp. Biol., 157: 143-160.

Head, E.J.H., 1980a. NADP-dependent isocitrate dehydrogenase from the mussel Mytilus edulis L. I. Purification and characterisation. Eur. J. Biochem., 111: 575-579.

Head, E.J.H., 1980b. NADP-dependent isocitrate dehydrogenase from the mussel Mytilus edulis L. Eur. J. Biochem., 111: 581-586.

Head, E.J.H. and Gabbott, P.A., 1980. Properties of NADP-dependent isocitrate dehydrogenase from the mussel Mytilus edulis L. Comp. Biochem Physiol., 66B: 285-289.

Head, E.J.H. and Gabbott, P.A., 1981. Kinetic control of NADP-dependent isocitrate dehydrogenase activity in the mussel Mytilus edulis L. Comp. Biochem. Physiol., 68B: 383-388.

Hemminga, M. A., Maaskant, J. J. and Joosse, J., 1985. Direct effects of the hyperglycemic factor of the freshwater snail Lymnaea stagnalis on isolated glycogen cells. Gen. Comp. Endocrinol., 58: 131-136.

Herlin-Houtteville, P., 1974. Contribution à l'étude cytologique et expérimentale du cycle annuel du tissu de réserve du manteau de Mytilus edulis L. Thèse de 3ème cycle, Université de Caen, France.

Hochachka, P.W. and Mommsen, T.P., 1983. Protons and anaerobiosis. Science, 219: 1391-1397.

Holland, D.L., 1978. Lipid reserves and energy metabolism in the larvae of benthic marine invertebrates. Biochem. Biophys. Perspect. Mar. Biol., 4: 85-123.

Holwerda, D.A. and Zwaan, A. de, 1979. Fumarate reductase of Mytilus edulis L. Mar. Biol. Lett., 1: 33-40.

Holwerda, D.A. and Zwaan. A. de, 1980. On the role of fumarate reductase in anaerobic carbohydrate catabolism of Mytilus edulis L. Comp. Biochem. Physiol., 67B: 447-453.

Holwerda, D.A., Kruitwagen, E.C.J. and Bont, A.M.T. de, 1981. Regulation of pyruvate kinase and phosphoenolpyruvate carboxykinase activity during anaerobiosis in Mytilus edulis L. Mol. Physiol., 1: 165-171.

Holwerda, D.A., Veenhof, P.R., Heugten, H.A.A. van and Zandee, D.I., 1983. Modification of mussel pyruvate kinase during anaerobiosis and after temperature acclimation. Mol. Physiol., 3: 225-234.

Holwerda, D.A., Veenhof, P.R. and Zwaan, A. de, 1984. Physiological and biochemical investigations of the ecological relevance of anaerobiosis in bivalves. I. The changes in activity of mussel adductor muscle and mantle pyruvate kinase during aerial exposure and reimmersion. Mar. Biol. Lett., 5: 185-190.

Holwerda, D.A., Veldhuizen-Tsoerkan, M., Veenhof, P.R. and Evers, E., 1989. In vivo and in vitro studies on the pathway of modification of mussel pyruvate kinase. Comp. Biochem. Physiol., 92B: 375-380.

Horst D.J. van der, 1974. In vivo biosynthesis of fatty acids in the pulmonate land snail *Cepaea nemoralis* (L.) under anoxic conditions. Comp. Biochem. Physiol., 47B: 181-187.

Hummel, H., Wolf, L. de, Zurburg, W., Apon, L., Bogaards, R.H. and Ruitenburg, M. van, 1989. The glycogen content in stressed marine bivalves: the initial absence of a decrease. Comp. Biochem. Physiol., 94B: 729-733.

Ibarguren, I., Vazquez-Illanes, M.D. and Ramos-Martinez, J.I., 1990. Seasonal variations in glycolysis in *Mytilus galloprovincialis* Lmk. Comp. Biochem. Physiol., 97B: 279-283.

Idler, D.R. and Wiseman, P., 1971. Sterols of molluscs. Int. J. Biochem., 2: 516-528.

Idler, D.R. and Wiseman, P., 1972. Molluscan sterols: A review. J. Fish. Res. Board Can., 29: 385-398.

Illanes-Bucher, J., 1979. Recherches cytologiques et expérimentales sur la neurosécrétion de la moule *Mytilus edulis* L. (Mollusque, Lamellibranche). Thèse de 3ème cycle, Université de Caen, France.

Illanes-Bucher, J. and Lubet, P., 1980. Etude de l'activité neurosécrétrice au cours du cycle sexuel annuel de la moule (*Mytilus edulis* L.). Mollusque Lamellibranche. Bull. Soc. Zool. Fr., 105(1): 141-145.

Jamieson, D.D. and Rome, P. de, 1979. Energy metabolism of the heart of the mollusc *Tapes watlingi*. Comp. Biochem. Physiol., 63B: 399-405.

Jokumsen, A. and Fyhn, H.J., 1982. The influence of aerial exposure upon respiratory and osmotic properties of haemolymph from two intertidal mussels, *Mytilus edulis* L. and *Modiolus modiolus* L. J. Exp. Mar. Biol. Ecol., 61: 189-203.

Jong-Brink, M. de, Boer, H.H., Hommes, T.G. and Kodde, A., 1977. Spermatogenesis and the role of Sertoli cells in the freshwater snail *Biomphalaria glabrata*. Cell Tissue Res., 181: 37-58.

Joosse, J. and Geraerts, W.P.M. , 1983. Endocrinology. In: A.S.M. Saleuddin and K.M. Wilbur (Editors), The Mollusca, Physiology, Part 1, 4. Academic Press, New York, pp. 317-406.

Joseph, J.D., 1982. Lipid composition of marine and estuarine invertebrates. Part II. Prog. Lipid Res., 21: 109-154.

Kanazawa, A., Teshima, S. and Ono, K., 1979. Relationship between essential fatty acid requirements of aquatic animals and the capacity of bioconversion of linolenic acid to highly unsaturated fatty acids. Comp. Biochem. Physiol., 63B: 295-298.

Karam, G.A., Paynter, K.T. and Bishop, S.H., 1987. Ketoglutarate dehydrogenase from ribbed mussel gill mitochondria: modulation by adenine nucleotides and calcium ions. J. Exp. Zool., 243: 15-24.

Kargbo, A and Swift, M.L., 1983. NAD^+-dependent isocitrate dehydrogenase from the oyster, *Crassostrea virginica* Gmelin. Comp. Biochem. Physiol., 76B: 123-126.

Katz, J. and Wood, H.G., 1963.The use of $^{14}CO_2$ yields from glucose-1-and 6-C^{14} for the evaluation of the pathways of glucose metabolism. J. Biol. Chem., 238: 517-523.

Kelley, R.N., Ashwood-Smith, M.J. and Ellis, D.V., 1982. Duration and timing of spermatogenesis in a stock of the mussel *Mytilus californianus*. J. Mar. Biol. Ass. U.K., 62: 509-519.

Kluytmans, J.H., Veenhof, P.R. and Zwaan, A. de, 1975. Anaerobic production of volatile fatty acids in the sea mussel *Mytilus edulis* L. J. Comp. Physiol., 104: 71-78.

Kluytmans, J.H., Bont, A.M.T. de, Janus, J. and Wijsman, T.C.M., 1977. Time dependent changes and tissue specificities in the accumulation of anaerobic fermentation products in the sea mussel *Mytilus edulis* L. Comp. Biochem. Physiol., 58B: 81-87.

Kluytmans, J.H., Graft, M. van, Janus, J. and Pieters, H., 1978. Production and excretion of volatile fatty acids in the sea mussel *Mytilus edulis* L. J. Comp. Physiol., 123: 163-167.

Kluytmans, J.H., Zandee, D.I., Zurburg, W. and Pieters, H., 1980. The influence of seasonal changes on energy metabolism in *Mytilus edulis* (L.). III. Anaerobic energy metabolism. Comp. Biochem. Physiol., 67B: 307-315.

Kluytmans, J.H., Boot, J.H., Oudejans, C.H.M. and Zandee, D.I., 1985. Fatty acid synthesis in relation to gametogenesis in the mussel *Mytilus edulis* L. Comp. Biochem. Physiol., 81B: 959-963.

Kollmann, J.,1908. Die Bindesubstanz des Acephalen. Arch. Mikr. Anat., XIII, 558pp.

Kreutzer, U., Siegmund, B.R. and Grieshaber, M.K., 1989. Parameters controlling opine formation during muscular activity and environmental hypoxia. J. Comp. Physiol., 159: 617-628.

LaNoue, K.F. and Schoolwerth, A.C., 1979. Metabolite transport in mitochondria. Ann. Rev. Biochem., 48: 871-922.

Lenoir, F., 1989. Mise au point de techniques de dissociation, de purification et de culture cellulaires chez la moule *Mytilus edulis* L. Application à l'étude des régulations du métabolisme du glucose et du glycogène dans les cellules à glycogène (= cellules vésiculeuses). Thèse d'Université, Caen, France.

Lenoir , F. and Mathieu, M., 1986. Utilisation de cultures de cellules dissociées dans l'étude des contrôles exercés sur la gamétogénèse chez la moule *Mytilus edulis* L. C. R. Acad. Sci., Paris, 303 (III) 12: 523-528.

Lenoir, F., Robbins, I., Mathieu, M., Lubet, P. and Gabbott, P. A., 1989. Isolation, characterization and glucose metabolism of glycogen cells (= vesicular connective-tissue cells) from the labial palps of the marine mussel *Mytilus edulis*. Mar. Biol., 101: 495-501.

Leung, M. and Stephano, G.B., 1983. Isolation of Molluscan opioid peptides. Life Sci., 33: 77-80.

L–Fando, J. J., Garcia–Fernandez, M. C. and Candela, J. L., 1972. Glycogen metabolism in *Ostrea edulis* L. : Factors affecting glycogen synthesis. Comp. Biochem. Physiol., 43B: 807-814.

List, T., 1902. Fauna und Flora des Golfes von Neapel und der angrenzenden Meeres-Abschnitte. I. Die Mytiliden des Golfes von Neapel und der angrenzenden Meeres-Abschnitte. Mitt. Zool. Stn. Neapel, 27: 1-312.

Lindinger, M.I., Laurën, D.J. and McDonald, D.G., 1984. Acid-base balance in the sea mussel, *Mytilus edulis*. III. Effects of environmental hypercapnia on intra- and extracellular acid-base balance. Mar. Biol. Lett., 5: 371-381.

Livingstone, D.R., 1981. Induction of enzymes as a mechanism for the seasonal control of metabolism in marine vertebrates: glucose-6-phosphate dehydrogenases from the mantle and hepatopancreas of the common mussel *Mytilus edulis* L. Comp. Biochem. Physiol., 69B: 147-156.

Livingstone, D.R., 1982. Energy production in the muscle tissues of different kinds of molluscs. In: A.D.F. Addink and N. Spronk (Editors), Exogenous and endogenous influences on metabolic and neural control, 1. Invited lectures, Proc. 3rd Congr. Eur. Soc. Comp. Physiol. Biochem., Noordwijkerhout, Netherlands, 1981. Pergamon Press, Oxford, pp. 257-274.

Livingstone, D.R., 1983. Invertebrate and vertebrate pathways of anaerobic metabolism: evolutionary considerations. J. Geol. Soc. (Lond.), 140: 27-37.

Livingstone, D.R. and Clarke, K.R., 1983. Seasonal changes in hexokinase from the mantle tissue of the common mussel *Mytilus edulis* L. Comp. Biochem. Physiol., 74B: 691-702.

Livingstone, D.R., Widdows, J. and Fieth, P., 1979. Aspects of nitrogen metabolism of the common mussel *Mytilus edulis*: adaptation to abrupt and fluctuating changes in salinity. Mar. Biol., 53: 41-55.

Livingstone, D.R., Zwaan, A. de and Thompson, R.J., 1981. Aerobic metabolism, octopine production and phosphoarginine as sources of energy in the phasic and catch adductor muscles of the giant scallop *Placopecten magellanicus* during swimming and the subsequent recovery period. Comp. Biochem. Physiol., 70B: 35-44.

Livingstone, D.R., Zwaan, A. de, Leopold, M. and Marteijn, E., 1983. Studies on the phylogenetic distribution of pyruvate oxidoreductases. Biochem. Syst. Ecol., 11: 415-425.

Livingstone, D.R., Stickle, W.B., Kapper, M.A., Wang, S. and Zurburg, W., 1990. Further studies on the phylogenetic distribution of pyruvate oxidoreductase activities. Comp. Biochem. Physiol., 97B: 661-666.

Longcamp, D. de, Lubet, P. and Drosdowsky, M., 1974. The in vitro biosynthesis of steroids by the gonad of the mussel *Mytilus edulis*. Gen. Comp. Endocrinol., 22: 116-127.

Longo, F.J. and Dornfield, E.J., 1967. The fine structure of spermatid differentiation in the mussel *Mytilus edulis*. J. Ultrastruct. Res., 20: 462-480.

Loomis, S.H., Carpenter, J.F. and Crowe, J.H., 1988. Identification of strombine and taurine as cryoprotectants in the intertidal bivalve *Mytilus edulis*. Biochim. Biophys. Acta, 943: 113-118.

Lowe, D.M., Moore, M.N. and Bayne, B.L., 1982. Aspects of gametogenesis in the marine mussel *Mytilus edulis* L. J. Mar. Biol. Ass. U.K., 62: 133-145.

Lubet, P., 1955. Cycle neurosécrétoire de *Chlamys varia* L. et *Mytilus edulis* L. C. R. Acad. Sci. Paris, 24: 119-121.

Lubet, P., 1959. Recherches sur le cycle sexuel et l'émission des gamètes chez les Mytilidés et les Pectinidés. Rev. Trav. Off. (Scient. Tech.) de Pêch. Marit., 23 (4): 396-545.

Lubet, P., 1966. Essai d'analyse expérimentale des perturbations produites par les ablations de ganglions nerveux chez *Mytilus edulis* L. et *Mytilus galloprovincialis* Lmk. (Mollusques Lamellibranches). Ann. Endocrinol., 27: 353-365.

Lubet, P., 1971. Influence des ganglions cérébroides sur la croissance de *Crepidula fornicata* Phil. (Mollusque Mésogastéropode). C. R. Acad. Sci., Paris, 273 (III): 2309-2311.

Lubet, P., 1981. Action des facteurs internes sur la reproduction des mollusques lamellibranches. Seminarios de Biologia Marinha. Bol. Zool., 5: 121-139.

Lubet, P. and Mathieu, M., 1982. The action of internal factors on gametogenesis in Pelecypod Mollusks. Malacologia, 22 (1-2): 131-136.

Lubet, P. and Mathieu, M., 1991. Les régulations endocriniennes chez les Mollusques Bivalves. Année Biol., 29 (4): 235-252.

Lubet, P., Herlin, P., Mathieu, M. and Collin, F., 1976. Tissu de réserve et cycle sexuel chez les Lamellibranches. Haliotis, 7: 59-62.

Lubet, P., Albertini, L. and Robbins, I., 1986. Recherches expérimentales au cours de cycles annuels sur l'action gonadotrope exercée par les ganglions cérébroides sur la gamétogénèse femelle chez la moule *Mytilus edulis* L. (Mollusque bivalve). C. R. Acad. Sci., Paris, 303 (III): 575-580.

Mahon, A.C., Nambu, J.R., Taussig, R., Shyamala, M., Roach, A. and Scheller, R.H., 1985. Structure and expression of the egg-laying hormone gene family in *Aplysia*. J. Neurosci., 5: 1872-1880.

Malanga, C.J. and Aiello E.L., 1972. Succinate metabolism in the gills of the mussels *Modiolus demissus* and *Mytilus edulis*. Comp. Biochem. Physiol., 43B: 795-806.

Mane, U.H. and Nagabhushanam, R., 1975. Body distribution and seasonal changes in the biochemical composition of the estuarine mussel *Mytilus veridis* at Ratnagiri. Riv. Idrobiol., 14 (3): 163-175.

Mangum, C. and Winkle, W. van, 1973. The response of aquatic invertebrates to declining oxygen conditions. Am. Zool., 13: 529-541.

Martin, R., Haas, C. and Voigt, K.H., 1986. Opioid and related neuropeptides in Molluscan neurons. In: G.B. Stephano (Editor), CRC handbook of comparative opioid and related neuropeptide mechanisms. CRC Press Inc. Boca Raton, Florida, pp. 49-64.

Mathieu, M., 1985. Partial characterisation of aspartate transcarbamylase from the mantle of the mussel *Mytilus edulis*. Comp. Biochem. Physiol., 82B: 667-674.

Mathieu, M., 1987a. Utilization of ATCase activity in the study of neuroendocrine control of gametogenesis in *Mytilus edulis*. J. Exp. Zool., 241 (2): 247-252.

Mathieu, M., 1987b. Etude expérimentale des contrôles exercés par les ganglions nerveux sur la gamétogénèse et les processus métaboliques associés chez la moule *Mytilus edulis* L. (Mollusque Lamellibranche). Thèse de Doctorat d'Etat, Université de Caen, France.

Mathieu, M. and Lubet, P., 1980. Analyse expérimentale en cultures d'organes de l'action des ganglions nerveux sur la gonade adulte de la moule. Bull. Soc. Zool. Fr., 105: 149-153.

Mathieu, M. and Minnen, J. van, 1989. Mise en évidence par immunocytochimie de cellules neurosécrétrices peptidergiques dans les ganglions cérébroïdes de la moule *Mytilus edulis*. C. R. Acad. Sci. Paris, 308 (III): 489-494.

Mathieu, M., Bergeron, J.P. and Alayse-Danet, A.M., 1982. L'aspartate transcarbamylase, indice d'activité gamétogénétique chez la moule *Mytilus edulis* L. Int. J. Invertebr. Reprod., 5(6): 337-343.

Mathieu, M., Lenoir, F. and Robbins, I.,1988. A gonial mitosis-stimulating factor in cerebral ganglia and hemolymph of the marine mussel *Mytilus edulis* L. Gen. Comp. Endocrinol., 72: 257-263.

Matsutani, T. and Nomura, T., 1982. Induction of spawning by serotonin in the scallop, *Patinopecten yessoensis* (Jay). Mar. Biol. Lett., 3: 353-358.

Matsutani, T. and Nomura, T., 1986. Pharmacological observations on the mechanisms of spawing in the scallop, *Patinopecten yessoensis*. Bull. Jpn. Soc. Sci. Fish., 52 (9): 1589-1594.

Meinardus, G. and Gäde G., 1981. Anaerobic metabolism of the cockle, *Cardium edule*. IV. Time dependent changes of metabolites in the foot and gill tissue induced by anoxia and electrical stimulation. Comp. Biochem. Physiol., 70B: 271-277.

Michaelidis, B. and Storey, K.B., 1990. Phosphofructokinase from the anterior byssus retractor muscle of *Mytilus edulis*: modification of the enzyme in anoxia and by endogenous protein kinases. Int. J. Biochem., 22: 759-765.

Michaelidis, B. and Storey, K.B., 1991. Evidence for phosphorylation/dephosphorylation control of phosphofructokinase from organs of the anoxia-tolerant sea mussel *Mytilus edulis.* J. Exp. Zool., 257: 1-9.

Moore, M.N., Lowe, D.M. and Fieth, P.E.M., 1978. Lysosomal responses to experimentally injected anthracene in the digestive cells of *Mytilus edulis.* Mar. Biol., 48: 297-302.

Morse, D.E., Duncan, H., Hooker, N. and Morse, A., 1977. Hydrogen peroxide induces spawning in molluscs with activation of prostaglandin endoperoxide synthetase. Science, 196: 298-300.

Motavkine, P.A. and Varaskine, A.A., 1989. La reproduction chez les mollusques bivalves. Rôle du système nerveux et régulation. Rapp Sci. Tech. IFREMER, 10pp.

Murphy, D.J., 1977. Metabolic and tissue solute changes associated with changes in the freezing tolerance of the bivalve mollusc *Modiolus demissus.* J. Exp. Biol., 69: 1-12.

Niijima, L. and Dan, J. ,1965. The acrosome reaction in *Mytilus edulis* L. 1. Fine structure of the intact acrosome. 2. Stages in the reaction, observed in supernumerary and calcium-treated spermatozoa. J. Cell Biol., 25: 243-259.

Nomura, T. and Ogata, H., 1976. Distribution of prostaglandins in the animal kingdom. Biochim. Biophys. Acta, 431: 127-131.

Nuccitelli, R. and Heiple, J.M., 1982. Summary of the evidence and discussion concerning the involvement of pH in the control of cellular functions. In: Kroc Foundation series, Vol. 15, intracellular pH: its measurement, regulation and utilization in cellular functions, Alan R. Liss, New York, pp. 567-586.

Osada, M. , Nishikawa, M. and Nomura, T., 1989. Involvement of prostaglandins in the spawning of the scallop, *Patinopecten yessoensis.* Comp. Biochem. Physiol., 94C: 595-601.

Oudejans, R.C.H.M. and Horst, D.J., van der, 1974. Aerobic-anaerobic biosynthesis of fatty acids and other lipids from glycolytic intermediates in the pulmonate land snail *Cepaea nemoralis* (L.). Comp. Biochem. Physiol., 47B: 139-147.

Pamatmat, M.M., 1980. Facultative anaerobiosis of benthos. In: K.R. Tenore and B.C. Coull (Editors), Marine benthic dynamics. University of South Carolina Press, Columbia, pp. 69-90.

Papadopoulos, A.I. Gaitanaki, C.J. and Beis, I.D., 1990. Pyruvate kinase isoenzymes in marine invertebrates: a comparative study by the use of monoclonal antibodies. Comp. Biochem. Physiol., 96B: 229-234.

Paradis, M. and Ackman, R.G., 1977. Potential for employing the distribution of anomalous non-methylene-interrupted dienoic fatty acids in several marine invertebrates as part of food web studies. Lipids, 12: 170-176.

Paynter, K.T., Ellis, L.L. and Bishop, S.H., 1984a. Cellular location and partial characterization of the alanine aminotransferase in ribbed mussel gill tissue. J. Exp. Zool., 232: 51-58.

Paynter, K.T., Hoffmann, R.J., Ellis, L.L. and Bishop, S.H., 1984b. Partial characterization of the cytosolic and mitochondrial aspartate aminotransferase from ribbed mussel gill tissue. J. Exp. Zool., 231: 185-197.

Paynter, K.T., Karam, G.A., Ellis, L.L. and Bishop, S.H., 1985a. Pyruvate dehydrogenase complex from ribbed mussel gill mitochondria. J. Exp. Zool., 236: 251-257.

Paynter, K.T., Karam, G.A., Ellis, L.L. and Bishop, S.H., 1985b. Subcellular distribution of aminotransferases, and pyruvate branch point enzymes in gill tissue from four bivalves. Comp. Biochem. Physiol., 82B: 129-132.

Peek, K. and Gabbott, P.A., 1989. Adipogranular cells from the mantle tissue of *Mytilus edulis* L. I. Isolation, purification and biochemical characteristics of dispersed cells. J. Exp. Mar. Biol. Ecol., 126: 203-216.

Peek, K. and Gabbott, P.A., 1990. Seasonal cycle of lysosomal enzyme activities in the mantle tissue and ilsolated cells from the mussel *Mytilus edulis.* Mar. Biol., 104(3): 403-412.

Peek, K., Gabbott, P.A. and Runham, N.W., 1989. Adipogranular cells from the mantle tissue of *Mytilus edulis* L. II. Seasonal changes in the distribution of dispersed cells in a performed percoll density gradient. J. Exp. Mar. Biol. Ecol., 126: 217-230.

Philip, R.A., Newsholme, E.A. and Zammit V.A., 1976. Activities of citrate synthase and NAD$^+$-linked and NADP$^+$-linked isocitrate dehydrogenase in muscle from vertebrates and invertebrates. Biochem. J., 154: 689-700.

Pieters, H., Kluytmans, J.H., Zandee, D. and Cadee, G.C., 1980. Tissue composition and reproduction of *Mytilus edulis* dependent on food availability. Neth. J. Sea Res., 14: 349-361.

Pipe, R. K., 1985. Seasonal cycles and effects of starvation on egg development in *Mytilus edulis*. Mar. Ecol. Prog. Ser., 24: 121-128.

Pipe, R. K., 1987. Ultrastructural and cytochemical study on interactions between nutrient storage cells and gametogenesis in the mussel *Mytilus edulis*. Mar. Biol., 96: 519-528.

Pipe, R.K. and Moore, M.N., 1985. The ultrastructural localization of lysosomal acid hyrolases in developing oocytes of the common marine mussel *Mytilus edulis*. Histochem. J., 127: 939-949.

Plaxton, W.C. and Storey, K.B., 1986. Glycolytic enzyme binding and metabolic control in anaerobiosis. J. Comp. Physiol., B, 156: 635-640.

Pörtner, H.O., 1987. Contributions of anaerobic metabolism to pH regulation in animal tissues: theory. J. Exp. Biol., 131: 69-87.

Reis-Henriques, M.A. and Coimbra, J., 1990. Variations in the levels of progesterone in *Mytilus edulis* during the annual reproductive cycle. Comp. Biochem. Physiol., 95A, 3: 343-348.

Reis-Henriques, M.A., Le Guellec, D. , Remy-Martin J.P. and Adessi, G.L., 1990. Studies of endogenous steroids from the marine mollusc *Mytilus edulis* L. by gas chromatography and mass spectrometry. Comp. Biochem. Physiol., 95B: 303-309.

Robbins, I., Lenoir, F. and Mathieu, M., 1990. A putative neuroendocrine factor that stimulates glycogen mobilization in isolated glycogen cells from the marine mussel *Mytilusedulis*. Gen. Comp. Endocrinol., 79: 123-129.

Robbins, I., Lenoir, F. and Mathieu, M., 1991. Neuroendocrine factors affecting the glycogen metabolism of purified *Mytilus edulis* glycogen cells. Partial characterization of the putative glycogen mobilization hormone. Demonstration of a factor that stimulates glycogen synthesis. Gen. Comp. Endocrinol., 82: 45-52.

Rodegker, W. and Nevenzel, J.C., 1964. The fatty acid composition of three marine invertebrates. Comp. Biochem. Physiol, 11: 53-60.

Rodriguez-Segade, S., Carrion, A. and Freire, M., 1979. Isolation and purification of a regulating cofactor of the pentose-phosphate pathway. Biochem.Biophys. Res. Commun., 89: 148-154.

Rodriguez-Segade, S., Eguiraun, A. and Freire, M., 1980. Properties of the regulatory cofactor of the pentose phosphate pathway in mussel hepatopancreas (*Mytilus edulis* L.). Comp.Biochem.Physiol., 65B: 579-581.

Ruiz, R.A., Santos, M.J.H. and Ruiz-Amil, M., 1985. $NADP^+$-dependent isocitrate dehydrogenase from hepatopancreas of mussel (*Mytilus edulis* L.). Comp. Biochem. Physiol., 81B: 953-957.

Ryan, C.A. and King, T.E., 1962. Succinate dehydrogenase from the bay mussel *Mytilus edulis*. Biochim. Biophys. Acta, 62: 269-278.

Sailer, M., Reuzel-Selke, A. and Achazi, R.K., 1990. The calmodulin-protein-kinase system of *Mytilus edulis* catch muscle. Comp. Biochem. Physiol., 96B: 533-541.

Sato, M., Sato, Y. and Tsuchiya, Y., 1982a. Distribution of *meso*-α-iminodipropionic acid and D-α-iminopropioacetic acid in a variety of aquatic organisms. Nippon Suisan Gakkaishi, 48: 1411-1414.

Sato, M., Sato, Y. and Tsuchiya, Y., 1982b. Biosynthesis of *meso*-α-iminodipropionic acid and D-α-iminopropioacetic acid in scallop. Nippon Suisan Gakkaishi, 48: 1415-1419.

Sato, M., Takahara, M., Kanno, N., Sato, Y. and Ellington, W.R., 1987. Isolation of a new opine, α-alanopine, from the extracts of the muscle of the marine bivalve mollusc, *Scapharca broughtonii*. Comp. Biochem. Physiol., 88B: 803-806.

Schaefer, M., Piciotto, M.R., Kreiner, T., Kaldany, R.R., Taussig, R. and Scheller, R.H., 1985. *Aplysia* neurons express a gene encoding multiple FMRFamide neuropeptides. Cell, 41: 457-467.

Scharrer, B., 1935. Ueber das Hanströmsche Organ X bei Opistobranchiern. Pubbl. Stn. Zool. Napoli, 15: 132-142.

Schell, J., Montagu, M. van, Beukelaar, M. de, Block, M. de, Depicker, A., Wilde, M. de, Engler, G., Genetello, C., Hernalsteens, J.P., Holsters, M., Seurinck, J., Silva, B., Vliet, F. van and Villarroel, R., 1979. Interaction and DNA transfer between *Agrobacterium tumefaciens*, the Ti plasmid and the plant host. Proc. R. Soc. Lond., Ser. B, 204: 251-266.

Scheller, R.H., Jackson, J.F., McAllister, L.B., Schartz, J.H., Kandel, E.R. and Axel, R., 1982. A family of genes that codes for ELH, a neuropeptide eliciting a stereotyped pattern of behaviour in *Aplysia*. Cell, 28: 707-719.

Schot, L.P.C., Boer, H.H., Swaab, D.F. and Noorden, S. van, 1981. Immunocytochemical demonstration of peptidergic neurons in the central nervous system of the pond snail *Lymnaea stagnalis* with antisera raised to biologically active peptides of vertebrates. Cell Tissue Res., 216: 273-291.

Schulz, T.K.F. and Kluytmans, J.H., 1983. Pathway of propionate synthesis in the sea mussel *Mytilus edulis* L. Comp. Biochem. Physiol., 75B: 365-372.

Schulz, T.K.F., Kluytmans, J.H. and Zandee, D.I., 1982. In vitro production of propionate by mantle mitochondria of the sea mussel *Mytilus edulis* L. : overall mechanism. Comp. Biochem. Physiol., 73B: 673-680.

Schulz, T.K.F., Joosse, A. and Kluytmans, J.H., 1984. Propionate synthesis in the sea mussel *Mytilus edulis* L. : possible role of acyl-CoA transferase on the occurrence of a lag time. Mar. Biol. Lett., 5: 155-169.

Schulz, T.K.F., Duin, M. van and Zandee, D.I., 1983. Propionyl-CoA carboxylase from the sea mussel *Mytilus edulis* L. : some properties and its role in the anaerobic energy metabolism. Mol. Physiol., 4: 215-230.

Seed, R., 1975. Reproduction in *Mytilus* (Mollusca: Bivalvia) in European waters. Pubbl. Stn. Zool. Napoli, 39 (1): 317-334.

Shick, J.M. and Widdows, J., 1981. Direct and indirect calorimetric measurement of metabolic rate in bivalve molluscs during aerial exposure. Am. Zool., 21: 983.

Shick, J.M., Zwaan, A. de and Bont, A.M.T. de, 1983. Anoxic metabolic rate in the mussel *Mytilus edulis* L. estimated by simultaneous direct calorimetry and biochemical analysis. Physiol. Zool., 56: 56-63.

Shick, J.M., Gnaiger, E., Widdows, J., Bayne, B.L. and Zwaan, A. de, 1986. Activity and metabolism in the mussel *Mytilus edulis* L. during intertidal hypoxia and aerobic recovery. Physiol. Zool., 59: 627-642.

Shick. J. M., Widdows, J. and Gnaiger, E., 1988. Calorimetric studies of behavior, metabolism and energetics of sessile intertidal animals. Am. Zool., 28: 161-181.

Siebenaller, J. F., 1979. Regulation of pyruvate kinase in *Mytilus edulis*, by phosphorylation-dephosphorylation. Mar. Biol. Lett., 1: 105-110.

Siegmund, B. and Grieshaber, M.K., 1983. Determination of *meso*-alanopine and D-strombine by high pressure liquid chromatography in extracts from marine invertebrates. Hoppe-Seyler's Z. Physiol. Chem., 364: 807-812.

Smit, A.B., Vreugdenhil, E., Ebberink, R.M.N., Geraerts, W.P.M., Klootwijk, J. and Joosse, J., 1988. Growth-controlling molluscan neurons produce the precursor of an insulin-related peptide. Nature (Lond.), 331(6156): 535-538.

Srivastava, K.C. and Mustafa, T., 1984. Arachidonic acid metabolism and prostaglandins in lower animals. Mol. Physiol., 5: 53-60.

Srivastava, K.C. and Mustafa, T., 1985. Formation of prostaglandins and other comparable products during aerobic and anaerobic metabolism of [1-^{14}C]arachidonic acid in the tissues of sea mussel, *Mytilus edulis* L. Mol. Physiol., 8: 101-112.

Stanley-Samuelson, D.W., 1987. Physiological roles of prostaglandins and other eicosanoids in invertebrates. Biol. Bull., 173: 92.

Stokes, T.M. and Awapara, J., 1968. Alanine and succinate as end-products of glucose degradation in the clam *Rangia cuneata*. Comp. Biochem. Physiol., 25: 883-892.

Storey, K.B., 1985a. A re-evaluation of the Pasteur effect: new mechanisms in anaerobic metabolism. Mol. Physiol., 8: 439-461.

Storey, K.B., 1985b. Fructose 2,6-biphosphate and anaerobic metabolism in marine molluscs. FEBS Lett., 181: 245-248.

Storey, K.B., 1986a. Fructose-2,6-bisphosphate and anaerobic metabolism in marine molluscs. FEBS Lett., 181: 245-248.

Storey, K.B., 1986b. Aspartate activation of pyruvate kinase in anoxia tolerant molluscs. Comp. Biochem. Physiol., 83B: 807-812.

Storey, K.B. and Storey, J.M., 1988. Freeze tolerance in animals. Physiol. Rev., 68: 27-84.

Storey, K.B. and Storey, J.M., 1990. Metabolic rate depression and biochemical adaptation in anaerobiosis, hibernation and estivation. Q. Rev. Biol., 65: 145-174.

Storey, K.B., Miller, D.C., Plaxton, W.C. and Storey, J.M., 1982. Gas-liquid chromatography and enzymatic determination of alanopine and strombine in tissues of marine invertebrates. Analyt. Biochem., 125: 50-58.

Takanashi, M. and Mitshuhashi, T., 1968. Lipids of mussel (*Mytilus edulis*). Tokio Gdugei Daigaku Kiyo Ser. 20: 39-42.

Teshima, S.I. and Kanasawa, A., 1972. Comparative study on the sterol composition of marine mollusks. Bull. Jpn. Soc. Sci. Fish., 38: 1299-1304.

Teshima, S.I. and Kanasawa, A., 1974. Biosynthesis of sterols in abalone, *Haliotis gurneri* and mussel *Mytilus edulis*. Comp. Biochem. Physiol., 47B: 55-561.

Teshima, S.I. , Kanasawa, A.and Miyawaki, H., 1979. Metabolism of β-sitosterol in the mussel and the snail. Comp. Biochem. Physiol., 63B: 323-328.

Theede, H., 1972. Vergleichende Ekologisch-physiologische Untersuchungen zur zellulären Kelteresistenz mariner Evertebraten. Mar. Biol., 15: 160-191.

Theede, H., Ponat, A., Hiroki, K. and Schlieper, C., 1969. Studies on the resistance of marine bottom invertebrates to oxygen deficiency and hydrogen sulphide. Mar. Biol., 2: 325-337.

Thillart, G. van den and Vries, I. de, 1985. Excretion of volatile fatty acids by anoxic *Mytilus edulis* and *Anodonta cygnea*. Comp. Biochem. Physiol., 80B: 299-301.

Thillart, G. van den, Lieshout, G. van, Storey, K.B., Cortesi, P. and Zwaan, A. de, 1992. Influence of long-term hypoxia on the energy metabolism of the hemoglobin containing bivalve *Scapharca inaequivalvis*: critical pO_2 levels for metabolic depression. J. Comp. Physiol., in press.

Thompson, R.J., Ratcliffe, N.A. and Bayne, B.L., 1974. Effects of starvation on structure and function in the digestive gland of the mussel (*Mytilus edulis* L.). J. Mar. Biol. Ass. U.K., 54: 699-712.

Toullec, J.Y., 1989. Recherche de substances hormonales actives sur la croissance d'invertébrés marins d'intérêt aquacole. Thèse d'Université , Paris VI.

Toullec, J.Y., Lenoir, F., Wormhoudt, A. van and Mathieu, M., 1989. Approche expérimentale du contrôle de la croissance chez les Bivalves. Oceanis, 15: 511-517.

Toyama, Y. and Tanaka, T., 1956. Fatty oils of aquatic invertebrates XII. Fatty oils of *Buccinum perryi*, *Tegula argyrostoma sublaevis* and *Mytilus edulis* with a particular reference to their sterol components. Mem. Fac. Eng. Nagoya Univ., 8: 29-34.

Tsuneki, K.,1974. Histochemical study of the neurosecretion in the mussel *Mytilus edulis*. J. Fac. Sci. Univ. Tokyo, 4(13): 159-173.

Uki, N. and Kikuchi, S., 1974. On the effect of irradiated sea water with ultraviolet rays on inducing spawning of the scallop, *Patinopecten yessoensis* (Jay) (in Japanese, with English abstract). Bull. Tohoku Reg. Fish. Res. Lab., 34: 87-92.

Umiji, S., 1969. Neurosecretion in the mussel *Mytilus perna*. Bol. Fac. Filos. Cienc. Let., Univ. Sao Paulo, 26: 181-254.

Usuki, I., 1956a. A comparison of the effects of cyanide and azide on the ciliary activity of the oyster gill. Sci. Rep. Tohoku Univ., Ser. IV (Biol.), 22: 137-142.

Usuki, I., 1956b. Effects of malonate and 2,4-dinitrophenol on the ciliary activity of the oyster gill. Sci. Rep. Tohoku Univ., Ser. IV (Biol.), 28: 59-83.

Villamarin, J.A., Rodriguez-Torres, A.M., Ibarguren, I. and Ramos-Martinez, J.I., 1990a. Phosphofructokinase in the mantle of the sea mussel *Mytilus galloprovincialis* Lmk. J. Exp. Zool., 255: 272-279.

Villamarin, J.A., Vazquez-Illanes, M.D., Barcia. R. and Ramos-Martinez, J.I., 1990b. Effect of pH on the phosphofructokinase activity from the posterior adductor muscle of the sea mussel *Mytilus galloprovincialis* Lmk. Biochem. Int., 21: 77-84.

Voogt, P. A., 1972. Lipid and sterol components and metabolism in Mollusca. In: M. Florkin and B.T. Scheuer (Editors), Chemical Zoology, 7. Academic Press, New York, pp. 245-300.

Voogt, P.A., 1975a. Investigations of the capacity of synthesizing 3ß-sterols in Mollusca–XIII. Biosynthesis and composition of sterols in some bivalves (Anisomyaria). Comp. Biochem. Physiol., 50B: 499-504.

Voogt, P.A., 1975b. Investigations of the capacity of synthesizing 3ß-sterols in mollusca–XIV. Biosynthesis and composition of sterols in some bivalves (Eulamellibranchia). Comp. Biochem. Physiol., 50B: 505-510.

Vooys, C.G.N. de, 1980. Anaerobic metabolism in sublittoral-living *Mytilus galloprovincialis* Lamk. in the mediterranean. II. Partial adaptation of pyruvate kinase and phosphoenolpyruvate carboxykinase. Comp. Biochem. Physiol., 65B: 513-518.

Vooys, C.G.N. de and Holwerda, D.A., 1986. Anaerobic metabolism in sublittoral-living *Mytilus galloprovincialis* Lmk. in the mediteranean. III. The effect of anoxia and osmotic stress on some kinetic parameters of adductor muscle pyruvate kinase. Comp. Biochem. Physiol., 85B: 217-221.

Waldock, M.J. and Holland, D.L., 1979. Seasonal changes in the triglyceride fatty acids of the mantle tissue of *Mytilus edulis*. Biochem. Biophys. Acta, 7: 898-900.

Walsh, P.J., McDonald, D.G. and Booth, C.E., 1984. Acid-base balance in the sea mussel, *Mytilus edulis*. II. Effects of hypoxia and air-exposure on intracellular acid-base status. Mar. Biol. Lett., 5: 359-369.

Walton, M.J. and Pennock, J.F., 1972. Some studies on the biosynthesis of ubiquinone, isoprenoid alcohols, squalene and sterols by marine invertebrates. Biochem. J., 127: 471-479.

Watabe, S. and Hartshorne, D.J., 1990. Paramyosin and catch mechanism. Comp. Biochem. Physiol., 96B: 639-646.

Watanabe, T. and Ackman, R.G., 1974. Lipids and fatty acids of the american (*Crassostrea virginica*) and european flat (*Ostrea edulis*) oysters from a common habitat, and after feeding with *Dicrateria inornata* or *Isochrisis galbana*. J. Fish. Res. Board. Can., 31: 403-409.

Whittle, M.A., 1982. Glycogen synthetase in the mussel, *Mytilus edulis* L. Ph. D. Thesis, University of Wales, U.K.

Whittle, M. A. and Gabbott, P. A., 1986. Glycogen synthetase in the sea mussel *Mytilus edulis* L. III. Regulation by glucose in a mantle tissue slice preparation. Comp. Biochem. Physiol., 83B: 209-214.

Whittle, M. A., Mathieu, M., Gabbott, P. A. and Lubet, P., 1983. The effect of glucose and neuro-endocrine factors on the activity ratio of glycogen synthetase in organ cultures of the mantle of *Mytilus edulis*. In: J. Lever et H. H. Boer (Editors), Molluscan Neuroendocrinology. North-Holland, Amsterdam. p. 183.

Widdows, J., 1987. Application of calorimetric methods in ecological studies. In: A.M. James (Editor), Thermal and energetic studies of cellular biological systems. Wright, Bristol, pp. 182-215.

Widdows, J. and Shick, J.M., 1985. Physiological responses of *Mytilus edulis* and *Cardium edule* to aerial exposure. Mar. Biol., 85: 217-232.

Widdows, J. and Hawkins, A.J.S., 1989. Partitioning of rate of heat dissipation by *Mytilus edulis* into maintenance, feeding and growth components. Physiol. Zool., 62: 764-784.

Widdows, J., Newell, R.I.E. and Mann, R., 1989. Effects of hypoxia and anoxia on survival, energy metabolism and feeding of oyster larvae (*Crassostrea virginica* Gmelin). Biol. Bull., 177: 154-166.

Widdows, J., Bayne, B.L., Livingstone, D.R., Newell, R.I.E. and Donkin, P., 1979. Physiological and biochemical responses of bivalve molluscs to exposure to air. Comp. Biochem. Physiol., 62A: 301-308.

Wijsman, T.C.M., Bont, A.M.T. de and Kluytmans, J.H.F.M., 1977. Anaerobic incorporation of radioactivity from 2,3-^{14}C-succinic acid into citric acid cycle intermediates and related compounds in the sea mussel *Mytilus edulis* L. J. Comp. Physiol., 114: 167-175.

Worral, C. M. and Widdows, J., 1984. Investigations of factors influencing mortality in *Mytilus edulis* L. Mar. Biol. Lett., 5: 85-98.

Wourms, J.P., 1987. Oogenesis. In: A.C. Giese, J.S. Pearse and V.B. Pearse (Editors), Reproduction of marine invertebrates. IX. General aspects: seeking unity in diversity. Blackwell Scientific Publications and Boxwood Press, California, pp. 49-178.

Zaba, B.N., 1981. Glycogenolytic pathways in the mantle tissue of *Mytilus edulis* L. Mar. Biol. Lett., 2: 67-74.

Zaba, B.N., 1983. On the nature of oxygen uptake in two tissues of *Mytilus edulis*. Mar. Biol. Lett., 4: 59-66.

Zaba, B.N. and Davies, J.I., 1979. The contribution of the pentose phosphate cycle to the central pathways of metabolism in the marine mussel *Mytilus edulis* L. Biochem. Soc. Trans., 7: 900-902.

Zaba, B.N. and Davies, J.I., 1980. Glucose metabolism in an in vitro preparation of the mantle tissue from *Mytilus edulis* L. Mar. Biol. Lett., 1: 235-243.

Zaba, B.N. and Davies, J.I., 1981. Carbohydrate metabolism in isolated mantle tissue of *Mytilus edulis* L. Isotopic studies on the activities of the Embden-Meyerhof and pentose phosphate pathways. Mol. Physiol., 1: 97-112.

Zaba, B.N. and Davies, J.I., 1984. Glycogen metabolism and glucose utilisation in the mantle tissue of *Mytilus edulis*. Mol. Physiol., 5: 261-282.

Zaba, B.N., Bont, A.M.T. de and Zwaan, A. de, 1978. Preparation and properties of mitochondria from tissues of the sea mussel *Mytilus edulis* L. Int. J. Biochem., 9: 191-197.

Zaba, B.N., Gabbott, P.A. and Davies, J.I., 1981. Seasonal changes in the utilisation of ^{14}C- and ^{3}H-labelled glucose in a mantle tissue slice preparation of *Mytilus edulis* L. Comp. Biochem. Physiol., 70B: 689-695.

Zandee, D.I., Holwerda, D.A. and Zwaan, A. de, 1980a. Energy metabolism in bivalves and cephalopods. In: R. Gilles (Editor), Animals and environmental fitness. Pergamon Press, Oxford, pp. 185-206.

Zandee, D.I., Kluytmans, J.H, Zurburg, W. and Pieters, H., 1980b. Seasonal variations in biochemical composition of *Mytilus edulis* with reference to energy metabolism and gametogenesis. Neth. J. Sea Res., 14: 1-29.

Zammit, V.A. and Newsholme, E.A., 1976. The maximum activities of hexokinase, phosphorylase, phosphofructokinase, glycerol phosphate dehydrogenases, lactate dehydrogenase, octopine dehydrogenase, phosphoenolpyruvate carboxykinase, nucleoside diphosphatekinase, glutamate-oxaloacetate transaminase and arginine kinase in relation to carbohydrate utilization in muscles from marine invertebrates. Biochem. J., 160: 447-462.

Zange, J., Pörtner, H.O. and Grieshaber, M.K., 1989. The anaerobic energy metabolism in the anterior byssus retractor muscle of *Mytilus edulis* during contraction and catch. J. Comp. Physiol. B, 159: 349-358.

Zange, J., Pörtner, H.O., Jans, A.W.H. and Grieshaber, M.K., 1990. The intracellular pH of a molluscan smooth muscle during contraction-catch-relaxion cycle estimated by the distribution of [^{14}C] DMO and by ^{31}P-NMR spectroscopy. J. Exp. Biol., 150: 81-93.

Zs.Nagy, I. and Galli, C., 1977. On the possible role of unsaturated fatty acids in the anaerobiosis of *Anodonta cygnea* L. (Mollusca, Pelecypoda). Acta Biol. Acad. Sci. Hung., 28: 123-131.

Zurburg, W. and Kluytmans, J.H., 1980. Organ specific changes in energy metabolism due to anaerobiosis in the sea mussel *Mytilus edulis* (L.). Comp. Biochem. Physiol., 67B: 317-322.

Zurburg. W. and Ebberink, R.H.M., 1981. The anaerobic energy demand of *Mytilus edulis*. Organ specific differences in ATP-supplying processes and metabolic routes. Mol. Physiol., 1: 153-164.

Zurburg, W., Bont, A.M.T. de and Zwaan, A. de, 1982. Recovery from exposure to air and the occurrence of strombine in different organs of the sea mussel *Mytilus edulis*. Mol. Physiol., 2: 135-147.

Zwaan, A. de, 1977. Anaerobic energy metabolism in bivalve molluscs. Oceanogr. Mar. Biol. Annu. Rev., 15: 103-187.

Zwaan, A. de, 1983. Carbohydrate catabolism in bivalves. In: P.V. Hochachka (Editor), The Mollusca, Vol. 1, Academic Press, New York, pp. 137-175.

Zwaan, A. de, 1989. Physiological and biochemical approaches to the assessment of pollution of european estuarine and coastal systems. Progress Report of Research Contract EV4V.0122/NL(GDF), 43pp.

Zwaan, A. de, 1991. Molluscs. In: C. Bryant (Editor), Metazoan life without oxygen. Chapman and Hall, London, pp. 186-217.

Zwaan, A. de and Zandee, D. I., 1972. Body distribution and seasonal changes in the glycogen content of the common sea mussel *Mytilus edulis* L. Comp. Biochem. Physiol., 43A: 53-58.

Zwaan, A. de and Bont, A.M.T. de, 1975. Phosphoenolpyruvate carboxykinase from adductor muscle tissue of the sea mussel *Mytilus edulis* L. J. Comp. Physiol., 96: 85-94.

Zwaan, A. de and Wijsman, T.C.M., 1976. Anaerobic metabolism in Bivalvia (Mollusca). Characteristics of anaerobic metabolism. Comp. Biochem. Physiol., 54B: 313-324.

Zwaan, A. de and Zurburg, W., 1981. The formation of strombine in the adductor muscle of the sea mussel *Mytilus edulis* L. Mar. Biol. Lett., 2: 179-192.

Zwaan, A. de and Dando, P.R., 1984. Phosphoenolpyruvate-pyruvate metabolism in bivalve molluscs. Mol. Physiol., 5: 285-310.

Zwaan, A. de and Putzer, V., 1985. Metabolic adaptations of intertidal invertebrates to environmental hypoxia (a comparison of environmental anoxia to exercise anoxia). In: M.S. Lavarack (Editor), Physiological adaptations of marine animals, Vol. 39, Symposia of the Society of Experimental Biology. Company of Biologists Ltd., University of Cambridge, pp. 33-62.

Zwaan, A. de and Thillart, G. van den, 1985. Low and high power output modes of anaerobic metabolism: invertebrate and vertebrate strategies. In: R. Gilles (Editor), Circulation, respiration, and metabolism. Springer-Verlag, Berlin, pp. 167-192.

Zwaan, A. de, Bont, A.M.T. de and Kluytmans, J.H.F.M., 1975a. Metabolic adaptations on the aerobic-anaerobic transition in the sea mussel, *Mytilus edulis* L. Proc. 9th Eur. Mar. Biol. Symp., Oban, Scotland, 1984. Aberdeen University Press, Aberdeen, pp. 121-138.

Zwaan, A. de, Holwerda, D.A. and Addink, A.D.F., 1975b. The influence of divalent cations on allosteric behaviour of muscle pyruvate kinase from the sea mussel *Mytilus edulis* L. Comp. Biochem. Physiol., 52B: 469-472.

Zwaan, A. de, Holwerda, D.A. and Veenhof, P.R., 1981. Anaerobic malate metabolism in mitochondria of the sea mussel *Mytilus edulis* L. Mar. Biol. Lett., 2: 131-140.

Zwaan, A. de, Bont, A.M.T. de and Verhoeven, A., 1982. Anaerobic energy metabolism in isolated adductor muscle of the sea mussel *Mytilus edulis* L. J. Comp. Physiol., 149: 137-143.

Zwaan, A. de, Bont. A.M.T. de and Hemelraad, J., 1983a. The role of phosphoenolpyruvate carboxykinase in the anaerobic metabolism of the sea mussel *Mytilus edulis* L. J. Comp. Physiol., 153: 267-274.

Zwaan, A. de, Bont, A.M.T. de, Zurburg, W., Bayne, B.L. and Livinstone, D.R., 1983b. On the role of strombine formation in the energy metabolism of adductor muscle of a sessile bivalve. J. Comp. Physiol., 149: 557-563.

Zwaan, A. de, Cattani, O., Isani, G. and Cortesi, P., 1991a. Survival and anaerobic metabolism of the arcid clam *Scapharca inaequivalvis* under anoxia and oxia in the presence of sulphide and cyanide. In: P.L. Lutz (Editor), Cellular defence strategies to hypoxia. Second IUB Satellite Symposium, Noordwijkerhout, Netherlands, 1991. Abst., p.17

Zwaan, A. de, Cortesi, P., Thillart, G. van den, Roos, J. and Storey, K.B., 1991b. Differential sensitivities to hypoxia by two anoxia-tolerant marine molluscs: a biochemical analysis. Mar. Biol., 111: 343-351.

Chapter 7

GENETICS OF *MYTILUS*

ELIZABETH M. GOSLING

INTRODUCTION

Mussels (*Mytilus*) have been more extensively used in genetic studies than any other group of marine organisms. There are several reasons for this: the widespread nature of the genus lends itself particularly well to the study of populations—either of the same or different species—on different geographic scales. In addition, because mussels occupy a wide range of environments within a particular geographic area, genetic variability of populations can be compared between different habitats. Also, the two distinct stages in the life cycle of mussels—an extended dispersal stage and a sedentary adult phase—facilitate the study of the relative contributions of natural selection and gene flow in maintaining patterns of genetic variation in natural populations. High fecundity combined with ease of culture makes *Mytilus* particularly well-suited for laboratory-based genetic studies. By experimentally altering temperature, salinity or food ration, the role of selection can be more easily assessed in laboratory cultures, as opposed to wild populations. Mussels also are being increasingly used as model organisms to experimentally address current genetics phenomena, such as heterozygosity/growth correlations and heterozygote deficits.

Notwithstanding the great amount of published information on the genetics of *Mytilus*, there have been only two reviews of the topic, and these deal exclusively with population genetics (Levinton and Koehn, 1976; Koehn, 1991). This chapter, therefore, presents an all-embracing review on *all* aspects of *Mytilus* genetics. The chapter is divided into three main sections. The first and largest section, population genetics, deals first with macro and microgeographic genetic variation in *Mytilus*, then with the adaptive significance of such variation, and lastly deals with the phenomena of heterozygote deficits and correlations between heterozygosity and fitness traits. The second section covers chromosomal genetics, while the third examines the role of genetics in aquaculture.

POPULATION GENETICS

Genetics in general concerns the genetic constitution (genotype) of organisms and the laws governing the transmission of this hereditary information, contained in the genes, from one generation to the next. Population genetics is that branch of genetics concerned with heredity in groups of individuals i.e. in populations; and population geneticists study the genetic constitution of populations and how this genetic constitution changes from generation to generation. Hereditary changes through generations underlie the evolutionary process. Therefore, population geneticists may also be considered as evolutionary geneticists.

The central problems facing population geneticists over the past 60 years or so have been the characterization of hidden genetic variation and the mechanisms maintaining genetic variation in natural populations. Lewontin (1974) states " It is clear that descriptions of the genetic variation in populations are the fundamental observations on which evolutionary genetics depends." The study of population genetics therefore involves furnishing an adequate description of genetic variation in populations, which necessarily involves a description of the statistical distribution of genotypes.

Electrophoresis and the Hardy–Weinberg Principle

Up to the mid 1950s most of the scientific literature documenting magnitudes of genetic variation in populations consisted of morphological and cytological studies on *Drosophila*. These studies indicated that large amounts of genetic variation were present in natural populations, signifying ample opportunity for evolutionary change. But how many gene loci were polymorphic (variable) and what proportion of all gene loci were heterozygous in a typical individual of the population, were questions which were only answered through the discoveries of molecular genetics, and the development of an analytical method of protein separation called gel electrophoresis (see Ferguson (1980) for details). Briefly, a piece of tissue from an individual organism is ground up to disrupt cells, centrifuged to remove insoluble material and the resulting protein solution is inserted into narrow slots in a gel of starch or polyacrylamide. An electric current is applied for a fixed time and different proteins migrate at different rates, depending on their charge and configuration. The gel is treated with appropriate chemical solutions to visualize the position of proteins or specific enzymes, which appear as discrete zones or bands on the gel.

The usefulness of the method lies in the fact that the genotype at the gene locus coding for the enzyme can be inferred for each individual in the sample. Single bands of identical mobility in all individuals are assumed to be products of a single gene locus. Such loci are referred to as monomorphic loci. When variation is present i.e. where the locus is polymorphic, patterns are generally consistent with those expected from simple models of single locus Mendelian inheritance i.e. homozygotes exhibit single-banded phenotypes, while heterozygotes are either double, triple or quintuple-banded phenotypes, depending on whether the enzyme is monomeric, dimeric or tetrameric, respectively. By counting all the different homozygotes and heterozygotes in a large sample of individuals one can estimate the number of alternative genes (alleles) coding for a particular protein, and also the frequency of the different alleles.

By determining the proportion of randomly selected enzymes which show electrophoretic variation in a population the proportion of loci which are polymorphic can be estimated. Alternatively, a better and more preferred measure of genetic variation is the average frequency of heterozygous individuals per locus or, simply, the heterozygosity of the population. This is calculated by first obtaining the frequency of heterozygous individuals at each locus and then averaging these frequencies over all loci to get the mean observed heterozygosity per locus ($\bar{H}_o$)

The electrophoretic technique makes it possible to compare allele frequencies and levels of genetic variability within and between different populations of a species, between different species, and so on. The degree of genetic differentiation between different taxa can be quantified using Nei's (1972) index of genetic identity (I) and distance (D). Genetic identity is calculated for any pair-wise combination of populations, or species using:

$$I_{xy} = \frac{\Sigma P_{ix} P_{iy}}{\sqrt{\Sigma P_{ix}^2 P_{iy}^2}}$$

where P_{ix} is the frequency if allele i in population x and P_{iy} is the corresponding frequency in population y. At a locus the genetic identity value may range from zero, where no alleles are shared between populations, to one, where the same alleles are present at identical frequencies in both populations. D_{xy} is calculated from I using the formula:

$$D_{xy} = - \ln I_{xy}$$

Mean genetic identity ($\bar{I}$) and genetic distance ($\bar{D}$) are the mean values over all analysed loci, including monomorphic ones.

Observed genotype frequencies can be tested against frequencies expected according to the Hardy–Weinberg formulation. This states that in the absence of migration, inbreeding, mutation or selection, gene and genotype frequencies in a sexually outbreeding population remain constant from generation to generation. If A and a represent a pair of alleles at a given gene locus with frequencies of p and q and with p+q = 1, the frequency of the genotypes AA, Aa and aa is given by the expansion of $(p+q)^2$ i.e. p^2, 2pq, and q^2 respectively. Hardy and Weinberg showed that in subsequent generations the distribution of genotypes in the population does not change. Deviations from Hardy–Weinberg expectations signify that one or more of the necessary requirements for equilibrium is absent. The degree of deviation is usually formulated using an index; the most commonly used one for molluscs is that of Selander (1970):

$$D = (H_o - H_e)/H_e$$

where H_o is the observed number of heterozygotes, and H_e is the number of heterozygotes expected from the estimated allele frequencies. A negative value of D indicates a deficiency of heterozygotes, while a positive value indicates an excess.

Over the last 30 years, with the aid of electrophoretic techniques, large amounts of genetic variation have been observed in most natural populations of animal and plant species. Levels of genetic variability, measured as $\bar{H}_o$ values (see above), are generally higher in invertebrate than vertebrate species. Detailed lists of these may be found in Ferguson (1980). For *Mytilus edulis* Ahmad et al. (1977) have estimated a $\bar{H}_o$ value of 0.095, which is close to the estimated value of 0.112 for invertebrate species in general (Hedrick, 1985), but lower than the values estimated for other species of marine bivalves (Beaumont and Zouros, 1991). How relevant such values are in terms of the adaptive capabilities of a species is far from clear and has been the subject of much controversy—the well-known selectionist/neutralist controversy (see Kimura, 1983; Endler,1986; Futuyama, 1986 for detailed discussion). Also, it must be remembered that because $\bar{H}_o$ values are based on very few gene loci, usually less than 20, they may not be a true reflection of *overall* genomic variability.

Population genetic studies on *Mytilus* tend to fall into two main groups: those dealing with surveys of genetic variation in mussels over large (macrogeographic) distances i.e. on a scale of hundreds of kilometres, and those analysing genetic heterogeneity over small (microgeographic) distances e.g. in small bays and estuaries and at different tidal levels.

Macrogeographic Differentiation in *Mytilus*

Most studies on macrogenetic differentiation in *Mytilus* have been concerned with elucidating taxonomic relationships within the genus.   Prior to the use of electrophoresis, about nine distinct species of *Mytilus* were recognized—although there was by no means a general concensus on this: *Mytilus edulis* from northern temperate latitudes, *Mytilus galloprovincialis* from the Mediterranean Sea, *Mytilus trossulus* from the Pacific coast of North America, *Mytilus coruscus* from Japan and China, *Mytilus californianus* from the Pacific coast of North America, *Mytilus chilensis* from Chile, *Mytilus platensis* from Argentina, *Mytilus planulatus* from Australia and *Mytilus desolationis* from the Kerguelen Islands in the southern Indian Ocean.

Much of the confusion in mussel taxonomy has arisen because of the emphasis placed on shell morphological characters (see Chapter 1).   Such characters are enormously plastic being subject to a wide range of environmental influences. Therefore, systematic information that is relatively free of environmently-induced changes is highly desirable.   The study of allozyme variation in different populations of *Mytilus* has gone some way in helping to resolve the systematics of the genus.

Genetic variability in allopatric *Mytilus edulis*

Genetic variability has been surveyed in allopatric populations of *M. edulis* from over a hundred locations in the northern and southern hemispheres (Levinton and Koehn, 1976 and references therein; Ahmad et al., 1977; Gartner–Kepkay et al., 1980, 1983; Gosling and Wilkins, 1981; Skibinski et al., 1983; Koehn et al., 1984; Bulnheim and Gosling, 1988; Varvio et al., 1988; Johannesson et al., 1990; McDonald et al., 1990, 1991).   The most studied loci include the following: peptidase-II (*Aap*, EC 3.4.11.-), leucine aminopeptidase-II (*Lap-2*, often referred to as aminopeptidase-I (EC 3.4.11.-) (Young et al., 1979)), aminopeptidase (*Ap*, EC 3.4.-.-), esterase-D (*Est-D*, EC 3.1.1.1), glucose-6-phosphate isomerase (*Gpi*, EC 5.3.1.9), octopine dehydrogenase (*Odh*, EC 1.5.1.11), phosphoglucomutase (*Pgm*, EC 5.4.2.2) and mannose-6-phosphate isomerase (*Mpi*, EC 5.3.1.8).   These are not in fact a random sample of gene loci but have been chosen principally on the basis of their usefulness in distinguishing between different mussel types within the genus.

The first large-scale survey was carried out on samples of *M. edulis* collected along the east coast of North America (Koehn et al., 1976).   Genetic variation at the *Lap*, *Gpi* and *Ap* loci was investigated.   Five alleles, two of them rare, were detected at the *Lap* locus.   The three common alleles are designated by numbers—*Lap*[94], *Lap*[96] and *Lap*[98]— which reflects their relative electrophoretic mobility.   *Lap* allele frequency was homogeneous in oceanic samples collected throughout the large geographic area (

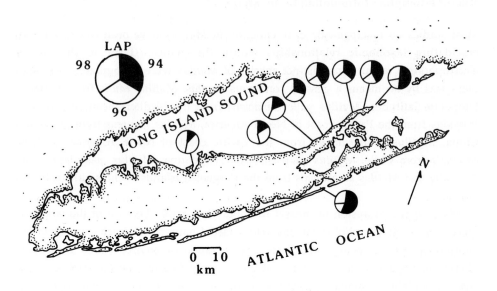

Fig. 7.1. The frequency of *Lap⁹⁴* among populations of *Mytilus edulis* in the Long Island Sound area. (After Hilbish, 1985).

about 800km) between Virginia and Cape Cod (Koehn et al., 1976) However, the frequency of Lap^{94} progressively declined on both the northern and southern shores of Long Island Sound, from 0.55 in oceanic populations to 0.15 in the sound—a distance of only some 30km (Fig. 7.1). This decline in frequency also occurred throughout the Gulf of Maine, as well as in samples from the Gulf of St. Lawrence and northern Newfoundland (see later p.334–336). Each year mussel larvae, exhibiting *Lap* allele frequencies typical of oceanic populations, emigrate into the sound. Following recruitment, genotype-dependent mortality among these immigrants results in the establishment in adult mussels of the allele frequency cline, which is both temporally and spatially stable (Koehn et al., 1980b; Hilbish, 1985). Results from detailed biochemical investigations (see p.344–346 for details) have suggested that decreasing salinity along the sound is the selective factor responsible for the decreasing frequency of Lap^{94}.

At the *Gpi* locus up to nine alleles have been observed in populations of *M. edulis* on the east coast of North America. The frequency of the three most common alleles was linearly dependent upon the latitude from which the samples were taken (Koehn et al., 1976), and more recent studies have demonstrated a correlation between genetic variability at this locus and temperature (Hall, 1985). In the same survey *Ap* allele frequency was homogeneous from Virginia to southern Nova Scotia. However, samples from the Canadian Maritimes differed significantly from those further south, mainly due to an increase in one of the common *Ap* alleles from 0.32 to 0.52. This change was accompanied by a large significant deficiency of heterozygotes, which was partially attributed to population mixing (Wahlund effect); a deficiency of heterozygotes is often observed at a locus when a sample is a mixture of two or more populations characterized by different allele frequencies.

In contrast to *M. edulis*, little variation in either *Lap* or *Gpi* allele frequency was observed in *M. californianus* samples collected over a large geographic distance (~3000km) along the west coast of North America (Levinton and Suchanek, 1978); the difference between the two species was attributed to the wider variety of habitats occupied by *M. edulis*. But, a decade later there has been little additional genetic information on *M. californianus* to either substantiate or refute the claim.

Geographic or clinal variation in the frequency of allozyme variants have also been observed in European populations of *Mytilus* (Theisen, 1978; Skibinski and Beardmore, 1979; Gosling and Wilkins, 1981). In North America (Koehn et al., 1976) and in the Baltic (Theisen, 1978) geographic variation has been initially viewed as occurring within a single species, *M. edulis*, and has been interpreted in terms of natural selection acting at individual allozyme loci; temperature and salinity have been identified as possible selective factors (Koehn et al., 1980a; Hall, 1985). However, in Ireland and the U.K. geographic variation has been interpreted as resulting from the mixing of *M. edulis* with the Mediterranean mussel, *M. galloprovincialis* (Gosling and Wilkins, 1977; Skibinski and Beardmore, 1979; Gosling and Wilkins, 1981). There is now good evidence that some of the anomalous results obtained by Koehn et al. (1976) and Gartner-Kepkay et al. (1980, 1983) for the Canadian Maritimes (see later), and by Theisen (1978) for the Baltic, were due to authors sampling in these areas not one, but a mixture of two forms of mussels—*M. edulis* and the recently rediscovered *M. trossulus* (Koehn et al., 1984; Bulnheim and Gosling, 1988; McDonald and Koehn, 1988; Varvio et al., 1988). Therefore, with the exception of the *Lap* cline in *M. edulis*, most of the gene frequency clines observed over large or small geographic distances are now known to involve contact (and hybridization) between different mussel types within the genus (see Fig.1.6 Chapter 1).

In areas e.g. North-west Europe and the east coast of the U.S.A., south of Cape Cod, where only pure populations of *M. edulis* have been analysed, allele frequencies within each region are remarkably homogeneous over large geographic distances (Ahmad et al., 1977; Gosling and Wilkins, 1981; Skibinski et al., 1983; Koehn et al., 1984; Bulnheim and Gosling, 1988; Varvio et al., 1988; Johannesson et al., 1990; McDonald et al., 1990; Väinölä and Hvilsom, 1991). However, if allele frequencies (Table 7.1) on the two sides of the Atlantic are compared with each other there is marked transoceanic differentiation, most notably at the *Lap*, *Ap*, *Gpi* and *Pgm* loci.

In the southern hemisphere, mussels on the west and east coasts of South America have, until recently, been regarded as *M. chilensis* and *M. platensis* respectively. However, McDonald et al. (1991) in a study on the taxonomy and distribution of *Mytilus* have suggested, from data based on allozymic analysis of 8 loci, and 18 shell characters (see Chapter 1), that South American populations should tentatively be included in *M. edulis*, as should the Kerguelen Islands mussel, *M. desolationis*—up to recently regarded as a semispecies of *M. edulis* (Blot et al., 1988). The tentative nature of this suggestion stems from the fact that mussels on the coasts of South America do not in fact closely resemble northern hemisphere *M. edulis*, but appear to fall somewhere in between the electrophoretically distinguishable *M. edulis* and *M. galloprovincialis* clusters of the northern hemisphere (Table 7.1 and Fig. 7.2). Also, at the majority of loci analysed by McDonald et al. (1991) there are substantial genetic differences between mussels from the west and east coasts of South America and between these and the Kerguelen Islands sample. Unfortunately, it was not possible to test the significance of this genetic differentiation because sample sizes were small (22–25 individuals), and consequently there were too few numbers, often as low as one, in many of the allelic classes at a locus.

Genetic variability in allopatric *Mytilus galloprovincialis*

Pure populations of *M. galloprovincialis* have been analysed electrophoretically at much the same loci as *M. edulis*, but from a smaller number of sites (approximately 50 to date). Similar to *M. edulis*, allele frequencies are fairly homogeneous over large geographic distances e.g. the Mediterranean Sea, the east coast of China and the U.S.S.R. (Skibinski et al., 1980, 1983; Wilkins et al., 1983; Grant and Cherry, 1985; Varvio et al., 1988; Beaumont et al., 1989a; McDonald et al., 1990, 1991; Sanjuan et al., 1990). For example, samples from the Mediterranean coast of Spain and South Africa are genetically very similar with a genetic distance value, based on frequencies at 23 loci, of 0.010 ± SE 0.005 (Grant and Cherry, 1985). Comparisons of populations within the Mediterranean give similar values of D (Skibinski et al., 1980), and are in the range expected for conspecific populations (0.0–0.05; Ferguson, 1980). In contrast, Beaumont et al. (1989a) have observed significant differences in allele frequency at the *Est-D, Mpi* and *Odh* loci between samples of *M. galloprovincialis* from two sites in South-west England situated less than 5km apart!

Until recently, southern hemisphere mussels from the Australian and New Zealand coasts have been regarded as *M. planulatus*. However, McDonald et al. (1991) have demonstrated that mussels from these two areas fall into the same electrophoretically distinguishable cluster as northern hemisphere *M. galloprovincialis* (see Fig 7.2) and, therefore, should now be regarded as *M. galloprovincialis*. An exam

Table 7.1. Allele percentages in representative northern and southern hemisphere samples of *Mytilus*. Sample sizes are 23–25 individuals, except for the Newfoundland sample, which numbers about 60 individuals. Data from Varvio et al. (1988), McDonald et al. (1990) and McDonald et al. (1991).

Locality	Aap										Lap				
	80	85	90	95	100	105	110	115	120	125	92	94	96	98	100
N. Hemisphere															
M. edulis															
N.W. Europe — Aarhus, Denmark	0	0	2	12	76	10	0	0	0	0	0	8	70	22	0
East coast U.S.A. — Shinnecock, N.Y.	0	0	4	8	84	4	0	0	0	0	0	44	19	35	2
M. galloprovincialis															
Mediterranean — Venice	0	0	0	2	4	10	26	52	6	0	0	0	46	42	12
West coast U.S.A. — Los Angeles	0	0	0	0	4	4	52	40	0	0	0	0	62	36	2
East coast U.S.S.R. — Fal'shivyi Island	0	0	2	2	4	12	36	40	4	0	8	2	64	24	2
M. trossulus															
Baltic — Tvärminne	0	0	8	92	0	0	0	0	0	0	0	20	16	2	62
N.E. Canada — Newfoundland	0	6	36	48	0	0	0	0	0	0	9	45	38	8	0
West coast U.S.A. — Tillamook, Oregon	2	6	34	34	6	0	2	0	0	0	6	50	28	12	4
N.W. U.S.S.R. — Magadan	4	26	34	34	2	0	0	0	0	0	12	56	26	0	6
S. Hemisphere															
S. America (west coast) — Chiloe, Chile	0	0	2	26	67	2	2	0	0	0	2	13	81	4	0
S. America (east coast) — Marde Plata (Arg)	0	0	0	52	48	0	0	0	0	0	2	26	65	7	0
Kerguelen Islands — –	0	0	0	76	24	0	0	0	0	0	0	10	81	9	0
W. Australia — Albany	0	0	0	2	4	4	8	64	18	0	0	2	89	9	0
Tasmania — R. Huon Estuary	0	0	0	0	2	0	15	54	25	4	0	12	79	8	0
New Zealand — Wellington	0	0	0	4	16	10	36	2	8	6	2	4	74	10	10

Table 7.1. continued

	Ap								Est-D				
Locality	90	95	100	105	108	117	120	125	80	90	95	100	110
N. Hemisphere													
M. edulis													
N.W. Europe Aarhus, Denmark	4	4	64	24	0	0	0	0	0	4	0	94	2
East coast U.S.A. Shinnecock, N.Y.	16	6	38	40	0	0	0	0	2	2	0	92	4
M. galloprovincialis													
Mediterranean Venice	0	0	18	42	22	10	4	4	0	100	0	0	0
West coast U.S.A. Los Angeles	0	0	22	38	18	6	12	4	4	96	0	0	0
East coast U.S.S.R. Fal'shivyi Island	0	0	26	40	14	14	2	0	0	100	0	0	0
M. trossulus													
Baltic Tvärminne	10	0	84	6	0	0	0	0	0	78	22	0	0
N.E. Canada Newfoundland	17	7	45	30	1	0	0	0			0		
West coast U.S.A. Tillamook, Oregon	2	8	62	22	4	2	0	0	4	90	6	0	0
N.W. U.S.S.R. Magadan	0	8	74	16	0	0	0	0	2	98	0	0	0
S. Hemisphere													
S. America (west coast) Chiloe, Chile	0	0	52	33	11	4	0	0	0	59	0	41	0
S. America (east coast) Marde Plata (Arg)	0	6	58	30	6	0	0	0	0	0	0	100	0
Kerguelen Islands –	0	0	86	12	2	0	0	0	0	0	0	100	0
W. Australia Albany	0	0	6	14	56	18	4	2	2	74	0	24	0
Tasmania R. Huon Estuary	0	0	19	23	56	2	0	0	2	48	0	48	2
New Zealand Wellington	0	6	54	22	18	0	0	0	0	100	0	0	0

Table 7.1. continued

Locality		Gpi										Odh				
		86	89	93	96	98	100	102	105	107	110	80	90	98	100	110
N. Hemisphere																
M. edulis																
N.W. Europe	Aarhus, Denmark	0	2	2	16	4	14	8	0	44	10	0	0	0	82	18
East coast U.S.A.	Shinnecock, N.Y.	0	4	2	12	0	54	4	0	24	0	0	4	0	84	12
M. galloprovincialis																
Mediterranean	Venice	0	0	0	0	0	92	0	6	2	0	0	6	0	34	60
West coast U.S.A.	Los Angeles	0	0	0	0	0	86	0	10	4	0	0	6	0	22	72
East coast U.S.S.R.	Fal'shivyi Island	0	0	2	0	2	92	0	2	2	0	0	10	0	12	78
M. trossulus																
Baltic	Tvärminne	0	5	4	0	94	0	2	0	0	0	0	4	0	88	8
N.E. Canada	Newfoundland	0	5	15	1	65	7	1	0	4	2	0	7	8	79	6
West coast U.S.A.	Tillamook, Oregon	0	6	24	0	56	2	10	2	0	0	2	10	26	50	12
N.W. U.S.S.R.	Magadan	28	4	26	0	32	4	4	2	0	0	2	4	14	74	6
S. Hemisphere																
S. America (west coast)	Chiloe, Chile	0	0	0	2	85	13	0	0	0	0	0	7	0	0	93
S. America (east coast)	Marde Plata (Arg.)	0	10	0	16	32	42	0	0	0	0	0	2	0	70	28
Kerguelen Islands	–	0	4	0	6	52	38	0	0	0	0	0	16	0	25	59
W. Australia	Albany	0	2	0	0	0	98	0	0	0	0	0	36	0	8	56
Tasmania	R. Huon Estuary	2	4	0	6	0	85	0	2	2	0	2	57	7	0	35
New Zealand	Wellington	0	0	0	6	2	92	0	0	0	0	0	4	20	4	72

Table 7.1. continued

Locality	Mpi								Pgm					
	84	90	92	94	96	100	104	110	89	93	100	106	111	114
N. Hemisphere														
M. edulis														
N.W. Europe Aarhus, Denmark	0	6	0	0	0	94	0	0	8	37	45	10	0	0
East coast U.S.A. Shinnecock, N.Y.	0	10	0	0	0	88	0	2	4	14	82	0	0	0
M. galloprovincialis														
Mediterranean Venice	0	0	100	0	0	0	0	0	6	14	50	28	2	0
West coast U.S.A. Los Angeles	0	0	100	0	0	0	0	0	8	16	44	24	8	0
East coast U.S.S.R. Fal'shivyi Island	0	0	100	0	0	0	0	0						
M. trossulus														
Baltic Tvärminne	10	0	0	90	0	0	0	0	0	0	4	2	86	8
N.E. Canada Newfoundland									0	0	14	25	58	3
West coast U.S.A. Tillamook, Oregon	6	0	0	92	0	0	2	0	0	0	10	32	52	6
N.W. U.S.S.R. Magadan	0	0	0	98	0	0	2	0						
S. Hemisphere														
S. America (west coast) Chiloe, Chile	0	4	20	0	0	76	0	0	0	0	80	20	0	0
S. America (east coast) Marde Plata (Arg)	0	2	4	0	0	88	0	6	0	6	56	38	0	0
Kerguelen Islands –	0	0	2	0	2	96	0	0	0	0	90	10	0	0
W. Australia Albany	0	0	100	0	0	0	0	0	10	6	62	22	0	0
Tasmania R. Huon Estuary	0	0	96	0	0	4	0	0	0	29	69	2	0	0
New Zealand Wellington	0	0	94	0	0	6	0	0	0	22	72	6	0	0

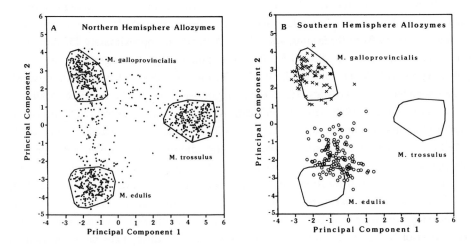

Fig. 7.2. (A) First and second principal component allozyme data for eight loci for individuals from 45 locations in the northern hemisphere illustrating the discrimination among the three *Mytilus* taxa: *Mytilus edulis, Mytilus galloprovincialis* and *Mytilus trossulus*. To aid comparison with southern hemisphere results, lines have been drawn subjectively around clusters of individuals in A, and are repeated in B. Individuals outside of clusters are from locations in contact areas between two taxa. (B) Same as A, but only individuals from nine southern hemisphere locations sampled. (o: mussels from Chile, Argentina, the Falkland Islands and the Kerguelen Islands; x: mussels from Australia and New Zealand). (Details in McDonald et al., 1991, from which this figure was redrawn).

-ination of the data in Table 7.1., however, indicates that there are striking differences in allele frequencies between mussels from Australia and New Zealand. For the same reason stated above, it was not possible to test this statistically.

Genetic variability in allopatric *Mytilus trossulus*

Because of its recent rediscovery (McDonald and Koehn, 1988) there have been few allopatric populations of *M. trossulus* analysed to date. Unlike the situation pertaining to *M. edulis* and *M. galloprovincialis*, populations of *M. trossulus* have been identified entirely on the basis of allozyme data. Of the three mussel types, the geographic range of this mussel is more restricted, being confined (as far as we know) to the northern hemisphere and having, within that region, a northerly, disjunct distribution (see Fig.1.6., Chapter 1). Allele frequencies, within single geographic areas tend to be homogeneous (Bulnheim and Gosling, 1988; McDonald and Koehn, 1988;

McDonald et al., 1990, 1991). For example, for comparisons between three samples of mussels collected in the Baltic Sea (Bulnheim and Gosling, 1988) genetic identity values, albeit based on only six loci, were very close to one (0.97). Also, *M. trossulus* samples collected from three sites, spanning a distance of approximately 1500km along the north-east coast of the U.S.S.R. were genetically homogeneous at seven polymorphic loci (McDonald et al., 1990). On the west coast of North America, *M. trossulus* collected at two sites, Tillamook, Oregon—believed to be the type locality of *M. trossulus* Gould 1850—and Peterburg, Alaska, situated about 2000km away, exhibited similar allele frequencies at eight polymorphic loci (McDonald and Koehn, 1988). In neither of these studies were genetic identity values given for pair-wise comparisons of *M. trossulus* populations. Although *M. trossulus* has also been identified from several locations in the Canadian Maritimes (McDonald et al., 1991) no allele frequencies were presented for that area in the paper. When *M. trossulus* frequencies are compared between regions e.g. the Baltic Sea (Tvärminne) and the west coast of North America (Tillamook) there are large differences between them at most of the analysed loci (Table 7.1). Marked allele frequency differences (no significance values given) between Baltic and Canadian Maritime *M. trossulus* have also been observed at four out of the five loci analysed by Varvio et al. (1988).

In summary, allele frequencies within each mussel type (*M. edulis, M. galloprovincialis* and *M. trossulus*) for a particular distinct geographic region are generally homogeneous. However, when allopatric populations of a single type are compared from *different* geographic regions, e.g. *M. edulis* from the east versus west sides of the North Atlantic Ocean, or *M. trossulus* from the Baltic versus the west coast of North America, there are invariably substantial differences between them.

Genetic distance (D) for comparisons between allopatric populations of *M. edulis* (U.K.) and *M. galloprovincialis* (Italy) was found to be 0.172, based on allele frequency data from 16 loci (Skibinski et al., 1980). In a comparison between *M. edulis* from Denmark and *M. galloprovincialis* from South Africa, Grant and Cherry (1985) observed a similar D value (0.162), based on frequency data from 23 loci. When allopatric populations of *M. edulis* (North Sea) and *M. trossulus* (Baltic Sea) were compared at 22 loci, Väinölä and Hvilsom (1991) observed a D value of 0.28 ± 0.12, while the Baltic–North Sea distance from *M. galloprovincialis* (North-west Spain) was 0.25. All of these D values are very similar and are in the range expected for comparisons between subspecies of invertebrate taxa (D= 0.228; Ferguson, 1980). Blot et al. (1988), have used Euclidian distance (Rogers, 1986) to ascertain genetic relationships between populations of *M. desolationis* (Kerguelen Islands), *M. galloprovincialis* and *M. edulis*. Euclidian distances are more suited to comparisons of closely related taxa and, in contrast to Nei's index, need no biological assumptions, such as absence of selection or migration. Euclidian distance values, which range between 0 (identical frequencies between two populations) and 4.47 (no electromorph in common), for the *M. desolationis* /*M. edulis* (Long Island) and *M. desolationis* /*M.* *galloprovincialis* (Venice) comparisons were 1.22 and 1.27 respectively, while the *M.* *edulis* /*M. galloprovincialis* value was 1.27. If, on the basis of these values, *M. edulis* and *M. galloprovincialis* are considered as subspecies within the genus *Mytilus*, then the same taxonomic rank should be accorded to *M. desolationis*.

The taxonomic value of individual loci

As stated earlier, the principal loci used in *Mytilus* genetics studies are those which have proved most useful in descriminating between various mussel types within the genus. In this section each of the loci mentioned above will be discussed in terms of its value in differentiating between *M. edulis, M. galloprovincialis, M. trossulus* and where appropriate, *M. desolationis*. In the case of *M. californianus* there is no need for diagnostic allozyme markers, since this species is easily distinguished from the others by means of the radiating ribs on the shell. In discussing the various loci the nomenclature of McDonald et al. (1991) has been followed.

Peptidase-II (*Aap*): This locus has been valuable in descriminating between the three different mussel types in the northern hemisphere (Table 7.1). However, in the southern hemisphere, where the distribution of *M. edulis* and *M. galloprovincialis* has been mapped, based on morphological and allozymic comparisons with the northern forms (Fig. 7.2. and Fig.1.4., Chapter 1), *Aap* allele frequencies for putatative *M. edulis* vary considerably between the different geographic areas (Table 7.1), although in the case of putatitive *M. galloprovincialis* the locus maintains its diagnostic value. Both Blot et al. (1988) and McDonald et al. (1991) have observed large allele frequency differences at this locus between *M. desolationis* from the Kerguelen Islands and *M. edulis* from both sides of the North Atlantic Ocean—of about the same magnitude as the differences observed between *M. galloprovincialis* and *M. edulis* in the northern hemisphere.

Aminopeptidase-I (*Lap-2*): This locus has no discriminatory power in differentiating between the different *Mytilus* forms, which is not unexpected in view of the well-documented evidence for the selective effect of salinity on *Lap* allele frequencies in *M.* *edulis* (discussed later). One interesting feature of this locus is that *Lap*[100], which is generally absent or at low frequency in populations of *Mytilus* (Blot et al., 1988, 1989; Bulnheim and Gosling, 1988; McDonald and Koehn, 1988; Varvio et al., 1988; McDonald et al., 1990, 1991; Väinölä and Hvilsom, 1991) is at high frequency in Baltic *M. trossulus*.

Aminopeptidase (*Ap*): This locus has no discriminatory power, chiefly because of the strong differentiation within each of the three mussel types in the northern hemisphere. Like *Lap*, *Ap* allele frequency may be determined by the direct effect of some selective factor on the locus itself, and thus the genetic differentiation observed *within* a single mussel type, or the genetic similarity *between* different types, may be attributed to population genetic mechanisms rather than systematic differentiation. An alternative explanation is that the contact zones of the different taxa do not effectively inhibit interlineage gene flow at this locus (Varvio et al., 1988).

Esterase-D (*Est-D*): In the northern hemisphere this locus has good discriminatory value in differentiating between *edulis* and *galloprovincialis*, or *edulis* and *trossulus*, but not between *galloprovincialis* and *trossulus*. In the southern hemisphere the diagnostic value of the locus breaks down (Table 7.1).

Glucose-6-phosphate isomerase (*Gpi*): This was one of the first loci to be used in population genetic studies on mussels (see reviews by Levinton and Koehn, 1976 and Gosling, 1984). In the northern hemisphere the *Gpi* locus is reasonably good in differentiating between the three mussel types. However, Gpi^{98} which is at low frequency in northern *M. edulis* varies between 0.28 and 0.92 in southern hemisphere putative *M. edulis*, including *M. desolationis* (Table 7.1), and Gpi^{107} which is characteristic of northern *edulis* is generally absent from the southern form (McDonald et al., 1991)!

Octopine dehydrogenase (*Odh*): In the northern hemisphere the frequency of Odh^{110} can be used to separate both *M. edulis* and *M. trossulus* from *M. galloprovincialis* However, in the southern hemisphere the frequency of Odh^{110} in putative *M. galloprovincialis* falls within the range observed for putative *M. edulis* (Table 7.1).

Mannose-6-phosphate isomerase (*Mpi*): This locus has only recently been used in genetic studies on *Mytilus* and has been cited by Varvio et al. (1988) as being "virtually diagnostic" for the three types of mussels. For the majority of populations studied to date this locus is indeed a good discriminator of the three different mussel types. It is probably on the basis of *Mpi* (and *Est-D*) frequencies that McDonald et al. (1991) have suggested that mussels from South America and the Kerguelen Islands (*M. desolationis*) should tentatively be considered as *M. edulis*, and that those from Australia and New Zealand should be included in *M. galloprovincialis*.

Phosphoglucomutase (*Pgm*): In the northern hemisphere this locus can be used to differentiate *M. trossulus* from either *M. edulis* or *M. galloprovincialis*; but since there is little difference in allele frequency at this locus between *M. edulis* and *M.*

galloprovincialis in the northern hemisphere, the locus plays little part in helping to differentiate *edulis* from *galloprovincialis* in the southern hemisphere.

It is quite clear that none of the above loci are truly diagnostic (a locus is considered to be diagnostic if an individual can be assigned to the correct species, or form, with a probability >0.99 (Avise, 1974). The search for diagnostic loci, where allele frequency distributions do not overlap, has to date been unsuccessful. Of the eight polymorphic loci, only six (*Aap, Est-D, Gpi, Pgm, Odh* and *Mpi*) have proved useful in differentiating between the three different mussel types in the northern hemisphere. Some of the loci are more useful than others e.g. *Mpi, Gpi* and *Aap* can discriminate between all three types, whereas *Odh, Pgm* and *Est-D* can only discriminate between pairs (Fig. 7.3). In the southern hemisphere only two loci, *Mpi*, and to a lesser extent *Aap*, retain their diagnostic value. One locus, which has received little attention, but which may prove useful in future taxonomic studies, is a peptidase locus using leucine-glycine-glycine as substrate. Grant and Cherry (1985) found this locus to be almost completely diagnostic for *M. edulis* and *M. galloprovincialis*.

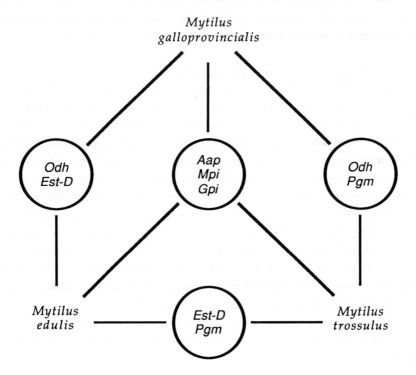

Fig. 7.3. Diagram illustrating the best diagnostic loci for differentiating between the different forms of *Mytilus* in the northern hemisphere.

It should be pointed out, as mentioned earlier (p.314–315), that the *Gpi* locus may have limited taxonomic usefulness, since there is good evidence that clinal variation in the frequency of the common *Gpi* alleles along the east coast of the U.S.A. at least, is maintained by the selective effect of temperature (Hall, 1985). This highlights the problems of using allozyme frequency differences, which may be caused by differential selection in different environments (either on a micro or macrogeographic scale), as indicators of taxonomic ranks.

The systematic status of *M. edulis, M. galloprovincialis* and *M. trossulus*

Nearly all the most illuminating investigations of speciation have been those where a number of techniques have been employed. In this section information from hybridization, cytological, immunological and mitochondrial DNA (mtDNA) studies on *Mytilus* will therefore be used, in conjunction with the morphometric (see Chapter 1) and electrophoretic data already presented, to try and reach some sort of concensus on the systematic status of various forms within the genus. Since populations of *M. trossulus* have been identified solely on the basis of allozyme data most of this additional information is pertinent only to *M. edulis* and *M. galloprovincialis*.

Hybridization in wild and laboratory populations:
There is now abundant evidence from morphometric, electrophoretic (see review by Gosling, 1984) and mtDNA data (see below) that hybridization and introgression occurs to varying degrees in mixed populations of *M. edulis* and *M. galloprovincialis*. The extent of introgression in hybrid zones can be assessed by measuring deviations from Hardy–Weinberg expectations and genotypic correlations between 'diagnostic' loci. Three statistics have been used by Skibinski and Beardmore (1979) to study mixed populations of *M. edulis* and *M. galloprovincialis* in the British Isles: F (Wright, 1951), which measures deviation from Hardy–Weinberg equilibrium; R1 which measures the strength of association of homozygote genotypes, and R2 which measures the excess of double heterozygotes expected (for a locus pair). General trends in the values of these statistics occur as a result of interbreeding: on hybridization F decreases, R2 increases and R1 remains unaltered and if intergradation proceeds all three measures decrease at different rates to zero when the process is complete. The analysis is complicated by the fact that deviations from Hardy–Weinberg equilibrium, in the form of heterozygote deficits, are a common occurrence in mussel populations (see p.349–352).

In some areas such as South-west England hybridization but little intergradation is occurring between the two forms of mussel, but at other localities e.g. parts of Scotland, North-east England and at exposed sites on the Atlantic coasts of Ireland, intergradation between them is extensive (Skibinski and Beardmore, 1979; Gosling and Wilkins, 1981). The evidence so far indicates that the amount of genetic mixing between the two forms is such as to strongly suggest that there is no clear evidence of an absolute reproductive barrier between them. This conclusion is supported by the results from artificial hybridization studies, which indicate that there is little evidence of genetic incompatibility between the two forms, although differences in sperm morphology have been reported by Hodgson and Bernard (1986) and Crespo et al. (1990). Reciprocal crosses between allopatric *M. edulis* from France and *M. galloprovincialis* from Spain produced fertile F_1 individuals and there was no indications of reduced viability or growth or increased mortality among the F_1 'hybrids', when compared to the parental crosses (Lubet et al., 1984). Backcrosses of hybrid F_1 individuals to *M. edulis* F_1 and *M. galloprovincialis* F_1 were also performed, but these, having reached a length of about 1cm, were unfortunately lost in a storm. More recently, reciprocal crosses have been performed both between sympatric *M. edulis* and *M. galloprovincialis* from Rock in South-west England, and between allopatric Mediterranean *M. galloprovincialis* and Irish Sea *M. edulis* (Beaumont, Matin and Seed, unpublished data). In laboratory trials hybrid veliger larvae showed significantly ($P<0.05$) reduced survival rates compared to pure *edulis* or *galloprovincialis* veligers. However, neither normality during early development, nor veliger growth rates were significantly different ($P<0.05$) in hybrids compared to pure lines.

In contrast to North-west Europe, there has been little detailed electrophoretic analysis carried out in the areas of contact between *M. edulis* and *M. trossulus*, or *M. galloprovincialis* and *M. trossulus*. However, the indications are that in every case where there is contact between two mussel types, they are hybridizing (McDonald and Koehn, 1988; Koehn, 1991; Väinölä and Hvilsom, 1991). The zones vary in size depending on location. For example, in North-west Europe the width of the hybrid zone between *M. edulis* and *M. galloprovincialis* is large, extending from northern Spain to the Shetland and Orkney Islands off the north of Scotland, a distance of some 1400km; while the hybrid zone between Baltic *M. trossulus* and North Sea *M. edulis* is narrow, hybridization occurring over a relatively short distance of about 150km in the Danish Belt Sea (Väinölä and Hvilsom, 1991). Hybridization between *M. galloprovincialis* and *M. trossulus* on the west coast of North America occurs in Central California, over a distance of less than 300km (McDonald and Koehn, 1988), while in North-east Canada the hybrid zone between *M. edulis* and *M. trossulus*—though poorly studied—probably occurs in the Gulf of St. Lawrence, Newfoundland and Nova Scotia (see Fig. 1.6., Chapter 1). Hybrid zones are generally spatially complex, containing a mixture of pure, hybrid and introgressed individuals; and, as in the case of *M. edulis* and *M. galloprovincialis*, the amount of introgression can vary even from place to place within the hybrid zone. For a full discussion on the analysis, characteristics and dynamics of hybrid zones readers are referred to: Gosling (1984) and references therein; Barton and Hewitt (1985, 1989) and references therein; Edwards and Skibinski (1987) and Väinölä and Hvilsom (1991).

Cytological studies:
The karyotypes of *M. edulis* and *M. galloprovincialis* (and also *M. desolationis*) are very similar (Thiriot-Quiévreux, 1984b and p.357–360 for details) with the exception of chromosome pairs number two and three: number two is metacentric and number three subtelocentric in *M. edulis*, while both chromosome pairs are metacentric in *M. desolationis* and subtelocentric in *M. galloprovincialis*. Thiriot-Quiévreux concludes, however, that the differences observed are not sufficient evidence yet for a major taxonomic separation between *M. edulis* and *M. galloprovincialis* (and presumably *M. desolationis*). Other authors (Dixon and Flavell, 1986; Pasantes et al., 1990) have found none of the major chromosomal differences reported by Thiriot-Quiévreux and colleagues. It is obvious that there is a certain amount of confusion regarding the actual karyotype of *Mytilus*, and that further work is necessary before the problem can be resolved.

If the differences are found to be real and repeatable how typical are they of molluscan species-level differences? White (1968, 1978) has suggested that certain types of chromosomal rearrangements e.g. inversions and translocations, of various kinds may play a determining role in the speciation process in many groups of animals (but see Zouros, 1982). Such inversions could perhaps provide the raw material for the forces of speciation which would eventually convert these polymorphisms within the ancestral population into a system of fixed inversions in two lineages. While this may be an attractive hypothesis to explain possible chromosomal differences between the different mussel types it must be kept in mind that polymorphisms for pericentric inversions have been observed within and between populations of both *M. edulis* and *M. galloprovincialis* (Ahmed and Sparks, 1970; Dixon and Flavell, 1986; Pasantes et al., 1990). It is obvious that more comprehensive sampling of populations within each *Mytilus* form is needed. In addition karyotypes of individuals collected at sites where *M. edulis* and *M. galloprovincialis* (or *M. trossulus*) and their natural hybrids are found together would be useful.

Immunological studies:
Serological methods are relatively free of environmentally-induced changes, being based on the quantitative precipitation of proteins. Moreover, when the precipitations are performed in gel-immunodiffusion or immunoelectrophoresis the sensitivity of serological methods is enhanced, and even the rate of cross-reactivity between individuals very close on the zoological scale can be appreciated.

Table 7.2. Number of antigens precipitated by antibodies against *Mytilus edulis* (Denmark) and *Mytilus galloprovincialis* (S. France) in different populations of *Mytilus*. The samples from Ireland are a mixture of *M. edulis* and *M. galloprovincialis*. Samples marked with an asterisk are most likely *Mytilus trossulus* . (From Brock, 1985).

Antigen	Locality	Antibodies against M. galloprovincialis Number of reactions	Antibodies against M. edulis Number of reactions
M. edulis*	Baltic Sea	27	0
M. edulis	Denmark	27	0
Mixed	Ireland	27	0
M. galloprovincialis	South France	27	0
M. edulis*	Nova Scotia, Canada	27	0
M. edulis*	Oregon, U.S.A	27	0
M. californianus	Oregon, U.S.A	25	0
Musculus sp.	Suez, Egypt	19	0

Brock (1985) has used crossed immunoelectrophoresis on a large number of *Mytilus* populations but observed no type-specific reactions (Table 7.2). It is difficult to evaluate these results in view of the lack of information on immunological comparisons between conspecific populations, congeneric species or species of different genera of molluscs. Two reports, however, are worth mentioning. Davis and Fuller (1981) have carried out an impressive immunological survey on 52 species belonging to 27 genera of freshwater bivalves. Cross comparisons between species pairs of the same genus indicated that they differed by one to six antigens, with most species pairs differing by more than two antigens. In a later report, Brock (1987) observed seven species-specific antigen-antibody reactions between two closely related cockle species, *Cardium edule* and *C. glaucum*. Although the relationship between immunoelectrophoretic genetic distance and taxonomic hierarcy (usually based on comparative morphology) is not a simple one, one might tentatively suggest at this point that a minimum of 1-2 antigen differences are expected in comparisons between species pairs. This would mean that from the immunological evidence to date, there is no valid reason for considering *M. edulis*, *M. galloprovincialis* and *M. trossulus* as separate species. It is interesting to note that, as expected, *M. californianus* showed a lower number (25) of common reactions with antibodies against the three other *Mytilus* forms (Table 7.2).

Mitochondrial DNA studies:
There are several features which distinguish mtDNA from nuclear DNA: (1) the mode of mtDNA inheritance is through the maternal line and therefore all individuals with a particular mtDNA genotype probably belong to the same maternal clone; since inheritance is maternal the male and female parents of naturally occurring hybrids can be easily identified; (2) most changes in mtDNA are believed to be neutral (Brown, 1983), and (3), there is evidence that mtDNA evolves about 5–10 times faster than nuclear DNA (Brown et al., 1979). The analysis of mtDNA variation has provided valuable information in taxonomic studies of vertebrates at the specific or subspecific level (Moritz et al., 1987), but has only recently been applied to marine invertebrates (Edwards and Skibinski, 1987).

MtDNA can be cleaved into short sections by site-specific proteolytic enzymes called endonucleases, and the fragments produced by these enzymes can be separated on electrophoretic gels on the basis of size. Closely related species show a greater number of mtDNA bands of equivalent mobility than species which are less closely related. MtDNA variation has been studied in several pure and mixed populations of *M. edulis* and *M. galloprovincialis* from Britain (Skibinski, 1985; Edwards and Skibinski, 1987). Significant differences in mtDNA genotype frequencies were observed between allopatric populations of *M. edulis* (Swansea, Wales) and *M. galloprovincialis* (Padstow, South-west England), but none of the mtDNA genotypes were perfectly diagnostic. Also, when the genetic distance value was computed it was within the range commonly found for conspecific comparisons. Thus, there is no indication that mtDNA variation provides greater overall diagnostic power than allozyme variation in differentiating between the two forms of mussel. Mixed populations of the two forms of mussel contained genotypes found at both Swansea and Padstow, which confirms the hybrid nature of these populations. The base sequence divergence between mtDNA from the two forms is about 3–4% (Skibinski, unpublished results). Assuming a 2% rate of change in mtDNA per million years (Brown et al., 1979) the divergence time between these two mussel types is consistent with the palaeontological evidence (Barsotti and Meluzzi, 1968). *Mytilus* mtDNA has now been cloned by Skibinski and colleagues and Hoeh et al. (1991) have recently used it as a probe to show that the mitochondrial genome in *Mytilus* is actually *biparentally* inherited. Whether this also applies to other bivalve taxa remains to be seen.

In conclusion, evidence from morphological (Chapter 1), electrophoretic (nuclear and mtDNA), immunological and cytological studies on *M. edulis, M. galloprovincialis, M. trossulus* and *M. desolationis* indicate that they are very similar to one another, with no single morphological or genetic character being clearly diagnostic. In addition, the different forms appear to be genetically compatible with one another, since everywhere two of them come in contact there is evidence of hybridization; and introgression occurs to varying extents. Thus, with the evidence to date, there would seem to be little reason for regarding the different forms as distinct species. There is substantial agreement with this point of view (Skibinski et al., 1983; Gosling, 1984 and references therein; Brock, 1985; Skibinski, 1985; Hodgsen and Bernard, 1986; Edwards and Skibinski, 1987; Blot et al., 1988; Johannesson et al., 1990; Tedengren et al., 1990; Väinölä and Hvilsom, 1991). In fact, Johannesson et al. (1990) and Tedengren et al. (1990) believe that the allozymic differences observed between Baltic and North Sea populations are due to differential selection occurring over the steep salinity gradient between the two regions (W. Baltic 100°/oo to Kattegat 20°/oo, and they suggest that the mussel in the Baltic is a specialized brackish water ecotype of North Sea *M.edulis* i.e. that the two taxa originate from the same evolutionary lineage.

Others however, (McDonald and Koehn, 1988; Koehn, 1991; McDonald et al., 1991), consider *M. edulis*, *M. galloprovincialis* and *M. trossulus* to be distinct species because, despite hybridization and a massive potential for dispersal, each of the three taxa maintains a distinct set of alleles, with *fairly* (my italics) homogeneous allele frequencies across vast distances" (McDonald et al., 1991). Perhaps the difference between the two viewpoints is simply a question of emphasis? Koehn and colleagues have tended to study widely separated allopatric populations of the three forms of mussel, and are therefore struck by the amount of genetic *differences* between the different taxa—differences which they feel are large enough to warrant taxonomic recognition at the species level. Other workers, however, have concentrated on areas of mixing between pairs of mussel types e.g. North-west Europe, or the Baltic/North Sea areas; in such areas, because of hybridization and varying amounts of introgression between *M. edulis* and *M. galloprovincialis*, or between *M. edulis* and *M. trossulus*, workers are impressed by the amount of genetic *similarity* between the different taxa, and are therefore reluctant to consider them as separate species.

The biological species concept states that species are "groups of interbreeding natural populations that are reproductively isolated from other such groups" (Mayr, 1970). This concept, which appears to have survived despite frequent criticism over the past 30 years (see Avise and Ball, 1990 for recent review), places heavy reliance on the presence of barriers to gene exchange between species pairs. According to Mayr's definition, the extent of interbreeding among the three mussel taxa is, therefore, clearly too high to warrant full specific status for each. Yet, on the other hand, there are plenty of examples, both in plants and animals, of stable hybrid zones between what are generally recognized to be distinct species. This indicates that a certain amount of gene flow *can* take place and all the while the parental forms can still maintain their genetic integrity on each side of the zone. However, unlike the situation in *Mytilus,* such zones are invariably narrow, due to a stable balance between dispersal (gene flow) and selection (see Barton and Hewitt, 1985 for review). For gene flow to be significantly reduced across a hybrid zone hybrids must be substantially less fit than nonhybrids, and the number of genes which make the barrier must be so large that most other genes are closely linked to one or another of the loci under selection (Johannesson et al., 1990).

Clearly the solution now lies in deciding how much hybridization is permissible before two taxa can no longer be regarded as separate species? If this can be done then we are quite some way along the road to acquiring an agreed operational definition of biological species. In the aquatic environment at least, and for *Mytilus* in particular, this is long overdue.

Meanwhile, the controversy concerning the systematic status of *M. edulis* and *M. galloprovincialis,* which has raged for nearly 150 years, continues. The recognition of a third, and perhaps a fourth form of mussel i.e. *M. trossulus* and *M. desolationis* respectively, further complicates the situation. There is now increasing awareness that a multidisciplanary approach to the problem is needed. Promising areas of further research should include:

•More extensive electrophoretic analysis of *Mytilus,* with particular emphasis on areas where few samples have been collected e.g. the northern and eastern coasts of the U.S.S.R., eastern China, Greenland, the northern coasts of North America and all over the southern hemisphere—but in particular the South American coasts, where mussels are somewhat intermediate in both morphological and electrophoretic characteristics to *M. edulis* and *M. galloprovincialis* (McDonald et al., 1991), and may represent a different lineage.

•Detailed investigations on the distribution of genes and barriers to gene flow in areas where the ranges of taxa overlap. This presumably would lead to some understanding of the mechanisms that are keeping the different taxa distinct over large geographic distances.

•A extension of the mtDNA studies, initiated by Skibinski and colleagues on British populations of *Mytilus,* to both pure and hybrid populations in other areas.

•Investigations on the reproductive cycles of the different forms, particularly in areas where there is overlap between two types.

•Detailed measurements of growth and survival in hybrid populations—along the lines of Skibinski and Roderick (1989) for South-west England.

•Laboratory crosses between the various mussel forms under controlled conditions, with details on larval abnormalities, survival and growth rates of F_1 and backcross individuals.

•More karyological investigations on *Mytilus* (and to include *M. trossulus*) from pure and hybrid populations.

•In particular, reciprocal transplants of nonsympatric populations e.g. *M. gallo-provincialis* from southern California and *M. trossulus* from Oregon, or *M. edulis* from the North Sea and *M. trossulus* from the Baltic, in order to show whether the genetic differences observed between types reflect distinct evolutionary backgrounds, or are due to differential selection.

•Finally, as mentioned already, there is a real need for a more objective definition of 'biological species'.

Until then, we are still left with the problem of what to call the different taxa. In order to avoid unnecessary confusion, the most prudent scenario would seem to be to continue referring to the taxa as: *M. edulis*, *M. galloprovincialis*, *M. trossulus* and *M. desolationis*; while at the same time recognizing—with the morphological and genetic information we have at present—that there is a considerable lack of agreement on their exact taxonomic status.

Microgeographic Differentiation in *Mytilus*

As seen above, the variation observed in populations sampled over large geographic distances, with the single exception of the *Lap* cline along the east coast of the U.S.A., reflects differences in the distributions of different mussel types within the genus, each having its own characteristic allele frequencies at several partially diagnostic loci. These loci (presumably linked to loci responsible for the systematic identity of the different lineages), are therefore seen as neutral markers, whose frequencies are not regulated by environmental selection pressures. Within each lineage allele frequencies for a particular distinct geographic region are generally homogeneous. In contrast, microgeographic studies, which involve sampling mussels over small geographic distances e.g. within bays and estuaries, at different levels in the intertidal zone, often reveal significant genetic differentiation between samples collected less than one metre apart. This variation is superimposed on the macrogeographic variation described above. How do such differences arise?

In order to answer this question the life cycle of *Mytilus* must be considered. The extended pelagic life of mussels—often as long as 6mo (Lane et al., 1985)—ensures a wide dispersal and substantial mixing of different larval populations. The result is that larvae recruiting to a particular shore may be genetically heterogeneous i.e. they may be an assemblage of larvae from different populations. Once settled, further genetic differentiation can occur in spat and adults through the selective effects of such factors as temperature, salinity and exposure. It is clear, therefore, that both natural selection and interpopulational migration i.e. gene flow, are important in maintaining intrapopulational genetic variation in mussels. But the relative effects of each on spatial patterns of variation are difficult to estimate. The evidence to date would suggest, however, that it is selection rather than migration which is primarily responsible (see below; also Koehn ,1975; Koehn et al.,1976).

Variation within the intertidal zone

Microgeographic studies in the early 1970s were mainly concerned with genetic variation in samples of mussels, usually *M. edulis*, collected along one or more transects in the intertidal area of the shore, or at different locations along an estuary. The results of these studies have been comprehensively reviewed by Levinton and Koehn (1976) and may be briefly summarized as follow: in general, (1) small mussels exhibit heterozygote deficits, which decrease with increasing length (age) of mussel; (2) there is a negative correlation between heterozygosity and intertidal height, and (3) significant positive correlations have been observed between the frequency of particular alleles at a locus (e.g. *Lap*) and shell length or tidal position. All these results have been ascribed to selective mortality, favouring heterozygotes, or particular alleles or genotypes at a locus (but see Tracey et al., 1975 and Levinton and Suchanek, 1978).

More recently, Blot et al. (1987, 1989) have analysed samples of *M. desolationis* at a number of loci (*Lap-1*, *Lap-2*, *Pgm*, 6-Phosphogluconate (*Pgd*, EC 1.1.1.49) and Acid phosphatase (*Acp-2*, EC 3.1.3.2)) from several geographic regions in the isolated Kerguelen Islands. Significant heterogeneity, analysed using correspondance analysis, was observed both between and within the different geographic regions. This was later confirmed from mtDNA analysis of these populations (Blot et al., 1990). Genetic differentiation between regions is probably due to reduced gene flow related to local hydrographic conditions. Within each geographic group the most important environmental factor influencing spatial structure was salinity, with the maximum contribution from the *Lap-2* locus. This agrees well with the results of Koehn and colleagues for this locus in *M. edulis* from Long Island Sound (p.344–345). Additional variation in Kerguelen samples was related to mussel size—mainly due to *Pgm* and *Pgd*—and to a lesser extent to shore exposure.

Similar type studies have been carried out on *Mytilus* populations in North-east Canada. Mussels were sampled from four sites (Fig. 7.4), St. Margaret's Bay (SMB) and Bedford Basin (BB), two closely situated sites on the Atlantic coast of Nova Scotia, and Bideford River estuary (PEI), on Prince Edward Island and Apple River (BF), a small inlet in the Bay of Fundy (Gartner–Kepkay et al., 1980) On the basis of allele frequencies at four loci (*Lap-1*, *Pep-2*, *Gpi* and *Pgm*) the sampling sites fell into two

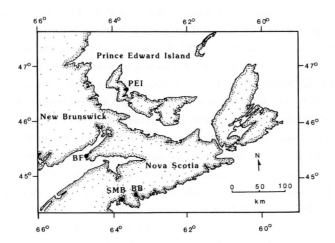

Fig. 7.4. The Canadian Maritime area indicating the *Mytilus* sampling locations of Gartner–Kepkay et al. (1980, 1983).

pairs, SMB–BB and PEI–BF. These authors concluded that the strong pairing was based on a similarity of environmental parameters, despite the wide geographical separation of sample locations. Mussels at the PEI and BF sites experience a greater range and variability in both salinity and temperature than those from the two Atlantic coast sites. A more detailed study of the SMB site (Gartner–Kepkay et al., 1983), has given similar results in that populations within the bay are differentiated into two groups, those at the head of the bay, where ambient conditions fluctuate widely and those at the mouth of the bay, where conditions are more uniform. In both studies large deficiencies of heterozygotes were observed for all loci and at all sampling locations; this was tentatively attributed to selection against heterozygotes. Also, the geographic grouping for the *Lap-1* locus (equivalent to *Lap-2* of Koehn et al., 1976) was consistent with the results of Koehn for this locus i.e. a decrease in the frequency of *Lap*[94] with decreasing ambient salinity. Gartner–Kepkay et al. (1983) suggested that two of the other loci, *Pep-2* (*Lap-1* of Koehn et al., 1984) and *Pgm* are subject to much the same, or closely associated, environmental selective factors as affect *Lap-2*. (Interestingly, results from the additional locus *Ap* analysed by Gartner–Kepkay et al. (1983) showed no differences among localities, indicating that environmental factors appeared to have little influence on this locus). Thus, the authors contend that despite extensive gene flow between the different sites within the bay, clustering of populations occurs as a genetic response to common environmental selective factors. This of course implies that rather strong differential mortality is operating on different genotypes (mostly heterozygotes?) within a common gene pool in order to maintain this spatial differentiation. Koehn et al. (1984) have suggested, however, that the results of Gartner–Kepkay et al. (1980, 1983) may be explained if these authors sampled not one but two forms of mussels in the Nova Scotia region. An examination of figure 2 in Koehn et al. (1984) does indeed indicate that the SMB and BB sites lie in a *trossulus* area, while PEI and BF appear to be in an *edulis* area. Yet allele frequencies at the SMB and BB sites do not correspond to the frequencies observed by Koehn et al. (1984) for Nova Scotian *M. trossulus*. Furthermore, there are large differences even in *edulis* frequencies between the two studies! The comparison is complicated by the fact that Gartner–Kepkay et al. (1980) observed many fewer alleles at each locus than Koehn et al. (1984). Also, while the latter have stated that *Pep-2* is virtually monomorphic in North America, Gartner–Kepkay et al. (1980) have observed four alleles at this locus in all of the populations they sampled. The question still remains: is the spatial differentiation observed by Gartner–Kepkay and colleagues the result of environmental selective factors operating on a single panmictic population of *M. edulis* (or *M. trossulus* in the case of the SMB study), or is it due to these authors having sampled a mixture of two genetically different taxons, *M. edulis* and *M. trossulus*? It is surprising, in view of the fundamental differences between the two competing hypotheses, that further analysis of mussel populations in the Canadian Maritimes has not yet been undertaken.

Variation in intertidal *Mytilus* in the British Isles

In western Europe genetic analysis of mussels in bays and estuaries is complicated by the fact that many sites contain not one but two forms of mussels intermixed together in varying proportions. Over the past 15 years the genetic structure of such mixed populations of *M. edulis* and *M. galloprovincialis* from Britain and Ireland has been studied in detail, using morphometric and electrophoretic data (Ahmad and Beardmore, 1976; Gosling and Wilkins, 1977; Skibinski et al., 1978a, b; Skibinski and Beardmore, 1979; Gosling and Wilkins, 1981; Skibinski, 1983; Skibinski et al., 1983; Gosling, 1984; Gardner and Skibinski, 1988, 1990, 1991; Skibinski and Roderick, 1989, 1991; Gosling and McGrath, 1990).

Seed (1974) was the first to note the presence of the *M. galloprovincialis* form in Ireland. In a morphological survey of mussels at 43 sites he found *M. galloprovincialis* intermixed with *M. edulis* on the Atlantic coasts of Ireland, but found no evidence for its occurrence on the Irish Sea coast (see Fig. 1.5., Chapter 1). Its absence from Irish Sea sites is believed to be chiefly related to the hydrographic conditions prevailing at the entrances to this sea (Gosling, 1977). At all Atlantic sites there was a considerable degree of overlap in morphological characters, and the greatest number of intermediate forms were encountered on exposed shores.

The results from electrophoretic surveys of mussel populations from about 20 sites on Irish coasts (Gosling and Wilkins, 1977, 1981) did indeed show that mussels on the Irish Sea coast exhibited allele frequencies at three loci (*Gpi*, *Pgm* and *Lap-2*) which were very similar to populations of 'pure' *M. edulis* from the coasts of France (Gosling and Wilkins, 1981) and Britain (Ahmad et al., 1977). Allele frequencies were homogeneous over localities, with no significant deviations from Hardy–Weinberg expectations; deviations might be expected as a result of population mixing—the Wahlund effect. All the evidence suggested that mussels in the Irish Sea area constituted a single panmictic population of *M. edulis* alone.

On the Atlantic coasts of Ireland there were marked differences in allele frequency at all loci between exposed and sheltered shore mussels and large significant deficiencies of heterozygotes at the *Gpi* locus were observed at all but one of the nine exposed sites, while the majority of the sheltered shore samples did not exhibit such deficiencies. Heterozygote deficits were not correlated with absolute or relative size of individual mussel, nor with position of the mussels on the shore. The indications were—from analyses of *M. galloprovincialis* from the Mediterranean and northern France (Gosling and Wilkins, 1977, 1981)—that the exposed shore samples, and to a much lesser extent the sheltered shore samples, constituted a mixture of interbreeding *M. galloprovincialis* and *M. edulis*. Skibinski et al. (1983), using the more powerful diagnostic locus *Est-D*, have confirmed our observations for Irish sites.

How universal this partitioning is for the British Isles in general is impossible to say, since a similar analysis has not been carried out for British sites. But, preliminary results indicate that the *M. galloprovincialis* form predominates at exposed sites in the Shetland and Orkney Islands to the north of Scotland (Gosling, unpublished data). It should, however, be pointed out at this stage that exposure score alone cannot be used to predict allele frequency, since at several sheltered sites on the Atlantic coasts of Ireland (Gosling and Wilkins, 1981 and Gosling, unpublished data) allele frequency and genotype proportions resemble those observed for exposed shore mussels. Also, mussels at the single exposed site sampled on the Irish Sea coast did not differ appreciably in allele frequency from the sheltered shore samples in the same area, and did not exhibit deficiencies of heterozygotes at the *Gpi* locus. Skibinski et al. (1983) have also observed low frequencies of $Est\text{-}D^{90}$ (a diagnostic allele for *M. galloprovincialis*) at exposed sites on the Irish Sea coast. The results for Irish sites are interesting in the light of an earlier paper by Murdock et al. (1975), which reported a significant correlation between allele frequencies at a *Lap* locus and exposure in samples of *Mytilus* from Irish coasts (their *Lap* frequencies do not correspond to either *Lap-1* or *Lap-2* frequencies in Irish populations). This paper has been much cited as evidence for selection in mussel populations. However, the observed correlation may not in fact be a correlation with exposure *per se* but may reflect the higher incidence of *M. galloprovincialis* alleles at the more exposed sites investigated by Murdock.

Further analysis of exposed shore mussels (Gosling and McGrath, 1990) has revealed that there are significant differences in the genetic composition of adult *Mytilus* from different tidal heights. Significant differences in allele frequencies were observed at the diagnostic *Odh* (but not *Est-D*) locus between samples analysed from different tidal levels. These differences were repeatable whether one was analysing replicate samples at a single point in time i.e. in different years, or samples from different exposed shores. Mussels higher up the shore tended to be more *galloprovincialis*-like than mussels lower down (Table 7.3). A similar result has also been reported by Skibinski (1983) for populations of *Mytilus* in South-west England. However, our results differ from those of Skibinski (1983) and Gardner and Skibinski (1988), in that they have observed differences between high- and low-shore mussels at *several* of the partially diagnostic loci, and have also observed a marked size-dependent variation in allele frequency; larger mussels having a higher frequency of *galloprovincialis* alleles than smaller ones. Gosling and McGrath (1990) have also analyzed samples of mussels recruiting to artificial substrates set out for a period of one month at different tidal levels at one exposed site. These were also observed to be genetically different: mussels higher up the shore exhibited higher frequencies of *galloprovincialis* alleles at *both* the *Odh* and *Est-D* loci than those settling lower down (Table 7.3). Hence, the genetic differences observed in adult mussels are much more exaggerated in juveniles and are already apparent, at least in Irish populations, within the first month of benthic life. Whether these differences derive from differences in the genetic composition of recruits at attachment (see Gosling and Wilkins, 1985), or from site-specific genotypic selection during benthic existence is unclear at present. Selection of suitable sites by different genotypes has been observed in the sedentary polychaete *Spirorbis borealis* (Doyle, 1974). In mussels, however, there is no evidence to date of differential settlement of larvae on the shore.

In South-west England, where the genetic composition of mussels has been extensively studied by Skibinski and colleagues over the past number of years, sympatric populations of *edulis* and *galloprovincialis* exhibit a strong positive correlation between shell length and allele frequency at three diagnostic loci, *Odh*, *Est-D* and *Lap-1* (equivalent to *Aap* locus). In addition, as mentioned above, high-shore samples are more *galloprovincialis*-like than low-shore samples. A further study on samples sof mussels collected from 17 localities within hybrid zones of the two forms of mussels in South-west and North-east England confirmed these observations (Skibinski and Roderick, 1991). The shape of the relationship between frequency and size appeared to be fairly consistent across localities, although the size of greatest frequency increase varied between localities (Fig. 7.5). The higher frequency of *galloprovincialis* alleles in larger mussels could be due to differential growth, a historical change in the genetic composition of mussel populations in these areas, or to differential viability. Growth comparisons involving mussels (pure *M. edulis* and

Table 7.3. Allele frequencies at the *Odh* and *Est-D* loci in samples of *Mytilus* collected from pads and adult beds from the upper, mid and lower tidal levels of the exposed shore at Ballynahown, Galway, Ireland. Alleles labelled 'E' and 'G' are compound alleles obtained by pooling those alleles at highest frequency in *M. edulis* (E) and *M. galloprovincialis* (G) respectively; *** represents significant (P<0.001) deviations from Hardy–Weinberg expectations; N is sample size. (From Gosling and McGrath, 1990).

Allele	Pads			Adult Beds	
	Upper	Mid	Lower	Mid	Lower
Odh					
94G	0.448	0.324	0.219	0.452	0.456
100E	0.379	0.560	0.714	0.387	0.456
105	0.172***	0.116	0.067	0.160***	0.088
N	29	117	91	156	80
Est-D					
89G	0.571	0.487	0.360	0.663	0.724
100E	0.429	0.513	0.640	0.330	0.276
N	28	119	89	156	76

M. galloprovincialis and mixed populations) collected from five different localities in Britain were made after they were held for varying periods of time at three different transplant localities (Skibinski and Roderick, 1989). There was no clear evidence of a growth difference between *M. edulis* and *M. galloprovincialis*, nor of growth differences between allozyme genotypes within populations. Allozyme heterozygotes had growth rates intermediate between allozyme homozygotes (Fig. 7.5), and thus there was no evidence for overdominance with respect to growth rate in this study (see p.352–356 on heterozygosity/growth correlations). In order to test whether a historical change in allele frequencies is occurring within hybrid populations, samples were collected at Croyde Bay and Whitsand Bay in South-west England in 1980–81 and 1986–87 (Gardner and Skibinski, 1988). Allele frequencies at the *Est-D* and *Odh* loci showed little change at these sites within the 6-year period. It is clear that there is no strong support for the hypothesis that *edulis* is gradually replacing *galloprovincialis*. The size-dependent genetic variation is best explained by differential viability. The decrease in *edulis* allele frequencies is very similar in both intra- and intersite comparisons, occurs all over the intertidal distribution of the mussels, and is constant from year to year.

In the light of the results of Skibinski and colleagues for hybrid mussel populations in Britain, it is likely that differential viability of the two forms is responsible for the marked differences in the genetic composition of mussels at different tidal levels in Irish exposed shore populations.

This does imply, however, that in Ireland at least, very strong selection pressures must be operating on mussels, within the first month of benthic life, to bring about the large and significant differences observed between upper and lower shore pad mussels (Table 7.3). Such differences might be related to increased tolerance to aerial exposure by the more southern *galloprovincialis* form. The presence of the *galloprovincialis* form on exposed sites in Ireland, together with its virtual absence from sheltered localities, does suggest that this mussel type has an advantage in exposed regions. At Whitsand the frequency of *edulis* alleles on the high shore begins to decrease at a smaller size than on the low shore. This is consistent with more intense selection against the *edulis* form higher up the shore, and suggests that *galloprovincialis* may be better able to withstand longer periods of emersion than *edulis*. In addition, shell shape in the *galloprovincialis* form may be particularly well-suited for high-shore exposed locations: the hooked appearance of the anterior end of the shell, together with the ventral flattening and posterior expansion of the shell are believed to be adaptations for maintaining firm attachment to substrates and utilization of space (see Chapter 2). A recent paper by Gardner and Skibinski (1991) has indeed shown that when mussel strength of attachment to the substrate (SOA) was tested at Croyde and Whitsand that *edulis*-like mussels of all lengths had a significantly lower SOA than *galloprovincialis*-like mussels. This factor could be responsible for the genotype-dependent mortality of *edulis* observed at British (and perhaps Irish exposed) sites, thus explaining the higher incidence of the *galloprovincialis* form in exposed environments. Why this form is virtually absent from sheltered locations, where the *edulis* form predominates, is an intriguing question and one not easily answered. It cannot be a problem of limited recuitment of the *galloprovincialis* form to sheltered sites, given the long larval life-span of mussels, and the geographical proximity of such sites, particularly in Ireland, to exposed locations. One reason might be that there are differences in the timing of spawning in the *edulis* and *galloprovincialis* forms. If, in general, *edulis* spawns earlier than *galloprovincialis*, then recruitment of the latter to sheltered shores could

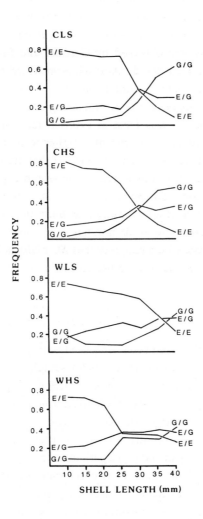

Fig. 7.5. Frequencies of the compond genotypes E/E, E/G and G/G (averaged over five loci, *Ap, Est-D, Mpi,*
Odh and *Gpi*) plotted against shell length for *Mytilus* collected from Croyde low shore (CLS), Croyde
high shore (CHS), Whitsand low shore (WLS) and Whitsand high shore (WHS). At a given locus the
compound allele E is obtained by pooling those alleles which are at highest frequency in *M. edulis*, and
the compound allele G is obtained by pooling those alleles at highest frequency in *M. galloprovincialis*.
(After Gardner and Skibinski, 1988).

be severely limited, because suitable settling surfaces would have been already colonized by *edulis* spat. But, although Seed (1971) has shown that at Rock in South-west England, *edulis* spawns some weeks earlier than *galloprovincialis*, no major differences in spawning cycles were observed by Gardner and Skibinski (1990) for mixed populations at Croyde and Whitsand. Also widespread hybridization between the two forms argues against different spawning cycles.

In hybrid populations *galloprovincialis* is not replacing *edulis*. This is surprising in view of the higher frequency of this form at exposed locations and the recent evidence that *M. galloprovincialis* has a consistently higher fecundity than *M. edulis* (Gardner and Skibinski, 1990). Because of this, and the fact that mortality of hybrids is intermediate to pure forms of the two mussel types, one would expect elimination of *M. edulis* by directional selection. However, Skibinski and Roderick (1991) have shown, from an analysis of spat and adult mussels at Bude in South-west England, that the frequency of *edulis* alleles in adult mussels is too low to account for its relatively high frequency in spat, which suggests that *M. edulis* is being maintained through the immigration of *edulis* spat from other localities.

Estuarine variation

Most estuarine studies on *Mytilus* have been concerned with genetic variation at the *Lap-2* locus. Milkman (cited in Koehn et al., 1976) sampled mussels inside and outside an estuary on the north shore of Cape Cod, U.S.A., and found that small individuals (4–5mm shell length) displayed identical frequencies of Lap^{94} while significant divergence occurred in larger size classes. These authors concluded that newly-established juveniles had initially the same allele frequency, but that subsequent divergence took place in older (larger) individuals along the estuarine gradient. Boyer (1974) sampled *M. edulis* at two separate estuarine localities and found larger upstream mussels to have a lower frequency of Lap^{94} than larger downstream mussels, with small mussels exhibiting intermediate frequencies of this allele. Similar results have been reported for several populations of *M. trossulus* from the Oregon coast, U.S.A. (McDonald and Siebenaller, 1989). So, it would appear that a reduction in salinity—at least for North American populations of *M. edulis* and *M. trossulus*—is always accompanied by a reduction in the frequency of Lap^{94}. In addition to these size-related differences great interlocality heterogeneity in allele frequency at the *Lap-2*, *Gpi* and *Ap* loci has been observed within estuaries, probably reflecting the greater environmental variation of such habitats. An example which serves to illustrate this is furnished by Koehn and Mitton's (1972) survey of sympatric intertidal populations of *M. edulis* and *Modiolus demissus* (= *Guekensia demissa*), where in collections made at four separate sites along an estuary, the two bivalves showed a similar pattern of three common and two rare alleles at the *Lap-2* locus, although their alleles differed in electrophoretic mobility. Different allele frequencies were seen at each site, but within any single locality the frequencies of *Lap* alleles were similar for both species. The correlated response of these two species to certain patterns of environmental heterogeneity lead the authors to conclude that both bivalves were responding in a similar manner to selection pressures at the *Lap* locus.

In the light of these results it is remarkable that so little additional information has been published on estuarine variation in *Mytilus* over the past 15 years. The single exception is of course the detailed investigation by Koehn and colleagues on the *Lap* polymorphism in *M. edulis* from Long Island Sound (see below p.344–345).

In summary, despite an extended dispersal period, the genetic composition of mussels often varies sharply over very short distances and/or among different size categories. These differences are most pronounced when samples are taken from different localities, which differ in environmental parameters such as salinity (e.g. the *Lap* polymorphism in *M. edulis*) and/or receive larval immigrants from different parental populations (e.g. *M. edulis/galloprovincialis* in the British Isles, or *M. edulis/trossulus* in the Canadian Maritimes).

Interactions between loci

There have been only two published papers on inter-locus interactions in *Mytilus*. Mitton et al. (1973) examined the distribution of di-locus genotypes at the *Lap-2* and *Ap*, two functionally related but unlinked loci, in a sample of *M. edulis* from Long Island Sound. They observed that genotypes at the two loci were not distributed independantly of one another, and they concluded that epistatic selection appeared to be the basis for the disequilibrium. However, a subsequent study on populations of *M. edulis* from the U.K., found no such association when these, or additional loci (*Gpi* and *Pgm*) were examined (Ahmad and Hedrick, 1985). These conflicting results were explained in terms of the large differences in allele frequenciy at the *Lap* and *Ap* loci between American and European *M. edulis* (see Table 7.1), and the strong possibility that the selection pressures, which gave rise to the nonrandom association in American *M. edulis*, may not be operative in British *M. edulis*.

The Adaptive Significance of Genetic Variation

It is now abundantly clear from the preceding section that the genetic characteristics of natural populations of *Mytilus* can change over time and space. The inference usually is that such changes occur in wild populations as a response to some environmental selective factor, acting either directly on the locus itself, or on an associated locus or loci. Support for this comes from laboratory-based studies, where significant changes in genetic composition can be produced by specific experimental conditions (see p.348–349). But how important these changes are in terms of the physiological adaptation of the organisms concerned is poorly understood (see Chapter 5). ' "The problem of understanding the physiological consequences of genetic variation, or stated oppositedly, the genetic basis of physiological variation, is a problem in identifying those genes, or genotypes, that produce specific physiological phenotypes. In short, it is a problem in 'mapping' phenotypic variation of physiological or biochemical characters to specific genetic states."(Koehn, 1985).

The *Lap* polymorphism
Koehn and colleagues have adopted such an approach in their investigation of the aminopeptidase–1(*Lap–2*) polymorphism in *M. edulis* on the east coast of the U.S.A. and Canada, and this is the only case—at least for the marine environment—where the biochemical and physiological consequences of genetic variation are thoroughly understood. The *Lap* polymorphism has been comprehensively reviewed by Koehn (1983, 1985) and Koehn and Hilbish (1987). Therefore, what follows is a brief summary of the major findings.

Three common (*Lap*^{94}, *Lap*^{96} and *Lap*^{98}) and two rare alleles have been detected at the *Lap-2* locus in populations of mussels from the eastern Atlantic Ocean (p.313–314). In this region the *Lap*^{94} allele has a lower frequency in estuarine populations than in oceanic populations. For example, *Lap*^{94} is approximately 0.55 in oceanic waters south of Cape Cod but declines abruptly to about 0.15 to the north of Cape Cod, and at the entrance to all estuaries south of Cape Cod (Koehn, 1985). Despite the long pelagic dispersal phase in the life cycle of mussels, strong selective forces, possibly related to environmental salinity, must be acting on the *Lap* locus to bring about these sharp discontinuities in allele frequency. It should be stated at this point that outside of North America the frequency of *Lap*^{94} is not correlated with environmental salinity (Bulnheim and Gosling, 1988; Varvio et al., 1988; Väinölä and Hvilsom, 1991).

Mussels osmoregulate by changing the concentration of intracellular amino acids in response to changing environmental salinity. The aminopeptidase–1 enzyme— biochemically characterized by Young et al. (1979)—is involved in the degradation of intracellular protein to supply the cytosolic free amino acid pool. Mussel populations which encounter different salinities display differences in the activity of aminopepti-dase–1 and other lysosomal enzymes involved in protein degradation (Moore et al., 1980; Koehn et al., 1980a; Koehn and Immermann, 1981); for example, oceanic popula-tions exhibit about 50% greater levels of specific activity than estuarine populations (Koehn and Hilbish, 1987). There are significant differences in aminopeptidase–1 ac-tivity between the various *Lap* genotypes (Koehn and Immermann, 1981; Koehn, 1985 and references therein). Genotypes that possess the Lap^{94} allele, either in homozygous or heterozygous form, have significantly greater specific activities than genotypes without this allele. During hypo-osmotic stress (under field and laboratory conditions), genotypes with the Lap^{94} allele excrete ammonia and amino acids at rates nearly twice those of alternate genotypes (Deaton et al., 1984; Hilbish and Koehn, 1985a, b). Also, this allele is observed at its highest frequency in those populations experiencing fully marine conditions. As mentioned already (p.313–314), each summer mussel larvae exhibiting Lap allele frequencies typical of oceanic populations emigrate into the sound. Following recruitment, immigrants with the Lap^{94} allele, either in homozygous or heterozygous state (Hilbish and Koehn, 1985c), are selected against, because of the higher rate with which Lap^{94} genotypes expend energy reserves, thereby reducing the energy available to carriers of this allele for other vital functions (Koehn and Hilbish, 1987). This selection is most intense in the autumn among individuals <20mm shell length (Hilbish, 1985). Selective mortality in newly-settled mussels results in the establishment in adult mussels of the allele frequency cline, which is both temporally and spatially stable (Koehn et al., 1980b; Hilbish, 1985).

This detailed biochemical and physiological study of the Lap polymorphism in *Mytilus* is unique in that it is the only study (apart from the Adh polymorphism in *Drosophila* (van Delden, 1982) in which the frequency of major alleles are clearly determined by an environmental factor. It is significant, in view of the continuing rhetoric on heterozygote superiority (see later), that Koehn and colleagues found no evidence of overdominance (where the heterozygote for two alleles, e.g. Aa, is superior to either corresponding homozygote AA or aa), only dominance, at this locus i.e. individuals carrying the Lap^{94} allele in either homozygous or heterozygous form appeared to have an equal selective advantage in the marine environment.

There is only one other, albeit incomplete, study which has attempted to establish the adaptive importance of allelic variation at a locus. In a kinetic study on Gpi allozymes in *M. edulis* Hall (1985) observed that, while the two major allozymes, Gpi^{100} and Gpi^{96} were catalytically similar at low temperatures (5–10°C), Gpi^{100} diverged to become significantly more efficient than Gpi^{96} at high temperatures (15–25°C). These differences are consistent with the latitudinal distribution of these alleles in populations of *M. edulis* along the east coast of the U.S.A. (see p.313–315) and the author tentatively suggested that environmental temperature may be the selective agent maintaining the cline. However, further investigations, along the lines of Koehn and colleagues for the Lap locus, have not been undertaken.

So it would appear that spatial differentiation in allele frequency can occur between different populations of mussels as a result of natural selection acting in combination with gene flow to adjust the frequency of specific genotypes to local conditions. In view of the evidence there is a surprising lack of follow-up experimental genetic studies on either wild or laboratory-based populations of mussels.

Reciprocal transplant experiments

Reciprocal transplant experiments have been used by several authors to find out the extent to which differences between populations occupying different habitats are environmentally induced (phenotypic adaptation), or inherited (genotypic adaptation). Several such studies have been recently carried out on mussels in the North and Baltic Seas. Populations from these seas are morphologically (Kautsky et al., 1990 and references therein) and physiologically distinct (Tedengren et al., 1990 and references therein), but these differences are largely due to the very different salinity regimes experienced by mussels in the two areas. As mentioned earlier substantial differences in allele frequency have also been observed between North Sea and Baltic *Mytilus*. To test whether the observed differences are systematic in nature, or whether they are due to environmentally induced selection, Johannesson et al. (1990) carried out reciprocal transplants of mussels between a North Sea (20–30°/oo) and Baltic (6–7°/oo) site. Large mortalities occurred in transplanted mussels at both sites, and the survivors were very similar in *Gpi* frequency to native mussels; this was also true for surviving *Pgm* genotypes of Baltic mussels transplanted to the North Sea, but not in the opposite direction. Results of a similar nature have also been reported by other workers for conspecific populations of *M. edulis* (Theisen, 1978), *M. desolationis* (Blot et al., 1989) and *Chlamys varia* (Gosling and Burnell, 1988). Johannesson et al. (1990) concluded from their results that mussels in the North and Baltic Seas are conspecific and that the observed genetic differences between the populations reflect differential selection, rather than distinct evolutionary backgrounds. Väinölä (1990) however, rejects selection and states that an obvious explanation for their results is that the survivors represented contamination of transplanted batches by local mussels; (surprisingly, Johannesson et al. (1990) did not use tagged animals in their study). He argues his case convincingly for the *Gpi* locus by showing that the observed number of survivors could not have been present in random samples of the given sizes from the sites of origin, or even in samples many times larger. However, in the case of the North Sea to Baltic transplant *Pgm* allele frequencies deviate from *both* native Baltic and North Sea populations—a result which is incompatible with the contamination hypothesis. Therefore, neither selection nor contamination alone can fully explain the results of Johannesson et al. (1990). The burning question still remains: are *Mytilus* populations in these areas genetically different because each represents a different evolutionary lineage, or are they different as a result of environmentally induced selection? Well-designed reciprocal tranplant experiments should certainly provide the answer, not just for the Baltic/North sea situation, but also for other areas such as the Canadian Maritimes referred to earlier (p.334–336).

Pollution studies

Additional indirect evidence for selection comes from the results of a few studies on the genetic effects of pollution on mussels. Battaglia et al. (1980a, b) analysed samples of *M. galloprovincialis* along a gradient of increasing hydrocarbon pollution in the Lagoon of Venice, Italy. Seven polymorphic loci were studied (*Ap, Lap, Gpi, Pgm, Pgd* and isocitrate dehydrogenase (*Idh$_s$, Idh$_m$*, EC 1.1.1.42), and four of these (*Ap, Lap, Idh$_s$* and *Idh$_m$*) showed a common trend: higher frequency of the more common allele in the more polluted areas, together with a significant drop in the frequency of heterozygotes. Preliminary results by Rodino (1973) for the same area had earlier indicated a similar trend for the *Lap* locus. Also, the results from a later study from the east coast of Italy (de Matthaeis et al., 1983) have shown that a sample of *M. galloprovincialis*, from a site polluted with human sewerage, had a lower mean heterozygosity (measured over 31 loci) than a sample from a cleaner site nearby. However, Fevolden and Garner (1986) found no evidence of a reduction in heterozygosity (based on data from 30 loci) in samples of *M. edulis* from experimental oil-polluted basins in the Oslofjorden, Norway.

Electrophoresis has also been used to screen populations of *M. edulis* for the presence of rare variants; these are alleles not normally present in populations at frequencies >0.01. At several British sites (Beardmore, 1980) a positive correlation has been observed between the concentration of heavy metals in mussel flesh and the frequency of rare variants. However, studies of this type have not been followed up probably because, in order to detect rare variants, large samples of mussels must be analysed, and also results from such studies are usually inferrential (see below).

It is obvious that such studies report correlations of observed gene frequency *after* the operation of many unknown and complex factors. They can neither specifically pinpoint the selective agent nor demonstrate cause-effect relationships between the pollutants and observed gene frequency. To quote Nevo et al. (1983) "They are essentially statistical associations that always leave room for ambiguities and spuriousness, hence skepticism." Therefore, a preferential approach is the use of controlled laboratory-based experiments, where the cause-effect influence of various pollutants such as heavy metals (Hg, Cd, Pu, Cu, Zn), detergents or crude oil can be tested. In controlled laboratory experiments Nevo and colleagues (see Nevo et al., 1983 for review and references) have produced ample evidence that in many different species of marine organisms different allozymes display differential tolerance to specific pollutants. However, only two such studies have been performed in *Mytilus* and both of these involved copper toxicity testing. Hawkins et al. (1989b) found that when *M. edulis* individuals were exposed to 0.15-ppm dissolved copper until death that *Gpi* heterozygotes survived significantly longer than homozygotes. This is in agreement with the findings from field studies described above. Hvilsom (1983) exposed samples of *M. edulis* to different concentrations of copper (0.1-, 0.2- and 0.5-ppm) until 50% mortality had been reached. Significant differences were observed in *Gpi* (but not *Pgm*) allele frequency between dead and surviving mussels exposed to 0.1-ppm, but not to 0.2- and 0.5-ppm; no information was presented on the survival of homozygotes versus heterozygotes. Hvilsom explains that at lower copper exposure, mussels are open and therefore under copper stress, while at higher concentrations the mussels are tightly closed, and when they eventually are forced to open they die immediately. However, there may in fact be an alternative explanation for Hvilsom's results; the mussels used in the experiment were a mixture of individuals originating in the Kattegat and Baltic Sea. In other words, the experimental animals were not genetically homogeneous, but constituted a mixture of *M. edulis, M. trossulus* and their hybrids (see p.327). This serves to illustrate how important it is to have *prior knowledge* of the genetic identity of experimental animals, since it is likely that genetically differentiated populations might have different susceptibilities to pollutants. This applies not just to pollution studies but also to studies on growth, mortality, disease, parasitic infection, physiology and biochemistry.

The evidence to date, albeit sparse, corroborates the results of Nevo for other marine organisms. Further genetic analyses at specifically polluted sites in the sea may confirm the patterns found in the laboratory, and biochemical investigations (similar to those carried out for the *Lap* polymorphism) may uncover the molecular basis of the selective mechanisms involved.

Restricted pair matings

A different type of laboratory-based study has been carried out by Beaumont and colleagues over the past number of years. In a series of experiments using single families and a mass mating of *M. edulis* they have investigated the effects of salinity and temperature on genotype-specific mortalities at a number of loci (*Lap, Gpi, Pgm, Est-D, Odh*, phosphoglycerate kinase (*Pgk*, EC 2.7.2.3) and β-n-acetyl glucose amidase (*Hex*, EC 3.2.1.30)), at both the late spat and juvenile stages of development (Beaumont et al., 1988, 1989b, 1990). Significant deviations from expected genotype frequencies were observed at most loci in many of the single family cultures, but, with the exception of the *Lap* locus, these deviations were unrelated to the temperature or salinity at which the cultures had been reared. The authors concluded that many of the loci may be acting as markers for other loci on the same chromosome which are under selection. In single families a particular genotype at a marker locus is associated with a very restricted range of genetic backgrounds in the offspring, while in wild populations the offspring are the products of many chromosomes and, therefore (because of recombination), particular genotypes are not as likely to act as markers for such loci (Beaumont et al., 1990). Support for this comes from their observation that such extensive and consistent genotype-specific mortalities were not observed in the mass matings.

Their data for the *Lap* locus corroborated the findings of Koehn, in that the Lap^{94} allele was selected against at low salinity; but selection only occurred during the postlarval to juvenile stage of development. In addition, in contrast to wild populations (see p.345), the laboratory cultures showed a significant excess of *Lap* heterozygotes.

The advantage in using single families is that inbreeding, population mixing (Wahlund effect) and null alleles (see below) can be diregarded as causative agents in deviations from Hardy–Weinberg equilibrium. However, "...the extrapolation of genetic data obtained in this way must be undertaken with caution." (Beaumont et al., 1989b), in view of the more restricted genetic background of laboratory crosses and the fact that the selective factors operating, either directly or indirectly on the analysed loci, are probably very different from those in wild populations.

Heterozygosity in Wild and Laboratory-Reared *Mytilus*

Heterozygote deficiency

A deficit of heterozygotes—when the observed number of heterozygous individuals at a polymorphic locus is significantly lower than Hardy–Weinberg expectations—is a common phenomenon in marine bivalves and has been well-documented since the early 1970s (see Zouros and Foltz, 1984a; Volkaert and Zouros, 1989; and Beaumont, 1991 for relevant references). More recently, studies have also demonstrated the existence of a relationship between heterozygosity at a small number of polymorphic loci and fitness traits, such as growth, fertility and viability in bivalve species. Many explanations for the two phenomena have been suggested (Zouros and Foltz, 1984a, b; Zouros, 1987; Zouros and Foltz, 1987; Zouros et al., 1988; Zouros and Mallet, 1989; Beaumont, 1991) and these will be briefly discussed in the following section.

Although heterozygote deficits are a general characteristic of natural populations of marine bivalves the degree of heterozygote deficiency varies with age, and is not randomly spread over loci. For example, the results of several studies have shown that deficits are more pronounced in smaller size classes i.e. in younger individuals, (Koehn et al., 1973; Tracey et al., 1975; Koehn et al., 1976; Diehl and Koehn, 1985) and that some loci, e.g. *Lap, Odh* or *Pgm* in *Mytilus* are most likely to show marked deficiencies whereas others, e.g. *Gpi* or *Ap* are not.

Zouros and Foltz (1984a) have put forward several hypotheses that might help to explain heterozygote deficits in bivalve populations. Some of these involve selection—selective models—while others are nonselective and include such factors as inbreeding, the Wahlund effect, null alleles and aneuploidy. Which of these could be responsible for the persistent deviations from Hardy–Weinberg equilibrium observed in mussel populations will now be briefly examined. For a more thorough treatment of the topic the reader is referred to the papers of Zouros and Foltz (1984a, b) and Singh and Green (1984).

Inbreeding increases the frequency of homozygotes and decreases the frequency of heterozygotes relative to the expectations from random mating. There are two factors which argue against inbreeding as the causative agent for heterozygote deficits in natural populations of mussels: the nature of the life cycle i.e. external fertilization and an extended larval dispersal phase; and the fact that heterozygote deficits are not observed at all loci.

The Wahlund effect is observed when a sample comprises a mixture of individuals from two or more populations which differ in allele frequencies at a locus. It is manifest as an overall deficiency of heterozygotes compared to the number predicted by the Hardy–Weinberg equilibrium. The extent of the deficit at a particular locus depends on the variance in allele frequencies at that locus across the adult populations. While the Wahlund effect may explain some cases of heterozygote deficiency i.e. where a sample is a mixture of two genetically differentiated forms with partial hybridization e.g. the *M. edulis/M. galloprovincialis* situation in South-west England, it cannot account for the marked heterozygote deficits observed in areas where allele frequencies at a locus e.g. *Lap*, are homogeneous over large distances.

Null allele is the term used for an allele which produces a nonfunctional protein. Heterozygotes for a null and active allele will not be distinguished from homozygotes for an active allele, and therefore the frequency of heterozygotes will be underestimated, thus creating an apparent heterozygote deficiency in a sample. Several factors militate against null alleles as a cause of heterozygote deficits in natural populations. Null homozygotes i.e. nonstaining individuals on gels have not been observed in several large surveys (see Zouros and Foltz, 1984b for references); but this absence could be explained if the null allele is lethal or nearly lethal in homozygotes but advantageous in heterozygotes. This however requires large mutation rates (but see Skibinski et al., 1983), or unrealistically high heterozygote advantage (Zouros et al., 1980; Zouros and Foltz, 1984a).

Aneuploidy results from nondisjunction of chromosome pairs at meiosis and therefore the resulting zygotes will be 2n+1 or 2n-1. High levels of aneuploidy have been reported in embryos of *M. edulis* (Dixon, 1982; see also p.361) and in another marine bivalve, *Mulinia lateralis* (Gaffney et al., 1990). Because monosomic (2n-1) individuals will be scored as homozygotes there will be an apparent heterozygote deficiency in a population (Zouros and Foltz, 1984a). Also, accepting that aneuploid individuals are less viable than normal individuals, this might also explain why heterozygote deficits tend to decrease with increasing age in many bivalve populations. However, aneuploidy would still not explain the large heterozygote deficits which persist in many adult bivalve populations. To date, there have been no reports on the incidence of aneuploidy in adult individuals.

Selection has been the factor most widely cited as the cause of heterozygote deficits in natural populations of marine bivalves. Zouros and Foltz (1984a) have examined various models which might account for the phenomenon in wild populations. One such model examines presettlement and postsettlement viabilities to determine if heterozygote inferiority (underdominance) at the presettlement stage is compensated by heterozygote superiority (overdominance) in adults. This model is consistent with reports of heterozygote deficits in wild mussels collected soon after settlement i.e. within 2–4 weeks (Gosling and Wilkins, 1985; Gosling and McGrath, 1990; Fairbrother and Beaumont, unpublished data) and also in laboratory cultures of juvenile mussels (Beaumont et al., 1983; Hvilsom and Theisen, 1984; Mallet et al., 1985; Beaumont et al., 1988, 1989b, 1990). (As mentioned already, the use of single families or controlled mass mating in laboratory cultures allows stochastic effects such as inbreeding, population mixing, null alleles or genotype–dependent spawning time (see later) to be discounted as possible causes of heterozygote deficiencies). The model is also consistent with the observation that heterozygote deficits tend to diminish with increasing age; and several studies (see later) have demonstrated a heterozygous advantage in postlarval populations of bivalves.

Zouros and Foltz (1984b) considered another model where viability selection is confined to the larval stage, and is compensated by overdominance for fertility in the adult stage. However, both selection models require large selection pressures, a reversal in the direction of selection between the larval and adult stage and, for large heterozygote deficiencies, underdominance for larval viabilities (Zouros and Foltz, 1984a, b). Because of the problems associated with purely selective models, Zouros and Foltz (1984a) have proposed a mixed model which assumes that heterozygotes in the wild have a fitness advantage as adults, primarily because of their faster growth (see later), and that heterozygotes deficiencies are generated in the population as a result of nonrandom fertilization—a nonselective mechanism. Zouros and Foltz (1984a) have suggested that one possible source of nonrandom fertilization could be variation in the time of spawning of heterozygotes and homozygotes, due for example to differences in body size. But, in a later paper Zouros et al. (1988) found no evidence that stage of gonad maturation in M. edulis was related to degree of heterozygosity. To date, Hilbish and Zimmerman (1988) provide the only direct evidence of genotype–dependent spawning time in bivalves. These authors observed that in a population of M. edulis from Long Island Sound the spawning of Lap^{94} genotypes was delayed by as much as six weeks compared to genotypes lacking this allele; they add, however, that this is unlikely to be of general importance in generating Lap heterozygote deficits outside of the Long Island Sound situation.

Thus, for laboratory cultures selective mortality clearly provides the most likely explanation for heterozygote deficits. In the wild, however, observed deficits are probably due to a combination of selective and stochastic processes. There is as yet little information on what these selective factors might be. One thing is certain, they are unlikely to be the same for wild and laboratory-reared animals.

Heterozygosity and fitness traits

Positive correlations between multiple locus heterozygosity and fitness-related characters have been observed in a variety of marine molluscs. Growth rate has been found to be positively associated with heterozygosity in a variety of bivalves including mussels (Singh and Zouros, 1978; Zouros et al., 1980, 1983, 1988; Fujio, 1982; Garton et al., 1984; Koehn and Gaffney, 1984; Diehl and Koehn, 1985; Diehl et al., 1986; Koehn et al., 1988; Alvarez et al., 1989; Gaffney, 1990; Gaffney et al., 1990). Positive correlations have also been observed between multiple locus heterozygosity and viability, fecundity, locomotory activity, fluctuating asymmetry and metabolic parameters such as, decreased O_2 consumption, ability to withstand starvation, and more efficient protein synthesis (Koehn and Shumway, 1982; Leary et al., 1983; Garton, 1984; Garton et al., 1984; Rodhouse and Gaffney, 1984; Diehl et al., 1985; Hawkins et al., 1986, 1989; Rodhouse et al., 1986; Volkaert and Zouros, 1989; Gaffney, 1990). In addition, several studies have demonstrated that heterozygous individuals exhibit lower levels of morphological variance than do more homozygous individuals (Zouros et al., 1980; Koehn and Gaffney, 1984; Mitton and Koehn, 1985).

The positive correlation between multiple-locus heterozygosity and certain fitness traits is believed to be due to a greater average efficiency of basal or standard metabolism in more heterozygous individuals as reflected by reduced respiratory rates (Diehl et al., 1985 and references therein). Hawkins et al. (1986, 1989) have shown that in *M. edulis* energy saved by heterozygotes resulted in increased ingestion and absorption efficiency, which in turn resulted in an increase in the net energy available for growth and reproduction (see also Chapter 5). However, heterozygosity explains only about 2–4% of the variation in size among individuals (Zouros et al., 1980; Koehn and Gaffney, 1984; Alvarez et al., 1989).

Several studies have failed to show a positive growth/heterozygosity correlation, but in the majority of cases such studies involved cohorts derived from restricted matings (see Beaumont 1991 and relevant references). Limited parentage results in a decrease in heterozygosity, not just at the marker loci themselves, but also in genetic background due to increased linkage disequilibrium. Some studies which have failed to show the relationship have, however, been on outbred populations where limited parentage was not in question (Foltz and Zouros, 1984; Beaumont et al., 1985; Gosling, 1989; Skibinski and Roderick, 1989; Volkaert and Zouros, 1989). Factors such as age

(Beaumont et al., 1985), environmental stress (Gentili and Beaumont, 1988), number and type of analysed loci and sample size (Zouros, 1987) have been cited as being responsible for the lack of a correlation. However, even in studies where the same organism and same set of loci were used, conflicting results have been obtained. For example, Koehn and Gaffney (1984) reported a significant positive correlation between heterozygosity and growth rate in a cohort of juvenile *M. edulis* from Long Island Sound. In a similar study on *M. edulis* from the west coast of Ireland, where the experimental design closely approximated that of Koehn and Gaffney (1984), and where the same loci (with one exception) were analysed, Gosling (1989) observed no correlation between multiple heterozygosity and size (Fig. 7.6), although there was a significant effect of *Odh* heterozygosity on size.

The genetic basis for heterozygosity/fitness traits correlations

Several hypotheses have been put forward to explain the heterozygosity/fitness traits correlations. The two main ones will be briefly discussed in the following section. For a more detailed treatment of the topic interested readers are referred to Zouros (1987), Koehn et al. (1988), Zouros et al. (1988), Zouros and Mallet (1989), Gaffney et al. (1990), Koehn (1990), Zouros (1990) and Beaumont (1991).

The first, the direct involvement hypothesis, proposes that the scored loci themselves are directly responsible for the correlation. This requires either true overdominance at the single-locus level, where heterozygotes are superior to their associated homozygotes, or multiple-locus dominance, where heterozygotes at all or many of the scored loci are *on average* superior to homozygotes. The second hypothesis, termed the inbreeding or associative overdominance hypothesis, is based on the assumption that electrophoretic homozygosity is an index of homozygosity for recessive alleles having deleterious effects on the fitness-related character in question. Therefore, individuals scored as multiple homozygotes will have a lower fitness compared to multiple-locus heterozygotes because of homozygosity for deleterious genes at loci in linkage disequilibrium (Zouros and Mallet, 1989). This effect will be particularly pronounced in highly homozygous inbred populations, but may also be present when multiple-locus homozygotes are sampled in outbred populations. The important difference between the two hypotheses, as Beaumont (1991) succinctly points out, "revolves around the question of whether the heterozygosity/growth correlation is caused by some intrinsic advantage of being heterozygous (principally at the scored loci but possibly also at linked loci: 'overdominance') or by some disadvantages of being homozygous (principally at linked loci: 'associative overdominance')."

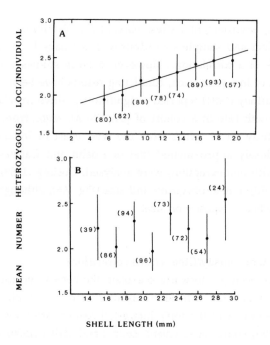

Fig. 7.6. Relationship between shell length (mean ± 95% confidence intervals) and mean individual heterozygosity within size categories. Data in (A) from Koehn and Gaffney (1984) (r = 0.952, df = 6, P<0.001) and (B), from Gosling, 1989 (r = 0.431, df = 6, ns). Sample sizes in parentheses.

Support for the direct involvement hypothesis comes from a study of the growth/heterozygosity correlation in the clam *Mulinia lateralis* (Koehn et al., 1988). The correlation was independantly determined for each of 15 loci, and only those loci involved in protein catabolism and glycolysis had large and significant effects on growth rate. As mentioned earlier, Hawkins et al. (1986, 1989) have shown that in *M. edulis* more heterozygous animals have decreased energy expenditure through greater efficiencies of protein synthesis than less heterozygous individuals, and that the energy thus saved is used for increased ingestion, thereby enhancing growth. The results of the two studies are clearly complementary and led Koehn et al. (1988) to conclude that protein catabolic (and glycolytic?) enzymes are directly implicated as *causative* in their effects upon energetic status and thereby, upon growth rate. However, other studies, in which significant positive correlations have also been reported between heterozygosity and growth, have not reported locus-specific effects (Singh and Zouros, 1978; Garton, 1984; Garton et al., 1984; Koehn and Gaffney, 1984; Hawkins et al., 1986; Zouros, 1987; Zouros et al., 1988). If the findings of Koehn et al. (1988) are corroborated by subsequent studies on other bivalve species, then this would clearly provide strong evidence for the direct involvement hypothesis.

Table 7.4. Heterozygosity and growth rate in natural populations of marine bivalves. N, sample size; k, number of loci assayed; D, index of heterozygote excess (+) or deficiency (–) in the population; r, correlation between degree of heterozygosity and growth rate (+ = positive correlation; – = negative correlation; 0 = no correlation). (Modified from Zouros, 1987).

Species	N	k	Age	D	r	Reference
Crassostrea virginica	1500	7	12mo	–	+	Zouros et al., 1980
Macoma balthica	150	6	adult	–	+	Green et al., 1983
Mytilus edulis	640	5	2mo	–	+	Koehn and Gaffney, 1984
Mytilus edulis	480	5	4mo	0/–	–	Diehl and Koehn, 1985
Mytilus edulis	460	5	9mo	0	0	Diehl and Koehn, 1985
Mytilus edulis	550	5	5mo	–	0	Gosling, 1989
Mytilus edulis	400	7	adult	–	+	Zouros et al., 1988
Mulinia lateralis	110	6	adult	0	+	Garton et al., 1984
Mulinia lateralis	1900	15	6mo	–	+	Gaffney et al., 1990
Placopecten magellanicus	640	6	adult	0/–	0	Foltz and Zouros, 1984
Placopecten magellanicus	310	6	5mo	0/–	0	Volkaert and Zouros, 1989
Pecten maximus	110	5	adult	0	0	Beaumont et al., 1985

Several lines of evidence have been put forward by Zouros (1987) and Zouros and Mallet (1989) as supporting the associative overdominance hypothesis. In natural populations the heterozygosity/growth correlation is normally associated with an excess of homozygosity (see Gaffney et al., 1990 for detailed discussion and Table 7.4 for relevant references), the strength of which diminishes with increasing age of the cohort of individuals examined (but see Zouros et al., 1988). In addition, heterozygote deficits tend to be larger at those loci where locus-specific effects of heterozygosity on growth have been reported (Gaffney, 1990; Gaffney et al., 1990).

Zouros and colleagues maintain that the co-occurrence of the growth/heterozygosity correlation with heterozygote deficiency is expected from models of associative overdominance. Any process causing increased homozygosity in a population will also increase the degree of disequilibria between deleterious genes and marker loci, and hence will make the noncausal correlation between electrophoretic homozygosity and traits affected by homozygosity for deleterious genes easier to detect (Zouros et al., 1988). A striking feature of the heterozygosity/growth correlation is its absence in cohorts derived from restricted matings. This demonstrates the dependence of the heterozygosity/growth correlation on the breeding structure of the population (Gaffney and Scott, 1984). Zouros and Mallet (1989) see the absence of a correlation in the progeny of pair matings as supporting the associative (inbreeding) overdominance hypothesis. In restricted pair matings deleterious recessives linked to scored loci will be rare because of the small number of parental chromosomes sets involved in the production of the cohort. Because of this there will be less difference in fitness-related traits such as growth rate between multiple-locus homozygous and heterozygous individuals, and thus little reason to expect a heterozygosity/growth correlation. Beaumont (1991) has suggested that more than 40 contributing parents would be necessary before such a correlation is likely to be detected in laboratory cultures. In that same paper Beaumont points out that, in contrast to natural populations, heterozygote deficits and the heterozygosity/growth correlation do not co-occur within data sets from single pair or restricted matings, indicating that there is no single cause for the two phenomena. In an earlier study on *Mulinia lateralis* Gaffney et al. (1990) had reached the same conclusion.

Another observation which lends support to the associative overdominance hypothesis is that whenever a positive association between heterozygosity and growth rate is observed there is no consistency in the relative effects of particular loci across populations. Using the same species, *M. edulis*, and the same five loci employed by Koehn and Gaffney (1984), Zouros et al. (1988) found that when the loci from these two studies were ranked according to their effect on growth there was no evidence for a matching of the two rankings. This result is anticipated under the associative overdominance hypothesis since linkage associations between marker loci and deleterious genes are expected to vary from population to population. Perhaps this also explains the results of Gosling (1989) mentioned above.

In summary, while there is a lot of evidence, albeit indirect, to support the associative overdominance hypothesis, there is also indirect evidence (Koehn et al., 1988) that some loci, especially those involved in protein catabolism (e.g. *Ap* and *Lap*), are causative agents of the heterozygosity/growth correlation. Zouros and Mallet (1989) aptly describe the status quo: "...there is no consensus about the genetic mechanism underlying the growth/heterozygosity correlation or any other correlation of heterozygosity with a phenotypic trait. This is not because we cannot choose among competing hypotheses that can explain the observations, but rather because we cannot arrive at a hypothesis that will be consistent with all (or even the bulk) of the data."

CYTOGENETICS

Chromosome Morphology and Inter- and Intraspecific Variation

The study of chromosome numbers has been used as a valuable complement to biochemical methods of identifying species, hybrids and more rarely populations, and has been of particular importance in the application of polyploidy (Thorgaard and Allen, 1987; see below). Both chromosome numbers and chromosome morphology in *Mytilus* have been described by Ahmed and Sparks (1970), Ieyama and Inaba (1974), Thiriot-Quiévreux and Ayraud (1982) and Thiriot-Quiévreux (1984a); and karyotypes, including measurements, by Ieyama (1983), Moynihan and Mahon (1983), Ieyama (1984), Thiriot-Quiévreux (1984b), Dixon and Flavell (1986), Pasantes et al. (1990). There is general agreement that all species of *Mytilus* examined to date exhibit a diploid (2n) complement of 28 chromosomes, and there is no evidence of sex chromosomes in *Mytilus*, or indeed in any other bivalve species to date (Dixon and McFadzen, 1987; C. Thiriot-Quiévreux, personal communication, 1991). There has been a recent suggestion, however, that mtDNA may play a role in sex determination in *Mytilus* (Fisher and Skibinski, 1991).

Chromosomes are classified according to the position of the centromere—the nonstaining, constricted region of the chromosome, which attaches the chromosome onto the spindle at cell division. The position of the centromere, which is fixed for particular chromosome pairs, can be terminal (telocentric) or virtually terminal (acrocentric), median (metacentric) or somewhere in between (subtelocentric, submetacentric); the number of different types of chromosomes in an organism constitutes its karyotype. In mussels there is a considerable amount of inconsistency relating to the morphological composition of the karyotype. Confusion is largely due to the different methods of tissue preparation, duration of colchicine treatment, slide preparation and measurement techniques employed in the separate studies (see Moynihan and Mahon, 1983); although it is also possible that mussel chromosomes may display a greater degree of plasticity of form than is usual for mammalian chromosomes.

Ahmed and Sparks (1970), examining *M. edulis* and *M. californianus* from the west coast of North America, observed three pairs of acrocentric and a variable number of metacentric and submetacentric chromosome pairs. They also observed structural polymorphism within populations of the two species, which they attributed to the presence of pericentric inversions or centromere shifts. (It must be noted that *M. edulis* is absent from the west coast of North America and that the authors were probably sampling either *M. galloprovincialis* or *M. trossulus*, or both). Moynihan and Mahon (1983) observed five to six metacentric pairs, six to eight submetacentric pairs and zero to three subtelocentric pairs in the karyotypes of 16 cells in a sample of three individuals of *M. edulis* from Galway, Ireland. They attributed the variation in the number of chromosomes in the different subgroups, both within and between individuals, to differential contraction of chromosomes, rather than to any structural changes which could affect centromere position. e.g. pericentric inversions or nonreciprocal translocations.   Samples of *M. edulis* (French Atlantic coasts), *M. galloprovincialis* (Mediterranean) and *M. desolationis* (Kerguelen Islands) were analysed by Thiriot-Quiévreux (1984b), using an improved technique on her earlier *Mytilus* studies.   She observed six metacentric, four submetacentric and four subtelocentric chromosome pairs in *M. edulis*. The karyotype of the three *Mytilus* taxa were very similar, with the exception of chromosome pairs number two and three: number two was metacentric and number three subtelocentric in *M. edulis*, while both chromosome pairs were metacentric in *M. desolationis* and subtelocentric in *M. galloprovincialis*.   Although Thiriot-Quiévreux and Ayraud (1982) concluded that the difference in the position of the centromere between *M. edulis* and *M. galloprovincialis* was strong evidence for regarding these two forms as separate species, in a later paper Thiriot-Quiévreux (1984b) concluded that the differences observed were not sufficient evidence yet for a major taxonomic separation between the two.   In South-west England Dixon and Flavell (1986) have analysed samples of *M. edulis* from Whitsand Bay and Robin Hood's Bay (North-east England) and a sample of *M. galloprovincialis* from Venice (Italy). The karyotype for both consisted of six metacentric pairs and a variable number of submetacentric and subtelocentric pairs.   There was considerable intra-and inter-individual variation in the number of subtelocentric pairs recorded, but no evidence of the *consistent* major chromosomal differences reported by Thiriot-Quiévreux and colleagues. This is doubly surprising when one considers that mussels at both of the British sites sampled by Dixon and Flavell (1986) are in fact a mixture of *M. edulis* and *M. galloprovincialis* (Skibinski and Roderick,1991).   A recent report by Pasantes et al. (1990) also indicates that these authors have not found any difference in the morphology of chromosome pair number between *M. galloprovincialis* from Galicia in North-west Spain and Thiriot's or Dixon's *M. edulis* samples.   However, interpopulational comparisons of *M. galloprovincialis* from Galicia, Italy and France (Pasantes et al., 1990) did reveal large differences both in the number of chromosome groups and in the composition of each group.   Table 7.5. presents a summary of the above results, and highlights the lack of agreement by the different workers on the precise karyotype of *Mytilus*.

There are many unanswered questions. Are the differences observed between the various *Mytilus* species real differences or are they artifacts due to poor chromosomal preparations or differing measurement techniques, with consequent misidentification and mispairing of chromosomes? If the differences are found to be real and repeatable how typical are they of molluscan species-level differences? It is obvious that more comprehensive sampling of populations within each *Mytilus* taxon is

Table 7.5. Karyotypes in different populations of *Mytilus*. T = telocentric or acrocentric, S = subtelocentric, SM = submetacentric, M = metacentric, v = variable number.

† some chromosomes in this group are SM/ST.

Species	Location	Centromeric Position				Remarks	Author
		T	ST	SM	M		
M. edulis	North	3		v	v	Intrapopulation	Ahmed and Sparks,
M. californianus	America	3		v	v	polymorphism	1970
						for pericentric	
						inversions	
M. edulis	Ireland		0–3	6–8	5–6	Variation	Moynihan and
						ascribed to	Mahon, 1983
						differential	
						chromosome	
						contraction	
M. edulis	England	v	v		6	Considerable	Dixon and Flavell,
M. galloprovincialis	Italy	v	v		6	intra-and inter-	1986
						individual	
						variation	
M. edulis	France	4	4		6	Variation	Thiriot-Quiévreux,
M. galloprovincialis	France	3	4		7	interpreted as	1984b
M. desolationis	Kerguelen Is.	5	4		5	real taxonomic	
						differences	
M. galloprovincialis	Spain	3	5†		6		Pasantes et al., 1990

needed. In addition karyotypes of individuals collected at sites where *M. edulis, M. galloprovincialis* (or *M. trossulus*) and their natural hybrids are found together would provide useful information.

An improvement in the production of mitotic cells for well-spread metaphases could be achieved by: upgrading laboratory rearing techniques to enhance growth and division, assaying mitotic agents or growth accelerating hormones, and developing cell or tissue culture methods (Thiriot-Quiévreux, 1991). In addition, the development of classic banding techniques, and now new banding techniques such as the restriction endonuclease method, which has been used very successfully by Lloyd and Thorgaard (1988) on rainbow trout chromosomes, would appear worthwhile. Finally, the adoption of a standard technique, for the preparation and interpretation of chromosome spreads would greatly facilitate the comparison of results between investigators.

Heterochromatin

Heterochromatin refers to areas on the chromosome which are highly condensed and genetically inactive (either because they lack genes, or contain genes that are re-pressed). These areas can be identified by means of various chromosomal banding techniques.

A special type of heterochromatin, called constitutive heterochromatin, found on centomeric regions of chromosomes and the telomeres (ends of chromosomes), contains DNA sequences which are not organized into genes. The role of constitutive heterochromatin in sex determination has been reported for a variety of vertebrate and invertebrate species (Bull, 1983). In some species which lack heteromorphic sex chromosomes a dimorphism in constitutive heterochromatin has been observed (Bull, 1983), which makes it possible to sex individuals on the basis of the number of C-bands in their interphase nuclei. Dixon and McFadzen (1987) have reported significant dimorphism in the quantity of C-band heterochromatin between male and female *M. edulis*. Males on average had one extra small block of C-band heterochromatin in their nuclei; but although the difference was significant between the sexes there was also a significant difference in the number of C-bands between individuals of the *same* sex. Therefore, for mussels at least, there does not appear to be a chromosomal mechanism of sex determination. It was suggested by Dixon and McFadzen (1987) that the observed dimorphism could be a reflection of rate differences in e.g. rRNA synthesis, which could be under the influence of the reproductive cycle. But how can the differences between individuals of the same sex be explained? Perhaps the five animals of each sex were not all pure *M. edulis*. The site sampled by Dixon and McFadzen in the estuary of the River Lynher (southeastern Cornwall, U.K.) is very close (10–20km) to the mixed population of *M. edulis* and *M. galloprovincialis* at Whitsand Bay.

Chromosome Damage by Mutagens and Various Pollutants

Mussels because they are common, sessile, filter feeders and have been shown to be effective concentrators of trace elements and hydrocarbons (Chapters 8 and 9 and references therein) are now widely used as test animals for coastal monitoring of trace toxic substances.

Field and laboratory studies have shown that the chromosomes of *Mytilus* and other marine molluscs are sensitive to damage inflicted by certain types of chemical pollutants in the marine environment. There are two recognized classes of chromosomal abnormalities: numerical aberrations such as aneuploidy, and structural defects such as chromatid exchange and micronuclei induction.

Aneuploidy

Aneuploidy refers to the condition of a cell, tissue or organism in which one or more whole chromosomes of a normal set are absent or are present more than once. The majority of aneuploidies are lethal and usually exert their effect early in embryonic life. Apart from a single report by Dixon (1982), there is little information on the effects of chemical pollutants on the frequency of aneuploidy in mussels. In that paper Dixon demonstrated that mussel embryos from a site with high levels of aromatic hydrocarbons (King's Dock, Swansea, South Wales) showed a significant increased incidence of aneuploidy (26% aneuploid nuclei) compared with the value (8%) from a clean open coastal site (Whitsand Bay, South-west England). He suggested that this high level of aneuploidy may be partly due to the residual effects of accumulated toxic compounds in the lipid reserves of the embryo. At the clean site the value of 8% is very similar to the 5–10% value of abnormal chromosome complements observed by Ahmed and Sparks (1970) for *Mytilus* populations from unpolluted sites on the west coast of North America. There is a need for further studies of this type, in order to determine the causative agents responsible for increased aneuploidy. Dixon (1982) does point out, however, that the increased frequency may not be directly due to high levels of hydrocarbons or heavy metals. Confirmation will depend on the results of laboratory dose-effect experiments with the presumed causative agents.

Sister chromatid exchange

Sister chromatid exchanges (SCEs) are events which involve breaks in both chromatids at coincident locations with subsequent interchange and repair (Latt, 1974). Techniques have been developed which distinguish between sister chromatids. These involve exposing cells to bromodeoxyuridine (BrdU) for two rounds of replication so that chromosomes on the second cell cycle possess one chromatid unifilarly substituted with BrdU with its sister chromatid bifilarly substituted. Such chromatids stain differentially with Giemsa or other fluorescent dyes; the dye bound to chromosomes containing DNA with little BrdU substitution fluoresces with highest efficiency, while that bound to chromosome regions containing DNA with both chains substituted fluoresces most weakly. It is at this stage that SCEs can be scored to provide a sensitive assay for the mutagenicity of environmental agents. It should be emphasized that since BrdU is itself mutagenic some minimal number of SCEs are present under baseline conditions.

For some time now SCE analysis has been used for both in vitro and in vivo studies on a wide range of vertebrate and invertebrate species (see Kadim, 1990 for relevant references). In *Mytilus* the effect of mutagens such as mitomycin C (MMC), $HgCl_2$, nitrilotriacetic acid (NTA) and methylmethanesulphate (MMS) on the frequency of SCE has been investigated, using various life cycle stages (eggs, larvae, adults) in both natural and laboratory populations (Dixon and Clarke, 1982; Harrison and Jones, 1982; Brunetti et al., 1986). In general, for laboratory-based studies, there is a linear relationship between the frequency of SCEs and the concentration of the mutagen. For example, in *M. galloprovincialis* Brunetti et al. (1986) observed the mean number of SCEs per metaphase in newly fertilized eggs to be significantly ($P<0.001$) higher with respect to the control (only BrdU-treated) after treatment with $HgCl_2$ (Table 7.6). Doubling the concentration of BrdU did not increase the frequency of SCEs in the various treatments. Brunetti et al. (1986) also compared the frequency of SCEs in developing eggs from mussels collected at two sites, which were geographically close, but which were characterized by very different degrees of eutrophication and hydrocarbon pollution. The mean number of SCEs at the polluted site was more than double that at the clean site. Their results however show a lower frequency of SCEs when compared with the results from the laboratory-based studies of Dixon and Clarke (1982) on adult gill cells, and those of Harrison and Jones (1982) on day-old *M. edulis* larvae. A more recent study by Jones and Harrison (1987) has shown that unexposed larvae from different laboratory crosses showed striking differences in the frequency of SCEs (0.02–0.41/chromosome), and they suggest that exposure of adult mussels to genotoxic agents, prior to the induction of spawning, could be responsible for the occurrence of elevated SCE frequencies i.e. SCE-inducing agents could have accumulated in the eggs of females and been released as a result of metabolism during the larval period spanned by their experiments. However, Dixon (personal communication, 1991) disagrees, and believes that their observations simply reflect differences in egg quality at the time of collection. Dixon has found that in SCE analysis egg quality is a very important factor, and that any hydrolysis due to partial resorption of effete gametes results in highly variable baseline SCE frequencies.

Sister chromatid exchange has also been shown to have potential as an extremely sensitive indicator of stimulation of the microsomal detoxification system in *M.*

Table 7.6. Induction of SCE by NTA and HgCl₂ in developing eggs of *M. galloprovincialis* after exposure to BdrU in the laboratory. SCE frequency is also presented for two different sites on the coast of Tuscany (Italy). (From Brunetti et al., 1986).

Treatment	Mean number SCE/metaphase	Variance
BdrU (10^{-3}M)	2.30	0.75
NTA (5mg/L)	2.40	0.88
HgCl₂ (0.03mg/L)	4.65	1.61
HgCl₂(0.03mg/L)+NTA (ratio 1:1)	4.05	0.47
unpolluted site	2.15	1.11
polluted site	4.74	3.12

edulis. Dixon et al. (1985) have observed that cyclophosphamide (CPA), a water soluble promutagen significantly increased SCE frequency in adult mussel gill.

When adults and larvae were pretreated with phenobarbital (PB)—an inducer of the microsomal detoxication system in mammals—there was a two-fold greater increase in the frequency of SCEs. PB appears to increase the rate at which CPA is transformed to metabolites capable of SCE induction—a reminder of the complexity of organismal responses to the mixtures of chemicals found in natural environments (see also p.446–448 Chapter 9).

Although only a small number of studies on SCE frequency in mussels have been published to date, it is a difficult task to cross compare results. This is because in laboratory studies different workers have employed different concentrations and different incubation times of BrdU, different test mutagens, different exposure times to individual mutagenic agents, and different *Mytilus* taxa. Since there appears to be a high variability in the frequency of SCEs between different batches of larvae within a species—suggesting that there may also be a genetic component to the incidence of SCE—it is even more likely that the response of individual species to mutagenic agents will be different. In spite of these limitations and others cited by Kadim (1990), SCE remains one of the most sensitive quantitative methods for assaying chromosomal mutagenicity of environmental agents.

Micronuclei

Micronuclei (MN) are chromosome(s) fragments which have become dislodged from the mitotic spindle and remain behind in the cytoplasm as blobs of DNA at the next interphase stage of cell division. The micronucleus test is another reliable assay for detecting mutagenic damage. It has the advantage over SCE in that it is applicable under field conditions (the SCE method requires that cells are exposed to BrdU for two replication cycles before a response can be detected), and does not suffer from the inability of SCE to detect certain chemicals e.g. benzene, or ionizing radiation.

Significant increases in the number of micronuclei have been detected in the gill cells of adult *M. galloprovincialis* exposed to different concentrations of heavy metals (Gola et al., 1986) and MMC (Majone et al., 1987). A recent paper by Scarpato et al. (1990) reported a significant linear increase in the frequency of micronucleated cells in gill tissue of adult *M. galloprovincialis* when treated in the laboratory with different concentrations of vincristine and benzo[a]pyrene. Mussels, collected from a clean site (control) and transferred to two polluted sites on the coast of Tuscany (Italy), were also analysed on a weekly basis to compare their MN frequencies to the basal level of the control sample (assessed at the start of the experiment). Thus, mutagenic activity in several environments could be compared by using animals from the same reference site. The frequency of MN rose significantly over the first few weeks of exposure, but declined somewhat over the next two months, suggesting to the authors the possibility that mussels subject to long-term exposure may reduce their MN frequency to a lower level than that observed during the first weeks of exposure. This indeed is a feature common to tests such as MN and SCE and others, that are dependent upon cell division to gain expression. One effect of exposure to toxicants is usually to modify cell turnover rates and at high concentrations toxicants may completely inhibit cell turnover, reducing response and leading to an underestimate of the effect of the toxicant (D. Dixon, personal communication, 1991).

Ploidy Manipulation

In recent years techniques have been applied to fish and shellfish species which result in the production of triploid (3n), tetraploid (4n) and diploid gynogenomic individuals (see Thorgaard, 1983 and Beaumont and Fairbrother, 1991 for reviews). In shellfish species triploids are produced by applying a thermal, pressure or chemical shock during the maturation divisions of the egg. When such shocks are applied during the late stages of either meiosis I or II they block the formation of the first or second polar bodies by allowing karyokinesis (chromosome division), but not cytokinesis (cytoplasmic division), to occur. This produces an egg with a diploid female pronucleus which, when united with a normal haploid sperm pronucleus, gives a triploid zygote.

Triploids are essentially sterile due to the inability of homologous chromosomes to synapse in the formation of gametes. From a commercial point of view sterility is advantageous in that metabolic energy normally utilized in gonadal development may instead be available for increased somatic growth. There is now increasing evidence, from studies on several commercially important oyster and scallop species, that triploid individuals do indeed grow faster than diploid individuals (see Beaumont and Fairbrother, 1991 for references).

To date there has been little interest in the production of triploid mussels, since mussels grow well in the wild, and do not command a high price compared to other shellfish species. However, as Beaumont and Kelly (1989) point out, triploid mussels could serve as a useful research tool to address current genetic phenomena such as heterozygosity/growth correlations and heterozygote deficits in marine bivalves. In the induction of triploidy the genetic outcome of targeting meiosis I or meiosis II is different (Stanley et al., 1984). When targeting meiosis I homologous chromosomes, which normally separate from each other at this stage of cell division, are prevented from doing so, and all of the heterozygosity of the female parent is retained in the egg. The addition of the male haploid chromosome set further increases heterozygosity beyond that of meiosis II induced triploids, or diploids.

Triploidy has been induced in *M. edulis* by both chemical and thermal shock. High percentages of triploids (70–98%) have been produced by Yamamoto and Sugawara (1988) using temperatures between 28°C and 35°C, with highest yields at 32°C. A more recent paper by Yamamoto et al. (1990) reported increased yields of triploid *M. edulis* using a combination of heat shock (29°C) and caffeine (15mM). Beaumont and Kelly (1989) targeting meiosis I and meiosis II using cytochalasin B— the most commonly used induction method—produced between 26% and 67% triploidy, with best results at 0.5mg L^{-1} concentration. In contrast to the results of Yamamoto and Sugawara (1988), these authors observed low yield of triploids (25% and less) using heat shock (25°C). Beaumont and Kelly (1989) also reported a significant increase in mean shell length of 36-day-old larvae derived from eggs treated at meiosis I, compared to either meiosis II or control larvae; this they ascribed to the predicted higher heterozygosity of meiosis I triploids. Further studies along these lines are in progress (A. Beaumont, personal communication, 1991).

Theoretically, tetraploids can be produced by the suppression of first cleavage. A tetraploid female should produce only diploid eggs, which when fertilized by normal haploid sperm should produce 100% triploid offspring. Although techniques for the production of tetraploids have been successfully applied to fish (Thorgaard and Allen, 1987), these have so far proved unsuccessful in shellfish.

Ultraviolet light can be used to destroy the DNA material in the sperm without however inhibiting its ability to activate the egg. This, together with suppression of either meiosis I, meiosis II or first cleavage, can produce diploid gynogenomes. Such individuals contain only chromosomes derived from the egg and, depending on which cell division is targeted, will exhibit either reduced heterozygosity or complete homozygosity. Diploid gynogenesis can therefore be used to provide highly homozygous individuals for genetic study. In addition, crossing gynogenetic inbred lines should result in individuals displaying 'hybrid vigour'. Once again, while the technique has been successfully applied to fish (Thorgaard and Allen, 1987), there is no published information on its use in shellfish. No doubt it is only a matter of time before this situation is rectified.

GENETICS AND MARICULTURE

Quantitative Genetics and Selective Breeding

Polygenic characters

The application of genetic techniques, such as selective breeding, to mariculture is many years behind such applications in agriculture (Newkirk, 1980a; Wilkins, 1981). In fish species, notably salmonids, tilapia (*Oreochromis* sp.) and catfish (*Ictalurus* sp.), specific strains have been selected and their performance characteristics elucidated (see Wilkins and Gosling, 1983; Gall and Busack, 1986; Gjedrem, 1990). In marine species, such as bivalves, very little effort has been made in this direction, although we have had control of the life cycle of all the commercially important species for a number of years.

The traits of most interest, from a production viewpoint, are polygenic i.e. they are controlled by a large number of genes, and include such characteristics as growth rate, survival, meat yield and disease resistance. Genetic analysis of such traits involves working with means and variances, trying to partition the variance of the trait (V_p) into genetic (V_g) and environmental (V_e) components. The genetic component can itself be partitioned into additive and nonadditive components:

$$V_g = V_a + V_d + V_i$$

Additive effects (V_a) result from the cumulative contribution of alleles at all the loci governing a quantitative trait, and as such are important because they contribute to the breeding value of individuals, and are passed on to progeny in a predictable manner. Nonadditive genetic effects are due to dominance and epistasis. Dominance effects (V_d) result from interactions among alleles at the same locus, while epistatic (V_i) effects are due to interactions among loci. Neither of these are passed on to progeny, due to segregation of parental alleles at meiosis.

The ratio of the additive genetic variance to the total phenotypic variance for a trait is called the heritability, denoted by h^2.

$$h^2 = V_a / V_p$$

Values of h^2 may theoretically range between zero, where V_p is entirely due to environmental effects, and one, where all the variance is due to additive genetic effects. The additive genetic variance is estimated from the analysis of half-sib families, while nonadditive variance is included in h^2 estimates from the analysis of full-sib families along with the additive genetic variance. The difference between estimates from half-and full-sib families gives an indication of the amount of nonadditive variance (Newkirk, 1980a). The response to selection of a particular trait can be predicted from the heritability estimate and the phenotypic variance, and selection methods are chosen on the basis of these values (see Falconer, 1981 for details).

In contrast to other shellfish species there have been few studies on h^2 estimates in *Mytilus*. This is because the cultivation of this species is entirely dependent on the natural environment; therefore, h^2 values would be of little practical importance. However, the well-studied biology of *Mytilus*, combined with its high fecundity (~25 million eggs/season) makes it particularly suitable as a model for other shellfish species. The earliest quantitative study was carried out by Innes and Haley (1977a) on *M. edulis* larvae. Families of larvae were raised at different salinities, ranging from 11°/oo to 30°/oo . There was substantial genetic variation in larval length, measured 16 days after fertilization, as well as significant genetic interaction with salinity. Although no h^2 values were given in that paper, Newkirk (1980a) estimated a half-and full-sib h^2 of about 0.16 and 0.29 respectively. Newkirk (unpublished results) in further experiments on the same population of *M. edulis* has observed a h^2 value from half-and full-sib analysis of 0.12 and 0.62 respectively. These, and additional results of Newkirk et al. (1981), indicate that there is some additive genetic variance for larval growth rate and considerable nonadditive variance in this population. In a more recent paper by Mallet et al. (1986) h^2 estimates for growth in larval, juvenile and mature *M. edulis* individuals from a population not previously subjected to artificial selection, were moderate to large (0.11–0.92). Strømgren and Nielsen (1989) also reported high h^2 values for shell growth in larval and juvenile *M. edulis*. However, some families with slow larval growth developed into very fast-growing juveniles and *vice versa*. This suggests that, although heritability is high, the physiological mechanisms associated with growth are affected differently, depending on the developmental stage of the mussel. The results from both these studies indicate that rapid phenotypic changes in the shell length of *M. edulis* could be achieved in a few generations of selection.

Single-gene characters

The majority of published papers discussing genetics in aquaculture deal with genetic variation in single-gene characters (an example of nonadditive genetic variation), such as blood group and protein polymorphisms. However, the number of single-gene characters having an easily detectable phenotypic expression, are few. This may of course reflect the fact that few have been looked for. Innes and Haley (1977b) proposed a single locus 2 allele mechanism for the black and brown-striped colour morphs observed in *M. edulis*, while Newkirk (1980b), using a much larger number of families, confirmed these results. However, because of departures from the clear-cut ratios expected from a single locus 2 allele model, Newkirk suggested that there may also be one or more additional loci influencing shell colour and striping.

Mitton (1977) has demonstrated a latitudinal cline in the frequency of colour morphs in *M. edulis* samples collected along the east coast of North America. In Maine the frequency of striped individuals was 7%, which rose to 40% in Virginia. Mitton suggests that the variation is adaptive: in the north, the darker-shelled individuals in the intertidal region of the shore would have a reduced risk of freezing, since they absorb more radiant energy, while in more southerly latitudes the absorption of too much sun would cause overheating and mortality—thus, the lighter-striped individuals would be better adapted. Mitton also observed a similar cline in colour morph frequency along an intertidal transect at Stony Brook, New York. Here the percentage of striped individuals was 24.3% in lower shore mussels, and differed significantly from the value of 19.6% at the upper edge of the distribution. The pattern was reversed in samples of dead shells along the transect, suggesting that the frequency differences were generated by differential mortality.

However, Newkirk (1980b) presents results which are opposite to the macro- and microgeographic trends observed by Mitton (1977). At Ostrea Lake, a warm water area on the coast of Nova Scotia (Canada) the frequency of black shells was greater than 90% in all three year classes collected at this site, while at Spanish Ship Bay, a low temperature site nearby, the frequency fell to 50%. In addition, experimental transplants from Ostrea Lake to Spanish Ship Bay did not show a change in colour morph frequency. Perhaps one of the reasons for the observed discrepancy in their results is that Mitton's observations were on intertidal animals of unknown age, whereas Newkirk was working with mussels of known age and grown in suspended culture, so that incident radiant energy would have little influence on the temperature of the animals.

In that same study Newkirk (1980b) demonstrated that within each population and each year class, striped individuals were significantly smaller (10–20%) than black individuals, and this association held true even when mussels were transplanted from the warmer environment of Ostrea Lake to a colder site further north. Newkirk suggests that there may be a physiological link between the gene or genes determining shell colour and growth rate. In view of the commercial implications of these results e.g. choice of seed source for cultivation, it is indeed surprising that further work on the possible association between shell colour and growth has not been established for other mussel poulations.

Characterization of Stocks

Mussel cultivation is an expanding industry, particularly in temperate latitudes (see Chapter 10). A critical factor in the development of the industry is the availability of seed mussels. Due to uneven distribution in mussel spatfall some areas are better seed collecting sites than others. Thus growers may collect seed at one site and transfer it to another more productive site for ongrowing. Some of the questions facing mussel farmers today is whether the quality of seed varies between different collecting sites and if so, how can this quality be measured? Is it possible that there are sufficient genetic differences between stocks of mussels to cause variability in seed quality, as seen from the growers' point of view?

Genetic differences have indeed been shown to exist between different populations or stocks of mussels. Such differences, detected using gel electrophoresis, have been observed at a number of specific enzyme loci between different populations of mussels separated by large distances i.e. on a scale of hundreds of kilometers, short distances i.e. on a scale of several kilometers or even meters, and also between individual cohorts of mussels over a single settlement season (see earlier sections: Macrogeographic and Microgeographic differentiation). When genetic differences have been observed between two stocks can we assume that there may also be genetic variation between them for the traits of interest to the mussel grower, such as growth rate, survival and disease resistance? To date, it has not been possible to demonstrate any direct connection between an electrophoretic locus and such a trait. In other words, these small number of loci are simply a random sample of genes of unknown relation to the fitness of the individual, which provide some idea of the genetic variation within and between wild populations of a species.

The experience of agricultural breeders suggests that the traits of interest to the mussel grower are determined by a relatively large number of genes with considerable environmental influence (see below). In a series of experiments carried out on the Nova Scotia coast, workers have tried to estimate how much of the variation in growth rate and mortality between different populations of mussels is genetically determined, and how much is determined by environmental conditions at the growing site. In the first of these experiments Dickie et al. (1984) observed that reciprocal transfer of M. edulis seed from three different stocks among three rearing

Table 7.7. Percent of the mortality variance explained by various main effects and their interactions in populations of *Mytilus edulis* from Nova Scotia, Canada. (From Mallet et al., 1987a).

Source	Percent of variance
Stock	16.7
Site	0.0
Season	0.4
Stock x site	22.2
Stock x season	14.2
Site x season	4.1

Table 7.8. Percent of the total variance (%) for shell length, shell weight and tissue weight explained by site and stock in populations of *Mytilus edulis* from Nova Scotia, Canada. (From Mallet et al., 1987b).

	Shell length	Shell weight	Tissue weight
Source	%	%	%
Site	40.0	47.0	81.0
Stock	27.0	24.0	6.6

sites resulted in marked changes in growth, mortality and maximum biomass. For example, highest average stock growth was of the order of 1.5 times that of the lowest stock growth, and highest average stock mortality rate was over six times that of the lowest. Allozyme analysis had previously shown that there were significant differences in genetic constitution between the stocks (Gartner-Kepkay et al., 1983; Koehn et al., 1984 and p.335); moreover recent evidence now indicates that all of the transfer experiments have been carried out in an area of mixing between *M. edulis* and *M. trossulus* (Varvio et al., 1988; Koehn, 1991; McDonald et al., 1991). Dickie et al. (1984) found that site differences were major determinants of the growth effects, while stock differences were chiefly responsible for the mortality effects. The overall effect on yield was about equally attributable to site and stock influences.

Further more extensive experiments have involved the reciprocal transfer of 11 populations of juvenile mussels to nine sites on the coasts of Nova Scotia for varying periods of time. In the first of several papers Mallet et al. (1987a) have attempted to quantify the effects of stock, site and season and their interactions on mortality over a one year period (Table 7.7). Stock, stock by site and stock by season together accounted for 53% of the variance in mortality, whereas site and season together explained less than 1% of the variance. Mallet et al. (1987a) used the mean cumulative mortality of all 11 stocks at each site as an index for ranking the sites according to their level of environmental stress; high cumulative mortality being indicative of high stress. When mean cumulative mortality in each stock was regressed against mean cumulative mortality at each site, the two variables were negatively correlated, suggesting that stocks originating from unfavourable environments tended to have lower overall mortalities than stocks originating from favourable environments. Newkirk et al. (1980) and Dickie et al. (1984) had also come to a similar conclusion i.e. the best seed stock comes from the 'toughest' environment.

Mussels, as do bivalves in general, show reduced growth rate in winter, primarily due to low temperature and low food levels. However, mussel producers have reported that considerable growth *can* take place in some populations over the winter months, suggesting that stocks may differ in their ability to adapt to low temperatures. To evaluate the importance of genetics in the physiological adaptation of *M. edulis* to low temperature Mallet et al. (1987b) carried out experiments, employing much the same set-up as before. After a six-month period over the winter months of November to April, increases in shell length, shell weight and tissue weight were recorded. Both site and stock differences were important in explaining the variance in shell growth (both length and weight), but site alone accounted for most of the variance in tissue growth (Table 7.8), probably explained by differences in food levels, due to ice coverage, between the various sites.

Summer mortality is a significant problem in Canadian and American populations of *Mytilus* but does not appear to be disease-related. Mallet et al. (1990) have examined the relative importance of stock, age and environmental conditions in determining mortality levels during the summer months using reciprocal transfers of three stocks of juvenile and adult mussels among sites on Prince Edward Island in eastern Canada. Stock accounted for 32% of the variance in survival, while stock x age explained 43%, with juvenile mussels showing higher survival rates than adults. Viability is known to be highly heritable (Mallet et al., 1986), which suggests that the variation in mortality among stocks is indicative of genetic differences among populations. Site and its interactions accounted for less than 2% of the variance. The importance of stock supports the earlier findings of Mallet et al. (1987a), where stock and its interactions accounted for 53% of the variance in mortality over a period of a year (see above). Results from further trials by Mallet and Carver (1989) on growth and mortality in *M. edulis*—using the same experimental design as before (Mallet et al., 1987a)—substantiate their earlier findings: stock and its interactions explain most of the variance in mortality, while site and its interactions are more important in explaining the variance in growth.

It is clear from this series of papers that the source of mussel stocks i.e. genetic constitution, as well as ongrowing sites have a practical impact on production. However, a cautionary note should be added at this point: the genetic constitution of seed at a particular site is not necessarily constant from year to year, since recruitment to that locality may be from *different* parental gene pools each generation (Newkirk et al., 1980). Also, although genetic differences between stocks are important in explaining variation in growth, nongenetic factors e.g. environmental conditions to which the juvenile mussels have been exposed to prior to transplantation, may also play a part. One way of testing this would be to use individuals of different stocks which had a similar pre-implantation environment, such as hatchery-produced seed. In addition, the environmental conditions at a particular site may change from year to year so that the results obtained in these experiments, while they hold true for the particular time-span of the experiment, may not be repeatable in subsequent trials.

The Future Role of Genetics in Mussel Cultivation

Mussel cultivation involves the collection of natural spat or seed for relaying on the ground or in suspension in mid water. This method is entirely dependent on the natural environment and therefore responsive only to the simplest stock management concepts, such as the use of naturally occurring advantageous stocks. Therefore, the main contribution of genetics to mussel mariculture would seem to be the *identification* of advantageous stocks. How can this be achieved? One approach is that of Mallet and colleagues described above, where testing of stocks for growth and survival is carried out by reciprocal transplants. Their results have indeed shown that, at least in Maritime Canada, stocks do differ in growth and survival rates, and that genetic constitution makes a significant contribution to survival and, to a lesser extent, growth of a stock. These results are corroborated by heritability estimates for larval growth and survival in hatchery-produced mussels.

There is ample evidence from electrophoretic surveys that there are differences in the level of genetic variability between stocks or populations of mussels. Can we assume that genetic variability (heterozygosity) at a small number of gene loci, as measured by electrophoresis, is indicative of genetic variability at loci governing traits of interest to the aquaculturist? Results from several studies (see Zouros, 1987, for references) have shown that the degree of heterozygosity at enzyme loci is practically uncorrelated with variability at sets of loci affecting a particular phenotypic trait. In those cases where a positive association has been observed between multiple locus heterozygosity and a fitness trait e.g. growth in *M. edulis* (Koehn and Gaffney, 1984), heterozygosity explains only about 4% of the total variance in shell length, the rest being attributable to both genetic and environmental background effects on growth rate. The approach adopted by Mallet and co-workers is more likely to yield beneficial results to the industry, than one involving the electrophoretic screening of mussel populations.

Mussels are unique in that, in contrast to most commercially important shellfish species, controlled hatchery production has never been necessary; they are widely distributed, grow well, mature early, have a high fecundity and are relatively disease-free. Therefore, genetic manipulation techniques such as artificial selection, intraspecific hybridization of different races, production of inbred lines, triploidy induction, which have been applied to other shellfish species, would seem to have little relevance to the mussel growing industry. For the future, therefore, efforts would be best spent in examining mussel production under different environmental conditions to evaluate the relative contribution of stock, age, season and growing site on fitness traits such as growth, survival and fecundity. At the same time it should be kept in mind, however, that mussels, by virtue of their many advantages, serve as valuable research tools in genetic studies.

ACKNOWLEDGEMENTS

I wish to thank Andrew Beaumont, Drs. David Dixon, John McDonald, Raymond Seed and Catherine Thiriot–Quiévreux for helpful comments on this chapter; thanks also to Dr. John McDonald for kind permission to use figures 3b and 3c from McDonald et al., 1991.

REFERENCES

Ahmad, M. and Beardmore, J.A., 1976. Genetic evidence that the "Padstow Mussel" is *Mytilus galloprovincialis*. Mar. Biol., 35: 139-147.

Ahmad, M. and Hedrick, P.W., 1985. Electrophoretic variation in the common mussel *Mytilus edulis*: random association of alleles at different loci. Heredity, 55: 47-51.

Ahmad, M., Skibinski, D.O.F. and Beardmore, J.A.,1977. An estimate of the amount of genetic variation in the common mussel *Mytilus edulis*. Biochem. Genet., 15: 833-846.

Ahmed, M. and Sparks, A.K., 1970. Chromosome number, structure and autosomal polymorphism in the marine mussel *Mytilus edulis* and *Mytilus californianus*. Biol. Bull., 138 (1): 1-13.

Alvarez, G., Zapata, C., Amaro, R. and Guerra, A., 1989. Multilocus heterozygosity at protein loci and fitness in the European oyster, *Ostrea edulis* L. Heredity, 63: 359-372.

Avise, J.C. 1974. The ststematic value of electrophoretic data. Syst. Zool., 23: 465-481.

Avise, J.C. and Ball, R.M. Jr., 1990. Principles of genealogical concordance in species concepts and biological taxonomy. Oxf. Surv. Evol. Biol., 7: 45-67.

Barsotti, G. and Meluzzi, C., 1968. Osservazioni su *Mytilus edulis* L. e *Mytilus galloprovincialis* Lamarck. Conchiglie (Milan), 4: 50-58.

Barton, N.H. and Hewitt, G.M., 1985. Analysis of hybrid zones. Annu. Rev. Ecol. Syst., 16: 113-148.

Barton, N.H. and Hewitt, G.M., 1989. Adaptation, speciation and hybrid zones. Nature (Lond.), 341: 497-503.

Battaglia, B., Bisol, P.M. and Rodino, E., 1980a. Experimental studies on some genetic effects of marine pollution. Helgoländer Wiss. Meeresunters., 33: 587-595.

Battaglia, B., Bisol, P.M., Fossato, V.U. and Rodino, E., 1980b. Studies on the genetic effects of pollution in the sea. Rapp. P.-V. Reun. Cons. Int. Explor. Mer, 179: 267-274.

Beardmore, J.A., 1980. Genetical considerations in monitoring effects of pollution. Rapp. P.-V. Reun. Cons. Int. Explor. Mer, 179: 258-266.

Beaumont, A.R., 1991. Genetic studies of laboratory reared *Mytilus edulis*: heterozygote deficiencies, heterozygosity and growth. Biol. J. Linn. Soc., 44(3): 273-285.

Beaumont, A.R. and Kelly, K.S. 1989. Production and growth of triploid *Mytilus edulis* larvae. J. Exp. Mar. Biol. Ecol., 132: 69-84.

Beaumont, A.R. and Fairbrother, J.E. 1991. Ploidy manipulation in molluscan shellfish: a review. J. Shellfish Res., 10(1): 1-18.

Beaumont, A.R. and Zouros, E. 1991. Scallops: biology, ecology and aquaculture. In: S.E. Shumway (Editor), Developments in Aquaculture and Fisheries Science, Vol. 21. Elsevier Science Publishers, B.V., Amsterdam, pp. 585-623.

Beaumont, A.R., Beveridge, C.M. and Budd, M.D., 1983. Selection and heterozygosity within single families of the mussel *Mytilus edulis* (L.). Mar. Biol. Lett., 4: 151-151.

Beaumont, A.R., Gosling, E.M., Beveridge, C.M., Budd, M.D. and Burnell, G.M., 1985. Studies of heterozygosity and size in the scallop, *Pecten maximus* (L.). In: P.E. Gibbs (Editor), Proc. 19th Eur. Mar. Biol. Symp., Plymouth, England, 1984. Cambridge University Press, Cambridge, pp. 443-454.

Beaumont, A.R., Beveridge, C.M., Barnet, E.A., Budd, M.D. and Smyth-Chamosa, M., 1988. Genetic studies of laboratory reared *Mytilus edulis*. I. Genotype specific selection in relation to salinity. Heredity, 61: 389-400.

Beaumont, A.R., Seed, R. and Garcia-Martinez, P., 1989a. Electrophoretic and morphometric criteria for the identification of the mussels *Mytilus edulis* and *M. galloprovincialis*. In: J. Ryland and P.A.Tyler (Editors), Proc. 23rd Eur. Mar. Biol. Symp., Swansea, U.K., 1988. Olsen and Olsen, Fredensborg, Denmark, pp. 251-258.

Beaumont, A.R., Beveridge, C.M., Barnet, E.A. and Budd, M.D., 1989b. Genetic studies of laboratory reared *Mytilus edulis*. II. Selection at the leucine amino peptidase (*Lap*) locus. Heredity, 62: 169-176.

Beaumont, A.R., Beveridge, C.M., Barnet, E.A. and Budd, M.D., 1990. Genetic studies of laboratory reared *Mytilus edulis*. III. Scored loci act as markers for genotype-specific mortalities which are unrelated to temperature. Mar. Biol., 106: 227-233.

Blot, M., Soyer, J. and Thiriot-Quiévreux, C., 1987. Preliminary data on the genetic differentiation of *Mytilus desolationis* Lamy 1936 and *Aulacomya ater regia* Powell 1957 (Bivalvia, Mytilidae) in the Kerguelen Islands (Terres Australes et Antarctiques Francaises). Polar Biol., 7: 1-9.

Blot, M., Thiriot-Quiévreux, C. and Soyer, J., 1988. Genetic relationships among populations of *Mytilus desolationis* from Kerguelen, *M. edulis* from the North Atlantic and *M. galloprovincialis* from the Mediterranean. Mar. Ecol. Prog. Ser., 44: 239-247.

Blot, M., Thiriot-Quiévreux, C. and Soyer, J., 1989. Genetic differences and environments of mussel populations in Kerguelen Islands. Polar Biol., 10: 167-174.

Blot, M., Legendre, B. and Albert, P., 1990. Restriction fragment length polymorphism of mitochondrial DNA in subantarctic mussels. J. Exp. Mar. Biol. Ecol., 141: 79-86.

Boyer, J. F., 1974. Clinal and size-dependent variation at the *Lap* locus in *Mytilus edulis*. Biol. Bull., 147: 535-549.

Brock, V. 1985. Immuno-electrophoretic studies of genetic relations between populations of *Mytilus edulis* and *M. galloprovincialis* from the Mediterranean, Baltic, East and west Atlantic, and east Pacific. In: P.E. Gibbs (Editor), Proc. 19th Eur. Mar. Biol. Symp., Plymouth, England, 1984. Cambridge University Press, Cambridge, pp. 515-520.

Brock, V., 1987. Genetic relations between the bivalves *Cardium* (*Cerastoderma*) *edule*, *Cardium lamarki* and *Cardium glaucum*, studied by means of crossed immunoelectrophoresis. Mar. Biol., 93: 493-498.

Brown, W.M., 1983. Evolution of animal mitochondrial DNA. In: M. Nei and R. K. Koehn (Editors), Evolution of genes and proteins. Sinauer Associates, Sunderland, Massachusetts, pp. 62-88.

Brown, W.M., George, M. Jr. and Wilson, A.C., 1979. Rapid evolution of animal mitochondrial DNA. Proc. Natl. Acad. Sci. (U.S.A.), 76: 1967-1971.

Brunetti, R., Gola, I. and Majone, F., 1986. Sister-chromatid exchange in developing eggs of *Mytilus galloprovincialis* Lmk. (Bivalvia). Mutat. Res., 174: 207-211.

Bull, J.J., 1983. Evolution of sex determining mechanisms. Benjamin/Cummings Publishing Company, California, 316pp.

Bulnheim, H.-P. and Gosling, E.M., 1988. Population genetic structure of mussels from the Baltic Sea. Helgoländer. Wiss. Meeresunters., 42: 113-129.

Crespo, C.A., Garcia–Caballero, T., Beiras, A. and Espinosa, J., 1990. Evidence from sperm ultrastructure that the mussel of Galician estuaries is *Mytilus galloprovincialis* Lamarck. J. Molluscan Stud., 56: 127–128.

Davis, G.M. and Fuller, S.L.H., 1981. Genetic relationships among recent Unionacea (Bivalvia) of North America. Malacologia, 20: 217-253.

Deaton, L.E., Hilbish, T.J. and Koehn, R.K., 1984. Protein as a source of amino nitrogen during hyperosmotic volume regulation in the mussel *Mytilus edulis*. Physiol. Zool., 57(6): 609-619.

Delden, W. van, 1982. The alcohol dehydrogenase polymorphism in *Drosophila melanogaster*: Selection at an enzyme locus. Evol. Biol., 15: 187-222.

Dickie, L.M., Boudreau, P.R. and Freeman, K.R., 1984. Influence of stock and site on growth and mortality in the Blue Mussel (*Mytilus edulis*). Can. J. Fish. Aquat. Sci., 41: 134-140.

Diehl, W.J. and Koehn, R.K., 1985. Multiple-locus heterozygosity, mortality, and growth in a cohort of *Mytilus edulis*. Mar. Biol., 88: 265-271.

Diehl, W.J., Gaffney, P.M., McDonald, J.H. and Koehn, R.K., 1985. Relationship between weight-standardised oxygen consumption and multiple–locus heterozygosity in the mussel, *Mytilus edulis*. In: P.E. Gibbs (Editor), Proc. 19th Eur. Mar. Biol. Symp., Plymouth, England, 1984. Cambridge University Press, Cambridge, pp. 529-534.

Diehl, W.J., Gaffney, P.M. and Koehn, R.K., 1986. Physiological and genetic aspects of growth in the mussel *Mytilus edulis*. I. Oxygen consumption, growth and weight loss. Physiol. Zool., 59(2): 201-211.

Dixon, D.R., 1982. Aneuploidy in mussel embryos (*Mytilus edulis* L.) originating from a polluted dock. Mar. Biol. Lett., 3: 155-161.

Dixon, D.R. and Clarke, K.R., 1982. Sister chromatid exchange: a sensitive method for detecting damage caused by exposure to environmental mutagens in the chromosomes of adult *Mytilus edulis*. Mar. Biol. Lett., 3: 163-172.

Dixon, D.R. and Flavell, N., 1986. A comparative study of the chromosomes of *Mytilus edulis* and *Mytilus galloprovincialis*. J. Mar. Biol. Ass. U.K., 66: 219-228.

Dixon, D.R. and McFadzen, I.R.B., 1987. Heterochromatin in the interphase nuclei of the common mussel *Mytilus edulis* L. J. Exp. Mar. Biol. Ecol., 112: 1-9.

Dixon, D.R. Jones, I. M and Harrison, F.L., 1985. Cytogenetic evidence of inducible processes linked with metabolism of a xenobiotic chemical in adult and larval *Mytilus edulis*. Sci. Total Environ., 46: 1-8.

Doyle, R.W., 1974. Choosing between darkness and light: the ecological genetics of photic behaviour in the planktonic larvae of *Spirorbis borealis*. Mar. Biol., 25: 311-317.

Edwards, C.A. and Skibinski, D.O.F., 1987. Genetic variation of mitochondrial DNA in mussel (*Mytilus edulis* and *M. galloprovincialis*) populations from South West England and South Wales. Mar. Biol., 94: 547-556.

Endler, J.A. 1986. Natural Selection in the wild. Princeton University Press, Princeton, New Jersey, 336pp.

Falconer, D.S., 1981. Introduction to Quantitative Genetics. Longman, London, 340pp.

Ferguson, A., 1980. Biochemical Systematics and Evolution. Blackie, Glasgow, 194pp.

Fevolden, S.E. and Garner, S.P., 1986. Population genetics of *Mytilus edulis* (L.) from Oslofjorden, Norway, in oil-polluted and non oil-polluted water. Sarsia, 71: 247-257.

Fisher, C. and Skibinski, D.O.F., 1991. Sex-biassed mitochondrial DNA heteroplasmy in the marine mussel *Mytilus*. Proc. R. Soc. Lond., Ser. B, 242: 149-156.

Foltz, D.W. and Zouros, E., 1984. Enzyme heterozygosity in the scallop *Placopecten magellanicus* (Gmelin) in relation to age and size. Mar. Biol. Lett., 5: 255-263.

Fujio, Y., 1982. A correlation of heterozygosity with growth rate in the Pacific oyster, *Crassostrea gigas*. Tohoku J. Agric. Res., 33: 66-75.

Futuyama, D.J. 1986. Evolutionary Biology. Sinauer Associates, Sunderland, Massachusetts, 600pp.

Gaffney, P. M., 1990. Enzyme heterozygosity, growth rate and viability in *Mytilus edulis*: another look. Evolution, 44(1): 204-210.

Gaffney, P.M. and Scott, T.M., 1984. Genetic heterozygosity and production traits in natural and hatchery populations of bivalves. Aquaculture, 42: 289-302.

Gaffney, P.M., Scott, T.M., Koehn, R.K. and Diehl, W.J., 1990. Interrelationships of heterozygosity, growth rate and heterozygote deficiencies in the coot clam, *Mulinia lateralis*. Genetics, 124: 687-699.

Gall, G.A.E. and Busack, C.A., (Editors), 1986. Genetics in Aquaculture II. Proc. 2nd Int. Symp. Gen. Aquacul., Davis, California, 1985. Aquaculture, 57: 386pp.

Gardner, J.P.A. and Skibinski, D.O.F., 1988. Historical and size-dependent genetic variation in hybrid mussel populations. Heredity, 61: 93-105.

Gardner, J.P.A. and Skibinski, D.O.F., 1990. Genotype-dependant fecundity and temporal variation of spawning in hybrid mussel (*Mytilus*) populations. Mar. Biol., 105: 153–162.

Gardner, J.P.A. and Skibinski, D.O.F., 1991. Biological and physical factors influencing genotype-dependant mortality in hybrid mussel populations. Mar. Ecol. Prog. Ser., 71(3): 235-244.

Gartner-Kepkay, K.E., Dickie, L.M., Freeman, K.R. and Zouros, E., 1980. Genetic differences and environments of mussel populations in the Maritime Provinces. Can. J. Fish. Aquat. Sci., 37: 775-782.

Gartner-Kepkay, K.E., Zouros, E., Dickie, L.M. and Freeman, K.R., 1983. Genetic differentiation in the face of gene flow: a study of mussel populations from a single Nova Scotia embayment. Can. J. Fish. Aquat. Sci., 40: 443-451.

Garton, D.W., 1984. Relationship between multiple locus heterozygosity and physiological energetics of growth in the estuarine gastropod *Thais haemostoma*. Physiol. Zool., 57: 530–543.

Garton, D.W., Koehn, R.K. and Scott, T.M., 1984. Multiple-locus heterozygosity and the physiological energetics of growth in the coot clam, *Mulinia lateralis*, from a natural population. Genetics, 108: 445–455.

Gentili, M.R. and Beaumont, A.R., 1988. Environmental stress, heterozygosity, and growth rate in *Mytilus edulis* L. J. Exp. Mar. Biol. Ecol., 120: 145-153.

Gjedrem, T. (Editor), 1990. Genetics in Aquaculture III. Proc. 3rd. Int. Symp. Gen. Aquacul., Trondheim, Norway, 1988. Aquaculture, 85: 340pp.

Gola, I., Brunetti, R., Majone, F. and Levis, A.G., 1986. Applications of the micronucleus test to a marine organism treated with NTA and insoluble heavy metals. Atti Assoc. Genet. Ital., 32: 95-96.

Gosling, E.M., 1977. Genetic variation in marine mussels (*Mytilus*) in western Europe. Ph. D. Thesis, National University of Ireland, Galway.

Gosling, E.M., 1984. The systematic status of *Mytilus galloprovincialis* in western Europe: a review. Malacologia, 25(2): 551-568.

Gosling, E.M., 1989. Genetic heterozygosity and growth rate in a cohort of *Mytilus edulis* from the Irish coast. Mar. Biol., 100: 211-215.

Gosling, E.M. and Wilkins, N.P., 1977. Phosphoglucosisomerase allele frequency data in *Mytilus edulis* from Irish coastal sites: its ecological implications. In: B.F. Keegan, P. O Céidigh and P. S. Boaden (Editors), Biology of Benthic Organisms. Proc. 11th Eur. Mar. Biol. Symp., Galway, Ireland, 1976. Pergamon Press, London, pp. 297-309.

Gosling, E.M. and Wilkins, N.P., 1981. Ecological genetics of the mussels *Mytilus edulis* and *M. galloprovincialis* on Irish coasts. Mar. Ecol. Prog. Ser., 4: 221-227.

Gosling, E. M. and Wilkins, N.P., 1985. Genetics of settling cohorts of *Mytilus edulis* (L.): preliminary observations. Aquaculture, 44: 115-123.

Gosling, E. M. and Burnell, G.M. 1988. Evidence for selective mortality in *Chlamys varia* transplant experiments. J. Mar. Biol. Ass. U.K., 68: 251-258.

Gosling, E.M. and McGrath, D.M., 1990. Genetic variability in exposed–shore mussels, *Mytilus* sp. along an environmental gradient. Mar. Biol., 104: 413–418.

Gould, A.A., 1850. Shells from the United States Exploring Expedition. Proc. Boston Soc. Nat. Hist., 3: 343-348.

Grant, W.S. and Cherry M.I., 1985. *Mytilus galloprovincialis* Lmk. in southern Africa. J. Exp. Mar. Biol. Ecol., 90: 179-191.

Green, R.H., Singh, S.M., Hicks, B. and McCuaig, J., 1983. An arctic intertidal population of *Macoma balthica* (Mollusca, Pelcypoda): genotypic and phenotypic components of population structure. Can. J. Fish. Aquat. Sci., 40: 1360-1371.

Hall, J.G. 1985. Temperature-related kinetic differentiation of Glucosephosphate-isomerase alleloenzymes isolated from the Blue Mussel *M. edulis*. Biochem. Genet., 23: 705– 728.

Harrison, F.L. and Jones, I.M., 1982. An in vivo sister chromatid exchange assay in the larvae of the mussel *Mytilus edulis*: response to 3 mutagens. Mutat. Res., 105: 235-242.

Hawkins, A.J.S., Bayne, B.L. and Day, A.J., 1986. Protein turnover, physiological energetics and heterozygosity in the blue mussel, *Mytilus edulis*: the basis of variable age-specific growth. Proc. R. Soc. Lond., Ser. B, 229: 161-176.

Hawkins, A.J.S., Bayne, B.L., Day, A.J. Rusin, J. and Worrall, C.M., 1989a. Genotype-dependent interrelations between energy metabolism, protein metabolism and fitness. In: J. Ryland and P.A.Tyler (Editors), Proc. 23rd Eur. Mar. Biol. Symp., Swansea, U.K., 1988. Olsen and Olsen, Fredensborg, Denmark, pp. 283-292.

Hawkins, A.J.S., Rusin, J., Bayne, B.L. and Day, A.J., 1989b. The metabolic/physiological basis of genotype-dependent mortality during copper exposure in *Mytilus edulis*. Mar. Environ. Res., 28: 253-257.

Hedrick, P.W., 1985. Genetics of populations. Jones and Bartlett Inc., Boston, 629pp.

Hilbish, T.J., 1985. Demographic and temporal structure of an allele frequency cline in the mussel *Mytilus edulis*. Mar. Biol., 86: 163-171.

Hilbish, T. J., and Koehn, R.K., 1985a. Genetic variation in nitrogen metabolism in *Mytilus edulis*: contribution of the *Lap* locus. In: P. E. Gibbs (Editor), Proc. 19th Eur. Mar. Biol. Symp., Plymouth, England, 1984. Cambridge University Press, Cambridge, pp. 497-504.

Hilbish, T.J. and Koehn, R.K., 1985b. The physiological basis for selection at the *Lap* locus. Evolution, 39(6): 1302-1317.

Hilbish, T.J. and Koehn, R.K., 1985c. Dominance in physiological phenotypes and fitness at an enzyme locus. Science, 229: 52-54.

Hilbish, T.J. and Zimmermann, K.M.,1988. Genetic and nutritional control of the gametogenic cycle in *Mytilus edulis*. Mar. Biol., 98: 223-228.

Hodgson, A.N. and Bernard, R.T.F., 1986. Observations on the ultrastructure of the spermatozoon of two Mytilids from the south-west coast of England. J. Mar. Biol. Ass. U.K., 66: 385-390.

Hoeh, W. R., Blakley, K.H. and Brown, W.M., 1991. Heteroplasmy suggests limited biparental inheritance of *Mytilus* mitochondrial DNA. Science, 251: 1488-1490.

Hvilsom, M., 1983. Copper-induced differential mortality in the mussel *Mytilus edulis*. Mar. Biol., 76: 291-295.

Hvilsom, M. and Theisen, B.F., 1984. Inheritance of allozyme variations through crossing experiments with the blue mussel, *Mytilus edulis*. Heriditas, 101: 1-7.

Ieyama, H., 1983. Karyotype in *Mytilus edulis* (Bivalvia, Mytilidae). Mem. Fac. Educ. Ehime Univ., Series III, 3: 23–26.

Ieyama, H., 1984. Karyotype in eight species of the Mytilidae (Bivalvia: Pteriomorphia). Venus, 43(3): 240-254.

Ieyama, H. and Inaba, A., 1974. Chromosome numbers of ten species in four families of Pteriomorphia (Bivalvia). Venus, 33(3): 129-137.

Innes, D.J. and Haley, L.E., 1977a. Genetic aspects of larval growth under reduced salinity in *Mytilus edulis*. Biol. Bull., 153: 312-321.

Innes, D.J. and Haley, L.E., 1977b. Inheritance of a shell-color polymorphism in the mussel. J. Hered., 68: 203-204.

Johannesson, K., Kautsky, N. and Tedengren, M., 1990. Genotypic and phenotypic differences between Baltic and North Sea populations of *Mytilus edulis* evaluated through reciprocal transplantations. II. Genetic variation. Mar. Ecol. Prog. Ser., 59: 211-219.

Jones, I.M. and Harrison, F.L., 1987. Variability in the frequency of sister-chromatid exchange in larvae of *M. edulis*: implications for field monitoring. J. Exp. Mar. Biol. Ecol., 113: 283-288.

Kadim, M.A., 1990. Methodologies for monitoring the genetic effects of mutagens and carcinogens accumulated in the body tissue of marine mussels. Rev. Aquat. Sci., 2(1): 83-107.

Kautsky, N., Johannesson, K. and Tedengren, M., 1990. Genotypic and phenotypic differences between Baltic and North Sea populations of *Mytilus edulis* evaluated through reciprocal transplantations. I. Growth and morphology. Mar. Ecol. Prog. Ser., 59: 203-210.

Kimura, 1983. The neutral theory of molecular evolution. Cambridge University Press, Cambridge, 367pp.

Koehn, R.K.,1975. Migration and population structure in the pelagically dispersing marine invertebrate, *Mytilus edulis*. In: C. Markert (Editor), Proceedings of the Third International Congress on Isozymes. Academic Press, New York, pp. 945-959.

Koehn, R.K., 1983. Biochemical genetics and adaptation in molluscs. In: A.S.M. Saleuddin and K.M. Wilbur (Editors), The Mollusca, 6. Academic Press, New York, pp. 305-330.

Koehn, R.K., 1985. Adaptive aspects of biochemical and physiological variability. In: P. E. Gibbs (Editor), Proc. 19th Eur. Mar. Biol. Symp., Plymouth, England, 1984. Cambridge University Press, Cambridge, pp. 425-441.

Koehn, R.K., 1990. Heterozygosity and growth in marine bivalves: Comments on the paper by Zouros, Romero–Dorey and Mallet (1988). Evolution, 44: 213-216.

Koehn, R.K., 1991. The genetics and taxonomy of species in the genus *Mytilus*. Aquaculture, 94(2/3): 125-145.

Koehn, R.K. and Mitton, J.B., 1972. Population genetics of marine pelecypods. I. Ecological heterogeneity and evolutionary strategy at an enzyme locus. Am. Nat., 106: 47-56.

Koehn, R.K. and Immermann, F.W., 1981. Biochemical studies of aminopeptidase polymorphism in *Mytilus edulis*. I. Dependence of enzyme activity on season, tissue and genotype. Biochem. Genet., 19: 1115–1142.

Koehn, R.K. and Shumway, S.E., 1982. A genetic/physiological explanation for differential growth rate among individuals of the American oyster, *Crassostrea virginica* (Gmelin). Mar. Biol. Lett., 3: 35–42.

Koehn, R.K. and Gaffney, P.M., 1984. Genetic heterozygosity and growth rate in *Mytilus edulis*. Mar. Biol., 82: 1-7.

Koehn, R.K. and Hilbish, T.J., 1987. The adaptive importance of genetic variation. Am. Sci., 75: 134-141.

Koehn, R.K., Turano, F.J. and Mitton, J.B., 1973. Population genetics of marine pelecypods. II. Genetic differences in microhabitats of *Modiolus demissus*. Evolution, 27: 100-105.

Koehn, R.K., Milkman, R. and Mitton, J.B., 1976. Population genetics of marine pelecypods. IV. Selection, migration and genetic differentiation in the blue mussel *Mytilus edulis*. Evolution, 30: 2-32.

Koehn, R.K., Bayne, B.L., Moore, M.N. and Siebenaller, J.F., 1980a. Salinity related physiological and genetic differences between populations of *Mytilus edulis*. Biol. J. Linn. Soc., 14(3/4): 319-334.

Koehn, R.K., Newell, R.I.E. and Immermann F., 1980b. Maintenance of an aminopeptidase allele frequency cline by natural selection. Proc. Natl. Acad. Sci. (U.S.A.), 77(9): 5385-5389.

Koehn, R.K., Hall, J.G., Innes, D.J. and Zera, A.J., 1984. Genetic differentiation of *Mytilus edulis* in eastern North America. Mar. Biol., 79: 117-126.

Koehn, R.K., Diehl. W.J. and Scott, T.M., 1988. The differential contribution by individual enzymes of glycolysis and protein catabolism to the relationship between heterozygosity and growth rate in the coot clam, *Mulinia lateralis*. Genetics, 118: 121-130.

Lane, D.J.W., Beaumont, A.R. and Hunter, J.R., 1985. Byssus drifting and the drifting threads of the young post-larval mussel *Mytilus edulis*. Mar. Biol., 84: 301–308.

Latt, S.A., 1974. Sister chromatid exchanges, indices of human chromosome damage and repair: detection by fluorescence and induction by mitomycin C. Proc. Natl. Acad. Sci. (U.S.A), 71(8): 3162-3166.

Leary, R.F., Allendorf, F.W. and Knudsen, K.L. 1983. Developmental stability and enzyme heterozygosity in rainbow trout. Nature (Lond.), 301: 71–72.

Levinton, J.S. and Koehn, R.K., 1976. Population genetics of mussels. In: B.L. Bayne (Editor), Marine mussels: their ecology and physiology. Cambridge University Press, Cambridge, pp. 357-384.

Levinton, J.S. and Suchanek, T.H., 1978. Geographic variation, niche breath and geographic differentiation at different geographic scales in the mussels *Mytilus californianus* and *M. edulis*. Mar. Biol., 49: 363-375.

Lewontin, R.C., 1974. The Genetic Basis of Evolutionary Change. Columbia University Press, New York, 346pp.

Lloyd M. and Thorgaard, G. H., 1988. Restriction endonuclease banding of rainbow trout chromosomes. Chromosoma, 96: 171-177.

Lubet, P., Prunus, G., Masson, M. and Bucaille, D., 1984. Recherches experimentales sur l'hybridisation de *Mytilus edulis* L. et *Mytilus galloprovincialis* Lmk. Bull. Soc. Zool. Fr., 109: 87-98.

Majone, F., Brunetti, R., Gola, I. and Levis, A.G., 1987. Persistence of micronuclei in the marine mussel *M. galloprovincialis*, after treatment with mitomycin C. Mutat. Res., 191: 157-161.

Mallet, A.L. and Carver, C.E.A., 1989. Growth, mortality, and secondary production in natural populations of the blue mussel, *Mytilus edulis*. Can. J. Fish. Aquat. Sci., 46: 1154-1159.

Mallet, A.L., Zouros, E., Gartner-Kepkay, K.E., Freeman, K.R. and Dickie, L.M., 1985. Larval viability and heterozygote deficiency in populations of marine bivalves: evidence from pair matings of mussels. Mar. Biol., 87: 165-172.

Mallet, A.L., Freeman, K.R. and Dickie, L.M., 1986. The genetics of production characters in the blue mussel *Mytilus edulis*. I. Preliminary analysis. Aquaculture, 57: 133-140.

Mallet, A.L., Carver, C.E.A. Coffen, S.S. and Freeman, K.R.,1987a. Mortality variations in natural populations of the Blue Mussel (*Mytilus edulis*). Can. J. Fish. Aquat. Sci., 44: 1589-1594.

Mallet, A.L., Carver, C.E.A. Coffen, S.S. and Freeman, K.R., 1987b. Winter growth of the blue mussel *Mytilus edulis* L: importance of stock and site. J. Exp. Mar. Biol. Ecol., 108: 217-228.

Mallet, A.L., Carver, C.E.A. and Freeman, K.R.,1990. Summer mortality of the blue mussel in Eastern Canada: spatial, temporal, stock and age variation. Mar. Ecol. Prog. Ser., 67: 35-41.

Matthaeis, E. de, Pagnotta, R. and Sbordoni, V. 1983. Variabilita genetica in *M. galloprovincialis* come indicatore biologico della qualita delle acque marine antistanti la foce del Tevere. In: Instituto di Ricerca sulle Acque (Editor), L'esperimento Tevere. Influenza di un fiume sull'ecosistema marino prospiciente la sua foce. Quad. Ist. Ric. Acque, 66: 329-336.

Mayr, E., 1970. Populations,Species and Evolution. Belknap Press, Cambridge, Massachusetts, 476pp.

McDonald, J.H. and Koehn, R.K., 1988. The mussels *Mytilus galloprovincialis* and *M. trossulus* on the Pacific coast of North America. Mar. Biol., 99: 111-118.

McDonald, J.H. and Siebenaller, J.S., 1989. Similar geographic variation in the *Lap* locus in the mussels *Mytilus trossulus* and *M. edulis*. Evolution, 43: 228-231.

McDonald, J.H., Koehn, R.K., Balakirev, E.S., Manchenko, G.P., Pudovkin, A.I., Sergiyevskii, S.O. and Krutovskii, K.V. 1990. Species identity of the "common mussel" inhabiting the Asiatic coasts of the Pacific Ocean. Biol. Morya, 1990 (1): 13–22.

McDonald, J.H., Seed, R. and Koehn, R.K., 1991. Allozyme and morphometric characters of three species of *Mytilus* in the Northern and Southern hemispheres. Mar. Biol., 111: 323-335.

Mitton, J.B., 1977. Shell colour and pattern variation in *Mytilus edulis* and its adaptive significance. Chesapeake Sci., 18: 387-390.

Mitton, J. B. and Koehn, R.K., 1985. Shell shape variation in the blue mussel, *Mytilus edulis* L. and its association with enzyme heterozygosity. J. Exp. Mar. Biol. Ecol., 90: 73–80.

Mitton, J.B., Koehn, R.K. and Prout, T., 1973. Population genetics of marine pelecypods. III. Epistasis between functionally related isoenzymes of *Mytilus edulis*. Genetics, 73: 487-496.

Moore, M.N., Koehn, R.K. and Bayne, B.L., 1980. Leucine aminopeptidase (aminopeptidase-1), N-acetyl-ß-hexosamidase and lysosomes in the mussel *Mytilus edulis* L., in salinity changes. J. Exp. Zool., 214: 239–249.

Moritz, C., Dowling, T.E. and Brown, W.M., 1987. Evolution of animal mitochondrial DNA: relevance for population biology and systematics. Annu. Rev. Ecol. Syst., 18: 269-292.

Moynihan, E. P. and Mahon, G.A.T., 1983. Quantitative karyotype analysis in the mussel *Mytilus edulis* L. Aquaculture, 33: 301–309.

Murdock, E.A., Ferguson, A. and Seed, R. 1975. Geographical variation in leucine aminopeptidase in *Mytilus edulis* from the Irish coasts. J. Exp. Mar. Biol. Ecol., 19: 33-41.

Nei, M., 1972. Genetic distance between populations. Am. Nat., 106: 283–292.

Nevo, E., Lavie, B. and Ben-Shlomo, R. 1983. Selection of allelic isozyme polymorphisms in marine organisms: pattern, theory and application. Curr. Top. Biol. Med. Res., 10: Alan R. Liss, New York, 69–92.

Newkirk, G.F.,1980a. Review of the genetics and the potential for selective breeding of commercially important bivalves. Aquaculture, 19: 209-228.

Newkirk, G.F., 1980b. Genetics of shell colour in *Mytilus edulis* L. and the association of growth rate with shell color. J. Exp. Mar. Biol. Ecol., 47: 89-94.

Newkirk, G.F., Freeman, K.R. and Dickie, L.M., 1980. Genetic studies of the Blue Mussel, *Mytilus edulis*, and their implication for commercial culture. Proc. World Maricul. Soc., 11: 596-604.

Newkirk, G.F., Haley, L.E. and Dingle, J., 1981. Genetics of the blue mussel *Mytilus edulis* (L.): nonadditive genetic variation in larval growth rate. Can. J. Genet. Cytol., 23: 349-354.

Pasantes, M.J., Martinez Exposito, A., Martinez Lage, A. and Mendez, J., 1990. Chromosomes of Galician mussels. J. Molluscan Stud., 56(1): 123-126.

Rodhouse, P.G. and Gaffney, P.M., 1984. Effect of heterozygosity on metabolism during starvation in the American oyster *Crassostrea virginica*. Mar. Biol., 80: 179-187.

Rodhouse, P.G., McDonald, J.H., Newell. R.I.E. and Koehn. R.K., 1986. Gamete production, somatic growth and multiple-locus heterozygosity in *Mytilus edulis*. Mar. Biol., 90: 209-214.

Rodino, E., 1973. Polimorfismo della *Lap* (Leucina-Amino-Peptidasi) in *Mytilus galloprovincialis* della Laguna di Venezia. Atti V Congr. Soc. Ital. Biol. Mar., 5: 179-180.

Rogers, J.S. 1986. Deriving phylogenetic trees from allele frequencies: a comparison of nine genetic distances. Syst. Zool., 35(3): 297-310.

Sanjuan, A., Quesada, H., Zapata, C. and Alvarez, G., 1990. On the occurrence of *Mytilus galloprovincialis* Lmk. on NW coasts of the Iberian Peninsula. J. Exp. Mar. Biol. Ecol., 143: 1–14.

Scarpato, R., Migliore, L., Alfinito–Cognetti, G. and Barale, R., 1990. Induction of micronuclei in gill tissue of *M. galloprovincialis* exposed to polluted marine waters. Mar. Pollut. Bull., 21 (2): 74–80.

Seed, R., 1971. A physiological and biochemical approach to the taxonomy of *Mytilus edulis* (L.) and *M. galloprovincialis* (Lmk.) from S.W. England. Cah. Biol. Mar., 12: 291-322.

Seed, R., 1974. Morphological variations in *Mytilus* from the Irish coasts in relation to the occurrence and distribution of *Mytilus galloprovincialis* (Lmk.). Cah. Biol. Mar., 15: 1-25.

Selander, R.K., 1970. Behaviour and genetic variation in natural populations. Am. Zool., 10: 53-66.

Singh, S.M. and Zouros, E., 1978. Genetic variation associated with growth rate in the American oyster (*Crassostrea virginica*). Evolution, 32: 342-353.

Singh, S.M. and Green, R.H., 1984. Excess of allozyme homozygosity in marine molluscs and its possible biological significance. Malacologia, 25(2): 569-581.

Skibinski, D.O.F., 1983. Natural selection in hybrid mussel populations. In: G. S. Oxford and D. Rollinson (Editors), Protein polymorphism: adaptive and taxonomic significance. Academic Press, London, pp. 283-298.

Skibinski, D.O.F., 1985. Mitochondrial DNA variation in *Mytilus edulis* L. and the Padstow mussel. J. Exp. Mar. Biol. Ecol., 92: 251-258.

Skibinski, D.O.F. and Beardmore, J.A., 1979. A genetic study of intergradation between *Mytilus edulis* and *M. galloprovincialis*. Experientia, 35: 1442-1444.

Skibinski, D.O.F. and Edwards, C. A., 1987. Mitochondrial DNA variation in marine mussels *Mytilus*. In: K. Tiews (Editor), Symp. on Selection, Hybridisation and Genetic Engineering in Aquaculture, Bordeaux, 1986, Vol. I. Heenemann, Berlin, pp. 210-226.

Skibinski, D.O.F. and Roderick, E.E., 1989. Heterozygosity and growth in transplanted mussels. Mar. Biol., 102: 73-84.

Skibinski, D.O.F. and Roderick, E.E., 1991. Evidence of selective mortality in favour of the *Mytilus galloprovincialis* Lmk. phenotype in British mussel populations. Biol. J. Linn. Soc., 42(3): 351-366.

Skibinski, D.O.F., Beardmore, J.A. and Ahmad, M., 1978a. Genetic aids to the study of closely related taxa of the genus *Mytilus*. In: B. Battaglia and J.A. Beardmore (Editors), Marine Organisms: Genetics, Ecology and Evolution. Plenum Press, London, pp. 469-485.

Skibinski, D.O.F., Ahmad, M. and Beardmore, J.A., 1978b. Genetic evidence for naturally occurring hybrids between *Mytilus edulis* and *Mytilus galloprovincialis*. Evolution, 32: 354-364.

Skibinski, D.O.F., Cross, T.F. and Ahmad, M., 1980. Electrophoretic investigations of systematic relationships in the marine mussels *Modiolus modiolus* L., *Mytilus edulis* L. and *Mytilus galloprovincialis* Lmk. Biol. J. Linn. Soc., 13: 65-73.

Skibinski, D.O.F., Beardmore, J.A. and Cross, T.F., 1983. Aspects of the population genetics of *Mytilus* (Mytilidae: Molluscs) in the British Isles. Biol. J. Linn. Soc., 19: 137-183.

Stanley, J.G., Hidu, H. and Allen, S.K., 1984. Growth of American oysters increased by polyploidy induced by blocking meiosis I but not meiosis II. Aquaculture, 37: 147-155.

Strømgren, T. and Nielsen, M.V., 1989. Heritability of growth in larvae and juveniles of *Mytilus edulis*. Aquaculture, 80: 1-6.

Tedengren, M., Andre, C., Johannesson, K. and Kautsky, N., 1990. Genotypic and phenotypic differences between Baltic and North Sea populations of *Mytilus edulis* evaluated through reciprocal transplantations. III. Physiology. Mar. Ecol. Prog. Ser., 59: 221-227.

Theisen, B.F., 1978. Allozyme clines and evidence of strong selection in three loci in *Mytilus edulis* (Bivalvia) from Danish waters. Ophelia, 17: 135-142.

Thiriot-Quiévreux, C., 1984a. Les caryotypes de quelques Ostreidae et Mytilidae. Malacologia, 25(2): 465-476.

Thiriot-Quiévreux, C., 1984b. Chromosome analysis of three species of *Mytilus* (Bivalvia: Mytilidae). Mar. Biol. Lett., 5: 265-273.

Thiriot-Quiévreux, C., 1991. Future of cytogenetic research in marine and estuarine animals. In: H. Hummel (Editor), The Genetics of Marine and Estuarine Organisms. Proceedings of a workshop, Delta Institute, Yerseke, The Netherlands, 1990, pp. 51–55.

Thiriot-Quiévreux, C. and Ayraud, N., 1982. Les caryotypes de quelques especes de bivalves et de gasteropodes marin. Mar. Biol., 70: 165-172.

Thorgaard, G. H., 1983. Chromosome set manipulation and sex control in fish. In: W.S. Hoar, D.J. Randall and E.M. Donaldson (Editors), Fish Physiology, Vol. 9(B). Academic Press, London, pp. 405-434.

Thorgaard, G. H. and Allen, S.K. 1987. Chromosome manipulation and markers in fishery management. In: N. Ryman and F. Utter (Editors), Population genetics and fishery management. University of Washington Press, Seattle, pp. 319-332.

Tracey, M.L., Bellet, N.F. and Gravem, C.D., 1975. Excess allozyme homozygosity and breeding population structure in the mussel *Mytilus californianus*. Mar. Biol., 32: 303-311.

Väinölä, R., 1990. Allozyme differentiation between Baltic and North Sea *Mytilus* populations: a reassessment of evidence from transplantation. Mar. Ecol. Prog. Ser., 67: 305–308.

Väinölä, R. and Hvilsom, M.M., 1991. Genetic divergence and a hybrid zone between Baltic and North Sea *Mytilus* populations (Mytilidae; Mollusca). Biol. J. Linn. Soc., 43(2): 127-148.

Varvio, S.-L., Koehn, R.K. and Vainola, R., 1988. Evolutionary genetics of the *Mytilus edulis* complex in the North Atlantic region. Mar. Biol., 98: 51-60.

Volckaert, F. and Zouros, E., 1989. Allozyme and physiological variation in the scallop *Placopecten magellanicus* and a general model for the effects of heterozygosity on fitness in marine molluscs. Mar. Biol., 103: 51-61.

White, M. J. D., 1968. Models of speciation. Science, 159: 1065-1070.

White, M. J. D., 1978. Modes of Speciation. Freeman, San Francisco, 455pp.

Wilkins, N.P., 1981. The rationale and relevance of genetics in aquaculture: an overview. Aquaculture, 22: 209–228.

Wilkins, N.P. and Gosling, E.M. (Editors), 1983. Genetics in Aquaculture I. Developments in Aquaculture and Fisheries Science, Vol. 12. Elsevier Science Publishers, B.V., Amsterdam, 426pp.

Wilkins, N.P., Fujino, K. and Gosling, E.M., 1983. The Mediterranean mussel *Mytilus galloprovincialis* Lmk. in Japan. Biol. J. Linn. Soc., 20: 365-374.

Wright, S., 1951. The genetical structure of populations. Ann. Eugen., 15: 323–354.

Yamamoto, S. and Sugawara, Y., 1988. Induced triploidy in the mussel, *Mytilus edulis*, by temperature shock. Aquaculture, 72: 21-29.

Yamamoto, S. Sugawara, Y. and Nomura, T., 1990. Chemical and thermal control of triploid production in Pacific oysters and mussels, with regard to controlling meiotic maturation. In: M. Hoshi and O. Yamashita (Editors), Advances in Invertebrate Reproduction. Elsevier Science Publishers, B.V., Amsterdam, pp. 455-460.

Young, J.P., Koehn, R.K. and Arnheim, N., 1979. Biochemical characterisation of "Lap", a polymorphic aminopeptidase from the blue mussel *Mytilus edulis*. Biochem. Genet., 17: 305-323.

Zouros, E., 1982. On the role of chromosomal inversions in speciation. Evolution, 36: 414-416.

Zouros, E., 1987. On the relation between heterozygosity and heterosis: an evaluation of the evidence from marine mollusks. In: M.C. Rattazzi, J.C. Scandalios and G.S. Whitt (Editors), Isozymes: Curr. Top. Biol. Med. Res., 15. Alan R. Liss, New York, pp. 255-270.

Zouros, E., 1990. Heterozygosity and growth in marine bivalves: response to Koehn's remarks. Evolution, 44(1): 218-221.

Zouros, E. and Foltz, D.W., 1984a. Minimal selection requirements for the correlation between heterozygosity and growth, and for the deficiency of heterozygotes, in oyster populations. Dev. Genet., 4: 393-405.

Zouros, E. and Foltz, D.W., 1984b. Possible explanations of heterozygote deficiency in bivalve molluscs. Malacologia, 25(2): 583-591.

Zouros, E. and Foltz, D.W., 1987. The use of allelic isozyme variation for the study of heterosis. In: M.C. Rattazi, J.G. Scandalios and G.S.Whitt (Editors), Isozymes: Curr. Top. Biol. Med. Res., 13. Alan R. Liss, New York, pp. 1-59.

Zouros, E. and Mallet, A.L., 1989. Genetic explanations of the growth/heterozygosity correlation in marine molluscs. In: J. Ryland and P.A.Tyler (Editors), Proc. 23rd Eur. Mar. Biol. Symp., Swansea, U.K., 1988. Olsen and Olsen, Fredensborg, Denmark, pp. 317-324.

Zouros, E., Singh, S.M. and Miles, H.E., 1980. Growth rate in oysters: an overdominant phenotype and its possible explanation. Evolution, 31: 856–867.

Zouros, E., Singh, S.M., Foltz, D. W. and Mallet, A.L., 1983. Post-settlement viability in the American oyster (*Crassostrea virginica*): an overdominant phenotype. Genet. Res., 41: 259–270.

Zouros, E., Romero-Dorey, M. and Mallet, A.L., 1988. Heterozygosity and growth in marine bivalves: further data and possible expalanations. Evolution, 42(6): 1332-1341.

Chapter 8

MUSSELS AND ENVIRONMENTAL CONTAMINANTS: BIOACCUMULATION AND PHYSIOLOGICAL ASPECTS

JOHN WIDDOWS AND PETER DONKIN

INTRODUCTION

Growth in industrial activity during the twentieth century has resulted in a rapid increase in inputs of chemicals, either mobilized or synthesized by man, into the estuarine and coastal environments. Many of these chemicals are bioaccumulated within the tissues of biota to concentrations significantly above ambient levels in the environment. Furthermore, some of these environmental contaminants may also be present at toxic levels, and thus induce adverse biological effects.

This chapter reviews the 'Mussel Watch' concept and the role of mussels in monitoring environmental pollution. It summarizes some of the factors affecting bioaccumulation of contaminants, the recorded levels of environmental contaminants in mussel tissues in different regions of the world, and the deleterious efects on physiological responses of mussels that are induced by environmental pollution. Molecular and cellular responses of mussels to environmental contamination are considered in Chapter 9.

'MUSSEL WATCH' CONCEPT

In the mid 1970s it was recognized that the geographical extent and severity of marine environmental contamination and the associated biological impact was largely unknown and undocumented. Goldberg (1975) thus proposed the establishment of a 'Mussel Watch' monitoring programme to assess the spatial and temporal trends in chemical contamination in estuarine and coastal areas of North America, using mussels as 'sentinel' organisms. This then led to the establishment of similar local or regional 'Mussel Watch' programmes in many countries of the world (e.g. U.K., France, Canada, Australia, Japan, Taiwan, India, Mediterranean, South Africa, U.S.S.R.). The primary reasons for assessing the level of chemical contaminants in coastal waters were to protect human health by estimating exposure via the dietary route back to man, and to protect valuable living natural resources.

The 'Mussel Watch' programme was initially concerned with analyzing trace metals, radionuclides associated with the nuclear fuel cycle, and organic chemical pollutants such as polychlorinated biphenyls (PCBs) and DDT (Farrington et al., 1987). However, during the past 15 years analytical capabilities have advanced, thus extending the range of chemicals that can now be analyzed routinely in environmental samples, including mussels (e.g. hydrocarbons, other classes of organochlorines and alkyltins).

ADVANTAGES OF USING MUSSELS FOR MONITORING ENVIRONMENTAL CONTAMINATION

The following attributes have led to the use of bivalves, particularly mussels, as 'sentinel' or 'indicator' organisms in environmental monitoring programmes throughout the world (NRC, 1980; Phillips, 1980; Widdows 1985; Farrington et al., 1987):

(1) Bivalves, such as mussels, are dominant members of coastal and estuarine communities and have a wide geographical distribution. This minimizes the problems inherent in comparing data for markedly different species.

(2) They are sedentary and are therefore better than mobile species as integrators of chemical contamination in a given area.

(3) They are relatively tolerant of (but not insensitive to) a wide range of environmental conditions, including moderately high levels of many types of contaminants.

(4) They are suspension-feeders that pump large volumes of water (several litres per hour) and concentrate many chemicals in their tissues, by factors of 10 to 10^5, relative to the concentration in seawater. This often makes measurement of trace contaminants easier to accomplish in their tissues than in seawater.

(5) The measurement of chemicals in bivalve tissue provides an assessment of biological availability which is not apparent from measurement of contaminants in environmental compartments (water, suspended particulates and sediment).

(6) In comparison to fish and crustacea, bivalves have a very low level of activity of those enzyme systems capable of metabolizing organic contaminants, such as aromatic hydrocarbons and PCBs (see Chapter 9). Therefore, contaminant concentrations in the tissues of bivalves more accurately reflect the magnitude of environmental contamination.

(7) Mussel populations are relatively stable and can be sufficiently large for repeated sampling, thus providing data on short- and long-term temporal changes in contaminant levels.

(8) They can be readily transplanted and maintained in cages to sites of interest, either in the intertidal zone or subtidally on moorings, where populations would not normally grow because of a lack of a suitable substrate.

(9) Mussels are a commercially important seafood species on a worldwide basis and measurement of chemical contamination is of interest for public health considerations.

FACTORS AFFECTING BIOACCUMULATION OF CHEMICAL CONTAMINANTS IN MUSSELS

In monitoring programmes using mussels primarily to assess environmental contamination, it is important to understand the factors which determine the relationships between contaminant levels in the environment (water, suspended particulates and sediments) and contaminant levels in the mussel tissue. However, in those programmes more concerned with assessing contaminant levels in mussels and the associated biological (ecological) effects, this aspect becomes less important. In such cases, it is the bioaccumulated (i.e. bioavailable) and potentially toxic contaminants which are of primary concern.

The degree to which inorganic and organic contaminants are accumulated by mussels depends upon both abiotic factors (e.g. the physicochemical properties of contaminants and their speciation) and biotic factors (e.g. pumping activity, growth, biochemical composition, reproductive condition and metabolism/elimination). These abiotic and biotic factors are in turn affected by environmental variables, such as temperature and salinity. All these factors can influence the rates of dynamic processes concerned with uptake, disposition and depuration, which together determine the degree of bioaccumulation.

Physicochemical Parameters

Aqueous solubility/hydrophobicity
The degree to which organic contaminants are accumulated from water and concentrated in biota is largely dependent on their aqueous solubility and hydrophobicity. Solubility limits the quantity of contaminant available to the mussels in a given volume of water, and for many organic compounds this is less than $10\mu g\ L^{-1}$

(Miller et al., 1985). However, reduced aqueous solubility reflects increased hydrophobicity and an increasing tendency to partition into the tissues of organisms; a phenomenon which partially offsets the effect of limited supply (Hawker and Connell, 1986). This partitioning behaviour can be expressed as the bioconcentration factor (BCF), which is the contaminant concentration in the organism (wet weight basis) relative to its concentration in the water. For mussels in field and experimental situations, the BCF of a range of structurally diverse organic compounds can be precisely related to physicochemical characteristics of the compound such as aqueous solubility [log S], or the log octanol-water partition coefficient [log K_{ow}] (Donkin and Widdows, 1990). For example, from Geyer et al. (1982):

$$\log BCF = -0.682 \log S + 4.94 \ (n = 16, r^2 = 0.889)$$

and

$$\log BCF = \ 0.858 \log K_{ow} - 0.808 \ (n = 16, r^2 = 0.912; \text{see also Fig. 8.3})$$

Such relationships indicate that uptake is generally by diffusion across the external membranes of mussels until a steady state between rates of uptake and depuration is attained. The time required to achieve such a steady state can range from hours for more soluble organic compounds (log K_{ow} <3; e.g. toluene, Hansen et al., 1978) to weeks or even months for very hydrophobic molecules (log K_{ow} >7; e.g. hexachloro-biphenyl, Hawker and Connell, 1986). High BCFs (e.g. 100,000) characteristic of hydrophobic molecules, therefore, enable considerable body burdens to be accumulated from very dilute solutions; though the maximum achievable by partitioning declines with increasing log K_{ow} because contaminant solubility falls more rapidly than the increase in partitioning into the hydrophobic phase (Hawker and Connell, 1986).

Although hydrophobicity-driven partitioning satisfactorily explains many of the observations of organic contaminant bioaccumulation in bivalve molluscs, there are anomalies. For example, the bioconcentration factor of ca. 30,000 for tributyltin (TBT) is >100-fold higher (Widdows et al., 1990; Salazar and Salazar, 1992) than would be predicted by the published BCF–log K_{ow} relationships for hydrophobic organic compounds (Donkin et al., 1989; see Fig. 8.3). This enhanced BCF reflects the strong interactions between TBT and biological materials, particularly protein binding (Saxena, 1987).

Pruell et al. (1986) noted that the BCF of PCBs into mussels was always higher than that of hydrocarbons of equivalent log K_{ow}, and suggested that this difference may be due to a more rapid metabolism of the hydrocarbons. However, Shaw and Connell (1984) have advanced the hypothesis that bioaccumulation of PCBs is strongly influenced by their stereochemistry.

Since water solubility and bioconcentration are inversely correlated, physical factors which influence solubility (e.g. temperature, pH and salinity) might be expected to affect bioaccumulation, but at present there is insufficient information available with which to assess their influence on bioaccumulation in mussels.

Most metals occur in forms which are highly soluble in seawater, so direct uptake from solution is a quantitatively important process in mussels (Lead (Pb), Schulz-Baldes, 1974; Cadmium (Cd), Riisgård et al., 1987; Mercury (Hg), King and Davies, 1987; Vanadium (V), Miramand et al., 1980; Americium (Am) and Plutonium (Pu), Bjerregaard et al., 1985)). Although Carpene and George (1981) provided evidence that Cd enters isolated gills by passive diffusion, the mechanism by which mussels accumulate heavy metals directly from water is unclear (Viarengo, 1989). Metals in saline waters occur either as free ion or as inorganic and organic complexes, some of which are uncharged (Davenport and Redpath, 1984; Viarengo, 1989). Simkiss (1983, 1984) has suggested that metal uptake is the result of partitioning of these relatively hydrophobic complexes into the lipid membrane, a mechanism analogous to the uptake of hydrophobic organic contaminants.

Metals can also be absorbed directly from seawater and incorporated into mussel shells, thus providing the potential for long-term records of changes in environmental contamination. Metals such as uranium (U) (Hamilton, 1980), V (Miramaund et al., 1980) and Pb (Sturesson, 1976) are highly concentrated in the periostracum covering the shell. While there is no apparent relationship between metal levels in the whole shells and bivalve tissues (Bryan and Uysal, 1978), primarily due to shell surface adsorption of metals, recent studies have shown significant relationships between Pb in the inner shell nacreous layer (i.e. the layer most recently deposited) and mussel tissue concentrations (Bourgoin, 1990).

Dissolved organic matter

Saline waters contain dissolved organic material (DOM) composed of small molecules and macromolecules of natural origin. DOM concentrations are highest in estuaries (Mantoura and Woodward, 1983) and can influence the solution behaviour of both metals and organic contaminants, with concomitant effects on bioavailability (Farrington, 1989; Suffet and MacCarthy, 1989).

Boehm and Quinn (1976) showed that a reduction in DOM (from 1.8 to 0.6mg DOC L^{-1} expressed as dissolved organic carbon) more than doubled the uptake of the very hydrophobic hydrocarbon, hexadecane (log K_{ow} 8.3), into *Mercenaria mercenaria* during 8h exposure, but had little effect on the accumulation of the more polar compound phenanthrene (log K_{ow} 4.6). Similarly, a change from 3.1 to 0.7mg DOC L^{-1} increased the uptake of No.2 fuel oil by between 5 and 17 times, depending upon the hydrophobicity of the oil fraction considered (Boehm and Quinn, 1976). DOM in seawater appears to increase the 'solubilization' of aliphatic hydrocarbons and thus reduces their presence in the dispersed 'particulate' state. In addition to the reduction in uptake via filter-feeding, the binding to DOM also makes them less available for uptake via adsorption onto the gill surface.

The influence of DOM on the bioaccumulation of metals by bivalve molluscs appears more complex, both increases and decreases having been reported. For example, chelators of both large and small molecular weight (e.g. humic acid and EDTA, respectively) increase the bioaccumulation of Cd by mussels (*Mytilus edulis*, George and Coombs, 1977; Pempkowiak et al., 1989), which, compared to dissolved organic Cd complexes, possibly enhance the rate of Cd exchange with membrane ligands. In contrast, structurally similar dissolved organic compounds decrease the uptake of Cu by the oyster, *Crassostrea virginica*, due to its strong binding to DOM which makes the Cu less bioavailable (Zamuda and Sunda, 1982; Zamuda et al., 1985).

Particulates/food

Mussels are efficient filter feeders capable of removing and concentrating particles from the water column. This is therefore a potentially important route of entry of contaminants into the animal. Organic contaminants in the environment can be associated with many different types of particles varying from digestible, bioavailable materials such as algae, to combustion particulates (Farrington, 1989) and coal (Bender et al., 1987), from which hydrocarbons are not readily removed. The significance of particulates in the bioaccumulation of organic contaminants therefore varies widely, depending on the nature of the contaminant and the particle.

Algal food has been found to enhance the rate of uptake of crude-oil derived aromatic hydrocarbons by *M. edulis* in a manner suggesting a role for dietary uptake (Widdows et al., 1982). Furthermore, in the deposit feeding bivalve *Abra nitida* (Ekelund et al., 1987), bioaccumulation of the organochlorine pesticide lindane (solubility ca. 2 to 9mg L^{-1}) was little influenced by the presence of particulates, whereas sediments enhanced the accumulation of the less soluble (ca. 5 to 9µg L^{-1}) hexachlorobenzene.

Pruell et al. (1986), however, concluded that the uptake of aromatic hydrocarbons and PCBs by *M. edulis* exposed to environmentally contaminated sediments was primarily from the aqueous phase, although the sediments were clearly the source of the dissolved material. These observations are consistent with those of McLeese and Burridge (1987), which indicated that sediment did not enhance the uptake of aromatic hydrocarbons into mussels, when compared to uptake from water.

Similarly for metals, the significance of particulates in bioaccumulation is dependent on both the nature of the contaminant and the particulates. Uptake from food appears to contribute little to the bioaccumulation of Cd by *M. edulis* in experimental systems (Borchardt, 1983; Riisgård et al., 1987), but contributes significantly to the uptake of Am and Pu (Bjerregaard et al., 1985) and Pb (Schulz-Baldes, 1974). In the environment, particulates may be the major source of bioaccumulated Pb (Loring and Prossi, 1986; Bourgoin, 1990), and are also of paramount importance in the process of iron accumulation (George et al., 1976).

In experimental studies, the relative bioavailability of inorganic mercury (Hg) in water, phytoplankton and sediment is in the approximate ratio of 10:5:1 (King and Davies, 1987). However, because of differences in their relative abundance within a mercury-contaminated estuary, mussels accumulated mainly inorganic Hg from suspended particulates at the seaward end, whereas both dissolved inorganic Hg and methylmercury compounds became more important further up the estuary. Although this study showed that Hg in highly organic particles (such as phytoplankton) is readily available, the bioavailability of this metal to the deposit feeding bivalves *Scrobicularia plana* and *Macoma balthica* is reduced with increasing organic content of the sediment (Langston, 1982).

Biotic Factors

Pumping rate
Since contaminant uptake occurs either from solution or from particles, any factor (e.g. temperature, salinity, anoxia, toxicant) which markedly reduces the processing of water and particles (i.e. pumping/feeding activity) by *Mytilus*, will thus reduce the rate of uptake and time to equilibrium, but not necessarily the steady state tissue concentration.

For example, the rate of accumulation of hexachlorobenzene by the clam *Macoma nasuta* (Boese et al., 1988), and of water-accommodated aliphatic petroleum hydrocarbons (Widdows et al., 1982) and Cd (Riisgård et al., 1987) by *M. edulis*, can be related to ventilation rate.

Biochemical composition/physiological condition
Once absorbed, the physiological condition and the biochemical composition of the mussel are important in determining the contaminant's tissue distribution and whole-animal retention.

Generally, when organic contaminant flux between a lipid-rich animal and its environment approaches a steady state, the distribution in the tissues can be correlated with lipid concentration within the tissue. Therefore, factors which affect lipid levels, such as seasonal storage cycles in digestive and reproductive tissues and the development of lipid-rich eggs (Pieters et al., 1980; Zandee et al., 1980; Lubet et al., 1985, 1986), can potentially affect bioaccumulation and the relative tissue distribution of contaminants. In addition, spawning of lipid-rich eggs can represent a major route of loss for hydrophobic organics (Hummel et al., 1990). In practice, however, there is often little or no correlation between the concentration of hydrophobic contaminants and the relatively low lipid content of mussels (i.e. ca. 5-8%), and therefore expressing contaminants on a lipid basis fails to reduce variability (Sericano et al., 1990; Widdows et al., 1990).

Metal levels are actively controlled by physiological and biochemical factors, and consequently there can be considerable individual animal variability in bioaccumulation (Lobel, 1986). The tissue disposition of metals can also be modified as a result of seasonal reproductive cycles and major alterations in biochemical composition (Cossa et al., 1980; Simpson, 1989; Coimbra and Carraça, 1990). The concentration of metal within the whole body can be strongly influenced by growth and weight loss (Simpson, 1979; Cossa et al., 1980; Fischer, 1988; Tušnik and Planinc, 1988; Borchardt et al., 1989).

Metabolism/storage/elimination

The influence of metabolism on the bioaccumulation of organic contaminants in mussels is discussed in Chapter 9. While it is unlikely that metabolism will have a significant effect on the accumulation of many aromatic compounds, it has been suggested as a possible reason for the unexpectedly rapid elimination of some alkyl phenanthrenes (Farrington, 1989) and noncyclic alkanes (Clement et al., 1980). Compounds with functional groupings are metabolized faster (Chapter 9), although even in these cases bioaccumulation of the untransformed contaminant is primarily a function of partitioning (Ekelund et al., 1990).

Metabolism is not the only active means of contaminant elimination. Widdows et al. (1983) showed that pre-exposure of mussels to unlabelled naphthalene enhanced the elimination of [^{14}C] naphthalene in a manner suggesting active excretion of the unmetabolized compound, particularly by the kidney. Palmork and Solbakken (1981) have also suggested that the kidney of the mussel, *Modiolus modiolus* may have an active role in hydrocarbon excretion.

The fate of metals absorbed by mussels is largely determined by a complex set of binding and sequestration processes, involving specialized proteins (metallothioneins and ferritin), lysosomes and inorganic granules (Bootsma et al., 1988; Viarengo, 1989; see also p.442-443 Chapter 9). The distribution of these agents of metal ion homeostasis within tissues influences their capacity to accumulate and remobilize metals, as well as their bioavailability at sites of toxic action within the tissues.

For metals such as Hg, the initial tissue distribution (determined by route of uptake) can be maintained for long periods; Hg concentrations are notably elevated in the gill when uptake is from water, and in the digestive gland when uptake is from particulates (King and Davies, 1987; see also section: Feeding and digestion, p.449–452

Chapter 9). However, some mobilization of metals between tissues clearly occurs (e.g. Cd, Theede and Jung, 1989).

Two tissues, the kidney (see p.452 Chapter 9) and the byssal gland/threads have been shown to be of considerable importance in the storage and elimination of metals. For example, zinc (Zn) (Lobel, 1986) and Cd (George and Coombs, 1977) levels are influenced by the activity of the kidney, whereas byssal threads appear to be an important route for the elimination for Fe (George et al., 1976), U (Hamilton, 1980) and arsenic (As) (Ünlü and Fowler, 1979).

Depuration curves for both metals and organic contaminants often approximate to a biphasic relationship, indicating accumulation into, and release from, at least two 'compartments', often termed 'fast' and 'slow' compartments. Support for the multicompartment model of organic bioaccumulation comes from the observations that the longer the exposure, the slower the depuration kinetics (Farrington, 1989; Livingstone, 1991), and that prolonged exposure of mussels to some contaminants results in a rapid initial phase of uptake, followed by relatively slow uptake, with no establishment of a steady state (Widdows et al., 1982; Laughlin et al., 1986). The more stable 'slow' compartment probably includes organelles and lipid stores (Widdows et al., 1983; Farrington, 1989).

Possible 'slow' release compartments for metals include specific metal-binding proteins (metallothioneins), lysosomes and inorganic granules (Viarengo, 1989). In the case of metallothioneins, their importance can increase with time since they are inducible (Roesijadi, 1982; Viarengo et al., 1985). Depuration half lives for metals can be short (10 days for Cu; Viarengo et al., 1985), intermediate (32 days for As; Ünlü and Fowler, 1979), or long (>120 days for Cd; Borchardt, 1983; Viarengo et al., 1985).

Contaminant Interactions

In the environment contaminants are generally present in the form of complex mixtures. However, the nature of contaminant interactions is poorly understood, with most observations being derived from single contaminant exposure experiments.

In general, the bioaccumulation (and toxicity) of mixtures of organic compounds appears to be additive, with body burdens reflecting environmental levels (Page et al., 1987; Widdows and Donkin, 1991). However, antagonistic interactions between naphthalene, PCBs and benzo[a]pyrene, with respect to their accumulation in the oyster, Crassostrea virginica, have been reported by Fortner and Sick (1985), but no mechanism was proposed.

There have been several reports of interactions between bioaccumulating metals. Mussels exposed to mixtures of Cu, Cd and Pb at 20μg L^{-1} showed a reduction in the uptake of Cd and Pb (Theede and Jung, 1989), probably as a result of the relatively rapid rate of uptake of Cu and its toxic effects on ventilation rate (Manley, 1983). Exposure of mussels to Cu or Cd can induce metallothioneins which then serve to 'protect' the animal during subsequent exposure to mercury (Roesijadi and Fellingham, 1987; and p.442 Chapter 9). In addition, Cu can displace Zn from the thionein pool, thus reducing the tissue concentration of the latter element. Zinc can also alter the bioaccumulation of cadmium when the Zn/Cd ratio falls outside the range typical of 'natural coastal environments' (Fischer, 1988).

Factors Affecting Bioaccumulation: Implications for Monitoring Programmes

For most contaminants, it is apparent that bioaccumulation in mussels adequately reflects the changing levels in the environment (Phillips, 1980; Farrington et al., 1987; Cossa, 1988; Fowler, 1990). Typically, abiotic and biotic factors can modify bioaccumulation by ca. 2-fold, and consequently they are of less importance when detecting and comparing marked (i.e. >5–10-fold) spatial and temporal differences in contaminant concentrations in mussels (Phillips, 1980). If a more precise discrimination is required, the influence of factors affecting bioaccumulation must be considered during the design of sampling strategy and interpretation of results.

The uptake and depuration half-lives of different contaminants range from hours to months, and this will influence both the frequency of oscillation in environmental levels that can be tracked by mussels and the appropriate sampling frequency. Variations in bioaccumulation resulting from individual physiological/biochemical differences can be overcome by choosing a large sample size (e.g. pools of >20 individuals). The influence of body size and seasonal changes in body composition can be minimized by sampling a 'standard body size' and sampling at the same time of year, preferably outside the spawning period.

CONTAMINANT LEVELS IN MUSSELS: 'SENTINELS' OR 'INDICATORS' OF MARINE ENVIRONMENTAL QUALITY

Due to the rapid growth and the worldwide application of the 'Mussel Watch' approach it is not possible to provide a thorough review of all chemical monitoring programmes published in the scientific literature; only the results of some of the more comprehensive regional studies are presented. Table 8.1 provides the ranges of concentrations for chemical contaminants recorded in mussels collected from different coastal regions of the world. The lower end of the range represents the 'background' level for the different contaminants (metals, organo-metals and organic compounds), while the upper end of the range represents the elevated levels found in the most densely populated and heavily industrialized areas, which are often located near major river estuaries.

'Mussel Watch' programmes have not only quantified the degree of contamination in the estuarine and coastal zone, but they have also identified 'unexpected contaminant hot-spots'. For example, the U.S. Mussel Watch programme identified very high concentrations of PCBs in mussels from New Bedford, Massachusetts; whereas other classes of contaminants did not occur at such elevated levels. These data, along with subsequent analyses of a more detailed spatial and temporal nature, were used to convince local governing officials that the area was severely contaminated with PCBs (Farrington et al., 1987). An example of the dangers of mis-interpretation of data, and the need to be aware of large-scale biogeochemical processes, comes from observations of elevated levels (10-fold) of plutonium (^{239}, ^{240}Pu) recorded in mussels on the Pacific coast of the U.S.A. (Farrington et al., 1987). The Pu was initially thought to come from leakage of radioactive waste dumped offshore in this region. However, more careful interpretation showed that the Pu maximum was due to upwelling of mid-depth waters from the North Pacific Ocean, which had been contaminated with fallout from nuclear weapons tests in the 1960s. However, the Mussel Watch programme did detect a release of radioactive material from a nuclear power reactor on the east coast of the U.S.A. in 1976, but further sampling showed that this was an isolated event.

In a few cases the concentrations recorded in monitoring programmes have exceeded the international concentration limits for edible food, which have then raised some concern for human health (Fowler, 1990).

There is now evidence of widespread contamination by persistent contaminants (e.g. organochlorine residues) far removed from known sources, and more recently reports of highly toxic dioxins in coastal mussels. Furthermore, there is a suggestion of a long-term shift in environmental contamination by organochlorine pesticides from the north towards the tropics and the southern hemisphere (Fowler, 1990). 'Mussel Watch' monitoring programmes may, therefore, continue to provide quantitative evidence of the global scale of distribution and contamination.

The overall conclusion from the two International Mussel Watch Workshops (NRC, 1980; cited Farrington et al., 1987) was that bivalve sentinel organisms have proved to be useful tools in identifying variation in chemical contamination between sites, and have contributed to an understanding of trends in coastal contamination. However, the participants stressed the need for: (a) knowledge of biogeochemical

Table 8.1. Summary of whole body tissue concentrations of organic and inorganic contaminants in mussels and oysters. (Reported concentrations ranging from 'clean' reference sites to industrial sites; for details of analysis and quantification see references).

A. Organics (Hydrocarbons µg g^{-1} dry wt)

Region	Location	Species	Hydrocarbon fraction	Concentration	Reference
North Atlantic	Norwegian fjord	M. e.	PAH (3–5 ring)	3.7–149.5	Bjorseth et al., 1979
	Shetland (oil terminal)	M. e.	PAH (2–3 ring)	0.3–7.6	Widdows et al., 1987a
	Scotland	M. e.	Total PAH	0.3–14.2	Mackie et al., 1980
	France	M. e.	Total PAH	0.1–303	Claisse, 1989
	Spain (Rias)	M. g.	Selected PAHs	3.6–110	Soler et al., 1989
			Selected Alkanes	4.6–220	
	USA, Boston	M. e.	Selected PAH	0.05–3.6	Farrington et al., 1983
			F_1 UCM	5–300	
South Atlantic	Brazil	A. b. †	Total PAH	2.5–82.5	Porte et al., 1990
			F_1 UCM	2.5–650	
	South Africa	M. e.	Total HC	10–5000	Mason, 1988
Pacific	USA, Alaska	M. e.	Total (F_1+F_2)	20–936	Shaw et al., 1986
	USA, San Francisco Bay	M. e.	Total PAH	ND–375	Di Salvo et al., 1975
	USA, California	M. c.	Total F_1	8–98	Martin and Castle, 1984
			Total F_2	70–1040	
	South-east Australia	M. e.	Total F_1+F_2	40–1975	Burns and Smith, 1981

Table 8.1. continued

B. Organics (Organochlorines, Pesticides in ng g⁻¹ dry wt)

Region	Location	Species	DDTs	HCH	Dieldrin	PCBs	PCDDs	PCDFs	PCTs	Reference
N. Atlantic	Norway	M. e.	–	–	–	70–300	–	–	–	Klungsøyr et al., 1988
	Norway	M. e.	–	–	–	–	0.137–1.51	0.68–8.48	–	Oehme et al., 1989
	Scotland	M. e.	6–380	12–76	6–2430	60–7100	–	–	–	Cowan, 1981
	England	M. e.	20–430	15–50	10–220	110–300	–	–	–	Franklin, 1987
	France	M. e.	10–1000	–	–	270–10000	–	–	–	Claisse, 1989
	Netherlands	M. e.	–	–	–	200–2000	–	–	–	Hummel et al., 1990
	Spain	M. g.	–	–	–	40–630	–	–	–	Soler et al., 1989
	USA, East coast	M. e.	4–44	–	–	15–630	–	–	–	Goldberg et al., 1978
	Chesapeake Bay	C. v.	–	–	–	–	–	–	35000	Hale et al., 1990
	Gulf of Mexico	C. v.	3.02–3570	0.25–9.06	0.25–51.6	3.60–1740	–	–	–	Sericano et al., 1990
Mediterranean	France/Italy	M. g.†	–	–	–	100–8130	–	–	–	Geyer et al., 1984
S. Atlantic	Nigeria	C. ga.†	24–760	ND–9	–	180–1435	–	–	–	Osibanjo and Bamgbose, 1990
Indian	South-east Asia	P. v.†	14–190	20–80	–	5–35	–	–	–	Ramesh et al., 1990
Pacific	USA, West coast	M. c.	35–550	1–9	2–20	10–360	–	–	–	Martin and Castle, 1984
	USA, San Francisco Bay	M. c.	42–22470	–	–	56–1500	–	–	–	Phillips and Spies, 1988
	Hong Kong	P. v.	86–2040	<30–210	–	20–3150	–	–	–	Phillips, 1985
	Japan	M. e.	–	–	–	17–2680	–	–	–	Tanabe et al., 1987
	Japan	M. e.	–	–	–	–	0.038–1.25	–	–	Watanabe et al., 1987
										Miyata et al., 1987

Table 8.1. continued

C. Organo-metals (Tri- and dibutyltin in μg g⁻¹ dry wt)

Region	Location	Species	TBT	DBT	Reference
North Atlantic					
	UK	M. e.	0.20–2.08	0.03–1.00	Page and Widdows, 1991
	Sweden	M. e.	0.20–10.2	0.57–2.68	Linden, 1987
	France	M. e.	4.1	–	Page and Widdows, 1991
	USA, Maine	M. e.	0.04–4.3	0.04–3.28	Page (unpublished data)
	USA, East coast	M. e.	0.12–0.8	0.04–0.40	Uhler et al., 1989
	Bermuda	A. z.	0.20–1.1	0.06–0.32	Widdows et al., 1990
Pacific					
	USA, West coast	M. e.	0.09–3.3	0.04–1.71	Uhler et al., 1989
	USA	M. e.	0.20–11	–	Grovhoug et al., 1986
					Wade et al., 1988
	USA, Honolulu, Hawaii	M. e.	7.4	–	Grovhoug et al., 1986
	New Zealand	C. g.	0.08–6.3	–	King et al., 1989

D. Radionuclides (dpm kg⁻¹ dry wt)

Region	Location	Species	239,240 Pu	Reference
Atlantic				
	USA, West coast	M. e.	0.16–5.0	Goldberg et al., 1983
Pacific				
	USA, West coast	M. c.	0.03–7.5	Goldberg et al., 1983

Table 8.1. continued

E. Metals (µg g⁻¹ dry wt)

Region	Location	Species	Ag	Cd	Co	Cr	Cu	Fe	Hg	Mn	Ni	Pb	Zn	Reference
Atlantic	England	M.e.	–	0.9-36.5	–	–	4.8-18.1	–	0.13-1.04	–	–	2.5-30.7	68-443	Franklin, 1987
	Scotland	M.e.†	–	1-3	–	–	4.9-12.3	–	0.15-1.15	–	–	2.2-13	105-196	Davies and Pirie, 1980
	France	M.e.	–	0.1-36.2	–	–	–	–	0.02-0.83	–	–	0.1-21.4	–	Claisse, 1989
	USA, East coast	M.e.	0.04-0.70	0.8-6.2	–	–	4.3-35	–	–	–	0.4-2.9	0.2-15.6	45-320	Goldberg et al., 1983
	Bermuda	A.z.	–	2.9-4.1	–	–	5.7-9.8	–	–	–	–	1.7-7.3	64-86	Widdows et al., 1990
Mediterranean	France/Italy	M.g.	0.10-1.89	0.4-5.9	0.5-7.4	0.5-28.8	2.4-154	149-2220	–	3.3-69.8	0.9-14.1	2.7-117	97-644	Fowler and Oregioni, 1976
Pacific	USA, West coast	M.c.	0.02-10.5	0.5-20.2	–	–	3.6-11	–	–	–	0.7-5.5	0.2-10	60-280	Goldberg et al., 1983
	Hong Kong	P.v.	–	0.1-1.44	–	–	8.5-278	–	<0.11-0.14	–	–	1.4-60.5	77-164	Phillips, 1985
	SE Australia	M.e.	–	6.7-111	–	–	4.8-54	568-1099	–	9.2-20.8	–	7.5-30.5	198-505	Fabris et al., 1986

† % water content not stated; therefore, concentration in terms of wet weight converted to dry weight using a standard conversion factor of 5.

M. e.: Mytilus edulis; M. g.: Mytilus galloprovincialis; M. c.: Mytilus californianus; A. z.: Arca zebra; P. v.: Perna viridis; A. b.: Anomalocardia brasiliana; C. g.: Crassostrea gigas; C. ga.: Crassostrea gasar; C. v.: Crassostrea virginica.

UCM: Unresolved complex mixture; F_1: Alkanes fraction; F_2: Aromatic fraction; PAH: Polyaromatic hydrocarbons; Total HC: Total hydrocarbons; HCH: Hexachlorocyclohexane (or Lindane); PCB: Polychlorinated biphenyl; PCT: Polychlorinated terphenyls; PCDF: Polychlorinated dibenzofuran; PCDD: Polychlorinated dibenzo-p-dioxin; – : not determined; ND: not detected.

cycles and related processes in the coastal zone in order to interpret the chemical data, and (b) coupling of chemical and biological effects measurements.

Some Limitations of Chemical Monitoring Programmes

A practical problem for analytical chemists involved in such surveys is the number of potential environmental contaminants, and their degradation products, which could be accumulated by mussels (e.g. metals, organo-metals and >10,000 organic contaminants). Consequently, chemical monitoring programmes are usually selective, often focusing on those classes of contaminants that can be easily measured at relatively low cost (e.g. metals), or those of 'known' concern in a particular area.

Even for a widespread and important class of organic environmental contaminants, such as petroleum-derived hydrocarbons, extraction, identification and quantification can be difficult because of the complex mixture of structurally diverse compounds. For example, the routine extraction of animal tissues and subsequent analytical procedures can lose a significant proportion of the more volatile toxic compounds (e.g. log K_{ow}<3; Farrington et al., 1988), and the unresolved complex mixture (UCM) component of chromatograms, which contains many potentially toxic compounds, is often not quantified (or semiquantified), due to the lack of appropriate standards for purposes of calibration.

Intercalibration exercises, such as those organized by the International Council for the Exploration of the seas (ICES) and the Intergovernmental Oceanographic Commission (IOC), are essential before results can be directly compared between analytical laboratories. Analyses of metal and organo-metals (e.g. alkyltins) have demonstrated a high degree of accuracy and precision (Topping, 1983). A corresponding intercomparison exercise for organic (hydrocarbon) analyses highlighted a lower level of accuracy and precision (ca. 10% CV within laboratory but 100% CV between laboratories), due in part to the diverse/complex nature of the contaminant mixture, and the variety of methodologies used. Farrington et al. (1988) therefore concluded that published hydrocarbon concentrations from different laboratories should be compared with caution. Consequently, it is often useful to focus on selected groups of compounds (e.g. 2 and 3 ring aromatic hydrocarbons which can be analyzed reliably) as 'indicators' of the total petroleum derived toxic load, and to establish appropriate conversion factors (Widdows et al., 1987a).

It is generally recognized that there is no single or simple method of assessing environmental pollution; it must ultimately be a combination of physicochemical (cause) measurements and biological (effect) measurements.

BIOLOGICAL RESPONSES OF MUSSELS FOR ASSESSING THE EFFECTS OF POLLUTION

The term 'pollution' implies a biological effect, whereas 'contamination' is a physical-chemical phenomenon resulting from the discharge to the environment of compounds in excess of normal concentrations. Consequently, the assessment of pollution and environmental quality must ultimately be in terms of biological measurements, preferably in concert with appropriate measurements of chemical contaminants.

Biological effects measurements for assessing and monitoring environmental pollution should ideally fulfil most of the following criteria:

(1) They should be sensitive to environmental levels of pollutants, and have a large scope for response throughout the range from optimal to lethal conditions.

(2) They should reflect a quantitative and predictable relationship with toxic contaminants (i.e. pollutants).

(3) They should have a relatively short response time, in the order of hours to weeks, so that pollution impact can be detected in its incipient stages.

(4) The technique should be applicable to both laboratory and field studies in order to relate laboratory-based concentration-response relationships to field measurements of spatial and temporal changes in environmental quality.

(5) They should not only provide an integrated response to the 'total pollutant load', thus providing a measure of the overall impact, but also provide insight into the underlying cause and the mechanism of toxicity.

(6) The biological response should have ecological relevance and be shown to reflect deleterious effects on growth, reproduction or survival of the individual, the population and ultimately the community.

Many of the biological reponses suggested as potential techniques for monitoring the effects of marine environmental pollution (McIntyre and Pearce, 1980; NRC, 1980; Bayne et al., 1985) have been applied to bivalves, particularly mussels, due to their established role as 'sentinel organisms'. In this chapter we will discuss only those physiological responses that are measured at the level of the whole animal. Specific biochemical and cellular responses to toxicants are discussed in Chapter 9.

The major whole animal physiological responses which have been used to measure the effects of specific toxicants and the effects of environmental pollution are listed below:

- Mortality (larvae and adults)
- Shell valve gape (adults)
- Shell growth (larvae and adults)
- Tissue growth (adults)
- Scope for growth (adults) involving components of the energy budget

Table 8.2 compares the relative sensitivities of lethal and sublethal responses of larval and adult mussels to a representative metal (Cu), organometal (tributyltin, TBT) and petroleum hydrocarbons.

Lethal Responses

The most common acute toxicity test is the 96h LC_{50}, which determines the toxicant concentration resulting in a 50% lethal response over a period of 96h exposure. Such lethality tests provide a relatively rapid method of estimating the concentration of test materials that cause direct and irreversible harm to the test organism. These toxicity tests are widely used because they are inexpensive, rapid and the response is easily quantified. However, application of the acute lethal response to adult bivalves is impractical because of their ability to close their valves and thus isolate themselves from extreme (i.e. ultimately lethal) environmental conditions for long periods of time (i.e. days). Consequently, LC_{50} (Table 8.2) values give a false impression of high tolerance to toxicants.

Use of the pelagic larval stage of aquatic invertebrates, particularly bivalves, for routine toxicity testing was first proposed by Dimick and Breese (1965). Since then the larval bioassay, using oyster or mussel larvae, has been adopted as a method for testing the toxicity of chemicals and assessing the 'water quality' of samples taken directly from the environment, and particularly to test the toxicity of dilutions of effluents, sewage sludge/dredge spoils/sediment extracts (ASTM, 1980; Martin et al., 1981; Klöckner et al., 1985). In this bioassay the embryos are simply incubated in the water samples at a constant temperature for a given period of time, and the number of larvae developing and surviving to the easily identifiable 'D' stage (first prodissoconch shelled larvae) are counted. Such a larval bioassay is therefore measuring a lethal response. The toxicities of ten metals, expressed in terms of the concentration inducing abnormal development in 50% of *M. edulis* embryos after 48h exposure (48h EC_{50}), are summarized in Table 8.3 (Martin et al., 1981).

A feature often attributed to the larval bioassay is its sensitivity (Connor, 1972). However, there is now a growing body of evidence which demonstrates that the early larval stages of bivalves are not necessarily more sensitive than adults to a wide range of pollutants (see Table 8.2, Beaumont et al., 1987 and Butler et al., 1990). These studies show, not only sublethal stress responses of adult and larval mussels to be more sensitive than lethal responses, but also adults to be more sensitive than larvae, whether comparing % survival, growth or feeding rate. Although, mortality in adults may not take effect for many months, due to their extensive body reserves, long-term lethal effects can be predicted from measurements of negative energy balance or scope for growth (e.g. mesocosm studies, Widdows et al., 1987b; Widdows and Johnson, 1988). Generally, adults appear to be >10-fold more sensitive than larvae with respect

Table 8.2. Comparison of lethal and sublethal responses of larval (L), juvenile (J) and adult (A) *Mytilus edulis* when exposed to selected toxicants (copper, tributyltin and petroleum hydrocarbons).

Toxicant	Biological response (50% of control)	Water concentration ($\mu g\ L^{-1}$)	Tissue concentration ($\mu g\ g^{-1}$ dry wt)	Life stage	Reference
Cu					
	Lethal (15 days)	400		L	Beaumont et al., 1987
	Lethal (15 days)	50		A	Martin, 1979
	Lethal (39 days)	20	59	A	Martin, 1979
	Shell growth	150		L	Beaumont et al., 1987
	Shell growth	5		A	Redpath, 1985; Strømgren, 1986
	Valve movement	10-20		A	Davenport and Manley, 1978
	Clearance rate/ Scope for growth	12		A	Widdows (unpublished data)
TBT					
	Lethal and shell growth (15 days)	0.4		L	Beaumont and Budd, 1984
	Lethal (5 days)		20	A	Page and Widdows, 1991
	Valve movement	10		A	Kramer et al., 1989
	Shell growth	6		J	Strømgren and Bongard, 1987
	Scope for growth	0.1	4	A	Page and Widdows, 1991
Hydrocarbons (crude oil)	Lethal (6h)	$>10 \times 10^3$		L	Craddock, 1977
(crude oil)	Lethal (4 days)	$1–10 \times 10^3$		A	Craddock, 1977
(diesel oil)	Lethal (4mo)	125		A	Widdows et al., 1987b
(crude oil)	Valve movement	6×10^3		A	Kramer et al., 1989
(crude oil)	Shell growth	$1.5 \times 10^{3\dagger}$		A	Strømgren, 1986
(crude oil)	Clearance rate/	30	150		
	Scope for growth			A	Widdows et al., 1987b
(aliphatics and aromatics log K_{ow} <5)	Clearance rate		95	A	Donkin et al., 1989

$\dagger$Total concentration in microcapsules and water.

Environmental ranges:

Cu: In water: 1 to 30$\mu g\ L^{-1}$ (Davenport and Redpath, 1984)

In tissue: 2.4 to 95$\mu g\ g^{-1}$ dry wt (Fowler and Oregioni, 1976)

TBT: In water: 0.005 to 0.5$\mu g\ L^{-1}$ (Salazar and Salazar, 1992)

In tissue: 0.1 to >10$\mu g\ g^{-1}$ dry wt (Page and Widdows, 1991)

Petroleum hydrocarbons: In water: 1 to 74$\mu g\ L^{-1}$ (Law, 1981); 19 to 560$\mu g\ L^{-1}$ (12-351 days after oil spill; Blackman and Law, 1980)

In tissue: Aliphatic UCM 1 to 270$\mu g\ g^{-1}$ dry wt

Aromatic UCM <3 to 111$\mu g\ g^{-1}$ dry wt

Selected PAH <0.02 to 17.5$\mu g\ g^{-1}$ dry wt (Farrington et al., 1982, 1983)

Table 8.3. Concentration inducing abnormal development in 50% of *Mytilus edulis* embryos after 48h incubation with toxicant (i.e. 48h EC_{50}) (Martin et al., 1981).

	Toxicant									
	Ag	As	Cd	Cr	Cu	Hg	Ni	Pb	Se	Zn
Water concentration										
$\mu g\, L^{-1}$	14	>3000	1200	4469	5.8	5.8	891	476	>10000	175

to Cu (Beaumont et al., 1987), petroleum hydrocarbons (Widdows et al., 1987b) and sewage sludge (Butler et al., 1990), and ca. 4-fold for TBT (Page and Widdows, 1991). The explanation for this lower sensitivity in the early larval stage may be a combination of their reliance on energy reserves provided by the parent rather than on direct feeding, and the absence of a developed nervous system, which is an important site of toxic action.

These findings therefore suggest that LC_{50} measurements will not provide the necessary sensitivity either to 'protect the environment' by adequately screening toxic materials prior to release, or to identify pollutant effects in all but the most extreme and acute environmental pollution incidents.

Shell Valve Gape

Valve movement response of mussels has been proposed as a biological monitoring tool by Sloof et al. (1983) and Kramer et al. (1989). The rationale is that mussels under optimal conditions remain 'open' for most of the time in order to pump water for respiration and feeding; whereas in response to environmental stress and many toxicants mussels close their shells for extended periods of time. Bivalves therefore offer the possibility of providing a continuous and automated detection system for monitoring acute changes in water quality.

The monitoring device described by Kramer et al. (1989) is based on electromagnetic induction sensors which detect the opening and closing of eight mussels mounted on a system which records, integrates and analyses their activity. Kramer et al. (1989) also present data on the detection limits (i.e. sensitivity) of the mussel shell closure response to several toxicants (Table 8.2). However, most of these values are very high relative to the concentrations of toxicants found in polluted environments, and this suggests that valve closure response may not be sufficiently sensitive for monitoring all but the most acute pollution incidents.

If valve closure represents an extreme response to very high levels of toxicants (i.e. lethal levels), then the degree of valve gape might supposedly be a more appropriate sublethal response if it is correlated with the pumping rate or clearance rate of mussels. Such a relationship has been recorded under 'optimal' conditions by Jørgensen et al. (1988). However, a simple relationship between valve gape and pumping rate is not maintained when mussels are exposed to a range of toxicants (with different mechanisms of toxicity). For example, exposure to hydrocarbons (inducing narcosis), TBT (causing uncoupling of oxidative phosphorylation and neurotoxic action) and pentachlorophenol (causing uncoupling and narcosis) all induce a wide shell gape at concentrations that inhibit pumping rates (Widdows and Donkin, 1991; Widdows, unpublished observations). Consequently, further research is required before the shell valve gape response can be routinely applied in pollution monitoring.

Shell and Tissue Growth

The rate of growth is a fundamental measure of physiological fitness/performance and therefore growth has often been used as a measure of environmental quality and pollution effects. However, bivalve growth is often difficult to quantify and interpret, especially in relation to pollution, because (a) a large proportion of the total production can be lost in the form of gametes (either in a single complete release or by gradual/continual spawning); (b) there is no tight coupling between shell growth and other growth components such as somatic or gonad growth (Hilbish, 1986; Salkeld and Widdows, unpublished data; and see Fig. 4.7 Chapter 4) and (c), in the field it is difficult to separate and distinguish between 'nutrition effects' and sublethal 'toxicant effects'. Temporal and spatial variation in food quantity and quality are clearly important factors determining growth (Smaal and van Stralen, 1990), but it is difficult to provide an integrated measure of food availability over a time-scale of weeks. This will therefore contribute to an increased variance in any growth data, and tend to mask any important but sublethal effects of pollutants (Salazar and Salazar, 1992).

These factors are clearly less important in controlled short-term experiments using larvae or juvenile mussels. Consequently, there have been many laboratory studies in which shell and tissue growth have been used to assess the toxic effects of specific chemicals. Early studies by Calabrese et al. (1977) not only demonstrated reduced growth rates of larval oysters (*Crassostrea virginica*) and clams (*Mercenaria mercenaria*) in response to metals, but also showed that metal tolerances of bivalve embryos and larvae can be markedly different. Table 8.2 includes EC_{50} values for growth of mussel larvae in response to Cu (EC_{50} of 150µg Cu L^{-1}; Beaumont et al., 1987) and TBT (EC_{50} of 0.4µg TBT L^{-1}; Beaumont and Budd, 1984).

Table 8.4. Water concentration inducing a 50% reduction in shell growth rate of juvenile *Mytilus edulis*.

Toxicant	Water concentration ($\mu g\ L^{-1}$)	Reference
Cu	4	Strømgren, 1982
Cd	100	Strømgren, 1982
Hg	0.4	Strømgren, 1982
Zn	60	Strømgren, 1982
Pb	>200	Strømgren, 1982
Ni	>200	Strømgren, 1982
TBT	0.2	Strømgren and Bonard, 1987
Oil	1500	Strømgren, 1986

In the environment, the planktonic larval stage is accompanied by high mortality rates (ca. 15% day^{-1}; Jørgensen, 1981) due to predation, disease and dispersion. Therefore, any factor, including toxicants, that reduces larval growth rate, and thus extends the period of larval development in the water column, will have a major effect on mortality and the chances of survival to the settlement stage and beyond (Calabrese et al., 1977; Widdows, 1991).

The effect of metals, TBT and oil on the growth in shell length of juvenile mussels has been measured by a laser diffraction technique (Strømgren, 1982, 1986; Strømgren and Bongard, 1987). Table 8.4 summarizes the water concentrations inducing a 50% reduction in shell growth rate. These EC_{50} values, however, suggest that shell growth rate does not exhibit the same degree of sensitivity as tissue growth and energetic measurements (Table 8.2).

Mussel tissue growth has been used to assess the environmental effects of TBT in flow-through microcosms (Salazar and Salazar, 1987) and in field studies (e.g. San Diego Bay, California; Salazar and Salazar, 1992). These studies recorded significant reductions in the growth rate of juvenile mussels (ca. 10mm) at concentrations of >0.2μg TBT L^{-1} or >2μg TBT g^{-1} wet tissue (ca. 10μg TBT g^{-1} dry wt). Salazar and Salazar (1992) found that at concentrations <0.1μg TBT L^{-1}, environmental factors (including nutrition, temperature, suspended silt and other contaminants) modified or masked any effects of TBT (see earlier comments). Juvenile mussels were used because their growth was not affected by gametogenesis and they grew faster than adults, thus providing a greater range for detecting significant differences. Salazar and Salazar (1992), however, reported that frequent (i.e. weekly) disturbance and measurement resulted in a significant inhibition of growth compared to 'undisturbed' mussels and those measured biweekly.

The form and shape of bivalve shells has also been used as a 'specific indicator' of TBT, which originates from antifouling paints. Shell deformities in oysters (*Crassostrea gigas*), characterized by the production of a series of cavities within the shell and causing extreme thickening of the valves, were first linked to TBT by Alzieu et al. (1982) and Waldock and Thain (1983). Subsequently, other workers have shown a similar relationship between environmental levels of TBT and shell deformities in mussels (Page, Dassanayake, Gilfillan and Kresja, unpublished data).

The body condition index, which relates tissue wet weight to shell volume or total wet weight, is a measure traditionally used by shellfisheries biologists to assess bivalve condition, mainly in relation to seasonal and spatial changes (Widdows, 1985; see also p.107-108 Chapter 4). On occasions when this index has been used to assess pollution effects, either no significant changes were detected (Widdows and Johnson, 1988) or significant effects were limited to the spring (McDowell Capuzzo et al., 1989).

Scope for Growth

Determination of the energy available for growth and reproduction, based on the physiolgical analysis of the energy budget rather than direct measurement of growth itself, has proved to be particularly useful in assessing the biological effects of pollution. It provides an instantaneous measurement of the energy status of the animal, as well as insight into the underlying components which effect changes in growth rate (see details in Chapter 5). Furthermore, there is good agreement between indirect estimates of growth (based on the energy budget) and direct measurement of tissue and shell growth (Riisgård and Randløv, 1981), and also determination of production based on detailed population size-class analysis (Gilfillan and Vandermeulen, 1978; Bayne and Worrall, 1980).

The energy budget of an animal represents an integration of the basic physiological responses such as feeding, food absorption, respiration, excretion and production. Each component is converted into energy equivalents ($J\ h^{-1}$) and alterations in the amount of energy incorporated into growth and production can be described by the balanced energy equation of Winberg (1960):

$$C = P + R + E + F$$

where C = total consumption of food energy, P = total production of shell, somatic tissue and gametes, R = respiratory energy expenditure (i.e. costs of maintenance, feeding, digestion and growth), E = energy lost in excreta, and F = faecal energy loss.

The absorbed ration (A) is the product of energy consumed (C) and the efficiency of absorption of energy from the food. Production may then be expressed as:

$$P = A - (R + E)$$

When production (P) is estimated from the difference between the energy gains (energy absorbed) and the energy losses (energy expenditure via respiration and excretion), it is referred to as 'scope for growth' (Warren and Davis, 1967). Scope for growth (SFG) can range from maximum positive values under optimum conditions, but declining to negative values when the animal is severely stressed and utilizing its body reserves for maintenance. Details of the methods for measuring physiological energetic responses and SFG are presented in Widdows (1985) and Widdows and Johnson (1988). Measurements should ideally be made during the summer period of active growth after the spawning season, and preferably using intertidal mussels which are adapted to a more rigorous environment and which recover rapidly from any disturbance or handling stress.

While there is no single biological effects measurement that can satisfy all environmental situations, SFG fulfils most of the criteria outlined above. The relative sensivities of SFG and other sublethal and lethal responses of mussels (*M. edulis*), at different life stages, to environmentally important toxic contaminants such as Cu, TBT and petroleum hydrocarbons are compared in Table 8.2. It shows that (a) the physiological components of the energy budget, which determine growth, are responsive to environmental levels of pollutants, (b) the physiological energetic and growth responses are considerably more sensitive than lethal responses, and (c) the larval stages are apparently less sensitive to some contaminants than adult mussels. In addition, mussels and their physiological responses appear to have greater or equal sensitivity to environmental toxicants when compared to other aquatic animals (hydrocarbons, Donkin et al., 1989; Cu, Davenport and Redpath, 1984; TBT, Widdows and Page, 1991). One notable exception is the very sensitive and toxicant-specific 'imposex' response of the gastropod, *Nucella lapillus*, to TBT. Imposex occurs at 0.005µg TBT L^{-1} or 0.5µg g^{-1} dry wt (Gibbs et al., 1987), and is therefore an order of magnitude more sensitive to TBT than the growth response of mussels.

ROLE OF COMBINED MEASUREMENT OF PHYSIOLOGICAL ENERGETICS AND TISSUE RESIDUE CHEMISTRY IN ECOTOXICOLOGY

The ultimate aim of environmental toxicology is to both predict and diagnose the causes of biological/ecological effects resulting from exposure to chemicals and other stressors in the environment. To meet this objective it is necessary to establish (a) relationships between the concentrations of chemical contaminants in the environment and in the tissues of biota (BCFs), and (b) cause-effect relationships between tissue contaminant concentration and the resultant biological effects, based on an understanding of their mode of toxic action. The combination of chemical analysis of contaminant levels in the body tissues of mussels, and the measurement of biological effects in terms of physiological energetics, is ideally suited to such a 'cause-effect' framework (Fig. 8.1; Widdows and Donkin, 1989, 1991).

Quantitative Concentration-Response Relationships and QSARs

Tissue concentration-physiological response relationships derived from controlled laboratory and mesocosm studies, not only facilitate the identification of chemical contaminants causing effects recorded in the environment, but also enable biological effects to be predicted from environmental levels of contaminants in water and body tissues. Such toxicological research therefore provides the information necessary to establish an appropriate database for the toxicological interpretation of tissue residue data derived from chemical monitoring programmes (e.g. Mussel Watch).

While it may be feasible to determine tissue concentration-response relationships (Fig. 8.2) for individual contaminants with specific mechanisms of toxicity, for example Cu (Redpath, 1985), TBT (Widdows and Page, 1991) and selected aromatic hydrocarbons (Fig. 8.4; Widdows et al., 1987a), it is clearly unrealistic to examine the sublethal effects of more than ten thousand individual organic contaminants which enter the environment. Consequently, the application of a Quantitative Structure-Activity Relationship (QSAR) approach (Hermens, 1986), which facilitates the prediction of toxicological properties of organic compounds from their chemical/structural properties, provides a means of overcoming this problem (Donkin et al., 1989; Donkin and Widdows, 1990). Furthermore, QSARs provide a unified approach to modelling and predicting the environmental behaviour, fate and effects of structurally diverse organic contaminants from their physicochemical properties.

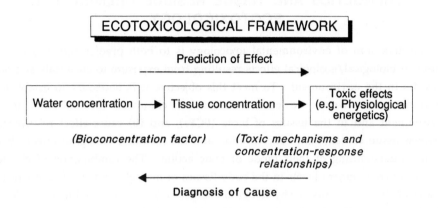

Fig. 8.1. Ecotoxicological framework enabling the prediction and diagnosis of toxic effects resulting from contaminant exposure.

A QSAR approach based on sensitive physiological responses in relation to toxicant concentrations in the tissues is distinct from that adopted by other aquatic toxicologists, who establish relationships between LC_{50} or EC_{50} values for growth and aqueous toxicant concentrations only. As a result, laboratory-derived relationships between sensitive responses and toxicant tissue concentrations, are more relevant to environmental situations where animals are typically exposed to sublethal and varying contaminant concentrations in the water, which are accumulated and time-integrated by the body tissues.

A number of important features concerning QSARs and the sublethal toxic effects of a range of hydrophobic organic compounds are illustrated in Figure 8.3:

(1) There is an inverse linear relationship between the hydrophobicity of organic compounds (measured in terms of log K_{ow}, the octanol-water partition coefficient) and the log concentration in the water inducing a 50% reduction in the feeding (clearance) rate of *M. edulis*.

(2) The log bioconcentration factor (BCF) increases linearly with log K_{ow}.

(3) Changes in bioconcentration account for most of the differences in the water concentration-based expression of toxicity. Consequently, hydrophobic organic compounds, such as aromatic and aliphatic hydrocarbons with log K_{ow} values <5 and 6 respectively, have equal toxicity when expressed on the basis of toxicant concentration in the tissues (i.e. horizontal line).

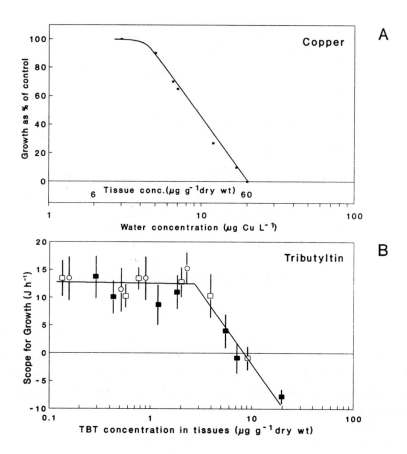

Fig. 8.2. (A) Relationship between the water/tissue concentration of copper and the growth of *Mytilus edulis*. (From Redpath, 1985; Widdows and Johnson, 1988). (B) Relationship between TBT concentration in the tissues and scope for growth of *Mytilus edulis* (mean ± 95% C.I.; results from experiments in October (■), June (□) and July (O). (From Widdows and Page, 1991).

(4) A single QSAR line implies a common mechanism of toxicity and a simple concentration additive effect when present as complex mixtures (Könemann, 1980; Widdows and Donkin, 1991). Conversely, QSARs can be used to identify outliers, which then indicate different modes of toxic action or mechanisms of defence against xenobiotics.

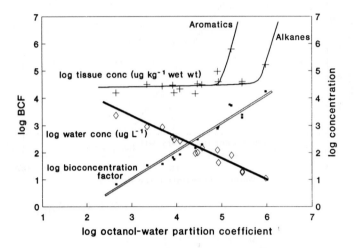

Fig. 8.3. Quantitative structure-activity relationships between log octanol-water partition coefficient (log K_{ow}) of various hydrophobic hydrocarbon compounds, and the log of their bioconcentration factor into *Mytilus edulis*, the concentration of these compounds in the water and their concentrations in the mussel tissues which reduce clearance rates by 50%. (For details see Donkin et al., 1989).

(5) There exists a molecular weight cut-off in the QSAR line occurring at a log K_{ow} of 5 for aromatic and 6 for aliphatic hydrocarbons; a well-known phenomenon associated with compounds inducing narcosis. Compounds with a log K_{ow} >5 are accumulated in the tissues, but induce little or no effect on feeding rate. This 'cut-off' identifies the molecular/structural range over which compounds inhibit feeding rate, and demonstrates that, while many hydrocarbons may be detected in mussels sampled from contaminated environments, not all produce adverse effects on feeding rate.

Once established, a QSAR between a physicochemical descriptor (e.g. K_{ow}) and a biological response can then play an important role in: (a) predicting the toxicity of untested but structurally related compounds, (b) identifying potentially toxic environmental contaminants, which need to be incorporated into chemical monitoring programmes, (c) the systematic comparison of relative sensitivities of different organisms to classes of toxicants, thus enabling extrapolation from a test organism (e.g. mussels) to other species and (d), providing a sound basis for mathematical models on the fate and effects of organic chemicals in the aquatic environment. Current and future research is concerned with establishing QSARs for other major classes of organic compounds with different physicochemical properties, and for other biological responses reflecting different mechanisms of toxicity.

Mechanisms of toxicity reflected in physiological energetics

Physiological energetics and the disturbance of the mussel's energy balance provides insight into, and integration of, some of the primary mechanisms of toxicity that are both biologically and environmentally important. Major mechanisms of toxicity reflected in physiological energetics include:

•nonspecific narcosis affecting the ciliary feeding activity of bivalves (e.g. hydrocarbons, Donkin et al., 1989);

•neurotoxic effects on the neural control of gill cilia (e.g. dinoflagellate toxins, Widdows, et al., 1979; Cu, Howell et al., 1984; TBT, Snoeij et al., 1987);

•uncoupling of oxidative phosphorylation causing an increase in respiration rate (e.g. TBT, Snoeij et al., 1987; phenols, Buikema et al., 1979);

•inhibition of oxidative metabolism thus reducing respiration rate (e.g. DBT, Snoeij et al., 1987; hypoxia, Widdows et al., 1989);

•toxic effects on membrane structure and function affecting processes of food digestion and absorption (e.g. hydrocarbons, Widdows et al., 1987b).

While QSARs provide a means of predicting the additive effects of complex mixtures of structurally related toxicants, physiological energetics provides a biological integration of the consequences of multiple mechanisms of toxicity (Widdows and Donkin, 1991). Many toxicants induce effects via more than one mechanism of toxicity. Pentachlorophenol, for example, is an uncoupler of oxidative phosphorylation, which results in enhanced oxygen consumption, and simultaneously induces narcosis which reduces ciliary feeding activity in mussels (Widdows and Donkin, 1991). In the case of TBT, the primary mechanism of toxicity is via uncoupling, although at high concentrations (i.e. >3μg TBT g^{-1} dry wt) it also induces neurotoxic effects on the ciliary feeding mechanism of the gill (Widdows and Page, 1991).

Furthermore, studies of interactions between structurally unrelated toxicants, such as petroleum hydrocarbons and Cu (Strømgren 1986; Widdows and Johnson, 1988), have found combined effects on mussel shell growth and clearance rate to be simply additive on a proportional basis (Widdows and Donkin, 1991); contrary to the conclusions drawn by Strømgren (1986).

In contrast to the additive effects shown by structurally related and some unrelated toxicants (e.g. hydrocarbons and Cu), recent laboratory studies have demonstrated a significant antagonism between TBT and hydrocarbons in terms of their effects on the clearance rate of *Mytilus* (Widdows and Donkin, 1991). At present there is no mechanistic explanation for this antagonism, but there is circumstantial evidence in support of these findings. For example, mussels are found surviving with very high levels of TBT in their tissues (e.g. 5–10µg g⁻¹ dry wt) at sites which are also likely to be contaminated by hydrocarbons (e.g. Honolulu Harbor, Hawaii; Norfolk Harbor, Virginia, U.S.A.; Page and Widdows, 1991).

Physiological energetics, therefore provides a means of integrating the resultant effects of different mechanisms of toxicity, as well as providing a valuable tool in future research concerned with understanding and predicting complex toxicant interactions occurring in the environment.

From mechanistic interpretation to ecological relevance

Responses measured at the organism level generally have the advantage of being readily interpreted as being beneficial or deleterious. In addition, energetics offers a common currency (energy) enabling the consequences of primary toxic mechanisms at the cellular level to be translated into effects on growth, reproduction and survival at the individual and population levels. The ultimate effects at the higher levels of biological organization are thus more readily interpreted and understood. Field and mesocosm studies have provided confirmation that the long-term consequences to growth and survival of individuals and the population can be predicted from measured effects on energy balance observed at the individual level (Gilfillan and Vandermeulen, 1978; Widdows et al., 1987b).

While the well-being of populations and communities are often the ultimate concern, indices of community change need to be complemented by more sensitive, sublethal measures that (a) are predictive (anticipatory) of likely population effects, (b) allow toxicological interpretation of contaminant levels, and (c) enable identification of causality through mechanistic understanding of readily established tissue-concentration relationships and QSARs.

Assessment of environmental pollution using combined chemical and physiological energetic measurements: case studies

The assessment of environmental pollution using physiological energetic measurements of bivalves in conjunction with analysis of contaminants in their tissues began with field studies by Gilfillan et al. (1977), Widdows et al. (1981) and Martin and Severeid (1984). These initial studies, carried out in the U.S.A., examined the biological effects of an oil spill in Maine, and pollution gradients in Narragansett Bay and San Francisco Bay. Since these early field studies, methodological development has continued and laboratory/mesocosm studies have enhanced toxicological understanding of contaminant tissue concentration-response relationships. Moreover, recent studies have shown that direct comparisons can be made between mussels from different sites that are separated by considerable distances; by adopting standard conditions and procedures, using intertidal mussels, and confining measurement to the summer period of growth. For instance, mussel transplantation experiments over distances >1000km have shown that physiological responses and growth reflect environmental rather than genetic differences (Kautsky et al., 1990; Widdows and Salkeld, unpublished data).

Recent applications of this approach and the environmental questions addressed are illustrated with reference to a field monitoring programme in the vicinity of the North Sea oil terminal at Sullom Voe in the Shetlands (for details see Widdows et al., 1987a), and two practical Biological Effects Workshops which investigated pollution gradients in subtropical and northern temperate environments (Bermuda, Widdows et al., 1990; Oslo, Widdows and Johnson, 1988).

Effects of Sullom Voe Oil Terminal:
Combined measurement of SFG and the concentration of two and three ring aromatic hydrocarbons (a major toxic component of oil) in the tissues of M. edulis during the period from 1982 to 1989 has provided an assessment of spatial and temporal impact of oil pollution in the vicinity of the Sullom Voe oil terminal (1982–1985, reported in Widdows et al., 1987a). A unique feature of this case study has been the quantification of (1) environmental inputs of oil, (2) subsequent hydrocarbon accumulation in the mussel tissues, and (3) the resultant biological effects. Mussels living near the source of oil inputs, the tanker loading areas, accumulate hydrocarbons to concentrations typically 10-fold higher than those at the 'clean reference' site, and this results in a significant reduction in the SFG of mussels at these contaminated sites. The relationship between SFG and the log concentration of two and three ring aromatic hydrocarbons indicates that an order of magnitude increase in the tissue concentration can account for an approximately 50% reduction in the growth potential of M. edulis (Fig. 8.4).

Annual variations in SFG and hydrocarbon residues in the tissues of mussels sampled near the oil terminal indicate that there has been neither a gradual increase in the level of hydrocarbon contamination of the water column, nor a gradual deterioration in water quality within Sullom Voe, since the oil terminal became operational. The year-to-year variation in hydrocarbon levels in the mussels is closely correlated (r=0.89; 1984–1989) with the total amount of oil spilt during the month preceding the annual field sampling (in July). These observations therefore suggest

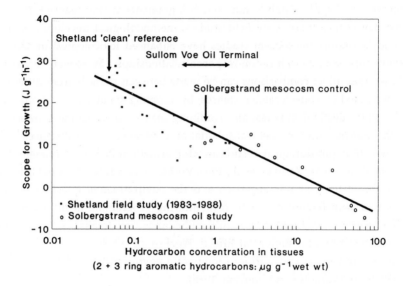

Fig. 8.4. Relationship between scope for growth and the concentration of 2 and 3 ring aromatic hydrocarbons in the tissues of *Mytilus edulis*. Combination of data from Sullom Voe (Shetland oil terminal; Widdows et al., 1987a) and Solbergstrand (Norway) mesocosm studies (long-term oil exposure and recovery experiments; Widdows et al., 1987b).

that the system can readily respond to, and recover from, any transient increases in oil inputs.

For purposes of environmental management, it is necessary not only to quantify the degree of contamination and the resultant deleterious biological effects, but also to address such questions as: "how bad or good are the conditions?" and "at what level of impact should managers start to be concerned and take remedial action?". This requires the presentation and interpretation of data in a wider geographical and ecological context. Figure 8.4 shows a synthesis of data derived from the Sullom Voe study and from mesocosm oil experiments (Widdows et al., 1987b). The semilogarithmic relationship (r= –0.87) illustrates that:

(1) There is an inverse relationship between SFG and the log concentration of aromatic hydrocarbons in the tissues of *M. edulis* over three orders of magnitude and without an apparent threshold effect; this suggests a simple mode of toxicity based on loading of body tissues, and the absence of any significant physiological adaptation.

(2) Mussels from the 'reference' site near Sullom Voe have a very low level of hydrocarbon contamination compared to other sites studied (see below).

(3) At Sullom Voe mussels at the tanker loading jetties nearest the source of pollution are only moderately contaminated, but to no greater extent than many urban estuaries (e.g. River Tamar, S.W. England) and sites such as Solbergstrand on

the Oslofjord (ca. 35km south of Oslo) in a region that is not in the vicinity of any point sources of industrial or urban pollution.

(4) In comparison with other sites, mussels in Sullom Voe experience only moderate levels of pollution and appear to have sufficient capacity (i.e. SFG) to grow, reproduce, and thus maintain a viable population.

Contaminant Gradient in Bermuda:
At the Bermuda Workshop, effects on the physiological energetics of transplanted mussels (*Arca zebra*) were quantified and related to contaminant gradients in Hamilton Harbour, Bermuda, a 'non-industrialized' region receiving relatively low levels of contaminant inputs, and at a 'dump site' in Castle Harbour, Bermuda (Widdows et al., 1990). The results highlight several important features:

(1) In Hamilton Harbour, *Arca* accumulated petroleum hydrocarbons, their polar oxygenated derivatives, PCBs, TBT and Pb in their tissues, and these were all significantly correlated with reductions in SFG. However, toxicological interpretation of the tissue residue data extended the analysis beyond statistical correlations and indicated that the lower molecular weight hydrocarbons (+ polar oxygenated derivatives) and TBT could account for the observed decline in SFG.

(2) In addition, the overall reduction in SFG at the various sites could be 'partitioned', such that at the most contaminated site TBT accounted for 21%, and hydrocarbons for 79%, of the observed effect.

(3) The most conspicuous 'dump site' caused only slight contamination of the marine environment and the recorded effects were not significant.

(4) Any reduction in SFG caused by relatively moderate levels of pollution is likely to shift the energy balance of a suspension feeding bivalve, living in a food limited subtropical environment, closer to the limit of growth and reproduction. Consequently, the effects of pollution are likely to be more marked in oligotrophic environments.

Contaminant Gradient in Langesundfjord:
In contrast, the Oslo Workshop examined a pollution gradient (in Langesundfjord) resulting from an 'industrialized' environment (Widdows and Johnson, 1988; Widdows and Donkin, 1989). Here the results showed that:

(1) The 'cleanest/outermost' site in Langesundfjord was significantly contaminated by hydrocarbons and TBT, and was therefore not an appropriate 'reference' site.

(2) Use of a well-established 'uncontaminated reference' site in the Shetlands enabled the contaminant levels and effects at all sites in Langesundfjord to be placed in a broader context.

(3) There were marked reductions in SFG of *M. edulis* particularly at the inner northern end, nearest to the source of contaminants.

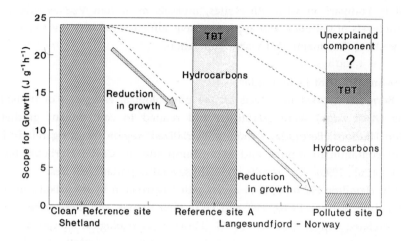

Fig. 8.5. Effect of pollution on scope for growth (SFG) of *Mytilus edulis* from Langesundfjord in comparison to a 'clean reference site' in the Shetlands. Partitioning and relative contribution of hydrocarbons and TBT towards observed decline in SFG at Langesundfjord. (SFG measured under 'standardized conditions' and calculated for a ration level of 0.4mg Particulate Organic Matter L^{-1}; data from Widdows and Johnson, 1988).

(4) Observed effects at the most polluted site were far greater than could be explained by the tissue concentrations of contaminants analyzed (metals, hydrocarbons and TBT), thus indicating toxic effects resulting from both known (PCBs, dioxins and mercury), and as yet unidentified toxicants (Fig. 8.5).

CONCLUSIONS

It is apparent that the degree and extent of pollution in the marine environment can only be assessed if combined chemical and biological effects monitoring programmes are more widely adopted by agencies concerned with environmental protection and management. While the toxicological approach outlined in this chapter provides the possibility of extrapolation from laboratory to environmental conditions, with respect to bivalve sentinel organisms, knowledge at present does not allow extrapolation from the health of bivalves to the health of ecosystems. Further research is necessary to provide information on the relative sensitivities of different species which form components of the ecosystem, as well as on the nature of the relationship between responses of sentinel organisms and responses at the community/ecosystem level.

While significant progress has been made towards the long-term objective of 'predicting effects and interpreting the causes of observed effects', further research is required to provide a more comprehensive toxicological interpretation of contaminant residues in body tissues. Future research within this toxicological framework needs to establish (a) QSARs for other important classes of contaminants, (b) relationships for other important biological responses (e.g. reproductive processes), (c) relationships enabling extrapolation to other key species and (d), further understanding of the combined effects of complex contaminant mixtures.

REFERENCES

Alzieu, C., Héral, M., Thiband, Y., Dardignac, M.J. and Feuillet, M., 1982. Influence des peintures antisalissures à base d'organostanniques sur la calcification de la coquille de l'huitre *Crassostrea gigas*. Rev. Trav. Inst. Pêches. Marit., 45: 101-116.

ASTM (American Society for Testing and Materials) 1980. Standard practice for conducting static acute toxicity tests with larvae of four species of bivalve mollusc. Annual book of ASTM standards, E724-80.

Bayne, B.L. and Worrall, C.M., 1980. Growth and production of mussels (*Mytilus edulis*) from two populations. Mar. Ecol. Prog. Ser., 3: 317-328.

Bayne, B.L., Brown, D.A., Burns, K., Dixon, D.R., Ivanovici, A., Livingstone, D.R., Lowe, D.M., Moore, M.N., Stebbing, A.R.D. and Widdows, J. (Editors), 1985. The Effects of Stress and Pollution on Marine Animals. Praeger Press, New York, 384pp.

Beaumont, A.R. and Budd, M.D., 1984. High mortality of the larvae of the common mussel at low concentrations of tributyltin. Mar. Pollut. Bull., 15: 402-405.

Beaumont, A.R., Tserpes, G. and Budd, M.D., 1987. Some effects of copper on the veliger larvae of the mussel, *Mytilus edulis* and the scallop, *Pecten maximus*, (Mollusca, Bivalvia). Mar. Environ. Res., 21: 299-309.

Bender, M.E., Roberts, M.H., Jr. and Fur, P.O. de, 1987. Unavailability of polynuclear aromatic hydrocarbons from coal particles to the eastern oyster. Environ. Pollut., 44: 243-260.

Bjerregaard, P., Topçuoglu, S., Fisher, N.S. and Fowler, S.W., 1985. Biokinetics of americium and plutonium in the mussel *Mytilus edulis*. Mar. Ecol. Prog. Ser., 21: 99-111.

Bjørseth, A., Knutzen, J. and Skie, J., 1979. Determination of polycyclic aromatic hydrocarbons in sediments and mussels from Sandafjord, W. Norway, by glass capillary chromatography. Sci. Total Environ., 13: 71-86.

Blackman, R.A.A. and Law, R.J., 1980. The Eleni V oil spill: Fate and effects of the oil over the first twelve months. Mar. Pollut. Bull., 11: 199-204.

Boehm, P.D. and Quinn, J.G., 1976. The effect of dissolved organic matter in sea water on the uptake of mixed individual hydrocarbons and Number 2 fuel oil by a marine filter-feeding bivalve (*Mercenaria mercenaria*). Est. Coast. Mar. Sci., 4: 93-105.

Boese, B.L., Lee II, H. and Specht, D.T., 1988. Efficiency of uptake of hexachlorobenzene from water by the tellinid clam, *Macoma nasuta*. Aquat. Toxicol., 12: 345-356.

Bootsma, N., Macey, D.J., Webb, J., and Talbot, V., 1988. Isolation and characterisation of ferritin from the hepatopancreas of the mussel *Mytilus edulis*. Biol. Metals, 1: 106-111.

Borchardt, T., 1983. Influence of food quality on the kinetics of cadmium uptake and loss via food and seawater in *Mytilus edulis*. Mar. Biol., 76: 67-76.

Borchardt, T., Karbe, L., Burchert, S., Haar, E., Hablizel, H. and Zeitner, R., 1989. Influence of size and body condition on metal concentrations in mussels. Implications for biological monitoring programs. In: R.Z. Klekowski, E. Styczynska-Jurewicz and L. Falkowski (Editors), Proc. 21st Eur. Mar. Biol. Symp., Gdansk, Poland, 1988. Ossolineum, Gdansk, pp. 531-541.

Bourgoin, B.P., 1990. *Mytilus edulis* shell as a bioindicator of lead pollution: considerations on bioavailability and variability. Mar. Ecol. Prog. Ser., 61: 253-262.

Bryan, G.W. and Uysal, H., 1978. Heavy metals in the burrowing bivalve *Scrobicularia plana* from the Tamar estuary in relation to environmental levels. J. Mar. Biol. Ass. U.K., 58: 89-108.

Buikema, A.L., McGinnis, M.J. and Cairns, J., 1979. Phenolics in aquatic ecostystems: A selected review of recent literature. Mar. Environ. Res., 2: 87-181.

Burns, K.A. and Smith, J.L., 1981. Biological monitoring of ambient water quality: the case for using bivalves as sentinel organisms for monitoring pertroleum pollution in coastal waters. Est. Coast. Shelf Sci., 13: 433-443.

Butler, R., Roddie, B.D. and Mainstone, C.P., 1990. The effects of sewage sludge on two life-history stages of *Mytilus edulis*. Chem. Ecol., 4: 211-219.

Calabrese, A., MacInnes, J.R. Nelson, D.A. and Miller, J.E., 1977. Survival and growth of bivalve larvae under heavy-metal stress. Mar. Biol., 41: 179-184.

Carpene, E. and George, S.G., 1981. Absorption of cadmium by gills of *Mytilus edulis* (L.). Mol. Physiol., 1: 23-24.

Claisse, D., 1989. Chemical contamination of French coasts. Mar. Pollut. Bull., 20: 523-528.

Clement, L.E., Stekoll, M.S. and Shaw, D.G., 1980. Accumulation, fractionation and release of oil by the intertidal clam *Macoma balthica*. Mar. Biol., 57: 41-50.

Coimbra, J. and Carraça, S., 1990. Accumulation of Fe, Zn, Cu, and Cd during the different stages of the reproductive cycle of *Mytilus edulis*. Comp. Biochem. Physiol., 95C: 265-270.

Connor, P.M., 1972. Acute toxicity of heavy metals to some marine larvae. Mar. Pollut. Bull., 3: 190-192.

Cossa, D., 1988. Cadmium in *Mytilus* spp. : worldwide survey and relationship between seawater and mussel content. Mar. Environ. Res., 26: 265-284.

Cossa, D., Bourget, E., Pouliot, D., Piuze, J. and Chanut, J.P., 1980. Geographical and seasonal variations in the relationship between trace metal content and body weight in *Mytilus edulis*. Mar. Biol., 58: 7-14.

Cowan, A.A., 1981. Organochlorine compounds in mussels from Scottish coastal waters. Environ. Pollut., 2: 129-143.

Craddock, D.R., 1977. Acute toxic effects of petroleum on arctic and subarctic marine organisms. In: D.C. Malins (Editor), Effects of petroleum on arctic and subarctic marine environments and organisms. Vol. II. Biological effects. Academic Press, New York, pp. 1-93.

Davenport, J. and Manley, A.R., 1978. Detection of heightened seawater copper concentrations by the mussel, *Mytilus edulis*. J. Mar. Biol. Ass. U.K., 58: 843-850.

Davenport, J. and Redpath, K.J., 1984. Copper and the mussel *Mytilus edulis* L. In: L. Bolis, J. Zadunaisky and R. Gilles (Editors), Toxins, Drugs and Pollutants in Marine Animals. Springer–Verlag, Berlin, pp. 176-189.

Davies, I.M. and Pirie, J.M., 1980. Evaluation of a 'Mussel watch' project for heavy metals in Scottish coastal waters. Mar. Biol., 57: 87-93.

Dimick, R.E. and Breese, W.P., 1965. Bay mussel embryo bioassay. Proc. 12th Pacific Northwest Industrial Waste Conference, College of Engineering, Univ. of Washington, pp. 165-175.

Di Salvo, L.H., Guard, H.E. and Hunter, L., 1975. Tissue hydrocarbon burden of mussels as potential monitor of environmental hydrocarbon insult. Environ. Sci. Technol., 9: 247-251.

Donkin, P. and Widdows, J., 1990. Quantitative structure—activity relationships in aquatic invertebrate toxicology. Rev. Aquat. Sci., 2: 375-398.

Donkin, P., Widdows, J., Evans, S.V., Worrall, C.M. and Carr, M., 1989. Quantitative structure—activity relationships for the effect of hydrophobic organic chemicals on rate of feeding by mussels (*Mytilus edulis*). Aquat. Toxicol., 14: 277-294.

Ekelund, R., Granmo, Å., Berggren, M., Renberg, L. and Wahlberg, C., 1987. Influence of suspended solids on bioavailability of hexachlorobenzene and lindane to the deposit-feeding marine bivalve, *Abra nitida* (Müller). Bull. Environ. Contam. Toxicol., 38: 500-508.

Ekelund, R., Bergman, Å, Granmo, Å. and Berggren, M., 1990. Bioaccumulation of 4-nonylphenol in marine animals—a re-evaluation. Environ. Pollut., 64: 107-120.

Fabris, G.T., Gibbs, C.F. and Brown, P., 1986. Cadmium and other heavy metals in mussels (*Mytilus edulis planulatus* L.) from Corio Bay, Victoria. Tech. Rep. Mar. Sci. Lab., 49: 1-12.

Farrington, J.W., 1989. Bioaccumulation of hydrophobic organic pollutant compounds. In: S.A. Levin, M.A. Harwell, J.R. Kelly and K.D. Kimball (Editors), Ecotoxicology: Problems and Approaches. Springer-Verlag, New York. pp. 279-313.

Farrington, J.W., Goldberg, E.D., Risebrough, R.W., Martin, J.H. and Bowen, V.T., 1983. U.S. 'Mussel Watch' 1976-1978: An overview of the trace-metal, DDE, PCB, hydrocarbon and artificial radionuclide data. Environ. Sci. Technol., 17: 490-496.

Farrington, J.W., Davis, A.C., Tripp, B.W., Phelps, D.K. and Galloway, W.B., 1987. 'Mussel Watch'— Measurements of chemical pollutants in bivalves as one indicator of coastal environmental quality. In: T.P. Boyle (Editor), New Approaches to Monitoring Aquatic Ecosystems, ASTM STP 940. American Society for Testing and Materials, Philadelphia. pp. 125-139.

Farrington, J.W., Davis, A.C., Frew, N.M. and Knap, A., 1988. ICES/IOC Intercomparison exercise on the determination of petroleum hydrocarbons in biological tissues (mussel homogenate). Mar. Pollut. Bull., 19: 372-380.

Fischer, H., 1988. *Mytilus edulis* as a quantitative indicator of dissolved cadmium. Final study and synthesis. Mar. Ecol. Prog. Ser., 48: 163-174.

Fortner, A.R. and Sick, L.V., 1985. Simultaneous accumulations of naphthalene, a PCB mixture, and benzo[a]pyrene by the oyster, *Crassostrea virginica*. Bull. Environ. Contam. Toxicol., 34: 256-264.

Fowler, S.W., 1990. Critical review of selected heavy metal and chlorinated hydrocarbon concentrations in the marine environment. Mar. Environ. Res., 29: 1-64.

Fowler, S.W. and Oregioni, B., 1976. Trace metals in mussels from the N.W. Mediterranean. Mar. Pollut. Bull., 7: 26-29.

Franklin, A. 1987. The concentrations of metals, organochlorine pesticides and PCB residues in marine fish and shellfish: results from MAFF fish and shellfish monitoring programme 1977-1984. Aquatic Environment Monitoring Report, No 16. MAFF Directorate of Fisheries Research, Lowestoft, 38pp.

George, S.G. and Coombs, T.L., 1977. The effects of chelating agents on the uptake and accumulation of cadmium by the mussel (*Mytilus edulis*). Mar. Biol., 39: 261-268.

George, S.G., Pirie, B.J.S. and Coombs, T.L., 1976. The kinetics of accumulation and excretion of ferric hydroxide in *Mytilus edulis* (L.) and its distribution in the tissues. J. Exp. Mar. Biol. Ecol., 23: 71-84.

Geyer, H., Sheehan, P., Kotzias, D., Freitag, D. and Korte, F., 1982. Prediction of ecotoxicological behaviour of chemicals: relationship between physico-chemical properties and bioaccumulation of organic chemicals in the mussel *Mytilus edulis*. Chemosphere, 11: 1121-1134.

Geyer, H., Freitag, D. and Korte, F., 1984. Polychlorinated biphenyls (PCBs) in the marine environment, particularly in the Mediterranean. Ecotoxicol. Environ. Saf., 8: 129-151.

Gibbs, P.E., Bryan, G.W., Pascoe, P.L. and Burt, G.R., 1987. The use of the dog-whelk, *Nucella lapillus*, as an indicator of tributyltin (TBT) contamination. J. Mar. Biol. Assoc. U.K., 67: 561-569.

Gilfillan, E.S., Mayo, D.W., Page, D.S., Donovan, D. and Hanson, S., 1977. Effects of varying concentrations of petroleum hydrocarbons in sedimants on carbon flux in *Mya arenaria*. In: F.J.Vernberg, A. Calabrese, F.P. Thurberg, and W.B. Vernberg (Editors), Physiological responses of marine biota to pollutants. Academic Press, New York. pp. 299-314.

Gilfillan, E.S. and Vandermenlen, J.H. 1978. Alterations in growth and physiology of soft shell clams, *Mya arenaria*: chronically oiled with Bunker C from Chedabucto Bay, Nova Scotia, 1970-76. J. Fish. Res. Board Can., 35: 630-636.

Goldberg, E.D., 1975. The Mussel Watch - A first step in global marine monitoring. Mar. Pollut. Bull., 6: 111.

Goldberg, E.D., Bowen. V.T., Farrington, J.H., Harvey, G., Martin, J.H., Parker, P.L., Riseborough, R.W., Robertson, W., Schneider, E. and Gamble, E., 1978. The Mussel Watch. Environ. Conserv., 5: 1-25.

Goldberg, E.D., Koide, M., Hodge, V., Flegal, A.R. and Martin, J., 1983. U.S. Mussel Watch: 1977-1978 results on trace metals and radionuclides. Est. Coast. Shelf Sci., 16: 69-93.

Grovhoug, J.G., Seligman, P.F., Vafa, G. and Fransham, R.L., 1986. Baseline Measurements of butyltin in U.S. harbours and estuaries, In: Proceedings of the Organotin Symposium of the Oceans 86 Conference, Washington DC, 1986. IEEE Piscataway, NJ and Marine Technology Society, Washington, DC, pp. 1283-1288.

Hale, R.C., Greaves, J., Gallagher, K. and Vadas, G.G., 1990. Novel chlorinated terphenyls in sediments and shellfish of an estuarine environment. Environ. Sci. Technol., 24: 1727-1731.

Hamilton, E.I., 1980. Concentration and distribution of uranium in *Mytilus edulis* and associated materials. Mar. Ecol. Prog. Ser., 2: 61-73.

Hansen, N., Jensen, V.B., Appelquist, H. and Mørch, E., 1978. The uptake and release of petroleum hydrocarbons by the marine mussel *Mytilus edulis*. Prog. Water Technol., 10: 351-359.

Hawker, D.W. and Connell, D.W., 1986. Bioconcentration of lipophilic compounds by some aquatic organisms. Ecotoxicol. Environ. Saf., 11: 184-197.

Hermens, J.L.M., 1986. Quantitative structure—activity relationships in aquatic toxicology. Pestic. Sci., 17: 287-296.

Hilbish, T.J., 1986. Growth trajectories of shell and soft tissue in bivalves: Seasonal variation in *Mytilus edulis* L. J. Exp. Mar. Biol. Ecol., 96: 103-113.

Howell, R., Grant, A.M. and Maccoy, N.E.J., 1984. Effect of treatment with reserpine on the change in filtration rate of *Mytilus edulis* subjected to dissolved copper. Mar. Pollut. Bull., 15: 436-439.

Hummel, H., Bogaards, R.H., Nieuwenhuize, J., De Wolf, L. and Liere, J.M. van, 1990. Spatial and seasonal differences in the PCB content of the mussel *Mytilus edulis*. Sci. Total Environ., 92: 155-163.

Jørgensen, C.B., 1981. Mortality, growth and grazing impact of a cohort of bivalve larvae, *Mytilus edulis* L. Ophelia, 20: 185-192.

Jørgensen, C.B., Larsen, P.S., Møhlenberg, F. and Riisgård, H.U., 1988. The mussel pump: properties and modelling. Mar. Ecol. Prog. Ser., 45: 205-216.

Kautsky, N., Johanneson, K. and Tedengren, M., 1990. Genotypic and phenotypic differences between Baltic and North Sea populations of *Mytilus edulis* evaluated through reciprocal transplantations. I. Growth and morphology. Mar. Ecol. Prog. Ser., 59: 203-210.

King, D.G. and Davies, I.M., 1987. Laboratory and field studies of the accumulation of inorganic mercury by the mussel *Mytilus edulis* (L). Mar. Pollut. Bull., 18: 40-45.

King, N., Miller, M. and De Mora, S., 1989. Tributyltin levels for sea water, sediment, and selected marine species in coastal northland and Auckland, New Zealand. N. Z. J. Mar. Freshw. Res., 23: 287-294.

Klöckner, K., Rosenthal, H. and Willführ, J., 1985. Invertebrate bioassays with North Sea water samples. I. Structural effects on embryos and larvae of serpulids, oysters and sea urchins. Helgoländer. Wiss. Meeresunters., 39: 1-19.

Klungsøyr, J. Wilhelmsen, S., Westrheim, K., Saetredt, E. and Palmork, K.H., 1988. The GEEP Workshop: Organic chemical analyses. Mar. Ecol. Prog. Ser., 46: 19-26.

Könemann, H., 1980. Structure-activity relationships and additivity in fish toxicities of environmental pollutants. Ecotoxicol. Environ. Saf., 4: 415-421.

Kramer, K.J.M., Jenner, H.A. and de Zwart, D., 1989. The valve movement response of mussels: a tool in biological monitoring. Hydrobiologia, 188/189: 433-443.

Langston, W.J., 1982. The distribution of mercury in British estuarine sediments and its availability to deposit-feeding bivalves. J. Mar. Biol. Ass. U.K., 62: 667-684.

Laughlin, R.B., French, W. and Guard, H.E., 1986. Accumulation of bis(tributyltin) oxide by the marine mussel *Mytilus edulis*. Environ. Sci. Technol., 20: 884-890.

Law, R.J., 1981. Hydrocarbon concentrations in water and sediments from U.K. marine waters, determined by fluorescence spectroscopy. Mar. Pollut. Bull., 12: 153-157.

Linden, O., 1987. The scope of the organotin issue in Scandinavia. In: Proceedings of the International Organotin Symposium of the Oceans 87 Conference, Nova Scotia, Canada, 1987. IEEE Piscataway, NJ and Marine Technology Society, Washington, DC, pp. 1320-1323.

Livingstone, D.R., 1991. Organic xenobiotic metabolism in marine invertebrates. In: R. Gilles (Editor), Adv. Comp. Environ. Physiol., 7: 45-185.

Lobel, P.B., 1986. Role of kidney in determining the whole soft tissue zinc concentration of individual mussels (*Mytilus edulis*). Mar. Biol., 92: 355-359.

Loring, D.H. and Prossi, F., 1986. Cadmium and lead cycling between water, sediment, and biota in an artificially contaminated mud flat on Borkum (FRG). Water Sci. Technol., 18: 131-139.

Lubet, P., Brichon, G., Besnard, J.Y. and Zwingelstein, G., 1985. Composition and metabolism of lipids in some tissues of the mussel *Mytilus galloprovincialis* L. (Moll. Bivalvia)—in vivo and in vitro incorporation of 1(3)-[^{3}H]-glycerol. Comp. Biochem. Physiol., 82B: 425-431.

Lubet, P., Brichon, G., Besnard, J.Y. and Zwingelstein, G., 1986. Sexual differences in the composition and metabolism of lipids in the mantle of the mussel *Mytilus galloprovincialis* Lmk. (Mollusca: Bivalvia). Comp. Biochem. Physiol., 84B: 279-285.

Mackie, P.R., Hardy, R., Whittle, K.J., Bruce, C. and McGill, A.S., 1980. Polynuclear aromatic hydrocarbons: In: A. Bjørseth and A.J. Dennis (Editors), Chemistry and Biological Effects. Battelle Press, Columbus, Ohio, pp. 379-393.

Manley, A.R., 1983. The effects of copper on the behaviour, respiration, filtration and ventilation activity of Mytilus edulis. J. Mar. Biol. Ass. U.K., 63: 205-222.

Mantoura, R.F.C. and Woodward, E.M.S., 1983. Conservative behaviour of riverine dissolved organic carbon in the Severn Estuary: chemical and geochemical implication? Geochim. Cosmochim. Acta, 47: 1293-1309.

Martin, J.L.M., 1979. Schema of lethal action of copper on mussels. Bull. Environ. Contam. Toxicol., 21: 808-814.

Martin, M. and Castle, W., 1984. Petrowatch: Petroleum hydrocarbons, synthetic organic compounds, and heavy metals in mussels from the Monterey Bay area of central California. Mar. Pollut. Bull., 15: 259-266.

Martin, M., Osborn, K.E., Billig, P. and Glickstein, N., 1981. Toxicities of ten metals to Crassostrea gigas and Mytilus edulis embryos and Cancer magister larvae. Mar. Pollut. Bull., 12: 305-308.

Martin, M. and Severeid, R., 1984. Mussel Watch monitoring for the assessment of trace toxic constituents in California marine waters. In: H.H. White (Editor), Concepts in Marine Pollution Measurements. Maryland Sea Grant College, University of Maryland, pp. 291-323.

Mason, R.P. 1988. Hydrocarbons in mussels around the Cape Peninsula, South Africa. S. Afr. J. Mar. Sci., 7: 139-151.

McDowell Capuzzo, J., Farrington, J.W., Rantamaki, P., Hovey Clifford, C., Lancaster, B.A., Leavitt, D.F. and Jia, X., 1989. The relationship between lipid composition and seasonal differences in the distribution of PCBs in Mytilus edulis L. Mar. Environ. Res., 28: 259-264.

McIntyre, A.D. and Pearce, J.B. (Editors) 1980. Biological effects of marine pollution and problems of monitoring. Rapp. P.-V. Reun. Cons. Int. Explor. Mer.,179:

McLeese, D.W. and Burridge, L.E., 1987. Comparative accumulation of PAHs in marine invertebrates. In: J.M. Capuzzo and D.R. Kestor (Editors), Oceanic Processes in Marine Pollution, Vol. 1. Biological Processes and Wastes in the Ocean. Krieger, Malabar, Florida, pp. 109-117.

Miller, M.M., Wasik, S.P., Huang, G-L., Shiu, W-Y. and Mackay, D., 1985. Relationship between octanol-water partition coefficient and aqueous solubility. Environ. Sci. Technol., 19: 522-529.

Miramand, P., Guary, J.C. and Fowler, S.W., 1980. Vanadium transfer in the mussel Mytilus galloprovincialis. Mar. Biol., 56: 281-293.

Miyata, H., Takayama, K., Ogaki, J., Kashimote, T., and Fukushima, S., 1987. Polychlorinated dibenzo-p-dioxins in blue mussel from marine coastal waters in Japan. Bull. Environ. Contam. Toxicol., 39: 877-883.

NRC. 1980. The International Mussel Watch: Report of a Workshop. Washington DC, U.S. National Academy of Sciences. National Research Council, Publications Office, 248pp.

Oehme, M., Manø, S., Brevik, E.M. and Knutzen, J., 1989. Determination of poly-chlorinated dibenzofuran (PCDF) and dibenzo-p-dioxin (PCDD) levels and isomer patterns in fish, crustacea, mussel and sediment samples from a fjord region polluted by Mg production. Fresenius Z. Anal. Chem., 335: 987-997.

Osibanjo, O., and Bamgbose, O., 1990. Chlorinated hydrocarbons in marine fish and shellfish of Nigeria. Mar. Pollut. Bull., 21: 581-586.

Page, D.S., Foster, J.C. and Gilfillan, E.S., 1987. Kinetics of aromatic hydrocarbon depuration by oysters impacted by the Amoco Cadiz oil spill. In: J. Kuiper and W.J. van den Brink (Editors), Fate and Effects of Oil in Marine Ecosystems. Nijhoff, Dordrecht, pp. 243-252.

Page, D.S. and Widdows, J., 1992. Temporal and spatial variation in levels of alkyltins in mussel tissues: A toxicological interpretation of field data. In: 3rd International Organotin Symposium, April 1990, Monaco. Mar. Environ. Res., 32: in press.

Palmork, K.H. and Solbakken, J.E., 1981. Distribution and elimination of [9-14C]phenanthrene in the horse mussel (Modiolus modiolus). Bull. Environ. Contam. Toxicol., 26: 196-201.

Pempkowiak, J., Bancer, B., Legezynska, E. and Kulinski, W., 1989. The accumulation and uptake of cadmium by four selected Baltic species in the presence of marine humic substances. In: R.Z. Klekowski, E. Styczynska-Jurewicz and L. Falkowski (Editors), Proc. 21st Eur. Mar. Biol. Symp, Gdansk, Poland, 1988. Ossolineum, Gdansk, pp. 599-608.

Phillips, D.J.H., 1980. Quantitative aquatic biological indicators. Applied Science Publishers Ltd., London. 488pp.

Phillips, D.J.H., 1985. Organochlorines and trace metals in green-lipped mussels *Perna viridis* from Hong Kong waters: a test of indicator ability. Mar. Ecol. Prog. Ser., 21: 251-258.

Phillips, D.J.H. and Spies, R.B., 1988. Chlorinated hydrocarbons in the San Francisco estuarine ecosystem. Mar. Pollut. Bull., 19: 445-453.

Pieters, H., Kluytmans, J.H., Zandee, D.I. and Cadée, G.C. 1980. Tissue composition and reproduction of *Mytilus edulis* in relation to food availability. Neth. J. Sea Res., 14: 349-361.

Porte, C., Barcelo, D., Tavares, T.M., Rocha, V.C. and Albaiges, J., 1990. The use of the Mussel Watch and Molecular Marker concepts in studies of hydrocarbons in a tropical bay (Todos os Santos, Bahia, Brazil). Arch. Environ. Contam. Toxicol., 19: 263-274.

Pruell, R.J., Lake, J.L., Davis, W.R. and Quinn, J.G., 1986. Uptake and depuration of organic contaminants by blue mussels (*Mytilus edulis*) exposed to environmentally contaminated sediment. Mar. Biol., 91: 497-507.

Ramesh, A., Tanabe, S., Subramanian, A.N., Mohan, D., Venugopalan, V.K. and Tatsukawa, R., 1990. Persistent organochlorine residues on green mussels from coastal waters of South India. Mar. Pollut. Bull., 21: 587-590.

Redpath, K.J., 1985. Growth inhibition and recovery in mussels (*Mytilus edulis*) exposed to low copper concentrations. J. Mar. Biol. Ass. U.K., 65: 421-431.

Riisgård, H.U. and Randløv, A. 1981. Energy budgets, growth and filtration rates in *Mytilus edulis* at different algal concentrations. Mar. Biol., 61: 227-234.

Riisgård, H.U., Bjørnestad, E. and Møhlenberg, F., 1987. Accumulation of cadmium in the mussel *Mytilus edulis*: kinetics and importance of uptake via food and sea water. Mar. Biol., 96: 349-353.

Roesijadi, G., 1982. Uptake and incorporation of mercury into mercury-binding proteins of gills of *Mytilus edulis* as a function of time. Mar. Biol., 66: 151-157.

Roesijadi, G. and Fellingham, G.W., 1987. Influence of Cu, Cd, and Zn pre-exposure on Hg toxicity in the mussel *Mytilus edulis*. Can. J. Fish. Aquat. Sci., 44: 680-684.

Salazar, M.H. and Salazar, S.M., 1987. TBT effects on juvenile mussel growth. In: Proceedings of the International Organotin Symposium of the Oceans 87 Conference, Nova Scotia, Canada, 1987. IEEE Piscataway, NJ and Marine Technology Society, Washington, DC, pp. 1504-1510.

Salazar, M.H. and Salazar, S.M., 1992. Mussels as bioindicators: Effects of TBT on survival, bioaccumulation and growth under natural conditions. In: M.A. Champ and P.F. Seligman (Editors), Tributyltin: Environmental fate and effects, Part III. Elsevier Science Publishers, B.V., New York, in press.

Saxena, A.K. 1987. Organotin compounds: Toxicology and biomedical applications. Appl. Organometallic Chem., 1: 39-56.

Schulz-Baldes, 1974. Lead uptake from sea water and food, and lead loss in the common mussel *Mytilus edulis*. Mar. Biol., 25: 177-193.

Sericano, J.L., Atlas, E.L., Wade, T.L. and Brooks, J.M., 1990. NOAA's status and trends mussel watch program: Chlorinated pesticides and PCBs in oysters (*Crassostrea virginica*) and sediments from the Gulf of Mexico, 1986-1987. Mar. Environ. Res., 29: 161-203.

Shaw, D.G., Hogan, T.E. and McIntosh, D.J., 1986. Hydrocarbons in bivalve mollusks of Port Valdez, Alaska: Consequences of five years permitted discharge. Est. Coast. Shelf Sci., 23: 863-872.

Shaw, G.R. and Connell, D.W., 1984. Physicochemical properties controlling polychlorinated biphenyl (PCB) concentrations in aquatic organisms. Environ. Sci. Technol., 18: 18-23.

Simkiss, K., 1983. Lipid solubility of heavy metals in saline solutions. J. Mar. Biol. Ass. U.K., 63: 1-7.

Simkiss, K., 1984. Effects of metal ions on respiratory structures. In: L. Bolis, J. Zadunaisky and R. Gilles (Editors), Toxins, Drugs, and Pollutants in Marine Animals. Springer-Verlag, Berlin, pp. 137-146.

Simpson, R.D., 1979. Uptake and loss of zinc and lead by mussels (*Mytilus edulis*) and relationships with body weight and reproductive cycle. Mar. Pollut. Bull., 10: 74-78.

Sloof, W., de Zwart, D. and Marquenie, J.M., 1983. Detection limits of a biological monitoring system for chemical water pollution based on mussel activity. Bull. Environ. Contam. Toxicol. 30: 400-405.

Smaal, A.C., and Stralen, M.R. van, 1990. Average annual growth and condition of mussels as a function of food source. Hydrobiologia, 195: 179-188.

Snoeij, N.J., Penninks, A.H. and Seinen, W., 1987. Biological activity of organotin compounds—an overview. Environ. Res., 44: 335-353.

Soler, M., Grimalt, J.O and Albaigés, J., 1989. Distribution of aliphatic, aromatic and chlorinated hydrocarbons in mussels from the Spanish Atlantic coast (Galicia). An assessment of pollution partameters. Chemosphere, 19: 1489-1498.

Strømgren, T. 1982. Effects of heavy metals (Zn, Hg, Cu, Cd, Pb, Ni) on the length growth of Mytilus edulis. Mar. Biol., 72: 69-72.

Strømgren, T. 1986. The combined effect of copper and hydrocarbons on the length growth of Mytilus edulis. Mar. Environ. Res., 19: 251-258.

Strømgren, T. and Bongard, T., 1987. The effects of tributyltin oxide on growth of Mytilus edulis. Mar. Pollut. Bull., 18: 30-31.

Sturesson, U., 1976. Lead enrichment in shells of Mytilus edulis. Ambio, 5: 253-256.

Suffet, I.H. and MacCarthy, P., 1989. Aquatic Humic Substances. Influence on Fate and Treatment of Pollutants. American Chemical Society, Washington, DC, 864pp.

Tanabe, S., Tatsukawa, R., and Phillips, D.J., 1987. Mussels as bioindicators of PCB pollution: A case study of uptake and release of PCB isomers and congeners in green-lipped mussels (Perna viridis) in Hong Kong waters. Environ. Pollut., 47: 41-62.

Theede, H. and Jung, C.T., 1989. Experimental studies of the effects of some environmental factors on the accumulation and elimination of cadmium by the mussel Mytilus edulis. In: R.Z. Klekowski, E. Styczynska-Jurewicz and L. Falkowski (Editors), Proc. 21st Eur. Mar. Biol. Symp, Gdansk, Poland, 1988. Ossolineum, Gdansk, pp. 615-624.

Topping, G., 1983. The analysis of trace metals in biological reference materials: A discussion of the results of the intercomparison studies conducted by the International Council for the Exploration of the Sea. In: C.S. Wong, E. Boyle, K.W. Bruland, J.D. Burton and E.D. Goldberg (Editors), Trace Metals in seawater. Plenum Press, New York, pp. 155-173.

Tušnik, P. and Planinc, R., 1988. Concentrations of the trace metals, (Hg, Cd) and its seasonal variations in Mytilus galloprovincialis. Biol. Vestn., 36: 1-82.

Uhler, A.D., Coogan, T.H., Davis, K.S., Durell, G.S., Steinhauer, W.G., Freitas, S.Y. and Boehm, P.D., 1989. Findings of tributyltin, dibutyltin and monobutyltin in bivalves from selected U.S. coastal waters. Environ. Toxicol. Chem., 8: 971-979.

Ünlü, M.Y. and Fowler, S.W., 1979. Factors affecting the flux of arsenic through the mussel Mytilus galloprovincialis. Mar. Biol., 51: 209-219.

Viarengo, A., 1989. Heavy metals in marine invertebrates: mechanisms of regulation and toxicity at the cellular level. Rev. Aquat. Sci., 1: 295-317.

Viarengo, A., Palmero, S., Zanicchi, G., Capelli, R., Vaissiere, R., and Orunesu, M., 1985. Role of metallothioneins in Cu and Cd accumulation and elimination in the gill and digestive gland cells of Mytilus galloprovincialis Lmk. Mar. Environ. Res., 16: 23-36.

Wade, T., Garcia-Romero, B., and Brooks, J.M., 1988. Tributyltin contamination in bivalves from United States coastal estuaries. Environ. Sci. Technol., 22: 1488-1493.

Waldock, M. J. and Thain, J.E., 1983. Shell thickening in Crassostrea gigas: Organotin antifouling or sediment induced? Mar. Pollut. Bull., 14: 411-415.

Warren, C.E. and Davis, G.E., 1967. Laboratory studies on the feeding, bioenergetics and growth in fish. In: S.D. Gerking (Editors), The biological basis of freshwater fish production. Blackwell Scientific Pubs. Ltd., Oxford. pp. 175-214.

Watanabe, I., Kashimoto, T. and Tatsukawa, R., 1987. Polybrominated biphenyl esters in marine fish, shellfish and river and marine sediments in Japan. Chemosphere, 16: 2389-2396.

Widdows, J., 1985. Physiological procedures. In: B.L. Bayne, D.A. Brown, K. Burns, D.R. Dixon, A. Ivanovici, D.R. Livingstone, D.M. Lowe, M.N. Moore, A.R.D. Stebbing and J. Widdows (Editors), The Effects of Stress and Pollution on Marine Animals. Praeger Press, New York, pp. 161-178.

Widdows, J., 1991. Physiological ecology of mussel larvae. Aquaculture, 94(2/3): 147-164.

Widdows, J. and Johnson, D., 1988. Physiological energetics of Mytilus edulis: Scope for growth. Mar. Ecol. Prog. Ser., 46: 113-121.

Widdows, J., and Donkin, P., 1989. The application of combined tissue residue chemistry and physiological measurements of mussels (Mytilus edulis) for the assessment of environmental pollution. Hydrobiologia, 188/198: 455-461.

Widdows, J. and Donkin, P. 1991. Role of physiological energetics in ecotoxicology. Comp. Biochem. Physiol., 100C: 69-75.

Widdows, J., and Page, D.S., 1992. Effects of tributyltin and dibutyltin on the physiological energetics of the mussel, *Mytilus edulis*. Mar. Environ. Res., in press.

Widdows, J., Moore, M.N., Lowe, D.M. and Salkeld, P.N., 1979. Some effects of a dinoflagellate bloom (*Gyrodinium aureolum*) on the mussel, *Mytilus edulis*. J. Mar. Biol. Assoc. U.K., 59: 522-524.

Widdows, J., Phelps, D.K. and Galloway, W., 1981. Measurement of physiological condition of mussels transplanted along a pollution gradient in Narragansett Bay. Mar. Environ. Res., 4: 181-194.

Widdows, J., Bakke, T., Bayne, B.L., Donkin, P., Livingstone, D.R., Lowe, D.M., Moore, M.N., Evans, S.V. and Moore, S.L., 1982. Responses of *Mytilus edulis* on exposure to the water-accommodated fraction of North Sea oil. Mar. Biol., 67: 15-31.

Widdows, J., Moore, S.L., Clarke, K.R. and Donkin, P., 1983. Uptake, tissue distribution and elimination of [1-^{14}C] naphthalene in the mussel *Mytilus edulis*. Mar. Biol., 76: 109-114.

Widdows, J., Donkin, P., Salkeld, P.N. and Evans, S.V., 1987a. Measurement of scope for growth and tissue hydrocarbon concentrations of mussels (*Mytilus edulis*) at sites in the vicinity of Sullom Voe:—A case study. In: J. Kuiper and W.J. van den Brink (Editors), Fate and effects of oil in marine ecosystems. Nijhoff, Dordrecht, pp. 269-277.

Widdows, J., Donkin, P., and Evans, S.V., 1987b. Physiological responses of *Mytilus edulis* during chronic oil exposure and recovery. Mar. Environ. Res., 23: 15-32.

Widdows, J., Newell, R.I.E. and Mann, R., 1989. Effects of hypoxia and anoxia on survival, energy metabolism and feeding of oyster larvae (*Crassostrea virginica*, Gmelin). Biol. Bull., 177: 154-166.

Widdows, J., Burns, K.A., Menon, N.R., Page, D.S. and Soria, S., 1990. Measurement of physiological energetics (scope for growth) and chemical contaminants in mussels (*Arca zebra*) transplanted along a contamination gradient in Bermuda. J. Exp. Mar. Biol. Ecol., 138: 99-117.

Winberg, G.G., 1960. Rate of metabolism and food requirements of fishes. Transl. Ser. Fish. Res. Board Can., 194: 1-202.

Zamuda, C.D. and Sunda, W.G., 1982. Bioavailability of dissolved copper to the American oyster *Crassostrea virginica*. 1. Importance of chemical speciation. Mar. Biol., 66: 77-82.

Zamuda, C.D., Wright, D.A. and Smucker, R.A., 1985. The importance of dissolved organic compounds in the accumulation of copper by the American oyster *Crassostrea virginica*. Mar. Environ. Res., 16: 1-12.

Zandee, D.I., Kluytmans, J.H., Zurburg, W. and Pieters, H., 1980. Seasonal variations in biochemical composition of *Mytilus edulis* with reference to energy metabolism and gametogenesis. Neth. J. Sea Res., 14: 1-29.

Chapter 9

MUSSELS AND ENVIRONMENTAL CONTAMINANTS: MOLECULAR AND CELLULAR ASPECTS

DAVID R. LIVINGSTONE AND RICHARD K. PIPE

INTRODUCTION

Information is presented on the metabolism and toxicity of metal and organic contaminants, so-called xenobiotics (foreign compounds), at the molecular and cellular levels. Data are largely for mussels, but information for other bivalve and molluscan species are included where relevant. Recent major reviews include Viarengo (1989), Renwrantz (1990) and Livingstone (1991a) on, respectively, heavy metals, internal defence systems and organic xenobiotics in molluscs. The uptake and discharge of xenobiotics, including tissue concentrations of metals and organic pollutants accumulated in the field, in particular data from the Mussel Watch monitoring programmes, are described in detail in Chapter 8. However, aspects which are of relevance to the metabolism and effects of xenobiotics are also summarized below.

Organic Xenobiotics

Rates of metabolism of organic xenobiotics are dependent on the tissue concentration of the compound (see p.435–440). Therefore, factors which affect uptake or discharge of the xenobiotic will also affect its metabolism. Lipophilic compounds such as organic xenobiotics are readily taken up into the tissues of bivalves and concentrated to levels greatly above those of the surrounding seawater. The bioaccumulation of a vast range of compounds by mollusc species has been recorded in field and laboratory studies. The xenobiotics include aliphatic hydrocarbons and polynuclear aromatic hydrocarbons (PAHs), polychlorobiphenyls (PCBs), polychlorinated dibenzo-p-dioxins and dibenzofurans, organochlorines such as 1,1-*bis* [4-chlorophenyl]-2,2,2-trichloroethane (DDT) and aldrin and dieldrin, aromatic amines, nitroaromatics, phthalate esters, organophosphorous pesticides and organometallics such as phenylmercuric acetate and *bis*(tributyltin)oxide (Livingstone, 1991a). Xenobiotics are bioaccumulated from the water-column, and when associated with either abiotic particulate material, or associated with biota such as detritus and phytoplankton.

The qualitative and quantitative patterns of bioaccumulation and depuration depend on many variables, including bioavailability and route of uptake, exposure time, and various physicochemical and biological factors. The bioaccumulation of most organic xenobiotics involves an initial linear rate of uptake followed by the eventual attainment of a maximal tissue equilibrium concentration, both parameters increasing with the exposure concentration of the xenobiotic. In some cases biphasic or multiphasic uptake has been observed, and this has been interpreted in terms of xenobiotics entering a multicompartment system. Rates of uptake are much lower from sediments, reflecting the lower bioavailability of the associated chemicals. Examples of approximate rates of uptake in relation to xenobiotic exposure concentration, for a range of compounds and species, are given in Livingstone (1991a) and p.435–440.

Uptake is thought to be essentially a passive process, involving movement and equilibrium of the chemical between aqueous (external) and biotic (organism) compartments. The tendency to bioaccumulate increases with increasing hydrophobicity of the chemical, allowing prediction of bioaccumulation from the octanol/water partition coefficient or water solubility of the compound. In some cases the observed bioaccumulation exceeds the predicted, e.g. a x10 excess of *bis*(tributyl)tin oxide taken up by the common mussel, *Mytilus edulis* (Laughlin et al., 1986), whereas in others chemical structure appears to be important, e.g. non-*ortho* coplanar congeners were taken up much more slowly than many other PCB isomers by the mussel *Perna viridis* (Kannan et al., 1989). Possible mechanisms for selective bioaccumulation include membrane penetration phenomena, specific binding sites, macromolecular adduct formation and localization in major subcellular sites of uptake such as the lysosomes (Kurelec and Pivčević, 1989, 1991; Livingstone, 1991a). Important factors in determining tissue distribution and disposition of xenobiotics are lipid levels and route of uptake. Bioaccumulation generally increases with increasing tissue lipid levels, either in individual organisms of the same species, or in different tissues, resulting in the hepatopancreas (digestive gland), which is also high in lysosomes, being a major site of xenobiotic uptake. Gills are an initial site of uptake and can be a major one for particular xenobiotics, e.g. the uptake of *o*-toluidine (Knezovich and Crosby, 1985) and *bis*(tributyl)tin oxide (Laughlin and French, 1988) by *M. edulis*. Differences in disposition occur that are not obviously attributable to either lipid levels or route of uptake, viz. in *M. edulis*, whereas the mantle is a major site of uptake of naphthalene, proportional to lipid levels (Widdows et al., 1983), the bioaccumulation of *bis*(tributyl)tin oxide by this tissue is similar to that of the adductor muscle (Laughlin and French, 1988). Patterns of bioaccumulation are also

Table 9.1. Relationship between exposure time and depuration half-life for the elimination of various hydrocarbons from mussels and other bivalves[1]

Hydrocarbon	Species	Exposure time	Depuration half-life	Reference
N-alkanes	*Mytilus edulis*	2 days	0.2–0.8 days	Farrington et al., 1982
Paraffins/ naphthalene	*M. edulis*	16 days	~6 days	Broman and Ganning, 1986
Naphthalene	*M. edulis*	4h	2.2–5.1h	Widdows et al., 1983
Alkyl- naphthalene	*M. edulis*	2 days	0.9 days	Farrington et al., 1982
Phenanthrene	*Modiolus modiolus*	2 days	~1–2 days	Palmork and Solbakken, 1981
Methyl- phenanthrenes	*M. edulis*	2 days	1.7 days	Farrington et al., 1982
4-ring PAH[2]	*M. edulis*	40 days	14–30 days	Pruell et al., 1986
2–4 ring PAH	*Macoma balthica*	180 days	>60 days	Clement et al., 1980
Benzo[a]pyrene	*Rangia cuneata*	1 day	6–10 days	Neff et al., 1976
	M. edulis	40 days	15.4 days	Pruell et al., 1986
	M. edulis	Field[3]	16 days	Dunn and Stitch, 1976

[1] Half-lives are direct from reference or calculated from body-burden data.
[2] Polynuclear aromatic hydrocarbons (PAH).
[3] Field animals of unspecified exposure time.

affected directly, or indirectly (e.g. via lipid levels), by reproductive condition and/or seasonality.

Depuration, like uptake, is thought to be essentially a passive process, although some active excretion has been proposed, e.g. loss of naphthalene from gills and kidneys of *M. edulis* (Widdows et al., 1983). Exponential depuration curves are observed, which for many xenobiotics may be markedly affected by the duration of pre-exposure to the chemical. Thus, whereas elimination of the xenobiotic, following short-term exposure, is usually rapid and complete, long-term exposure results in slower and often incomplete elimination. This results in an increase in the half-life of depuration with increasing time of pre-exposure, and the relationship is particularly evident for petroleum hydrocarbons (Table 9.1). The mechanisms underlying such phenomena are unknown, but have been interpreted in terms of the xenobiotic entering a more stable molecular or cellular compartment, with a lower

Table 9. 2. Approximate depuration half-lives of polychlorobiphenyls from mussels and other bivalves[1]

Species	PCB[2]	Exposure time (days)	Depuration half-life (days)	Reference
Cardium edule	Cl_3	10	~7	Langston, 1978
	Cl_4		7–14	
	Cl_5		21	
	Cl_6		>21	
Perna viridis	Cl_2	17	0.5–2.5	Tanabe et al., 1987
	Cl_3		0.5–6.5	
	Cl_5		4.9–8.3	
Mytilus edulis	Cl_3	40	16.3	Pruell et al., 1986
	Cl_5		27.9	
	Cl_6		37–46	
Mytilus edulis	Cl_{2-4}	89	4.6–9.1	Calambokidis et al., 1979
	Cl_{5-7}		20–50	

[1]Half-lives are direct from reference or calculated from body-burden data.
[2]Number of chlorine atoms as subscript.

rate of xenobiotic turnover. Chemical structure, either directly or through hydrophobicity, also affects elimination. Thus, higher molecular weight, less water-soluble compounds are generally eliminated at a lower rate, this being particularly evident for increasingly chlorinated PCB congeners (Table 9.2). The position of chlorine atoms in PCB congeners also affects elimination, those with most *ortho*-substituted chlorines being retained least by the cockle *Cardium edule* and the clam *Macoma balthica* (Langston, 1978). Patterns of elimination are also affected by other factors such as season (probably via lipid levels) and temperature.

Metals

Aspects of the mechanisms of uptake and disposition of metals in bivalves have been the subject of several reviews (Simpkiss et al., 1982; Simpkiss and Mason, 1984; George and Viarengo, 1985; Fowler, 1987; Viarengo, 1989). Mussels are excellent bioaccumulators of both 'essential' metals (e.g. Zn, Cu, Co) and those for which no biological function has yet been discerned (e.g. Cd, Hg). Other metals accumulated include Ag, Al, Cr, Fe, Mn, Ni and Pb, and radionuclides such as uranium and the transuranium elements [239,240]Pu, [238]Pu and [241]Am. The bioaccumulation of metals is dependent on chemical speciation, routes and mechanisms of uptake, intracellular compartmentation and other aspects of cellular metal homeostasis.

For most metals uptake is proportional to the concentration of the metal in the external medium, and for the soluble form of heavy metals is indicated to be mainly a passive-transport process. Although the gill is the most important tissue for soluble metal uptake, the digestive gland is the major site for particulate-bound metal via endocytosis, an active-transport mechanism requiring ATP; the endocytotic vesicles subsequently fuse with primary lysosomes. The kidney is another major site of metal accumulation. Major sites of localization of radionuclides in *M. edulis* include the byssal threads, periostracum and pericardial gland. Metal gradients across membranes are maintained, and metals retained within the tissues, by their binding to various specific and nonspecific reactive ligands within the cell (see section on heavy metal metabolism p.442–443). Metal accumulation also varies with many biological and physical factors, including season/reproductive state, salinity and water-depth.

The movement and discharge of metals has been studied mainly in relation to mechanisms of metal homeostasis. Release into the external medium is mainly from lysosomes via exocytosis of residual bodies. Indicated or hypothesized release into the haemolymph is from insoluble granules and possibly also thiol-containing cytosolic compounds.

MOLECULAR ASPECTS

Biotransformation of Organic Xenobiotics

Enzymes

Metabolism of organic xenobiotics by organisms employs a wide range of biotransformation enzymes and is largely divisible into two phases: phase I (functionalization) and phase II (conjugative) metabolism (see Livingstone, 1991a for details of reaction types and enzymes). The enzymes of phase I metabolism catalyze virtually every possible chemical reaction that a compound can undergo, i.e. oxidation, reduction, hydrolysis, hydration and others, and introduce a reactive functional group (–OH, –NH$_2$, –COOH etc.) into the xenobiotic, so preparing it for phase II metabolism. Phase II metabolism attaches a variety of polar molecules to the reactive groups producing water-soluble, excretable conjugates, and is, therefore, the main detoxication pathway. A phase III metabolism is also known in higher organisms, involving hydrolysis of the conjugates and remetabolism of the product. Many of the biotransformation enzymes are also involved in the normal metabolism of endogenous compounds, such as steroids, vitamins and prostaglandins, and the presence of xenobiotics can alter or interfere with these processes. Of key importance in determining the metabolic fate of a xenobiotic are the levels of the biotransformation enzymes, and their subcellular and tissue distribution, properties (substrate specificity, isozymes) and regulation (induction). Also of importance for mussels are seasonal changes in activities of the biotransformation enzymes, which will affect their capacity to metabolize xenobiotics.

Most known phase I and II enzymes have been detected in vitro, or indicated from in vivo studies, viz. phase I: the cytochrome *P*-450 monooxygenase or mixed-function oxygenase (MFO) system, the flavoprotein monooxygenase system (EC 1.14.13.8), other oxidases, epoxide hydratase (EC 4.2.1.64), reductases, esterases, sulphatases, phosphatases and deacetylases; phase II: glutathione *S*-transferases, uridine diphosphate (UDP)-glucuronytransferases (EC 2.4.1.17), UDP-glucosyltransferases, acetylases, sulphotransferases, methylases and formylases. No information is available on the existence of amino acid conjugases. Specific activities are given in Table 9.3. Not surprisingly, the activities of biotransformation enzymes are generally lower than in fish and other vertebrates, but are in the same proportion to the levels of nonbiotransformation enzymes, e.g. MFO activities and MFO system components compared to enzymes of intermediary metabolism (Livingstone and Farrar, 1984). Depending on the substrates employed, some marine invertebrate biotransformation enzyme activites have been suggested to be particularly high, e.g. glutathione *S*-transferases in molluscs (Kurelec and Pivčević, 1991). The specific activities of molluscan biotransformation enzymes are generally similar to, or possibly slightly lower, than those observed for echinoderms and crustaceans (Livingstone, 1990, 1991a).

MFO system and cytochrome P-450:
The MFO system has been detected or indicated in 23 species of bivalves and other molluscs (reviewed in Livingstone et al., 1989a, 1990a; Livingstone, 1990). Cytochrome *P*-450 and the associated components and oxidative activities of the MFO system (Table 9.3) are localized primarily in the digestive gland, but are also present or indicated in the gills and blood cells. Also found is a 'low wavelength haemoprotein' (so-called '416-' or '418-peak') which is thought either to be derived from cytochrome *P*-450 (equivalent to cytochrome *P*-420, the denatured form of *P*-450), or to be an independent functional protein, possibly a peroxidase.

Most information available on the catalytic nature and functioning of cytochrome *P*-450 and the MFO system is for digestive gland microsomes of *M. edulis* and other mytilids. Substrates binding to any cytochrome *P*-450 give typical difference (binding) spectra, depending upon whether they bind to the apoprotein (type I compounds and type I spectra), or to the haem part (type II compounds and type II spectra) of the

Table. 9.3. Phase I and II biotransformation enzymes in mussels and other molluscs. (From Livingstone, 1991a).

Enzyme	Source and number of species[1]		Units (mg^{-1} protein)	Range
A. PHASE I				
A.1. Mixed function oxidase system	D/M			
Cytochrome P-450	(11)		pmol	8–134
Cytochrome b$_5$	(7)		pmol	26–160
Cytochrome P-450 reductase[2]	(8)		nmol min^{-1}	4–22
Benzo[a]pyrene hydroxylase	(4)		pmol min^{-1}	3–35
7-ethoxycoumarin O-deethylase	(2)		pmol min^{-1}	0.1–3.2
Aldrin epoxidase	(1)		pmol min^{-1}	1–5
Dimethylaniline N-demethylase	(2)		nmol min^{-1}	0.6–3.4
Aminopyrine N-demethylase	(2)		nmol min^{-1}	0.05–0.67
A. 2. Others				
Flavoprotein monooxygenase system	D/M	(2)	nmol min^{-1}	0.2–1.0
Monoamine oxidase	D/M	(1)	nmol min^{-1}	0.1–0.3
Diamine oxidase	D/M	(1)	nmol min^{-1}	0.3
Epoxide hydratase	W/M	(4)	nmol min^{-1}	0.1–13.2
Azoreductase	D/M	(1)	nmol min^{-1}	2.9
	D/C	(1)	nmol min^{-1}	0.4
Nitroreductase	D/M	(1)	nmol min^{-1}	0.4
	D/C	(1)	nmol min^{-1}	0.2
Organophosphate acid anhydrase-glucuronidase	D/C	(2)	nmol min^{-1}	20–100
	D/S	(1)	nmol min^{-1}	62,000
B. PHASE II				
Glutathione S-transferase (epox.)[3]	W/C	(4)	nmol min^{-1}	0.1–15.8
Glutathione S-transferase (others)	W/C	(2)	nmol min^{-1}	21–91
	D/C	(3)	μmol min^{-1}	1.2–10.7
UDP-glucuronyl-transferase	D/M	(1)	nmol min^{-1}	38
	D/S	(1)	nmol min^{-1}	7.6
UDP-glucosyl-transferase	D/M	(1)	nmol min^{-1}	0.002

[1] D: digestive gland; W: whole animal; M: microsomes; C: cytosol; S: supernatant; number of species in brackets.

[2] NADPH-cytochrome c reductase activity.

[3] Epoxide substrate.

enzyme (see Schenkman et al., 1982 for explanation of binding spectra). In the case of *M. edulis*, whereas type II binding compounds (clotrimazole, ketoconazole, miconazole, metyrapone and pyridine) gave type II difference spectra, type I compounds (7-ethoxycoumarin, α-naphthoflavone, SKF-525A and testosterone) gave apparent reverse type I difference spectra. The reason for the latter unusual observation is unknown, but may be due to the presence of endogenous substrates bound to the mussel cytochrome *P*-450, which are subsequently displaced by the added substrates, causing the observed spectral changes. The results differ from those obtained for several crustacean species which showed type I spectra with type I compounds (hexobarbital, phenobarbital, aminopyrine, ethylmorphine, benzphetamine and SKF-525A) (see Livingstone, 1991a).

Other unusual observations, related to the nature and catalytic mechanism of cytochrome *P*-450 action, have also been obtained for mussels compared to crustaceans or echinoderms (Livingstone, 1991a). The typical cytochrome *P*-450-catalyzed MFO reaction requires molecular oxygen and a source of reducing equivalents, usually reduced nicotinamide adenine dinucleotide phosphate (NADPH), the stoichiometry of the reaction being:

$$RH + NADPH + H^+ + O_2 = R\text{--}OH + NADP^+ + H_2O$$

where RH is the xenobiotic or endogenous substrate and R–OH is the hydroxylated (or other) product. The mechanism of this reaction usually proceeds by what is termed two-electron monooxygenation. A second type of reaction can also be catalyzed by cytochrome *P*-450, which utilizes a peroxide, such as cumene hydroperoxide, as the source of activated oxygen (the oxygen has to be activated to react with the organic substrate), and which does not require NADPH or molecular oxygen. The stoichiometry of this peroxidation reaction is:

$$RH + POOH = ROH + POH$$

where POOH is the peroxide. Although this type of reaction has been studied extensively in vitro, the extent to which it has physiological significance and occurs in vivo (employing say a lipid hydroperoxide as substrate) is unknown. Another reaction mechanism that can occur, other than two-electron monooxygenation, once the cytochrome *P*-450/activated oxygen molecular species has been formed, is termed one-electron oxidation. This proceeds via the formation of a cation radical of the xenobiotic substrate, and recent studies in mammalian systems indicate that this mechanism may be of importace in vivo. Only certain PAHs with particular structural features (high ionization potential), such as benzo[a]pyrene (BaP), will react via one-electron oxidation, and their major fate in vitro is covalent binding to DNA, protein and other macromolecules (macromolecular adduct formation), and the formation of quinones, viz. the 1,6-, 3,6- and 6,12-quinones of BaP.

A number of MFO activities catalyzed by digestive gland microsomes of *M. edulis* or *Mytilus galloprovincialis*, in particular BaP hydroxylase (BPH), 7-ethoxycoumarin

O-deethylase (ECOD) and N,N-dimethylaniline N-demethylase (DMAD), occur in vitro in the absence of added NADPH. This contrasts with the situation for vertebrates, echinoderms, arthropods and polychaetes, which have an absolute requirement for NADPH (or NADH) for MFO activity. The source of activated oxygen for the mussel MFO reactions is unknown, but is presumed to be either molecular oxygen and a source of unidentified endogenous reducing equivalents, or a hydroperoxide such as lipid hydroperoxide. The NADPH-independent BPH activity (total metabolites of BaP measured radiometrically using ^{3}H-BaP; N.B. BaP can be metabolized to many products, including dihydrodiols, quinones and phenols), and the ECOD and DMAD activities, are inhibited by reducing agents, including the addition of NADPH itself. The addition of reducing agents could inhibit a one-electron oxidation mechanism by converting the cation radical intermediate of the xenobiotic substrate (lacks an electron) back to the original substrate molecule. The major metabolites of mussel microsomal BaP metabolism are quinones (indicative of one-electron oxidation—see previous paragraph) (contrasted with mainly phenols for crustaceans and vertebrates) and one-electron oxidation has therefore been proposed as a possible mechanism of molluscan cytochrome P-450 catalytic action. Additionally, because, unlike quinone formation, the putative formation of phenols from BaP by mussel microsomes (fluorometric assay) is stimulated by NADPH, it is postulated that more than one process may be involved in the microsomal metabolism of BaP, viz. one-electron oxidation and two-electron monooxygenation giving rise to, respectively, quinones and phenols. Putative protein adducts of BaP were formed in incubations with digestive gland microsomes, possibly providing further evidence for one-electron oxidation (Livingstone et al., 1990a). MFO activities have been observed towards a number of other substrates, viz. epoxidase (substrate: aldrin), hydroxylase (biphenyl, antipyrine), O-dealkylase (p-nitroanisole) and N-dealkylase (benzphetamine, aminopyrine, p-chloro-N-methylaniline) (see also Table 9.3).

Multiple forms of cytochrome P-450 are indicated (N.B. a classification system now exists for different P-450 gene families and subfamilies based on the degree of similarity of gene and protein sequence information—(e.g. P-450 IA1 where I is the family, A is the subfamily and 1 is the individual member)—see Nebert et al., 1989). Partial purification studies revealed at least two forms in digestive gland microsomes of M. edulis. Transcript sequences showing similarity to rat cDNA probes to cytochrome P-450 IVA have been detected in the total RNA from the digestive gland of M. edulis; cytochrome P-450 IVA is thought to be an ancient cytochrome P-450, involved in fatty acid metabolism, originating some 800 or more million years ago. Activity specifically associated with hydrocarbon-inducible cytochrome P-450 IA (7-ethoxyresorufin O-deethylase) (involved in the metabolism of many organic xenobiotics, such as PAHs and particular PCB congeners) has only rarely been detected.

Seasonal variations in cytochrome P-450 isozyme composition are indicated for digestive gland of *M. edulis* (Kirchin et al., 1992): BPH activity in the tissue varied seasonally and was highest in autumn.

Other phase I enzymes:
The flavoprotein monooxygenase system has been detected or indicated in seven species of mollusc, including *M. edulis* and *M. galloprovincialis* (Livingstone, 1991a; Table 9.3). Metabolism of a wide range of nitrogen- and sulphur-containing xenobiotics has been demonstrated or indicated e.g. *o*- toluidine, *N,N*-dimethylaniline (DMA), aminofluorene (AF), *N*-acetylaminofluorene (AAF), aminoanthracene (AA), 4-amino-*trans*-stilbene, phenylhydrazine, 1,1-dimethylhydrazine and methimazole. Enzyme activity was present in digestive gland microsomes of *M. edulis* (Kurelec, 1985) and indicated in all tissues, but predominantly in the digestive gland of the clam *Mercenaria mercenaria* (Anderson and Döös, 1983). Aspects of reaction stoichiometry have been examined for DMA. Whereas *N*-oxide formation by digestive gland microsomes of *M. edulis* was NADPH-dependent and likely catalyzed by the flavoprotein monooxygenase system, formaldehyde formation was in part, or totally, NADPH-independent and possibly catalyzed by cytochrome P-450 (Livingstone et al., 1990a).

Epoxide hydratase activity towards styrene-7,8-oxide, octene-1,2-oxide and BaP-4,5-oxide has been detected variously in four species of bivalves (Table 9.3; Bend et al., 1977; Galli et al., 1988). Activity was similar in whole body and digestive gland microsomes of *M. galloprovincialis* (Suteau and Narbonne, 1988). Azoreductase activity (substrate: 1,2-dimethyl-4-(*p*-carboxyphenylazo)-5-hydroxybenzene) (Hanzel and Carlson, 1974) and nitroreductase activity (substrate: *p*-nitrobenzoic acid) (Carlson, 1972) have been detected in cytosol and microsomes of digestive gland of *M. mercenaria* (Table 9.3). Nitroreductase activity was also found in lower activities in mantle, gill, foot and gonadal tissue. Nitroreductase activity towards 4-nitroquinoline *N*-oxide has recently been demonstrated in cytosol and microsomes of digestive gland of *M. edulis* (Garcia Martinez et al., 1992).

Hydrolases, including esterases, acid- and alkaline-phosphatases (respectively, EC 3.1.3.2 and 3.1.3.1), organophosphate acid anhydrases, β-glucuronidases (EC 3.2.1.31) and hexoseaminidases are widespread in molluscs, and present in a number of tissues (Moore et al., 1989; Livingstone, 1991a; see also Table 9.3). The identified substrates are mostly endogenous, but activities towards xenobiotics and xenobiotic conjugates (phase III metabolism) have been detected or indicated. A number of the enzymes are polymorphic, e.g. nonspecific esterases in *M. edulis*. Esterase activity towards polyethoxylate fatty acid ester-containing commercial oil spill dispersants was present in digestive gland of the scallop *Chlamys islandicus* (Payne, 1982). Putative glucuronides formed from the aromatic amines AA and AAF by postmitochondrial fractions of digestive gland of *M. galloprovincialis* were hydrolyzed by added mammalian β-glucuronidase (Kurelec et al., 1986). Deacetylase activity towards *N*-hydroxy-AAF was absent in digestive gland microsomes of *M. galloprovincialis* (Kurelec and Krča, 1987), but present in those of *M. edulis* (Marsh et al., 1992). Deaceytlase activity towards AAF and *N*-acetyl-*o*-toluidine has been indicated in *M. edulis* (Knezovich and Crosby, 1985; Knezovich et al., 1988).

Phase II enzymes:
Glutathione *S*-transferases have been detected in 18 species of mollusc (Livingstone, 1991a; Table 9.3). Activity has been observed towards a range of substrates, including 1-chloro-2,4-dinitrobenzene (CDNB), 1,2-dichloro-4-nitrobenzene, ethacrynic acid, *p*-nitrobenzylchloride, methyl iodide 1,2-epoxy-3-(*p*-nitrophenoxy)propane. The enzyme has a wide tissue distribution, including high activities in the digestive gland, and is mainly cytosolic. Glutathione *S*-transferase isoenzymes have been detected or indicated in a number of gastropod species (Livingstone, 1991a), but as yet no information is available for mussels or other bivalves. Less is known of other phase II enzymes. Glucuronidation of phase I metabolites of AAF, but not of BaP, was indicated in postmitochrondrial supernatants of digestive gland of *M. galloprovincialis* (Kurelec et al., 1986). Sulphotransferase activity (substrate: pentachlorophenol (PCP)) is present in short-necked or 'ascari' clams (*Tapes* or *Ruditapes philippinarum*) and is mainly cytosolic (Kobayashi, 1985). Sulphotransferase activities have also been indicated from mutagenic and in vivo metabolic studies, e.g. in *M. edulis* (substrate: 1-naphthol) and *M. galloprovincialis* (activation of hydroxy-AAF, AAF and AF). Such studies have also indicated the presence of *N,O*-acetyltransferases and paraoxon-sensitive cytosolic enzyme in *M. galloprovincialis*, *N*-acetylases (substrate: *p*-toluidine, aniline and AF) in *M. edulis*, and *N*-methylases and, unusually, *N*-formylases (substrate: *o*-toluidine) in both *M. edulis* and the oyster, *Crassostrea gigas*.

In vivo metabolism

The metabolism of a wide variety of organic xenobiotics by different mollusc species has been observed (Livingstone, 1991a). In the case of mussels and other bivalves, the species and compounds include *p*-toluidine, aniline, AF, *N*-acetyl-*o*-toluidine, AAF, PCP, chlorinated paraffins and 1-naphthol (metabolized in *M. edulis*); *o*-toluidine (*M. edulis* and *C. gigas*); picric acid (2,4,6-trinitrophenol), picramic acid (2-amino-4,6-dinitrophenol), dibutylphthalate, di(2-ethylhexyl)phthalate and *bis*(tributyltin)oxide (the oyster *Crassostrea virginica*); *p*-nitroanisole and antipyrine (*Mytilus californianus*); aldrin (*M. californianus* and the freshwater bivalve, *Anodonta* sp.); dieldrin (the freshwater bivalve, *Sphaerium corneum*); and naphthalene (the clam, *Macoma inquinata* and the oyster, *Ostrea edulis*). Other xenobiotics metabolized by various gastropods and other molluscs include 2,6-diethylaniline, benzidine, benzoic acid, nitrobenzene, anisole, *p*-nitrophenetole, methoxychlor, ethoxychlor, chlorobenzene, hexachlorobenzene, DDT, vinyl chloride, chlordane, phthalic anhydride, biphenyl and phenylmercuric acetate.

Metabolism of organic xenobiotics usually produces multiple end-products, involving the catalytic action of several phase I and phase II enzymes, e.g. *o*-toluidine is converted to four metabolites in *M. edulis* by methylation, formylation, hydroxylation and oxidation (Knezovich and Crosby, 1985). Most enzymes identified in vitro are indicated to be functional in vivo, although not always on the same substrate, e.g. aromatic ring and alkyl side-chain oxidations were not detected in the metabolism of aromatic amines by *M. edulis* and *C. gigas*, indicating that the observed *N*-oxidations are likely catalyzed by the flavoprotein monooxygenase rather than the MFO system (Knezovich and Crosby, 1985; Knezovich et al., 1988). A number of cytochrome *P*-450-catalyzed reactions are evident, however, e.g. aldrin epoxidation, antipyrine hydroxylation and *p*-nitroanisole *O*-demethylation in *M. californianus* (Krieger et al., 1979), and dealkylation of *bis*(tributyl)tin oxide in *C. virginica* (Lee, 1986). Glucosidic and sulphated conjugates have been identified in vivo, but not as yet those of glucuronic acid or glutathione, e.g. no indication of glucuronide formation from AF or other xenobiotics was obtained in *M. galloprovincialis* (Kurelec and Krča, 1989), and glutathione conjugates of dieldrin could not be detected in the bivalve *Sphaerium corneum* (Boryslawskyj et al., 1988). In contrast, two isomeric methylthioheptachlorostyrenes were isolated from *M. edulis* exposed to octachlorostyrene, which was indicative of the existence of mercapturic acid pathways (Bauer et al., 1989). Conjugates and other metabolites are released by molluscs into the surrounding water, but formation of the former does not particularly appear to predominate as an excretion mechanism, e.g. 1-naphthylsulphate, formed from 1-naphthol, was seen to be retained by *M. edulis* (Ernst, 1979b). Temperature appeared to have little effect on sulphated PCP formation by *M. edulis* (Ernst, 1979a). In some cases, no metabolism has apparently been detected at all e.g. *p*-nitroanisole in *M. edulis* (Landrum and Crosby, 1981).

Information on the metabolism of hydrocarbons, the first step of which would involve oxidation by cytochrome *P*-450 and the MFO system, is limited. Metabolites or conjugates were not detected in the bivalves *M. edulis* (Lee et al., 1972), *Macoma inquinata* (Augenfeld et al., 1982), *Macoma nasuta* (Varanasi et al., 1985) and *Modiolus modiolus* (Palmork and Solbakken, 1981) exposed to variously naphthalene, phenanthrene, BaP and other hydrocarbons. In contrast, 1- and 2-naphthols were formed from naphthalene in the oyster, *Ostrea edulis* (Riley et al., 1981), and unidentified metabolites and conjugates were detected in the freshwater snail, *Physa* sp. exposed to BaP (Lu et al., 1977). A number of possible reasons have been put forward to explain the absence, or low levels, of hydrocarbon metabolism, including generally or seasonally low MFO activities, and inadequate analytical methodologies for the levels of metabolites produced (Moore et al., 1989). However, given that the levels of microsomal cytochrome *P*-450 and MFO activities such as BPH are not markedly lower in molluscs than in crustaceans or polychaetes (Livingstone, 1990), but that in vivo hydrocarbon metabolism is much more readily detectable in the latter two (see later), another possible explanation, or contributory factor, could be fundamental differences in cytochrome *P*-450 catalytic action (see section on enzymes p.429–435). Specifically, cytochrome *P*-450-catalyzed one-electron oxidation in vivo could result in a major fate of certain PAHs being covalently bound to protein and other macromolecules, rather than to polar metabolite and conjugate formation (Livingstone et al., 1989a, 1990a). Macromolecular adducts of hydrocarbons were indicated in *M. inquinata* (Augenfeld et al., 1982) and *Physa* sp. (Lu et al., 1977), but were not examined for in the other hydrocarbon studies. In digestive gland of *M. edulis*, putative protein adducts of BaP were formed in in vitro incubations with microsomes (Livingstone et al., 1990a), and more recently, in vivo binding of BaP to protein and DNA has been demonstrated (Marsh et al., 1992). Chlorinated paraffins were incorporated into the protein fraction of *M. edulis* (Renberg et al., 1986), which is interesting given that a suggested catalytic function of (rat liver) cytochrome *P*-450 IVA1 is hydroxylation of straight-chain alkanes (Lock et al., 1987). Changes in tissue profiles of accumulated aliphatic hydrocarbons, which could also be due to factors other than metabolism, have been observed for several bivalve species, e.g. in *M. edulis* (Widdows et al., 1982).

The formation of macromolecular adducts could be partly responsible for the observed increase in depuration half-life for hydrocarbons following increased periods of exposure (Table 9.1). The process of adduct formation, and presumably excision and release, could therefore contribute to the more stable cellular compartment, with a lower rate of xenobiotic turnover that has been postulated (Stegeman and Teal, 1973). Similarly, it could contribute to other observations on bioaccumulation, such as aspects of the specific patterns of PCB congener uptake and elution, and the greater than predicted accumulations of xenobiotics such as *bis*(tributyltin)oxide (see section on organic xenobiotics p.425–428 and Table 9.2). Macromolecular adduct formation in bivalves has also been demonstrated for the metabolism of AAF by *M. edulis* (Knezovich et al., 1988) and picric and picramic acids (Burton et al., 1984) and phthalate esters by *C. virginica* (Wofford et al., 1981).

Quantitative analyses of in vivo metabolism have been carried out using pooled literature data involving a large number of xenobiotics and molluscan species (Livingstone, 1991a, 1992). The compounds were divided into two groups: (a) those already containing functional groups (–OH, –NH$_2$ etc.), such as nitroaromatics, aromatic amines, esters and phenols, and termed 'functional group compounds',

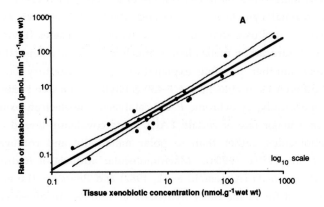

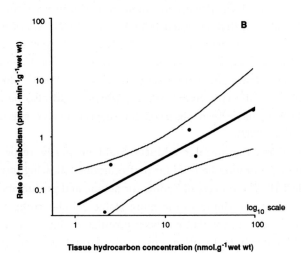

Fig. 9.1. Relationship between rate of in vivo metabolism and tissue xenobiotic concentration for the pooled data of a variety of functional group compounds (A) and hydrocarbons (B) and molluscan species. The former include nitroaromatics, aromatic amines, phenols, phthalate esters and others. The latter include naphthalene, benzo[a]pyrene, chlorinated paraffins and hexachlorobenzene, but not studies in which metabolism was not detected. The species include *Mytilus edulis, Crassostrea gigas, Crassostrea virginica, Macoma inquinata, Ostrea edulis, Cryptochiton stelleri* and *Physa* sp. The correlation coefficients for (A) and (B) were respectively 0.96 (n=20) and 0.87 (n=5) and the 95% confidence limits are shown on the plots. (Figures from Livingstone, 1992 with permission; data from Livingstone, 1991a, 1992).

Table 9.4. Theoretical comparative rates of in vivo organic xenobiotic metabolism for molluscs and crustaceans calculated for a tissue xenobiotic concentration of 10 nmol per gram wet weight[1]

Animal group	Type of compound	Regression equation[2]	Rate of metabolism ($pmol\ min^{-1}\ g^{-1}$ wet wt)[3]
Crustacean	Functional group	$log_{10}R = -0.46 + 1.01\ log_{10}T$	3.55 (2.82 – 4.47)
Mollusc	Functional group	$log_{10}R = -0.51 + 0.94\ log_{10}T$	2.69 (1.91 – 3.80)
Crustacean	Hydrocarbon	$log_{10}R = -0.90 + 0.93\ log_{10}T$	1.01 (0.83 – 1.38)
Mollusc	Hydrocarbon	$log_{10}R = \quad 1.31 + 0.92\ log_{10}T$	0.41 (0.09 – 1.91)

[1] From Livingstone (1991a, 1992).

[2] The regression equations relate rate of in vivo metabolism to tissue substrate concentration for the pooled data of a large number of organic xenobiotic metabolism studies in the literature; R and T are, respectively, rate of metabolism of xenobiotic in pmol per min per gram wet weight, and tissue xenobiotic concentration in nmol per gram wet weight.

[3] Mean plus, in parenthesis, the range for 2 standard errors.

Crustaceans include species of crabs, lobsters, crayfish, copepods, amphipods and zooplankton. Functional group compounds include aromatic amines, nitroaromatics, phthalate esters, phenols and others. Hydrocarbons include aliphatic and aromatic compounds. Mollusc equations are derived from Figure 9.1.

which can be directly metabolized by both phase I and II enzymes; and (b) hydrocarbons which require oxidation by the MFO system, before phase II enzymes can act on them. For functional group compounds, calculated whole body rates of xenobiotic metabolism linearly increased with tissue concentrations of parent compound, over four orders of magnitude of the two parameters (Fig. 9.1A). A similar relationship was indicated for the much smaller data set for hydrocarbons (Fig. 9.1B). However, in both cases it is important to realize that the treatments represent generalizations, and exceptions which fall outside the derived relationships will exist with respect to both particular compounds and species. Increased rates of metabolism with increased tissue concentrations of xenobiotic have also been seen in studies on single compounds, viz. sulphated PCP formation by *M. edulis* (Ernst, 1979a). Using the regression equations of Figures 9.1A and 9.1B, and data similarly derived for crustaceans (Livingstone, 1991a, 1992), it is possible to calculate rates of in vivo metabolism for particular tissue concentrations of xenobiotics, allowing comparison between types of compounds and animals. Thus for molluscs, rates of metabolism are lower for hydrocarbons than for functional group compounds (Table 9.4), presumably reflecting either that several enzymes can act on the latter substrates at once, or that the action of the MFO system is rate-limiting. A similar difference is observed between functional group compounds and hydrocarbons for crustaceans, and the rates of metabolism of both groups of compounds are higher than in molluscs (Table 9.4).

The rates of uptake and final tissue equilibrium concentrations of organic xenobiotics, including hydrocarbons, are both mainly determined by the exposure concentration of xenobiotic in the water-column (see section on organic xenobiotics p.. 425–428). A relationship therefore exists between the former two parameters, allowing determination of one from the other. The tissue xenobiotic concentration determines the in vivo rate of metabolism (see above), and therefore, it is possible to compare rates of uptake and metabolism for a given tissue concentration of xenobiotic. Thus, using pooled literature data on the uptake of hydrocarbons by molluscs, it has been shown that rates of uptake generally exceed rates of metabolism by an order of magnitude or more (Livingstone, 1991a, 1992), thus accounting for the marked bioaccumulation of these compounds by molluscs, and the process being explainable on the basis of a simple lipid/water equilibrium model (Burns and Smith, 1981). The fundamental reason for this phenomenon is presumably that whereas uptake is mainly passive and determined by physicochemical principles, metabolism (i.e. enzyme activities) is intrinsic and largely determined (limited) by endogenous considerations, i.e. energy costs and endogenous functions of the biotransformation enzymes. Other factors that undoubtedly contribute to the long residence times of organic xenobiotics in molluscs, and other marine invertebrates, are a slow release of metabolites into the seawater, and the covalent binding of xenobiotics to macromolecules (Livingstone, 1991a, 1992).

Induction studies
A fundamental feature of most biotransformtion enzymes in mammals and other vertebrates is that the enzymes are induced by exposure of the organism to the xenobiotic. However, differences in gene regulation and induction are seen between different animal groups, e.g. cytochrome *P*-450 IIB1 is inducible in mammals but not apparently in fish (Stegeman, 1989). The amount of information available for molluscs is limited, but generally indicates that the enzymes, particularly the MFO system, are less responsive, and the increases in enzyme activities are less than in vertebrates. Responses of biotransformation enzymes to xenobiotic exposure have recently been reviewed (Livingstone, 1991b). Increases in MFO system components and activities, including BPH, occur with experimental exposure to organic xenobiotics (3-methylcholanthrene (3MC), BaP, PCBs, hydrocarbon mixtures (diesel oil

and crude oil), β-naphthoflavone), but the results are variable, including absence of responses. Increases in microsomal cytochrome P-450 specific content of digestive gland have been reasonably consistently observed with exposure of M. edulis and M. galloprovincialis to PAH and other hydrocarbons. Changes in cytochrome P-450 isozyme composition have been indicated for M. edulis. With one exception (Galli et al., 1988), no responses have been seen with exposure to phenobarbital or related compounds. Similar variability has been seen in field comparisons of molluscs from clean and polluted sites, e.g. no differences were evident in M. edulis from around Cape Cod, U.S.A. and the Shetland Islands, U.K., but increases in digestive gland cytochrome P-450 content were seen in Langesundfjord, Norway. Increases with pollution have been observed for the 418-peak of digestive gland microsomes of M. edulis, C. edule and the periwinkle, Littorina littorea, which in some cases have been associated with low or undetectable levels of cytochrome P-450. It is postulated that the haemoprotein responsible for the 418-peak may be some sort of microsomal breakdown product, possibly derived from cytochrome P-450. Elevated NADPH-cyctochrome c (P-450) reductase activity and levels of cytochrome P-450 and the 418-peak were seen in digestive gland microsomes of Mytilus sp. from polluted sites on the Catalan coast, Spain (Porte et al., 1991). In a study of M. galloprovincialis from sites around the Mediterranean Sea, whole body microsomal BPH activity (total metabolites) was correlated with sediment levels of PAH (Garrigues et al., 1990).

The flavoprotein monooxygenase system has been shown not to be inducible, at least in M. galloprovincialis (Britvić and Kurelec, 1986). Epoxide hydratase activity of whole body microsomes of M. galloprovincialis was increased following exposure to 3MC-type inducers (BaP, 3,4,3',4'-tetrachlorobiphenyl (TCBP), diesel oil emulsion), and to the phenobarbital-type inducer 2,4,5,2',4',5'-hexachlorobiphenyl (Suteau et al., 1988a, b). In contrast, decreases in activity only have been observed in field mussels with increasing body burdens of PAHs and PCBs (Suteau et al., 1988b). Indirect evidence for increases in epoxide hydratase activity in M. galloprovincialis was seen in the increased ability of digestive gland postmitochondrial fractions to detoxify styrene oxide in a yeast genotoxicity test following pre-exposure of the mussels to phenobarbital (Galli et al., 1988).

Glutathione S-transferase activity (enzyme substrate: CDNB) was increased in whole body of S. corneum after exposure of the bivalve to dieldrin or lindane (Boryslawskj et al., 1988), and in digestive gland of the freshwater mussel, Anodonta cygnea after exposure to diesel oil or polluted river water (Kurelec and Pivčević, 1989). With styrene oxide as the substrate for the enzyme, glutathione S-transferase activity in whole body of M. galloprovincialis increased in animals exposed to TCBP (Suteau et al., 1988a), but not in those exposed to diesel oil emulsion or in mussels from polluted field sites (Suteau et al., 1988b). Sulphotransferase activity in T. philippinarum increased following exposure to PCP; increases in activity also occurred with exposure to resorcinol, *o*-cresol, *p*-chlorophenol and *p*-nitrophenol but not phenol (Kobayashi, 1985). Changes have also been seen in antioxidant enzymes with exposure to xenobiotics (see section on oxyradical metabolism p.443–446).

Heavy Metal Metabolism

On entering cells, heavy metals are primarily complexed by thiol-containing molecules such as amino acids, glutathione and, in particular, the metal-binding detoxication proteins, metallothioneins. In addition, part of the metal is compartmentalized in the lysosomal vacuolar system, or trapped in different types of specialized inorganic granules.

Metallothioneins

Metallothioneins are a class of low molecular weight, soluble (generally cytosolic), thiol-rich (high cysteine content) proteins with a high heavy metal content (Viarengo, 1989). They contain Zn^{2+} and Cu^{2+}, but can also bind xenobiotic metals such as Hg^{2+}, Cd^{2+}, Au^{2+} and Ag^{2+}. Amongst other identified or putative physiological roles, e.g. Cu/Zn homeostasis and oxyradical scavenging, metallothioneins mainly function to maintain low levels of free heavy metal cations in cells, initially through the displacement of Zn in existing metallothionein and the binding of the xenobiotic metal, and then through induction by the synthesis of increased amounts of metallothionein.

Metallothioneins have a wide tissue distribution in mussels, e.g. gills, mantle and digestive gland in *M. galloprovincialis* (Viarengo et al., 1981). Copper-binding (Viarengo et al., 1984), Cd-binding (George et al., 1979) and Hg-binding (Roesijadi and Hall, 1981) proteins (molecular weight 10,000 to 15,000 daltons, or more) have been isolated and characterized from *M. edulis* or *M. galloprovincialis*. Copper displaced Zn (but not Cd) from Cd, Zn-thioneins in digestive gland of *M. galloprovincialis* (Viarengo, 1989). Elevation of metallothionein levels have been seen in various tissues of mussels exposed to Cu (Viarengo et al., 1981), Cd (Noel–Lambot, 1976) and Hg (Roesijadi, 1982), and in other molluscs (Viarengo, 1989). Enhanced Hg-tolerance in *M. edulis* was related to the induction of Hg-binding proteins (Roesijadi et al., 1982a). Mercury-tolerance was also enhanced by pre-exposure to other heavy metals capable of inducing metallothioneins (Roesijadi and Fellingham, 1987). Metallothionein levels were elevated in the larvae of Hg-exposed *M. edulis* (Roesijadi et al., 1982b).

Lysosomes

Metal accumulation in lysosomes in mussels is well-documented, although information on the molecular processes involved is limited (Moore et al., 1989; Viarengo, 1989). Lysosomes are present in considerable numbers in digestive gland and kidney (see also p.449–452). Tertiary lysosomes accumulate undegradable end-products of lipid peroxidation (oxidized lipid and protein polymers), so-called lipofuschin. In kidneys, lipofuschin granules have been shown to bind metals in two ways, viz. (a) metals weakly bound by acidic groups in the outer region of the granules, and thus able to dissociate and be in equilibrium with cations in the cytoplasm; and (b), metals sterically 'trapped' in a nontoxic form in the centre of the developing granules (George, 1983). Active excretion of these residual bodies by exocytosis leads to metal elimination. A second method of metal elimination has been indicated for Cu in the digestive gland, involving the accumulation of Cu-rich thionein-like proteins in lysosomes, followed again by elimination of residual bodies (Viarengo, 1989; Viarengo et al., 1989). Neither of these two biochemical elimination pathways appear to play a major role in Cd removal in *Mytilus* sp., with the result that the biological half-life for Cu is nine days compared to seven months for Cd (Viarengo, 1989).

Inorganic granules and vesicles

Two major types of metal-containing granules, involved in heavy metal detoxification, are found, viz. Cu-sulphur-containing granules and calcium-containing granules. An association of Cu with sulphur has been seen in membrane-limited vesicles of oyster granular amoebocytes (Viarengo, 1989). Heavy metal cations are trapped in calcium-insoluble concretions as ortho- and pyrophosphates, e.g. in the kidneys of scallops (Fowler, 1987). Kinetic relationships are likely to exist between these concentrations of metals, and those of lysosomes and metallothioneins.

Toxic Effects

Oxyradical metabolism

Oxyradical generation, oxidative damage and antioxidant defenses in mussels and other molluscs have been reviewed (Viarengo, 1989; Livingstone et al., 1990b). The normal fate of molecular oxygen is tetravalent reduction to water, coupled to oxidative phophorylation and the production of energy. However, small amounts are continually partially reduced by endogenous and xenobiotic-stimulated processes to reactive oxygen species, so-called oxyradicals. These oxyradicals include the superoxide anion radical (O_2^-; univalent reduction), hydrogen peroxide (H_2O_2; bivalent reduction) and the highly reactive hydroxyl radical (•OH; trivalent reduction). Oxyradicals are implicated in oxidative tissue damage and free radical pathology. Deleterious molecular effects include enzyme inactivation, lipid peroxidation and DNA damage, the latter being implicated in carcinogenesis. Combating oxyradicals are antioxidant defences which include various free radical scavengers and specific antioxidant enzymes. Only when these defences are overcome, either through their depletion, or a marked increase in oxyradical production, are pro-oxidant processes thought to result in significant molecular damage.

NADH-dependent and/or NADPH-dependent generation of O_2^-, H_2O_2 and •OH have been observed variously in digestive gland microsomes of *M. edulis* (Winston et al., 1990), *R. cuneata* and the mussel, *Geukensia demissa* (Wenning and Di Giulio, 1988a). Redox cycling metals such as Fe (and probably Cu) are necessary for •OH generation to catalyze the Haber–Weiss reaction (Winston et al., 1990), viz.

$$O_2^- + Fe^{3+} = O_2 + Fe^{2+}$$
$$H_2O_2 + Fe^{2+} = \bullet OH + OH^- + Fe^{3+}$$

Net (Haber–Weiss reaction): $O_2^- + H_2O_2 = O_2 + \bullet OH + OH^-$

The Haber–Weiss reaction which produces •OH from O_2^- and H_2O_2 is thermodynamically favourable but kinetically very slow, and therefore in biological systems the production of •OH is thought to be dependent on, and site specific to, the presence of a suitable chelated metal catalyst (probably Fe or Cu), of which nothing is known in molluscs.

Microsomal NAD(P)H-dependent oxyradical production is stimulated by the redox cycling compounds menadione (2-methyl-1,4-naphthoquinone) (Livingstone et al., 1989b), nitrofurantoin (*N*-(5-nitro-2-furfurylidene)-1-aminohydantoin) (Garcia Martinez et al., 1989) and paraquat (1,1'-dimethyl-4,4'-bipyridinium dichloride) (Wenning and Di Giulio, 1988b). More recently, 4-nitroquinoline *N*-oxide has been observed to stimulate NAD(P)H-oxyradical production by microsomes and cytosol of digestive gland of *M. edulis* (Garcia Martinez et al., 1992). Likely loci for the oxyradical production are cytochrome *P*-450 reductase and NADH-dependent microsomal flavoprotein reductases (autoxidation, redox cycling), and cytochrome *P*-450 (autoxidation, oxidase activity). Menadione-stimulated NADPH-dependent oxyradical production was indicated to be limited by the presence of a microsomal DT-diaphorase (EC 1.6.99.2) catalyzing the two-electron reduction of menadione to the hydroquinone, so preventing one-electron reduction to the semiquinone and redox cycling. Oxyradical production was also supported by the cytosolic fraction of M. edulis digestive gland (Winston et al., 1990).

Antioxidant enzymes have been detected in a number of bivalves, viz. superoxide dismutase (SOD; EC 1.15.1.1), catalase (EC 1.11.1.6), NAD(P)H-dependent DT-diaphorases and selenium-dependent (EC 1.11.1.9) and selenium-independent glutathione peroxidases (GPX) variously in, for example, M. edulis (Goldfarb et al., 1989; Winston et al., 1990; Livingstone et al., 1992) and the giant clam, Tridacna maxima (Shick and Dykens, 1985). A wide tissue distribution is indicated, with activities related to the potential for oxyradical production, e.g. high in digestive gland and gills. CuZnSOD (cytosolic form) and MnSOD (mitochondrial form) were found in G. demissa, R. cuneata (Wenning and Di Giulio, 1988a), the clams, Calyptogena magnifica, M. mercenaria (Blum and Fridovich, 1984), and in M. edulis (Livingstone et al., 1992). Small molecular weight free radical scavengers, such as glutathione, vitamins C (ascorbic acid), A (retinol) and E (α-tocopherol), and carotenoids, including β-carotene, have been detected variously in M. edulis (Ribera et al., 1991; Viarengo et al., 1991a), M. galloprovincialis (Ribera et al., 1989), G. demissa and R. cuneata (Wenning and Di Giulio, 1988a). Levels of antioxidant defenses vary seasonally in digestive gland of Mytilus spp. (Viarengo et al., 1991b).

Changes in antioxidant defences have been observed with exposure to xenobiotics. Catalase and SOD activities were transiently elevated in digestive gland of G. demissa exposed to paraquat: the increases were maximal after 6–12h, but had declined to control levels by 24h (Wenning et al., 1988). Minimal increases were seen or indicated in catalase, SOD, GPX and DT-diaphorase activities in digestive gland of M. edulis exposed to BaP and menadione (Livingstone et al., 1990b). Catalase and SOD, but not GPX, activities were elevated in digestive gland of M. edulis from polluted sites on the Catalan coast, Spain (Porte et al., 1991). GPX activities in digestive gland of A. cygnea were unaffected by exposure of mussels to polluted river water or diesel oil (Kurelec and Pivčević, 1989).

Glutathione occasionally increased in digestive gland of G. demissa exposed to paraquat (Wenning et al., 1988), whereas no change or decreases were seen in whole body M. edulis exposed to diesel oil/copper mixture, or to PAH and PCBs in the field (Suteau et al., 1988b). Glutathione in the digestive gland and gills of M. edulis decreased with exposure to Cu^{2+}, but increased slightly with exposure to Cd^{2+} or Zn^{2+} (Viarengo et al., 1988). Changes in the status of glutathione and vitamins A and E were observed variously in digestive gland, gills and remaining tissues of M. edulis exposed to menadione, BaP or carbon tetrachloride (Ribera et al., 1991). Total carotenoids increased in M. galloprovincialis exposed to a mineral oil/hypoxia condition (Karnaukhov et al., 1977).

Lipid peroxidation has been observed in vitro and in vivo. Enzyme-mediated lipid peroxidation was demonstrated in mantle microsomes of M. edulis (Musgrave et al.,

1987). Lipid peroxidation (malonaldehyde equivalents) increased in digestive gland of *G. demissa* exposed to paraquat (Wenning et al., 1988), in gill and digestive gland of *M. edulis* exposed to Cu^{2+} (Viarengo et al., 1988), and in various tissues of *M. edulis* exposed to menadione or BaP (Livingstone et al., 1990b; Ribera et al., 1991). Other toxic aldehydes such as 4-hydroxyalkenals were also detected in *M. edulis* (Viarengo et al., 1988). Lipofuschin formation increased in digestive gland of *M. edulis* following exposure to organic pollution in the field (Moore, 1988). Increased levels of lipid peroxide have been correlated with reduced antioxidant defenses in digestive gland of *Mytilus* spp. in relation to age (animal size) (Viarengo et al., 1991a) and seasonal variability (Viarengo et al., 1991b). Copper is considered to be the most likely metal peroxidative agent in the field, with seasonal variations in membrane lipid composition, and antioxidant defenses also being important (Viarengo, 1989).

Genotoxicity

The primary role of biotransformation enzymes is to detoxify organic xenobiotics by converting them to polar excretable products. Paradoxically, however, some of the biotransformations result in products that are more mutagenic or carcinogenic than the parent compound. Although information in molluscs is limited, such processes are indicated and genetic effects with exposure to xenobiotics have been observed.

Sister chromatid exchange (SCE) frequencies in developing eggs of *M. galloprovincialis* were higher in mussels from a polluted compared to a clean, field site (Brunetti et al., 1986; see also p.361–363 Chapter 7). Increased incidence of chromosomal aberrations were found in embryos of *M. edulis* (aneuploidy) (Dixon, 1982) and gill cells of *M. galloprovincialis* (aberrant metaphase) (Al-Sabti and Kurelec, 1985) in animals from polluted field sites. The percentage of aberrant metaphases in *M. galloprovincialis* was increased in a dose-dependent manner by exposure to BaP (Al-Sabti and Kurelec, 1985). In contrast, neoplastic diseases are found in bivalves but no convincing association with environmental pollution could be demonstrated (Mix, 1986; see also p.453–454).

Mutagenic chemicals have been found in the tissues of *M. edulis* and other molluscs, and a relationship is seen with pollution, season and the digestive gland (Moore et al., 1989). The identity of the mutagenic chemicals was not established, and therefore, it is not known to what extent they were bioaccumulated or subsequently produced through biotransformation. However, mutagenic activity of material extracted from the tissues with concentrated nitric acid was increased by incubation with mussel microsomes (Parry et al., 1981).

The digestive glands of *M. edulis*, *M. mercenaria* and *C. virginica* produced the proximate carcinogen BaP-7,8-dihydrodiol in in vitro incubations (Anderson, 1985; Stegeman, 1985), but showed only minimal or no mutagenic activation of BaP or 3MC in Ames bacterial (*Salmonella typhimurium*) tests (Anderson and Döös, 1983; Britvić and Kurelec, 1986; Marsh et al., 1992), consistent with the predominance of quinones in the in vitro metabolism of BaP by cytochrome *P*-450. Similarly, using ^{32}P-postlabelling analysis of DNA adducts, incubation of digestive gland homogenate of *M. galloprovincialis* with BaP either showed no adduct, or a very weak adduct spot (1 adduct per 1–4 x 10^9 nucleotides) (Kurelec et al., 1988). In contrast to PAH, mutagenic activation was seen with aromatic amines such as AA, AF, AAF, *N*-hydroxy-2AAF, aminobiphenyl and 4-amino-trans-stilbene, and the properties of the activating ability were consistent with microsomal flavoprotein monooxygenase being the principal enzyme responsible (Anderson and Döös, 1983; Kurelec et al., 1986; Britvić and Kurelec, 1986; Marsh et al., 1992). Digestive gland homogenates of *M. galloprovincialis* incubated with AF showed one major and one minor DNA-adduct in the range of 1 to 4 x 10^8 nucleotides (Kurelec et al., 1988). Inhibitor studies established that *N,O*-acetyltransferase, sulfotransferase and paraoxon sensitive cytosolic enzyme (deacetylase) are also involved in the mutagenic activation (*S. typhimurium*) of *N*-hydroxy-2AAF, AF and AAF in digestive gland cytosol of *M. galloprovincialis* (Kurelec and Krča, 1987). The nitroaromatics 4-nitroquinoline *N*-oxide (Garcia Martinez et al., 1992) and 1-nitropyrene (Marsh et al., 1992) were activated to bacterial mutagens by subcellular fractions of digestive gland of *M. edulis*.

Mechanisms of xenobiotic activation have also been investigated using SCE and alkaline unwinding as endpoints of genetic damage. The promutagen cyclophosphamide (cytochrome *P*-450-activated), increased frequencies of SCE in both adult and larval *M. edulis* (Dixon et al., 1985). Pre-exposure to phenobarbital resulted in an increase in the levels of SCE produced by cyclophosphamide, possibly indicating induction of enzymes such as cytochrome *P*-450 or epoxide hydratase. In contrast, 9,000g supernatant of digestive gland of *M. galloprovincialis* did not activate cyclophosphamide in a yeast genotoxicity test (Galli et al., 1988). The frequency of SCE was not affected in eggs of *M. galloprovincialis* (Brunetti et al., 1986) and larvae of *M. edulis* (Dixon and Prosser, 1986) by exposure to, respectively, nitriloacetic acid and *bis*(tributyl)tin oxide.

The effects of BaP and 4-nitroquinoline *N*-oxide on damage to DNA in haemolymph of *M. galloprovincialis* were examined using the alkaline elution technique which measures alkali-labile sites and single strand breaks in the DNA (Bihari et al., 1990). Following injection of the xenobiotics into the pallial cavity, dose-dependent DNA damage was observed for both compounds 1.5h after exposure. However, the effects decreased after two days for BaP and five days for 4-nitroquinoline *N*-oxide, indicating the presence of significant DNA repair mechanisms in mussel haemolymph. The formation of peroxides and/or radicals was postulated as the mechanism of genotoxic action of BaP.

Damage to DNA (oxidation) can also be effected by enhanced oxyradical generation. Thus, redox cycling, formation of quinones from PAHs (BaP), and elevation of cytochrome *P*-450 content and cytochrome *P*-450 reductase activity represent possible mechanisms for linking pollution exposure to DNA and other damage in the digestive gland of molluscs (Livingstone et al., 1990b).

Little is known of metal-mediated genotoxicity (Viarengo, 1989), although a role in enhanced oxyradical generation is possible. Copper accumulated in appreciable amounts in the nuclei of exposed *M. galloprovincialis*. Although mRNA syntheis was increased in nuclei isolated from digestive gland and subsequently exposed to Cu, the activities of nuclear polymerases I and II (involved in, respectively, rRNA and mRNA synthesis) were decreased. In contrast, nuclear polymerase activities were unaffected in whole mussels exposed to Cu or Hg.

Other effects

Many other molecular and subcellular toxic interactions are possible. Identified ones include those on intermediary metabolism, lysosomes, mitochondria, endoplasmic reticulum, and calcium homeostasis. Effects on the adenosine phosphate system are variable, with decreases, increases and no change in adenylate energy charge (AEC) being recorded: similar variability has been seen for the different enzymes of intermediary metabolism (Livingstone, 1985; Viarengo, 1989). Lysosomal membranes are readily destablized by Cu, Cd and various organic xenobiotics such as PAHs (Moore et al., 1989; Viarengo, 1989). Decreases in cytochrome *P*-450 content and BPH activity occurred in *M. edulis* exposed to Cd (Livingstone, 1988). Calcium levels were elevated in tissues of *Mytilus* spp. exposed to Cu, Cu/PAHs mixture, or from polluted field sites, possibly affecting its key role as a secondary messenger: a mechanistic basis for this, involving oxidative damage and impared Ca^{2+} pumping (Ca, Mg-ATPases), has been proposed (Viarengo, 1989).

CELLULAR ASPECTS

It is possible to detect structural and functional alterations at the cellular level resulting from exposure to environmental contaminants at an early stage of the stress response. The tissues and cell types investigated generally reflect potentially sensitive indicators of an adaptive response in the mussel with research concentrated on the cells associated with the processes of feeding and digestion, reproduction and defence. Studies on contaminant induced cellular alterations can be broadly divided into three categories: (a) descriptive histopathology, where pathological changes in cell and tissue morphology are recorded; (b) quantitative histology, where subtle changes in cell (or organelle) size and numerical ratios are recorded, and (c) cytochemistry, where changes relating morphology to cellular biochemistry are recorded. The first two categories are described together under headings for feeding and digestion, reproduction, and internal defence. Cytochemistry which investigates subcellular organelles and can generally be applied to various cell and tissue types is dealt with separately in the final subsection.

Feeding and Digestion

Laboratory studies have been made on the acute effects of a wide range of potential pollutants on the gills of mussels (see Fig. 9.2 and Chapter 2 for details on gill structure). An inflammatory reaction consisting of enlargement of the postlateral cells, dilation of the blood spaces and invasion by granular haemocytes resulted from exposure to a number of heavy metals, diesel oil and N-nitroso compounds (Rasmussen, 1982; Sunila, 1988; Auffret, 1988; Hietanen et al., 1988). In addition exposure to high concentrations of Zn caused swelling and degeneration of mucus secretory cells and necrosis of haemocytes within the gill (Hietanen et al., 1988). Lead and N-nitroso compounds caused loss of lateral cilia and sloughing of lateral cells from the chitinous rod, while treatment with Ag resulted in vacuolation of endothelial cells (Rasmussen, 1982; Sunila, 1988). Uncoupling of the interfilamentar junctions of the gill resulted from exposure to Cu and Cd (Sunila and Lindstrom, 1985); in addition Cu caused fusion of the gill filaments (Sunila, 1986a). Field studies, taking samples from polluted sites on the southern coast of Finland, and at dredge spoil dumpsites in New Haven, U.S.A., demonstrated similar effects in gill tissues (Arimoto and Feng, 1983; Sunila, 1987); however, no chemical analysis, to determine the levels or types of contaminant present at the sites or in the mussel tissues, was carried out.

Few studies have investigated contaminant effects on the cells of the stomach or digestive gland ducts. In contrast, however, there is a considerable body of work on the digestive diverticula of mussels (see Chapter 2 for details on structure of digestive organs). Mussels exposed to Cu and a mixture of diesel oil and Cu showed cytoplasmic erosion and invasion by clusters of brown cells of the epithelial cells lining the stomach wall; vacuolization, loss of cilia and cytoplasmic erosion of the ciliated columnar cells in the digestive gland ducts were also noted (Calabrese et al., 1984; Auffret, 1988). Mussels dosed with Ag accumulated it as black deposits in the basement membranes and connective tissue around the stomach, intestine and digestive diverticula (Calabrese et al., 1984; George et al., 1986).

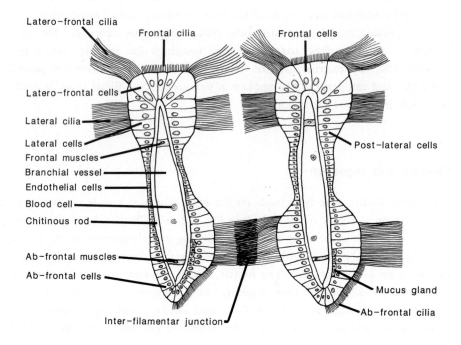

Fig. 9.2. Diagram of transverse section across adjacent gill filaments of *Mytilus edulis* showing histological structure under normal conditions.

The response of the digestive diverticula of mussels to contaminant exposure has been well-documented from both laboratory and field studies. Changes vary from gross pathology involving atrophy of cells to more subtle alterations in phasic activity of the tubules or variations in the lysosomal-vacuolar system. Mussels challenged with a wide range of contaminants develop some degree of atrophy of the digestive tubules (see Fig. 9.3); in extreme cases, such as injection with *N*-nitroso compounds, this involves severe necrosis of the digestive epithelium and replacement with collagenous scars (Rasmussen, 1982; Rasmussen et al., 1983a, b, 1985a). The degree of tubule thinning can be quantified using stereological techniques (Lowe et al., 1981). Other more sophisticated measurements, such as the ratio of lumen surface area to digestive cell volume can also be estimated using stereology (Lowe and Clarke, 1989). The breakdown of the digestive epithelium appears to be a generalized stress response, resulting not only from exposure to a wide range of contaminants (Pipe and Moore, 1986; Sunila, 1986b; Moore et al., 1987; Lowe, 1988; Lowe and Clarke, 1989), but also physiological extremes such as increased salinity and starvation (Thompson et al., 1974; Pipe and Moore, 1985a). Associated with the epithelial breakdown is an increase

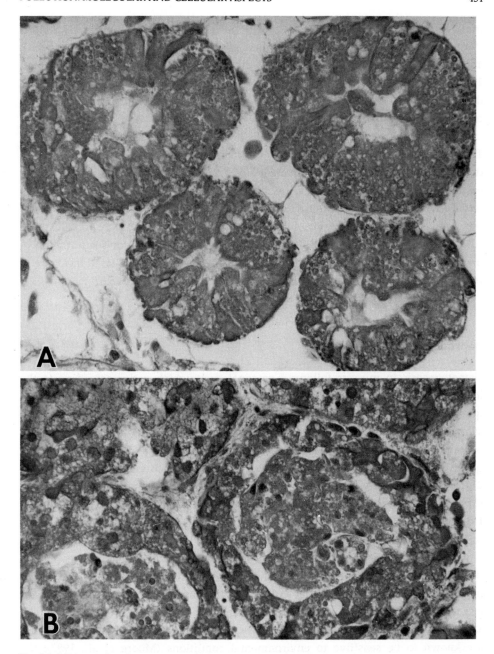

Fig. 9.3. Micrographs of transverse sections across the digestive gland of *Mytilus edulis* showing the structure of the digestive diverticulae under (A) normal and (B) oil-exposed conditions; notice thinning of the tubules and accumulation of material within the lumen. (Unpublished micrographs courtesy of D.M. Lowe).

in lipid within the cells (Wolfe et al., 1981; Pipe and Moore, 1986; Lowe, 1988; Moore, 1988; Lowe and Clarke, 1989), and an enlargement of the secondary lysosomes which are the organelles responsible for internal digestion of food (Lowe et al., 1981; Lowe, 1988; Moore, 1988). Experimental evidence suggests that the breakdown of the digestive cells is mediated through release of hydrolytic enzymes of pre-lysosomal origin (Pipe and Moore, 1986).

The excretory system in mussels consists of paired kidneys and pericardial glands. These organs are also the principal sites of accumulation of heavy metals. Metals are detoxified by intracellular compartmentation in tertiary lysosomes and eventually excreted in the urine (George and Pirie, 1979; George et al., 1982; George, 1983). These tertiary lysosomes or lipofuschin granules are found in large numbers in the kidney; they are composed of highly insoluble lipoprotein pigments and may contain up to 10% metal (George et al., 1982). While the kidney forms a site for sequestration of metals, it is not itself immune to the toxic effects, and shows an increased incidence of renal cysts in mussels subjected to long term Cu exposure (Sunila, 1989).

Reproduction

Gametogenesis in mussels follows a seasonal cycle which varies according to the location of the population (see Chapter 4). The reproductive state of mussels can be quantified in terms of the cellular composition of the mantle tissue using stereological techniques (Lowe et al., 1982; Pipe, 1985); this enables the subtle effects of contaminant exposure, such as slowing down of gamete development, to be quantified in addition to the gross pathological changes such as oocyte degeneration. The use of stereological analysis has demonstrated that low-level hydrocarbon exposure can reduce the levels of stored reserves in the mantle tissue of *M. edulis*. However, the indications are that this effect is reversible following depuration (Lowe and Pipe, 1986, 1987).

A frequently observed phenomenon in mussels is gamete atresia and resorption; this may be brought about by a wide range of environmental conditions (Pipe, 1987a, b). Exposure to a number of metal and organic contaminants induces enhanced levels of gamete breakdown (Maung Myint and Tyler, 1982; Rasmussen et al., 1983a; Sunila, 1984, 1986b; Lowe and Pipe, 1986, 1987; Lowe, 1988). Molluscan lysosomal membranes are known to be sensitive to environmental conditions (Moore et al., 1987); the susceptibility of the gametes to breakdown may therefore be due, in part, to their considerable lysosomal compartment (Pipe and Moore, 1985b; Pipe, 1987a; Pipe and Da Silveira, 1989). An inflammatory response often accompanies the contaminant induced breakdown of gametes, with granular haemocytes infiltrating both male and female follicles (Sunila, 1984).

The suppression of gamete development has been reported following exposure of mussels to Cu, Cd and Zn (Maung Myint and Tyler, 1982; Calabrese et al., 1984; Kluytmans et al., 1988). This may well result from a lack of nutrients caused by damage to the gills and digestive tract. A number of contaminants, including metallic and N-nitroso compounds, have also been shown to stimulate spawning activity (Rasmussen, 1982; Sunila, 1986b; Kluytmans et al., 1988).

The mode of action of heavy metals, in terms of respiration rate, has been investigated (Akberali et al., 1984, 1985; Earnshaw et al., 1986). The results show that addition of Cu to unfertilized eggs from *M. edulis* results in an initial stimulation of respiration. However, pre-incubation with Cu and Zn inhibits the respiration rate of both eggs and sperm (Akberali et al., 1984, 1985). In addition, both metals cause a decrease in sperm motility, with Zn also causing some mitochondrial damage (Earnshaw et al., 1986).

Internal Defence

The internal defence of mussels relies upon circulating haemocytes which play a key role during inflammation, wound repair and phagocytosis of pathogens. The circulating haemocytes of *M. edulis* have been classified into at least two morphologically distinct types: granular and agranular cells (Moore and Lowe, 1977; Cheng, 1981; Rasmussen et al., 1985b; Pipe, 1990a). The granular haemocytes have been shown recently to be heterogeneous in terms of the lectin binding characteristics of their granules, indicating the possibility that a number of cell types or different developmental stages are present (Pipe, 1990a). The mechanisms available for destroying invading pathogens include release of hydrolytic enzymes (Pipe, 1990b) and the generation of reactive oxygen metabolites (Pipe, 1992). Further aspects of the internal defence system of *M. edulis* have been reviewed recently (Renwrantz, 1990).

There have been relatively few studies concerned with the direct effects of contaminants on the cells of the immune system in mussels. Oil emulsions and Cu and Cd have been shown to cause fluctuations in haemocyte numbers and, in addition, it has been demonstrated that oil affects their phagocytic ability (Livingstone, 1984; Sunila, 1984; McCormick-Ray, 1987). Silver has been shown to accumulate within the vacuoles of connective tissue macrophages (which are also responsible for phagocytosis), where it was associated with sulphur (George et al., 1986). The haemocyte response to contaminant exposure has been noted in numerous field and experimental studies; a general response to a wide range of contaminants is haemocytic infiltration of tissues (Rasmussen, 1982, Sunila, 1984, 1988; Auffret, 1988; Hietanen et al., 1988). Another non-neoplastic response to a wide range of environmental pollutants is the formation of granulocytomas (for definition see p.543 Chapter 12). This condition was first described by Lowe and Moore (1979) in the digestive gland and mantle tissues of *M. edulis*. Rasmussen et al. (1985a) were able to induce a similar pathological effect following chronic chemical exposure; more recently it has been described in mussel tissues exposed to chemical contamination under field and experimental conditions (Auffret, 1988). A virus-associated granulocytoma has also been described in *M. edulis* (Rasmussen, 1986).

Proliferative blood cell disorders have been described in *M. edulis* from different locations worldwide (Mix, 1986). Early work (Lowe and Moore, 1978), implicated environmental contamination, in particular PAHs, as potential causative agents of the haematopoietic neoplasms. Subsequent studies have demonstrated that mussels inhabiting highly contaminated environments did not display neoplastic disorders, while others have shown neoplasms from nonpolluted sites (Mix, 1988). Recent studies (Elston et al., 1988) have indicated that the haemic neoplasm can be transmitted allogenically and suggest a retrovirus as the causative agent (see p.557 Chapter 12). This does not preclude a possible role for environmental pollutants which may initiate the expression of an integrated viral genome.

Cytochemistry

A quantitative approach to the functional disturbances of subcellular organization resulting from contaminant exposure in mussels has concentrated on the use of cytochemical techniques. The lysosomal system and the endoplasmic reticulum are the organelles which have received most attention. Increases in permeability of lysosomal membranes, resulting from exposure to a wide range of contaminants, has been extensively reported, (Moore, 1991). The technique used to measure lysosomal permeability or stability is based on substrate penetrability (hydrolase latency), and bears a quantitative relationship to the magnitude of the stress imposed. Destabilization of the lysosomal membrane appears to lead to enhanced protein catabolism and cellular atrophy (Moore, 1985; Moore and Viarengo, 1987). In addition to quantifying effects in terms of membrane damage, the accumulation of compounds within lysosomes can also be measured. Changes in the lysosomal content of lipofuschin, neutral lipid and metallothionein have all been quantified following contaminant exposure (Viarengo et al., 1987; Moore, 1988). Accumulation of neutral lipid in enlarged lysosomes appears to be as a consequence of exposure to lipophilic xenobiotics such as PAHs (Moore et al., 1988; Moore, 1991).

The smooth endoplasmic reticulum forms the site of metabolism of many lipophilic organic xenobiotics. NADPH-ferrihaemoprotein reductase activity is associated with cytochrome *P*-450 reductase of the MFO system, and is measured cytochemically using tetrazolium salts (Altman, 1972; van Noorden and Butcher, 1986; Moore, 1988). Field and experimental studies have demonstrated stimulation of cytochemically determined NADPH-ferrihaemoprotein reductase following exposure to a range of organic xenobiotics (Moore et al., 1987; Moore, 1988).

POLLUTION MONITORING

The need for biological effects measurements in pollution monitoring has long been appreciated. The extensive use of mussels for such purposes has resulted in the development of various molecular and cellular indices of stress for application in environmental impact assessment (Moore et al., 1987; Livingstone et al., 1989c). The details of many such responses have already been described in the previous sections, and the main points of general interest are highlighted below.

Induction of metallothioneins (Viarengo, 1989) and the MFO system (Livingstone, 1991b) have been used as specific indicators of impact by, respectively, metals and organic pollutants. More success has been achieved with the former than the latter, and in the case of the MFO system a multiparameter approach has been suggested, e.g. measurement of cytochrome *P*-450, 418-peak, BPH and other parameters. In contrast to the variability, or lack of change, of biochemical measures of the MFO system, cytochemically measured NADPH-ferrihaemoprotein (NADPH-neotetrazolium) reductase activity has been used extensively as an indicator of impact by organic pollution. Other less specific, or general, molecular stress indices that have been considered include AEC, antioxidant enzymes, free radical scavengers, and end-points of biological damage such as lipid peroxidation and genotoxicity.

At the molecular/subcellular level, lysosomal latency has been routinely used as a general indicator of stress. Cellular measurements have included stereological quantification of gametogenic and digestive conditions.

Future directions in the development of molecular and cellular indices of biological effect—so-called biomarkers—for molluscs, are likely to include modern molecular biological and immunological techniques. The former is likely to focus on the expression of genes (measured at either the mRNA or protein levels) concerned with producing such molecules as biotransformation enzymes (e.g. cytochrome *P*-450 (Spry et al., 1989)) and multidrug resistance proteins (Kurelec and Pivčević, 1991). The extent to which a molecular biomarker for impact by organic pollution will be developed will depend upon the inducibility of the system concerned and the depth of understanding of its functioning in molluscs (Livingstone, 1991b). For example, the current extensive use of EROD activity, and other measurements of cytochrome *P*-450 IA1 expression, as a specific biomarker for impact by organic pollution in fish, is based on a thorough characterization of the enzyme and its functioning in this animal (Stegeman, 1989).

ACKNOWLEDGEMENTS

The authors gratefully acknowledge the stimulating discussions with colleagues, students and collaborators in the preparation of this chapter. Thanks also to Dr. P Lemaire for the preparation of Figure 9.1 and to D.M. Lowe for the unpublished micrographs used in Figure 9.3.

REFERENCES

Akberali, H.B., Earnshaw, M.J. and Marriott, K.R.M., 1984. The action of heavy metals on the gametes of the marine mussel, *Mytilus edulis* (L.). I. Copper-induced uncoupling of respiration in the unfertilized egg. Comp. Biochem. Physiol., 77C: 289-294.

Akberali, H.B., Earnshaw, M.J. and Marriott, K.R.M., 1985. The action of heavy metals on the gametes of the marine mussel, *Mytilus edulis* (L.). II. Uptake of copper and zinc and their effects on respiration in the sperm and unfertilized egg. Mar. Environ. Res., 16: 37-59.

Al-Sabti, K. and Kurelec, B., 1985. Induction of chromosomal aberrations in the mussel *Mytilus galloprovincialis*. Bull. Environ. Contam. Toxicol., 35: 660-665.

Altman, F.P., 1972. Quantitative dehydrogenase histochemistry with special reference to the pentose shunt dehydrogenases. Prog. Histochem. Cytochem., 4: 225-273.

Anderson, R.S., 1985. Metabolism of a model environmental carcinogen by bivalve molluscs. Mar. Environ. Res., 17: 137-140.

Anderson, R.S. and Döös, J.E., 1983. Activation of mammalian carcinogens to bacterial mutagens by microsomal enzymes from a pelecypod mollusc, *Mercenaria mercenaria*. Mutat. Res., 116: 247-256.

Arimoto, R. and Feng, S.Y., 1983. Histological studies on mussels from dredge spoil dumpsites. Est. Coast. Shelf Sci., 17: 535-546.

Auffret, M., 1988. Histopathological changes related to chemical contamination in *Mytilus edulis* from field and experimental conditions. Mar. Ecol. Prog. Ser., 46: 101-107.

Augenfeld, J.M., Morehouse, J.W., Riley, R.G. and Thomas, B.L., 1982. The fate of polyaromatic hydrocarbons in an intertidal sediment exposure system: bioavailability to *Macoma inquinata* (Mollusca : Pelecypoda) and *Abarenicola pacifica* (Annelida : Polychaeta). Mar. Environ. Res., 7: 31-50.

Bauer, I., Weber, K., Ernst, W. and Weigelt, V., 1989. Metabolism of octachlorostyrene in the blue mussel (*Mytilus edulis*). Chemosphere, 18: 1573-1579.

Bend, J.R., James, M.O. and Dansette, P.M., 1977. In vitro metabolism of xenobiotics of some marine animals. Ann. N.Y. Acad. Sci., 298: 505-517.

Bihari, N., Batel, R. and Zahn, R.K., 1990. DNA damage determination by the alkaline elution technique in the haemolymph of mussel *Mytilus galloprovincialis* treated with benzo[a]pyrene and 4-nitroquinoline-*N*-oxide. Aquat. Toxicol., 18:13-22.

Blum, J. and Fridovich, I., 1984. Enzymatic defenses against oxygen toxicity in the hydrothermal vent animals *Riftia pachyptila* and *Calyptogena magnifica*. Arch. Biochem. Biophys., 228: 617-620.

Boryslawskyji, M., Garood, A.C., Pearson, J.T. and Woodhead, D., 1988. Elevation of glutathione S-transferase activity as a stress response to organochlorine compounds in the freshwater mussel, *Sphaerium corneum*. Mar. Env. Res., 24: 101-104.

Britvić, S. and Kurelec, B., 1986. Selective activation of carcinogenic aromatic amines to bacterial mutagens in the marine mussel *Mytilus galloprovincialis*. Comp. Biochem. Physiol., 85C: 111-114.

Broman, D. and Ganning, B., 1986. Uptake and release of petroleum hydrocarbons by two brackish water bivalves, *Mytilus edulis* L. and *Macoma baltica* L. Ophelia, 25: 49-57.

Brunetti, R., Gola, I. and Majone, F., 1986. Sister-chromatid exchange in developing eggs of *Mytilus galloprovincialis* Lmk. (Bivalvia). Mutat.Res., 174: 207-211.

Burns, K.A. and Smith, J.L., 1981. Biological monitoring of ambient water quality: the case for using bivalves as sentinel organisms for monitoring petroleum pollution in coastal waters. Est. Coast. Shelf Sci., 13: 433-443.

Burton, D.T., Cooper, K.R., Goodfellow, W.L. Jr. and Rosenblatt, D.H., 1984. Uptake, elimination, and metabolism of ^{14}C-picric acid and ^{14}C-picramic acid in the American oyster (*Crassostrea virginica*). Arch. Environ. Contam. Toxicol., 13: 653-663.

Calabrese, A., MacInnes, J.R., Nelson, D.A., Greig, R.A. and Yevich, P.P., 1984. Effects of long-term exposure to silver or copper on growth, bioaccumulation and histopathology in the blue mussel *Mytilus edulis*. Mar. Environ. Res., 11: 253-274.

Calambokidis, J., Mowrer, J., Beug, M.W. and Herman, S.G., 1979. Selective retention of polychlorinated biphenyl components in the mussel, *Mytilus edulis*. Arch. Environ. Contam. Toxicol., 8: 299-308.

Carlson, G.P., 1972. Detoxification of foreign organic compounds by the quahog, *Mercenaria mercenaria*. Comp. Biochem. Physiol., 43B: 295-302.

Cheng, T.C., 1981. Bivalves. In: N.A. Ratcliffe and A.F. Rowley (Editors), Invertebrate blood cells 1. Academic Press, London, pp. 233-300.

Clement, L.E., Stekoll, M.S. and Shaw, D.G., 1980. Accumulation, fractionation   and release of oil by the intertidal clam *Macoma balthica*. Mar. Biol., 57: 41-50.

Dixon, D.R., 1982. Aneuploidy in mussel embryos (*Mytilus edulis* L.) originating from a polluted dock. Mar. Biol. Lett., 3: 155-161.

Dixon, D.R. and Prosser, H., 1986. An investigation of the genotoxic effects of an organotin antifouling compound [*bis*(tributyltin) oxide] on the chromosomes of the edible mussel, *Mytilus edulis*. Aquat. Toxicol., 8: 185-195.

Dixon, D.R., Jones, I.M. and Harrison, F.L., 1985. Cytogenetic evidence of inducible processes linked with metabolism of a xenobiotic chemical in adult and larval *Mytilus edulis*. Sci. Total Environ., 46: 1-8.

Dunn, B.P. and Stich, H.F., 1976. Release of the carcinogen benzo[a]pyrene from environmentally contaminated mussels. Bull. Environ. Contam. Toxicol., 15: 398-401.

Earnshaw, M.J., Wilson, S., Akberali, H.B., Butler, R.D. and Marriott, K.R.M., 1986. The action of heavy metals on the gametes of the marine mussel, *Mytilus edulis* (L.). III. The effect of applied copper and zinc on sperm motility in relation to ultrastructural damage and intracellular metal localisation. Mar. Environ. Res., 20: 261-278.

Elston, R.A., Kent, M.L. and Drum, A.S., 1988. Transmission of hemic neoplasia in the bay mussel, *Mytilus edulis*, using whole cells and cell homogenate. Dev. Comp. Immunol., 12: 719-727.

Ernst, W., 1979a. Factors affecting the evaluation of chemicals in laboratory experiments using marine organisms. Ecotoxicol. Environ. Saf., 3: 90-98.

Ernst, W., 1979b. Metabolic transformation of 1-(1-^{14}C)naphthol in bioconcentration studies with the common mussel, *Mytilus edulis*. Veroeff. Inst. Meeresforsch. Bremerhaven., 17: 233-240.

Farrington, J.W., Davis, R.C., Frew, N.M. and Rabin, K.S., 1982. No. 2 fuel oil compounds in *Mytilus edulis*. Retention and release after an oil spill. Mar. Biol., 66: 15-26.

Fowler, B.A., 1987. Intracellular compartmentation of metals in aquatic organisms: roles in mechanisms of cell injury. Environ. Health Perspect., 71: 121-128.

Galli, A., Del Chiero, D., Nieri, R. and Bronzetti, G., 1988. Studies on cytochrome P-450 in *Mytilus galloprovincialis*: induction by Na-phenobarbital and ability to biotransform xenobiotics. Mar. Biol., 100: 69-73.

Garcia Martinez, P., Hajos, A.K.D., Livingstone, D.R. and Winston, G.W., 1992. Metabolism and mutagenicity of 4-nitroquinoline-N-oxide by microsomes and cytosol of digestive gland of the mussel *Mytilus edulis* , in press.

Garcia Martinez, P., O'Hara, S., Winston, G.W. and Livingstone, D.R., 1989. Oxyradical generation and redox cycling mechanisms in digestive gland microsomes of the common mussel, *Mytilus edulis* L. Mar. Environ. Res., 28: 271-274.

Garrigues, P., Raoux, C., Lemaire, P., Ribera, D., Mathieu, A., Narbonne, J.F. and Lafaurie, M., 1991. In situ correlations between polycyclic hydrocarbons (PAH) and PAH metabolizing system activities in mussels and fish in the Mediterranean sea: preliminary results. Int. J. Environ. Anal. Chem., 38: 379-387.

George, S.G., 1983. Heavy metal detoxification in *Mytilus* kidney—an in vitro study of Cd- and Zn-binding to isolated tertiary lysosomes. Comp. Biochem. Physiol., 76C: 59-65.

George, S.G. and Pirie, B.J.S., 1979. The occurrence of cadmium in subcellular particles in the kidney of the marine mussel *Mytilus edulis*, exposed to cadmium: The use of electron probe microanalysis. Biochim. Biophys. Acta, 580: 125-143.

George, S.G. and Viarengo, A., 1985. A model for heavy metal homeostasis and detoxication in mussels. In: F.J. Vernberg, F.P. Thurberg, A. Calabrese and W.B. Vernberg (Editors), Marine pollution and physiology: recent advances. University of South Carolina Press, Columbia, pp. 234-244.

George, S.G., Carpene, E., Coombs, T.L., Overnell, J. and Youngson, A., 1979. Characterization of cadmium-binding protein from mussels, *Mytilus edulis* (L.), exposed to cadmium. Biochim. Biophys. Acta, 580: 225-233.

George, S.G. Coombs, T.L. and Pirie, B.J.S., 1982. Characterization of metal-containing granules from the kidney of the common mussel, *Mytilus edulis*. Biochim. Biophys. Acta, 716: 61-71.

George, S.G., Pirie, B.J.S., Calabrese, A. and Nelson, D.A., 1986. Biochemical and ultrastructural observations of long-term silver accumulation in the mussel, *Mytilus edulis*. Mar. Environ. Res., 18: 255-265.

Goldfarb, P., Spry, J.A., Dunn, D., Livingstone, D.R., Wiseman, A. and Gibson, G.G., 1989. Detection of mRNA sequences homologous to the human glutathione peroxidase and rat cytochrome P-450 IVA1 genes in *Mytilus edulis*. Mar. Environ. Res., 28: 57-60.

Hanzel, M.E. and Carlson, G.P., 1974. Azoreductase activity in the hard clam, *Mercenaria mercenaria* (L.). J. Exp. Mar. Biol. Ecol., 14: 225-229.

Hietanen, B., Sunila, I. and Kristoffersson, R., 1988. Toxic effects of zinc on the common mussel *Mytilus edulis* L. (Bivalvia) in brackish water. I. Physiological and histopathological studies. Ann. Zool. Fenn., 25: 341-347.

Kannan, N., Tanabe, S., Tatsukawa, R. and Phillips, D.J.H., 1989. Persistency of highly toxic coplanar PCBs in aquatic ecosystems: uptake and release kinetics of coplanar PCBs in green-lipped mussels (*Perna viridis* Linnaeus). Environ. Pollut., 56: 65-76.

Karnaukhov, V.N., Milovidova, N.Y. and Kargopolova, I.N., 1977. On the role of carotenoids in tolerance of sea molluscs to environmental pollution. Comp. Biochem. Physiol., 56A: 189-193.

Kirchin, M.A., Wiseman, A. and Livingstone, D.R., 1992. Seasonal and sex variation in the mixed-function oxygenase system of digestive gland microsomes of the common mussel, *Mytilus edulis* L. Comp. Biochem. Physiol., 101C: 81-91.

Kluytmans, J.H., Brands, F. and Zandee, D.I., 1988. Interactions of cadmium with the reproductive cycle of *Mytilus edulis* L. Mar. Environ. Res., 24: 189-192.

Knezovich, J.P. and Crosby, D.G., 1985. Fate and metabolism of o-toluidine in the marine bivalve molluscs *Mytilus edulis* and *Crassostrea gigas*. Environ. Toxicol. Chem., 4: 435-446.

Knezovich, J.P., Lawton, M.P. and Harrison, F.L., 1988. In vivo metabolism of aromatic amines by the bay mussel, *Mytilus edulis*. Mar. Environ. Res., 24: 89-91.

Kobayashi, K., 1985. The effect of the herbicide PCP on short-necked clams and the process of detoxification. In: T. Okutani, T. Tomiyama and T. Hibiya (Editors), Fisheries in Japan. Bivalves. Japan Marine Product Photo Material Association, pp. 161-165.

Krieger, R.I., Gee, S.J., Lim, L.O., Ross, J.H. and Wilson, A., 1979. Disposition of toxic substances of mussels (*Mytilus californianus*): preliminary metabolic and histologic studies. In: M.A.Q. Khan, J.J. Lech and J.J. Menn (Editors), Pesticide and xenobiotic metabolism in aquatic organisms. Am. Chem. Soc. Symp. Ser., 99: 259-277.

Kurelec, B., 1985. Exclusive activation of aromatic amines in the marine mussel *Mytilus edulis* by FAD-containing monooxygenase. Biochem. Biophys. Res. Commun., 127: 773-778.

Kurelec, B. and Krča, S., 1987. Metabolic activation of 2-aminofluorene, 2-acetylaminofluorene and *N*-hydroxy-acetylaminofluorene to bacterial mutagens with mussel (*Mytilus galloprovincialis*) and carp (*Cyprinius carpio*) subcellular preparations. Comp. Biochem. Physiol., 88C: 171-177.

Kurelec, B. and Krča, S., 1989. Glucuronides in mussel *Mytilus galloprovincialis* as a biomonitor of environmental carcinogens. Comp. Biochem. Physiol., 92C: 371-376.

Kurelec, B. and Pivčević, B., 1989. Distinct glutathione-dependent enzyme activities and a verapamil-sensitive binding of xenobiotics in a fresh-water mussel, *Anodonta cygnea*. Biochem. Biophys. Res. Commun., 164: 934-940.

Kurelec, B. and Pivčević, B., 1991. Evidence for a multixenobiotic resistance mechanism in the mussel *Mytilus galloprovincialis*. Aquat. Toxicol., 19: 291-301.

Kurelec, B., Britvić, S., Krča, S. and Zahn, R.K., 1986. Metabolic fate of aromatic amines in the mussel *Mytilus galloprovincialis*. Mar. Biol., 91: 523-527.

Kurelec, B., Chacko, M. and Gupta, R.C., 1988. Postlabelling analysis of carcinogen-DNA adducts in mussel, *Mytilus galloprovincialis*. Mar. Environ. Res., 24: 317-320.

Landrum, P.F. and Crosby, D.G., 1981. Comparison of several nitrogen-containing compounds in the sea urchin and other marine invertebrates. Xenobiotica, 11: 351-361.

Langston, W.J., 1978. Persistence of polychlorinated biphenyls in marine bivalves. Mar. Biol., 46: 35-40.

Laughlin, R.B. Jr. and French, W., 1988. Concentration dependence of *bis*(tributyl)tin oxide accumulation in the mussel, *Mytilus edulis*. Environ. Toxicol. Chem., 7: 1021-1026.

Laughlin, R.B. Jr., French, W. and Guard, H.E., 1986. Accumulation of *bis* (tributyltin) oxide by the marine mussel *Mytilus edulis*. Environ. Sci. Technol., 20: 884-890.

Lee, R.F., 1986. Metabolism of *bis*(tributyltin) oxide by estuarine animals. Mar. Technol. Soc. Washington D.C., Symp. Vol.4, pp. 1182-1188.

Lee, R.F., Sauerheber, R. and Benson, A.A., 1972. Petroleum hydrocarbons: uptake and discharge by the marine mussel *Mytilus edulis*. Science, 177: 344-346.

Livingstone, D.R., 1984. Biochemical differences in field populations of the common mussel *Mytilus edulis* L. exposed to hydrocarbons: Some considerations of biochemical monitoring. In: L. Bolis, J. Zadunaisky and R. Gilles (Editors), Toxins, drugs and pollutants in marine animals. Springer Verlag, Berlin, pp. 161-175.

Livingstone, D.R., 1985. Biochemical measurements. In: The effects of stress and pollution on marine animals. Praeger Publishers, New York, pp. 81-132.

Livingstone, D.R., 1988. Responses of microsomal NADPH-cytochrome c reductase activity and cytochrome *P*-450 in the digestive gland of *Mytilus edulis* and *Littorina littorea* to environmental and experimental exposure to pollutants. Mar. Ecol. Prog. Ser., 46: 37-43.

Livingstone, D.R., 1990. Cytochrome *P*-450 and oxidative metabolism in invertebrates. Biochem. Soc. Trans., 18: 15-19.

Livingstone, D.R., 1992. Persistent pollutants in marine invertebrates. In: C. Walker and D.R. Livingstone (Editors), Persistent pollutants in marine ecosystems. Pergamon Press, Oxford, in press.

Livingstone, D.R., 1991a. Organic xenobiotic metabolism in marine invertebrates. Adv. Comp. Environ. Physiol., 7: 45-185.

Livingstone, D.R., 1991b. Towards a specific index of impact by organic pollution for marine invertebrates. Comp. Biochem. Physiol., 100C: 151-155.

Livingstone, D.R. and Farrar, S.V., 1984. Tissue and subcellular distribution of enzyme activities of mixed-function oxygenase and benzo[a]pyrene metabolism in the common mussel *Mytilus edulis* L. Sci. Total Environ., 39:209-235.

Livingstone, D.R., Kirchin, M.A. and Wiseman, A., 1989a. Cytochrome *P*-450 and oxidative metabolism in molluscs. Xenobiotica, 19: 1041-1062.

Livingstone, D.R., Garcia Martinez, P. and Winston, G.W., 1989b. Menadione-stimulated oxyradical production in digestive gland microsomes of the common mussel, *Mytilus edulis* L. Aquat. Toxicol., 15; 213-236.

Livingstone, D.R., Dixon, D.R., Donkin, P., Lowe, D.M, Moore, M.N. and Widdows, J. 1989c. Molecular, cellular and physiological responses of the common mussel *Mytilus edulis* to pollution: uses in environmental monitoring and management. In: Ecotoxicology, Institute of Biology, UK., pp. 26-30.

Livingstone, D.R., Arnold, R., Chipman, K., Kirchin, M.A. and Marsh, J., 1990a. The mixed-function oxygenase system: metabolism, responses to xenobiotics, and toxicity. Oceanis, 16: 331-347.

Livingstone, D.R., Garcia Martinez, P., O'Hara, S., Michel, X., Narbonne, J.F., Ribera, D. and Winston, G.W., 1990b. Oxyradical production as a pollution-mediated mechanism of toxicity in the common mussel, *Mytilus edulis* L. and other molluscs. Funct. Ecol., 4: 415-424.

Livingstone, D.R., Lips, F., Garcia Martinez, P. and Pipe R. K., 1992. Antioxidant enzymes in digestive gland of the common mussel *Mytilus edulis* . Mar. Biol., in press.

Lock, E.A., Stonard, M.D. and Elcombe, C.R., 1987. The induction of ω- and β-oxidation of fatty acids and effect on a_{2u}globulin content in the liver and kidneys of rats administered 2,2,4-trimethylpentane. Xenobiotica, 17: 513- 522.

Lowe, D.M., 1988. Alterations in cellular structure of *Mytilus edulis* resulting from exposure to environmental contaminants under field and experimental conditions. Mar. Ecol. Prog. Ser., 46: 91-100.

Lowe, D.M. and Moore, M.N., 1978. Cytology and quantitative cytochemistry of a proliferative atypical hemocyte condition in *Mytilus edulis* (Bivalvia, Mollusca). J. Natl. Cancer Inst., 60: 1455-1459.

Lowe, D.M. and Moore, M.N., 1979. The cytology and occurrence of granulocytomas in mussels. Mar. Poll. Bull., 10: 137-141.

Lowe, D.M. and Pipe, R.K., 1986. Hydrocarbon exposure in mussels: a quantitative study of the responses in the reproductive and nutrient storage cell systems. Aquat. Toxicol., 8: 265-272.

Lowe, D.M. and Pipe, R.K., 1987. Mortality and quantitative aspects of storage cell utilization in mussels, *Mytilus edulis*, following exposure to diesel oil hydrocarbons. Mar. Environ. Res., 22: 243-251.

Lowe, D.M. and Clarke, K.R., 1989. Contaminant-induced changes in the structure of the digestive epithelium of *Mytilus edulis*. Aquat. Toxicol., 15: 345-358.

Lowe, D.M., Moore, M.N. and Clarke, K.R., 1981. Effects of oil on digestive cells in mussels: quantitative alterations in cellular and lysosomal structure. Aquat. Toxicol., 1: 213-226.

Lowe, D.M., Moore, M.N. and Bayne, B.L., 1982. Aspects of gametogenesis in the marine mussel *Mytilus edulis* L. J. Mar. Biol. Ass. U.K., 62: 133-145.

Lu, P.Y., Metcalf, R.L., Plummer, N. and Mandel, D., 1977. The environmental fate of three carcinogens benzo[a]pyrene, benzidene and vinyl chloride evaluated in laboratory model ecosystems. Arch. Environ. Contam. Toxicol., 6: 129-142.

Marsh, J.W., Chipman, J.K. and Livingstone, D.R., 1992. Activation of xenobiotics to reactive and mutagenic products by the marine invertebrates *Mytilus edulis*, *Carcinus maenas* and *Asterias rubens*. Aquat. Toxicol., in press.

Maung Myint, U. and Tyler, P.A., 1982. Effects of temperature, nutritive and metal stressors on the reproductive biology of *Mytilus edulis*. Mar. Biol., 67: 209-223.

McCormick-Ray, M.G., 1987. Hemocytes of *Mytilus edulis* affected by Prudoe Bay crude oil emulsion. Mar. Environ. Res., 22: 107-122.

Mix, M.C., 1986. Cancerous diseases in aquatic animals and their association with environmental pollutants: a critical literature review. Mar. Environ. Res., 20: 1-141.

Mix, M.C., 1988. Shellfish diseases in relation to toxic chemicals. Aquat. Toxicol., 11: 29-42.

Moore, M.N., 1985. Cellular responses to pollutants. Mar. Pollut. Bull., 16: 134-139.

Moore, M.N., 1988. Cytochemical responses of the lysosomal system and NADPH-ferrihemoprotein reductase in molluscan digestive cells to environmental and experimental exposure to xenobiotics. Mar. Ecol. Prog. Ser., 46: 81-89.

Moore, M.N., 1991. Cellular reactions to toxic environmental contaminants in marine molluscs. In: P.D. Abel and V. Axiak, (Editors), Ecotoxicology and the marine environment. Ellis Horwood, Chichester, pp. 157-175.

Moore, M.N. and Lowe, D.M., 1977. The cytology and cytochemistry of the hemocytes of *Mytilus edulis* and their response to experimentally injected carbon particles. J. Invertebr. Pathol., 29: 18-30.

Moore, M.N. and Viarengo, A., 1987. Lysosomal membrane fragility and catabolism of cytosolic proteins: evidence for a direct relationship. Experientia, 43: 320-323.

Moore, M.N., Livingstone, D.R., Widdows, J., Lowe, D.M. and Pipe, R.K., 1987. Molecular, cellular and physiological effects of oil-derived hydrocarbons on molluscs and their use in impact assessment. Philos. Trans. Roy. Soc. Lond. Ser. B, 316: 603-623.

Moore, M.N., Pipe, R.K. and Farrar, S.V., 1988. Induction of lysosomal lipid accumulation and fatty degeneration by polycyclic aromatic hydrocarbons in molluscan digestive cells. Mar. Environ. Res., 24: 252-253.

Moore, M.N., Livingstone, D.R. and Widdows, J., 1989. Hydrocarbons in marine molluscs: biological effects and ecological consequences. In: U. Varanasi (Editor), Metabolism of polycyclic aromatic hydrocarbons in the marine environment. CRC Press, Boca Raton, Florida, pp. 291-328.

Musgrave, M.E., Gould, S.P. and Ablett, R.F., 1987. Enzymatic lipid peroxidation in the gonadal and hepatopancreatic microsomal fraction of cultivated mussels (Mytilus edulis L.). J. Food Sci., 52: 609-612.

Nebert, D.W., Nelson, D.R., Adnesik, M., Coon, M.J., Estabrook, R.W., Gonzalez, F.J., Guengerich, F.P., Gunsalus, I.C., Johnson, E.F., Kemper, B., Levin, W., Phillips, I.R., Sato, R. and Waterman, M.R., 1989. The P-450 gene superfamily. Update on the naming of new genes and nomenclature of chromosomal loci. DNA, 8: 1-13.

Neff, J.M., Cox, B.A., Dixit, D. and Anderson, J.W., 1976. Accumulation and release of petroleum-derived aromatic hydrocarbons by four species of marine animals. Mar. Biol., 38: 279-289.

Noel-Lambot, F., 1976. Distribution of cadmium, zinc and copper in the mussel Mytilus edulis. Existence of cadmium-binding proteins similar to metallothioneins. Experientia, 32: 324-326.

Noorden, C.J.F. van, and Butcher, R.G., 1986. A quantitative histochemical study of NADPH-ferrihemoprotein reductase activity. Histochem. J., 18: 364-370.

Palmork, K.H. and Solbakken, J.E., 1981. Distribution and elimination of [9-^{14}]phenanthrene in the horse mussel (Modiolus modiolus). Bull. Environ. Contam. Toxicol., 26: 196-201.

Parry, J.M., Kadhim, M., Barnes, W. and Danford, N., 1981. Assays of marine organisms for the presence of mutagenic and/or carcinogenic chemicals. In: C.J. Dawe, J.C. Harshbarger, S. Kondo, T. Sugimura and S. Takayama (Editors), Phyletic approaches to cancer. Japan Sci. Soc., Tokyo, pp. 141-166.

Payne, J.F., 1982. Metabolism of complex mixtures of oil spill surfactant compounds by a representative teleost (Salmo gairdneri), crustacean (Cancer irroratus) and mollusc (Chlamys islandicus). Bull. Environ. Contam. Toxicol., 28: 277-280.

Pipe, R.K., 1985. Seasonal cycles in and effects of starvation on egg development in Mytilus edulis. Mar. Ecol. Prog. Ser., 24: 121-128.

Pipe, R.K., 1987a. Ultrastructural and cytochemical study on interactions between nutrient storage cells and gametogenesis in the mussel Mytilus edulis. Mar. Biol., 96: 519-528.

Pipe, R.K., 1987b. Oogenesis in the marine mussel Mytilus edulis: an ultrastructural study. Mar. Biol., 95: 405-414.

Pipe, R.K., 1990a. Differential binding of lectins to haemocytes of the mussel Mytilus edulis. Cell Tissue Res., 261: 261-268.

Pipe, R.K., 1990b. Hydrolytic enzymes associated with granular haemocytes of the marine mussel Mytilus edulis. Histochem. J., 22: 595-603.

Pipe, R.K., 1992. The generation of reactive oxygen metabolites by the haemocytes of the mussel Mytilus edulis. Dev. Comp. Immunol., in press.

Pipe, R.K. and Moore, M.N., 1985a. Ultrastructural changes in the lysosomal-vacuolar system in digestive cells of Mytilus edulis as a response to increased salinity. Mar. Biol., 87: 157-163.

Pipe, R.K. and Moore, M.N., 1985b. The ultrastructural localization of lysosomal acid hydrolases in developing oocytes of the common marine mussel Mytilus edulis. Histochem. J., 17: 939-949.

Pipe, R.K. and Moore, M.N., 1986. Arylsulphatase activity associated with phenanthrene induced digestive cell deletion in the marine mussel Mytilus edulis. Histochem. J., 18: 557-564.

Pipe, R.K. and Da Silveira, H.M.C., 1989. Arylsulphatase and acid phosphatase activity associated with developing and ripe spermatozoa of the mussel Mytilus edulis. Histochem. J., 21: 23-32.

Porte, C., Solé, M., Albaigés, J. and Livingstone, D.R., 1991. Responses of mixed-function oxygenase and antioxidase enzyme system of Mytilus sp. to organic pollution. Comp. Biochem. Physiol., 100C: 183-186.

Pruell, R.J., Lake, J.L., Davis, W.R. and Quinn J.G., 1986. Uptake and depuration of organic contaminants by blue mussels (Mytilus edulis) exposed to environmentally contaminated sediment. Mar. Biol., 91: 497-507.

Rasmussen, L.P.D., 1982. Light microscopical studies of the acute effects of N-nitrosodimethylamine on the marine mussel, Mytilus edulis. J. Invertebr. Pathol., 39: 66-80.

Rasmussen, L.P.D., 1986. Virus-associated granulocytomas in the marine mussel, Mytilus edulis, from three sites in Denmark. J. Inverebr. Pathol., 48: 117-123.

Rasmussen, L.P.D., Hage, E. and Karlog, O., 1983a. Light and electron microscopic studies of the acute and chronic toxic effects of N-nitroso compounds on the marine mussel, Mytilus edulis (L.). I. N-nitrosodimethylamine. Aquat. Toxicol., 3: 285-299.

Rasmussen, L.P.D., Hage, E. and Karlog, O., 1983b. Light and electron microscopic studies of the acute and chronic toxic effects of N-nitroso compounds on the marine mussel, Mytilus edulis (L.). II. N-methyl-N-nitro-N-nitrosoguadine. Aquat. Toxicol., 3: 301-311.

Rasmussen, L.P.D., Hage, E. and Karlog, O., 1985a. Light and electron microscopic studies of the acute and long-term toxic effects of N-nitrosodipropylamine and N-methylnitrosurea on the marine mussel Mytilus edulis. Mar. Biol., 85: 55-65.

Rasmussen, L.D.P., Hage, E. and Karlog, O., 1985b. An electron microscope study of the circulating leukocytes of the marine mussel Mytilus edulis. J. Invertebr. Pathol., 45: 158-167.

Renberg, L., Tarkpea, M. and Sundstrom, G., 1986. The use of the bivalve Mytilus edulis as a test organism for bioconcentration studies. II. The bioconcentration of two ^{14}C-labelled chlorinated paraffins. Ecotoxicol. Environ. Saf., 11: 361-372.

Renwrantz, L., 1990. Internal defence system of Mytilus edulis. In: G.B. Stefano (Editor), Neurobiology of Mytilus edulis. Manchester University Press, Manchester, pp. 256-275.

Ribera, D., Narbonne, J.F., Daubeze, M. and Michel, D., 1989. Characterisation, tissue distribution and sexual differences of some parameters related to lipid peroxidation in mussels. Mar. Environ. Res., 28: 279-283.

Ribera, D., Narbonne, J.F., Michel, X., Livingstone, D.R. and O'Hara, S., 1991. Responses of antioxidants and lipid peroxidation in mussels to oxidative damage exposure. Comp. Biochem. Physiol. 100C: 177-181.

Riley, R.T., Mix, M.C., Schaffer, R.L. and Bunting, D.L., 1981. Uptake and accumulation of naphthalene by the oyster Ostrea edulis, in a flow-through system. Mar. Biol., 61: 267-276.

Roesijadi, G., 1982. Uptake and incorporation of mercury into mercury-binding proteins of gills of Mytilus edulis as a function of time. Mar. Biol., 66: 151-157.

Roesijadi, G. and Hall, R.E., 1981. Characterization of mercury-binding proteins from the gills of marine mussels exposed to mercury. Comp. Biochem. Physiol., 70C: 59-64.

Roesijadi, G. and Fellingham, G.W., 1987. Influence of Cu, Cd and Zn pre-exposure on Hg toxicity in the mussel Mytilus edulis. Can. J. Fish. Aquat. Sci., 44: 680-684.

Roesijadi, G., Drum, A.S., Thomas, J.M. and Fellingham, G.W., 1982a. Enhanced mercury tolerance in marine mussels and relationship to low molecular weight, mercury-binding proteins. Mar. Pollut. Bull., 13: 250-253.

Roesijadi, G., Calabrese, A. and Nelson, D., 1982b. Mercury-binding proteins of M. edulis. In: W.B. Vernberg, A. Calabrese, F.P. Thurberg and F.J. Vernberg (Editors), Physiological mechanisms of marine pollutant toxicity. Academic Press, New York, pp. 75-87.

Schenkman, J.B., Sligar, S.G. and Cinti, D.L., 1982. Substrate interaction with cytochrome P-450. In: J.B. Schenkman and D. Kupfer (Editors), Hepatic cytochrome P-450 monooxygenase system. Pergamon Press, Oxford, pp. 587-615.

Shick, J.M. and Dykens, J.A., 1985. Oxygen detoxification in algal-invertebrate symbioses from the Great Barrier Reef. Oecologia (Berl.), 66: 33-41.

Simkiss, K. and Mason, A.Z., 1984. Cellular responses of molluscan tissues to environmental metals. Mar. Environ. Res., 14: 103-118.

Simkiss, K., Taylor, M. and Mason, A.Z., 1982. Metal detoxification and bioaccumulation in molluscs. Mar. Biol. Lett., 3: 187-201.

Spry, J.A., Livingstone, D.R., Wiseman, A., Gibson, G.G. and Goldfarb, P.S., 1989. Cytochrome P-450 gene expression in the common mussel Mytilus edulis. Biochem. Soc. Trans., 17: 1013-1014.

Stegeman, J.J., 1985. Benzo[a]pyrene oxidation and microsomal enzyme activity in the mussel (Mytilus edulis) and other bivalve molluscs species from the Western North Atlantic. Mar. Biol., 89: 21-30.

Stegeman, J.J., 1989. Cytochrome P-450 forms in fish: catalytic, immunological and sequence similarities. Xenobiotica, 19: 1093-1110.

Stegeman, J.J. and Teal, J.M., 1973. Accumulation, release and retention of petroleum hydrocarbons by the oyster Crassostrea virginica. Mar. Biol., 22: 37-44.

Sunila, I., 1984. Copper and cadmium-induced histological changes in the mantle of Mytilus edulis L. (Bivalvia). Limnologica, 15: 523-527.

Sunila, I., 1986a. Chronic histopathological effects of short-term copper and cadmium exposure on the gill of the mussel, Mytilus edulis. J. Invertebr. Pathol., 47: 125-142.

Sunila, I., 1986b. Histopathological changes in the mussel Mytilus edulis L. at the outlet from a titanium dioxide plant in Northern Baltic. Ann. Zool. Fenn., 23: 61-70.

Sunila, I., 1987. Histopathology of mussels (*Mytilus edulis* L.) from the Tvärminne area, the Gulf of Finland (Baltic Sea). Ann. Zool. Fenn., 24: 55-69.

Sunila, I., 1988. Acute histological responses of the gill of the mussel, *Mytilus edulis*, to exposure by environmental pollutants. J. Invertebr. Pathol., 52: 137-141.

Sunila, I., 1989. Cystic kidneys in copper exposed mussels. Dis. Aquat. Org., 6: 63-66.

Sunila, I. and Lindström, R., 1985. The structure of the interfilamentar junction of the mussel (*Mytilus edulis* L.) gill and its uncoupling by copper and cadmium exposures. Comp. Biochem. Physiol., 81C: 267-272.

Suteau, P. and Narbonne, J.F., 1988. Preliminary data on PAH metabolism in the marine mussel *Mytilus galloprovincialis* from Arcachon Bay, France. Mar. Biol., 98: 421-425.

Suteau, P., Migaud, M.L., Daubeze, M. and Narbonne, J.F., 1988a. Comparative induction of drug metabolizing enzymes in whole mussel by several types of inducer. Mar. Environ. Res., 24: p.119.

Suteau, P., Daubeze, M., Migaud, M.L. and Narbonne, J.F., 1988b. PAH-metabolizing enzymes in whole mussels as biochemical tests for chemical monitoring pollution. Mar. Ecol. Prog. Ser., 46: 45-49.

Tanabe, S., Tatsukawa, R. and Phillips, D.J.H., 1987. Mussels as bioindicators of PCB pollution: a case study on uptake and release of PCB isomers and congeners in green-lipped mussels (*Perna viridis*) in Hong Kong waters. Environ. Pollut., 47: 41-62.

Thompson, R.J., Ratcliffe, N.A. and Bayne, B.L., 1974. Effects of starvation on structure and function in the digestive gland of the mussel (*Mytilus edulis* L.). J. Mar. Biol. Ass. UK., 54: 699-712.

Varanasi, U., Reichert, W.L., Stein, J.E., Brown, D.W. and Sanborn, H.R., 1985. Bioavailability and biotransformation of aromatic hydrocarbons in benthic organisms exposed to sediment from an urban estuary. Environ. Sci. Technol., 19: 836-841.

Viarengo, A., 1989. Heavy metals in marine invertebrates: mechanisms of regulation and toxicity at the cellular level. Rev. Aquat. Sci., 1: 295-317.

Viarengo, A., Pertica, M., Mancinelli, G., Palmero, S., Zanicchi, G. and Orunesu, M., 1981. Synthesis of Cu-binding proteins in different tissues of mussels exposed to the metal. Mar. Pollut. Bull., 12: 347-350.

Viarengo, A., Pertica, M., Mancinelli, G., Zanicchi, G., Bouquegneau, J.M. and Orunesu, M., 1984. Biochemical characterization of copper-thioneins isolated from the tissues of mussels exposed to the metal. Mol. Physiol., 5: 41-52.

Viarengo, A., Moore, M.N., Mancinelli, G., Mazzucotelli, A., Pipe, R.K. and Farrar, S.V., 1987. Metallothioneins and lysosomes in metal toxicity and homeostasis in marine mussels: the effects of cadmium in the presence and absence of phenanthrene. Mar. Biol., 94: 251-257.

Viarengo, A., Pertica, M., Canesi, L., Biasi, F., Cecchini, G. and Orunesu, M., 1988. Effects of heavy metals on lipid peroxidation in mussel tissues. Mar. Environ. Res., 24: p. 355.

Viarengo, A., Pertica, M., Canesi, L. Mazzucotelli, A., Orunesu, M. and Bouquegneau, J.M., 1989. Purification and biochemical characterization of a lysosomal copper-rich thionein-like protein involved in metal detoxification in the digestive gland of mussels. Comp. Biochem. Physiol., 93C: 389-395.

Viarengo, A., Canesi, L., Pertica, M., Livingstone, D.R. and Orunesu, M., 1991a. Age-related lipid peroxidation in the digestive gland of mussels: the role of antioxidant defense systems. Experientia, 47: 454-457.

Viarengo, A., Canesi, L., Pertica, M. and Livingstone, D.R., 1991b. Seasonal variations in the antioxidant defense systems and lipid peroxidation of the digestive gland of mussels. Comp. Biochem. Physiol., 100C: 187-190.

Wenning, R.J. and Di Giulio, R.T., 1988a. Microsomal enzyme activities, superoxide production, and antioxidant defenses in ribbed mussels (*Geukensia demissa*) and wedge clams (*Rangia cuneata*). Comp. Biochem. Physiol., 90C: 21-28.

Wenning, R.J. and Di Giulio, R.T., 1988b. The effects of paraquat on microsomal oxygen reduction and antioxidant defenses in ribbed mussels (*Geukensia demissa*) and wedge clams (*Rangia cuneata*). Mar. Environ. Res., 24: 301-305.

Wenning, R.J., Di Giuilo, E.T. and Gallagher, F.P., 1988. Oxidant-mediated biochemical effects of paraquat in the ribbed mussel, *Geukensia demissa*. Aquat. Toxicol., 12: 157-170.

Widdows, J., Bakke, T., Bayne, B.L., Donkin, P., Livingstone, D.R., Lowe, D.M., Moore, M.N., Evans, S.V. and Moore, S.L., 1982. Responses of *Mytilus edulis* on exposure to water-accommodated fraction of North Sea oil. Mar. Biol., 67: 15-31.

Widdows, J., Moore, S.L., Clarke, K.R. and Donkin, P., 1983. Uptake, tissue distribution and elimination of [1-^{14}C]naphthalene in the mussel *Mytilus edulis*. Mar. Biol., 76: 109-114.

Winston, G.W., Livingstone, D.R. and Lips, F., 1990. Oxygen reduction metabolism by the digestive gland of the common marine mussel, *Mytilus edulis* L. J. Exp. Zool., 255: 296-308.

Wofford, H.W., Wilsey, C.D., Neff, G.S., Giam, C.S. and Neff, J.M., 1981. Bioaccumulation and metabolism of phthalate esters by oysters, brown shrimp, and sheepshead minnows. Ecotoxicol. Environ. Saf., 5: 202-210.

Wolfe, D.A., Clark, R.C., Foster, C.A., Hawkes, J.W. and MacLoad, W.D., 1981. Hydrocarbon accumulation and histopathology in bivalve molluscs transplanted to the Baie de Morlaix and the Rade de Brest. In: Amoco cadiz: Fates and effects of the oil spill. Proc. Int. Symp., Brest, France, 1979. pp. 599-616.

Chapter 10

MUSSEL CULTIVATION

ROBERT W. HICKMAN

INTRODUCTION

Marine mussels, perhaps of all the species of shellfish which are cultivated around the world, most readily demonstrate those characteristics that go to make up the 'ideal candidate for aquaculture'. The extremely wide distribution of this mollusc family, the Mytilidae, throughout the world is one of these characteristics, and one that has contributed to the premier position that mussels now occupy in the world statistics for aquaculture production. The total world production of mussels exceeded 1.1 million tonnes(t) in 1988 (FAO, 1990). The vast majority of this comes from aquaculture. Over 20 countries now report regular harvests of farmed mussels but world production is dominated by two countries, China, with almost 40% of the total, and Spain, with a further 20% (Table 10.1). The Netherlands, which traditionally rivalled Spain in mussels, has seen a dramatic decline in its production since the early 1980s. By far the majority of the farmed crop is the 'blue' type of mussel, the various *Mytilus* species which occur in the temperate waters of Europe and Asia, North and South America. Green mussels of the various *Perna* species are farmed in warmer waters, particularly in Thailand and the Philippines, but also in China and New Zealand.

AQUACULTURE CHARACTERISTICS OF MUSSELS

High fecundity and a mobile free-living larval phase are two characteristics which have contributed to the widespread distribution of the relatively few mussel species, and at the same time have greatly influenced the technology and practice of mussel farming. The natural abundance of mussel larvae, particularly in temperate coastal waters, is reflected in the dominant position that the settled larvae, or mussel seed, frequently attain on the hard substrate of intertidal and subtidal zones of many shorelines. Though much of a mussel farmer's effort may need to be expended to obtain sufficient seed to stock his farm, the natural availability of seed sources,

Table 10.1. World production (in metric tonnes) of mussels for 1988 (from FAO, 1990) showing species and methods used for cultivation.

Country	Production	Species	Culture Methods
China	429,675	*Mytilus edulis* *Perna viridis*	Longline, raft
Spain	209,687	*Mytilus galloprovincialis* *Mytilus edulis*	Raft
Italy	85,400	*Mytilus galloprovincialis*	Longline, hanging park
Netherlands	77,596	*Mytilus edulis*	On-bottom
Denmark	72,524	*Mytilus edulis*	On-bottom
France	54,873	*Mytilus edulis* *Mytilus galloprovincialis*	Bouchot, longline
U.S.A.	35,724	*Mytilus edulis* *Mytilus californianus*	On-bottom, longline, raft
Thailand	35,270	*Perna viridis* *Musculus senhauseni*	Pole, fish trap
Germany	30,865	*Mytilus edulis*	On-bottom
Korea	27,356	*Mytilus crassitesta*	Longline
Chile	21,910	*Mytilus chilensis* *Choromytilus chorus* *Aulacomya ater*	Raft, longline
New Zealand	18,000	*Perna canaliculus*	Longline
Philippines	17,553	*Perna viridis*	Pole
Ireland	16,000	*Mytilus edulis*	On-bottom, longline, raft

Table 10.1. continued.

Country	Production	Species	Culture Methods
Peru	9,083	*Aulacomya ater*	no data
Turkey	7,953	*Mytilus galloprovincialis*	no data
U.K.	6,949	*Mytilus edulis*	On-bottom, raft
Albania	3,107	*Mytilus galloprovincialis*	no data
Australia	2,508	*Mytilus planulatus*	Raft, longline
Canada	1,984	*Mytilus edulis*	Raft, longline
Yugoslavia	1,370	*Mytilus galloprovincialis*	Hanging park
Singapore	1,192	*Perna viridis*	Raft, longline
Argentina	1,100	*Mytilus platensis*	no data
Greece	1,100	*Mytilus galloprovincialis*	Raft, longline
U.S.S.R.	1,082	*Mytilus galloprovincialis*	Longline
Sweden	862	*Mytilus edulis*	Longline
Malaysia	709	*Perna viridis*	Raft
Venezuela	575	*Perna perna*	Raft
Morocco	488	*Mytilus galloprovincialis*	no data
Mexico	458	*Mytilus californianus*	no data
Uruguay	256	*Mytilus platensis*	no data

Table 10.1. continued

Country	Production	Species	Culture Methods
Portugal	107	*Mytilus galloprovincialis*	Raft
Bulgaria	100	*Mytilus galloprovincialis*	Longline
Norway	87	*Mytilus edulis*	Longline
French Polynesia	5	*Perna viridis*	Rack
India	– †	*Perna viridis* *Perna indica*	Raft, longline
South Africa	–	*Perna perna*	Longline
Ecuador	–	*Perna perna*	Raft
Egypt	–	*Mytilus galloprovincialis*	On-bottom
Indonesia	–	*Perna viridis*	Raft
Brazil	–	*Perna perna*	Raft
Western Samoa	–	*Perna viridis*	Raft

† No data on production available

without the need to resort to hatchery production, has been a significant positive factor in the development of mussel farming.

The prominent mussel carpets which are typical of the exposed rocky shore, provide evidence of two further features of mussel biology that predispose these animals for aquaculture. The rapid growth rate which enables wild mussels to compete successfully against other benthic organisms, also ensures that a commercial-sized product can be reared in a short time period (less than three years) under farming conditions. At the same time the mussel's natural propensity to live in dense beds in the wild makes it readily adaptable to the high population densities necessary for an economically viable farming system.

The ideal aquaculture candidate should also possess two attributes which are less immediately obvious but equally essential to the successful farming operation. The farmed species should be cheap to feed and resistant to disease. The mussel, as a filter feeder primarily utilizing the phytoplankton, requires only a continuous supply of high productivity seawater to grow and fatten. Disease is an ever present risk in any intensive livestock production and numerous pathogenic and parasitic organisms have been identified in mussels (Bower and Figueras, 1989; see also Chapter 12). However, mussels growing in high densities, either in the wild or in farming situations, seem to be relatively free from mass mortality diseases such as affect other molluscs, most notably oysters.

One final feature of the mussel which particularly equips it for farming, and which distinguishes it from the other cultured molluscs, is its ability to attach itself by means of byssus threads to any firm substrate. Even more importantly, it is able to reattach again and again, and to make minor adjustments to its orientation whilst attached. This characteristic has had perhaps the greatest influence on the types of mussel farming practice and technology that have developed throughout the world.

METHODS OF CULTIVATION

All the methods used in the cultivation of mussels can be assigned to one of two categories; they are either on-bottom cultivation or off-bottom cultivation. Examples of both, though not categorized as such, were included in Korringa's (1976) elegant and detailed account of commercial marine farming operations in the 1960s and early 1970s. This remains the classic treatise on mussel farming methods and practice throughout the world. Comparison of Korringa's account with recent reviews of current methodology in the Netherlands (Dijkema, 1988; Dijkema and van Stralen, 1989), France (Figueras, 1989), Spain (Figueras, 1989; Perez-Camacho et al., 1991), and the Philippines (Rosell, 1991) confirms that in most areas basic mussel farming practice has changed hardly at all in the last quarter of a century. Only one new method of cultivation has been developed in the last 20–25 years, namely the longline system of floating ropes and buoys.

A brief description of the basic features which distinguish the various methods used to farm mussels throughout the world is provided here, followed by a discussion of the general principles governing mussel farming. These sections update Mason's chapter on mussel cultivation in Bayne's (1976) review of the biology of marine mussels. Readers wishing to become familiar with details of farming procedures used in different countries, are referred to the above reviews and to pertinent chapters in three other mussel farming references, namely Lutz (1980a), which covers cultivation in North America, Vakily (1989), which concentrates on the *Perna* species and Menzel (1991), which has information on the culture techniques used in the U.S.A., Spain, Thailand, Chile, India, China, the Philippines and New Zealand. Other significant sources of technical data on some of the more recently developed mussel industries include Zhang (1984) for China, Herriot (1984) for Ireland, Jenkins (1985) and Hickman (1989a, b) for New Zealand, and Muise (1990) for Canada.

The discussion of general principles attempts to emphasize the differences in productivity which result from the different methods of farming mussels.

On-Bottom Cultivation

On-bottom or seabed cultivation is based on the principle of transferring mussels from areas where they have settled in great abundance, to culture plots, where they can be spread at lower density in order to obtain much better growth and fattening of the mussels. This method has been developed over the last 120 years in the Netherlands into a highly sophisticated and mechanized industry producing 50,000–100,000t per year (Dijkema and van Stralen, 1989). Much of the farming practice described by Korringa (1976), remains unchanged. However, major changes to the hydrology of the mussel growing areas, which resulted from the construction of a flood barrier scheme in the 1980s (Dijkema, 1988), have led to the modernization and expansion of both the cultivation and processing sectors of the industry (Dijkema and van Stralen, 1989). The centre of cultivation has shifted from the Oosterschelde in the south to the Waddenzee in the north.

Dredging for the 10–30mm one-year-old seed mussels occurs mainly in the Waddenzee. The seed is relayed either on intertidal plots, to produce the thick-shelled adults with strong adductor muscles that are preferred for the fresh domestic and export markets, or on subtidal plots. Here the mussels grow faster and develop the high meat yield and thin shell that is preferred for processing. The mussels remain on the culture plots for 18–24mo before being harvested by dredging. Modern ships are bigger and faster, dredge more efficiently and can carry up to 140–180t of mussels. These factors are important because, although most of the mussels are now grown in the Waddenzee, all marketing still takes place through one auction at Yerseke, on the Oosterschelde, some 200km to the south. An integral part of the Dutch bottom culture methodology is the 'rewatering' process, which immediately precedes the marketing of the mussels. The crop is spread on special plots for 10–14 days for the purpose of eliminating weak and damaged mussels. Modern practice is to complement the rewatering with a final desanding period in a rapid flow of UV irradiated water to ensure a grit and bacteria-free product. Dutch bottom-grown mussels are $2^1/_2$–3 years old when harvested, and have a meat yield of up to 30–40% (Dijkema and van Stralen, 1989). Dutch culture techniques produce about 22t live weight of mussels acre^{-1} year^{-1} (Hurlburt and Hurlburt, 1980) or about 5.5kg m^{-2} of mussel bed.

On-bottom culture is a well-established technique in Denmark and West Germany (Lutz et al., 1991), has more recently been developed in the U.S.A. (Lutz et al., 1991) and Ireland (Anon, 1990), and is the method for the production of striped horse mussel, *Musculus senhauseni* in Thailand (Lutz et al., 1991). The European and North American industries practice similar, though less mechanized, farming methods to those used in Holland. Seed is derived from natural beds. On the east coast of the U.S.A. particular attention is paid to achieving the optimum seeding density on the culture plots. This is critical for optimal growth, and is largely determined by the food available to the mussel ('seston flux') at any given site (Lutz et al., 1991). On-bottom farming of striped horse mussels in Thailand utilizes very small seed—about 5mm—to sow onto shallow plots for harvesting, only 2–3mo later, when the mussels are about 20mm. These small mussels are sold as duck food. *Musculus senhauseni* planted slightly deeper may reach 40mm and be suitable for human consumption, after they have been harvested by hand and trampled under foot to clean and declump them (Lutz et al., 1991).

Off-Bottom Cultivation

Off-bottom cultivation can be used to describe all other types of mussel farming, and it encompasses the whole spectrum from cultivation on stakes or poles set into the seabed, through to methods of utilizing ropes or lines suspended from the sea surface (Figs. 10.1 and 10.2). Off-bottom cultivation adds a third dimension, that of depth, to the essentially two dimensional on-bottom culture. The crop is spread over, and can utilize, a much greater proportion of the water column.

Pole cultivation

The cultivation of mussels on poles, and specifically the 'bouchot' style of culture developed in France, is considered to be the original method for farming mussels derived from the observations of a shipwrecked sailor in 1235 (Mason, 1976). France's bouchot culture, which produces about 50,000t annually (Figueras, 1989), has changed little since Korringa's (1976) account of a Pertuis Breton farming operation. Wooden poles are set into the seabed in rows, or bouchots. The poles, usually 20–30cm diameter oak tree trunks, protrude 2–3m above the seabed and are spaced 20–50cm

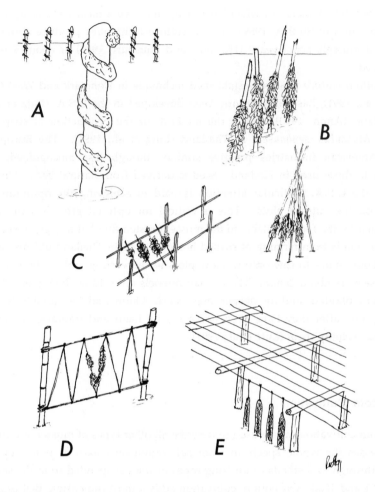

Fig. 10.1. Diagrammatic representation of various methods of fixed-suspended cultivation. A: bouchot; B: bamboo pole; C: rack and rod; D: rope-web; E: hanging park.

apart. A bouchot may contain about 125 poles and be up to 50m long. Bouchots are spaced 15–25m apart at right angles to the shoreline. Several series of bouchots may extend across the intertidal zone. The bouchots are used for both seed catching (on those deepest in the intertidal zone) and on-growing, although in recent years there has been increasing emphasis on catching the seed on horizontal ropes strung between poles, rather than on the vertical poles themselves. Farming involves stripping the seed from the catching poles and transferring it to mesh tubes. The

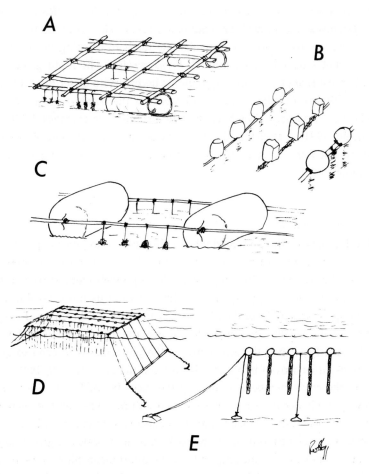

Fig. 10.2. Diagrammatic representation of various methods of floating-suspended cultivation. A: bamboo and oil drum raft; B: single longline; C: double longline; D: multi-longline; E: subsurface longline.

mussels are reattached by winding the tubes around the growing poles. Stripping, reattaching and harvesting by hand, are major operations in the labour intensive farming cycle. Seed caught on rope is transferred to the bouchots by winding short lengths (3m) around the poles and nailing each end. Again this is done manually.

Marketable sized mussels (over 4cm) are 12–18mo old and have a 35–50% meat yield. Bouchot culture techniques produce about 5t live weight of mussels per acre per year (Hurlburt and Hurlburt, 1980) or about 1kg per m² of seabed. An average of 25kg of mussels are harvested from each pole annually (Korringa, 1976).

Pole or stake cultivation of mussels is also conducted on a much less industrialized and mechanized level in Asian countries such as the Philippines (Rosell, 1991) and Thailand (Chalermwat and Lutz, 1989; Lutz et al., 1991). Here bamboo poles, set into the seabed in shallow bays, provide the substrate for both catching the seed and on-growing to market size, the green mussel, *Perna viridis*. The rapid growth of this species, to a harvest size of 5–10cm in 6–10mo, is necessary to offset the constraints of the monsoonal climate and the rapid deterioration of the bamboo poles, which both can lead to heavy crop losses. The harvest from a single pole averages 8–12kg of live mussels (Rosell, 1991). Perhaps the most extreme form of pole cultivation is represented by the harvesting of mussels from offshore oil drilling structures in California (Conte, 1990).

Fixed suspended cultivation
A logical development of the pole method of cultivation is to join the poles together with horizontal connections of rope or timber to achieve a fixed suspended culture (Mason, 1976). This is seen at its simplest in the rope-web method used for green mussel cultivation in the Philippines (Rosell, 1991), and at its most developed in the 'vivai' or hanging mussel parks of the Mediterranean (Ross, 1982). In fixed suspended culture systems, only usually used in shallow water, the main function of the poles is to support the structure. It is the area between the poles that is utilized to grow the mussels on ropes either woven into a matrix between the poles (Rosell, 1991), or hung vertically from the cross members connecting the poles (Ross, 1982). Seed collection and on-growing both occur *in situ* in rope-web culture, whereas the seed for the hanging parks in the Mediterranean may be collected elsewhere and seeded onto ropes or into mesh stockings to hang in the park. Green mussels are grown to market size in 6–10mo without any handling, but the Mediterranean blue mussels may be thinned and re-hung in the parks one or more times before being harvested from 18mo onwards.

The rack method of cultivation used for *Perna viridis* in Tahiti (Fig. 10.3) is a fixed suspended culture developed specifically for a high productivity lagoonal system with negligible tidal movement (Coeroli et al., 1984). The mussels are grown on metal rods or in mosquito netting trays (Vakily, 1989) laid on racks. The racks can be raised or lowered on strings so that the mussels can be positioned to avoid the extreme water temperature (up to 40°C) and salinity (down to 5‰) conditions that occasionally occur in these shallow (1–1.5m deep) lagoons.

An intertidal rack and rope method is used on the west coast of North America (Johnson, 1986). The seed are caught on collectors of coconut fibre twine inside nylon mesh tube (supported by a polypropylene rope), which are held on racks in the low

Fig. 10.3. Mussel farming methods have been adapted to suit particular environments, for example, the small-scale rack and rod system used in French Polynesia.

intertidal zone. Seed-laden collectors were transferred to racks higher up the intertidal to grow to marketable size.

Nylon mesh bags are used in fixed suspended culture of *Perna viridis* in India (Sreenivasan et al., 1988a). The 35 x 40cm bags of 15mm mesh are simply loaded with mussel seed and hung from a wooden framework in 1.5–2.0m of water. The mussels attached to the bag and to each other and increase from an average of 3kg of 27mm seed per bag to 18kg of 75mm mussels in 6mo.

Floating suspended cultivation

Although the Italian hanging parks may be constructed in water as deep as 11–12m (Korringa, 1976), culture systems suspended from the seabed are of necessity limited to shallow water. The development of floating suspended systems freed mussel culture from this constraint, allowing it to expand into deeper coastal waters and more effectively exploit the high primary productivity of these areas.

Rafts:

Mussel culture in the rias of Galicia in N.W. Spain is based exclusively on rafts as the means of flotation (Fig. 10.4). Spain's annual production of over 200,000t is derived from over 3,300 rafts (Figueras, 1989, 1990; Perez-Camacho et al., 1991). A great variety

Fig. 10.4. In the Spanish mussel industry the production unit is the raft, which is between 100 and 500 m^2 in total area, supports 200–700 culture ropes, and produces between 20 and 100 tonnes of *Mytilus edulis* per annum.

of raft designs are in existence in the Spanish mussel industry. The earlier rafts had a single, central flotation chamber with masts and guy ropes to support the wooden framework from which the mussel ropes were hung. Some of the earliest were, in fact, old boat hulls with outriggers attached. Four smaller wooden floats positioned towards the corners of the raft was the first development, and more recently the trend has been towards 4–6 cylindrical fibreglass or steel floats running the full width of the raft (Figueras, 1984). The useable area for hanging ropes increases as the number of floats increases, from 80% with one float to 90% with six floats (Perez-Camacho et al., 1991).

The well-established Spanish farming cycle relies on the collection of wild seed, which is either caught on collector ropes hung from the rafts themselves, or, more commonly, gathered from natural settlement on coastal rocks. The seed from rocks is wound onto ropes with rayon mesh; this job is usually done by hand. Seeded ropes are hung on the raft for 5–6mo, by which time they hold 100–200kg of 45–55mm mussels. At this stage they are thinned and redistributed onto 3–4 ropes to be rehung for a further 12–18mo until the mussels reach marketable size of 70–100mm length. Short wooden pegs inserted into the ropes every 30–45cm are necessary to prevent clumps of mussels sliding down and off the ropes. There is a trend towards mechanization of both the seeding and thinning operations. One-man operated machines can load ropes at about 1 metre every 10sec, or complete a 500-rope raft of 10m ropes in 14h, whereas a skilled person can only process about 1m of rope per minute by hand (Figueras, 1984). For harvesting, the ropes of mature mussels are lifted by crane while a basket is lowered beneath the rope. A vigorous shake detaches the mussels from the rope into the basket (Figueras, 1989). Annual production varies between 20 and 100t per raft, averaging 130kg live weight of mussels per m² of raft area.

The Spanish system provided the example for experimental and commercial raft culture of mussels all around the world including China (Zhang, 1984), Chile (Chanley and Chanley, 1991), Canada (Heritage, 1983), the U.S.A. (Lutz, 1980a), New Zealand (Jenkins, 1985), Australia (Maclean, 1972), Malaysia (Sivalingham, 1977), Singapore (Cheong and Lee, 1984), India (Nagabhushanam and Mane 1991), Venezuela (Mandelli and Acuna, 1975) and Ireland (Anon, 1990).

Submersible rafts have been proposed for utilizing more exposed situations in India (Rajan, 1980), and a self-buoyant bamboo raft submerged 1–1.5m below the surface is a standard culture method in the Philippines (Rosell, 1991). However, the more general approach to mussel farming in more exposed areas, or in areas where large floating structures, such as rafts, are considered aesthetically unacceptable, has been to adopt the longline method of cultivation.

Longlines:
The longline culture of mussels is based directly on the technology evolved for scallops and oysters in Japan (Imai, 1977). A series of floats connected together by horizontal lines support a large number of vertical ropes or mesh stockings on or in which the mussels are grown. Longlines are generally cheaper, easier to construct and maintain, and give slightly better mussel growth rates than equivalent raft systems.

They have been the preferred cultivation method for new mussel industries, particularly in developed countries, such as Ireland (Herriott, 1984), Canada (Muise, 1990) and New Zealand (Hickman, 1989a), where synthetic rope and float materials are readily available. Longlines have been used experimentally to evaluate mussel farming potential in at least 13 countries (Lutz et al., 1991). Even in Holland with its long tradition of bottom culture there have been recent trials to assess the potential for longline culture techniques (Gorski, 1991).

Longline farming practice is essentially the same as for raft cultivation. Seed collection or catching *in situ* is followed by a growth period, before thinning and reseeding onto the grow-out ropes or into stockings. The layout and construction of longlines is as variable as the countries in which this culture system is used. In China a 50–60m single longline is supported by 60–80 small (0.01m^3) glass or plastic floats and carries up to 100 growing ropes, each about 1.5m in length. All farming activities, including harvesting, are done manually to achieve an annual production per longline of about 1.5t of 7–8cm *Mytilus edulis* (Nie, 1991). At the other end of the spectrum, New Zealand's highly mechanized longline culture of *Perna canaliculus* is based on a 110m double longline supported by up to 50 large (0.30 m^3) plastic floats and carrying up to 440 vertical growing ropes, each 5–10m in length (Fig. 10.5). A high level of mechanization, from automated seeding through to custom-built harvesting vessels, is necessary to produce a crop of 20–25t per annum of 7.5–10.5cm green mussels per longline, or between 135kg and 170kg per m^2 of longline surface area (Hickman, 1989b).

Although subsurface longline systems have been proposed, particularly for oyster and scallop culture (Heral and Deslous-Paoli, 1991), they have only been used commercially for mussel culture in the Languedoc-Roussillon region of France, where production rates of 18–20t ha^{-1} year^{-1} have been reported (Vidal-Giraud, 1986). However, temporary submerging of mussel longlines may form part of the farming practice, as in New Zealand, for catching seed at 15–25m depth, or in Canada, to lower the crop below sea surface ice during the winter months (Muise, 1990).

Variations on the single or double longline system have been developed and used both experimentally and commercially. A triple longline system used in Italy was described by Ross (1982). Each 100m long unit was supported by eleven 1.75m long floats. The triple longline was able to carry up to 450 stockings, each about 3m long, and yielded a harvest, after 12–14mo, of about 10t of 70mm mussels. Ross (1982) suggested a future development might be to use smaller floats and only two headropes to make the system "more steady in a rough sea" and "far easier to work".

A multi-longline system has been developed in Sweden (Ackefors and Haamer, 1987). The unit consists of 7–9 headlines, either rope or wire, up to 200m in length, which are all attached at each end to a 10m length of rail, each rail being itself attached

Fig. 10.5. In the New Zealand mussel industry the production unit is the longline, which is 110m in length, supports up to 440 culture ropes and produces 20–40 tonnes of *Perna canaliculus* per annum.

to two 200kg anchors. The headlines are supported by 200L floats, the number of floats varying with the weight of the crop. The typical unit occupies a 1500m^2 area and can produce a crop of 80–120t of *M. edulis* in 16–17mo, or a production of about 400t ha^{-1} year^{-1}.

A similar system, developed for use in Prince Edward Island, Canada (O'Rourke, 1987), used 12m long, 500kg railway lines as the anchors for a unit of six 75m longlines. Two units placed end to end can be anchored by three lengths of railway iron and can support up to 1440 2.4m lengths of mussel stocking with the potential to produce approximately 20t of commercial size *M. edulis*, or about 17kg mussels per m^2 of surface area of longline unit.

FARM MANAGEMENT

The process of farming mussels can be separated into three phases; the initial seeding phase is followed by the grow-out phase, which culminates in the final harvesting phase.

Seeding Phase

During the seeding phase the farmer either catches a supply of seed by setting out suitable settlement substrate within his farm, or he collects seed from an area of natural settlement outside his farm. He attempts to establish and maintain his seed stock at densities which allow maximum growth and survival, but which minimize mortality or loss from overcrowding. Optimum seeding densities have been determined for several of the different farming methods, and for different sizes of seed mussels. In Dutch seabed cultivation of *M. edulis* 20t ha^{-1} of 'wheat grain' sized seed or 30–35t ha^{-1} of 'haricot bean' sized seed are considered optimal (Korringa, 1976). For on bottom culture in the U.S.A. seeding densities typically range between "500 and 1000 bushels acre^{-1} " (1 bushel = 8 American gallons) but may go "as high as 15,000 bushels acre^{-1} .... in highly productive areas" (Lutz et al., 1991), based on a general rule of thumb of seeding out at 10% of the density on natural seed beds. For rope culture suggested optimum densities are 6000 10–12mm seed or 300 40mm seed per metre of rope for *M. galloprovincialis* in Spain (Figueras, 1989), or 200 10mm seed per metre for *Perna canaliculus* in New Zealand (Jenkins, 1985). Seeding densities for *M. edulis* grown in mesh stockings in Irish longline culture are based on the weight of the small (2–8mm) seed. Recommended densities range from 0.5kg of seed per metre of stocking (Roantree, 1986) to 1.5kg m^{-1} (Herriot, 1984).

In practice it is difficult, if not impossible, to consistently obtain the optimum seeding density, particularly when using seed attached to a natural substrate such as seaweed, which may have up to 3000 seed per 90mm length of weed (Hickman, 1976), or eel grass, with up to 3000 seed per blade of *Zostera marina* (Newell et al., 1991). To overcome this difficulty the more highly developed cultivation methods, mainly those using rope culture, incorporate a reseeding operation into this phase of their farming (Fig. 10. 6). The ropes of juvenile mussels, derived either from *in situ*

Fig. 10.6. A wide variety of mesh stockings or tubes have been developed to hold both seed mussels (left) and the older mussels at reseeding (right).

catching or from an initial seeding, are completely stripped and the mussels, usually between 10 and 40mm, are reseeded onto several new ropes at the optimum density. Size grading can be included in the reseeding operation to facilitate maximum uniformity of the crop throughout the next phase of the farming operation.

Grow-Out Phase

The second or grow-out phase of mussel farming frequently involves little activity on the part of the farmer. In the fixed and floating cultivation systems used for tropical mussels, farm management throughout the 6–12mo required to produce a crop may involve no more than regular inspection, or it may simply include attempts to minimize crop losses resulting from the drop-off of oversized clusters of mussels, and the collapse of rotten stakes (Vakily, 1989). Even in the more sophisticated and mechanized cultivation systems used for temperate mussel species, minimal farm management activity during the grow-out phase seems to be the norm. It is, however, standard practice to thin and reattach the crop of mussels at least once during the grow-out phase in Spanish raft cultivation (Figueras, 1989). Thinning has also been shown to benefit both the growth rate and the production at harvest in French bouchot culture (Boromthanarat and Deslous-Paoli, 1988), and may be used in seabed cultivation in Holland to maintain the density on the beds below 8kg of mussels per m² (Korringa, 1976). On the other hand, thinning was shown to have an adverse effect on the economics of farming green mussels on rafts in Singapore, where labour costs have a major impact on profitability (Cheong and Loy, 1982).

The longline system of cultivation allows the farmer to vary the amount of flotation to maintain the minimum buoyancy required to safely support the maturing crop. By adding or removing floats during the grow-out phase, as well as by manipulating the spacing of the culture ropes, the farmer has the potential to make optimum use of his investment in farming equipment. At the same time he will be managing the production of a crop of uniform sized mussels in the most efficient manner (Jenkins, 1985).

During the grow-out phase cultivated mussels are subject to mortality, predation and a variety of other biological factors which can significantly influence the quantity and quality of the maturing crop, as well as vary the time taken for the crop to reach market size.

Mortality and predation

In principle, once the mussel farmer has secured a sufficient supply of seed to stock his farm all he has to do is husband the crop through to harvest. In practice, mussels being farmed are subject to many of the same pressures that keep wild mussel populations in check. Mussel culture, in fact, provides the mussels' natural predators with an abundance of food in their natural habitat, and thereby, the potential to proliferate greatly.

Crabs, starfish, and fish are the major predators on mussels and numerous species worldwide have been implicated in mortalities of farmed mussels (Table 10.2; see also section on predation, p.127–135 Chapter 4). Bottom culture is particularly vulnerable. Crabs predate mainly on seed mussels and, if present in large numbers, can completely denude newly seeded beds (Korringa, 1976). Starfish are capable of similar devastation and as well, in the case of the larger species, are able to consume mussels right up to commercial size (Dare, 1982). Other predatory bottom dwelling organisms,

Table 10.2. Organisms identified as being predators on cultivated mussels.

Predator	Type of Culture	Reference
Birds		
Somateria mollissma (eider duck)	On- and off-bottom	Incze and Lutz, 1980
Melanitta deglandi (white winged scooter)		
M. perspicillata (surf scooter)	On- and off- bottom	Waterstrat et al., 1980
Oidemia nigra (common scooter)		
Bucephala islandica (Burrow's goldeneye)		
Fish		
Rhabdosargus sarba (silver bream)	Off-bottom	Vakily, 1989
Rhacochilus vacca (pile perch)	Off-bottom	Waterstrat et al., 1980
Embiotoca lateralis (sea perch)		
Chrysophrys auratus (snapper)		
Pseudolabrus celidotus (spotty)	Off-bottom	Jenkins, 1985
Navodon scaber (leatherjacket)		
Trygon pastinacea (stingray)	Bouchot	Figueras, 1989
Diodan hystrix		
Arius spp.		
Arothron nigropunctatus	Raft	Nagabhushanam and Mane, 1991
Ostracion sp.		
Chaetodon sp.		
Acanthurus sp.		
Diplodus sargus	Raft	Figueras, 1990
Sparus aurata (golden mackerel)		
Crustaceans		
Carcinus maenas	On-bottom	Davies et al., 1980
Scylla serrata	Off-bottom	Vakily, 1989
Echinoderms		
Asterias rubens	On-bottom	Dare, 1982
A. vulgaris	Off-bottom	O'Neill et al., 1983
A. forbesi	Off-bottom	Incze and Lutz, 1980
Pisaster ochraceus	Off-bottom	Waterstrat et al., 1980
Evasterias troschelii	Off-bottom	Waterstrat et al., 1980
Marthasterias glacialis	Raft	Figueras, 1990
Paracentrotus lividus		
Molluscs		
Ocinebrina edwardsi	On-bottom	Curini Galletti and Galleni, 1984
Thais (Nucella) lapillus	On-bottom, raft	Dare, 1982; Figueras, 1990
Turbellarians		
Stylocus ellipticus	Off-bottom	Incze and Lutz, 1980

including gastropod molluscs and turbellarian flatworms, may be significant causes of seed mortality (Incze and Lutz, 1980; Curini Galletti and Galleni, 1984).

Suspended cultivation tends to eliminate the bottom dwelling predators but may accentuate the mussel's vulnerability to other species, notably fish and birds. Fish predation occurs throughout the complete farming cycle in New Zealand's longline mussel culture, with spotty and leatherjacket attacking the seed, and snapper feeding on mussels as large as 90mm in length (Jenkins, 1985). Silver bream completely destroyed an experimental rope culture in India, consuming even the adult *Perna viridis* (Vakily, 1989). Several other fish species are also reported to attack cultured mussels in India (Nagabhushanam and Mane 1991). Fish predation is somewhat ironic when one considers that mussel culture in several tropical countries, such as Thailand, started as opportunistic harvesting from the poles of fish traps (Chalermwat and Lutz, 1989). The stingray is an important mussel predator on the French bouchot culture (Figueras, 1989). Predation by birds is a major problem for mussel farmers in North America (Lutz, 1980) where eider ducks and scooters, in particular, cause major losses in both seabed and suspended mussel cultivation.

Methods to combat predation are many and varied, but generally they have only limited success. Special dredges, rollers and mops have been developed to remove starfish and crabs (Korringa, 1976) and enclosures have been designed to exclude them (Davies et al., 1980). The application of quicklime (calcium oxide) to culture plots may be successful in combating starfish (Shumway et al., 1988). Protection of suspended cultures is more difficult. Acoustic and mechanical scaring devices have generally proved ineffective against either fish or diving birds (Lutz et al., 1991), but nets to deter stingrays and fluttering plastic strips to scare birds, are used to protect the French bouchot mussels (Figueras, 1989).

Disturbance seems to be a critical factor in predation. Newly seeded ropes (Johns and Hickman, 1985) or culture plots (Korringa, 1976) are particularly susceptible. Careful farm management offers the best solution to minimize predation on the seed mussels. Jenkins (1985) makes several suggestions for countering persistent fish predation on longline cultured mussels. These include holding the seeded ropes close to the surface until they are well-established; isolating the seed mussels from the maturing crop, either on separate longlines or in distinct nursery areas known to be free of fish; using nets to protect the seed; or as a last resort, setting pots, nets or lines to fish out the predators.

Some degree of predation may be beneficial in certain mussel farming situations. O'Neill et al. (1983) showed that the culling of undersized *M. edulis* by *Asterias vulgaris* predation improved the yield in net culture of mussels in Canada. Seagulls and crabs are utilized in the Dutch industry to eliminate the weaker mussels when the crop is spread on intertidal beds prior to harvesting for sale (Korringa, 1976).

Major mortality in cultured mussels can result from adverse weather conditions. Gales can cause either dislodgement or smothering of whole beds of seed or adult mussels being cultured in shallow plots. This is a not infrequent problem for the Dutch mussel industry (Korringa, 1976; Anon., 1991c). Devastation of both the crop and the farming equipment can result from tropical storms, if growth from seed to harvest is not completed in the 6–9mo between the monsoon seasons in India (Nagabhushanam and Mane, 1991). Losses of as much as 40% of the crop can occur as a result of storms in the more exposed mussel farming areas in Spain (Perez-Camacho et al., 1991). Canadian mussel farmers combat the risk of severe crop losses from winter ice by sinking their longlines below the surface, and adopting special techniques for harvesting through the ice (Muise, 1990). Heavy rainfall has the potential to cause mortality of the uppermost mussels in suspended culture if a thick surface layer of water of severely reduced salinity is maintained for more than a few days.

Precise data on the extent of mortality in cultivated stocks of mussels is extremely scarce. A 33–36% mortality during the seed phase and 14–15% during the growth phase of thinned out mussels have been reported for Spanish raft culture methods (Roman and Perez, 1979). Natural wild populations of mussels are known to show highly variable mortality, which may be genetically related (Mallet et al., 1987; see also p.369–373 Chapter 7). It is likely that cultivated stocks would be similarly variable, but perhaps with the mortality moderated by good farm management. Mussel farmers would generally only be aware of the 'catastrophic' instances of mortality, such as a major loss of seed, which can only be combatted by provision of further stock for the farm.

Other biological factors

Mussel farming worldwide has been remarkably free of disease. Figueras (1989) reports "no significant mortalities or losses during the 50 years of mussel culture in Galicia" (Spain). Several disease organisms have, however, been described from farmed mussels (Figueras et al., 1991), and mass mortalities in European cultured mussels have been attributed to disease and parasites (Bower and Figueras 1989; see also Chapter 12). Potential mussel pathogens include rickettsial and chlamydial bacteria, haplosporidian protozoans (notably the oyster pathogen *Marteilia refringens*), digenean trematodes, polychaete annelids (particularly the shell-boring *Polydora* spp.), parasitic copepods (especially *Mytilicola intestinalis*) and pinnotherid pea crabs (Fig. 10.7). The pathenogenicity of the polychaetes (Kent, 1979), copepods (Paul, 1983) and pea crabs (Bierbaum and Ferson, 1986) is open to question, but it is apparent that mussels cultivated by off-bottom methods are generally less susceptible

Fig. 10.7. The pea crab is one of the potential disease organisms commonly associated with on-bottom cultivated mussels, but it occur less frequently in off-bottom cultivation.

Fig. 10.8. Cultivated mussels support an assemblage of fouling organisms (left), but when seeded at optimum density the mussels comprise the overwhelming majority of the biomass (right).

to infestations of these, and probably other pathogens, mainly because they are above the seabed, and because they grow to harvest size more rapidly (Pregenzer, 1983).

A significant side effect of the infection of mussels by the trematode parasite *Gymnophallus bursicola* is the formation of pearls. The high incidence of pearls in bottom grown mussels on the east coast of North America is a major problem in marketing wild mussels, but one which can be eliminated if mussels are grown to harvest, either on the bottom or in suspension, in less than five years (Lutz, 1980b).

Farmed mussels constitute an excellent substrate for the settlement of many other organisms which, from the perspective of the mussel farmer, are collectively termed fouling (Fig. 10.8). Floating suspended cultures are particularly vunerable to fouling, since they are continually submerged. They can exhibit a diverse assemblage of epifaunal and epifloral organisms, which has been described from the raft culture of *M. edulis* in N.W. Spain (Perez and Roman, 1979; Roman and Perez, 1979, 1982; Gonzalez, 1982). Almost 100 invertebrate species were identified from the mussel ropes, including crustaceans, gastropods and lamellibranch molluscs, polychaetes, ascidians, sponges and hydrozoans. The study concluded that the make-up of the epifauna was largely determined by the mussels themselves, which represent 95% of the biomass, through their efficient filtering activity and their high level of biodeposition. These two aspects of the mussel's physiology bias the epifauna towards predatory and detritivorous species, by the elimination of filter feeding competitor species.

Both green and brown algae of various species may form part of the fouling on cultured mussels (Nagabhushanam and Mane, 1991), but weed generally tends to be only a seasonal fouling problem. Oversettlement of the maturing crop by a subsequent cohort of seed may also represent a fouling problem. In New Zealand oversettlement of seeded ropes of the green mussel *Perna canaliculus* by blue mussels *M. edulis aoteanus* (= *M. galloprovincialis?*) is also considered a fouling problem, since only the greens are valued as the marketable crop (Hickman, 1989a).

Excessive fouling is a mussel farming problem mainly by virtue of the extra effort required to clean the crop for marketing. It may also have some influence on growth rate and productivity through competition for space, and reduction of water circulation. Again, farm management provides the best means of alleviating this problem—optimally seeded ropes tending to sustain less fouling—which in most mussel farming situations "seldom exceeds nuisance proportions" (Figueras, 1989).

Of far greater significance than fouling as a mussel farming problem is the activity of boring organisms. The teredinid and pholad boring molluscs can cause rapid destruction and disintegration of the timber used for mussel rafts, especially the untreated timber often used in tropical areas, and the speedy collapse of the bamboo stakes used in pole culture (Rosell, 1991).

Harvesting

The final phase of the mussel farming cycle, the harvesting phase, commences as soon as the mussels have reached marketable size. This varies for the different mussel species, for different countries and for mussels grown by different culture methods. Size must also be considered in relation to the particular form of the end product, whether it be fresh-in-the-shell, or one of the multitude of processed mussel product forms (Warwick, 1984). French bouchot mussels become marketable at only 40mm in length after 12–15mo on the poles (Figueras, 1989). Dutch seabed mussels reach the minimum harvesting size of 50mm after 18–24mo on the culture plots (Dijkema and van Stralen, 1989). Tropical mussels grown in suspension may reach the marketable size of 60–70mm in 6mo (Vakily, 1989). Rope-grown mussels are typically harvested at a larger size: 80–90mm, 15–20mo old *M. galloprovincialis* being the usual crop in Spain (Figueras, 1989) and 75–115mm, 12–18mo old *Perna canaliculus* being typical of New Zealand (Hickman, 1989a, b).

Harvesting methods are as diverse as are the situations in which mussels are farmed around the world. They range from the purely manual methods commonly used in tropical mussel producing countries, through to the fully mechanized dredging techniques practised in the Netherlands. Manual harvesting is time-consuming, laborious and labour intensive but, assisted by specialized tools for working the bouchots in France (Korringa, 1976), or by divers to raise the bamboo stakes in Thailand (Vakily, 1989), it has proved to be an effective method in both temperate and tropical mussel culture, even for an annual production of 30–50,000t.

The huge production of mussels which can be achieved through industrial scale on-bottom cultivation, made it essential to have mechanized harvesting techniques. The Netherlands has led the world in the technical improvements to its mussel dredging vessels (Fig. 10.9 upper) and in the development of mussel handling and processing equipment. The most modern Dutch mussel dredgers, operating four dredges, each capable of retaining over 0.5t of mussels, can harvest up to 130t in 4–5h. The dredges empty directly into on-board containers, which can subsequently be lifted ashore and connected to a supply of UV irradiated seawater, for desanding and post-harvest handling (Dijkema and van Stralen, 1989). Automated equipment for washing, declumping, grading and debyssing, followed by continuous pressure cooking and IQF freezing were also developed to assist the Dutch mussel industry in its modernization. Many of these innovations have been incorporated into other mussel industries around the world (Anon., 1991a, b).

Mechanization has been equally important to the development of floating suspended cultivation. Purpose-designed vessels and harvesting equipment facilitated the expansion of mussel culture in Spain (Korringa, 1976; Figueras, 1989) and New Zealand (Pooley, 1991b, c). In both countries harvesting requires the 5–15m

Fig. 10.9. Mechanization is necessary to harvest large quantities of mussels in both on-bottom and off-bottom culture. Four large dredges are operated on board the Dutch mussel boats (upper). Heavy lifting gear has been developed for harvesting from longlines in New Zealand (lower).

long culture ropes, which can each carry a crop of 100–200kg, to be lifted out of the water so that the mussels can be stripped off (Fig. 10. 9 lower).

Further innovation and experimentation may lead to improvements in mussel farming technique for all of the different culture methods. However, the general principle of establishing and maintaining the optimum stocking density for the particular farming situation will remain the key to the production of a harvestable crop in the shortest period of time, and therefore, the key to a profitable farming operation.

MUSSEL CULTURE PRODUCTIVITY AND ECONOMICS

Direct comparison of the productivity of the different mussel culture methods is difficult because of the variety of systems and equipment used. It is not realistic to directly relate the quantities produced, without reference to the volume or area of water that has provided the supply of food for the mussels. In most mussel farming situations the food source is a mixture of *in situ* primary production and in-flow from adjoining water bodies. The ratio between these sources is the result of interacting physical and biological factors, not the least of which is the current flow through the farm or over the mussel bed (Gibbs et al., 1991).

Farming technique also radically affects productivity. On a purely practical basis, without any consideration of possible limits on the ability of the area to support the mussels, the productivity of a farm can be manipulated by how many poles are set up in a given area, how many ropes are attached to a raft or longline, or what depth of water is utilized. In other words, what length of culture ropes or stockings are used.

Without full knowledge of all the parameters contributing to the overall production of a mussel culture operation, it is only realistic to compare productivity of on-bottom cultivation between different areas, and of off-bottom cultivation to a standard production unit—the most appropriate being a metre length of growing surface, whether it be pole, rope, stocking or mesh band (Table 10.3).

Korringa (1976) included an assessment of some economic aspects of the various mussel farming situations that he described. His assessment, however, was restricted to itemizing the labour and equipment requirements and the market value of the crop. He did not attempt detailed analyses of the profitability of mussel farming. Few such analyses are, in fact, available (Clifton, 1980; Cheong and Loy, 1982; Morris, 1984; Herriot, 1985; Bartlett and Hufflett, 1991) although cost/benefit analysis and economic models have been included in publications on other aspects of particular culture situations (Quasim et al., 1977; Yap et al., 1979; Herriot, 1984; Jenkins, 1985; Hickman, 1987; Vakily, 1989; Hickman et al., 1991)

Two detailed analyses of the economics of longline mussel farming (Morris, 1984; Bartlett and Hufflett, 1991) have identified many of the variables in the equipment costs, labour costs, method of financing, type of farm management, scale of operation, growth rate and yield at harvest, which can significantly vary the profitability of mussel farming, even within a relatively small and highly standardized industry such as exists in New Zealand. It is clearly not possible to generalize about the profitability

Table 10.3. Productivity (kg m⁻¹ substrate) of off-bottom cultivated mussels

Species	Country	Production kg m⁻¹	Substrate	Culture	Reference
Mytilus edulis	Ireland	5	Rope	Raft	Rodhouse et al., 1985
	France	6–11	Pole	Bouchot	Boromthanarat and Deslous-Paoli, 1988
	Italy	10–14	Pergolari	Longline	Ross, 1982
	Sweden	7	Mesh band	Longline	Ackefors and Haamer, 1987
Mytilus galloprovincialis	Spain	14.5	Rope	Raft	Perez-Camacho et al., 1991
Perna perna	Brazil	6–7	Rope	Raft	Marques et al., 1984
Perna canaliculus	New Zealand	20	Rope	Longline	Hickman, 1991
Perna indica	India	16	Rope	Raft	Kumaraswamy Achari and Thangavelu, 1980
Perna viridis	Singapore	10–12	Rope	Raft	Cheong and Loy, 1982

of mussel farming worldwide because of the huge variation in the economic conditions under which mussel farming is practised. However, Vakily's (1989) conclusion that "mussel farming is an economically viable alternative among the various aquaculture systems, if appropriate technology is applied" is certainly acceptable.

It is likely also that the observation, originating in the Dutch mussel industry (Korringa, 1976), that "the gradual evolution from a large number of small enterprises to a small number of units of greater size (constitutes a) favourable development" has universal application. This is particularly true when the consolidation occurs in combination with some degree of vertical integration of the growing, processing and marketing sectors of the industry.

CONSTRAINTS ON MUSSEL CULTIVATION

Mussel cultivation has frequently developed in response to unsustainable exploitation of wild stocks. This can be on either a noncommercial scale as, for example, in the subsistence gathering of intertidal *Perna perna* in South Africa (van Erkom Schurink and Griffiths, 1990), or on a commercial scale, as happened in the dredge fishery for *Perna canaliculus* in New Zealand (Hickman, 1989b). In either case the over-exploitation adversely affects the wild stocks in two ways. Firstly, the adult mussels are removed at a rate beyond which recruitment and growth can replace them, or in fishery terms beyond the maximum sustainable yield. Secondly, a major substrate for the settlement of mussel larvae, namely the shells and byssus threads of the adult mussels themselves (Bayne, 1965; Suchanek, 1979), is also removed. The presence of sustainable wild stocks of mussels and the abundance of seed which they can supply, are prerequisites and potential constraints on the development of mussel farming.

Mussel Seed Supply

Mussel farming can only be successful if the supply of mussel seed is both regular and reliable. Research into mussel settlement, on both natural and artificial surfaces, has focussed on these two characteristics, and has been conducted in all countries where experimental or commercial farming has been undertaken (Menzel, 1991), as well as in some countries where there is as yet no culture activity (Ekaratne, 1987).

Forecasting the timing of mussel settlement, which is the commercial objective of many of these studies, has proved difficult. Peaks of reproductive activity are characteristic of most mussel species both in temperate (King et al., 1989) and tropical (Vakily, 1989) regions. However, they do not necessarily correlate with subsequent settlements. In Spain, for example, where there are spring and autumn spawning peaks, the major mussel settlement occurs during the summer and hardly at all in autumn (Perez-Camacho et al., 1991). The timing of the settlement peaks can also vary from year to year (Cheong and Lee, 1984; King et al. 1989).

A variety of endogenous and exogenous factors influence the duration of the larval phase and therefore, also the larval mortality rate (Widdows, 1991; see also p.57–60 Chapter 3). The ability of the mussel larva to delay metamorphosis and settlement for up to seven weeks (Pechenik et al., 1990) further complicates the task of predicting mussel settlements. Even in areas where mussel settlement has been studied extensively over a large number of years, as for instance along the Yugoslavian coast (Hrs-Brenko, 1974, 1980, 1983), and where relationships between settlement and factors such as season, depth, temperature and salinity have been identified, environmental

Fig. 10.10. Just a handful of beach-stranded seaweed, heavily settled with mussel seed, can be sufficient to seed several metres of rope.

variability is generally still seen as a constraint on spatfall forecasting (Margus and Teskeredzic, 1986).

The phenomenon of primary and secondary settlement of mussel larvae (Bayne, 1964; see also p.61–63 Chapter 3 and p.108–112 Chapter 4) has received scant investigation with reference to commercial seed collection. It may, however, have indirectly influenced commercial practice through the development of specialized types of rope with improved spat catching and seed retaining characteristics. Palm fibre ropes used in China (Nie, 1991), coconut coir ropes in Singapore (Cheong and Lee, 1984) and the 'Christmas tree' rope used in New Zealand (Johns and Hickman, 1985) all utilize the similarity of these ropes to filamentous algae and hydroids to maximize the settlement of mussel seed.

Settlement of mussel seed directly onto seaweed (Fig. 10.10) has played a fundamental part in the rapid growth of *Perna canaliculus* culture in New Zealand (Hickman, 1989a, b). Beach stranded seaweed which washes ashore covered in prodigious quantities of minute *P. canaliculus* (Hickman, 1976) is the preferred source of seed for the industry despite its irregularity of supply. Settlement of *P. perna* on seaweed along the South African coast and its potential for mussel farming has been suggested as a factor to be considered in seaweed harvesting management (Beckley, 1979).

Species-specific settlement characteristics may influence commercial spat catching, particularly in areas where more than one mytilid species co-exist. The Atlantic coast of France where bouchot culture is practised is one such area. Here *Mytilus edulis* and *M. galloprovincialis* distributions overlap and hybridization occurs (see relevant references in Chapter 7). On the Oregon coast of North America, where *M. edulis* and *M. californianus* both occur, there is a pronounced settlement succession of the mytilids. *M. californianus* only settles and colonizes existing mussel beds (Suchanek, 1979 and Chapter 4).

Other behavioural characteristics may significantly affect commercial spat catching success. Attachment rates of *M. edulis* larvae has been shown to be 200–800% higher in agitated water as compared to static water (Eyster and Pechenik, 1987).

In purely practical terms the precise timing of settlement is often of less importance to the mussel farmer than is the postsettlement survival of the seed. Timing is of greater concern when spat catching involves setting up posts or laying out ropes, since a period of 'conditioning' of these surfaces in seawater is generally necessary before a successful spat catch can be expected (Figueras, 1989). It may be possible to eliminate the 'conditioning' period by applying 'mussel spawn attractant' to the settlement surface, as is done in the culture of *M. edulis* on offshore rigs in California (Kronman, 1986). When the seed source is from natural settlement on the rocky shore, as in Spain, or the seabed, as in Holland, the combined results of settlement, survival and early growth will determine the farmer's opportunity to obtain sufficient seed for his farming operation. He also has the choice of varying the timing of his seed collection in relation to the size of seed that he prefers to collect. Other aspects of the mussel farming cycle may determine a mussel farmer's seed collection activities. In the more industrialized culture operations space for seed ropes on the raft or longline must be balanced against the space requirements of the maturing crop (Perez-Camacho et al., 1991). On the other hand, in order to maintain production in tropical mussel culture areas such as India, where *Perna viridis* grows to marketable size in 5–6mo (Nagabhushanam and Mane, 1991) at least two seed collection periods may be needed.

Although seed supply has frequently been cited as a constraint on mussel culture development (Menzel, 1991) it appears to have been a theoretical rather than a practical constraint in most existing mussel farming industries. For example, even during a 2-year-period of complete postsettlement mortality of the locally caught spat in the New Zealand green mussel industry, seed supply was maintained by the use of beach stranded seed air-freighted to the farms from 600km away (Hickman, 1989b). The reliability of wild seed supply worldwide has largely precluded the need for hatchery rearing of mussels. Only two countries rely on hatcheries for commercial production of seed; China utilizes hatchery seed in its *M. edulis* farming (Nie, 1991) and the *Perna viridis* culture in Tahiti relies entirely on hatchery-produced seed (Coeroli et al., 1984). Hatchery rearing techniques have, nevertheless, been perfected for several species of mussels including *P. viridis* (Coeroli et al., 1984; Sreenivasan et al., 1988b), *P. indica* (Kuriakose, 1980), *M. edulis* (Nie et al., 1979; Body 1983; Zhang, 1984; Yamamoto and Sugawara, 1988), *Choromytilus chorus* and *Aulacomya ater* (Chanley and Chanley, 1991). The economics of hatchery seed production versus seed collection appear to strongly favour collection from the wild, except under particular circumstances. This is the situation on the Pacific island of Tahiti where no wild *P. viridis* seed are available. A case has also been made for hatchery production of *Choromytilus chorus* in Chile (Winter et al., 1984) to supply seed for aquaculture and repopulation of natural beds, whereas the authors state that "the high settlement densities normally found at various sites ... do not justify spat production by hatcheries" for *Mytilus chilensis*.

If hatchery production of seed for commercial mussel farming were to become the major source of supply for the various mussel industries there would be a requirement for further research into its genetic implications. Within-population and between-population genetic diversity has been well studied in transplantation experiments on natural mussel populations (Mallet et al., 1987; Skibinski and Roderick, 1989; Johannesson et al., 1990; see also p.346–347 Chapter 7), and under controlled conditions (Gentili and Beaumont, 1988; see also p.348–349 Chapter 7). The mussel farming implications of aspects such as the genetic differences between spring and autumn seed, or the higher level of homozygosity found in seed mussels as opposed to adults (Smith, 1988), have received much less attention.

Once a satisfactory supply of seed mussels for stocking the farm has been obtained and an effective farm management regime put into practice, the major constraint on the success and economic viability of all mussel farming operations is the adequacy of the nutrient supply to the mussels themselves.

Mussel Food Supply

The relationship between farmed mussels grown in dense concentrations and the quantity of food necessary for them to grow and fatten in culture conditions has received increasing research effort in recent years. Frequently, the aim of the research has been to improve the planning, use and management of coastal resources. Thus far, the research has achieved a high level of understanding of the many relationships which interact to determine 'carrying capacity' (Incze and Lutz, 1980) of mussel farming ecosystems. It has yet to achieve the level of sophistication or precision necessary for delineating which coastal waters are suited or not suited to marine farming, or for determining what constitutes sustainable marine farming production from specific areas of water.

Incze and Lutz (1980) used largely theoretical considerations in their carrying capacity model of a longline culture system. During the 1980s on-bottom, bouchot, raft and longline commercial farming operations have all been studied with respect to what has variously been termed their environmental physiology (Parulekar et al., 1982), energy flow (Rosenberg and Loo, 1983), resource allocation (Rodhouse et al., 1984), energy budget (Deslous-Paoli et al., 1990), or physiological energetics (Navarro et al., 1991). These studies have provided real data for modelling, as well as exposing further physiological parameters for consideration in the models. The scale of the models varies widely. The extensive ecosystem investigations on the Oosterschelde in Holland (Smaal et al., 1991) include mussel production and scope for growth as functions in models of the whole Delta system (Smaal, 1991). Mussel growth and condition have been related to food source for parts of the Delta (Smaal and van Stralen, 1990). Observations made in benthic tunnels have allowed the modelling of the nutrient fluxes at the level of individual mussel beds (Dame and Dankers, 1988; Prins and Smaal, 1990).

Similarly for suspended mussel culture, the scale of investigation on feeding relationships has varied from whole industry analyses (Hickman et al., 1991) through studies on a single coastal inlet (Carver and Mallet, 1990), to ones on individual rafts (Navarro et al., 1991), longlines (Rosenberg and Loo, 1983) or platforms (Page and Hubbard, 1987). Few studies have directly compared feeding energetics contemporaneously in suspended and on-bottom populations of mussels (Rodhouse et al., 1984a, b).

These studies on the dynamic aspects of mussel culture systems are complemented by a considerable volume of recent data on mussel growth and production (Table 10.4; see also Chapter 5 and p.122–124 and p.137 Chapter 4), covering seabed cultivation (Newell, 1990), bouchots (Boromthanarat and Deslous-Paoli, 1988), suspended parks (Ceccherelli and Barboni, 1988), rafts (Rodhouse et al., 1985), longlines (Loo and Rosenberg, 1983), subsurface longlines (Fabi et al., 1989) and manipulations of wild populations (Behrens Yamada and Peters, 1988; Raubenheimer and Cook, 1990).

The Incze and Lutz (1980) carrying capacity model for a longline culture system attempted to determine the decrease in seston concentration in the water passing through the farm as a function of the filtering activity of the mussels. The authors recognized the need for field data on size-specific filtration rates and energy requirements of mussels in culture conditions, and also on seasonal variations in the quantity and quality of the seston. These aspects have been addressed in subsequent studies, particularly on suspended cultures, but with varying emphasis placed on different elements in the ecosystem by different workers. Rosenberg and Loo (1983) utilized the energy values (kJ m^{-2} day^{-1}) of the biomass, production, ingestion, assimilation, respiration, and gamete and faeces production components of a *M. edulis* longline culture to determine an optimum yield, after 17–18mo, of 36kg

Table 10.4. Growth of various species of cultivated mussels

Species	Country	Growth	Method of Cultivation	Reference
Mytilus edulis	Holland	72mm/3yr	Seabed	Figueras, 1989
	France	46–54mm/12mo	Bouchot	Boromthanarat and Deslous-Paoli, 1988
	Sweden	60mm/18mo	Longline	Loo and Rosenberg, 1983
	Norway	46mm/3yr	Longline	Wallace, 1983
	British Columbia	50mm/12–14 mo	Raft	Heritage, 1983
	U.S.A.	43–50mm/12–15mo	Raft	Lutz, 1980
	Canada	55mm/18–24mo	Longline	Muise, 1990
	China	70–80mm/12mo	Longline	Nie, 1991
Mytilus galloprovincialis	Spain	80–90mm/12–18mo	Raft	Figueras, 1989
Mytilus chilensis	Chile	54mm/14mo	Raft	Winter et al., 1984
Perna viridis	India	93mm/12mo	Raft	Parulekar et al., 1982
	Tahiti	90mm/10mo	Rack	Coeroli et al., 1984
	Philippines	50–100mm/6–10mo	Poles	Rosell, 1991
Perna canaliculus	New Zealand	70–115mm/12mo	Raft/longline	Hickman, 1991
	New Zealand	113–165mm/2yr	Raft/longline	Hickman, 1991
Choromytilus chorus	Chile	60mm/2yr	Raft	Chanley and Chanley, 1991
		120mm/6yr	Raft	Winter et al., 1984
Aulacomya ater	Chile	70mm/2yr	Raft	Winter et al., 1984

mussels per m² for a 4500m² culture area. Based on carbon and nitrogen flows through the same components, Rodhouse et al. (1985) derived figures of 8.1kg C m^{-2} and 1.65 kg N m^{-2} for the harvest of mussels, after 14mo of raft culture. Carver and Mallet (1990) concentrated on the relationship between available food supply (g particulate organic matter (POM) per week) and food demand (g POM kg mussels^{-1} week^{-1}) to obtain values of 200–600t of market size (> 50mm) mussels as the carrying capacity of a 0.5km² coastal inlet.

The Carver and Mallet (1990) study shows how the modelling approach can provide usable predictive data for planning the use and commercial development of a discrete unit of coastal resource. However, the authors stress the importance to the model of site-specific data on the quantitative and qualitative aspects of the mussel food supply, and on the physiology of the mussels under commercial culture conditions. Site-specific characteristics can be responsible for variation in mussel feeding physiology (Navarro et al., 1990) and condition index (Hickman et al., 1991). Carver and Mallet also stress the need to update the model as mussel farming at the site expands, and in particular as 50% depletion of the available food supply is approached. The figure of 50% depletion has been used as a criterion of acceptable and sustainable shellfish production (Lutz, 1980; Rodhouse and Roden, 1987).

The relative importance of the various physical (abiotic) environmental parameters in the energetics of mussel culture systems is frequently confused by correlations existing between them, as for example, between water temperature and salinity in estuarine situations (Hickman et al., 1991). Cross-correlations may also exist between one or more of the abiotic and the biotic parameters, as is exemplified by the monsoonal influence on the food supply, and on the mussel's reproductive cycle in tropical regions (Parulekar et al., 1982). In both these situations the physical parameters can appear to be the dominant factors determining mussel production. Under more stable environmental conditions both water temperature and salinity exert less influence than food supply on the mussel energetics (Ceccherelli and Barboni, 1983; Page and Hubbard. 1987; see also Chapter 5), and food supply is generally considered to be the principal limiting factor in either suspended (Navarro et al., 1991; Perez-Camacho et al., 1991) or seabed (Smaal and van Stralen, 1990) culture systems. Empirical evidence for food limitation is seen in the quite pronounced difference in growth (and scope for growth) between mussels from the front and the back of a Spanish raft (Cabanas et al., 1979; Navarro et al., 1991). A similar affect occurs on mussel beds, where the mussels at the edge of a patch grow significantly larger than those in the middle (Newell, 1990). On a broader scale, the food limitation is evident in the well-documented difference between the production and biomass characteristics of suspended and bottom cultures (Smaal, 1991) or the four-fold difference between the production rates of raft cultured *M. edulis* and wild mussels from the west coast of Ireland (Rodhouse et al., 1984a).

Food limitation, in the dynamic environment of a mussel culture system, may result from either quantitative or qualitative depletion of the food source, which may be a reduction in the total particulate matter availability, or a change in the ratios of phytoplankton to detritus, or organics to inorganics. Equally, the limitation may result from an altered rate of supply of the food to the mussels. Despite its importance as a factor in food supply, current flow has received relatively little attention in relation to suspended cultivation of mussels. Authors have generally used averaged

Table 10.5. Current flows recorded at mussel cultivation sites

Current Speed (m sec⁻¹⁾		Reference
On-bottom Culture:		
< 0.03	(low)	⎫
0.03–0.10	(medium)	⎬ Newell, 1990
> 0.10	(high)	⎭
0.15–0.60		Lutz et al., 1991
Off-bottom Culture:		
0.05	Raft	Cabanas et al., 1979
0.17–0.35	Raft	Cheong, 1982
0.02–0.03	Longline	Rosenberg and Loo, 1983
1.0–1.5	Longline	Heritage, 1983
0.10	Raft	Rodhouse et al., 1985
0.05–0.25	Longline	Hickman et al., 1991

data derived from a few observations in the vicinity of the culture site (Rodhouse et al., 1985), or used hypothetical values for their calculations (Incze and Lutz, 1980; Loo and Rosenberg, 1983). Carver and Mallet (1990) incorporated tidal exchange into their model for a semi-enclosed inlet in place of current flow. By contrast, the hydrodynamics of mussel beds have been investigated extensively in connection with the effects of the storm surge barrier on water exchange over the Dutch mussel beds (Smaal et al., 1986; Dijkema 1988; Smaal and van Stralen, 1990). Food limitation in the boundary layer immediately above a mussel bed can depress the growth rate of the mussels (Fréchette and Bourget, 1985a) if the current speed is insufficient to prevent seston depletion (Wildish and Kristmanson, 1984, 1985). This does not occur in areas of high mixing energy such as the Oosterschelde mussel beds (Klepper and van de Kamer, 1988). However, the reduction in water flow and therefore food supply caused by the boundary layer effect (Fréchette and Bourget, 1985b), could be a significant factor in determining the food available to the densely packed mussels in double longline culture systems such as those used in Ireland and New Zealand, where the mussel growing ropes may be only 0.5m apart and the twin headlines separated by only 1.3m (Jenkins, 1985; Hickman, 1989a). Current measurements reported from mussel culture sites are typically in the range 0.02–0.10m sec⁻¹ (Table 10.5).

Fig. 10.11. Overcrowding of mussels can be used to slow the growth rate, but there is a risk of loss of part of the crop from drop-off.

Mussel density, of itself, can be a food limiting factor and manipulation of density provides a major tool for the mussel farmer to increase commercial production. This can be effected in on-bottom culture by varying the seeding density (Dijkema and van Stralen, 1989), the patch density (Behrens Yamada and Peters, 1988), or the patch size (Newell, 1990); in bouchot culture, by thinning and transplanting (Boromthanarat and Deslous-Paoli, 1988); or in suspended culture, by varying the spacing between the growing ropes or the number of longlines in the culture area (Rosenberg and Loo, 1983). Deliberate overcrowding can also be used by the mussel farmer to slow growth rate for specific purposes, such as to maintain a continuous supply of mussels of the right size for marketing, or to spread out the thinning and reseeding operations (Fig. 10.11).

Mussel farming, particularly off-bottom culture, tends to be sited in shallow, sheltered coastal areas, frequently in partially enclosed embayments. Many of the site-specific characteristics of these areas result from, or are manifest in, their water current patterns. Better understanding of water movement through and around the culture system, and any alterations to the pattern that result from changes in mussel density, mussel size, longline or raft density, may hold the key to optimizing mussel production for either the individual farmer or the mussel industry as a whole.

Mussel Supply to the Market

Mussel farming has evolved as a means of improving or augmenting mussel fishing activities (Dijkema and van Stralen, 1989; Hickman, 1989b), or has been developed to exploit a hitherto unutilized resource (Rosell, 1991). In either case, an established 'market' for the wild product pre-existed. Exploitation of wild mussel stocks is almost universal where they are abundant on rocky shorelines, and in many places has a history extending back over thousands of years (Chanley and Chanley, 1991). Mussels have come to occupy a low value, subsistence food niche for coastal populations in many parts of the world.

This perception of mussels as the poor man's oyster has provided mussel farming with a major dilemma, and is a continuing constraint on its development in certain areas. Any method of cultivation of mussels adds to the value of the end product, or, seen from the consumer's viewpoint, makes the product more expensive *vis a vis* the wild mussel. Minimizing farming costs is fundamental to all mussel farming methods. It is the purpose behind the industrialization and mechanization to reduce labour costs in the Dutch industry, and the reason for very limited technology and automation in the Thailand and Philippine industries, where labour is abundant and inexpensive. The mussel market varies markedly in different parts of the world. A huge and unsatisfied commodity market exists in Europe. This is supplied by the mussel farming industries of Spain, France, Holland, Belgium, Denmark and Germany. Within this market there are niches for particular mussel types or product forms which allow, for example, New Zealand green mussels or Irish rope-grown mussels a small stake. Development of specialized product forms can enable mussel producers to open up new markets, as has happened with New Zealand's frozen-on-the-half-shell mussel in Japan (Warwick, 1984). On the other hand, to produce mussels for local consumption in the less developed countries of Asia or South America, requires low cost, high volume production combined with cheap processing or preserving (Vakily, 1989).

Water quality deterioration from pollution is likely to impose constraints on expansion of existing mussel farming operations in Europe (Smaal, 1991), while population pressures may deter developments in tropical areas (Silas, 1980). The apparent increase in toxic algal blooms worldwide could be the cause of further constraint on either the siting of mussel farms, or the harvesting of the farmed product (Shumway 1989, 1990; see also p.520–521 Chapter 11). The association of wild mussels with paralytic shellfish poisoning already provides a hurdle to marketing farmed mussels in some countries.

THE FUTURE OF MUSSEL FARMING

In his review of cultivation in the 1970s Mason (1976) concluded that mussel farming was "more likely to provide a business opportunity in the advanced countries of the world than to provide a means of helping to overcome the protein deficiency ... in less well-developed countries". During the 1980s these business opportunities were taken up. In Ireland a suspended culture industry for *M. edulis* was developed, which by 1990 was producing 3500t (Corish, 1991). In New Zealand *P. canaliculus* farming expanded from an annual production of 300t in 1977 to over 35,500t in 1990 (Kingsbury, 1991). At the same time huge injections of finance went into modernizing the existing industries in Spain (Figueras, 1989) and Holland (Dijkema and van Stralen, 1989). Mussel farming in Thailand, on the other hand, has been beset by problems of deteriorating water quality, rising costs and depressed markets (Chalermwat and Lutz, 1989) with the result that production has fallen during the 1980s from 60–70,000t to around 35,000t (FAO, 1990).

Mussel farming, as with any other farming or fishing operation, must be economically viable. The farmer must be able to produce and sell his crop profitably, but at a price that the consumer is prepared to pay. The market must be sufficiently consistent and stable to encourage the farmer to continue producing. The high quality of cultivated mussels by comparison with wild ones is the premium on which the farmer must capitalize, but which the consumer must be prepared to pay.

The rapid growth and high meat yield that are characteristic of cultivated mussels contribute significantly to their quality. Cultivated mussels, in fact, allocate less than half as much of their energy budget to reproduction as do wild mussels (Rodhouse and Roden, 1984), to achieve their rapid growth. Food limiting factors in the mussel farm ecosystem are therefore important in determining the extent to which a farm or an industry can develop. Hence the interest in all aspects of 'carrying capacity' of mussel farming areas during the 1980s, and the need for continued research in this field.

Expansion of mussel farming in developed nations will occur at the expense of, or in competition with, other uses of the coastal resource. If it is to be allowed to expand into new areas, or even to significantly increase production in existing areas, mussel farming will need to be able to show that it is a benign user of the coastal water. Recent research using *in situ* measurement techniques on mussel beds (Dame and Dankers, 1988; Prins and Smaal, 1990) show that the release of inorganic nutrients from the mussel bed may compensate for the uptake of particulate material, estimated at between 170–600g suspended particulate matter m^{-2} day^{-1} (Smaal et al., 1986), and the consequent biodeposition. Selectivity in the feeding process of mussels (Shumway et al., 1985; Newell et al., 1989) and the influence of ingredients such as silt in the diet (Bayne et al. 1987) will have an effect on the environmental impact of mussel cultivation.

Further research is needed to establish the benign nature of suspended mussel culture, which may have different impacts from those of bottom culture (Smaal, 1990), and so far has only been looked at in terms of biodeposition (Dahlback and Gunnarsson 1981; Tenore et al., 1982) and nitrogen cycling (Kaspar et al., 1985). The relationship between the reduction in phytoplankton, caused by the filtering activity of a population of farmed mussels, and the potential increase in primary production, resulting from nutrient release by the mussels (Asmus and Asmus, 1991), is another important area for future research.

In less developed countries the challenge is not so much one of establishing mussel farming's credibility as of establishing its viability. Using the low technology approach, and the abundant labour force readily available in many tropical countries, mussel farming has the potential to become established as a village occupation providing a continuing food supply for local consumption. The ability of mussel farming in these areas to progress beyond the subsistence level and become a cash crop industry, will depend on the availability of a supporting infrastructure such as freezing, preserving or processing facilities, and reliable transport. Vakily (1989) summed up this step in the challenge as "a seemingly insolvable contradiction: to keep the price of the end product low and thus affordable to the poor, the mussel farmer's profit must be minimal".

The 1990s will see an increasing emphasis on industrialization, mechanization and automation in the mussel cultivation industries of the developed nations in an effort to increase production, to supply established market demand in Europe and North America, and the rapidly expanding markets in Asia. The dichotomy between high technology farming for high value markets and low technology techniques for low value commodity markets will increase.

Improvements in equipment, particularly in the types of rope and netting used in suspended culture (Anon, 1983) are inevitable. Improvements in farming techniques will also occur as farmers endeavour to offset rising material and labour costs. One such improvement can be seen in the change to continuous seeding of full coils of rope (220m) instead of individual 5–10m lengths, which has enabled seeding rates in the New Zealand longline mussel culture to increase from 800m day^{-1} to 15,000m day^{-1} (Pooley, 1991). However, it seems unlikely that the basic methods used for farming mussels during the 1970s and 80s will undergo radical change.

The world statistics (FAO, 1990) show a steadily increasing production of mussels from 0.97 million metric tonnes in 1985 through 1.00 million in 1986, 1.07 million in 1987 to 1.17 million in 1988. Most, but not all, of this production comes from aquaculture. However, closer examination of the figures (Table 10.6) shows that, in

Table 10.6. World mussel production in metric tonnes (t) for 1985 and 1988 (countries producing > 10,000t). (From FAO, 1990)

	1985	1988	Change (%)
China	128,860	429,675	+ 233
Spain	245,655	209,687	- 14
Italy	75,984	85,400	+ 12
Netherlands	116,252	77,596	- 33
Denmark	84,077	72,524	- 14
France	60,763	54,873	- 10
U.S.A.	15,798	35,724	+126
Thailand	68,964	35,270	- 49
Germany	23,108	30,865	+ 34
Korea	58,014	27,356	- 53
Chile	18,266	21,910	+ 20
New Zealand	10,860	18,000	+ 66
Philippines	26,159	17,553	- 33
Ireland	10,358	16,000	+ 54
The World	968,669	1,173,515	+ 21

fact almost all of the increase has been contributed by one country—China—which only started mussel farming in the 1970s (Zhang, 1984). Other nations that also developed their industries from about this time show increased production figures, but in the more traditional mussel farming countries, with longer established mussel industries, production has typically declined during the 1980s.

Is there perhaps a message to be taken from the observation that the one country that makes significant use of hatchery-produced spat in its mussel cultivation methodology (Zhang, 1984; Nie, 1991), is also the country that has achieved the most spectacular increase in mussel production in the last decade?

REFERENCES

Ackefors, H. and Haamer, J., 1987. A new Swedish technique for culturing blue mussel. Mariculture Ctte, Ref. Shellfish Ctte, Int Council Exp. Sea, C.M. 1987/K:36. 7pp.

Anon., 1983. Low stretch ropes could hold a lot more mussels. Fish Farming Int., 10 (1): 7.

Anon., 1990. Mussels. Aquacult. Ir., No. 45: 36-37.

Anon., 1991a. Dutch firm sells mussel plant to South Korea. Fish Farming Int., 18(3): 2.

Anon., 1991b. First half-shell machine sold to New Zealand. Fish Farming Int., 18(4): 20

Anon., 1991c. Gales and ducks deplete Dutch mussels. Fish Farming Int., 18(4): 20-21.

Asmus, R.M. and Asmus, H., 1991. Mussel beds: limiting or promoting phytoplankton? J. Exp. Mar. Biol. Ecol., 148: 215-232.

Bartlett, S. and Hufflett, P., 1991. Mussel farm economic study. An analysis of the costs of mussel production in the Marlborough Sounds and the Coromandel. New Zealand Fishing Industry Board, Wellington, 34pp.

Bayne, B.L., 1964. Primary and secondary settlement in Mytilus edulis L. J. Anim. Ecol., 33: 513-523.

Bayne, B.L., 1965. Growth and delay of metamorphosis of the larvae of Mytilus edulis (L). Ophelia, 2: 1-47.

Bayne, B.L. (Editor), 1976. Marine mussels: their ecology and physiology. Cambridge University Press, Cambridge, 506pp.

Bayne, B.L., Hawkins, A.J.S. and Navarro, E., 1987. Feeding and digestion by the mussel Mytilus edulis L. (Bivalvia: Mollusca) in mixtures of silt and algal cells at low concentrations. J. Exp. Mar. Biol. Ecol., 111: 1-22.

Beckley, L.E., 1979. Primary settlement of Perna perna (L.) on littoral seaweeds on St. Croix Island. S. Afr. J. Zool., 14: 171.

Behrens Yamada, S. and Peters, E.E., 1988. Harvest management and the growth and condition of submarket-size sea mussels, Mytilus californianus. Aquaculture, 74: 293-299.

Bierbaum, R.M. and Ferson, S., 1986. Do symbiotic pea crabs decrease growth rate in mussels? Biol. Bull., 170: 51-61.

Body, A., 1983. Hatchery-reared spat could boost NSW mussel industry. Aust. Fish., 42 (12): 11-13.

Boromthanarat, S. and Deslous-Paoli, J.M., 1988. Production of Mytilus edulis L. reared on bouchots in the bay of Marennes-Oleron: comparison between two methods of culture. Aquaculture, 72: 255-263.

Bower, S.M. and Figueras, A.J., 1989. Infectious diseases of mussels, especially pertaining to mussel transplantation. World Aquaculture, 20 (4): 89-93.

Cabanas, J.M., Gonzalez, J.J., Marino, J., Perez-Camacho, A. and Roman, G., 1979. Estudio del mejillon y de su epifauna en los cultivos flotantes de la Ria de Arosa. III. Observaciones previas sobre la retenion de particulas y la biodeposicion de una batea. Bol. Inst. Esp. Oceanogr., 5(1): 45-50.

Carver, C.E.A. and Mallet, A.L., 1990. Estimating the carrying capacity of a coastal inlet for mussel culture. Aquaculture, 88: 39-53.

Ceccherelli, V.U. and Barboni, A., 1983. Growth, survival and yield of Mytilus galloprovincialis Lmk. on fixed suspended culture in a bay of the Po River delta. Aquaculture, 34: 101-114.

Chalermwat, K. and Lutz, R.A., 1989. Farming the green mussel in Thailand. World Aquaculture, 20 (4): 41-46.

Chanley, M.H., and Chanley, P., 1991. Chilean mussel culture: Mytilus edulis chilensis (Hupé, 1854), Choromytilus chorus (Molina, 1782), Aulacomya ater (Molina, 1782). In W. Menzel (Editor), Estuarine and Marine Bivalve Mollusk Culture. CRC Press, Boca Raton, Florida, pp. 135-143.

Cheong, L., 1982. Country report, Singapore. In: F.B. Davy and M. Graham (Editors), Bivalve Culture in Asia and the Pacific: Proceedings of a Workshop, Singapore, 1982. IDRC, Ottawa, Canada, pp. 69-71.

Cheong, L. and Lee, H.B., 1984. Mussel farming. SAFIS Extension Manual No. 5. Southeast Asian Fisheries Development Centre, Bangkok, 51pp.

Cheong, L. and Loy, W.S., 1982. An analysis of farming green mussels in Singapore using rafts. Aquaculture Economics Research in Asia. Proceedings of a Workshop, Singapore, June 1981. IDRC, Ottawa, Canada, pp. 65-74.

Clifton, J.A., 1980. Some economics of mussel culture and harvest. In: R.A. Lutz (Editor), Mussel Culture and Harvest: A North American Perspective. Elsevier Science Publishers, B.V., Amsterdam, pp. 312-338.

Coeroli, M., Gaillande, D. de, Landret, J.P., and Aquacop. (Coatanea, D.), 1984. Recent innovations in cultivation of molluscs in French Polynesia. Aquaculture, 39: 45-67.

Conte, F.S., 1990. California aquaculture. World Aquaculture, 21 (3): 33-44.

Corish, C., 1991. The production cycle of a longline mussel farm, Fastnet Mussels Ltd, followed by a discussion of the state of mussel production in Ireland. Unpublished MS, Department of Zoology, University College, Cork, Ireland.

Curini Galletti, M. and Galleni, L., 1984. Mussel beds of the coast of Livorno. II. Mussels and their predators. Oebalia, 10: 117-131.

Dahlback, B. and Gunnarsson, L.A.H., 1981. Sedimentation and sulphate reduction under a mussel culture. Mar. Biol., 63: 269-275.

Dame, R.F. and Dankers, N., 1988. Uptake and release of materials by a Wadden Sea mussel bed. J. Exp. Mar. Biol. Ecol., 118: 207-216.

Dare, P.J., 1982. Notes on the swarming behaviour and population density of *Asterias rubens* L. (Echinodermata: Asteroidea) feeding on the mussel *Mytilus edulis*. J. Cons., 40: 112-118.

Davies, G., Dare, D.B. and Edwards, D.B., 1980. Fenced enclosures for the protection of seed mussels (*Mytilus edulis* L.) from predation by shore crabs (*Carcinus maenas* (L.)). Minist. Agric. Fish. Food, Fish. Res. Tech. Rep., 56: 1-14.

Deslous-Paoli, J.M., Boromthanarat, S., Heral, M., Boromthanarat, W. and Razet, D., 1990. Energy budget of a *Mytilus edulis* L. population during its first year on bouchots in the Bay of Marennes-Oleron. Aquaculture, 91: 49-63.

Dijkema, R., 1988. Shellfish cultivation and fishery before and after a major flood barrier construction project in the southwestern Netherlands. J. Shellfish Res., 7: 241-252.

Dijkema, R. and Stralen, M. van, 1989. Mussel cultivation in the Netherlands. World Aquaculture, 20 (4): 56-62.

Ekaratne, S.U.K., 1987. Availability of spat and exploitability of the marine intertidal mussel *Perna perna* in Sri Lanka. In: S. Chang, K Chan, and N.Y.S. Woo (Editors), Recent Advances in Biotechnology and Applied Biology. The Chinese University of Hong Kong, Hong Kong, pp. 353-362.

Erkom Schurink, C. van, and Griffiths, C.L., 1990. Marine mussels of southern Africa—their distribution, patterns, standing stocks, exploitation and culture. J. Shellfish Res., 9: 75-85.

Eyster, L.S. and Pechenik, J.A., 1987. Attachment of *Mytilus edulis* L. larvae on algal and byssal filaments is enhanced by water agitation. J. Exp. Mar. Biol. Ecol., 114: 99-110.

Fabi, G., Fiorentini, L. and Giannini, S., 1989. Experimental shellfish culture on an artificial reef in the Adriatic Sea. Bull. Mar. Sci., 44: 923-933.

FAO, 1990. FAO Yearbook Fishery Statistics. Catches and Landings Vol. 66, 1988. FAO, Rome, Italy, 503pp.

Figueras, A. J., 1984. Mussels—Spain updates an industry. Wld. Fishg., 33 (11): 23.

Figueras, A.J., 1989. Mussel culture in Spain and France. World Aquaculture, 20 (4): 8-17.

Figueras, A. J., 1990. Mussel culture in Spain. Mar. Behav. Physiol., 16: 177-207.

Figueras, A. J., Jordon, C.F. and Caldas, J.R., 1991. Diseases and parasites of rafted mussels (*Mytilus galloprovincialis* Lmk.): preliminary results. Aquaculture, 99: 17-33.

Fréchette, M. and Bourget, E., 1985a. Energy flow between the pelagic and benthic zones: factors controlling particulate organic matter available to an intertidal mussel bed. Can. J. Fish. Aquat. Sci., 42: 1158-1165.

Fréchette, M. and Bourget, E., 1985b. Food-limited growth of *Mytilus edulis* L. in relation to the benthic boundary layer. Can. J. Fish. Aquat. Sci., 42: 1166-1170.

Gentili, M.R. and Beaumont, A.R., 1988. Environmental stress, heterozygosity, and growth rate in *Mytilus edulis* L. J. Exp. Mar. Biol. Ecol., 120: 145-153.

Gibbs, M.M., James, M.R., Pickmere, S.E., Woods, P.H., Shakespeare, B.S., Hickman, R.W. and Illingworth, J., 1991. Hydrodynamic and water column properties at six stations associated with mussel farming in Pelorus Sound, 1984–85. N. Z. J. Mar. Freshw. Res., 25: 239-254.

Gonzalez, R., 1982. Estudio de la epifauna de la semilla de mejillon en la Ria de Arosa. Bol. Inst. Esp. Oceanogr., 7(1): 51-57.

Gorski, Z., 1991. Overview of suspended mussel culture systems over the world. Aquacult. Eur., 16(1): 6-10.

Heral, M., and Deslous-Paoli, J.M., 1991. Oyster culture in European countries. In: W. Menzel (Editor), Estuarine and Marine Bivalve Mollusk Culture. CRC Press, Boca Raton, Florida, pp. 153-190.

Heritage, G.D., 1983. A blue mussel, (*Mytilus edulis* Linnaeus), culture pilot project in south coastal British Columbia. Can. Tech. Rep. Fish. Aquat. Sci., No. 1174, 27pp.

Herriott, N., 1984. A guide to longline mussel cultivation. Aquacult. Tech. Bull. Ir., No. 9, 97pp.

Herriott, N., 1985. One way to profitable farming. Aquacult. Ir., 21(8): 14-15.

Hickman, R.W., 1976. Potential for the use of stranded seed mussels in mussel farming. Aquaculture, 9: 287-293.

Hickman, R.W., 1987. Growth potential and constraints in the New Zealand mussel farming industry. Proc. N. Z. Soc. Anim. Prod., 47: 131-133.

Hickman, R.W., 1989a. Farming the green mussel in New Zealand. Current practice and potential. World Aquaculture, 20 (4): 20-28.

Hickman, R.W., 1989b. Mussel farming in New Zealand; how much more can it grow? In N. De Pauw, E. Jaspers, H. Ackefors and N. Wilkins (Editors), Aquaculture—A Biotechnology in Progress. European Aquaculture Society, Bredene, Belgium, pp. 321-325.

Hickman, R.W., 1991. Perna canaliculus (Gmelin) in New Zealand. In: W. Menzel (Editor), Estuarine and Marine Bivalve Mollusk Culture. CRC Press, Boca Raton, Florida, pp. 325-334.

Hickman, R.W., Waite, R.P., Illingworth, J., Meredyth-Young, J.L. and Payne, G., 1991. Relationship between farmed mussels Perna canaliculus and available food in Pelorus-Kenepuru Sound, New Zealand, 1983-1985. Aquaculture, 99: 49-68.

Hrs-Brenko, M., 1974. The settlement of mussel larvae (Mytilus galloprovincialis Lmk.) in Limski channel in the northern Adriatic Sea. Rapp. Comm. Int. Mer Mediter., 26 (6): 51-52.

Hrs-Brenko, M., 1980. The settlement of mussels and oysters in the northern Adriatic Sea. Nova Thall., 4, Suppl.: 67-85.

Hrs-Brenko, M., 1983. Knowledge of the settlement of mussels in Kukuljina-Tivatski Bay. Stud. Mar., 13/14: 267-273.

Hurlburt, C.G. and Hurlburt, S.W., 1980. European mussel culture technology and its adaptability to North American waters. In R.A. Lutz (Editor), Mussel Culture and Harvest: A North American Perspective. Elsevier Science Publishers, B.V., Amsterdam, pp. 69-98.

Imai, T. (Editor), 1977. Aquaculture in shallow seas: progress in shallow sea culture. Amerind Publishing Company, New Delhi, 615pp.

Incze, L.S. and Lutz, R.A., 1980. Mussel culture: an east coast perspective. In: R.A. Lutz (Editor), Mussel Culture and Harvest: A North American Perspective. Elsevier Science Publishers, B.V., Amsterdam, pp. 99-140.

Jenkins, R.J., 1985. Mussel Cultivation in the Marlborough Sounds (New Zealand). New Zealand Fishing Industry Board, Wellington, 2nd Edition, 77pp.

Johannesson, K., Kautsky, N. and Tedengren, M., 1990. Genotypic and phenotypic differences between Baltic and North Sea populations of Mytilus edulis evaluated through reciprocal transplantations. II. Genetic variation. Mar. Ecol. Prog. Ser., 59: 211-219.

Johns, T.G. and Hickman, R.W., 1985. A manual for mussel farming in semi-exposed coastal waters; with a report on mussel research at Te Kaha, eastern Bay of Plenty, New Zealand, 1977-82. N. Z. Fish. Res. Div. Occ. Publ., No. 50, 28pp.

Johnson, T., 1986. Mussels: an up and coming cash crop. Natl. Fisherman 67 (4): 4-5.

Kaspar, H.F., Gillespie, P.A., Boyer, I.C. and McKenzie, A.L., 1985. Effects of mussel aquaculture on the nitrogen cycle and benthic communities in Kenepuru Sound, New Zealand. Mar. Biol., 85: 127-136.

Kent, R.M.L., 1979. The influence of heavy infestations of Polydora ciliata on the flesh content of Mytilus edulis. J. Mar. Biol. Ass. U.K., 59: 289-297.

King, P.A., McGrath, D. and Gosling, E.M., 1989. Reproduction and settlement of Mytilus edulis on an exposed rocky shore in Galway Bay, west coast of Ireland. J. Mar. Biol. Ass. U.K., 69: 355-365.

Kingsbury, R., 1991. Mussel industry grows, but is it sustainable? The Christchurch Press, New Zealand, 15th June 1991.

Klepper, O. and Kamer, H., van de, 1988. A definition of consistency of the carbon budget of an ecosystem and its application to the Oosterschelde estuary, S.W. Netherlands. Ecol. Model., 42: 217-232.

Korringa, P., 1976. Farming marine organisms low in the food chain; a multidisciplinary approach to edible seaweed, mussel and clam production. Elsevier Science Publishers, B.V., Amsterdam, 264pp.

Kronman, M. 1986. Mariculture in southern California is experiencing some growing pains. Natl. Fisherman, 66 (12): 18-19.

Kumaraswamy Achari, G.P. and Thangavelu, R., 1980. Mussel culture—its problems and prospects. Symposium on Coastal Aquaculture, Cochin, India, 1980. Mar. Biol. Ass. India, p. 101.

Kuriakose, P.S. 1980. Development of the brown mussel Perna indica. Symposium on Coastal Aquaculture, Cochin, India, 1980. Mar. Biol. Ass. India, p. 100.

Loo, L.O. and Rosenberg, R. 1983. Mytilus edulis culture: growth and production in western Sweden. Aquaculture, 35: 137-150.

Lutz, R.A. (Editor), 1980a. Mussel Culture and Harvest: A North American Perspective. Elsevier Science Publishers, B.V., Amsterdam, 305pp.

Lutz, R.A., 1980b. pearl incidence: mussel culture and harvest implications. In: R.A.Lutz (Editor), Mussel Culture and Harvest: A North American Perspective. Elsevier Science Publishers, B.V., Amsterdam, pp. 193-222.

Lutz, R.A., Chalermwat, K., Figueras, A.J., Gustafson, R.G. and Newell, C., 1991. Mussel aquaculture in marine and estuarine environments throughout the world. In: W. Menzel (Editor), Estuarine and Marine Bivalve Mollusk Culture. CRC Press, Boca Raton, Florida, pp. 57-97.

Maclean, J.L., 1972. Mussel culture: methods and prospects. Aust. Fish. Pap., No. 20, 13pp.

Mallet, A.L., Carver, C.E.A., Cotton, S.S. and Freeman, K.R., 1987. Mortality variations in natural populations of the blue mussel *Mytilus edulis*. Can. J. Fish. Aquat. Sci., 44: 1589-1594.

Mandelli, E.F. and Acuna, A.C., 1975. The culture of the mussel, *Perna perna*, and the mangrove oyster, *Crassostrea rhizophorae*, in Venezuela. Mar. Fish. Rev., 37: 15-18.

Margus, D. and Teskeredzic, E., 1986. Settlement of mussels (*Mytilus galloprovincialis* Lamarck) on rope collectors in the estuary of the River Krka, Yugoslavia. Aquaculture, 55: 285-296.

Marques, H.L.A., Pereira, R.T.L., Ostini, S. and Scorvo Filho, J.D., 1984. Obsrevacoes preliminares sobre o cultivo experimental do mexilhao *Perna perna* (Linnaeus, 1758) na regiao de Ubatuba (23°32'S e 45°04'W) estado de Sao Paulo, Brasil. Bull. Inst. Pesca, 12 (4): 23-34.

Mason, J., 1976. Cultivation. In: B.L. Bayne (Editor), Marine Mussels: their ecology and physiology. Cambridge University Press, Cambridge, pp. 385-410.

Menzel, W., 1991 (Editor), Estuarine and Marine Bivalve Mollusk Culture. CRC Press Boca Raton, Florida, 362pp.

Morris, C., 1984. An analysis of mussel farming viability in the Marlborough Sounds. N. Z. Fish. Ind. Board, Wellington, 39pp.

Muise, B. 1990. Mussel culture in eastern Canada. World Aquaculture, 21 (2): 12-23.

Nagabhushanam, R. and Mane, V.H., 1991. Mussels in India. In: W. Menzel (Editor), Estuarine and Marine Bivalve Mollusk Culture. CRC Press, Boca Raton, Florida, pp. 191-200.

Navarro, E., Iglesias, J.I.P., Perez-Camacho, A., Labarta, V. and Beiras, R., 1991. The physiological energetics of mussels (*Mytilus galloprovincialis* Lmk) from different cultivation rafts in the Ria de Arosa (Galicia, N.W. Spain). Aquaculture, 94 (2/3): 197-212.

Newell, C.R., 1990. The effects of mussel (*Mytilus edulis*, Linnaeus, 1758) position in seeded bottom patches on growth at subtidal lease sites in Maine. J. Shellfish Res., 9: 113-118.

Newell, C.R., Shumway, S.E., Cucci, T.L. and Selvin, R., 1989. The effects of natural seston particle size and type on feeding rates, feeding selectivity and food resource availability for the mussel *Mytilus edulis* Linnaeus, 1758 at bottom culture sites in Maine. J. Shellfish Res., 8: 187-196.

Newell, C.R., Hidu, H., McAlice, B.J. Podniesinski, G., Short, F. and Kindblom, L., 1991. Recruitment and commercial seed procurement of the blue mussel *Mytilus edulis* in Maine. J. World Aquacult. Soc., 22(2): 134-152.

Nie, Z-Q., 1991. The culture of marine bivalve molluscs in China. In: W. Menzel (Editor), Estuarine and Marine Bivalve Mollusc Culture. CRC Press, Boca Raton, Florida, pp. 261-276.

Nie, Z.Q., Wang, Z.Y., Niu, X.D., Ji, F.M., Shen, J.F., Chen, W.H., and Liu, L.H., 1979. Artificial rearing of *Mytilus edulis* on a large scale. Acta Oceanogr. Sinica, 1: 138.

O'Neill, S.M., Sutterlin, A.M. and Aggett, D., 1983. The effect of size-selective feeding by starfish (*Asterias vulgaris*) on the production of mussels (*Mytilus edulis*) cultured on nets. Aquaculture, 35: 211-220.

O'Rourke, T.D., 1987. Intensive longline culture systems for mussels (*Mytilus edulis*) on Prince Edward Island. Proc. Annu. Meet. Aquacult. Ass. Can., 1: 12-13.

Page, H.M. and Hubbard, D.M., 1987. Temporal and spatial patterns of growth in mussels *Mytilus edulis* on an offshore platform: relationships to water temperature and food availability. J. Exp. Mar. Biol. Ecol., 111: 159-179.

Parulekar, A.H., Dalal, S.G., Ansari, Z.A. and Harkantra, S.N., 1982. Environmental physiology of raft-grown mussels in Goa, India. Aquaculture, 29: 83-93.

Paul, J.D., 1983. The incidence and effects of *Mytilicola intestinalis* on *Mytilus edulis* from the rias of Galicia, North West Spain. Aquaculture, 31: 1-10.

Pechenik, J.A., Eyster, L.S., Widdows, J. and Bayne, B.L., 1990. The influence of food concentration and temperature on growth and differentiation of blue mussel larvae, Mytilus edulis L. J. Exp. Mar. Biol. Ecol., 136: 47-64.

Perez, A. and Roman, G., 1979. Estudio del mejillon y de su epifauna en los cultivos flotantes de la Ria de Arosa. II. Crecimiento, mortalidad y produccion del mejillon. Bol. Inst. Esp. Oceanogr., 5(1): 21-42.

Perez-Camacho, A., Gonzalez, R., and Fuentes, J., 1991. Mussel culture in Galicia (N.W. Spain). Aquaculture, 94 (2/3): 263-278.

Pooley, R., 1991a. Mussel farming—reseeding. N. Z. Prof. Fisherman 5 (4): 16-17.

Pooley, R., 1991b. Mussel farming—harvesting. N. Z. Prof. Fisherman 5 (5): 44-46.

Pooley, R., 1991c. Mussel farming—the boats. N. Z. Prof. Fisherman 5 (6): 12-13.

Pregenzer, C., 1983. Survey of metazoan symbionts of Mytilus edulis (Mollusca: Pelecypoda) in southern Australia. Aust. J. Mar. Freshw. Res., 34: 387-396.

Prins, T.C. and Smaal, A.C., 1990. Benthic pelagic coupling: the release of inorganic nutrients by an intertidal bed of Mytilus edulis. In: M. Barnes and R.N. Gibson (Editors), Trophic Relationships in the Marine Environment. Aberdeen University Press, Aberdeen, pp. 89-103.

Quasim, S.Z., Parulekar, A.H., Harkantra, S.N., Ansari, Z.A. and Nair, A., 1977. Aquaculture of green mussel Mytilus viridis L. : cultivation on ropes from floating rafts. Indian J. Mar. Sci., 6: 15-25.

Rajan, S.J., 1980. Experiments on submerged rafts for open sea mussel culture. In: K.N. Nayar, K. Mahadevan, K. Alagarswami and P.T. Meenakshisundaram (Editors), Coastal Aquaculture: Mussel Farming. Progress and Prospects. Bull. Cent. Mar. Fish. Res. Inst., India, 29: 46-51.

Raubenheimer, D. and Cook, P., 1990. Effects of exposure to wave action on allocation of resources to shell and meat growth by the subtidal mussel Mytilus galloprovincialis. J. Shellfish Res., 9: 87-93.

Roantree, V., 1986. Plastic drums and tractor tyres. Aquacult. Ir., No. 24, 8-9.

Rodhouse, P.G. and Roden, C.M., 1987. Carbon budget for a coastal inlet in relation to intensive cultivation of suspension-feeding bivalve molluscs. Mar. Ecol. Prog. Ser., 36: 225-236.

Rodhouse, P.G., Roden, C.M., Hensey, M.P. and Ryan, T.H., 1984a. Resource allocation in Mytilus edulis on the shore and in suspended culture. Mar. Biol., 84: 27-34.

Rodhouse, P.G., Roden, C.M., Burnell, G.M., Hensey, M.P., McMahon, T., Ottway, B. and Ryan, T.H., 1984b. Food resource, gametogenesis and growth of Mytilus edulis on the shore and in suspended culture: Killary Harbour, Ireland. J. Mar. Biol. Ass. U.K., 64: 513-529.

Rodhouse, P.G., Roden, C.M., Hensey, M.P. and Ryan, T.H., 1985. Production of mussels, Mytilus edulis, in suspended culture and estimates of carbon and nitrogen flow: Killary Harbour, Ireland. J. Mar. Biol. Ass. U.K., 65: 55-68.

Roman, G. and Perez, A., 1979. Estudio del mejillon y de su epifauna en los cultivos flotantes de la Ria de Arosa. I. Estudios preliminaires. Bol. Inst. Esp. Oceanogr., 5(1): 9-19.

Roman, G. and Perez, A., 1982. Estudio del mejillon y de su epifauna en los cultivos flotantes de la Ria de Arosa. IV. Evoluccion de la comunidad. Bol. Inst. Esp. Oceanogr., 7(2); 279-296.

Rosell, N.C., 1991. The green mussel (Perna viridis) in the Philippines. In: W. Menzel (Editor), Estuarine and Marine Bivalve Mollusk Culture. CRC Press, Boca Raton, Florida, pp. 298-305.

Rosenberg, R. and Loo, L.O., 1983. Energy flow in a Mytilus edulis culture in western Sweden. Aquaculture, 35: 151-167.

Ross, R., 1982. Report on: Intensive mussel culture in the Italian Adriatic. Aquacult. Tech. Bull. Ir., No. 5, 43pp.

Shumway, S.E., 1989. Toxic algae. A serious threat to shellfish aquaculture. World Aquaculture, 20 (4): 65-74.

Shumway, S.E., 1990. A review of the effects of algal blooms on shellfish and aquaculture. J. World Aquacult. Soc., 22 (2): 65-104.

Shumway, S.E., Cucci, T.L., Newell, R.C. and Yentsch, C.M., 1985. Particle selection, ingestion and absorption in filter feeding bivalves. J. Exp. Mar. Biol. Ecol., 91: 77-92.

Shumway, S.E., Card, D., Getchell, R. and Newell, C. R., 1988. Effects of calcium oxide (quicklime) on non-target organisms in mussel beds. Bull. Environ. Contam. Toxicol., 40: 503.

Silas, E.G., 1980. Mussel culture in India—constraints and prospects. In: K.N. Nayar, K. Mahadevan, K. Alagarswami and P.T. Meenakshisundaram (Editors), Coastal Aquaculture: Mussel Farming. Progress and Prospects. Bull. Cent. Mar. Fish. Res. Inst., India, 29: 51-56.

Sivalingam, P.M., 1977. Aquaculture of the green mussel, Mytilus viridis Linnaeus in Malaysia. Aquaculture, 11: 297-312.

Skibinski, D.O.F. and Roderick, E.E., 1989. Heterozygosity and growth in transplanted mussels. Mar. Biol., 102: 73-84.

Smaal, A.C., 1991. The ecology and cultivation of mussels: new advances. Aquaculture, 94 (2/3): 245-261.

Smaal, A.C. and Stralen, M.R. van, 1990. Average annual growth and condition of mussels as a function of food source. Hydrobiologia, 195: 179-188.

Smaal, A.C., Verhagen, J.H.G., Coosen, J., and Haas, H.A., 1986. Interaction between seston quantity and quality and benthic suspension feeders in the Oosterschelde, the Netherlands. Ophelia, 26: 385-399.

Smaal, A.C., Knoester, M., Nienhuis, P.H. and Meire, P.M., 1991. Changes in the Oosterschelde ecosystem induced by the Delta works. In: M. Elliot and J.P. Ducrotoy (Editors), Estuaries and Coasts: Spatial and Temporal Inter-comparisons. Olsen and Olsen, Fredensborg, Denmark, pp. 375-384.

Smith, P.J., 1988. Biochemical-genetic variation in the green-lipped mussel *Perna canaliculus* around New Zealand and possible implications for mussel farming. N. Z. J. Mar. Freshw. Res., 22: 85-90.

Sreenivasan, P.V., Thangavelu, R. and Poovannan, P., 1988a. Potentialities of Muttukadu mariculture farm for green mussel culture. Indian Marine Fisheries Information Service T. and E. Series, No. 81: 10-12.

Sreenivasan, P.V., Satyanarayan Rao, K., Poovannan, P. and Thangavelu, R., 1988b. Growth of larvae and spat of the green mussel *Perna viridis* (Linnaeus) in hatchery. Indian Marine Fisheries Information Service T. and E. Series, No. 79: 23-26.

Suchanek, T.J., 1979. The *Mytilus californianus* community: studies on the composition, structure, organisation, and dynamics of a mussel bed. Ph. D. Thesis, University of Washington, U.S.A.

Tenore, K.R., Bayer, L.F., Cal, R.M., Carrol, J., Garcia-Fernandez, C., Gonzalez, N., Gonzalez-Gurriaran, E., Hanso, R.B., Iglesias, J., Krom, M., Lopez-Jamar, E., McClain, J., Pomatmat, M.M., Perez, A., Rhoads, D.C., Santiago, G. de, Tietjen, J., Westruck, J. and Windom, H.L., 1982. Coastal upwelling in the Rias Bajas, N.W. Spain: contrasting the benthic regimes of the Rias de Arosa and de Muros. J. Mar. Res., 40: 701-772.

Vakily, J.M., 1989. The biology and culture of mussels of the genus *Perna*. ICLARM Stud. Rev. 17: 1-63.

Vidal-Giraud, B., 1986. Etat actuel de la conchyliculture en mer en Longuedoc-Roussillon et perspectives de developpement. In: Report of the technical consultation on open sea shellfish culture in association with artificial reefs. FAO Fisheries Report, No. 357, pp. 78-83.

Wallace, J.C., 1983. Spatfall and growth of the mussel *Mytilus edulis edulis*, in hanging culture in the Westfjord area (68°5'N), Norway. Aquaculture, 31: 89-94.

Warwick, J., 1984. A code of practice for mussel processing. Publ. N. Z. Fish. Ind. Board, Wellington, 35pp.

Waterstrat, P., Chew, K. Johnson, K. and Beattie, J.H., 1980. Mussel culture: a west coast perspective. In: R.A.Lutz (Editor), Mussel Culture and Harvest: A North American Perspective. Elsevier Science Publishers, B.V., Amsterdam, pp. 141-165.

Widdows, J., 1991. Physiological ecology of mussel larvae. Aquaculture, 94 (2/3): 147-163.

Wildish, D.J. and Kristmanson, D.D., 1984. Importance to mussels of the benthic boundary layer. Can. J. Fish. Aquat. Sci., 41: 1618-1625.

Wildish, D.J. and Kristmanson, D.D., 1985. Control of suspension feeding bivalve production by current speed. Helgoländer. Wiss. Meeresunters., 39: 237-243.

Winter, J.E., Toro, J.E., Navarro, J.M., Valenzuela, G.S. and Chaparro, O.R., 1984. Recent developments, status and prospects of molluscan aquaculture on the Pacific coast of South America. Aquaculture, 39: 95-134.

Yamamoto, S. and Sugawara, Y., 1988. Induced triploidy in the mussel, *Mytilus edulis*, by temperature shock. Aquaculture, 72: 21-29.

Yap, W.G., Young, A.L., Orano, C.E.F. and Castro, M.T. de, 1979. Manual on mussel farming. Aquaculture Extension Manual No. 6. Southeast Asian Development Centre, Aquaculture Department, Iloilo, Philippines, 17pp.

Zhang, F.S., 1984. Mussel culture in China. Aquaculture, 39: 1-10.

Chapter 11

MUSSELS AND PUBLIC HEALTH

SANDRA E. SHUMWAY

INTRODUCTION

"As at present carried on" (i.e....., the shellfish trade) "nobody who realizes the meaning of clean food could possibly recommend shellfish for eating, and I venture to think that did the buying public fully realize the conditions the consumption of this commodity would fall considerably." (Holden, 1925, in Dodgson, 1928).

Dodgson went on to comment that shellfish are, "at best, a dirty food, and may be, and frequently are, a dangerous one". Fortunately, conditions have improved in most areas and shellfish are now a popular and relatively safe commodity although, by virtue of their estuarine and near-coastal habitats, they are commonly exposed to sewage and land run-off. While clams and oysters have been the major concern of health agencies, the filtration process generally lacks specificity and selectivity; thus, rendering all filter-feeding bivalves as potential vectors for infection from water-borne agents (including bacteria, viruses, pesticides, industrial chemical and radioactive wastes, toxic metals and hydrocarbon derivatives of oil spills).

We have come a long way since the 1900s, when Asiatic cholera and typhoid fever (enteric fever) were commonly associated with the consumption of polluted shellfish; however, identifiable diseases such as typhoid, cholera and infectious hepatitis, along with gastroenteritis of unknown etiology, are all still associated with consumption of contaminated shellfish. It has long been recognized that 'mussel poisoning' includes at least three pathological conditions: allergic, infectious (bacterial and viral food poisoning (fulminant infection)) and toxic (paralytic shellfish poisoning) (Dodgson, 1928; Acres and Gray, 1978; Eastaugh and Shephard, 1989). To this list we can now add diarrhetic and amnesic shellfish poisoning, all of which have been associated with mussel consumption (see Shumway, 1990 for review; Sindermann, 1990).

The following review focuses on public health issues specifically related to consumption of mussels. It must be remembered, however, that mussels are likely to serve as vectors of *any* water-borne disease or contaminant, and these public health problems will continue to play an important role in molluscan aquaculture development and product marketing.

DISEASES TRANSMITTED BY MUSSELS

Bacterial Infections

Marine mussels are often subject to faecal contamination from domestic sewage discharges, which typically contain pathogenic bacteria such as *Salmonella* sp., *Shigella* sp., *Clostridium* and nonpathogenic *Escherichia coli* (Wood, 1972). These bacteria are concentrated to levels far in excess of the surrounding water by filter-feeding shellfish such as mussels (Ayres et al., 1975). The pathogens are then transmitted to humans after consumption of the shellfish.

While some diseases associated with consumption of contaminated shellfish such as typhoid fever and paratyphoid associated with *Salmonella* are no longer common, the presence of other harmful organisms remains a problem for public health officials worldwide. Even with today's relatively high standards of sewage treatment, significant quantities of potentially harmful micro-organisms may be discharged into areas of shellfish harvest and/or culture.

Most bacteriological diseases can be avoided through depuration and/or adequate cooking or marinating of shellfish (van den Broek et al., 1979). Although mussels are usually eaten cooked, it is the custom in some countries to consume them raw, which increases the potential for disease transmission. It is also a common practice in some areas to cook mussels only until the shell valves are open, which is not adequate to kill all bacteria present.

Because of the associated disease hazards to humans due to bacterial contamination, *Mytilus edulis* has been investigated by bacteriologists for over a hundred years (Dodgson, 1928; Al-Jebouri and Trollope, 1981). Estimates of accumulation of contaminants (bacteria and viruses) are based on indicator organisms, most notably the nonpathogenic *Escherichia coli*, the standard accepted indicator of faecal contamination (Escherich, 1885; Bernard, 1973, 1989). The unreliability of this standard as an indicator of other pathogens, especially viruses, will be discussed later.

Numerous species of bacterial and viral contaminants have been identified from various species of mussels (Table 11.1). This list is not intended to be all-inclusive and it should be remembered that *any* water-borne contaminant is a potential hazard. Brisou et al. (1962) isolated 44 strains of vibrios from *Mytilus galloprovincialis* from the Algerian coast, only some of which are considered pathogenic, and there is clear evidence that some strains of naturally-resident aquatic bacteria are capable of causing gastroenteritis, systemic infections and intoxications in man.

Contamination from pathogens associated with terrestrial soil, fresh and marine waters include bacteria of the genus *Vibrio*. The most important of these are *V. vulnificus*, *V. cholerae* non-01 (NAG *Vibrio*) and *V. cholerae*) group 1 and other

Table 11.1. Known contaminants of mussels. Species names are as they appear in original publications.

Species	Contaminant	Disease	Location	Reference
Aulacomya ater *Mytilus edulis*	*Vibrio parahaemolyticus*	Gastroenteritis	U.S.A., Japan, Argentina, Netherlands	Kampelmacher et al. ,1972; Casellas et al.,1977; van den Broek et al. ,1979
Aulacomya ater *Mytilus platensis*	*Vibrio alginolyticus* *Vibrio parahaemolyticus*		Argentina	Casellas et al., 1977
Crenomytilus *grayanus*	*Photobacterium* *Aeromonas* *Vibrio* *Flavobacterium* *Pseudomonas* Enterobacteriaceae *Bacillus* Coryneforms Mycelial fungi; *Trichoderma, Penicillium*		U.S.S.R.	Mikhailov et al., 1988
Mytilus edulis	not isolated	Hepatitis A	England	Bostock et al., 1979
Mytilus edulis	Hepatitis–A virus	Acute viral hepatitis	Australia	Deinstag et al., 1976
Mytilus edulis	*Klebsiella pneumoniae* *Enterobacter cloccae* *Enterobacter agglomerans* *Escherichia coli* *Shigella dysenteriae* *Yersinia enterocolitica*		U.K.	Al–Jebouri and Trollope,1978
Mytilus edulis	*Citrobacter* *Enterobacter* *Escherichia coli* *Salmonella* *Yersinia* *Klebsiella*		Spain	Ledo et al., 1983
Mytilus edulis	*Proteus* *Serratia*			
Mytilus edulis	*Escherichia coli* *Streptococcus faecalis* *Salmonella anatum*		France	Plusquellec et al., 1990
Mytilus edulis	*Bacillus subtilis* *Serratia marcescens*		U.K.	Al–Salihi and Trollope, 1978
Mytilus edulis	*Aeromonas hydrophila*		U.K.	Trollope, 1984

Table 11.1. continued

Species	Contaminant	Disease	Location	Reference
Mytilus edulis	*Acinetobacter calcoaceticus* var. *lwoffi* Bacillus spp. Campylobacter jejuni Clostridium perfringens Corynebacterium sp. Escherichia coli Enterobacter cloacae Erwinia herbicola Faecal streptococci Flavobacterium sp. Klebsiella pneumoniae Kurthia sp. Micrococcus sp. Pasteurella spp. Pseudomonas spp. Salmonella hadar Serratia sp. Shigella dysenteriae Staphylococcus sp. Vibrio parahaemolyticus Yersinia enterocolitica			
Mytilus galloprovincialis	faecal coliform		Yugoslavia	Krstulović and Šolić, 1988
Mytilus galloprovincialis	Echo virus 5,6,8,12 Coxsackie virus A18		Italy	Bendinelli and Ruschi, 1969
Perna canaliculus	Coxsackie virus B4 CB5 virus Polio viruses 1,2 and 3		New Zealand	Lewis et al., 1986
Mussels	*Vibrio alginolyticus* *Vibrio parahaemolyticus*		Netherlands	Kampelmacher et al., 1972
Mussels	Faecal coliforms; *Escherichia coli*		cosmopolitan	see: Dodgson, 1928; Wood, 1957; Volterra and Tosti. 1983; Power and Collins, 1990
Mussels	Faecal coliforms; *Escherichia coli*		cosmopolitan	see: Dodgson, 1928; Wood, 1957; Volterra and Tosti, 1983; Power and Collins, 1990

Table 11.1. continued

Species	Contaminant	Disease	Location	Reference
Mussels	*Vibrio cholerae*	Cholera	Italy	Baine et al., 1974
Mussels	Polio virus 3		Italy	Petrilli and Crovari, 1965
Mussels	Echo virus 3,9 and 13		Italy	Bellelli and Leogrande, 1967
Mussels	Coxsackie virus A18		France	Denis, 1973
Shellfish	*Salmonella typhi* *Salmonella paratyphi* *Shigella* spp. *Vibrio cholerae* *Vibrio cholerae* O-group 1	Typhoid Dysentary Cholera	U.K. Europe	Ayres et al., 1975

Vibrio species (Eastaugh and Shepherd, 1989). The presence of these types of bacteria is not associated with faecal contamination from human or animal sources, and they are not detected by standard methods of monitoring for bacterial contamination. Furthermore, vibrios are not always eliminated from shellfish using standard commercial decontamination techniques (Richards, 1985; Eastaugh and Shepherd, 1989).

There have been few studies concerned with the uptake of bacteria by mussels. Obviously, mussels located nearest to sewage outfalls contain the highest numbers of bacteria; however, bacteria have been noted in animals considerable distances from outfall areas. Trollope and Al-Salihi (1984) further demonstrated that mussels from the seabed contained higher numbers of *E. coli* than did mussels immersed just below the surface. It has also been shown that *E. coli* seems to accumulate in *M. edulis* to a smaller extent than other coliforms (Webber, 1982). In general, the uptake of bacteria by mussels, held in areas subject to constant levels of water-borne sewage pollution, will be greater during the summer than the winter.

Bernard (1989) showed that rates of uptake and elimination of bacteria by bivalve molluscs are species-specific and temperature dependant and further, that there is little correlation with ambient loading. Of four bivalve species, *M. edulis* attained the highest accumulations of coliforms, at a higher rate than the other species, and also eliminated them more effectively (Fig. 11.1). In a more recent study, Plusquellec et al. (1990) followed bacterial contamination of *M. edulis* both in the laboratory and under natural conditions. They showed that concentration of bacteria by mussels is influenced by the bacterial species, particle density in the surrounding seawater and season (Fig. 11.2).

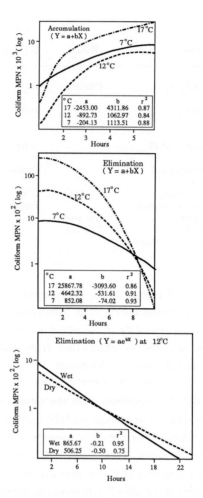

Fig. 11.1. Accumulation and elimination of *Escherichia coli* per g dry meat weight for *Mytilus edulis* as a function of time at three different temperatures and during wet and dry storage. Regression equations and constants are given in the figures; r^2 is the coefficient of determination; MPN = most probable number/100mL. (After Bernard, 1989).

It has been demonstrated that the bacterial load and retention values within mussels vary between tissues, with the digestive gland containing more than 75% of the bacteria (Al-Jebouri and Trollope, 1979). In a later study these authors showed that dissection of the bacteriologically rich digestive gland significantly increased the sensitivity of the *E. coli* detection (3–6 fold enhancement), even when lightly polluted mussels were used. Trollope and Webber (1977) reported a range of 10–87% retention values and a mean of 60 ± 25.35%. Retention by individual tissues also varies:

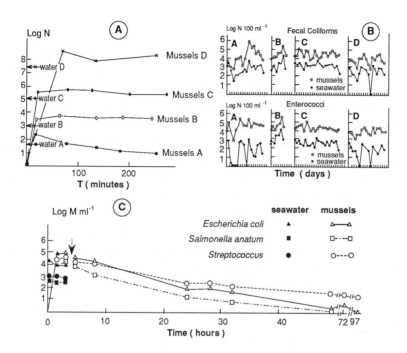

Fig. 11.2. 1: Influence of *Escherichia coli* density in seawater on mussel flesh contamination. (A, B, C, D).

2: Results of daily counts of faecal coliforms and Enterococci in seawater and mussel flesh. A: May–June 1984; B: October–November 1984; C: February 1985; D: May–June 1985.

3: Elimination of bacteria from contaminated mussel tissue. ↓ indicates the transfer to pure running seawater. (After Plusquellec et al., 1990).

intestine, > stomach, > residual tissue, > gill, ≅ mantle, ≅ labial palp, > foot (Al-Jebouri and Trollope, 1978) and Minet et al. (1987) also demonstrated that greater amounts of bacteria are always found in the hind gut. While this differential distribution may provide interesting data for microbiologists, it is of little consequence to the consumer since mussels are consumed whole. In another effort to increase the applicability of *E. coli* tests to public health monitors, Krstulović and Šolić (1988) demonstrated no essential difference in faecal coliform concentration in shellfish flesh alone and those in flesh plus intervalvular fluid. They recommended the use of flesh plus fluid since (1) the coefficient of correlation with the growing water is slightly higher, especially in more polluted waters; (2) the method is simpler since the flesh does not have to be separated from the fluid, and (3), both flesh and fluid are normally consumed.

Viruses

Metcalf (1978) pointed out that at least 66 enteroviruses of human origin might be expected in shellfish-growing areas. These include 3 polioviruses, 24 coxsackievirus A types, 6 coxsackievirus B types and 33 echoviruses. A total of more than 100 viruses might be involved if enteroviruses of animal origin are included. Gerba and Goyal (1978) estimated the number to be more than 100 in human faeces alone including enteroviruses (polio, coxsackie, echo), reoviruses, adenoviruses, infectious hepatitis and rotavirus. Theoretically, any virus excreted in faeces or urine, and capable of producing infection when ingested, could be transmitted by inefficiently treated water (IAWPRC, 1983) and consequently, accumulated by filter-feeding shellfish. These viruses can cause such illnesses as fever, paralysis, meningitis, respiratory disease, diarrhoea and others, and can range from the trivial to the fatal (Feacham et al., 1982). The role of shellfish as vectors of human enteric diseases has been well-documented (Gerba and Goyal, 1978) and luckily, only a few viruses have been shown epidemiologically to be transmitted by shellfish. These include hepatitis A, non-A, non-B hepatitis, Norwalk, Snow Mount agent, astroviruses, coxsackievirus and small round viruses.

Hepatitis A and Norwalk viruses are of chief concern to public health officials, although many others are present and pose potential health hazards and are most commonly associated with oysters (Portnoy et al., 1975; Murphy et al., 1979; Noble, 1990). Nearly all outbreaks of disease associated with excreted viral contamination of shellfish are outbreaks of hepatitis A or viral gastroenteritis (Feacham et al., 1982). To date, no outbreaks of hepatitis B associated with shellfish have been reported, although shellfish in Maine (U.S.A.) waters have been found to carry hepatitis B antigen (Konno et al., 1982). Dienstag et al. (1976) confirmed (both epidemiologically and serologically) acute viral hepatitis from incompletely cooked mussels (*M. edulis*) in Australia. Again, while adequate cooking will destroy most viruses, the habit of eating raw shellfish increases the likelihood of disease outbreak.

Viruses can survive for extended periods of time outside an animal host (see Akin et al., 1975), and can remain infectious for several weeks or longer after discharge into receiving waters. Once inside a shellfish, their survival appears to be further prolonged (Metcalf and Stiles, 1965). Enteric viruses can survive from a few days to over 130 days in marine water (see Akin et al., 1975) and it has been demonstrated that viruses survived for longer periods when in raw sewage (Metcalf and Stiles, 1965). Virus survival in seawater is dependent on temperature, salinity, type of virus, bacterial antagonism, suspended solids and pollution (Gerba and Goyal, 1978); and survival of enteroviruses in sea water is generally reported to be shorter than in fresh water, but they do survive longer in seawater than do coliform bacteria (Feacham et al., 1982). Temperature appears to be the prime determining factor in viral survival in seawater, with increased inactivation in warmer waters.

Lack of correlation between the depuration of viruses and bacteria has been clearly demonstrated for several species of shellfish (Scotti et al., 1983). It has been calculated that enteric viruses can survive in mussel tissue 3–6 times longer than coliform bacteria, and, in several studies, enteroviruses have been isolated from shellfish otherwise having a satisfactory coliform index (Gerba and Goyal, 1978). Gerba et al. (1975, 1979) reported that no correlation could be found between the presence of entroviruses in marine waters and indicator bacteria; often enteroviruses were found in waters that met current bacteriological standards. In one instance, shellfish (oysters) harvested from a 'clean' area caused a hepatitis A outbreak (Portnoy et al., 1975).

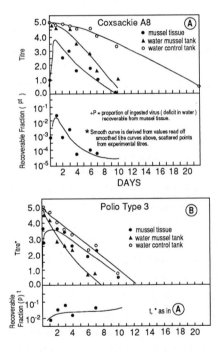

Fig. 11.3. Changes in titres for Coxackie A8 (A) and Polio Type 3 (B) in *Mytilus edulis aoteanus*. A: Titres are expressed as dose concentrations in suckling mice LD_{50s} per 0.03mL. B: Titres are expressed as dose concentrations in tissue culture ID_{50s} per 0.2mL. When virus was just recoverable but not titratable, the point is plotted just below the abscissa. (After Duff, 1967).

Uptake of viruses by mussels has been clearly demonstrated for three species: *M. edulis* (Dimmock, 1967), *M. galloprovincialis* (Milo, 1971), and *Mytilus edulis aoteanus* (Duff, 1967), now considered to be *M. galloprovincialis* (McDonald et al., 1991). Like bacteria, the majority of viruses are concentrated in the digestive system of the mussels. Duff (1967) demonstrated that polio type 3 virus was significantly more resistant to inactivation capacity of the mussel tissue than coxsackie A8 virus and that maximum infectivity occurred approximately 18-36h after pollution (see Fig. 11.3).

Finally, several authors have suggested that particle association significantly extends the survival capacity of viruses, and enhances their potential for interaction with local marine organisms (Landry et al., 1980, 1983; Liew and Gerba, 1980). Retention of viruses by gill structures is enhanced by adsorption of viruses to fine particulate matter (Duff, 1967). This may have special significance for mussels which are known to feed heavily on resuspended organic matter (Kiørboe et al., 1980; Lucas et al., 1987; Fréchette et al., 1989). The incidence of viral accumulations in the oyster, *Crassostrea virginica* and the clam, *Mercenaria mercenaria* was increased when the sediments (viruses) were resuspended in the water column (Landry et al., 1983). While increased viral accumulation in mussels has not yet been demonstrated in mussels exposed to resuspended sediments, it seems a likely cause for concern, and public health officials need to be more aware of the potential hazards associated with viral contamination of mussels.

Toxic Algal Blooms

Blooms of toxic algal species are common occurrences in shellfish growing areas worldwide and pose a severe threat to public health. The toxins associated with these algae are potent, and the filter-feeding shellfish accumulate the toxic cells during feeding, thus rendering them vectors in various forms of shellfish poisoning including: paralytic shellfish poisoning (PSP), diarrhetic shellfish poisoning (DSP) and amnesic shellfish poisoning (ASP). Of all shellfish consumed, mussels probably pose the greatest threat with regard to shellfish poisoning. Very early cases were noted but associated merely with 'toxic action on the nerves'. Many fatalitites were noted (see Dodgson, 1928) and the public was warned against taking mussels from 'foul and stagnant water' as they become 'intensely poisonous', and no association was made with plankton blooms. Both ASP and PSP can prove fatal, whereas DSP is easily confused with gastroenteritis and general stomach upsets associated with eating shellfish or contaminated shellfish. Sensitivity to PSP is highly variable and estimates of the lethal dose for humans range from 0.3–1.0mg of saxitoxin (Tennant et al., 1955; Schantz, 1973; Eastaugh and Shepherd, 1989). Severe symptoms have occurred with ingestion of as little as 124µg of toxin. Death has resulted from ingestion of only 456µg (Music et al., 1973).

Amnesic shellfish poisoning is a novel and devastating form of a previously unknown marine toxin. It has been attributed to a bloom of *Nitzschia pungens* f.

multiseries in Canada (Bates et al., 1988). Acute illness was characterized by gastrointestinal symptoms and unusual neurologic abnormalities, including loss of short-term memory. Some patients required intensive care due to seizures, coma, profuse respiratory secretions or unstable blood pressure (Perle et al., 1990). Domoic acid, which can act as an excitatory neurotransmitter, was identified as the causative agent, and has been described recently by Teitelbaum et al., (1990). An intense monitoring program was established in Canada after the first outbreak (Bates et al., 1988, 1989), and no new cases have occurred since December 1987.

All of these forms of shellfish poisoning have been associated with mussels and the topic has been recently reviewed by Shumway (1990). Table 11.2. summarizes the effects of both toxic and noxious algal blooms and their effects on mussels. No geographic area seems immune from possible blooms of toxic species, and many outbreaks of PSP cases occur in areas where there are no monitoring programmes, or amongst picnickers who ignore posted warnings. Government agencies, including food and drug, fishery and public health groups, have taken measures for many years to prevent toxic shellfish from getting into commercial markets and to warn the general public against collecting for their own use. Since the toxins are not inactivated by cooking and there are no known antidotes, mussels should only be eaten from areas known to be monitored regularly for the presence of toxins by an authorized and recognized public health agency.

Allergy

Halstead and Schantz (1984) described an allergic form of shellfish poisoning manifested by severe allergic reaction. The incubation period is usually short (a few hours) and the symptoms consist of "a diffuse erythema, swelling, and urticaria of the face and neck, but may involve the entire body (the rash may be accompanied by a severe itching); headache, sensation of warmth, conjunctivitis, coryza, gastric distress, dryness of the throat, swelling of the tongue and respiratory distress may be present". Patients usually recover within a few hours, but death may occur.

One of the best accounts of an allergic reaction to mussels is given by Dodgson (1928). The symptoms associated with an allergic reaction to mussels were so well-known that 'musselling' was a common term for any indisposition caused by the consumption of various foods. Symptoms were usually noted within a short period of time ranging from five minutes to several hours, and consisted of a red rash or 'nettle-rash', sometimes accompanied by intense itching. Some patients suffered breathing distress, vomiting and/or diarrhoea while others experienced little discomfort other than rash. Recovery was usually complete in less than 12 hours.

Table 11.2. A summary of toxic and noxious algal blooms and their effects on shellfish. Taxonomic nomenclature is as it appears in the original publications.

Shellfish species affected	Algal species	Notes	Location	Reference
Aulacomya ater *Mytilus chilensis*	*Gonyaulax catenella*	toxic	Chile	Guzman and Campodonico, 1978; Avaria, 1979
Modiolus auriculatus *Pinna* sp.	*Pyrodinium bahamense*	toxic shellfish; some human	New Guinea	Maclean, 1973, 1975; Worth et al., 1975
Modiolus sp.	*Pyrodinium bahamense*	toxic	Palau, Micronesia	Harada et al., 1982
Mytilus chilensis	*Amphidoma* sp.	mildly toxic	Chile	Campodonico and Guzman,1974; Avaria, 1979
Mytilus coruscus *Mytilus edulis*	*Protogonyaulax tamarensis* *Protogonyaulax catenella*	toxic	Japan	Anraku, 1984
Mytilus coruscus *Mytilus edulis*	*Dinophysis fortii* *Dinophysis acuminata* *Gymnodinium catenatum* *Protogonyaulax tamarensis*	toxic	Japan	Oshima et al., 1982; Anraku, 1984; Ikeda et al., 1989
Mytilus edulis	*Dinophysis acuminata* *Dinophysis acuta*	highly toxic	Netherlands	Kat, 1983, 1985, 1989
Mytilus edulis	*Dinophysis acuta*	DSP	Sweden	Edler and Hageltorn, 1990
Mytilus edulis	*Dinophysis* spp. including *acuta, acuminata, norvegica*	highly toxic; remained toxic for up to 7mo	Sweden, Norway, Denmark	Krogh et al., 1985; Underdal et al., 1985; Yndestad and Underdal, 1985; ICES, 1988
Mytilus edulis	*Dinophysis* spp. *Prorocentrum* sp.	DSP; first report from area	Wadden Sea, Germany	Meixner and Luckas, 1988
Mytilus edulis	*Prorocentrum micans*	40-50% mortality; probably due to low oxygen.	Brittany	Lassus and Berthome, 1988
Mytilus edulis	*Prorocentrum micans*	toxic; PSP	Portugal	Pinto and Silva, 1956
Mytilus edulis	*Gonyaulax tamarensis*	toxic	U.K.	Ingham et al., 1968

Table 11. 2. continued

Shellfish species affected	Algal species	Notes	Location	Reference
Mytilus edulis	Gonyaulax tamarensis Gymnodinium catenatum	toxic; PSP	Spain	Campos et al., 1982; Fraga et al., 1984; Blanco et al., 1985; Fraga and Sanchez, 1985
Mytilus edulis	Gonyaulax acatenella	several cases of PSP	British Columbia	Prakash and Taylor, 1966
Mytilus edulis	Gonyaulax excavata	toxic; shellfish mortalities	Faroe Is.	Mortenson, 1985; Dale et al., 1987; Gaard and Poulson, 1988
Mytilus edulis	Gonyaulax excavata	toxic	Argentina	Carreto et al., 1985
Mytilus edulis	Gonyaulax sp.	toxic	Uruguay	Davison and Yentsch, 1985
Mytilus edulis	Nitzschia pungens f. multiseries	highly toxic; over 106 illnesses and 3 human deaths	Prince Edward Is., Canada	Bates et al., 1988, 1989; Subba Rao et al., 1988; Addison and Stewart, 1989; Smith et al., 1990
Mytilus edulis	Dinophysis spp.	"probably source of DSP"	New York, U.S.A.	Freudenthal and Jijina, 1988
Mytilus edulis planulatus	Gymnodinium catenatum	toxic	Tasmania	Hallegraeff and Summer, 1986
Mytilus edulis Modiolus modiolus	Gonyaulax tamarensis (Protogonyaulax)	highly toxic	Gulf of Maine and E. Canada; Bay of Fundy; St. Lawrence regions	Prakash, 1963; Caddy and Chandler, 1968; Prakash et al., 1971; Hartwell, 1975; Hurst, 1975; Tufts, 1979; Shumway et al., 1988
Mytilus edulis Mytilus californianus	Gonyaulax catenella	toxic	California and Pacific coast states, U.S.A.	Sharpe, 1981; Nishitani and Chew, 1988
Mytilus edulis galloprovincialis Mytilus coruscus	not specified but probably Protogonyaulax spp.	toxic	Korea	Jeon et al., 1988
Mytilus sp.	Dinophysis sacculus Gymnodinium catenatum	DSP; first report from area; PSP outbreaks	Portugal	Franca and Almeida, 1989; Alvito et al., 1990

Table 11.2. continued

Shellfish species affected	Algal species	Notes	Location	Reference
Perna perna	*Protogonyaulax tamarensis Gonyaulax monilata*	toxic	Venezuela	Ferraz–Reyes et al., 1985
Perna perna	*Cochlodinium* sp.	symptoms similar to PSP; several fatalities; many illnesses	Venezuela	Reyes–Vasquez et al., 1979
Perna perna	*Gonyaulax tamarensis*	first record from Caribbean; 1 human fatality	Venezuela	Reyes–Vasquez et al., 1979
Perna viridis	*Protogonyaulax tamarensis*[1]	63 cases of PSP; 1 human fatality.	Pran Buri, S. Thailand	Tamiyavanich et al., 1985; Maclean, 1984
Perna viridis	*Pyrodinium bahamense*	highly toxic	Brunei, Philippines	Beales, 1976; Arafiles et al., 1984; Jaafar and Subramaniam, 1984; Gacutan et al., 1985; Gonzales et al., 1989
Perna viridis	*Pyrodinium bahamense*	several human fatalities; mostly juveniles.	Philippines	Estudillo and Gonzales, 1984
Mussels	*Alexandrium minutum*	PSP; first record from area.	France	Nezan et al., 1990
Mussels	*Alexandrium tamarensis Alexandrium acatenella*	toxic	Kamchatka, U.S.S.R.	Konovalova, 1989

[1] In a later study (Kodama, 1985) it was demonstrated that the strains of *Protogonyaulax tamarensis* in this area are nontoxic and that the toxicity exhibited by shellfish is due primarily to *P. cohorticula*.

DEPURATION

In many areas depuration is mandatory prior to marketing of shellfish, and in some areas only those shellfish which are cultured, or meant for export, are depurated. In others, depuration needs to be made a routine part of the cultivation process. Some regions are noted for unsanitary measures in the culture and harvesting of mussels, and this has resulted in lack of consumer demand. The two types of depuration processes available to cleanse shellfish contaminated with pathogenic viruses or bacteria are: (1) relaying to clean water and (2), treating with disinfectants, including ultraviolet light (UV), chlorine and ozone (see Blogoslawski, 1983, 1989, 1990; Blogoslawski and Stewart, 1983; Richards, 1988).

The simplest method of cleansing is to move the contaminated shellfish to unpolluted waters (relaying) or to maintain them in sterilized waters under controlled conditions (depuration). Details of plant construction and operation have been given by Furfari (1966), Canzonier (1984), the NSSP Manual, Part II (USDHHS, 1986; USPHS, 1988) and most recently by Howell and Howell (1989). As Dodgson (1928) so aptly put it, "The mussel is as diligent and successful in cleansing itself in favourable conditions as it is in polluting itself in unfavourable ones". Based on this ability to purge themselves, Dodgson established the first mussel purification plant at Conwy, Wales (U.K.), and showed that depuration is an effective method for reducing the microbial flora of contaminated shellfish. Since that time, relaying has been an accepted method for reducing the potential risk of public health hazards (Wood, 1969; Ledo et al., 1983), and this method remains one of the best for reducing public health risks associated with contaminated shellfish. Relaying or self-depuration technology has remained essentially unchanged since first established by Dodgson (Power and Collins, 1989).

Areas used for relaying (depuration in a natural setting (Richards, 1988)) are frequently closed by the regulating agency for various lengths of time until the relayed shellfish are deemed safe to harvest. This method uses waters from approved shellfish harvest areas, and has the advantage of being comparatively inexpensive, and the disadvantage of having a recovery rate of only 50% of the relayed shellfish. The major advantage of this method is that relayed shellfish are only required to meet open area shellfish bacterial standards. The major drawback to relaying is that these systems are labour-intensive and are thus frequently unfeasible (Blogoslawski, 1989). Canzonier (1988) discussed the many drawbacks associated with depuration which include: variable efficacy, unfeasibilty in the case of very heavy bacterial loads, virtual uselessness in reducing contaminants such as hydrocarbons and heavy metals, economic unfeasibility in some cases, lack of control over viral contaminants and potential conflicts with watermen.

Bacterial pathogens can be eliminated from seawater by treatment with UV light, and this system is used commonly to sterilize seawater for depuration of bivalve shellfish. Depuration in the U.S.A. is exclusively with UV light disinfection (Richards, 1988). In U.S. depuration plants the sterilized water must meet drinking water standards for bacteria, i.e. <1 MPN (most probable number)/100mL for coliforms. Water used for depuration in the U.S.A. must come from areas that do not exceed moderate pollution levels i.e. <700 coliforms/100mL. Chlorine has also been used to disinfect seawater which must then be dechlorinated before it can be used to depurate contaminated shellfish. Although this method is more costly than other forms, it is still the method of choice in many depuration facilities (France, Spain, England) because of its reliability. While artificial purification of mussels is widely practiced in Europe, mussels are not currently depurated in the U.S.A.

Table 11.3. Approximate times of contaminant retention for various species of mussels (represents time taken for levels to fall below either quarantine or detection levels).

Species	Source of contamination	Retention time	References
Choromytilus meridionalis	*Gonyaulax catenella*	3mo	Popkiss et al., 1979
Modiolus auriculatus	*Pyrodinium bahamense*	6 weeks	Worth et al., 1975
Modiolus modiolus	*Gonyaulax tamarensis*	up to 60 days[†]	Gilfillan et al., 1976
Mytilus californianus	*Gonyaulax catenella*	<1mo	Sommer and Meyer, 1937; Sharpe, 1981
Mytilus edulis	*Protogonyaulax tamarensis*	10 days–7 weeks up to 50 days	Oshima et al., 1982; Gilfillan et al., 1976; Prakash et al., 1971
	Gonyaulax acatenella	11 weeks 4 weeks	Quayle, 1965 Sharpe, 1981
	Gonyaulax excavata	2–3 weeks	Gaard and Poulsen, 1988
	Dinophysis spp.	1 week	Haamer et al., 1990
Mytilus edulis	*Escherichia coli*	<24h	Bernard, 1989
Mytilus edulis	Polio virus 2	48h	Crovari, 1958
Mytilus edulis	*Escherichia coli Salmonella anatum*	4 days	Plusquellec et al., 1990

[†] Dependant on initial level of toxicity

Ozone is another powerful disinfectant that does not leave harmful chemical residues, as does chlorine, and is now the depuration method of choice in major shellfish-cleansing stations in France. While ozone has been tried as a means of detoxifying shellfish exposed to paralytic shellfish toxins, it has not been shown to be effective and there are currently no commercial methods available for ridding shellfish of toxins (Blogoslawski, 1988). Heat processing has also been shown to reduce, but not eliminate, toxin levels (Medcof et al., 1947; Prakash et al., 1971). Mussels usually purge themselves of accumulated toxins after blooms of toxic algae subside. Time taken to reach quarantine levels varies between species and ranges from one week to three months (see Table 11.3).

Although mussels are possibly by volume the most frequently depurated shellfish, there have been few studies on the rate of elimination of bacteria by mussels. Wood (1957) showed that mussels continued to purge themselves of faecal coliforms at

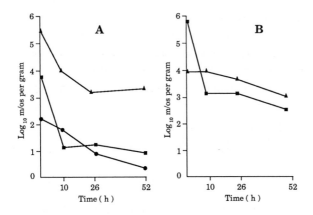

Fig. 11.4. (A) Elimination of poliovirus (plaque forming units (PFU) per gram) (•), *Escherichia coli* 4A (colony forming units (CFU) per gram) (■) and φA1-5a (PFU per gram) (▲). (B) Elimination of *E. coli* 4A (CFU per gram) (■) and φA1-5a (PFU per gram) (▲) from mussels during depuration in the laboratory-scale system. Environmental conditions during depuration were as follows: temperature, 15.5–19.5°C; salinity, 27–29.3°/oo; dissolved oxygen, >60% saturation; pH, 7.4–8.3. (After Power and Collins, 1989).

temperatures as low as 1–2°C, and Trollope and Webber (1976) demonstrated that immersion for 48h was necessary to ensure 100% removal of *E. coli* from mussels. Volterra and Tosti (1983) noted that streptococci were often in higher concentrations than coliforms, and it has been suggested (Wood, 1976) that faecal streptococci may be retained longer than coliforms. Kueh (1987) also demonstrated rapid uptake of bacteria and elimination via the faeces of *Perna viridis*. More recently, Plusquellec et al. (1990) showed that, while a four-day period is necessary to depurate down to undetectable levels of *E. coli* and *Salmonella anatum* in *M. edulis*, this period is not sufficient for a complete elimination of *Streptococcus faecalis*, and that these differences are more marked *in situ* than in laboratory studies.

Little is known of the depuration of viruses by mussels and the literature appears to be limited to three studies. Power and Collins (1989) monitored the elimination of sewage effluent-associated poliovirus, *E. coli* and 22nm icosahedral coliphage by *M. edulis* under both laboratory- and commercial-scale recirculating, UV depuration systems. Their results are summarized in Figure 11.4, and suggest that the organisms are eliminated from mussels by different mechanisms during depuration under stable conditions. The relative rates of elimination during depuration were: *E. coli* 4A >φ A1-5a > poliovirus type 1, regardless of experimental conditions. Spawning appeared to have little effect on the elimination of poliovirus. They also demonstrated slow elimination of viruses from nondigestive tract tissues. Again, these data indicate that conventional depuration practices are inappropriate for efficient elimination of viruses from mussels. Lewis et al. (1986) showed that no significant reduction in viral numbers had occurred after eight days of depuration in *Perna canaliculus*. They also showed no significant correlation between viral and faecal coliform numbers, supporting the contention of many others (Gerba and Goyal, 1978; Ellender et al., 1980; La Belle et al., 1981; Lewis et al., 1985) that faecal coliform numbers are unreliable indicators of the presence of human enteroviruses, i.e. the absence of faecal coliforms is not sufficient to ensure the safe consumption of shellfish (Power and Collins, 1986, 1989).

Properly used, depuration produces high-quality shellfish but lack of understanding of the depuration process has produced shellfish that have caused illnesses. While depuration can increase the marketability of shellfish (Blogoslawski, 1989), it must be remembered that viruses are not necessarily removed by purification procedures (Bryan, 1980, 1986; Power and Collins, 1987). With the increased market for shellfish worldwide, further studies are needed to develop rapid, sensitive analyses for viral contaminants in shellfish.

MONITORING AND REGULATIONS

With the continuing increase in international shipment of shellfish, especially mussels, strict attention must be paid to quality control. Constant surveillance by public health authorities is a necessity if the safe marketing of shellfish is to be assured. Most countries recognize the importance of quality control; however, inadequate sanitary and processing facilities are often the nemesis of shellfish operations, especially in developing countries (see Davy and Graham, 1982 and papers therein).

Any shellfish sanitation programme should include an infrastructure responsible for monitoring, culture and harvesting activities, and should provide adequate surveillance. This infrastructure is usually composed of public agencies. An administrative system, which coordinates the activities of the various agencies responsible to enjoin the prosecution of violators of the programme is the other major component.

Methods employed for sanitary control of molluscan shellfish have been reviewed by several authors (See Wood, 1972; AOAC, 1984; Richards, 1988) and references therein) and will not be dealt with here.

Many agencies work in cooperation with each other, and with the increased international trade in live shellfish, many countries have become involved in cooperative programmes. The United Nations Programme (UNEP) inititated the Regional Seas Programme in 1974. This programme was designed to assess the state of marine pollution, the sources and trends of the pollution and the impact of pollution on human health, marine ecosystems and amenities (UNEP/WHO, 1983a,

Table 11.4. Action levels, tolerances and other values for poisonous or deleterious substances in seafood (from NSSP Manual Part 1, Appendix C 1986)

Deleterious Substances	Level	Food Community	Reference
Aldrin/Dieldrin	0.30ppm	Fish and shellfish	CPG 7120.23–A
Chlordane	0.30ppm	Fish only	CPG 7120.23–C
DDT, DDE, TDE	5.00ppm	Fish only	CPG 7120.23–D
Endrin	0.30ppm	Fish and shellfish	CPG 7120.23–F
Heptachlor/Heptachlor Epoxide	0.30ppm	Fish and shellfish	CPG 7120.23–H
Kepone	0.30ppm	Fish and shellfish	CPG 7120.23–I
	0.40ppm	Crabmeat	CPG 7120.23–I
Mercury	1.00ppm	Fish and shellfish	CPG 7108.07
Mirex	0.10ppm	Fish only	CPG 7120.23–K
Paralytic shellfish poison	80μg/100g of meat	Fresh, frozen and canned clams, mussels and oysters	CPG 7108.20
Polychlorinated biphenyls (PCBs)	20ppm	Fish and shellfish	21 CR 109.30
Ptychodiscus brevis toxins	20 Mouse Units/100g	Shellfish	APHA Lab. Procedures (17)
Toxaphene	5.00ppm	Fish only	CPG 7120.23–L

b; 1988). A set of reference methods and guidelines for marine pollution studies (faecal coliform) have been developed and have been recommended for adoption to governments participating in the Regional Seas Programme .

The National Shellfish Sanitation Programme (NSSP) was established in the U.S.A. in 1925 to establish sanitary control mechanisms for prevention of further outbreaks of shellfish-borne typhoid fever and other diseases of bacterial origin (Hunt, 1972), and to insure that shellfish shipped interstate would not be the cause of communicable disease. In the 1940s, steps were taken to protect the public against paralytic shellfish poisoning, and in 1957 radionuclides were added to the list of possible contaminants of shellfish. In the 1960s and 1970s it became apparent that shellfish also concentrate other poisonous and/or deleterious substances, including metals, pesticides, hydrocarbons and others, to potentially unsafe levels (see Chapters 8 and 9). It is the responsibilty of each individual State to supervise the growing, harvesting, relaying and transportation of the shellfish. NSSP is a voluntary programme which encourages states to adopt shellfish sanitation regulations based on federal agency recommendations. Table 11.4. gives the action levels and tolerance values allowed for poisonous or deleterious substances for which standards exist in the U.S.A. (NSSP Manual, USPHS, 1988). At this time, bacterial standards for depurated mussels have not been set by NSSP. When a standard is established, it can be expected to be similar to those set for softshell clams, quahogs and oysters. Routine testing, or monitoring of viruses in shellfish or their waters, is not carried out or recommended due to the technical complexity, time required, high cost and limitations of the detection and recovery methods (USDHHS Manual, 1986; USPHS Manual, 1988). Conspicuous by its absence from this list is okadiac acid or diarrhetic shellfish poisoning (DSP). This is a recent phenomenon and individual countries have set their own tolerance levels and accepted methods of analysis (Table 11.5). Although DSP has not been positively identified from U.S. waters (Stamman et al., 1987), its presence is suspected (see Freudenthal and Jijina, 1988; Shumway, 1990). Acceptable methods for determinating levels of contamination have not as yet been arrived at.

The NSSP was reorganized in September 1982 and is now titled Interstate Shellfish Sanitation Conference (ISSC). The ISSC establishes guidelines for shellfish sanitation standards, which are published in a Manual of Operations (NSSP, 1989a, b), available to all interested parties. The Manual describes in detail how a programme should be operated in any member state, describes the interrelation of member state programmes, and cites the criteria to be applied by the Food and Drug Administration (FDA) in evaluating the programmes (Canzonier, 1988).

ISSC is a state, federal and industrial cooperative and includes shellfish sanitation control agencies in 22 States of the U.S.A., Canada, the Hiroshima Prefecture of Japan, and also shellfish industry organizations in these countries. It is administered by the FDA's Shellfish Sanitation Branch (Hunt, 1972).

Shellfish sanitation and the guarantee of a safe product for human consumption is an international problem. To this end, the U.S. Department of Health and Human Services, the Public Health Service and the FDA regularly publish an interstate certified shellfish shippers list. The shippers listed have been certified by regulatory authorities in the U.S.A., Australia, Canada, Japan, the Republic of Korea, Iceland, Mexico, England and New Zealand under the uniform sanitation requirements of the National Shellfish Programme, under the terms of the shellfish sanitations agreements with the governments of these countries (USPHS, 1990). Control measures of the states are evaluated by the FDA.

'Interstate Certified Shellfish Shippers List' (ISSN 0364–7048) is published monthly for the information of, and use by, food control officials, seafood industry and other interested persons (USPHS, 1990). The publication is distributed under authorities of the authorities of the Public Health Service Act and Food, Drug and Cosmetic Act by the U.S. Food and Drug Adminstration, 200 'C' Street, Washington, D.C. 20204.

Table 11.5. Diarrhetic shellfish poisoning (DSP) tolerance in various countries (from Krogh, 1992).

Country	Tolerance	Method of Analysis
Denmark	No detectable amount	Mouse bioassay: rat bioassay[a]
Germany	No detectable amount	Rat bioassay
France	0.2–0.4MU/g digestive glands	Mouse bioassay[b]
Ireland	No detectable amount	Rat bioassay (HPLC for confirmation)
Japan	5MU/100g soft tissue	Mouse bioassay[d]
Netherlands	No detectable amount	Rat bioassay
Norway	5–6MU/100g soft tissue	Mouse bioassay
Portugal	No detectable amount	Mouse bioassay
Spain	No detectable amount	Mouse bioassay[c]
Sweden	60µg/100g soft tissue	HPLC mouse bioassay[e]

[a] Employed for export commodities of shellfish to countries requiring this method of analysis

[b] A modified version, with shorter observation period, so the Mouse Units (MU) cannot be compared to those of the original Japanese method used in other countries

[c] No clean-up with either of the shellfish extract

[d] Okadaic acid + DTX-1

[e] Employed for export shellfish commodities

Shellfish programmes vary from state to state (see Broutman and Leonard, 1988; Leonard et al., 1989; Leonard and Slaughter, 1990) and country to country; however, the concern of all is the provision of a safe product. Any state which plans to ship their products interstate must conform to the ISSC (NSSP). Currently, the ISSC (NSSP) manual for interstate shipment stipulates that all growing areas must be certified, commercial shellfish harvesters must be licensed, and the processors and distributors be certified. All products can then be traced to their point of origin by either harvester number or distributor number on the shipping tag (Hungerford and Wekell, 1992). While the ISSC has generally provided a safe shellfish market there is still a need for more effective measures for monitoring the safety of shellfish products, particularly with reference to the presence of viruses (Subcommittee on Microbiological Criteria et al., 1985).

Table 11.6. Microbiological standards for shellfish and shellfish growing areas fixed by Italian regulation (DM, 1978) (from Bonadonna et al., 1990)

Areas	Shellfish		Water	Shellfish destination
	$E.\ coli$ mL^{-1}	*Salmonella*	$E.\ coli$ 100mL^{-1}	
Approved	4	absent–25 mL^{-1}	2 (10% of samples 7)	Purification treatment
Conditionally approved	39		34 (10% of samples 49)	Food preservation industry
Prohibited	>39		>34	

The Subcommittee on Microbiological Criteria, the Committee on Food Protection, the Food and Nutrition Board and the National Research Council (1985) have proposed that the Hazard Analysis Critical Control Point system (HACCP) be implemented to provide a "more specific and critical approach to the control of microbiological hazards in foods than that provided by traditional inspection and quality control approaches". First introduced at the 1971 National Conference on Food Protection (APHA, 1971), this system consists of three units: (1) identification and assessment of hazards associated with growing, harvesting, processing, marketing, preparation and use of a given raw material or food product; (2) determination of critical control points to control any identifiable hazard and (3), establishment of systems to monitor critical control points. Strong emphasis on the application of the HACCP system was given by the WHO Expert Committee on Microbiological Aspects of Food Hygiene (1976), and it has been suggested that discussion of the HACCP system be incorporated into appropriate WHO training programmes (WHO/ICMSF, 1985).

Other countries have establish their own criteria for shellfish water standards. Italian regulations (DM, 1978) distinguish three classes of areas where shellfish may be harvested and collected (Table 11.6). A microbiology network dates from April 1989 in France, and is focused on assessing the level and tendencies of bacteriological contamination in the marine environment as measured in shellfish used for integration and in consumer protection (Berthome, unpublished results). In addition, France has a monitoring and warning network for phytoplankton consisting of 100 sampling stations, 38 of which are sampled systematically throughout the year, the remainder being supplemental monitoring stations if toxic species occur (Berthome, unplublished results). France's extensive monitoring programme has been reviewed by Furfari and Hunt (1981).

Monitoring programmes for PSP, DSP and ASP are well-established in many countries. The United States' programme has been recently reviewed by Hungerford and Wekell (1992). Other countries with extensive monitoring programmes include France, Spain, Japan, Canada and most recently, Tasmania (see Shumway, 1990). More monitoring programmes are urgently needed, especially in developing countries, where primitive culture facilities are common and technical assistance may be lacking.

CONCLUSIONS

As the global culture and transportation of mussel species continues to grow, the public health aspects cannot be ignored. While improved inspection systems, laws and regulations governing the shellfishing industry, and monitoring programmes have all contributed to the decline in shellfish-borne illnesses, many outbreaks still occur. Cooking is not always sufficient to insure complete inactivation of infectious particles from heterogeneous and homogeneous populations (Milo, 1971), and it is impossible to be confident in our ability to certify filter-feeding molluscs which are to be consumed raw. In addition, viruses can be present in shellfish that have been certified 'clean' based on *E. coli* measurements. More outbreaks occur from home collections and consumption in private homes and beaches than from commercial suppliers, and no amount of regulation can stop these incidences. To prevent unnecessary outbreaks of shellfish-borne diseases, shellfish should be obtained only from approved, certified sources and never harvested from waters contaminated with raw sewage. The shellfish should be thoroughly cooked, not 'quick-steamed', as destruction of the viruses may not be complete otherwise. Increased public education and awareness are a must if disease outbreaks are to be curtailed.

As summarized by Richards (1988), more research is urgently needed to: (1) develop more sensitive, reliable and universally accepted assay techniques for virus anaylses; (2) delineate the role of environmental parameters on the depuration process; (3) reassess the usefulness of indicator organisms (e.g. *E. coli*) as predictors of overall shellfish quality and safety and (4), standardize depuration research to include internal viral and/or bacterial controls coupled with inter- and intralaboratory comparisons. To this list we would add implementation of the HACCP system to assure the quality of shellfish and provide a uniform means of testing and reporting contamination levels.

ACKNOWLEDGEMENTS

I am indebted to our librarian, P. Shephard–Lupo, without whom this review would not have been possible. I thank J. Rollins for preparing the figures and J. Barter for preparing the text for publication. J. Hurst provided many helpful discussions.

REFERENCES

Acres, J. and Gray, J., 1978. Paralytic shellfish poisoning. Can. Med. Assoc. J., 119 : 1195-1197.

Addison, R.F. and Stewart, J.E., 1989. Domoic acid and the eastern Canadian molluscan shellfish industry. Aquaculture, 77: 263-269.

Akin, E.W., Hill, W.F., Jr. and Clarke, N.A., 1975. Mortality of enteric viruses in marine and other waters. In: A.L.H. Gameson (Editor), Discharge of Sewage from Sea Outfalls. Pergamon Press, New York, pp. 227-235.

Al-Jebouri, M.M. and Trollope, D.R., 1978. The enumeration of enterobacteria from *Mytilus edulis* using CLED medium. Soc. Gen. Microbiol. Quart., pp. 29.

Al-Jebouri, M.M. and Trollope, D.R., 1979. The effects of season and site on the numbers of enteric bacteria from mussel organs: an analysis including multivariate analysis of variance. J. Appl. Bacteriol., 47: xi.

Al-Jebouri, M.M. and Trollope, D.R., 1981. The *Escherichia coli* content of *Mytilus edulis* from analysis of whole tissue or digestive tract. J. Appl. Bacteriol., 51: 135-142.

Al-Salihi, S. and Trollope, D.R., 1978. The uptake of *Serratia marcescens* and *Bacillus subtilus* var. *niger* by *Mytilus edulis* on a sewage-polluted shore. Soc. Gen. Microbiol. Quart., 6: 29.

Alvito, P., Sousa, I., Franca, S. and Sampayo, M.A. de, 1990. Diarrhetic shellfish toxins (DSP) in bivalve molluscs along the coast of Portugal. In: E. Graneli, D.M. Anderson, L. Edler and B.G. Sundstrom (Editors), Toxic Marine Phytoplankton. Elsevier Science Publishers, B.V., New York, pp. 443-448.

Anraku, M., 1984. Shellfish poisoning in Japanese waters. In: A.W. White, M. Anraku and K.-K. Hooi (Editors), Toxic Red Tides and Shellfish Toxicity in Southeast Asia, Proceedings of a Consultative Meeting, Singapore, 1984. Southeast Asian Fisheries Development Center and the International Development Research Centre, Singapore, pp. 105-109.

A.O.A.C. 1984. Official Methods of Analysis. In: S. William (Editor), Assoc. Offic. Anal. Chem., Arlington, VA., pp. 58-60.

APHA (American Public Health Association), 1971. Proceedings of the 1971 National Conference on Food Protection. Washington, D.C.: U.S. Department of Health, Education and Welfare, Public Health Service, Food and Drug Administration.

Arafiles, L.M., Hermes, R. and Morales, J.B.T., 1984. Lethal effect of paralytic shellfish poison (PSP) from *Perna viridis*, with notes on the distribution of *Pyrodinium bahamense* var. *compressa* during a red tide in the Philippines. In: A.W. White, M. Anraku and K.-K. Hooi (Editors), Toxic Red Tides and Shellfish Toxicity in Southeast Asia, Proceedings of a Consultative Meeting, Singapore, 1984. Southeast Asian Fisheries Development Center and the International Development Research Centre, Singapore, pp. 43-51.

Avaria, S.P., 1979. Red tides off the coast of Chile. In: L.T. Taylor and H.H. Seliger (Editors), Toxic Dinoflagellate Blooms. Elsevier Science Publishers, B.V., New York, pp. 161-164.

Ayres, P.A., Burton, H.W. and Cullum, M.L., 1975. Sewage pollution and shellfish. J. Hyg., 74: 51-62.

Baine, W.B., Mazzotti, M., Greco, D., Izzo, E., Zampieri, A., Angioni, G., Di Gioia, M., Gangarosa, E.J. and Pocchiari, F., 1974. Epidemiology of cholera in Italy in 1973. Lancet, 2(7893): 1370-1374.

Bates, S.S., Bird, C.J., Boyd, R.K., Freitas, A.S.W. de, Falk, M., Foxall, R.A., Hanic, L.A., Jamieson, W.D., McCulloch, A.W., Odense, P., Quilliam, M.A., Sim, P.G., Thibault, P., Walter, J.A. and Wright, J.L.C., 1988. Investigations on the source of domoic acid responsible for the outbreak of amnesic shellfish poisoning (ASP) in eastern Prince Edward Island. Atlantic Res. Lab. Tech. Rep., 57: 1-54.

Bates, S.S., Bird, C.J., deFreitas, A.S.W., Foxall, R.A., Gilgan, M., Hanic, L.A., Johnson, G.R.,McCulloch, A.W., Odense, P., Pocklington, R., Quilliam, M.A., Sim, P.G., Smith, J.C., Subba Rao, D.V., Todd, E.C.D., Walter, J.A. and Wright, J.L.C., 1989. Pennate diatom *Nitzschia pungens* as the primary source of domoic acid, a toxin in shellfish from eastern Prince Edward Island, Canada. Can. J. Fish. Aquat. Sci., 46: 1203-1205.

Beales, R.W., 1976. A red tide in Brunei's coastal waters. Brunei Mus. J., 3: 167-182.

Bellelli, E. and Leogrande, G., 1967. Ricerche batteriologiche e virologiche sui mitili. Ann. Sclavo, 9: 820-828.

Bendinelli, M. and Ruschi, A., 1969. Isolation of human enterovirus from mussels. J. Food Prot., 51: 218-251.

Bernard, F.R., 1973. Bacterial flora of positive coliform tests of Pacific oysters from polluted and clean regions of Vancouver Island. Fish. Res. Board Can. Tech. Rep. 421: 6pp.

Bernard , F.R., 1989. Uptake and elimination of coliform bacteria by four marine bivalve mollusks. Can. J. Fish. Aquat. Sci., 46: 1592-1599.

Blanco, J., Marino, J. and Campos, M.J., 1985. First toxic bloom of *Gonyaulax tamarensis* detected in Spain 1984. In: D.M. Anderson, A.W. White and D.G. Baden (Editors), Toxic Dinoflagellates. Elsevier Science Publishers, B.V., New York, pp. 79-84.

Blogoslawski, W.J., 1983. Influence of water quality on shellfish culture. Mariculture Ctte, Ref. Shellfish Ctte, Int Council Exp. Sea, Gothenburg, Sweden, ICES. C.M. 1983/F: 8, pp. 1-35.

Blogoslawski, W.J., 1988. Ozone depuration of bivalves containing PSP: Pitfalls and possibilities. J. Shellfish Res. 7: 702-705.

Blogoslawski, W.J., 1990. Depuration and clam culture. In: J.J. Manzi and M. Castagna (Editors), Clam mariculture in North America. Elsevier Science Publishers, B.V., Amsterdam, pp. 417-426.

Blogoslawski, W.J., 1991. Enhancing shellfish depuration. In: W.S. Otwell, G.E. Rodrick and R. Martin (Editors), Proceedings of the First International Conference on Molluscan Shellfish Depuration, CRC Press, Boca Raton, Florida, pp. 145-149.

Blogoslawski, W.J. and Stewart, M.E., 1983. Depuration and public health. J. World Maricul. Soc., 14: 535-545.

Bonadonna, L., Volterra, L., Aulicino, F.A. and Mancini, L., 1990. Accumulation power of some bivalve molluscs. Mar. Pollut. Bull., 21: 81-84.

Bostock, A.D., Mepham, P., Phillips, S., Skidmore, S. and Hambling, M.H., 1979. Hepatitis A infection associated with the consumption of mussels. J. Infect., 1: 171-177.

Brisou, J., Tysset, C., Mailloux, M. and Espinasse, S., 1962. Recherches sur les vibrion marins. A propos de 44 souches isolées de moules (*Mytilus galloprovincialis*) du littoral algerois. Bull. Soc. Pathol. Exot., 55: 260-275.

Broek, M.J.M. van den, Mossel, D.A.A. and Eggenkamp, A.E., 1979. Occurrence of *Vibrio parahaemoliticus* in Dutch mussels. Appl. Environ. Microbiol., 37: 438-442.

Broutman, M.A. and Leonard, D.L., 1988. The quality of shellfish growing waters in the Gulf of Mexico. National Oceanic and Atmospheric Administration, U.S. Dept. of Commerce, Rockville, Maryland, 43pp.

Bryan, F.L., 1980. Epidemiology of foodborne diseases transmitted by fish, shellfish and marine crustaceans in the United States, 1970-1978. J. Food Prot., 43: 859-876.

Bryan, F.L., 1986. Seafood-transmitted infections and intoxications in recent years. In: D.E. Kramer and J. Liston (Editors), Seafood Quality Determination. Elsevier Science Publishers B.V., Amsterdam, pp. 319-337.

Caddy, J.F. and Chandler, R.A., 1968. Accumulation of paralytic shellfish poison by the rough whelk (*Buccinum undatum* L.). Proc. Natl. Shellfish Assoc., 58: 46-50.

Campodonico, I. and Guzman, L., 1974. Marea Roja producida por *Amphidoma* sp. en el estrecho de Magallanes. Ans. Inst. Pat., Punta Arenas (Chile), V(1-2): 208-213.

Campos, M.J., Fraga, S., Marino, J. and Sanchez, F.J., 1982. Red tide monitoring programme in NW Spain. Report of 1977-1981. ICES. C.M. 1982/L: 27: 1-8.

Canzonier, W.J., 1984. Technical aspects of bivalve depuration plant operation: pipes, pumps and petri plates. In: A.J. O'Sullivan (Editor), Mussel Bound, Proceedings of an International Shellfish Seminar, Bantry, Ireland, 1982, pp. 68-96.

Canzonier, W.J., 1988. Public health component of bivalve shellfish production and marketing. J. Shellfish Res., 7: 261-266.

Carreto, J.I., Negri, R.M., Benavides, H.R. and Askelman, R., 1985. Toxic dinoflagellate blooms in the Argentine Sea. In: D.M. Anderson, A.W. White and D.G. Baden (Editors), Toxic Dinoflagellates. Elsevier Science Publishers, B.V., New York, pp. 147-152.

Casellas, J.M., Caria, M.A. and Gerghi, M.E., 1977. Aislamiento de *Vibrio parahaemolyticus*, a partir de cholgas y mejillones en Argentina. Rev. Asoc. Argent. Microbiol., 9: 41-53.

Crovari, P., 1958. Some observations on the depuration of mussels infected with poliomyelitis virus. Ig. Mod., 51: 22-32.

Dale, B., Baden, D.G., Bary, B.M., Edler, L., Fraga, S., Jenkinson, I.R., Hallegraeff, G.M., Okaichi, T., Tangen, K., Taylor, F.G.R., White, A.W., Yentsch, C.M. and Yentsch, C.S., 1987. The problems of toxic dinoflagellate blooms in aquaculture. Proceedings of an International Conference and Workshop, Sherkin Island Marine Station, Ireland, 1987.,62pp.

Davison, P. and Yentsch, C.M., 1985. Occurrence of toxic dinoflagellates and shellfish toxin along coastal Uruguay, South America. In: D.M. Anderson, A.W. White and D.G. Baden (Editors), Toxic Dinoflagellates. Elsevier Science Publishers, B.V., New York, pp. 153-158.

Davy, F.B. and Graham, M. (Editors), 1982. Bivalve Culture in Asia and the Pacific: Proceedings of a workshop held in Singapore, 1982. International Development Research Centre, Canada, 90pp.

Denis, F., 1973. Coxsackie virus group A in oysters and mussels. Lancet, 1(7814): 1262.

Dienstag, J.L., Gust, I.D., Lucas, C.R., Wong, D.C. and Purcell, R.H., 1976. Mussel-associated viral hepatitis, type A: Serological confirmation. Lancet, 1(7659): 561-564.

Dimmock, N.J., 1967. Differences between the thermal inactivation of opicornaviruses at 'High' and 'Low' temperatures. Virology, 31: 338-353.

D.M., 1978. Norme concernenti i requisiti microbiologici, biologici, chimici e fisici delle zone acquee sedi di banchi e di giacimenti naturali di molluschi eduli lamellibranchi e delle zone acquee destinate alla molluschicoltura ai fini della classificazione in approvate, condizionate e precluse. Gazzetta Ufficiale della Republica Italiana suppl. n. 125, 8/5/78, pp. 1-14.

Dodgson, R.W., 1928. Report on mussel purification. His Majesty's Stationery Office, London, England, 497pp.

Duff, M.F., 1967. The uptake of enteroviruses by the New Zealand marine blue mussel *Mytilus edulis aoteanus*. Am. J. Epidemiol., 85: 486-493.

Eastaugh, J. and Shepherd, S., 1989. Infectious and toxic syndromes from fish shellfish consumption. Arch. Intern. Med., 149: 1735-1740.

Edler, L. and Hageltorn, M., 1990. Identification of the causative organism of a DSP-outbreak on the Swedish west coast. In: E. Graneli, D.M. Anderson, Edler, L. and B.G. Sundstrom (Editors), Toxic Marine Phytoplankton. Elsevier Science Publishers, B.V., New York, pp. 345-349.

Ellender, R.D., Cook, D.W., Sheladia, V.L. and Johnson, R.A., 1980. Enterovirus and bacterial evaluation of Mississippi oysters. Gulf Res. Rep., 6: 371-376.

Escherich, T., 1885. Die Darmbakterien des Neugeborenen und Sauglings. Fortschr. Med. 3: 515-522.

Estudillo, R.A. and Gonzales, C.L., 1984. Red tides and paralytic shellfish poisoning in the Philippines. In: A.W. White, M. Anraku and K.-K. Hooi (Editors), Toxic Red Tides and Shellfish Toxicity in Southeast Asia. Southeast Asian Fisheries Development Center and the International Development Research Centre, Singapore, pp. 52-79.

Feacham, R., Garelick, H. and Slade, J., 1982. Enteroviruses in the environment. World Health Forum, 3: 170-180.

Ferraz-Reyes, E., Reyes-Vasquez, G. and Oliveros, A.L. de, 1985. Dinoflagellates of the genera *Gonyaulax* and *Protogonyaulax* in the Gulf of Cariaco, Venezuela. In: D.M. Anderson, A.W. White and D.G. Baden (Editors), Toxic Dinoflagellates. Elsevier Science Publishers, B.V., New York, pp. 69-72.

Fraga, S. and Sanchez, F.J., 1985. Toxic and potentially toxic dinoflagellates found in Galician rias (NW Spain). In: D. Anderson, A.W. White and D.G. Baden (Editors), Toxic Dinoflagellates. Elsevier Science Publishers, B.V., New York, pp. 51-54.

Fraga, S., Marino, J., Bravo, I., Miranda, A., Campos, M.J., Sanchez, F.J., Costas, E., Cabanas, J.M. and Blancos, J., 1984. Red tides and shellfish poisoning in Galicia (NW Spain). ICES special meeting on the causes, dynamics, and effects of exceptional marine blooms and related events, Copenhagen, Denmark, 1984. C:5, 10pp.

Franca, S. and Almeida, J.F., 1989. Paralytic shellfish poisons in bivalve molluscs on the Portuguese coast caused by a bloom of the dinoflagellate *Gymnodinium catenatum*. In: T. Okaichi, D.M. Anderson and T. Nemoto (Editors), Red tides: Biology, Environmental Science, and Toxicology. Elsevier Science Publishers, B.V., New York, pp. 93-96.

Fréchette, M., Butman, C.A. and Geyer, W.R., 1989. The importance of boundary-layer flows in supplying phytoplankton to the benthic suspension feeder, *Mytilus edulis* L. Limnol. Oceanogr., 34: 19-36.

Freudenthal, A.R. and Jijina, J.L., 1988. Potential hazards of *Dinophysis* to consumers and shellfisheries. J. Shellfish Res., 7: 695-701.

Furfari, S.A., 1966. Depuration plant design. U.S. Pub. Health Serv. Publ. 999-FP-7, 109pp.

Furfari, S.A. and Hunt, D.A., 1981. Sanitary control of shellfish in France, 1981. U.S. Food and Drug Administration, 171pp.

Gaard, E. and Poulsen, M., 1988. Blooms of the toxic dinoflagellate *Gonyaulax excavata* in a Faroese fjord. ICES, C.M. 1988/L: 6: 1-11.

Gacutan, R.Q., Tabbu, M.Y., Aujero, E. and Icatlo, F., Jr., 1985. Paralytic shellfish poisoning due to *Pyrodinium bahamense* var. *compressa* in Mati, Davao Oriental, Philippines. Mar. Biol., 87: 223-227.

Gerba, C.P. and Goyal, S.M., 1978. Detection and occurrence of enteric viruses in shellfish: a review. J. Food Prot., 41: 743-754.

Gerba, C.P., Wallis, C. and Melnick, J.L., 1975. Viruses in water: the problem, some solutions. Environ. Sci. Technol., 9: 1122-1126.

Gerba, C.P., Goyal, S.M., La Belle, R.L., Cech, I. and Bodgan, G.F., 1979. Failure of indicator bacteria to reflect the occurrence of enteroviruses in marine waters. Am. J. Public Health, 69: 1116-1119.

Gilfillan, E.S., Hurst, J.W., Jr., Hansen, S.A. and LeRoyer, C.P., III, 1976. Final report to the New England Regional Commission, 83pp.

Gonzales, C.L., Ordonez, J.A. and Maala, A.M., 1989. Red tide: the Philippine experience. In: T. Okaichi, D.M. Anderson and T. Nemoto (Editors), Red Tides: Biology, Environmental Science, and Toxicology. Elsevier Science Publishers, B.V., New York, pp. 45-49.

Guzman, L. and Campodonico, I., 1978. Red tides in Chile. Interciencia, 3: 144-150.

Haamer, J., Andersson, P.-O., Lange, S., Li, X.P. and Edebo, L., 1990. Effects of transplantation and reimmersion of mussels *Mytilus edulis* Linnaeus, 1758, on the contents of okadaic acid. J. Shellfish Res., 9: 109-112.

Hallegraeff, G.M. and Summer, C.E., 1986. Toxic phytoplankton blooms affect shellfish farms. Aust. Fish., 45: 15-18.

Halstead, B.W. and Schantz, E.J., 1984. Paralytic shellfish poisoning. World Health Organization, Geneva, 60pp.

Harada, T., Oshima, Y., Kamiya, H. and Yasumoto, T., 1982. Confirmation of paralytic shellfish toxins in the dinoflagellate *Pyrodinium bahamense* var. *compressa* and bivalves in Palau. Bull. Jpn. Soc. Sci. Fish., 48: 821-825.

Hartwell, A.D, 1975. Hydrographic factors affecting the distribution and movement of toxic dinoflagellates in the western Gulf of Maine. In: V.R. LoCicero (Editor), Proceedings of the First International Conference on Toxic Dinoflagellate Blooms, Boston, U.S.A., 1974. Massachusetts Science and Technology Foundation, Wakefield, Massachusetts, pp. 47-68.

Howell, T.L. and Howell, L.R., 1989. The controlled purification manual. New England Fisheries Development Association, Boston, MA., 77pp.

Hungerford, J.M. and Wekell, M.M., 1992. Control measures in shellfish and finfish industries: U.S.A. In: P. Krogh (Editor), Algal Toxins in Seafood and Drinking Water. Academic Press, New York, in press.

Hunt, D.A., 1972. Sanitary control of shellfish and marine pollution. In: M. Ruivo (Editor), Marine pollution and sea life. Fishing News (Books) Ltd., England, pp. 565-568.

Hurst, J.W. Jr, 1975. The history of paralytic shellfish poisoning on the Maine Coast: 1958-1974. In: V. LoCicero (Editor), Proceedings of the First International Conference on Toxic Dinoflagellate Blooms. Massachusetts Science and Technology Foundation, Wakefield, Massachusetts, pp. 525-528.

IAWPRC, 1983. The health significance of viruses in water. Water Res., 17: 121-132.

ICES, 1988. Report of the working group on harmful effects of algal blooms on mariculture and marine fisheries, Lisbon, Portugal, 1988. ICES C.M.1988/F: 33: 1-21.

Ikeda, T., Matsuno, S., Sato, S., Ogata, T., Kodama, M., Fukuyo, Y. and Takayama, H., 1989. First report on toxic shellfish poisoning caused by *Gymnodinium catenatum* Graham (Dinophyceae) in Japan. In: T. Okaichi, D.M. Anderson and T. Nemoto, (Editors), Red Tides: Biology, Environmental Science, and Toxicology. Elsevier Science Publishers, B.V., New York, pp. 411-414.

Ingham, H.R., Mason, J. and Wood, P.C., 1968. Distribution of toxin in molluscan shellfish following the occurrence of mussel toxicity in northeast England. Nature (Lond.), 220: 25-27.

Jaafar, M.H. and Subramaniam, S., 1984. Occurrences of red tide in Brunei Darussalam and methods of monitoring and surveillance. In: A.W. White, M Anraku, and K. -K. Hooi (Editors), Toxic Red Tides and Shellfish Toxicity in Southeast Asia. Southeast Asia Fisheries Development Research Centre, Singapore, pp. 17-24.

Jamieson, G.C., (Editor) 1989. Mussel Culture. World Aquaculture, 20: 112pp.

Jeon, J.-K., Yi, S.K. and Huh, H.T., 1988. Paralytic shellfish poison of bivalves in Korean waters. J. Oceanol. Soc. Korea, 23: 123-129.

Kampelmacher, E.H., Noorle Jansen, L.M. van, Mossel, D.A.A. and Groen, F.J., 1972. A survey of the occurrence of *Vibrio parahaemolyticus* and *V. alginolyticus* on mussels and oysters and in estuarine waters in the Netherlands. J. Appl. Bacteriol., 35: 431-438.

Kat, M., 1983. Diarrhetic mussel poisoning in the Netherlands related to the dinoflagellate *Dinophysis acuminata*. Antoine Leeuwenhoek J. Microbiol., 49: 417-427.

Kat, M., 1985. *Dinophysis acuminata* blooms, the distinct cause of Dutch mussel poisoning. In: D.M. Anderson, W. White and D.G. Baden (Editors), Toxic Dinoflagellates. Elsevier Science Publishers, B.V., New York, pp. 73-77.

Kat, M., 1989. Toxic and non-toxic dinoflagellate blooms on the Dutch coast. In: T. Okaichi, D.M. Anderson and T. Nemoto (Editors), Red tides: Biology, Environmental Science, and Toxicology. Elsevier Science Publishers, B.V., New York, pp. 73-76.

Kiørboe, T., Møhlenberg, F. and Nøhr, O., 1980. Feeding, particle selection and carbon absorption in *Mytilus edulis* in different mixtures of algae and resuspended bottom material. Ophelia, 19: 193-205.

Kodama, M., (Editor), 1985. Studies on paralytic shellfish poisoning occurring in the coastal water of Thailand, and its causative dinoflagellate. Report on surveys in 1985 supported by a Grant-in-Aid for Scientific Research (Overseas Scientific Survey) from the Ministry of Education, Science and Culture, Japan. No. 60041067 (1985) and No. 61043062 (1986), 97pp.

Konno, T., Suzuki, H., Ishida, N., Chiba, R., Mochizuki, K. and Tsunoda, A., 1982. Astrovirus-associated epidemic gastroenteritis in Japan. J. Med. Virol., 9: 11-17.

Konovalova, G.V., 1989. Phytoplankton blooms and red tides in the far east coastal waters of the USSR. In: T. Okaichi, D.M. Anderson and T. Nemoto (Editors), Red Tides: Biology, Environmental Science, and Toxicology. Elsevier Science Publishers, B.V., New York, pp. 97-100.

Krogh, P., 1992. Review of toxicology of diarrhoeic shellfish poisons. Toxicology, in press.

Krogh, P., Edler, L., Graneli, E. and Nyman, U., 1985. Outbreak of diarrhetic shellfish poisoning on the west coast of Sweden. In: D.M. Anderson, A.W. White and D.G. Baden (Editors), Toxic Dinoflagellates. Elsevier Science Publishers, B.V., New York, pp. 501-503.

Krstulović, N. and Šolić, M., 1988. Comparison of faecal coliform levels in mussel flesh together with intervalvular fluid. Acta Adriat., 29: 67-73.

Kueh, C.S.W., 1987. Uptake, retention and elimination of enteric bacteria in bivalve molluscs. Asian Mar. Biol., 4: 113-128.

La Belle, R.L., Gerba, C.P., Goyal, S.M., Melnick, J.L., Cech, I. and Bodgan, G.F., 1981. Relationships between environmental factors, bacterial indicators and the occurrence of enteric viruses in estuarine sediments. Appl. Environ. Microbiol., 39: 469-478.

Landry, E.F., Vaughn, J.M. and Vicale, T.C., 1980. Modified procedure for extraction of poliovirus from naturally-infected oysters using Cat-Floc and beef extract. J. Food Prot., 43: 91-94.

Landry, E.F., Vaughn, J.M., Vicale, T.J. and Mann, R., 1983. Accumulation of sediment-associated viruses in shellfish. Appl. Environ. Microbiol. 45: 238-247.

Lassus, P. and Berthome, J.P., 1988. Status of 1987 algal blooms in IFREMER. ICES/ annex III C.M. 1988/F:33A: 5-13.

Ledo, A., Gonzalez, E., Barja, J.L. and Toranzo, A.E., 1983. Effect of depuration systems on the reduction of bacteriological indicators in cultured mussels (*Mytilus edulis* Linnaeus). J. Shellfish Res., 3: 59-64.

Leonard, D.L. and Slaughter, E.A., 1990. The quality of shellfish growing waters on the west coast of the United States. U.S. Dept. of Commerce, National Oceanic and Atmospheric Administration, Rockville, Maryland, Administration. 51pp.

Leonard, D.L., Broutman, M.A. and Harkness, K., 1989. The quality of shellfish growing waters on the east coast of the United States. U.S. Dept. of Commerce, National Oceanic and Atmospheric Administration, 54pp.

Lewis, G., Loutit, M.W. and Austin, F.J., 1985. Human enteroviruses in marine sediments near a sewage outfall on the Otago Coast. N. Z. J. Mar. Freshw. Res., 19: 187-192.

Lewis, G., Loutit, M.W. and Austin, F.J., 1986. Enteroviruses in mussels and marine sediments and depuration of naturally accumulated viruses by green lipped mussels (*Perna canaliculus*). N. Z. J. Mar. Freshw. Res., 20: 431-437.

Liew, P.F. and Gerba, C.P., 1980. Thermostabilization of enteroviruses by estuarine sediment. Appl. Environ. Microbiol., 40: 305-308.

Lucas, M.I., Newell, R.C., Shumway, S.E., Seiderer, L.J. and Bally, R., 1987. Particle clearance and yield in relation to bacterioplankton and suspended particulate availability in estuarine and open coast populations of the mussel *Mytilus edulis*. Mar. Ecol. Prog. Ser., 36: 215-224.

Maclean, J.L., 1973. Paralytic shellfish poisoning in Papua New Guinea. Papua New Guinea Agric. J., 24: 131-138.

Maclean, J.L., 1975. Paralytic shellfish poisoning in various bivalves, Port Moresby, 1973. Pac. Sci., 29: 349-352.

Maclean, J.L., 1984. Indo-Pacific toxic red tide occurrences, 1972-1984. In: A.W. White, M. Anraku and K.-K. Hooi (Editors), Toxic Red Tides and Shellfish Toxicity in Southeast Asia. Southeast Asia Fisheries Development Center, Singapore, pp. 92-102.

McDonald, J.H., Seed, R. and Koehn, R.K., 1991. Allozyme and morphometric characters of three species of *Mytilus* in the Northern and Southern hemispheres. Mar. Biol., 111: 323-335.

Medcof, J.C., Leim, A.H., Needler, A.B., Needler, A.W.H., Gibbard, J. and Naubert, J., 1947. Paralytic shellfish poisoning on the Canadian Atlantic coast. Bull. Fish. Res. Board Can., 75: 32pp.

Meixner, R. and Luckas, B., 1988. On an outbreak of diarrhetic shellfish poisoning and determination of okadaic acid as a typical DSP-toxin in mussels. ICES C.M. 1988/K: 6: 1-9.

Metcalf, T.G., 1978. Indicators of viruses in shellfish. In: Berg, G. (Editor), Indicators of Viruses in Water and Food. Ann Arbor, Michigan, pp. 383-415.

Metcalf, T.G. and Stiles, W.C., 1965. Survival of enteric viruses in estuary waters and shellfish. In: G. Berg (Editor), Transmission of Viruses by the Water Route. Interscience Publishers, New York, pp. 439-447.

Mikhailov, V.V., Kochkin, A.V. and Ivanova, E.P., 1988. A comparison study of the microorganisms found in the mussel and its habitat. Mikrobiologiia, 57: 1-4.

Milo, G.E., 1971. Thermal inactivation of poliovirus in the presence of selective organic molecules (cholestrol, lecithin, collagen, and ß-carotene). Appl. Microbiol., 21: 198-202.

Minet, J., Barbosa, T., Prieur, D. and Cormier, M., 1987. The nature of the bacteria concentration process by the mussel *Mytilus edulis* (L.). C.R. Acad. Sci. Paris, 305 (III): 351-354.

Mortenson, A.M., 1985. Massive fish mortalities in the Faroe Islands caused by a *Gonyaulax excavata* red tide. In: D.M. Anderson, A.W. White and D.G. Baden (Editors), Toxic Dinoflagellates. Elsevier Science Publishers, B.V., New York, pp. 165-170.

Murphy, A.M., Grohmann, G.S., Christopher, P.J., Lopez, W.A., Davey, G.R. and Millsom, R.H., 1979. An Australia-wide outbreak of gastroenteritis from oysters caused by Norwalk virus. Med. J. Aust., 2: 329-333.

Music, S.I., Howell, J.T. and Brumback, C.L., 1973. Red tide: Its public health implications. J. Fla. Med. Assoc., 60: 27-29.

Nezan, E., C. Belin, P. Lassus, G. Piclet and J.P. Berthome, 1990. *Alexandrium minutum*: first PSP species occurrence in France. Programme Abstracts, Fourth International Conference on Toxic Marine Phytoplankton, Lund, Sweden, 1989, 111pp.

Nishitani, L. and Chew, K., 1988. PSP toxins in the Pacific coast states: monitoring programs and effects on bivalve industries. J. Shellfish Res., 7: 653-669.

Noble, R.C., 1990. Death on the half-shell: The health hazards of eating shellfish. Perspec. Biol. Med., 33: 313-320.

NSSP (National Shellfish Sanitation Program), 1989a. Manual of Operations, Part 1. Sanitation of Shellfish, Growing Areas. 1989 Revision. Food and Drug Administration, North Kingstown, RI., Shellfish Sanitation Branch, 115pp.

NSSP (National Shellfish Sanitation Program), 1989b. Manual of Operations, Part 2. Sanitation of the Harvesting, Processing and Distribution of Shellfish. Food and Drug Administration, North Kingstown, RI., Shellfish Sanitation Branch, 166pp.

Oshima, Y., Yasumoto, T., Kodama, M., Ogata, T. Fukuyo, Y. and Matsura, F., 1982. Features of shellfish poisoning in Tohoku district. Bull. Jpn. Soc. Sci. Fish., 48: 525-530.

Perl, T.M., Bernard, L., Kosatsky, T., Hockin, J.C., Todd, E.C.D. and Remis, R.S., 1990. An outbreak of toxic encephalopathy caused by eating mussels contaminated with domoic acid. New Engl. J. Med., 322: 1775-1780.

Petrilli, F. and Crovari, P., 1965. Aspetti dell'inquinamento delle acque marine con particolare riguardo alla situazione in Liguria. G. Ig. Med. Prev., 8: 269-311.

Pinto, J.D.S. and Silva, E.D.S., 1956. The toxicity of *Cardium edule* L. and its possible relation to the dinoflagellate *Prorocentrum micans*. Notas Estud. Inst. Biol. Marit. (Lisb.), 12: 1-20.

Plusquellec, A., Beucher, M., Prieur, D. and Le Gal, Y., 1990. Contamination of the mussel, *Mytilus edulis* Linnaeus, 1758, by enteric bacteria. J. Shellfish Res., 9: 95-101.

Popkiss, M.E., Horstman, D.A. and Harpur, D., 1979. Paralytic shellfish poisoning: a report of 17 cases in Cape Town. S. Afr. Med. J., 55: 107-123.

Portnoy, B.L., Mackowiak, P.A., Caraway, C.T., Walker, J.A., McKinley, T.W. and Klein, C.A., 1975. Oyster-associated hepatitis: failure of shellfish certification programs to prevent outbreaks. J. Am. Med. Assoc., 233: 1065-1068.

Power, U.F. and Collins, J.K., 1986. Evaluation of depuration as a means of rendering shellfish free from viral pathogens and bacterial indicators. Ir. J. Food Sci. Technol., 10: 159.

Power, U.F. and Collins, J.K., 1987. Conventional depuration techniques do not guarantee virus-free mussels for human consumption. Ir. J. Food Sci. Technol., 11: 189-190.

Power, U.F. and Collins, J.K., 1989. Differential depuration of poliovirus, *Escherichia coli*, and a coliphage by the common mussel, *Mytilus edulis*. Appl. Environ. Microbiol., 55: 1386-1390.

Power, U.F. and Collins, J.K., 1990. Tissue distribution of a coliphage and *Escherichia coli* in mussels after contamination and depuration. Appl. Environ. Microbiol., 56: 803-807.

Prakash, A., 1963. Source of paralytic shellfish toxin in the Bay of Fundy. J. Fish. Res. Board Can., 20: 983-996.

Prakash, A. and Taylor, F.J.R., 1966. A "red water" bloom of *Gonyaulax acatenella* in the Strait of Georgia and its relation to paralytic shellfish toxicity. J. Fish. Res. Board Can., 23: 1265-1270.

Prakash, A., Medcof, J.C. and Tennant, A.D., 1971. Paralytic shellfish poisoning in eastern Canada. Fish. Res. Board Can. Bull., 177. Fisheries Research Board of Canada, Ottawa, Canada, 87pp.

Quayle, D., 1965. Animal detoxification. Proceedings of Joint Sanitation Seminar on North Pacific Clams, Juneau, Alaska, U.S.A.,1965. U.S. Government Printing Office, Washington, D.C., pp. 7-8.

Reyes-Vasquez, G., Ferraz-Reyes, E. and Vasquez, E., 1979. Toxic dinoflagellate blooms in northeastern Venezuela during 1977. In: D.L. Taylor and H.H. Seliger (Editors), Toxic Dinoflagellate Blooms. Elsevier Science Publishers, B.V., New York, pp. 191-194.

Richards, G.P., 1985. Outbreaks of shellfish-associated enteric virus illness in the United States: requisite for development of viral guidelines. J. Food Prot., 48: 815-823.

Richards, G.P., 1988. Microbial purification of shellfish: a review of depuration and relaying. J. Food. Prot., 51: 218-251.

Schantz, E.J., 1973. Seafood toxicants. In: Committee Food Protection, Food and Nutrition Board and National Research Council (Editors), Toxicants Occurring Naturally in Foods. National Academy of Sciences, pp. 424-447.

Scotti, P.D., Fletcher, G.C., Buisson, D.H. and Fredericksen, S., 1983. Virus depuration of the Pacific oyster (*Crassostrea gigas*) in New Zealand. N. Z. J. Sci., 26: 9-13.

Sharpe, C.A., 1981. Paralytic shellfish poison, California–Summer 1980. State of California Dept. Health Services-Sanitary Engineering Section, 75pp.

Shumway, S.E., 1990. A review of the effects of algal blooms on shellfish and aquaculture. J. World Aquacult. Soc., 21: 65-104.

Shumway, S.E., Sherman-Caswell, S. and Hurst, J.W., Jr., 1988. Paralytic shellfish poisoning in Maine: monitoring a monster. J. Shellfish Res., 7: 643-652.

Sindermann, C., 1990. Shellfish diseases of public health significance. In: Principal Diseases of Marine Fish and Shellfish, Vol. 2, Disease of Marine Shellfish. Academic Press, California, pp. 457-487.

Smith, J.C., Cormier, R., Worms, J., Bird, C.J., Pocklington, R., Angus, R. and Hanic, L., 1990. Toxic blooms of the domoic acid containing diatom *Nitzschia pungens* in the Cardigan River, Prince Edward Island, in 1988. In: E. Graneli, D.M. Anderson, L. Edler and B.G. Sundstrom (Editors), Toxic Marine Phytoplankton. Elsevier Science Publishers, B.V., New York, pp. 227-232.

Sommer, H. and Meyer, K.F., 1937. Paralytic shellfish poisoning. Arch. Pathol., 24: 560-598.

Stamman, E., Segar, D.A. and Davis, P.G., 1987. A preliminary epidemiological assessment of the potential for diarrhetic shellfish poisoning in the Northeast United States. NOAA Tech. Mem. NOS OMA, 34: 1-18.

Subba Rao, D.V., Quilliam, M.A. and Pocklington, R., 1988. Domoic acid—a neurotoxic amino acid produced by the marine diatom *Nitzschia pungens* in culture. Can. J. Fish. Aquat. Sci., 45: 2076-2079.

Subcommittee on Microbiological Criteria, Committe on Food Protection, Food and Nutrition, and National Research Council, 1985. An Evaluation of the Role of Microbiological Criteria for Foods and Food Ingredients. National Academy Press, Washington, D.C., 436pp.

Tamiyavanich, S., Kodama, M. and Fukuyo, Y., 1985. The occurrence of paralytic shellfish poisoning in Thailand. In: D.M. Anderson, A.W. White and D.G. Baden (Editors), Toxic Dinoflagellates. Elsevier Science Publishers, B.V., New York, pp. 521-524.

Teitelbaum, J.S., Zatorre, R.J., Carpenter, S., Gendron, D., Evans, A.C., Gjedde, A. and Cashman, N.R., 1990. Neurologic sequelae of domoic acid intoxication due to the ingestion of contaminated mussels. New Engl. J. Med., 322: 1781-1787.

Tennant, A.D., Naubert, J. and Corbeil, H.E., 1955. An outbreak of paralytic shellfish poison. Can. Med. Assoc. J., 72: 436-439.

Trollope, D.R., 1984. Use of molluscs to monitor bacteria in water. In: J.M. Grainger and J.M. Lynch (Editors), Microbiological Methods for Environmental Biotechnology. Academic Press, London, pp. 393-408.

Trollope, D.R. and Webber, D.L., 1977. Shellfish bacteriology: coliform and marine bacteria in cockles (*Cardium edule*), mussels (*Mytilus edulis*) and *Scrobicularia plana*. In: A. Nelson-Smith and E.M. Bridges (Editors), Problems of a Small Estuary. Swansea University College, U.K., pp. 1-18.

Trollope, D.R. and Al-Salihi, S.B.S., 1984. Sewage-derived bacteria monitored in a marine water column by means of captive mussels. Mar. Environ. Res. 12: 311-322.

Tufts, N.R., 1979. Molluscan transvectors of paralytic shellfish poisoning. In: D.L. Taylor and H.H. Seliger (Editors), Toxic Dinoflagellate Blooms. Elsevier Science Publishers, B.V., New York, pp. 403-408.

Underdal, B., Yndestad, M. and Aune, T, 1985. DSP intoxication in Norway and Sweden, autumn 1984-spring 1985. In: D.M. Anderson, A.W. White and D.G. Baden (Editors), Toxic Dinoflagellates. Elsevier Science Publishers, B.V., New York, pp. 489-494.

UNEP/WHO, 1983a. Determination of faecal coliforms in sea water by the membrane filtration culture method. Reference Methods for Marine Pollution Studies. No. 3 Rev. 1, UNEP 1983, 23pp.

UNEP/WHO, 1983b. Determination of faecal coliforms in bivalves by multiple test tube method. Reference Methods for Marine Pollution Studies. No. 5 Rev. 1. UNEP 1983, 20pp.

UNEP/WHO, 1988. Guidelines for monitoring the quality of coastal recreation and shellfish growing areas. References Methods for Marine Pollution Studies. No. 1. Rev. 1. UNEP 1988, 36pp.

United States Department of Health and Human Services (NSSP), 1986. Sanitation of Shellfish Growing Areas. National Shellfish Sanitation Program (NSSP) Manual of Operations Part 1. U.S. Department of Health and Human Services, Washington, D.C., 93pp.

United States Public Health Service (NSSP), 1988. Sanitation of shellfish growing areas, National Shellfish Sanitation Program (NSSP) Manual of Operations, Interstate Shellfish Sanitation Conference (ISSC), and Food and Drug Administration, U.S. Department of Health and Human Services. ISSC, Phoenix, Arizona, 136pp.

United States Public Health Service, 1990. Interstate certified shellfish shippers list. U.S. Public Health Service. U.S. Government Printing Office, 46pp.

Volterra, L. and Tosti, E., 1983. Faecal pollution and shellfish hygienic condition. Ann. Ist. Super. Sanita, 19: 317-322.

Webber, D.L., 1982. The accumulation of faecal indicator bacteria by the mussel, *Mytilus edulis*. Ph. D. Thesis, University of Wales, U.K.

WHO (World Health Organization), 1976. Microbiological aspects of food hygiene. Report on a WHO Expert Committee with participation of FAO. Technical Report Series 578. Geneva, Switzerland: World Health Organization, 103pp.

WHO/ICMSF (World Health Organization/International Commission on Microbiological Specifications for Food), 1985. Report of the WHO/ICMSF meeting on hazard analysis: critical control point system in food hygiene Geneva, 1980. In: Subcommittee on Microbiological Criteria, Committee on Food Protection, Food and Nutrition Board and National Research Council (Editors), An Evaluation of the

Role of Microbiological Criteria for Foods and Food Ingredients. National Academy Press, Washington, D.C., pp. 399-419.

Wood, P.C., 1957. Factors affecting the pollution and self-purification of molluscan shellfish. J. Cons. Int. Explor. Mer, 22: 200-208.

Wood, P.C., 1969. The production of clean shellfish. Fish. Lab., MAFF., Burnham-on-Crouch, England. Laboratory leaflet (new series) No. 20, Ministry of Agriculture, Fisheries and Food, 16pp.

Wood, P.C., 1972. The principles and methods employed for the sanitary control of molluscan shellfish. In: M. Ruivo (Editor), Marine Pollution and Sea Life. Fishing News (Books) Ltd., England, pp. 560-565.

Wood, P.C., 1976. Guide to shellfish hygiene. Fish. Lab., MAFF., Burnham-on-Crouch, England. WHO Offset Publ., (no. 31), World Health Organization, Geneva, 80pp.

Worth, G.K., Maclean, J.L. and Price, M.J., 1975. Paralytic shellfish poisoning in Papua, New Guinea, 1972. Pac. Sci., 29: 1-5.

Yndestad, M. and Underdal, B., 1985. Survey of PSP in mussels (*Mytilus edulis* L.) in Norway. In: D.M. Anderson, A.W. White and D.G. Baden (Editors), Toxic Dinoflagellates. Elsevier Science Publishers, B.V., New York, pp. 457-460.

Chapter 12

DISEASES AND PARASITES OF MUSSELS

SUSAN M. BOWER

INTRODUCTION

In comparison to other cultured bivalves, such as oysters, little is known about the parasites and diseases of mussels. Also, epizootic diseases, like those that have devastated the oyster culture industry in some parts of the world, have not been encountered by the mussel culture industry. The relative lack of information on parasites and diseases in mussels may be attributed to the lesser economic importance of mussels, the shorter history of intensive mussel culture, and the comparatively fewer investigations into the causes of mussel mortalities (Bower and Figueras, 1989). Thus, the supposition that mussels have fewer parasites and diseases than other bivalves is probably incorrect.

With the current increase in mussel culture activity worldwide, disease problems, hitherto not recognized in wild stocks, have been encountered and several pathogenic organisms and potential agents of disease have been described. A brief summary of the parasites involved, geographic location, and recognized or suspected pathogens is presented below. Species names of the mussels are as they appear in the original publications. Ecological terminology follows the definitions of Margolis et al. (1982).

VIRUSES

To date, few viruses pathogenic to mussels have been described. However, a Picornaviridae-like virus was found associated with granulocytomas (non-neoplastic inflammatory lesions composed mainly of eosinophilic granular haemocytes, some of which contained large vacuoles and/or karyolysis) in the vesicular connective tissue of the digestive diverticula and mantle in 2.8% of the *Mytilus edulis* from Denmark (Rasmussen, 1986a). Histopathological examination of infected mussels indicated that the viral infection was progressive, but the virus was confined to haemocytes, and the majority of infected haemocytes were within granulocytomas.

BACTERIA

Excluding the bacteria that are ubiquitous problems in intensive culture, especially larval bivalve production, few bacteria have been noted as being pathogenic to adult mussels. Bacteria belonging to the family Vibrionaceae are common in bivalves including mussels (Prieur et al., 1985). Extracellular toxins found in culture fluids of various strains of *Vibrio* exhibited the ability to disaggregate excised gill tissue of *M. edulis* (Nottage and Birkbeck, 1987). Some strains of *Vibrio* inhibited filtration and thus feeding of *M. edulis*. But, *Vibrio* ingested before filtration ceased were degraded by the mussel in the same way as other Gram-negative bacteria (Birkbeck et al., 1987). Thus, most species or strains of *Vibrio* may only be pathogenic to mussels in situations where environmental conditions are poor.

A Gram-positive coccus was noted in 1.5% of *M. edulis* from the Gulf of Finland. This bacterium was observed in the epithelium of the intestine and had induced an inflammatory response (Sunila, 1987). On the east coast of the U.S.A., infections with Gram-negative, rod-shaped bacteria were observed to cause disease in the plicate organ of about 8% *M. edulis* (Farley, 1988).

Intracellular bacteria belonging to the Rickettsiae and Chlamydiae have been described from bivalves worldwide, including *M. edulis* and *Mytilus californianus* from the U.S.A. (Lauckner, 1983). Prevalences up to 50% have been observed in mussels from Galicia, Spain (Figueras and Montes, 1988a). On the west coast of Canada, light infections were observed in the epithelial cells of the digestive gland and gill (Fig. 12.1) of 35% and 20%, respectively, of *M. edulis* from one population. There was no evidence of a host response to these organisms. Since few cells were affected, the overall pathology was minimal, aside from the pathological changes observed in individual host cells. To date Rickettsiae and Chlamydiae are not known to cause mortality in mussels (Lauckner, 1983).

PROTOPHYTA

Two species of Protophyta, a lichen confined to the shell, and a microalga within the soft tissues, have been reported from mussels. The shell burrowing lichen, *Arthopyrenia sublitoralis*, was noted as being fairly common in living *M. edulis* from shallow waters in the Isefjord complex of Denmark (Rasmussen, 1973). The infection was usually confined to the eroded parts of the mussel shell near the umbo, and was easily confused with infections of the shell boring sponge *Cliona*. The endobiotic microalga, provisionally identified as a blue-green alga of the genus *Microcystis*, was reported to cause green spots in the mantle and adductor muscle of up to 68% of *M. edulis* from the western Baltic Sea (Meixner, 1984). Erosion of the dark periostracum

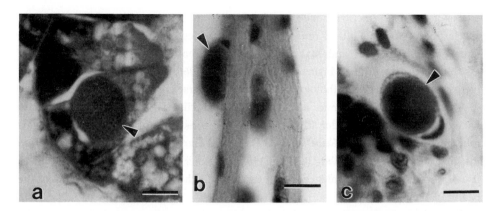

Fig. 12.1. Rickettsiae-like infection (arrows) in a secretory epithelial cell of the digestive gland (a); in a squamosal epithelial cell along the dense connective tissue of the gill (b); and in a cuboidal epithelial cell at the distal end of a gill filament (c) of *Mytilus edulis* from British Columbia, Canada. (Scale bars = 10μm; haematoxylin and eosin stain).

and blue calcite layers of the shell of infected mussels permitted sunlight to penetrate into the soft tissues through the remaining translucent shell. Heavily infected mussels were in poor physiological condition and the disease was thought to be terminal. The prevalence of infection increased with mussel size and seemed to follow a seasonal trend, with highest prevalences occurring in the late summer months.

PROTOZOA

Protozoa representing five phyla (according to the taxonomic classification of Levine et al., 1980) have been described from mussels. Although numerous species of the phyla Sarcomastigophora and Ciliophora inhabit mussels, especially the mantle cavity, most, under normal conditions, are not pathogenic. However, several potentially pathogenic species are noteworthy. Three species are motile ciliates belonging to the class Oligohymenophorea, order Scuticociliatida. *Peniculistoma mytili* on the foot epithelium and adjacent surfaces of *M. edulis* occurs in almost 100% of the mussels from various localities in the North and Baltic Seas (Lauckner, 1983). *Mytilophilus pacificae* was described from the foot and mantle of *M. californianus* from the Pacific coast of North America (Antipa and Dolan, 1985). *Mytilophilus pacificae* is suggested to occupy the same ecological niche as *Peniculistoma mytili*, and there is no evidence that either ciliate has deleterious effects on its host. *Ancistrum mytili,* with possible morphologically similar sympatric species, occurs on the gills of mussels and other pelecypods worldwide (Cheng, 1967; Lauckner, 1983; DaRos and Massignan, 1985). Although there is no indication of associated disease in most instances, Pauley et al. (1966) suggested that these ciliates were capable of causing pathology under certain adverse conditions.

The sessile ancistrocomid and sphenophryid ciliates, belonging to the class Kinetofragminophorea, order Rhynchodida, and consisting of species from at least four genera (*Crebricoma, Raabella, Isocomides,* and *Gargarius*), are found attached to the gills of mussels worldwide (Cheng, 1967; Lauckner, 1983). These attached ciliates usually occur in low prevalence and intensity and thus, are not likely to affect the health of their host.

Recently, an unusual ciliate has been observed within epithelial cells of the digestive gland of *M. edulis* from the east coast of North America, North-west Spain, and British Columbia (Figueras et al., 1991; Figueras and Montes, 1988a; McGladdery, 1990). Despite the intracellular location of this ciliate, there was no apparent damage to adjacent uninfected cells and no inflammatory reaction to its presence (Fig. 12.2). Thus, this parasite may not cause disease, even though prevalence can be high (between 60% and 100% in mussels from several locations in British Columbia during August), and intensities heavy in a few mussels.

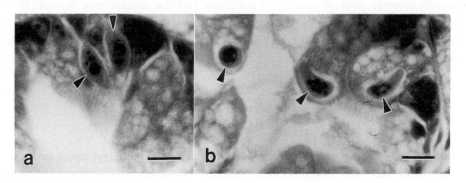

Fig. 12.2. Undescribed ciliate (arrows) apparently containing one large and several small nuclei (a), or containing several small nuclei (b), in the cytoplasm of the digestive gland epithelial cells of *Mytilus edulis* from British Columbia. (Scale bars = 10µm; haematoxylin and eosin stain).

In the phylum Microspora, one species, *Steinhausia mytilovum* (originally classified as a haplosporidian), has been described from the ova of *M. edulis* from the American Atlantic seaboard and Californian coast, and *M. galloprovincialis* from the Gulf of Naples and the northwestern Atlantic coast of Spain (Sprague, 1965; Lauckner, 1983; Sparks 1985; González et al., 1987; Farley, 1988; Hillman et al., 1988). The effect of this parasite on the fecundity of its host depended on the intensity of infection. In

Spain, the prevalence was about 10% at the time of maximum gonadal development and its presence was always accompanied by a strong haemocytic host response (Figueras and Montes, 1988a). Although infected eggs did not appear degenerate (Sparks, 1985), this parasite may have the potential for reducing the fecundity of infected mussels.

The phylum Ascetospora contains at least six representatives that infect mussels. Three are undescribed species of *Haplosporidium*. One has been reported in the digestive gland and kidney with associated 'tumefaction' (swellings) in 2.1% of *M. californianus* from California (Taylor, 1966). Another has been found within large lesions in the gonad of up to 34% of *M. edulis* from British Columbia (Quayle, 1978). The third was observed in the connective tissue of the digestive gland and gills of *M. edulis* from the coast of Maine (Figueras et al., 1991), and may be equivalent to the organism that was identified as *Minchinia* by Sherburne and Bean (1986) from the same locality.

The fourth ascetosporan, *Marteilia refringens*, which has caused high mortalities in flat oysters (*Ostrea edulis*) in Europe, was observed in *M. edulis* in France and Spain (Lauckner, 1983; Balouet and Poder, 1985; Figueras and Montes, 1988a, b). A closely related species, *Marteilia maurini*, was described from *M. galloprovincialis* and *M. edulis* from the Atlantic coasts of Spain and France and the Persian Gulf (Comps et al., 1981; Auffret and Poder, 1983; DaRos and Massignan, 1985; González et al., 1987). The availability of technology to obtain purified isolations of *Marteilia* from both mussels and oysters (*Ostrea edulis*) should facilitate the ultrastructural, biochemical, and physiological characterization of both forms of *Marteilia* in order to establish their taxonomic affinities (Mialhe et al., 1985). Nevertheless, properly controlled cross infection studies are essential in establishing the relationship between the two species.

Disease and mortality associated with the *Marteilia* spp. from mussels is poorly understood (Lauckner, 1983). However, high prevalences (37–70%) in *M. edulis* from the north coast of Brittany (Auffret and Poder, 1983), and cumulative mortalities of 46% over 19mo, with significant inhibition of gametogenesis in mussels from various localities in northern Spain that have high prevalences of this parasite (López et al., 1990; Villalba et al., 1990a, b), suggest that *Marteilia* spp. represent a risk for mussel culture.

The sixth ascetosporan is a *Bonamia*-like organism (microcell) that has been observed in *M. edulis* from New Jersey on the east coast of the U.S.A. (Figueras et al., 1991). Although *Bonamia ostreae* has devastated the flat oyster (*O. edulis*) industry in Europe (Grizel et al., 1988), and the *Bonamia*-like infections in mussels from New Jersey were associated with an intense haemocytic response, nothing is known about the pathogenicity of this microcell to mussels.

The protozoan phylum Apicomplexa includes representatives of *Nematopsis* found worldwide in mussels (Cheng, 1967; Lauckner, 1983). These are members of the

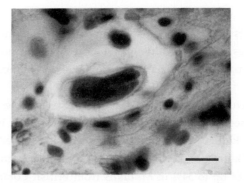

Fig. 12.3. Oocyst containing a sporozoite of an unidentified *Nematopsis* sp. in the connective tissue at the distal end of a gill filament of *Mytilus edulis* from British Columbia. (Scale bar = 10μm; haematoxylin and eosin stain).

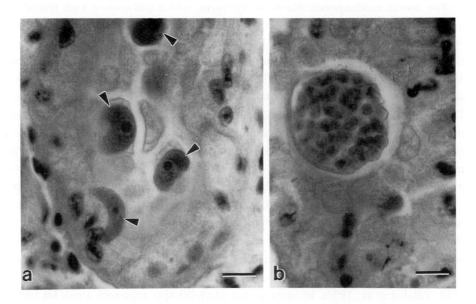

Fig. 12.4. Crescent shaped gamonts (arrows) (a) and a mature oocyst containing many sporocysts (b) of a *Pseudoklossia*-like coccidian in the kidney epithelium of *Mytilus edulis* from British Columbia. (Scale bars = 10μm; haematoxylin and eosin stain).

digenetic Eugregarinida, which complete their life cycles in the intestine of an arthropod final host. An example of an unidentified *Nematopsis* occurs mainly in the gills (and rarely in the connective tissue of the digestive gland, plicate membrane, and mantle) of up to 75% of the *M. edulis* from the south coast of British Columbia (Fig. 12.3). There is no evidence of an adverse effect on the mussel in British

Columbia by this particular *Nematopsis*. However, other reports concerning pathogenicity are inconclusive. For instance, the presence of *Nematopsis* oocysts in the mantle tissues of *M. edulis* in Europe has been associated with the production of pearls or calcareous deposits on the inner surface of the valves (Gotting, 1979; Lauckner, 1983). Another apicomplexan, but of the order Eucoccidiida and identified as *Pseudoklossia* sp., has been observed in the kidney cells of a few mussels from the east coast of the U.S.A. (Farley, 1988). A *Pseudoklossia*-like coccidian (Fig. 12.4) has been observed in about 10% of the mussels from two locations in British Columbia, with no associated pathological changes.

Finally, mortalities of *M. edulis* in Prince Edward Island (Canada) have been attributed to a poorly described protozoan (Li and Clyburne, 1979) that may be related either to *Perkinsus marinus* (phylum Apicomplexa), a pathogenic parasite of the oyster, *Crassostrea virginica* (Lauckner, 1983), or to *Labyrinthuloides haliotidis* (phylum Labyrinthomorpha), a recently described pathogenic parasite of juvenile abalone (Bower, 1987).

FUNGI

Fungi have been accused of causing malformation of shell and loss of condition in *M. edulis* in England and Spain (Lauckner, 1983; Figueras and Montes, 1988a). Degradation of byssal material of *M. galloprovincialis* in Italy has been attributed to an ascomycetes (Vitellaro-Zuccarello, 1973). However, all these instances were not fully investigated and the specific identity and pathogenicity of the fungi are still in question. *Leptolegnia marina*, normally a pathogen of pea crabs (*Pinnotheres pisum*) in South-west England, will invade the tissues of the mussel host of an infected pea crab (Sparks, 1985). It is unlikely that *Leptolegnia marina* has a significant impact on mussel populations.

MESOZOA

An orthonectid mesozoan belonging to the genus *Stoecharthrum* was observed in 15% of *M. edulis* from south Puget Sound, Washington on the west coast of the U.S.A. (Foster and Arnold, 1982). In moderate infections the orthonectid was usually confined to the mantle and gonads, but it also occurred in the digestive gland, adductor muscle, and gills of heavily infected mussels. Infection was associated with a reduction in gamete number and general emaciation of tissues. The geographical range and significance of *Stoecharthrum* spp. to *M. edulis* remains unknown.

CNIDARIA

Polyps of four species of hydroids have been described from the surface of tissues within the mantle cavity of mussels (Kubota, 1987). *Eugymnanthea inquilina* (= *Mytilhydra polimantii*) occurs in *M. galloprovincialis* along the coast of Italy (Lauckner, 1983; Kubota, 1989). Prevalences ranged from 29% in a lagoon to 8% in the Gulf of Naples, with large mussels harbouring many hundreds of polyps (Crowell, 1957). A second unnamed species of *Eugymnanthea*, originally thought to be a subspecies of the hydroid from Italy (*Eugymnanthea inquilina japonica*), occurs in three bivalves, including *M. edulis* from the Pacific coast of central Japan (Kubota, 1989). The third species *Eutima japonica* (=*Eugymnanthea cirrhifera*) occurs in at least 12 species of bivalves, including up to 70% *M. edulis* and *Mytilus coruscus* from the coast of Japan (Kubota, 1985a). From 11 to 2043 (mean 587 ± 497) polyps of *Eutima japonica* have been recorded in a single *M. edulis* (Kubota, 1983). The final species *Eucheilota intermedia* occurs in about 8% of *M. edulis* from central Japan (Kubota, 1985b).

In bivalve hosts, both *Eugymnanthea* and *Eutima* are strongly clustered, suggesting an initial invasion of the mantle cavity by one dispersal stage, followed by asexual reproduction by budding to form numerous hydroids (Crowell, 1957; Rees, 1967; Kubota, 1983; Lauckner, 1983). Regardless of the intensity of the infection, there is no invasion of the host tissue. The polyps are capable of slowly moving around (0.2mm h[-1]) on the ciliated epithelial surface of the mantle and foot, and some species survive well in the laboratory outside the host. However, the presence of a conspicuous basal disc used for attachment to the host, and the absence of the perisarc and stolons, are features adaptive to commensalism (Mattox and Crowell, 1951).

Although loss of cilia and the presence of granules in the epithelial cells of the mantle have been reported in *M. galloprovincialis* infected with *Eugymnanthea inquilina* in Italy, Kubota (1983) concluded that the bivalve host is not harmed by the commensal polyp. Rees (1967) proposed that mutualism may occur with the hydroid receiving some food and a sheltered environment in return for protecting the bivalve against other intruders with its nematocyst defences. In contrast, Lauckner (1983) suggested that the habitation of the mantle cavity by numerous hydroids may interfere with the normal filtering activity of the host. These conjectures have not been substantiated by experimental evidence.

PORIFERA

Three species of shell burrowing sponges of the class Demospongiae and family Clionidae have been reported in the shells of mussels. *Cliona celata* and *Cliona lobata* occurred in the shells of living and dead *M. edulis* from Isefjord in Denmark (Rasmussen, 1973). Infection with the more prevalent *C. lobata* usually started at the thickest part of the shell (umbo region) and a high percentage of infected shells were perforated. The thin sheet deposited by the mussel at the perforations provided little protection from crabs and other predators. The third species, *Cliona vastifica*, was identified in 60–100% of *M. galloprovincialis* from the Black Sea (Tkachuk, 1988). Heavily infected mussels had thicker shells (factor of 1.8x) and decreased body weight (factor of 0.7x). Thus, *Cliona* at high prevalence and intensity was detrimental to mussels in certain locations.

TURBELLARIA

Several species of Alloeocoela and Rhabdocoela Turbellaria were described from the mantle cavity and alimentary tract, respectively, of many species of Mytilidae. However, Turbellaria do not appear to be pathogenic to their hosts (Lauckner, 1983).

TREMATODA

Numerous species of digenean trematodes have been described from various species of mussels worldwide (Seed, 1976). In general, the trematodes that are the most pathogenic to mussels (as well as to other bivalves) are the species belonging to the families Bucephalidae and Fellodistomidae that utilize mussels as primary hosts. In such instances, the larval trematode life stages of sporocyst and cercaria occur within the mussel (Fig. 12.5). Although complete life cycles have not been elucidated for most species, in many cases, fishes are suspected to be the final host for these pathogenic trematodes of mussels. For species in which the complete life cycle is not known, larval stages have been described and arbitrarily placed in a collective group, treated as a genus under the name *Cercaria*. Thus, members of this collective group could belong to various families of Trematoda.

Members of the family Bucephalidae include two species from the eastern North Atlantic coast: *Prosorhynchus squamatus* in up to 38% of *M. edulis* and in less than 10% of *M. galloprovincialis*, and *Rudolphinus* (=*Prosorhynchus*) *crucibulum* in up to 60% *M. edulis*. A third species, *Cercaria noblei*, occurs in 0.42% of *M. californianus* from California and a fourth undescribed species was reported from *Mytilus platensis* in Argentina. The sporocysts of each of these species form a dense branching interwoven network that infiltrates most organs of its mussel host, especially the gonad (Giles, 1962; Matthews, 1973; Lauckner, 1983; Morris, 1983; Coustau et al., 1990).

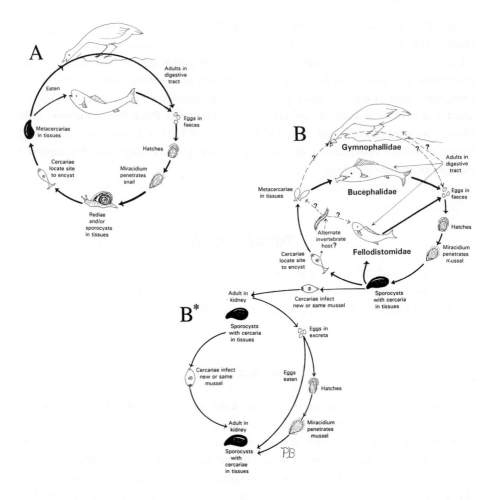

Fig. 12.5. Generalized life cycles of trematoda that infect mussels (*Mytilus* sp.). (A) represents the life cycle of the majority of trematode species in the marine environment, where the metacercariae encyst in, or on, mussels as well as other animals and/or plants. (B) represents the life cycle for members of the family Fellodistomidae, Bucephalidae, and possibly Gymnophallidae as indicated. (B*) represents the progenetic development of *Proctoeces maculatus*, which is capable of completing its entire life cycle in mussels. However, like other Fellodistomidae, adult forms also occur in the intestinal tract of fishes (usually tropical mollusc-eating fishes of the family Sparidae and Labridae) as illustrated in (B).

The resulting damage often led to castration (Seed, 1976; Coustau et al., 1990). In addition to castration, the bucephalid *Prosorhynchus squamatus* caused muscle weakness and gaping in *M. edulis* from northwestern Europe, which can be a problem for shipping and marketing (Lauckner, 1983; Coustau et al., 1990). Coustau et al. (1990) showed that *M. edulis* is more susceptible to *P. squamatus* than hybrids between *M. edulis* and *M. galloprovincialis* from Britany, France and suggested that it may be possible to select for a mussel stock that is resistant to this parasite.

In the family Fellodistomidae, *Proctoeces maculatus* (= *Cercaria milfordensis*) was reported from up to 46% of *M. edulis* and *M. galloprovincialis* on both sides of the North Atlantic Ocean and in the Mediterranean and Black Seas (Uzmann, 1953; Lang and Dennis, 1976; Mulvey and Feng, 1981; Shchepkina, 1982; Machkevski, 1985; Feng, 1988). In *M. galloprovincialis* from the Black Sea, up to 28,000 sporocysts per mussel, comprising 20% of the biomass of the soft tissues, were observed (Machkevski, 1985). Mussels may also serve as an alternate final host for *P. maculatus* in lieu of the usual fish final host (Stunkard and Uzmann, 1959; Lang and Dennis, 1976). As a result of this capability of progenesis (abbreviation of the developmental cycle by elimination of the final (definitive) host, Fig. 12.5), the current geographic range of *P. maculatus* could inadvertently be extended by movement of infected mussels, through for example, aquaculture activities.

Proctoeces maculatus is primarily a parasite of the vascular system in the mantle of mussels (Uzmann, 1953). Infection causes an alteration in haemolymph components, a sharp decrease in energy stores, a reduction in growth rate, and weakness with respect to valve closure and attachment to the substrate (Mulvey and Feng, 1981; Machkevski, 1982, 1988; Shchepkina, 1982). In heavily infected mussels, the numerous sporocysts developing in the mantle can seriously reduce glycogen content of the tissues and efficiency of the circulatory system, resulting in disturbances to gametogenesis and possibly castration and death (Uzmann, 1953; Mulvey and Feng, 1981; Machkevski and Shchepkina, 1985; Pascual et al., 1987; Feng, 1988). One species identified as *Cercaria tenuans* (family Fellodistomidae, and possibly homologous to *Proctoeces maculatus*), was suspected of causing extensive mortalities in cultured mussels in the Laguna Veneta, Italy. This parasite was thought to have been introduced via a depuration plant located on the shore of the lagoon (Munford et al., 1981).

Parasitic castration caused by either *Prosorhynchus squamatus* or *Proctoeces maculatus* was once thought to be beneficial to mussel culture. This was especially so in southern waters where mussels spawn more heavily, since parasitized mussels remained in good condition all summer. However, consumption of trematode-infested molluscs may be hazardous due to accumulation of toxic metabolites such as butyric and other short-chain fatty acids. These result from the degeneration of the host's neutral fats by parasite-secreted enzymes, and thus render mussels unsuitable for human consumption (Cheng, 1967; Lauckner, 1983).

Other species of trematodes utilize mussels as secondary or final hosts and are found worldwide in most mussel populations (Fig. 12.5). The encysted metacercarial stage is found in mussels serving as a secondary host, and gravid adult trematodes of some species will occasionally develop in mussels, thereby making them a final host (Lauckner, 1983). Except for *Proctoeces maculatus* (mentioned above), which in some situations will utilize mussels as a final host, most species of trematodes in these two categories are relatively nonpathogenic to mussels. However, infection can cause: (1) compression of adjacent tissues in *M. edulis* by encysted echinostomatid metacercariae of *Himasthla quissetensis* resulting in loss of normal organ architecture (Sparks, 1985), (2) reduced byssal production and impaired shell cleaning in young *M. edulis* infected with metacercariae of the bird trematode *Himasthla elongata* (family Echinostomatidae) from the North Sea and adjacent areas (Lauckner, 1984), and (3), induction of pearl formation by several species of bird gymnophallid metacercaria encysted between the mantle and shell of *M. edulis* on the European and American North Atlantic coasts, and of *M. galloprovincialis* in the Mediterranean and Black Seas, have been noted (Gotting, 1979; Lutz, 1980). Although pearl formation induced by metacercariae rarely causes mussel mortalities, the presence of pearls seriously affects the exploitation of some stocks. Large quantities of mussels from various areas along the Atlantic coast of North America, Denmark, England and France cannot be marketed because of the presence of numerous pearls (>1mm diameter) throughout the mantle epithelium (Lutz, 1980; see also p.487 Chapter 10).

CESTODA

Adult cestodes have not been reported from mussels. However, metacestodes of *Tylocephalum* spp. were observed in 11.4% *M. edulis* (as well as other bivalves) from one locality in Japan (Sakaguchi, 1973). Associated disease was not evident. The final host for this parasite is thought to be sharks that feed on molluscs.

NEMATODA

Parasitic nematodes have rarely been reported from mussels. However, in the North Atlantic, the 'codworm' *Phocanema decipiens* has been observed in *M. edulis*, which may serve as a paratenic host (i.e. accidental host for the transport of the larvae into the true teleost intermediate host) for this parasite (Lauckner, 1983).

ANNELIDA

The polychaete annelid *Polydora ciliata* is responsible for substantial mortalities and loss of market quality among mussels in European waters (Lauckner, 1983). The burrows excavated by *P. ciliata* in the mussel shell not only cause unsightly blisters containing compacted mud, but also result in a significant reduction in shell strength, thereby increasing susceptibility to predation by birds and shore crabs. This is particularly a problem for mussels over 6cm in shell length (Kent, 1981). Nacreous blisters produced by mussels in response to *P. ciliata*, which burrows into the region of the adductor muscle, can result in atrophy and detachment of muscle tissue and possible interference with gamete production, if the calcareous ridges press against the mantle (Lauckner, 1983).

In southern Australia, five species of polydorid polychaetes (*Polydora haswelli, Polydora hoplura, Polydora websteri, Boccardia chilensis*, and *Boccardia polybranchia*) were observed in up to 95% of *M. edulis*. Although the intensity of infection was generally low, about 15% of the mussels from two localities had serious shell damage attributed to polydorids. The most heavily infested mussels were from bottom samples (Pregenzer, 1983).

COPEPODA

Two species of parasitic copepods of the genus *Mytilicola* have been reported in mussels. *Mytilicola intestinalis* is found in *Mytilus edulis* and *M. galloprovincialis* as well as other bivalves, but only in European waters (Davey and Gee, 1976; Seed, 1976; Figueras and Figueras, 1981; Dethlefsen, 1985). Although this parasite has an annual cycle, the prevalence of infection in mussels from some localities only falls below 90% during the early summer months and intensity of infection often exceeds 30 copepods per mussel (Davey, 1989). The infective but free swimming copepodid stage generally occurs low in the water column. Thus, mussels growing close to the surface or in suspended culture are less likely to be as severely infected as those in bottom culture (Incze and Lutz, 1980; Sparks, 1985). Several workers have concluded that some of the periodic mass mortalities among cultured mussels in Europe were attributable to *M. intestinalis* (Sparks, 1985; Blateau et al., 1990). However, these conclusion were: (1) not substantiated by statistical analysis, (2) not supported by experimental evidence, and (3) did not rule out the possibility that microscopic pathogens were responsible for the mortalities (Lauckner, 1983). More recently, it has been suggested that *M. intestinalis* causes an alteration in the oocytes of *M. edulis* and thus, has an adverse affect on the fecundity as well as the condition of its host (Durfort et al., 1982; Theisen, 1987; Tiews, 1988). However, from the results of a 10-year study conducted in

Cornwall, England, Davey (1989) concluded that in every respect M. *intestinalis* exhibits the features of a commensal rather than those of a harmful parasite. Nevertheless, more work is required before the pest status of M. *intestinalis* is fully elucidated, especially with respect to its synergistic relations with other pathogens and/or pollutants (Davey and Gee, 1988).

The second species of parasitic copepod is *Mytilicola orientalis* which was thought to originate in *Mytilus crassitesta* (= *Mytilus coruscus*) from Japan. This parasite has now spread along the Pacific coast of North America, where it is found in up to 58% of M. *edulis*, and 65% of M. *californianus*, as well as in other bivalves (Chew et al., 1964; Bernard, 1969). *Mytilicola orientalis* was recently introduced into France with imported *Crassostrea gigas* and is now found in both *Mytilus edulis* and M. *galloprovincialis*, where dual infections with *Mytilicola intestinalis* occur (Lauckner, 1983). Apparently, M. *orientalis* is not pathogenic to mussels.

Several other species of parasitic copepods have been reported attached to the gills of mussels, but their prevalences and intensities were low (Cheng, 1967; Lauckner, 1983; Durfort et al., 1990; Poquet et al., 1990).

DECAPODA

Seven species of brachyuran crabs, all belonging to the family Pinnotheridae have been observed in the mantle cavities of mussels. *Pinnotheres pisum* occurs in M. *edulis* and M. *galloprovincialis* from European waters (Lauckner, 1983); *Pinnotheres maculatus* and *Pinnotheres ostreum* occur in M. *edulis* along the Atlantic coast of North and South America (Kruczynski, 1974; Lauckner, 1983); *Pinnotheres hickmani* occurs in *Mytilus planulatus* along the coast of southern Australia (Pregenzer, 1983); *Pinnotheres sinensis* and *Pinnotheres pholadis* occur in *Mytilus coruscus* (Lauckner, 1983) and M. *galloprovincialis* (Yoo and Kajihara, 1985), respectively, around the coast of Japan; and *Fabia subquadrata* occurs in M. *edulis* and M. *californianus* along the west coast of North America (Anderson, 1975). None of these crabs is host-specific. In general, the prevalence of infection can be locally high (often reaching 100%), with the highest prevalence in larger mussels and mussels from subtidal locations (Seed, 1969; Kruczynski, 1974; Anderson, 1975; Lauckner, 1983; Pregenzer, 1983; Yoo and Kajihara, 1985).

The relationship between the pinnotherid crabs and their hosts has been described as both commensal and parasitic. Apparently, the crabs do not feed on the host tissues, but upon the material collected by the gills of the host. Nevertheless, their presence within the mantle cavity may elicit a variety of host responses, ranging from slight irritation to structural alterations and pathology (Lauckner, 1983). Specific examples of these are gill damage, emaciation, shell thickening, delay in gonad maturation, and reduction of reproductive capacity (Seed, 1969; Anderson, 1975; Yoo and Kajihara, 1985).

PANTOPODA

Although records of associations between sea spiders and bivalves are rare, *Achelia chelata* has been reported from the mantle cavity of up to 50% of *Mytilus californianus* from two localities in California (Lauckner, 1983). Damage to the gills, gonadal tissues, visceral mass, foot, and palps in several mussels was reported (Benson and Chivers, 1960).

DISEASES OF UNKNOWN CAUSE

The final condition to be considered may result from a variety of causes and thus, may represent more than one disease. However, due to lack of information, the complex of proliferative blood cell disorders will be presented here as one disease under the common name of haemocytic (or haemic and occasionally called haematopoietic) neoplasia. The term neoplasia refers to new growth of abnormal tissue and is based strictly on histological criteria, recognizing that all necessary biological criteria (i.e. irreversible uncontrolled growth, metastases, transplantation, and host death) may not have been demonstrated (Bower, 1989). Haemocytic neoplasia is characterized by proliferative growth of abnormal haemocytes to a terminal stage in which these abnormal cells are found in overwhelming numbers that possibly occlude most vascular spaces (Fig. 12.6).

Prevalence of haemocytic neoplasia was usually less than 4% in mussels from the coast of Europe and the east coast of North America (Lowe and Moore, 1978; Green and Alderman, 1983; Gutiérrez and Sarasquete, 1986; Rasmussen, 1986b; Figueras et al., 1991). However, prevalences were much higher and have exceeded 40% in some localities on the west coast of North America (Cosson-Mannevy et al., 1984; Elston et al., 1988a; Bower, 1989; Elston et al., 1990). Haemocytic neoplasia in mussels from Puget Sound, Washington, and British Columbia, Canada, is progressive, impairs defence mechanisms, is lethal, and can be transmitted by cohabitation (Elston et al., 1988a, b; Bower, 1989; Kent et al., 1989). Although the etiological agent has yet to be identified, Elston et al. (1988b) suggested a retrovirus may be involved. Examples of remission were observed in some mussels.

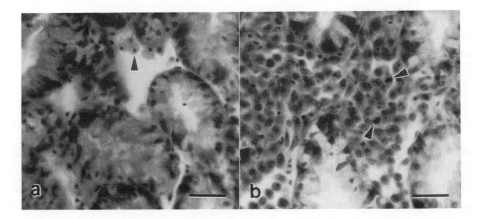

Fig. 12.6. Histological sections through the digestive gland of a normal *Mytilus edulis* (a), showing the low numbers of normal haemocytes (arrow) usually present in the spaces between the diverticula, and of a diseased *M. edulis* showing a heavy intensity of haemocytic neoplasia (b) with an abundance of abnormal haemocytes (each with a large nucleus containing two or more nucleoli and a small amount of peripheral cytoplasm) and haemocytes undergoing mitosis (arrows) between the diverticula. (Scale bars = 25μm; haematoxylin and eosin stain).

Mussels over 4cm in shell length from Puget Sound and British Columbia suffer annual cumulative mortalities often exceeding 75% (Heritage, 1983; Skidmore and Chew, 1985; Bower, 1989). Unexpectedly, there was poor correlation between the prevalence of haemocytic neoplasia and the cumulative mortalities of mussels from several localities in British Columbia (Bower, 1989). High mortalities may be related to reproductive stress resulting from maximum reproductive output at an early age (Emmett et al., 1987). Despite these high mortalities mussel recruitment is high, making mussels one of the major fouling pests of the salmon net-pen industry in British Columbia. This modification of the life cycle to early maturation in a circumpolar species may be induced by unidentified environmental factors in some localities. Another possibility is that these mussels are not *Mytilus edulis* as previously concluded (Bernard, 1983), but a different subspecies or species (i.e. *Mytilus trossulus*) (see McDonald and Koehn 1988; and Chapter 1). Regardless of the reason, these mussels seem to have a life cycle in which reproduction, followed by senescence and death for the majority of individuals, occurs at an earlier age and smaller body size than for *M. edulis* on the east coast of North America.

CONCLUSION

High mortalities caused by parasites or infectious diseases, comparable to those encountered by the oyster culture industry, have not yet been encountered in mussels. However, several parasites with the potential to cause similar epizootics have been observed in mussels. Possibly, mussels from localities where these parasites occur are resistant to the disease they can cause. But, the accidental introduction of these pathogens into new localities, through the movement of mussels, could initiate an epizootic (Bower and Figueras, 1989). In order to avoid this type of disaster, all movements of live shellfish must be conducted with caution. In the interim, it is essential that information on parasites and diseases of mussels be amassed so that the risk involved with past and impending movements can be accurately assessed.

ACKNOWLEDGEMENTS

I am grateful for the technical assistance of G. Meyer, J. Blackbourne and D. Nishimura.

REFERENCES

Anderson, G.L., 1975. The effects of intertidal height and the parasitic crustacean *Fabia subquadrata* Dana on the nutrition and reproductive capacity of the California sea mussel *Mytilus californianus* Conrad. Veliger, 17: 299-306.

Antipa, G.A. and Dolan, J., 1985. *Mytilophilus pacificae* n. g., n. sp. : A new mytilid endocommensal ciliate (Scuticociliatida). Trans. Am. Microsc. Soc., 104: 360-368.

Auffret, M. and Poder, M., 1983. Recherches sur *Marteilia maurini*, parasite de *Mytilus edulis* sur les côtes de Bretagne Nord. Rev. Trav. Inst. Pêches Marit., 47: 105-109.

Balouet, G. and Poder, M., 1985. Current status of parasitic and neoplastic disease of shellfish: a review. In: A.E. Ellis (Editor), Fish and Shellfish Pathology. Academic Press, London, pp. 371-380.

Benson, P.H. and Chivers, D.D., 1960. A pycnogonid infestation of *Mytilus californianus*. Veliger, 3: 16-18.

Bernard, F.R., 1969. The parasitic copepod *Mytilicola orientalis* in British Columbia bivalves. J. Fish. Res. Board Can., 26: 190-191.

Bernard, F.R., 1983. Catalogue of the living Bivalvia of the eastern Pacific Ocean: Bering Strait to Cape Horn. Can. Spec. Publ. Fish. Aquat. Sci., 61: 102pp.

Birkbeck, T.H., McHenery, J.G. and Nottage, A.S., 1987. Inhibition of filtration in bivalves by marines vibrios. Aquaculture, 67: 247-248.

Blateau, D., Le Coguic, Y., Mialhe, E., Grizel, H. and Flamion, G., 1990. Traitement des moules (*M. edulis*) contre le copepode *Mytilicola intestinalis*. In: A. Figueras (Editor), Abstracts, 4th Internat. Colloq. Pathol. Mar. Aquacul., Vigo, Spain, 1990. pp. 97-98.

Bower, S.M., 1987. *Labyrinthuloides haliotidis* n. sp. (Protozoa: Labyrinthomorpha), a pathogenic parasite of small juvenile abalone in a British Columbia mariculture facility. Can. J. Zool., 65: 1996-2007.

Bower, S.M., 1989. The summer mortality syndrome and haemocytic neoplasia in blue mussels (*Mytilus edulis*) from British Columbia. Can. Tech. Rep. Fish. Aquat. Sci., 1703: 65pp.

Bower, S.M. and Figueras, A.J., 1989. Infectious diseases of mussels, especially pertaining to mussel transplantation. World Aquacul., 20: 89-93.

Cheng, T.C., 1967. Marine molluscs as hosts for symbioses with a review of known parasites of commercially important species. In: F.S. Russell (Editor), Advances in Marine Biology, 5. Academic Press, New York, 424pp.

Chew, K.K., Sparks, A.K. and Katkansky, S.C., 1964. First record of *Mytilicola orientalis* in the California mussel *Mytilus californianus* Conrad. J. Fish. Res. Board Can., 21: 205-207.

Comps, M., Pichot, Y. and Papagianni, P., 1981. Recherche sur *Marteilia maurini* n. sp. parasite de la moule *Mytilus galloprovincialis* Lmk. Rev. Trav. Inst. Pêches Marit., 45: 211-214.

Cosson-Mannevy, M.A., Wong, C.S. and Cretney, W.J., 1984. Putative neoplastic disorders in mussels (*Mytilus edulis*) from southern Vancouver Island waters, British Columbia. J. Invertebr. Pathol., 44: 151-160.

Coustau, C., Combes, C., Maillard, C., Renaud, F. and Delay, B., 1990. *Prosorhynchus squamatus* (Trematoda) parasitosis in the *Mytilus edulis-Mytilus galloprovincialis* complex: specificity and host-parasite relationships. In: F.O. Perkins and T.C. Cheng (Editors), Pathology in Marine Science. Academic Press, New York, pp. 291-298.

Crowell, S., 1957. *Eugymnanthea*, a commensal hydroid living in pelecypods. Pubbl. Stn. Zool. Napoli, 30: 162-167.

DaRos, L. and Massignan, F., 1985. Indagine parassitologica su *Mytilus galloprovincialis* Lmk. allevato in Laguna di Venezia (Bacino di Chioggia). Oebalia. (N.S), 11: 809-811.

Davey, J.T., 1989. *Mytilicola intestinalis* (Copepoda: Cyclopoida): A ten year survey of infested mussels in a Cornish estuary, 1978-1988. J. Mar. Biol. Ass. U.K., 69: 823-836.

Davey, J.T. and Gee, J.M., 1976. The occurrence of *Mytilicola intestinalis* Steuer, an intestinal copepod parasite of *Mytilus*, in the south-west of England. J. Mar. Biol. Ass. U.K., 56: 85-94.

Davey, J.T. and Gee, J.M., 1988. *Mytilicola intestinalis*, a copepod parasite of blue mussels. Am. Fish. Soc. Spec. Publ., 18: 64-73.

Dethlefsen, V., 1985. *Mytilicola intestinalis*, parasitism. In: C.J. Sindermann (Editor), Identification Leaflets for Diseases and Parasites of Fish and Shellfish. Cons. Int. Explor. Mer, 4pp.

Durfort, M., Bargalló, B., Bozzo, M.G., Fontarnau, R. and López-Camps, J., 1982. Alterations des ovocytes de *Mytilus edulis* L. (Mollusca, Bivalvia) dues a l'infestation de la moule par *Mytilicola intestinalis*, Steuer (Crustacea, Copepoda). Malacologia, 22: 55-59.

Durfort, M., Sagrista, E., Bozzo, M.G., Poquet, M., Valero, J.G., Amor, M.J., Ferrer, J. and Ribes, E., 1990. Modified cilia in vibratil epithelia of mussels infected by *Mytilicola intestinalis* and *Modiolicola gracilis* (Crustacea, Copepoda). In: A. Figueras (Editor), Abstracts, 4th Internat. Colloq. Pathol. Mar. Aquacul., Vigo, Spain, 1990. pp. 108-109.

Elston, R.A., Kent, M.L. and Drum, A.S., 1988a. Progression, lethality and remission of hemic neoplasia in the bay mussel *Mytilus edulis*. Dis. Aquat. Org., 4: 135-142.

Elston, R.A., Kent, M.L. and Drum, A.S., 1988b. Transmission of hemic neoplasia in the bay mussel, *Mytilus edulis*, using whole cells and cell homogenate. Dev. Comp. Immunol., 12: 719-727.

Elston, R. A., Bonar, D., Brooks, K., Gee, A., Mialhe, E., Moore, J., Noel, D. and Stephens, L., 1990. Studies on pathogenesis and etiology of circulating sarcomas in *Mytilus*. In: A. Figueras (Editor), Abstracts, 4th Internat. Colloq. Pathol. Mar. Aquacul., Vigo, Spain, 1990. pp. 119.

Emmett, B., Thompson, K. and Popham, J.D., 1987. The reproductive and energy storage cycles of two populations of *Mytilus edulis* (Linné) from British Columbia. J. Shellfish Res., 6: 29-36.

Farley, C.A., 1988. A computerized coding system for organs, tissues, lesions, and parasites of bivalve mollusks and its application in pollution monitoring with *Mytilus edulis*. Mar. Environ. Res., 24: 243-249.

Feng, S.L., 1988. Host response to *Proctoeces maculatus* infection in the blue mussel, *Mytilus edulis* L. J. Shellfish Res., 7: 118.

Figueras, A. and Figueras, A.J., 1981. *Mytilicola intestinalis* Steuer, en el mejillón de la ría de Vigo. (NO de España). Inv. Pesq. (Barc.), 45: 263-278.

Figueras, A.J. and Montes, J., 1988a. Parasites and diseases of mussels (*Mytilus edulis* and *M. galloprovincialis*) cultivated on rafts in Galicia (NW Spain). In: F.O. Perkins and T.C. Cheng (Editors), Abstracts, 3rd Internat. Colloq. Pathol. Mar. Aquacul., Virginia, U.S.A., 1988. pp. 95-96.

Figueras, A.J. and Montes, J., 1988b. Aber disease of edible oysters caused by *Marteilia refringens*. Am. Fish. Soc. Spec. Publ., 18: 38-46.

Figueras, A.J., Jardon, C.F. and Caldas, J.R. 1991. Diseases and parasites of mussels (*Mytilus edulis* Linneaus 1758) from two sites on the east coast of the United States. J. Shellfish Res., 10: 89-94.

Foster, C.A. and Arnold, T.W., 1982. An orthonectid parasite in a new host, *Mytilus edulis*, and its relationship to seasonal mussel mortalities in south Puget Sound, Washington. J. Shellfish Res., 2: 118-119.

Giles, D.E., 1962. New bucephalid cercaria from the mussel *Mytilus californianus*. J. Parasitol., 48: 293-295.

González, P., Pascual, C., Quintana, R. and Morales, J., 1987. Parásitos del mejillón gallego cultivado: 1. Protozoos con especial referencia a *Marteilia maurini* y *Steinhausia mytilovum*. Alimentaria, 24: 37-44.

Götting, K.J., 1979. Durch parasiten induzierte perlbildung bei *Mytilus edulis* L. (Bivalvia). Malacologia, 18: 563-567.

Green, M. and Alderman, D.J., 1983. Neoplasia in *Mytilus edulis* L. from United Kingdom waters. Aquaculture, 30: 1-10.

Grizel, H., Mialhe, E., Chagot, D., Boulo, V. and Bachère, E., 1988. Bonamiasis: a model study of diseases in marine molluscs. Am. Fish. Soc. Spec. Publ., 18: 1-4.

Gutiérrez, M. and Sarasquete, M.C., 1986. Un caso de hemocitosarcoma hialino en el mejillón, *Mytilus edulis* L. (Pelecypoda: Mytilidae) de la costa NO de España. Inv. Pesq. (Barc.), 50: 265-269.

Heritage, G.D., 1983. A blue mussel, (*Mytilus edulis* Linnaeus), culture pilot project in south coastal British Columbia. Can. Tech. Rep. Fish. Aquat. Sci., 1174: 27pp.

Hillman, R.E., Boehm, P.D. and Freitas, S.Y., 1988. A pathology potpourri from the NOAA Mussel Watch Program. J. Shellfish Res., 7: 216-217.

Incze, L.S. and Lutz, R.A., 1980. Mussel culture: an east coast perspective. In: R.A. Lutz (Editor), Mussel culture and harvest: a North American perspective. Elsevier Science Publishers, B.V., New York, pp. 99-140.

Kent, M.L., Elston, R.A., Wilkinson, M.T. and Drum, A.S., 1989. Impaired defense mechanisms in bay mussels, *Mytilus edulis*, with hemic neoplasia. J. Invertebr. Pathol., 53: 378-386.

Kent, R.M.L., 1981. The effect of *Polydora ciliata* on the shell strength of *Mytilus edulis*. J. Cons. Int. Explor. Mer, 39: 252-255.

Kruczynski, W.L., 1974. Relationship between depth and occurrence of pea crabs, *Pinnotheres maculatus*, in blue mussels, *Mytilus edulis*, in the vicinity of Woods Hole, Massachusetts. Chesapeake Sci., 15: 167-169.

Kubota, S., 1983. Studies on life history and systematics of the Japanese commensal hydroids living in bivalves, with some reference to their evolution. J. Fac. Sci. Hokkaido Univ. Ser. VI, Zool., 23: 296-402.

Kubota, S., 1985a. Morphological variation of medusa of the northern form of *Eutima japonica* Uchida. J. Fac. Sci. Hokkaido Univ. Ser. VI, Zool., 24: 144-153.

Kubota, S., 1985b. Systematic study on a bivalve-inhabiting hydroid *Eucheilota intermedia* Kubota from central Japan. J. Fac. Sci. Hokkaido Univ. Ser. VI, Zool., 24: 122-143.

Kubota, S., 1987. The origin and systematics of four Japanese bivalve-inhabiting hydroids. In: J. Bouillon, F. Boero, F. Cicogna and P.F.S. Cornelius (Editors), Modern Trends in the Systematics, Ecology, and Evolution of Hydroids and Hydromedusae. Oxford University Press, Oxford, pp. 275-287.

Kubota, S., 1989. Systematic study of a paedomorphic derivative hydrozoan *Eugymnanthea* (Thecata-Leptomedusae). Zool. Sci. (Tokyo), 6: 147-154.

Lang, W.H. and Dennis, E.A., 1976. Morphology and seasonal incidence of infection of *Proctoeces maculatus* (Looss, 1901) Odhner, 1911 (Trematoda) in *Mytilus edulis* L. Ophelia, 15: 65-75.

Lauckner, G., 1983. Diseases of Mollusca: Bivalvia. In: O. Kinne (Editor), Diseases of Marine Animals. Volume II: Introduction, Bivalvia to Scaphopoda. Biologische Anstalt Helgoland, Hamburg, pp. 477-961.

Lauckner, G., 1984. Impact of trematode parasitism on the fauna of a North Sea tidal flat. Helgoländer Wiss. Meeresunters., 37: 185-199.

Levine, N.D., Corliss, J.O., Cox, F.E.G., Deroux, G., Grain, J., Honigberg, B.M., Leedale, G.F., Loeblich, A.R., Lom, J., Lynn, D., Merinfeld, E.G., Page, F.C., Poljansky, G., Sprague, V., Vavra, J. and Wallace, F.G., 1980. A newly revised classification of the Protozoa. J. Protozool., 27: 37-58.

Li, M.F. and Clyburne, S., 1979. Mortalities of blue mussel (*Mytilus edulis*) in Prince Edward Island. J. Invertebr. Pathol., 33: 108-110.

López, M.C., Carballal, M.J., Mourelle, S.G., Villalba, A. and Montes, J., 1990. Parasites and diseases of mussels, *Mytilus galloprovincialis* Lmk., from estuaries of Galicia. In: A. Figueras (Editor), Abstracts, 4th Internat. Colloq. Pathol. Mar. Aquacul., Vigo, Spain, 1990. pp. 105.

Lowe, D.M. and Moore, M.N., 1978. Cytology and quantitative cytochemistry of a proliferative atypical hemocytic condition in *Mytilus edulis* (Bivalvia, Mollusca). J. Natl. Cancer Inst., 60: 1455-1459.

Lutz, R.A., 1980. Pearl incidences: mussel culture and harvest implications. In: R.A. Lutz (Editor), Mussel Culture and Harvest: a North American Perspective. Elsevier Science Publishers, B.V., New York, pp.193-222.

Machkevski, V.K., 1982. Osobennosti razvitiya i biologii partenit Proctoeces maculatus (Trematoda) v chernomorskikh midiyakh (in Russian, with English abstract). Zool. Zh., 61: 1635-1642.

Machkevski, V.K., 1985. Some aspects of the biology of the trematode, *Proctoeces maculatus*, in connection with the development of mussel farms on the Black Sea. In: J.W. Hargis (Editor), Parasitology and Pathology of Marine Organisms of the World Ocean. U.S. Department of Commerce, pp. 109-110.

Machkevski, V.K., 1988. Vliyanie partenit trematody *Proctoeces maculatus* na rost chernomorskoy midii *Mytilus galloprovincialis* (in Russian, with English abstract). Parazitologiya (Leningr.), 22: 341-344.

Machkevski, V.K. and Shchepkina, A.M., 1985. Zarazhennot chernomorskikh midiy partenitami *Proctoeces maculatus* i ikh vliyanie na soderzhanie glikogena v tkaniyakh khozyaev (in Russian, with English abstract). Ekol. Moria., 20: 69-73.

Margolis, L., Esch, G.W., Holmes, J.C., Kuris, A.M. and Shad, G.A., 1982. The use of ecological terms in parasitology (Report of an *ad hoc* committee of the American Society of Parasitologists). J. Parisitol., 68: 131-133.

Matthews, R.A., 1973. The life-cycle of *Prosorhynchus crucibulum* (Rudolphi, 1819) Odhner, 1905, and a comparison of its cercaria with that of *Prosorhynchus squamatus* Odhner, 1905. Parasitology, 66: 133-164.

Mattox, N.T. and Crowell, S., 1951. A new commensal hydroid of the mantle cavity of an oyster. Biol. Bull., 101: 162-170.

McDonald, J.H. and Koehn, R.K., 1988. The mussels *Mytilus galloprovincialis* and *M. trossulus* on the Pacific coast of North America. Mar. Biol., 99: 111-118.

McGladdery, S., 1990. Shellfish parasites and diseases on the east coast of Canada. Bull. Aqua. Assoc. Can., 90(3): 14-18.

Meixner, R., 1984. On a microalgal infection of *Mytilus edulis*. ICES, C.M. 1984/K:30: 4pp.

Mialhe, E., Bachere, E., Le Bec, C. and Grizel, H., 1985. Isolement et purification de *Marteilia* (Protozoa: Ascetospora) parasites de bivalves marins. C. R. Hebd. Seances Acad. Sci. (III), Paris., 301: 137-142.

Morris, M.R., 1983. Estados larvales de trematodes digeneos en moluscos marinos *Mytilus platensis* d'Orb. y *Brachyodontes rodriguezi* d'Orb. Rev. Mus. La Plata (Secc. Zool.), 13 (135): 65-71.

Mulvey, M. and Feng, S.Y., 1981. Hemolymph constituents of normal and *Proctoeces maculatus* infected *Mytilus edulis*. Comp. Biochem. Physiol., 70A: 119-125.

Munford, J.G., DaRos, L. and Strada, R., 1981. A study on the mass mortality of mussels in the Laguna Veneta. J. World Maricul. Soc., 12: 186-199.

Nottage, A.S. and Birkbeck, T.H., 1987. The role of toxins in *Vibrio* infections of bivalve molluscs. Aquaculture, 67: 244-246.

Pascual, C., Quintana, R., Molares, J. and González, P., 1987. Parásitos del mejillón gallego cultivado: 2. Metazoos, con especial referencia a tremátodos y copépodos. Alimentaria, 24: 31-36.

Pauley, G.B., Sparks, A.K., Chew, K.K. and Robbins, E.J., 1966. Infection in Pacific coast mollusks by thigmotrichid ciliates. Proc. Natl. Shellfish. Assoc., 56: 8.

Poquet, M., Ribes, E., Amor, M.J., Bozzo, M.G., Garcia, J., Ferrer, J., Sagrista, E. and Durfort, M., 1990. Ultrastructural study of tegument of *Modiolicola gracilis*, parasitic Copepoda in the gills of *Mytilus edulis* from Delta de l'Ebre (Tarragona, Spain). In: A. Figueras (Editor), Abstracts, 4th Internat. Colloq. Pathol. Mar. Aquacul., Vigo, Spain, 1990. pp. 106-107.

Pregenzer, C., 1983. Survey of metazoan symbionts of *Mytilus edulis* (Mollusca: Pelecypoda) in Southern Australia. Aust. J. Mar. Freshw. Res., 34: 387-396.

Prieur, D., Barbosa, T. and Marhic, A., 1985. Les communautes bacteriennes des mollusques bivalves et du sediment en Rade de Brest. Oceanis, 11: 287-294.

Quayle, D.B., 1978. A preliminary report on the possibilities of mussel culture in British Columbia. Fish. Mar. Serv. Tech. Rep., 815: 37pp.

Rasmussen, E., 1973. Systematics and ecology of the Isefjord marine fauna (Denmark) with a survey of the eelgrass (Zostera) vegetation and its communities. Ophelia, 11: 1-507.

Rasmussen, L.P.D., 1986a. Virus-associated granulocytomas in the marine mussel, Mytilus edulis, from three sites in Denmark. J. Invertebr. Pathol., 48: 117-123.

Rasmussen, L.P.D., 1986b. Occurrence, prevalence and seasonality of neoplasia in the marine mussel Mytilus edulis from three sites in Denmark. Mar. Biol., 92: 59-64.

Rees, W.J., 1967. A brief survey of the symbiotic associations of Cnidaria with Mollusca. Proc. Malacol. Soc. Lond., 37: 213-231.

Sakaguchi, S., 1973. A larval tapeworm, Tylocephalum, in marine molluscan shellfishes. Bull. Nansei Reg. Fish. Res. Lab., 6: 1-8.

Seed, R., 1969. The incidence of the pea crab, Pinnotheres pisum in the two types of Mytilus (Mollusca: Bivalvia) from Padstow, south-west England. J. Zool. (Lond.), 158: 413-420.

Seed, R., 1976. Ecology. In: B.L. Bayne (Editor), Marine mussels: their ecology and physiology. Cambridge University Press, Cambridge, pp. 13-65.

Shchepkina, A.M., 1982. O patogennom vozdeistvii lichinok trematod na Mytilus galloprovincialis. In: R.N. Burukovskii (Editor), Problemy Ratsional'nogo Ispol'zovaniya Promyslovykh Bespozvonochnykh (in Russian). Tezisy Dokladov III Vsesoyuznoi Konferentsii, 1982. AtlantNIRO, Kaliningrad, pp. 229-231.

Sherburne, S.W. and Bean, L.L., 1986. A synopsis of the most serious diseases occurring in Maine shellfish. Fish Health Sec., Am. Fish. Soc. Newsletter., 14: 5.

Skidmore, D. and Chew, K.K., 1985. Mussel aquaculture in Puget Sound. Wash. Sea Grant Tech. Rep., WSG 85-4: 57pp.

Sparks, A.K., 1985. Synopsis of Invertebrate Pathology Exclusive of Insects. Elsevier Science Publishers B.V., New York, 423pp.

Sprague, V., 1965. Observations on Chytridiopsis mytilovum (Field), formerly Haplosporidium mytilovum Field, (Microsporida?). J. Protozool., 12: 385-389.

Stunkard, H.W. and Uzmann, J.R., 1959. The life-cycle of the digenetic trematode, Proctoeces maculatus (Looss, 1901) Odhner, 1911 (Syn. P. subtenuis (Linton, 1907) Hanson 1950), and description of Cercaria adranocerca n. sp. Biol. Bull., 116: 184-193.

Sunila, I., 1987. Histopathology of mussels (Mytilus edulis L.) from the Tvärminne area, the Gulf of Finland (Baltic Sea). Ann. Zool. Fenn., 24: 55-69.

Taylor, R.L., 1966. Haplosporidium tumefacientis sp. n., the etiologic agent of a disease of the California sea mussel, Mytilus californianus Conrad. J. Invertebr. Pathol., 8: 109-121.

Theisen, B.F., 1987. Mytilicola intestinalis Steuer and the condition of its host Mytilus edulis L. Ophelia, 27: 77-86.

Tiews, K., 1988. Die miesmuschel in der ökologie des Wattenmeeres. Inf. Fischwirtsch., 35: 110-112.

Tkachuk, L.P., 1988. Vrediteli chernomorskikh midiynykh plantatsiy. Ekol. Moria., 30: 60-64.

Uzmann, J.R., 1953. Cercaria milfordensis nov. sp., a microcercous trematode larva from the marine bivalve, Mytilus edulis L. with special reference to its effect on the host. J. Parasitol., 39: 445-451.

Villalba, A., López, M.C., Mourelle, S.G. and Carballal, M.J., 1990a. Assessment of the effects of the infection by Marteilia sp. on the reproduction of mussel, Mytilus galloprovincialis Lmk., in estuaries of Galicia (NW of Spain). In: A. Figueras (Editor), Abstracts, 4th Internat. Colloq. Pathol. Mar. Aquacul., Vigo, Spain, 1990. pp. 76.

Villalba, A., Mourelle, S.G., Carballal, M.J. and López, M.C., 1990b. Epizootiological study of the infection of mussel, Mytilus galloprovincialis Lmk., by the protistan parasite Marteilia sp. in estuaries of Galicia (NW. of Spain). In: A. Figueras (Editor), Abstracts, 4th Internat. Colloq. Pathol. Mar. Aquacul., Vigo, Spain, 1990. pp. 77.

Vitellaro-Zuccarello, L., 1973. Ultrastructure of the byssal apparatus of Mytilus galloprovincialis. I. Associated fungal hyphae. Mar. Biol., 22: 225-230.

Yoo, M.S. and Kajihara, T., 1985. The effect of the pea crab (Pinnotheres pholadis) on the reproductive capacity of the blue mussel (Mytilus edulis galloprovincialis). Bull. Korean Fish. Soc., 18: 581-585.

INDEX

Page numbers in boldface refer to a major text discussion of the entry